String-Math 2011

Proceedings of Symposia in
PURE MATHEMATICS

Volume 85

String-Math 2011

Jonathan Block
Jacques Distler
Ron Donagi
Eric Sharpe
Editors

American Mathematical Society
Providence, Rhode Island

STRING-MATH 2011,
UNIVERSITY OF PENNSYLVANIA, PHILADELPHIA, PA,
JUNE 6–11, 2011

with support from the National Science Foundation,
grants DMS-0963840 and DMS-0908487

2010 *Mathematics Subject Classification.* Primary 14-XX, 18-XX, 19-XX, 22-XX, 53-XX, 58-XX, 81-XX, 81Txx, 83Exx, 83E30.

Library of Congress Cataloging-in-Publication Data

String-Math (Conference) (2011 : Philadelphia, Pa.)
String-Math 2011 : June 6–11, 2011, University of Pennsylvania, Philadelphia, Pennsylvania / Jonathan Block, Jacques Distler, Ron Donagi, Eric Sharpe, editors.
pages cm. — (Proceedings of symposia in pure mathematics ; volume 85)
Includes bibliographical references.
ISBN 978-0-8218-7295-6 (alk. paper)
1. Geometry, Algebraic–Congresses. 2. Quantum theory–Mathematics–Congresses. I. Block, Jonathan, 1960–editor of compilation. II. Title.

QA564.S77 2011
516.3′5—dc23

2012025768

---

10 9 8 7 6 5 4 3 2 1 17 16 15 14 13 12

Contents

Preface

The conference 'String-Math 2011' was held June 6-11, 2011 at the University of Pennsylvania, as the first in a new annual series of large meetings exploring the interface of mathematics and string theory. This volume presents the proceedings of that conference.

The nature of interactions between mathematicians and physicists has been thoroughly transformed in recent years. String theory and quantum field theory have contributed a series of profound ideas which gave rise to entirely new mathematical fields and revitalized older ones. The influence flows in both directions, with mathematical techniques and ideas contributing crucially to major advances in string theory. By now there is a large and rapidly growing number of both mathematicians and physicists working at the string-theoretic interface between the two academic fields.

For mathematics, string theory has been a source of many significant inspirations, ranging from Seiberg-Witten theory in four-manifolds, to enumerative geometry and Gromov-Witten theory in algebraic geometry, to work on the Jones polynomial in knot theory, to recent progress in the geometric Langlands program and the development of derived algebraic geometry and n-category theory. In the other direction, mathematics has provided physicists with powerful tools, ranging from powerful differential geometric techniques for solving or analyzing key partial differential equations, to toric geometry, to K-theory and derived categories in D-branes, to the analysis of Calabi-Yau manifolds and string compactifications, to the use of modular forms and other arithmetic techniques. The depth, power and novelty of the results obtained in both fields thanks to their interaction is truly mind-boggling.

Despite these profound and wide-ranging interactions, there was no major conference bringing together the leading mathematicians and mathematically-minded physicists working in this interface. We initiated the String-Math series of conferences and organized String-Math 2011 to fill this gap. Our intention was that this conference would be a vehicle for promoting such interactions, giving attendees greater opportunities to cross cultural boundaries, learn aspects of other fields relevant for their research, and advertise important developments to audiences that might not otherwise hear of them or appreciate their importance.

The conference covered a wide array of topics at the interface of mathematics and high energy physics, including, in no particular order,

- Homological mirror symmetry
- Categorical constructions of topological field theories
- Mathematical string phenomenology
- F-theory

- Wall crossing formulas
- Geometric Langlands
- Arithmetic of strings
- Gromov-Witten theory and related enumerative geometry
- A-twisted Landau-Ginzburg models
- String topology
- Elliptic cohomology
- Heterotic mirror symmetry
- Topological T duality
- String measures
- Chiral de Rham complexes
- Noncommutative geometry

Altogether, this conference brought together approximately 140 mathematicians and physicists. There were 28 invited plenary talks given by leaders in both fields. Additionally, there were 36 contributed talks given in parallel sessions on the Wednesday of the meeting. All the talks are available at the conference web site: http://www.math.upenn.edu/StringMath2011/.

The conference also included a public lecture on 'Strings and Geometry', by Cumrun Vafa. There were three satellite and affiliated conferences:

- FRG Workshop on Holomorphic Symplectic Varieties, Courant Institute for Mathematical Science, NY, June 4-8, 2011, http://www.math.nyu.edu/~tschinke/.nyu/FRG11.html.
- Topological Heterotic Strings and (0,2) Mirror Symmetry, Schrödinger Institute for Mathematical Physics, Vienna, Austria, June 20-24, 2011, http://www.aei.mpg.de/~ilarion/ESIhet/ESIhet.html.
- Connections in Geometry and Physics: 2011, Fields Institute, Toronto, Canada, May 13-15, 2011. http://www.math.uwaterloo.ca/~gap/.

The math/strings collaboration is clearly here to stay, and we expect this conference series to continue as long the subject remains active and exciting. The venues and years of the first five conferences of the String-Math series were agreed in advance:

- String-Math 2011, Penn, June 6-11, 2011
- String-Math 2012, Bonn (Hausdorff Center for Mathematics), July 16 - 21, 2012
- String-Math 2013, Stony Brook (Simons Center for Geometry and Physics)
- String-Math 2014, Alberta (Pacific Institute of Mathematical Sciences)
- String-Math 2015, China (Sanya Conference Center)

We gratefully acknowlege support obtained from the following sources:

- The National Science Foundation, grants DMS 0963840 and DMS 0908487,
- University of Pennsylvania Research Foundation,
- University of Pennsylvania Provost's International Initiatives Funding,
- University of Pennsylvania School of Arts and Sciences,
- US Department of Energy.

We are also very grateful to Sergei Gelfand and Chris Thivierge of AMS for their help in various stages of preparing this volume, and to Arthur Greenspoon for his help in proofreading and improving many of the articles included here.

The organizers of String-Math 2011:

Jonathan Block

Mirjam Cvetic

Jacques Distler

Ron Donagi

Tony Pantev

Eric Sharpe

List of Participants

Mina Aganagic
University of California Berkeley

Murad Alim
Harvard University

Paolo Aluffi
Florida State University

Lara Anderson
University of Pennsylvania

Paul Aspinwall
Duke University

David Baraglia
Australian National University

Nicolas Behr
Max-Planck-Institute for Gravitational Physics

David Ben-Zvi
University of Texas Austin

Eric Bergshoeff
University of Groningen

Ali Bleibel
Lebanese University

Jonathan Block
University of Pennsylvania

Lev Borisov
Rutgers University

Vincent Bouchard
University of Alberta

Peter Bouwknegt
Australian National University

Christopher Braun
University of Leicester

Chris Brav
Leibniz University Hannover

Andrei Caldararu
University of Wisconsin

Nils Carqueville
LMU Munich

Oscar Chacaltana
University of Texas Austin

Isaac Chappell
University of Maryland

Shyamoli Chaudhuri
Virginia Tech

Dmitro Chebotarov
University of Southern California

Miranda Cheng
Harvard University

A. Egon Cholakian
IRDF Project Harvard / Columbia

Ming-Tao Chuan
Harvard University

Braxton Collier
University of Texas Austin

Kevin Costello
Northwestern University

Xiaoyi Cui
University of Minnesota

Mirjam Cvetic
University of Pennsylvania

Peter Dalakov
University of Massachusetts Amherst

Jennie D'Ambroise
University of Minnesota Morris

Rhys Davies
University of Oxford

Emanuel Diaconescu
Rutgers University

Colin Diemer
University of Miami

Jimmy Dillies
University of Utah

Tudor Dimofte
Institute for Advanced Study, Princeton

Jacques Distler
University of Texas Austin

Ron Donagi
University of Pennsylvania

Michael Douglas
Simons Center for Geometry and Physics

Moutuou El-Kaioum
University of Metz

Mboyo Esole
Harvard University

Bohan Fang
Columbia University

David Favero
University of Vienna

Fabio Ferrari Ruffino
University of Sao Paolo

Daniel Fleisher
NC State University

Dan Freed
University of Texas Austin

James Fullwood
Florida State University

Inaki Garcia Estebaria
University of Pennsylvania

Masoud Ghezelbash
University of Saskatchewan

Antonella Grassi
University of Pennsylvania

Sergei Grigorian
Simons Center for Geometry and Physics

Marco Gualtieri
University of Toronto

Sergei Gukov
Caltech

Sam Gunningham
Northwestern University

Masahashi Hamanaka
Nagoya University

Daniel Halpern-Leistner
University of California Berkeley

James Halverson
University of Pennsylvania

Chris Harris
University of Miami

Kentaro Hori
IPMU

Mary-Frances Jagod
University of Pennsylvania

Tyler Jarvis
Brigham Young University

Bei Jia
Virginia Tech

Xin Jin
Northwestern University

Benjamin Jurke
Max-Planck-Institute for Physics

Ning Kang
University of Texas Austin

Sheldon Katz
University of Illinois, Urbana-Champaign

Ludmil Katzarkov
University of Miami

Louis H. Kauffman
University of Illinois, Chicago

Michael Kay
LMU Munich

Takashi Kimura
Boston University

Albrecht Klemm
University of Bonn

Peter Koroteev
University of Minnesota

Peggy Kouroumalou
University of Athens

Thomas Kragh
Massachusetts Institute of Technology

Eric Korman
University of Pennsylvania

Woden Kusner
University of Pittsburgh

Seung-Joo Lee
Korean Institute for Advanced Study

Si Li
Harvard University

Wei Li
IPMU

Kevin Lin
University of California Berkeley

Melissa Liu
Columbia University

Yijia Liu
University of Miami

Zhentao Lu
University of Pennsylvania

Andreas Malmendier
Colby College

Andy Manion
Princeton University

Kishore Marathe
CUNY Brooklyn College

Fernando Marchesano
IFT-UAM/CSIC

Matilde Marcolli
Caltech

Takuo Matsuoka
Northwestern University

Stefan Mendez-Diez
University of Alberta

Sukhendu Mehrotra
University of Wisconsin

Nikolai Mnev
Steklov Math Institute, St. Petersburg

Samuel Monnier
LPTENS Paris

Greg Moore
Rutgers University

Andrew Neitzke
University of Texas Austin

Tony Pantev
University of Pennsylvania

Daniel Park
Massachusetts Institute of Technology

Krishna Kumar Pathak
Columbia Institute of Technology

Brent Pim
University of Toronto

Daniel Pomerleano
University of California Berkeley

Anatoly Preygel
Massachusetts Institute of Technology

You Qi
Columbia University

Thorsten Rahn
Max-Planck Institute of Physics

Daniel Robbins
Texas A&M University

Andy Royston
Rutgers University

Daniel Roggenkamp
Rutgers University

Jonathan Rosenberg
University of Maryland

Yongbin Ruan
University of Michigan Ann Arbor

Pavel Safronov
University of Texas Austin

Sakura Schafer-Nameki
King's College London

Yuan Shen
Northwestern University

Jihye Seo
McGill University

Eric Sharpe
Virginia Tech

Artan Sheshmani
University of Illinois, Urbana-Champaign

Paul Seidel
Massachusetts Institute of Technology

Nicolo Sibilla
Northwestern University

Yefeng Shen
University of Michigan

Ruifang Song
Harvard University

Yan Soibelman
Kansas State University

Masoud Soroush
LMU Munich

Piotr Sulkowski
Caltech

Matt Szczesny
Boston University

Yuji Tachikawa
IPMU

Meng-Chwan Tan
National University of Singapore

Hiro Tanaka
Northwestern University

Cumrun Vafa
Harvard University

Justin Vazquez-Poritz
New York City College of Technology

Janu Verma
Kansas State University

Martijn Wijnholt
LMU Munich

James Walbridge
University Paul Sabatier

Kevin Wray
University of California Berkeley

Junya Yagi
Chiba University

Masahito Yamazaki
Princeton University

Yuan Yao
University of Texas Austin

S.-T. Yau
Harvard University

Shilin Yu
Penn State University

Jie Zhao
University of Wisconsin

Ying Zhao
Massachusetts Institute of Technology

Jie Zhou
Harvard University

Plenary Talks

Proceedings of Symposia in Pure Mathematics
Volume **85**, 2012

Refined Chern-Simons Theory and Knot Homology

Mina Aganagic and Shamil Shakirov

ABSTRACT. The refined Chern-Simons theory is a one-parameter deformation of the ordinary Chern-Simons theory on Seifert manifolds. It is defined via an index of the theory on N M5 branes, where the corresponding one-parameter deformation is a natural deformation of the geometric background. Analogously with the unrefined case, the solution of refined Chern-Simons theory is given in terms of S and T matrices, which are the proper Macdonald deformations of the usual ones. This provides a direct way to compute refined Chern-Simons invariants of a wide class of three-manifolds and knots. The knot invariants of refined Chern-Simons theory are conjectured to coincide with the knot superpolynomials – Poincaré polynomials of the triply graded knot homology theory. This conjecture is checked for a large number of torus knots in S^3, colored by the fundamental representation. This is a short, expository version of arXiv:1105.5117, with some new results included.

1. Introduction

One of the beautiful stories in the marriage of mathematics and physics developed from Witten's realization [**1**] that three dimensional Chern-Simons theory on S^3 computes the polynomial invariant of knots constructed by Jones in [**2**] . While Jones constructed an invariant $J(K, \mathbf{q})$ of knots in three dimensions, his construction relied on projections of knots to two dimensions. This obscured the three dimensional origin of the Jones polynomial. The fact that Chern-Simons theory is a topological quantum field theory in three dimensions made it manifest that the Jones polynomial is an invariant of the knot, and independent of the two dimensional projection. Moreover, it also gave rise to new topological invariants of three-manifolds and knots in them. For any three-manifold M and a knot in it, the Chern-Simons path integral, with a Wilson loop observable inserted along the knot, gives a topological invariant that depends only on M, K and the representation of the gauge group. Moreover, Chern-Simons theory gives a whole family of invariants associated to M and K, by changing the gauge group G and the representation R on the Wilson line. The Jones polynomial $J(K, \mathbf{q})$ corresponds to $G = SU(2)$, and R the fundamental, two dimensional representation of G. Taking $G = SU(n)$ instead, one computes the HOMFLY polynomial $H(K, \mathbf{q}, \mathbf{a})$ [**3**] evaluated at $\mathbf{a} = \mathbf{q}^n$. The work in [**1**] was made even more remarkable by the fact that it explained how to solve Chern-Simons theory for any M and collection of knots in it.

Based on talks presented by M.A. at several conferences and workshops, including the String-Math 2011 Conference at the University of Pennsylvania.

A mystery left open by [**1**] is the integrality of the coefficients of the Jones and HOMFLY polynomials. They are both Laurent polynomials in $\mathbf{q}$, and in the latter case in $\mathbf{a}$, with integer coefficients. While Chern-Simons theory gives means of computing knot invariants, it gives no insight into the question of why the coefficients are integers. An answer to this question was provided by [**4**] . Khovanov associates a bi-graded (co)homology theory to a knot $H^{i,j}(K)$, in such a way that its Euler characteristic is the Jones polynomial,

$$J(K,\mathbf{q}) = \sum_{i,j} (-1)^i \mathbf{q}^j \dim H^{i,j}(K).$$

Interpreted in this way, the integrality of the coefficients is manifest, since they are counting dimensions of knot homology groups. This gives rise to a refinement of the Jones polynomial, where one computes the Poincaré polynomial instead,

$$Kh(K,\mathbf{q},\mathbf{t}) = \sum_{i,j} \mathbf{t}^i \mathbf{q}^j \dim H^{i,j}(K).$$

This depends on one extra parameter $\mathbf{t}$, and reduces to the Jones polynomial at $\mathbf{t} = -1$. Later, many generalizations of [**4**] were constructed. In particular, the generalization to $SU(N)$ knot invariants was constructed by Khovanov and Rozansky [**5, 6**]. A knot homology theory corresponding to refinement of the HOMFLY polynomial was conjectured to exist by [**7**]. The resulting three variable Poincaré polynomial was named the superpolynomial [**7**].

Chern-Simons theory arises naturally in the context of string theory, [**8**]. Namely, the $SU(N)$ Chern-Simons partition function on M is the same as the partition function of N M5 branes wrapping $M \times \mathbf{C} \times S^1$ in M-theory on $(T^*M \times \mathbb{C}^2 \times S^1)_q$ where, as one goes around the S^1, the two $\mathbb{C}$ planes get rotated by

$$(z_1, z_2) \to (qz_1, q^{-1}z_2).$$

The M-theory partition function is an index, $\mathrm{Tr}(-1)^F q^{S_1 - S_2}$, where S_1 and S_2 are the generators of rotations around the two complex planes. As explained in [**9**] it is expected that knot homologies arise from string theory by counting BPS states in M-theory where one keeps track of both quantum numbers S_1 and S_2 separately.

In [**10**], we argued that, provided that the three-manifold M and knots in it possess enough symmetry (this is the case when M is a Seifert manifold with Seifert knots), one can construct a refined index that keeps track of the spins S_1 and S_2 separately. This corresponds to the M-theory partition function in the background where z_1 and z_2 are allowed to rotate independently,

$$(z_1, z_2) \to (qz_1, t^{-1}z_2). \tag{1}$$

Here $q = e^{\epsilon_1}, t = e^{-\epsilon_2}$, in terms of the usual parameters $\epsilon_{1,2}$ defining the Nekrasov background [**11**] , and the refined topological string. The M-theory partition function is $\mathrm{Tr}(-1)^F q^{S_1 - S_r} t^{S_r - S_R}$, where S_r is the generator of the extra $U(1)$ symmetry present. Because of the additional grading, and since one is computing an index, the knot invariants of refined Chern-Simons theory need not be the same as the Poincaré polynomials of knot homology. However, as we will see in some cases, the BPS states have $S_r = 0$, there are no cancellations, and the two are equal.

This short note is devoted to an overview of the results of [**10**], and its organization is as follows. In section 2 we review the relation of the ordinary and the refined Chern-Simons theory. We explain how to compute the knot and the three-manifold invariants of the refined Chern-Simons theory. We discuss the conjecture relating refined Chern-Simons knot invariants to the superpolynomial of [**7**]. In section 3, we explain the M-theory definition of the refined Chern-Simons theory. We also discuss the relation of the present constructions to [**12**] . In section 4 we explain how to compute the S and the T matrices from M-theory. In section 5 we show that the large N dual of the refined Chern-Simons theory on S^3 is the refined topological string on $X = \mathcal{O}(-1) \oplus \mathcal{O}(-1) \to \mathbb{P}^1$. In section 6, we finish with comments on relations to previous work and interesting directions of generalization. We add an appendix, with the explicit refined Chern-Simons knot invariants for some more complicated torus knots.

2. Refined and Ordinary Chern-Simons Theory

In [**10**] we formulated a refinement of $SU(N)$ Chern-Simons theory, which we define on Seifert three-manifolds, with Seifert knots. Seifert three-manifolds are circle fibrations over a Riemann surface. They admit a (semi-)free $U(1)$ action, corresponding to rotating the S^1 fiber; the action of the $U(1)$ is free except that a discrete subgroup of $U(1)$ can act with fixed points. Seifert knots are knots wrapping the S^1 fiber, and projecting to points on the Riemann surface.

Refined Chern-Simons theory is a topological theory in three dimensions. In any such theory, all amplitudes, corresponding to any three-manifold, with arbitrary knots, can be written in terms of three building blocks: S, T and the braiding matrix B. In refined Chern-Simons theory, only a subset of amplitudes enter, as only those preserve the $U(1)$ symmetry. Moreover, the S and the T matrices of the refined Chern-Simons theory are a one parameter deformation of those in ordinary Chern-Simons theory. The S and the T matrices provide a unitary representation of the modular $SL(2,\mathbb{Z})$ group on the Hilbert space $\mathcal{H}_{T^2}$ of the theory on a torus: they satisfy

$$S^4 = 1, \qquad (ST)^3 = S^2. \tag{2}$$

A basis of the Hilbert space $\mathcal{H}_{T^2}$ can be obtained by taking a solid torus and placing Wilson lines in various representations in its interior in a particular way. More precisely, choosing a basis of $H_1(T^2)$ of the boundary torus, and taking the $(1,0)$ cycle of the T^2 to be contractible in the interior, one defines a state

$$|R_i\rangle \;\in\; \mathcal{H}_{T^2} \tag{3}$$

by the path integral on the solid torus with a Wilson line in representation R_i running along the $(0,1)$ cycle of the torus. Moreover, on the boundary of the T^2, one gets the action of $SL(2,\mathbb{Z})$ corresponding to the mapping class group of the torus. An element K of $SL(2,\mathbb{Z})$ acts on the basis states by

$$K|R_i\rangle = \sum_j K^j{}_i|R_j\rangle$$

simply corresponding to the fact that the Hilbert space is finite dimensional, and K acts on it. The representation of $SL(2,\mathbb{Z})$ acting on $\mathcal{H}_{T^2}$, is generated by S and T matrices,

$$S^i{}_j, \qquad T^i{}_j,$$

satisfying the defining relations of $SL(2,\mathbb{Z})$,

$$S^4 = 1, \qquad (ST)^3 = S^2$$

Let

$$K_{\bar{i}j} = \langle R_j | K R_i \rangle.$$

This has to be unitary,

$$K^*_{\bar{i}j} = K^{-1}{}_{\bar{j}i}, \tag{4}$$

since otherwise 3d general covariance would have been lost.[1] Topological invariance further constrains the representation. For example, the S matrix has to be symmetric and satisfy

$$S^{-1}_{\bar{i}j} = S^*_{\bar{i}j}. \tag{5}$$

This follows since $S_{\bar{i}j}$ is the amplitude of a Hopf link in S^3, obtained by gluing two solid tori, with Wilson lines corresponding to states $|R_i\rangle$ and $|R_j\rangle$ we defined before. Gluing these with an S transformation of the boundary we get the amplitude corresponding to two linked knots in S^3, the Hopf link:

$$S_{\bar{i}j} = \langle R_i | S | R_j \rangle.$$

The fact that we can smoothly re-arrange the link so that the roles of R_i and R_j get exchanged implies that S has to be symmetric, $S_{\bar{i}j} = S_{\bar{j}i}$. Unitarity then implies (5).

Indices are raised and lowered by a Hermitian metric g, defined by

$$g_{\bar{i}j} = \langle R_i | R_j \rangle = Z(S^2 \times S^1, \bar{R}_i, R_j). \tag{6}$$

This corresponds to taking two solid tori, with Wilson lines in representations R_i, R_j inside, and gluing together, with trivial identifications. The three-manifold one obtains is $S^2 \times S^1$, with two Wilson lines. The metric is Hermitian, since exchanging the roles of R_i and R_j corresponds to orientation reversal of the manifold,

$$g_{\bar{j}i} = \langle R_j | R_i \rangle = \langle R_i | R_j \rangle^* = g^*_{\bar{i}j},$$

where * denotes complex conjugation. We have, using the definitions,

$$K_{\bar{j}i} = \sum_k K^k{}_i \, g_{\bar{j}k}.$$

2.1. "Ordinary" Chern-Simons Theory. In the case of $SU(N)$ Chern-Simons theory, the basis of the Hilbert space $\mathcal{H}_{T^2}$ is provided by the conformal blocks on T^2 of the $SU(N)_k$ WZW model [**1**] . Only a subset of $SU(N)$ representations enter, those that correspond to integrable highest weight representations of the affine Lie algebra. This can be phrased in terms of a constraint on the corresponding Young diagram that $0 \leq R_1 \leq k$, where R_1 is the length of the first row. The basis of the Hilbert space (3) provided by the path integral on the solid torus with Wilson lines, is automatically orthonormal [**1**] :

$$g_{\bar{i}j} = \langle R_i | R_j \rangle = \delta^i{}_j. \tag{7}$$

The S and T matrices are given by

$$S_{\bar{i}j}/S_{00} = s_{R_i}(q^\rho) s_{R_j}(q^{\rho + R_i}),$$

where $s_{R_i}(x)$ are Schur functions, and

$$T^i_j = \delta^i{}_j q^{\frac{1}{2}(R_i + 2\rho, R_i) - \frac{k}{2N}(\rho,\rho)},$$

---

[1] This follows from $\langle K R_j | R_i \rangle = \langle R_j | K^{-1} R_i \rangle$ using $\langle K R_j | R_i \rangle = \langle R_i | K R_j \rangle^*$.

with

$$S_{00} = \frac{i^{N(N-1)/2}}{N^{\frac{1}{2}}(k+N)^{\frac{N-1}{2}}} \prod_{\alpha>0} (q^{-(\alpha,\rho)/2} - q^{(\alpha,\rho)/2})$$

where the product is over all positive roots α, and R_i denotes the highest weight of the corresponding representation R_i. When q is a root of unity, determined by the level k and the rank N of the Chern-Simons gauge group,

$$q = e^{\frac{2\pi i}{k+N}},$$

the matrices S and T provide a unitary representation of $SL(2,\mathbb{Z})$, in particular, satisfy the properties (2)(4)(5) .

2.2. Refined Chern-Simons Theory. The refined Chern-Simons theory depends on an extra parameter, but shares many of the same properties as the ordinary Chern-Simons theory. To specify the theory we need to specify the rank N and the level k, $N, k \in \mathbb{Z}_{>0}$, together with one additional, arbitrary, parameter β. The theory at level k has the Hilbert space $\mathcal{H}^2_T$ of the same dimension as ordinary Chern-Simons theory, with a basis labeled by integrable representations of $SU(N)_k$. The Hermitian metric is diagonal[2]

$$\langle R_i | R_j \rangle = g_{\bar{i}j} = g_i \delta^i{}_j.$$

Set

$$t = q^\beta.$$

While the theory makes sense for any β, the expressions are the simplest for $\beta \in \mathbb{Z}_{>0}$. We then get

$$g_i = \prod_{m=0}^{\beta-1} \prod_{\alpha>0} \frac{q^{-\frac{1}{2}(\alpha,R_i)} t^{-\frac{1}{2}(\alpha,\rho)} q^{-\frac{m}{2}} - q^{\frac{1}{2}(\alpha,R_i)} t^{\frac{1}{2}(\alpha,\rho)} q^{\frac{m}{2}}}{q^{-\frac{1}{2}(\alpha,R_i)} t^{-\frac{1}{2}(\alpha,\rho)} q^{\frac{m}{2}} - q^{\frac{1}{2}(\alpha,R_i)} t^{\frac{1}{2}(\alpha,\rho)} q^{-\frac{m}{2}}} \tag{8}$$

Setting

$$q = e^{\frac{2\pi i}{k+\beta N}}, \qquad t = e^{\frac{2\pi i \beta}{k+\beta N}}. \tag{9}$$

the metric vanishes for representations other than the those whose Young tableau fits in a box of width k, corresponding to representations of $SU(N)$ at level k. This is to be expected, since β is arbitrary and we can change it away from the unrefined value $\beta = 1$ adiabatically – in a finite dimensional Hilbert space, the states have nowhere to go to.

The S and the T matrices are given by

$$S_{\bar{i}j}/S_{00} = M_{R_i}(t^\rho) M_{R_j}(t^\rho q^{R_i}), \tag{10}$$

where $M_{R_i}(x)$ are Macdonald polynomials, and

$$T^i_j = \delta^i{}_j \, q^{\frac{1}{2}(R_i,R_i)} \, t^{(R_i,\rho)} t^{\frac{\beta-1}{2}(\rho,\rho)} q^{-\frac{k}{2N}(\rho,\rho)}, \tag{11}$$

where

$$S_{00} = \frac{i^{N(N-1)/2}}{N^{\frac{1}{2}}(k+\beta N)^{\frac{N-1}{2}}} \prod_{m=0}^{\beta-1} \prod_{\alpha>0} (q^{-m/2} t^{-(\alpha,\rho)/2} - q^{m/2} t^{(\alpha,\rho)/2})$$

Setting β to 1, q and t coincide, Macdonald polynomials become Schur functions, and the refined Chern-Simons amplitudes reduce to ordinary ones, as they should.

[2]We could have chosen a different normalization of $|R_i\rangle$, that would have made the basis orthonormal. We chose to stick to the conventions of **[10]** and not do that.

It is crucial, for the three dimensional interpretation of $S_{\bar{i}j}$ as the expectation value of the colored Hopf link, that it is symmetric. This fact is not obvious from the formula (10) , just like in the unrefined case. The symmetry of $S_{\bar{i}j}$ is in fact the content of one of the famous Macdonald conjectures [**13, 14**]. We will see that the M-theory derivation of $S_{\bar{i}j}$ will result in an expression for S that makes this symmetry manifest. Since the metric $g_{\bar{i}j}$ is not the identity, $S_{\bar{i}j}$ and $S^i{}_j$ are not the same, instead, $S_{\bar{i}j} = \sum_k g_{\bar{i}k} S^k{}_j = g_i S^i{}_j$, and similarly for T. In particular, while $S_{\bar{i}j}$ is symmetric, $S^i{}_j$, is not. The fact that S and T provide a unitary representation on $SL(2,\mathbb{Z})$ was proven in [**15, 16, 17**] (for integer β). The fact that this holds for arbitrary β can be checked in examples.

2.3. Verlinde Coefficients. One can define Verlinde coefficients N_{ijk} by the partition function of the theory on $S^2 \times S^1$ with Wilson lines in representations R_i, R_j and R_k inserted at three points on the S^2, and winding around the S^1.

$$N_{ijk} = Z(S^2 \times S^1, R_i, R_j, R_k) = \langle 0 | R_i R_j R_k \rangle. \tag{12}$$

It can be shown, by computing the same amplitude in two different ways, in the refined and unrefined case alike, that N_{ijk} satisfy the Verlinde formula [**18**]

$$S_{\bar{k}i}\, S_{\bar{k}j} / S_{\bar{k}0} = \sum_\ell N^\ell{}_{ij} S_{\bar{k}\ell} \tag{13}$$

or equivalently,

$$N_{ij\bar{k}} = \sum_\ell S_{\bar{\ell}i}\, S_{\bar{\ell}j}\, (S^*)_{\bar{k}}{}^{\bar{\ell}} / S_{\bar{\ell}0}. \tag{14}$$

The formula (13) can be proven by noting [**20**] that on the one hand, insertions of Wilson loops are realized by multiplication by Schur/Macdonald polynomials in N holonomy variables; and on the other hand, the S matrix is obtained by evaluating the Schur/Macdonald polynomials at special points. The proof, presented in [**10**] uses only a restricted notion of topological invariance that preserves the circle symmetry at every step. The one property of the Verlinde coefficients that is lost at $\beta \neq 1$ is integrality: the Verlinde coefficients are deformed to rational functions of q and t.

2.4. Three-Manifold Invariants from Refined Chern-Simons Theory. From S and T, by cutting and gluing, we can obtain invariants of knots and three-manifolds which preserve the $U(1)$ action. In particular, it is interesting to consider Seifert three-manifolds without knots. A compact Seifert three- manifold, fibered over the Riemann surface of genus g, carries labels (the description of the geometry is borrowed from [**21**])

$$(g, n; (\alpha_1, \beta_1), \ldots, (\alpha_r, \beta_r)),$$

where n is the degree of the circle bundle, and (α_i, β_i) are integers, parameterizing the type of special fibers that occur. The three-manifold invariant of this can be computed[3]

$$Z(g, n; (\alpha_1, \beta_1), \ldots, (\alpha_r, \beta_r)) = \sum_j T_j^n (g_j)^{g-1} (S_{0,j})^{2-r-2g} \Big(\prod_{i=1}^r (SK^{(\alpha_i,\beta_i)})_{0,j}\Big)$$

[3]See for example, [**22**] for details. Earlier work on three-manifold invariants of Seifert spaces includes [**23, 24, 25, 26, 27, 28, 29, 30**].

where $K^{(\alpha_i,\beta_i)}$ is an $SL(2,\mathbb{Z})$ matrix whose first column is (α_i, β_i). Each such matrix, as an element of $SL(2,\mathbb{Z})$ can be written as a product of S and T matrices. The $U(1)$ bundle over Σ_g has first Chern class $n - \sum_{i=1}^{r} \beta_i/\alpha_i$. In the presence of special fibers, Σ_g is an orbifold.

Written in terms of S and T, the three-manifold invariant no longer depends on the underlying topological field theory, but only on M. The dependence on the theory enters only through S and T. To get a three-manifold invariant corresponding to $SU(N)$ Chern-Simons theory on M, one would use the S and the T matrices of the $SU(N)_k$ WZW model. To get the path integral of the refined Chern-Simons theory on M instead, one uses the refined S and T matrices of the previous subsection.

In the simple case without special fibers (g,n), the three- manifold is a circle bundle over a smooth Riemann surface Σ_g of degree n, the corresponding partition function is

$$Z(g,n) = \sum_j T_j^n (g_j)^{g-1} (S_{0j})^{2-2g}.$$

We can obtain $(g, n; (\alpha_1, \beta_1), \ldots, (\alpha_r, \beta_r))$ from this by cutting out the neighborhoods of r knots wrapping the S^1 fibers over points on Σ, and gluing back corresponding solid tori by $SL(2,\mathbb{Z})$ transformation of their boundaries, corresponding to $K^{(\alpha_i,\beta_i)}$.

2.5. Knot Invariants in Operator Formalism and the Refined Chern-Simons Theory. The case of most interest for knot theory is $M = S^3$. S^3 can be viewed as a Seifert manifold in several different ways. The knots wrapping the Seifert fibers are torus knots in S^3. We view S^3 as a locus in $\mathbb{C}^2$, with coordinates z_1, z_2, where

$$|z_1|^2 + |z_2|^2 = 1 \tag{15}$$

An (n,m) torus knot $K_{n,m}$ is described by (15) together with the equation

$$z_1^n = z_2^m$$

This is invariant under the $U(1)$ action that takes $(z_1, z_2) \to (\zeta^m z_1, \zeta^n z_2)$, with $\zeta = e^{i\theta}$. This $U(1)$ action acts freely on S^3, except for a $\mathbb{Z}_m$ subgroup, generated by $\zeta = e^{2\pi i/m}$, that has fixed points at $z_2 = 0$, and a $\mathbb{Z}_n$ subgroup that similarly has fixed points at $z_1 = 0$. We get a knot provided n and m are relatively prime; otherwise, we get a link in S^3.

In any topological theory in three dimension, the partition function on S^3 with an (n,m) torus colored by representation R_i inserted, can be written as follows:

$$Z(S^3, K_{n,m}, R_i) = \sum_{j,k,\ell} K_{0k} N^k{}_{ij} (K^{-1})^j{}_\ell S^\ell{}_p, \tag{16}$$

Here, K is an element of $SL(2,\mathbb{Z})$ that takes the $(0,1)$ cycle to (n,m) cycle, i.e.

$$K = \begin{pmatrix} a & n \\ b & m \end{pmatrix} \in SL(2,\mathbb{Z}).$$

with $am - nb = 1$ since then the action of $\mathcal{O}_{R_i}^{(0,1)}$ on $|0\rangle$, and $\mathcal{O}_{R_i}^{(n,m)}$ on $K|0\rangle$ agree. Any such K can be written explicitly in terms of strings of S and T matrices

$$S = \begin{pmatrix} 0 & -1 \\ 1 & 0 \end{pmatrix} \in SL(2,\mathbb{Z}).$$

FIGURE 1. Two examples of torus knots, with winding numbers (2,3) and (4,5).

with the only ambiguity being the choice of framing $K \mapsto KT^f$ with integer f. For every particular (n, m), eq. (16) is fully explicit and allows one to compute straightforwardly the corresponding amplitude. For comparison with knot theory, it is most convenient to consider the normalized amplitude, where we set the expectation value of the unknot to 1,

$$\frac{Z(S^3, K, \square)}{Z(S^3, \bigcirc, \square)}$$

Introduce "knot theory" variables,

$$\mathbf{t} = \sqrt{t}, \quad \mathbf{q} = -\sqrt{q/t}, \quad \mathbf{a}^2 = t^N \sqrt{t/q}.$$

In terms of these, the $(2,3)$ torus knot (the trefoil) invariant, for example, is

$$\frac{Z(S^3, K_{2,3}, \square)}{Z(S^3, \bigcirc, \square)} = \mathbf{a}\mathbf{q}^{-1} + \mathbf{a}\mathbf{q}\mathbf{t}^2 + \mathbf{a}^2\mathbf{t}^3.$$

For the (4,5) torus knot one gets

$$\frac{Z(S^3, K_{4,5}, \square)}{Z(S^3, \bigcirc, \square)} =$$

$$= \mathbf{a}^3\mathbf{t}^4 + \mathbf{a}^3\mathbf{q}^3\mathbf{t}^6 + \mathbf{a}^3\mathbf{q}\mathbf{t}^4 + \mathbf{a}^3\mathbf{q}^{-1}\mathbf{t}^2 + \mathbf{a}^3\mathbf{q}^{-3} + \mathbf{a}^4\mathbf{q}^2\mathbf{t}^7 +$$
$$+\mathbf{a}^4\mathbf{q}\mathbf{t}^7 + \mathbf{a}^4\mathbf{t}^5 + \mathbf{a}^4\mathbf{q}^{-1}\mathbf{t}^5 + \mathbf{a}^4\mathbf{q}^{-2}\mathbf{t}^3 + \mathbf{a}^5\mathbf{t}^4.$$

Further examples are given in the Appendix.

2.6. Relation to Knot Superpolynomials. In the unrefined case, when $t = q$ and $\mathbf{t} = -1$, the refined amplitudes reduce to the amplitudes of ordinary Chern-Simons theory. Written in terms of $\mathbf{q}$ and $\mathbf{a}$, these are known to compute the HOMFLY polynomials. Therefore for generic $\mathbf{t} \neq -1$, refined Chern-Simons amplitudes provide a one-parameter generalization of HOMFLY polynomials, and a natural question arises: what is the corresponding object on the mathematical side?

We conjecture that refined Chern-Simons amplitudes (in the fundamental representation) compute the superpolynomial $\mathcal{P}(K)$ [**7**] of the knot homology theory categorifying the HOMFLY polynomial:

$$\frac{Z(S^3, K, \square)}{Z(S^3, \bigcirc, \square)} = \mathcal{P}(K)(\mathbf{a}, \mathbf{q}, \mathbf{t})$$

In [**10**] we have shown that the conjecture holds for all $(2, 2m+1)$ torus knots for any m, and for the $(3, 3m+1), (3, 3m+2)$ torus knots for $m = 1, 2$, computed previously in [**7, 31, 32**]. Our more recent calculations confirm that the conjecture holds for $(3, 8), (3, 10), (3, 11)$; $(4, 5), (4, 7), (4, 9), (4, 11)$; $(5, 6), (5, 8)$ and $(7, 8), (7, 9)$, by matching to the unpublished results of [**33**], made available to us by E. Gorsky.

We will give a physical explanation of these results, including the fact that $\frac{Z(S^3,K,\Box)}{Z(S^3,\bigcirc,\Box)}$ gives rise to polynomials with integer coefficients, in the next sections. We will first explain the physical origin of refined Chern-Simons theory itself, in terms of M-theory and the refined topological string.

3. M-Theory Definition of Refined Chern-Simons Theory

In this section we give a physical explanation of the above results using M-theory and the (refined) topological string. Indeed, ordinary Chern-Simons theory is known to be closely related to topological strings and also to M-theory on certain Calabi-Yau threefolds. In this section we will explain that one can turn this around and use M-theory to deduce the amplitudes of a three-dimensional topological field theory that refines Chern-Simons theory.

3.1. Chern-Simons Theory from Topological Strings. Let us start by summarizing the relation of ordinary Chern-Simons theory to the topological string. Topological string, or Gromov-Witten theory, on a Calabi-Yau threefold Y is given in terms of holomorphic maps $\varphi : \Sigma_g \to Y$, $\bar{\partial}\varphi = 0$ from a Riemann surface Σ_g into Y. It can be extended to open topological string, where the Riemann surface Σ_g possibly has boundaries, whose image under the map φ lies on a Lagrangian submanifold $M \subset Y$. Since M is a Calabi-Yau threefold, M is three-dimensional.

In [**8**] , Witten explained that, for any three-manifold M, the open topological string on

$$Y = T^*M,$$

with N topological D-branes on M is the same as $SU(N)$ Chern-Simons theory on M. The string coupling and the level of Chern-Simons get related via

$$g_s = \frac{2\pi i}{k+N}.$$

Since in this case $Y = T^*M$, there are no holomorphic curves of any kind, so only the degenerate maps φ can contribute. These degenerate maps of the open string theory on T^*M precisely reproduce the Feynman graphs of the underlying Chern-Simons theory on M. In particular, this implies that the Chern-Simons partition function on M, and the open topological string partition function on Y with N D-branes on M, $Z^{top}_{open}(T^*M)$ are the same:

$$Z_{CS}(M) = Z^{top}_{open}(T^*M).$$

Adding a knot K to M, in some representation R of the gauge group also has a topological string interpretation [**34**], of adding D-branes wrapping a non-compact Lagrangian L_K. L_K is a rank 2 bundle over the knot K, constructed as follows. Take a point on K and the vector V tangent to K at that point in M. One obtains a two-plane in the fiber of T^*M consisting of the cotangent vectors orthogonal to V in the pairing between the cotangent and tangent vectors provided by the symplectic form on Y (see [**34, 35**]). Such an L_K is topologically $\mathbb{R}^2 \times S^1$, and

$$L_K \cap M = K.$$

3.2. Chern-Simons Theory from M-Theory. For our purposes – to give a physical definition of the refined Chern-Simons theory – it is important that all the above constructions are further related to M-theory with M5 branes. While in the unrefined case all sides of this relation are well known, when we consider the refined case, the M-theory will provide the sole definition of the theory.

Consider M-theory on

$$(Y \times TN \times S^1)_q, \tag{17}$$

where Y is a Calabi-Yau manifold (for now Y is arbitrary) and TN is the Taub-NUT space. The Taub-NUT space is twisted along the S^1, in the sense that going around the circle, the complex coordinates z_1, z_2 of the TN rotate by

$$z_1 \to q z_1, \qquad z_2 \to q^{-1} z_2, \tag{18}$$

so the space is not a direct product. We denoted this twist by a subscript q in (17) . The M-theory partition function on this background is the same as the partition function of the closed topological string on Y [**36**], where one identifies $q = e^{g_s}$ with the string coupling g_s. To extend this to the open string [**37, 38, 39**], we add N M5 branes wrapping

$$(M \times \mathbb{C} \times S^1)_q$$

where M is a special Lagrangian 3-cycle in Y. The branes wrap a $\mathbb{C}$ subspace of TN space fixed by the rotations (18). We can take $\mathbb{C}$ to correspond to the z_1 plane. The partition function of the M5 branes on this background is

$$Z_{M5}(Y, M) = \mathrm{Tr}\,(-1)^F q^{S_1 - S_2}. \tag{19}$$

Here S_1 and S_2 are the generators of two $U(1)_{1,2}$ rotations in (18) , and $F = 2S_1$ measures the fermion number. The M5 brane partition function (19) is the same as the open topological string partition function on Y with N D-branes wrapping M,

$$Z_{M5}(Y, M, q) = Z^{top}_{open}(Y, M, g_s),$$

where $q = e^{g_s}$ in terms of g_s, the topological string coupling. In particular, in the case when $Y = T^*M$ with some three-manifold M, the partition function of M5 branes equals the $SU(N)$ Chern-Simons partition function on M:

$$Z_{M5}(T^*M, M, q) = Z^{top}_{open}(T^*M, M, g_s) = Z_{CS}(M)$$

We will now use M-theory in a slightly more general background to define what we mean by the refinement of the right hand side.

3.3. Refined Chern-Simons Theory from M-Theory. In certain cases, M-theory on Y can be used to define a refinement of the topological string [**40**]. Consider, as before, M-theory on $Y \times TN \times S^1$. We fiber TN over the S^1, so that going around the circle, the coordinates z_1,z_2 of the TN space are twisted by

$$z_1 \to q z_1, \qquad z_2 \to t^{-1} z_2. \tag{20}$$

We will denote the resulting space by

$$(Y \times TN \times S^1)_{q,t}.$$

If $t \neq q$ this alone breaks supersymmetry. However, if the Calabi-Yau Y is non-compact, M-theory on Y gives rise to a five dimensional gauge theory. This has an additional $U(1)_R \subset SU(2)_R$ symmetry, and supersymmetry of the theory can be preserved provided, as one goes around the S^1, one performs an additional R-symmetry twist.

Now consider adding N M5 branes on

$$(M \times \mathbb{C} \times S^1)_{q,t},$$

where $\mathbb{C}$ corresponds to the z_1 plane. The M5 brane configuration automatically preserves a $U(1)_1 \times U(1)_2$ symmetry rotating the z_1 and z_2 planes. For it to have the $U(1)_R$ symmetry as well, an additional requirement is needed. Let us focus for now on the main case of interest for us, when

$$Y = T^*M.$$

The theory has the $U(1)_R$ symmetry provided M admits a free $U(1)$ action. This $U(1)$ symmetry of M is itself not an R-symmetry; however, it can be used to construct one [**10**]. More precisely, the action only needs to be semi-free: this corresponds to allowing a discrete subgroup of $U(1)$ to act with fixed points. This implies that M is a Seifert three-manifold – an S^1 fibration over a genus g Riemann surface Σ_g,

$$S^1 \to M \to \Sigma_g$$

where the $U(1)$ action comes from the rotation of the fiber. The ordinary Chern-Simons theory on Seifert spaces was studied recently in [**21, 41, 42**].

With the additional $U(1)_R$ symmetry preserved, the M5 brane partition function on $M \times \mathbb{C} \times S^1$ defines an index,

$$Z_M(T^*M, q, t) = \mathrm{Tr}\,(-1)^F\, q^{S_1 - S_r}\, t^{S_r - S_2}. \tag{21}$$

Note that both S_2 and S_r are R-symmetries, as they correspond to rotations in the normal bundle to the M5 branes. Their difference $S_2 - S_r$ acts as a global symmetry, as the three dimensional $\mathcal{N} = 2$ theory at low energies has a unique $U(1)_R$ symmetry. A pair of the supercharges $Q, \bar{Q}$ preserved by the brane, whose (S_1, S_2, S_r) charges are $\pm(1/2, 1/2, 1/2)$ can be used to define the above index, see for example [**43**]. The index localizes on configurations that are annihilated by $Q, \bar{Q}$. Moreover, for $q = t$ it reduces to the unrefined index (19). We will take this index (21) as the definition of the refined $SU(N)$ Chern-Simons partition function on M:

$$Z_M(T^*M, q, t) \equiv Z_{CS}(M, q, t). \tag{22}$$

This is analogous to the unrefined case, where the N M5 brane partition function on M in T^*M equals the ordinary $SU(N)$ Chern-Simons partition function on M. Unlike the unrefined case, we now do not have an alternative definition of the theory.

We can also include knot observables in the refined Chern-Simons theory. As we explained above, in the ordinary topological string, including a Wilson loop K on a three-manifold M in Chern-Simons theory corresponds to adding D-branes on a special Lagrangian L_K in T^*M. To extend this to the refined case, both the knot K and the three-manifold M have to respect the $U(1)_R$ symmetry that is needed to define the theory. As explained in [**10**], this implies that the allowed knots and links in M are the Seifert knots [**42**], wrapping the S^1 fibers over Σ_g in M. In particular, in the simplest case $M = S^3$ where most of the actual calculations have been performed so far, the allowed knots and links are precisely the torus knots and links.

3.4. The extra $U(1)$ Symmetry. Given the importance of the existence of the extra $U(1)$ symmetry for the definition of the index, let us elaborate this point. The $U(1)$ symmetry we need is constructed as follows. M being a Seifert three-manifold has a nowhere vanishing vector field V. Thinking of M as an S^1 fibration over a Riemann surface, V acts by rotating the S^1 fiber. Using V, we can define a rank-two subbundle of the cotangent bundle T^*M to M. At each point in M the fiber of this bundle consists of those cotangent vectors that are orthogonal to V. The $U(1)$ action we want to use, whose generator we called S_R above, corresponds to rotation of the fibers of this bundle.

To be sure we can use this $U(1)$ action to define the refined index, we need to show that T^*M admits a metric that has this $U(1)$ action as an isometry, and moreover, that this metric preserves supersymmetry. We will now argue that such a metric exists[4]. Consider, to begin with, N M5 branes on a three-manifold of the special form, $M = S^1 \times \Sigma$, times the flat $\mathbb{R}^{2,1}$, inside $T^*M \times \mathbb{R}^{4,1} = T^*\Sigma \times T^*S^1 \times \mathbb{R}^{4,1}$. The theory has $SO(2)_1 \times SO(2)_2 \times SO(2)_R$ as a subgroup of the symmetry group. Here, $SO(2)_R$ rotates the fibers of $T^*\Sigma$, $SO(2)_1$ is a part of the Lorentz group of the brane, and $SO(2)_2$ rotates the normal directions to the brane in $\mathbb{R}^{4,1}$. $SO(2)_2$ and $SO(2)_R$ are both R-symmetries of the $\mathcal{N} = 4$ supersymmetric theory in three dimensions on the brane. It is helpful to reduce the $(2,0)$ theory along the S^1 (viewing this as the M-theory circle). Then, the theory we are discussing is obtained by compactifying five dimensional Yang-Mills theory on Σ, to obtain the three dimensional Yang-Mills theory on $\mathbb{R}^{2,1}$.

Now consider M5 branes on a Seifert three-manifold obtained by fibering the S^1 non-trivially over Σ. In this case, half of the supersymmetry of the theory is broken: the theory on $\mathbb{R}^{2,1}$ has $\mathcal{N} = 2$ supersymmetry in three dimensions. This also corresponds to the fact that T^*M is an honest Calabi-Yau threefold. Supersymmetries that get preserved by the background have their $SO(2)_2 \times SO(2)_R$ charges correlated. We can take the supercharges that survive to have $S_2 = S_R$. Thus, one combination of S_2 and S_R becomes an R symmetry of the theory. In addition, we potentially get a global $U(1)$ symmetry, corresponding to $S_2 - S_R$, if both S_2 and S_R survive as the symmetries. If this is the case, we can define the refined index (21) of the M5 brane theory on $M \times \mathbb{R}^2 \times S^1$.

What is the effect of fibering the S^1 over Σ on the M5 brane theory? We can answer this question from the perspective of the $(2,0)$ theory or better yet in its dimensionally reduced version – where we reduce on the S^1 fiber of the Seifert three-manifold, as we did above. The answer is that giving the S^1 bundle over Σ a degree p corresponds to turning on an $\mathcal{N} = 2$ Chern-Simons coupling on $\mathbb{R}^{2,1}$,

$$p \int_{\mathbb{R}^{2,1}} d^4\theta \, \mathrm{Tr}\mathcal{V}\Sigma(\mathcal{V}). \tag{23}$$

where $\mathcal{V}$ is the $\mathcal{N} = 2$ vector multiplet, and $\Sigma(\mathcal{V}) = \epsilon^{\alpha\beta}\bar{D}_\alpha D_\beta \mathcal{V}$ is the linear superfield [**44**]. The crucial point is that this coupling is neutral under both $SO(2)_2$ and $SO(2)_R$. Thus, turning it on breaks neither symmetry, and both S_2 and S_R survive as generators of R-symmetries in the theory on $\mathbb{R}^{2,1}$. To see that the $\mathcal{N} = 2$ Chern-Simons coupling preserves both symmetries, it suffices to note that one of its terms is the bosonic Chern-Simons coupling $p \int_{\mathbb{R}^{2,1}} \omega_{CS}(A)$, where A is the

[4]We thank E. Witten for discussions of this point.

gauge field on $\mathbb{R}^{2,1}$, clearly neutral under both of the symmetries, and $\omega_{CS}(A) = \mathrm{Tr}\ AdA - \frac{2}{3}A^3$ is the Chern-Simons three-form.

The origin of the Chern-Simons coupling is purely topological. The fastest way to see that is to recall that, viewing the circle fiber to M as the M-theory circle, the five dimensional YM theory is a theory on N D4 branes, and that has a term

$$\int_{\Sigma\times\mathbf{R}^{2,1}} F_{RR} \wedge \omega_{CS}(A) \tag{24}$$

in its action. The 11-dimensional interpretation of F_{RR} is the curvature of the circle bundle corresponding to the M-theory circle. In our present case, we would have

$$\int_\Sigma F_{RR} = p$$

resulting in the above coupling. The rest of (23) is fixed by $\mathcal{N}=2$ supersymmetry. It may seem that we are using some very particular facts about the couplings on the D4 branes to argue this. This is not the case. The origin of the term in (24) is a the fact that the $(2,0)$ theory has a self-dual two-form tensor on it. In any attempt to write down the action for the corresponding theory there is a peculiar Wess-Zumino type term that arises – albeit involving the metric on the six-manifold (see for example [**45, 46, 47**]). Using dimensional reduction to get to a five dimensional Yang-Mills the term (24) arises, with F_{RR} as the curvature of the S^1 bundle.

Thus, in the specific case when M is a Seifert three- manifold, the theory has, in addition to a $U(1)_R$ symmetry S_2, an additional $U(1)$ symmetry generated by S_R. Correspondingly, when M is a Seifert three-manifold, we can define the refined index (21) .

3.5. Related Work. In [**12**] a physical approach to knot homology was proposed, based on studying gauge theory on D4-branes wrapping a four-manifolds with a boundary on the three-manifold M, where the knots live, times a thermal S^1 (there were other duality frames studied in [**12**] as well, but we will focus on this one, as it is closest to us). The advantage of the approach initialized in [**12**] and developed further in [**48, 49**] is that it provides one a way to get at knot homology groups themselves, not relying on indices that exist when M is special. Nevertheless, it is important to note that the physical setting of [**12**] and the one we use here are related by a simple duality.

To define the refined Chern-Simons theory on a three-manifold M, we needed to study M-theory on $Y \times TN \times S^1$, where $Y = T^*M$ with N M5 branes on $M \times \mathbb{C} \times S^1$. Consider a dual description of this, by dimensionally reducing on the S^1 of the Taub-Nut space. Without M5 branes, we would obtain IIA string theory on the geometry,

$$Y \times \mathbb{R}^3 \times S^1$$

with a D6 brane wrapping $Y \times S^1$ and sitting at the origin of $\mathbb{R}^3$. Adding the N M5 branes on $M \times \mathbb{C} \times S^1$, we get IIA string theory with the addition of N D4 branes, wrapping $M \times S^1$ times a half-line $\mathbb{R}_+$ in $\mathbb{R}^3$, ending on the D6 brane. This is a D4 brane on a four manifold $\mathbb{R}_+ \times M \times S^1$, with the specific boundary condition imposed by the D6 brane. This setup is the same as that in [**12**] (see the discussion on the bottom of p. 13 of [**12**] and else where in the paper). Now consider how the symmetry generators map between the two pictures. From this we will deduce the index in IIA corresponding to the refined Chern-Simons partition function computed in M-theory, and recover in the unrefined limit, the index computation in [**12, 48, 49**] that gave rise to the Jones polynomial.

Before we add branes, the Taub-NUT geometry has an $SU(2)_\ell \times SU(2)_r$ isometry. We used the $U(1)_\ell \times U(1)_r$ subgroup of it in the definition of the index. The $U(1)_\ell \times U(1)_r$ act on the complex coordinates (z_1, z_2) of the TN space by $(e^{i(\theta_\ell+\theta_r)/2}z_1, e^{i(-\theta_\ell+\theta_r)/2}z_2)$. Asymptotically, Taub-NUT looks like $S^1 \times \mathbb{R}^3$ and the $U(1)_\ell$ isometry rotates the S^1, while the $SU(2)_r$ rotates the base geometry. So upon dimensional reduction, the charge under $U(1)_\ell$ becomes the D0-brane charge, while the charge under $SU(2)_r$ becomes the spin in the base $\mathbb{R}^3$. In addition to the this, IIA and M-theory have a common $SU(2)_R$ R-symmetry of a five-dimensional gauge theory.

The branes we add preserve the $U(1)_r$ subgroup of the $SU(2)_r$ rotation group, for any M. For any M, setting $q = t = q_0$, the partition function of the M5 brane theory (19) equals the partition function of the D4 brane theory in this background

$$Z_{D4}(T^*M, q_0) = \mathrm{Tr}\,(-1)^F\, q_0^{Q_0}$$

and both equal to the partition function of the ordinary $SU(N)$ Chern-Simons theory on M. In the D4 brane context, this was shown in [**12**] and studied further in [**48**] . In [**12, 48**] the Chern-Simons level arises due to non-zero value of the Wilson line of the RR 1-form potential C in IIA string theory, $\int_{S^1} C$. This couples to D0 brane charge. It is the same as $\log q_0$, the chemical potential for the D0 branes turned on in our setting.

When M is a Seifert three-manifold both the M5 brane and the D4 brane theories should also preserve a $U(1)$ subgroup of the $SU(2)_R$-symmetry group of the five dimensional background, by the duality. Then, we can define the refined index (21) , giving rise to the refined Chern-Simons theory, and depending on one more parameter. The refined partition function (21) becomes the partition function of the theory on N D4 branes in this background

$$Z_{D4}(T^*M, q_0, y) = \mathrm{Tr}\,(-1)^F\, q_0^{Q_0}\, y^{2J_3-2S_R}.$$

Here $q_0 = \sqrt{qt}$, $y = \sqrt{q/t}$, Q_0 is the D0 brane charge, and J_3 is the generator of the rotation group in $\mathbb{R}^3$, and S_R is the generator of the $U(1)$.

We could have also studied a different circle reduction, where we reduce to IIA on the thermal S^1 instead. In this case, we get IIA string theory on $Y \times TN$ with the four dimensional omega background. Adding N M5 branes, maps to adding N D-branes wrapping $M \times \mathbb{C}$. The D4 branes are nothing but surface operators in the four $\mathcal{N} = 2$ gauge theory in omega background (more precisely five dimensional gauge theory on $TN \times S^1$ where the data of the gauge theory is determined by the choice of Y in the usual way. The surface operators in 4d gauge theories were studied extensively in [**50, 51, 52, 53, 54, 55, 56, 57**].

4. Solving the Refined Chern-Simons Theory via M-Theory

Having used M-theory to define the refined Chern-Simons theory on Seifert three-manifolds M, we will explain how to solve it, and obtain the results quoted in section 2. The basic idea is to use topological invariance of the theory to solve it on simple pieces first, and recover the rest by gluing. More precisely, since the refined Chern-Simons theory makes sense only of Seifert manifolds, what one uses

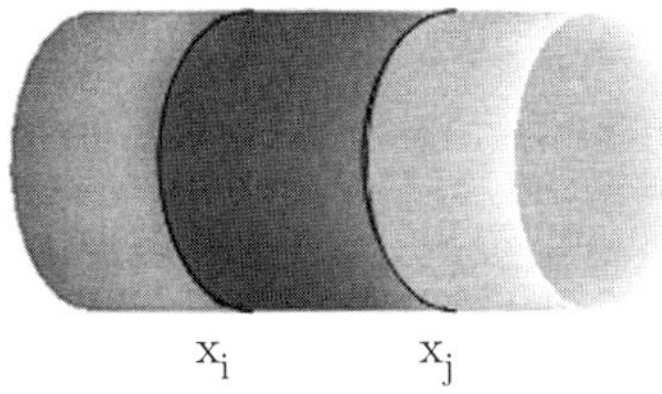

FIGURE 2. The D-branes wrap two S^1's on the cylinder $\mathbb{C}^*$ in Y_L. The open topological string is counting the maps to the annulus between them.

is a restricted notion of topological invariance that preserves the circle symmetry at every step; in essence, this is a combinatorial invariance[5], that governs how Seifert manifolds are cut up into tori.

The key amplitude we will obtain in this way corresponds to taking

$$M_L = S^1 \times \mathbb{R}^2$$

inside

$$Y_L = T^* M_L = T^*(S^1 \times \mathbb{R}^2) = \mathbb{C}^* \times \mathbb{C}^2.$$

The free $U(1)$ action on M_L corresponds to the rotation of the S^1, and the $U(1)_R$ symmetry one gets from it acts by rotating the fiber in $T^*\mathbb{R}^2$. The Calabi-Yau $Y_L = \mathbb{C}^* \times \mathbb{C}^2$ is essentially a flat space, hence non-compact; because of this, the $U(N)$ symmetry of the theory on M_L is a global symmetry. The partition function depends on the values of scalars x_I, $I = 1, \dots N$ of the background linear multiplets. These have interpretations as the positions of M5 branes along the fiber of T^*S^1 (this is a cylinder, and x_I parameterize the linear direction). We can, alternatively, view M_L as a solid torus $S^1 \times D$ where the path integral computes a wave function depending on the boundary conditions we impose – by topological invariance the two viewpoints are equivalent. With the branes widely separated, computing the partition function of the M5 brane theory on $(M_L \times \mathbb{C} \times S^1)_{q,t}$ inside $(Y_L \times TN \times S^1)_{q,t}$ amounts to counting BPS states of M2 branes stretching between them – i.e., cohomologies of the moduli of holomorphic curves embedded in Y and with boundaries on the M5 branes. As shown in detail in [**10**], this spectrum is very simple and consists precisely of two cohomology classes. (From the 3D perspective, the contributions to the index come from W-bosons and the adjoint scalar, parameterizing the position of the M5 branes along the $\mathbb{C}^2$ directions transverse to its world volume).

Because of this, one can compute the partition function explicitly as a function of the positions of the M5 branes. One finds

(25)
$$Z(S^1 \times \mathbb{R}^2)(q,t) = \prod_{1 \le I < J \le N} \prod_{n=1}^{\infty} \frac{(q^{-n/2} e^{(x_J - x_I)/2} - q^{n/2} e^{(x_I - x_J)/2})}{(t^{-1/2} q^{-n/2} e^{(x_J - x_I)/2} - t^{1/2} q^{n/2} e^{(x_I - x_J)/2})}$$

[5]We thank the referee for suggesting how to phrase this.

Note that this is consistent with the known result for the unrefined case: considering N branes and symmetrizing, for $q = t$ we recover

$$Z(S^1 \times \mathbb{R}^2)(q,q) = \prod_{1 \le I < J \le N} (e^{(x_J - x_I)/2} - e^{(x_I - x_J)/2}), \tag{26}$$

which is the Chern-Simons partition function on a solid torus [**58**] $S^1 \times \mathbb{R}^2$. To avoid dealing with infinite products, we can put

$$t = q^\beta, \qquad \beta \in \mathbb{N}, \tag{27}$$

where β is a positive integer. This specialization (27) is inessential – while it will simplify the intermediate formulas, it is easy to reinstate arbitrary β by inspection. At integer β, the partition function (25) can be written as a finite product:

(28)

$$Z(S^1 \times \mathbb{R}^2)(q,t) = \prod_{m=0}^{\beta-1} \prod_{1 \le I < J \le N} (q^{-m/2} e^{(x_J - x_I)/2} - q^{m/2} e^{(x_I - x_J)/2}) \equiv \Delta_{q,t}(x).$$

This is a partition function of M-theory on $T^*(S^1 \times \mathbb{R}^2)$. At the same time, by (22) , this should be regarded as the partition function of refined Chern-Simons theory on the solid torus. This simple result allows one to compute the partition functions for more complicated 3-manifolds M by cutting and gluing.

To consider gluing, it is useful to think of this in terms of the quantization of Chern-Simons theory. In this picture, as explained in Sec. 2, Chern-Simons theory on a manifold with T^2 boundary has a Hilbert space of states labeled by representations R of $SU(N)$, satisfying a certain condition. Then, the partition function eq. (26) is interpreted as the wavefunction of the vacuum state. Our results on refined Chern-Simons theory seem to suggest that at $q \neq t$, the theory has the same Hilbert space of states, but the wavefunction of the vacuum becomes (28) :

$$\langle 0 | x \rangle = \Delta_{q,t}(x)$$

Given the form of this vacuum, the most natural choice of basis in the Hilbert space is the one that has wavefunctions given by Macdonald polynomials:

$$\langle R_i | x \rangle = M_{R_i}(e^x) \Delta_{q,t}(x)$$

This is because Macdonald polynomials are the orthogonal polynomials associated with the $\Delta_{q,t}(x)$ measure, i.e. in Macdonald basis the identity operator has a diagonal matrix:

$$\langle R_i | R_j \rangle = \int dx_1 \cdots dx_N \Big| \Delta_{q,t}(x) \Big|^2 M_{R_i}(e^x) M_{R_j}(e^{-x}) = g_i 1_{ij}$$

where $g_i = ||M_{R_i}||^2$ is the Macdonald integral norm.

Now we can take two copies of T^2 with opposite orientation and glue them with trivial identification of the boundary, which gives the partition function of refined Chern-Simons theory on $S^1 \times S^2$ and M-theory on $T^*(S^1 \times S^2)$. This corresponds to taking the overlap of the two solid toruus wave functions, by integrating over $x_1, \ldots, x_N$:

$$Z(S^1 \times S^2) = \langle 0 | 0 \rangle = \int dx_1 \cdots dx_N \Big| \Delta_{q,t}(x) \Big|.^2$$

Instead of considering $S^1 \times S^2$, one can consider more general fibrations with the first Chern class p, by multiplying one of the two wave functions by

$$e^{-p\sum_i \frac{x_i^2}{2}}$$

before gluing. For example, S^3 is a Hopf fibration, with $p = 1$. Accordingly, the partition function of refined Chern-Simons theory on S^3 is given by the integral

$$Z(S^3) = \int dx_1 \cdots dx_N \left|\Delta_{q,t}(x)\right|^2 e^{-\sum_i \frac{x_i^2}{2}} = \langle 0|TST|0\rangle \tag{29}$$

i.e., the vacuum matrix element of the operator TST. Similarly, one deduces that other matrix elements of this operator are given by integrals

$$\langle R_i|TST|R_j\rangle = \int dx_1 \cdots dx_N \left|\Delta_{q,t}(x)\right|^2 e^{-\sum_i \frac{x_i^2}{2}} M_{R_i}(e^x) M_{R_j}(e^{-x})$$

The fact that this takes the explicit form that we gave before (10) , (11) was proven in the 90's in the context of proving Macdonald Conjectures.

Other examples of gluing include the Verlinde numbers that compute the amplitude on $S^1 \times S^2$ with three Wilson lines wrapping the S^1:

$$Z(S^1 \times S^2|R_i, R_j, R_k) = \int dx_1 \cdots dx_N \left|\Delta_{q,t}(x)\right|^2 M_{R_i}(e^x) M_{R_j}(e^x) M_{R_k}(e^x) = N_{ijk}$$

The fact that they satisfy the Verlinde formula (14) then follows from three-dimensional topological invariance – see the Appendix to [**10**] for derivation.

5. Large-N Dual of Refined Chern-Simons Theory

Ordinary Chern-Simons theory gets related, by string dualities, to other problems in mathematics and physics. Particularly important is that $SU(N)$ Chern-Simons theory has a large N dual description in terms of the refined topological string on a different manifold. Among other applications, this large N duality was used in [**34, 9**] to physically explain the integrality of coefficients of knot invariants. In this section we will first review this duality in the ordinary unrefined case, and then explain how it extends to the refined theory.

5.1. Large N Duality in Ordinary Chern-Simons Theory. Recall, from section 2, that $SU(N)$ Chern-Simons theory on S^3 is the same as the open topological string on

$$Y = T^* S^3,$$

with N D-branes on the S^3. Gopakumar and Vafa showed that this has a large N dual, the ordinary topological string theory on

$$X = \mathcal{O}(-1) \oplus \mathcal{O}(-1) \to \mathbb{P}^1.$$

The duality is a large N duality in the sense of 't Hooft [**59**]. The topological string coupling g_s is the same on both sides – it is related to the level k in Chern-Simons theory by

$$g_s = \frac{2\pi i}{k+N}.$$

The latter is the effective coupling constant of Chern-Simons theory, due to the famous shift of $k \to k+N$, generated by quantum effects. The rank N of the gauge theory is related to the area of the $\mathbb{P}^1$ in X by $\lambda = e^{-\mathrm{Area}(\mathbb{P}^1)}$ by

$$\lambda = q^N, \tag{30}$$

where $q = e^{g_s}$. This is a large N duality, since when the geometry of X is classical, λ is continuous, and this is only true in the limit of large N. The duality in this case also has a beautiful geometric interpretation: it is a geometric transition that shrinks the S^3 and grows the $\mathbb{P}^1$ at the apex of the conifold, thereby taking Y to X [**60**]. The duality has been checked, at the level of partition functions, to all orders in the $1/N$ expansion [**60**] .

As we explained in section 2, adding knots on the S^3 corresponds to introducing non-compact branes on a Lagrangian L_K in Y, intersecting the S^3 along a knot K,

$$K = L_K \cap S^3.$$

The geometric transition affects the interior of X and Y, but not their asymptotics, which are the same. A non-compact Lagrangian L_K on Y, goes through the transition to give a Lagrangian L_K on X. The correspondence of Chern-Simons theory with knots and the open topological string on X has been proven for the case of the unknot and the Hopf link, colored by arbitrary representations [**34, 61, 62, 63**]. This was also recently extended to torus knots in [**64**].

5.2. Large N Duality in Refined Chern-Simons Theory. The large N duality relating open topological string on $Y = T^*S^3$, i.e., Chern-Simons theory on S^3 to topological string on $X = \mathcal{O}(-1)\oplus\mathcal{O}(-1) \to \mathbb{P}^1$ was expected to extend to the refined topological string [**70, 9, 65**], where we view refinement as a deformation[6]. With the refined Chern-Simons theory in hand, we can offer compelling evidence that the refined theory indeed inherits the large N duality. The simplest test of the duality is to compare partition functions (without knot observables) on the two sides of the duality.

On X, we are computing the partition function of the refined closed topological string, which is well known. Let λ parameterize the size of the $\mathbb{P}^1$ in X as $\lambda = e^{-\mathrm{Area}(\mathbb{P}^1)}$, were by the area we really mean the mass of a BPS state wrapping the $\mathbb{P}^1$. The partition function of the closed topological string in this background is [**70**]

$$Z^{top}(X,\lambda,q,t) = \exp\Big(-\sum_{n=0}^{\infty}\frac{\lambda^n}{n(q^{n/2}-q^{-n/2})(t^{n/2}-t^{-n/2})}\Big), \tag{31}$$

up to simple factors that correspond to classical intersection numbers on the Calabi-Yau, and are ambiguous. Here q and t are the parameters defining the omega-background, $q = e^{\epsilon_1}$, $t = e^{-\epsilon_2}$. On Y, we are computing the partition function of N M5 branes, which is given by the Macdonald-type integral in (29) ,

$$Z(S^3,N,q,t) = \int dx_1\cdots dx_N \big|\Delta_{q,t}(x)\big|^2 e^{-\sum_i \frac{x_i^2}{2g_s}} \tag{32}$$

[6]We do not expect the large N duality to extend to the full superstring theory as in the unrefined case. Instead, we expect it to be the case for the topological, i.e. BPS sector of the theory, captured by the index.

The value of the integral is known explicitly for arbitrary finite N:

$$Z(S^3, N, q, t) = \prod_{m=0}^{\beta-1} \prod_{i=1}^{N-1} (1 - t^{N-i} q^m)^i, \tag{33}$$

where $q = e^{g_s}$ and $t = q^\beta$. Taking the large N limit of this exact finite-N expression, we find (see [**10**] for a derivation) that (32) and (31) agree,

$$Z(S^3, N, q, t) = Z^{top}(X, \lambda, q, t),$$

provided we identify

$$\lambda = t^N \sqrt{t/q}. \tag{34}$$

This generalizes the relation between the λ and N in the unrefined case. The fact that N ϵ_1-branes on the S^3 back-react on the geometry in such a way to open up a $\mathbb{P}^1$ of size $N\epsilon_2$ is as expected on general grounds, as discussed recently in detail in [**66**] . By ϵ_1 branes we mean the branes wrapping the z_1 plane rotated by the parameter $q = e^{\epsilon_1}$, and by ϵ_2 branes the branes wrapping the z_2 plane, rotated by $t = e^{\epsilon_2}$. The factors $\sqrt{t/q}$, which vanish in the unrefined theory at $q = t$, correspond to quantum shifts of the moduli by $(\epsilon_2 - \epsilon_1)/2$, and are typical [**65, 66**].

5.3. Large N Duality and Integrality of Knot Invariants. Two parallel approaches were developed to integrality of knot invariants. The one coming from mathematics, due originally to Khovanov, introduced the idea of knot homology, revealing the fact that Chern-Simons knot invariants are actually Euler characteristics of certain complexes of vector spaces [**4**]. On the physical side, nearly simultaneously with Khovanov's work, an explanation for this phenomenon was put forward in [**34**] .

The physical explanation from [**34**] was based on the two duality relations, the large N duality relating $SU(N)$ Chern-Simons theory on the S^3 to the topological string on $X = \mathcal{O}(-1) \oplus \mathcal{O}(-1) \to \mathbb{P}^1$, and the duality of the topological string on X with M-theory on $(X \times TN \times S^1)_q$. In [**34**] the authors showed that, taken together, the two dualities imply that computing the HOMFLY invariant of a knot K in S^3 is related to the index (19) of M5 branes wrapping $(L_K \times \mathbb{C} \times S^1)_q$,

$$Z_{CS}(K, S^3, N, q) = Z_M(L_K, X, \lambda, q). \tag{35}$$

The right hand side is the partition function of M5 branes wrapping the Lagrangian L_K in X, associated to the knot K. This partition function is computed by the index [7]

$$Z_M(L_K, X, \lambda, q) = \mathrm{Tr}_{\mathcal{H}_{BPS}} (-1)^F q^{S_1 - S_2} \lambda^Q, \tag{36}$$

counting the BPS states of M2 branes ending on the M5 branes on L_K. The grading λ of HOMFLY gets interpreted as the Kähler parameter λ, of the large N dual (30).

The fact that the knot invariants get related to a problem of counting BPS particles explains their integrality.

A natural further conjecture, made in [**9**] , is that the spaces of BPS states of M2 branes ending on L_K in X, where one keeps track of the spins S_1 and S_2 separately, are precisely the homologies of the knot K. Later, in [**7**] , motivated by this conjecture, a candidate for the right knot homology theory to match the physics

[7]For simplicity, we focus on knots colored by fundamental representation, so as to suppress an additional chemical potential one would have needed otherwise, to keep track of representations.

was put forward. There, the authors proposed existence of a triply graded knot homology theory categorifying the HOMFLY polynomial. The Poincaré polynomial of this theory has been named the superpolynomial in [**7**] . In terms of M-theory, the [**9**] conjecture implies that the superpolynomial of [**7**] is computed by counting the dimensions of spaces of BPS states on X, keeping track of their (S_1, S_2) spins: by refining the counting in (36) to

$$\mathrm{Tr}_{\mathcal{H}_{BPS}}(-1)^F q^{S_1} t^{-S_2} \lambda^Q \tag{37}$$

Note that, since F, the fermion number also equals $2S_1$, the $(-1)^F$ can be absorbed into q^{S_1}, to make all the coefficients in (37) positive. It is important to emphasize that, while one can define (37) for any knot K, this is not the partition function of the M5 branes on L_K, in any way of defining it. The partition function would need to extend over the whole Hilbert space $\mathcal{H}$, a subspace of which is the space of BPS states, $\mathcal{H}_{BPS}$. This would receive contributions from non-BPS states as well, since it is not an index. Equivalently, there should be no path integral way of computing (37) , analogously to what Chern-Simons theory does for the Jones polynomial[8].

If however, the theory has an additional $U(1)_R$ symmetry, as when K is a torus knot, than one can define a refined index:

$$Z_M(L_K, X, q, t) = \mathrm{Tr}_{\mathcal{H}_{BPS}}(-1)^F q^{S_1 - S_R} t^{S_R - S_2} \lambda^Q, \tag{38}$$

computed by the partition function of M5 branes on $(L_K \times \mathbb{C} \times S^1)_{q,t} \subset (X \times TN \times S^1)_{q,t}$, with the $U(1)_R$ symmetry twist as one goes around the S^1, to preserve supersymmetry. The index (38) is counting BPS states of M2 branes on X, and should equal the refined Chern-Simons knot invariant if the large N duality holds,

$$Z_{CS}(K, S^3, N, q, t) = Z_M(L_K, X, \lambda, q, t).$$

The results we presented in section two provide support for the conjecture of [**9**] and large N duality. Recall, we showed that, starting with the refined Chern-Simons invariant of a torus knot K, and setting

$$\lambda = -t^N \sqrt{t/q},$$

we obtained the corresponding superpolynomial of K, at least in a large number of cases. The change of variables is exactly what corresponds to the Káhler modulus of X, by large N duality. Moreover, the refined index on X and the Poincaré superpolynomial agree, assuming S_R in (38) vanishes (or is otherwise correlated with Q) for all BPS states of M2 branes ending on L_K in X, where K is a toric knot. In this particular case, the Poincaré polynomial and the index agree, as we found in section 2.5.

Moreover, the results of [**7**] provide a mathematical formulation of the index refined Chern-Simons theory is computing at finite N. In the categorification of [**7**] an important role is played by the so called d_N differential. The d_N operator is crucial in relating the HOMFLY knot homology of [**7**] and the better established SL_N knot homology of Khovanov and Rozansky [**5**][**6**], that categorifies $SU(N)$ Chern-Simons knot invariants: as explained in [**7**] the two are related in a non-trivial way at low enough N. At low enough N the HOMFLY homology is "too big": the SL_N homology emerges from the HOMFLY knot homology only upon taking the cohomology with respect to d_N.

[8]We are grateful to Edward Witten for discussions and explanations tied to this point.

On the other hand, to relate the superpolynomial $\mathcal{P}$, i.e. the Poincaré polynomial of the HOMFLY knot homology to the refined Chern-Simons knot invariants, we had to evaluate it at $\lambda = -t^N\sqrt{t/q}$ (this is equivalent to $\lambda = -\mathbf{q}^N/\mathbf{t}$, in knot theory variables). But, this is nothing but the index of the d_N differential: In the present language, d_N has (S_1, S_2, Q) gradings $(1/2, 1/2 + N, 1)$, so evaluating $q^{S_1}t^{-S_2}\lambda^Q$ at $\lambda = -t^N\sqrt{t/q}$, it anti-commutes with d_N, turning the superpolynomial into the corresponding index. This also explains how it is that refined Chern-Simons theory, which a priory computes an index, ends up having enough information to compute the Poincaré polynomial: provided N is sufficiently large, there are no cancellations in the index; the Poincaré polynomial and the index agree. On the other hand, both the index and the Poincaré polynomial are polynomials in λ or t^N, so to knowing them at sufficiently large N suffices to determine them for all N.

6. Conclusion

To summarize, the Poincaré polynomial of the theory of N M5 branes wrapping M in $Y = T^*M$,

$$\mathrm{Tr}_{\mathcal{H}_{BPS}}(-1)^{2S_1}q^{S_1}t^{-S_2} \tag{39}$$

counting BPS states along $\mathbb{C}\times S^1$ inside $TN\times S^1$ is expected [**12**] to compute the dimensions of $SU(N)$ knot homology groups, when one introduces knots K in the theory (by, for example, adding additional M5 branes wrapping $L_K\times\mathbb{C}\times S^1$, or). One can study the theory either directly, as in [**12**] or at large N, as originally proposed in [**9**] . This is defined for an arbitrary knot K and arbitrary three-manifold M. However, it is hard to compute in general, as one needs to know the space of BPS states $\mathcal{H}_{BPS}$.

By contrast, the index of the M5 brane theory,

$$\mathrm{Tr}_{\mathcal{H}}(-1)^{2S_1}q^{S_1-S_R}t^{S_R-S_2} \tag{40}$$

is simple to compute when it exists: when M and knots in it admit a semi-free $U(1)$ action, and the theory has an extra grading, S_R associated to it. The index receives contributions only from the space of BPS states, $\mathcal{H}_{BPS}\subset\mathcal{H}$. But, since the partition function of the theory, is computable simply by cutting and gluing, from S and T matrices of the refined Chern-Simons theory given in section 2.

As explained in [**12**] in the context of the Poincaré polynomial (39) , and in [**67**] in the context of the index (40) , both the index and the Poincaré polynomial have a natural generalization to other groups of ADE type, by replacing M5 branes by more general ADE type $(2,0)$ theory in the definition of (39) and (40) . In the case of the index, the corresponding generalization results in ADE type refined Chern-Simons theories. These were constructed and solved in [**67**], showing that, as expected [**10**] the connection to earlier work of I. Cherednik [**15**] on ADE type Macdonald Polynomials and $SL(2,\mathbb{Z})$ representations emerges.

From the mathematical perspective, we predict that, when M and knots in it admit a semi-free U(1) action, the knot homology groups admit an additional grading,

$$\mathcal{H}_{ij} = \oplus_k\mathcal{H}_{ijk}.$$

This allows one to define a refined index (40) , written in terms of knot theory variables abstractly as,

$$\sum_{i,j,k}(-1)^k \mathbf{q}^i \mathbf{t}^{j+k} \mathrm{dim}\mathcal{H}_{ijk}. \tag{41}$$

This has more information about knot homology than the Euler characteristic computed by the ordinary Chern-Simons theory of ADE type. The index (41) reduces to the Euler characteristic only upon setting $\mathbf{t} = -1$.

While the Poincaré polynomial of the knot homology theory

$$\sum_{i,j,k} \mathbf{q}^i \mathbf{t}^j \mathrm{dim}\mathcal{H}_{ijk} = \sum_{i,j} \mathbf{q}^i \mathbf{t}^j \mathrm{dim}\mathcal{H}_{ij}. \tag{42}$$

has yet more information than the index (41) , computing it is hard. The index can, by contrast, be obtained simply by cutting and gluing from S and T matrices of Macdonald type, described in the $SU(N)$ case in Sec. 2, and in the general ADE case, in [**67**].

Moreover, the index (40)(41) is interesting for another reason. Its robustness implies that string dualities can be used to relate it to other problems in physics and mathematics. For example, large N dualities should relate refined Chern-Simons invariants to partition functions of five dimensional $\mathcal{N} = 1$ gauge theories in omega background, analogous to [**63**] in the unrefined case. Moreover, the refined topological vertex of [**70**] should be derivable from the refined Chern-Simons theory, similarly to the way the unrefined topological vertex emerged from this [**68**] . Some evidence in this direction is presented in [**69**], where it was shown that the S matrix of the refined Chern-Simons theory also arises from the refined topological vertex formalism [**70, 71**]. This also proves that the Hopf link knot homologies of [**71**] and those arising from the refined Chern-Simons theory are equivalent. Moreover, we expect that it should be possible to define, at least for toric Calabi-Yau, a refined version of mirror symmetry (related to a five dimensional, or q-version of AGT correspondence [**72, 73**]). Chern-Simons theory is expected to play a role in this.

ATorus Knot Invariants From Refined Chern-Simons Theory

In this section we list several examples of the refined Chern-Simons theory knot invariants $Z(S^3, K, \Box)/Z(S^3, \bigcirc, \Box)$, colored by the fundamental representation, and written in terms of

$$\mathbf{t} = \sqrt{t}, \quad \mathbf{q} = -\sqrt{q/t}, \quad \mathbf{a}^2 = t^N \sqrt{t/q},$$

for various torus knots K. To the best of our knowledge, they agree with the corresponding superpolynomials $\mathcal{P}(K)(\mathbf{a}, \mathbf{q}, \mathbf{t})$, provided to us by E. Gorsky,

$$Z(S^3, K, \Box)/Z(S^3, \bigcirc, \Box) = \mathcal{P}(K)(\mathbf{a}, \mathbf{q}, \mathbf{t}),$$

We choose to normalize the invariant in such a way, that $\mathcal{P} = 1 + O(\mathbf{a}, \mathbf{q}, \mathbf{t})$.

$\mathcal{P}(K_{3,7}) = 1+\mathbf{q}^4\mathbf{t}^2+\mathbf{q}^6\mathbf{t}^4+\mathbf{q}^8\mathbf{t}^4+\mathbf{q}^{10}\mathbf{t}^6+\mathbf{q}^{12}\mathbf{t}^6+\mathbf{q}^{12}\mathbf{t}^8+\mathbf{q}^{14}\mathbf{t}^8+\mathbf{q}^{16}\mathbf{t}^8+\mathbf{q}^{18}\mathbf{t}^{10}+ \mathbf{q}^{20}\mathbf{t}^{10}+\mathbf{q}^{24}\mathbf{t}^{12}+(\mathbf{q}^2\mathbf{t}^3+\mathbf{q}^4\mathbf{t}^5+\mathbf{q}^6\mathbf{t}^5+2\mathbf{q}^8\mathbf{t}^7+\mathbf{q}^{10}\mathbf{t}^7+\mathbf{q}^{10}\mathbf{t}^9+2\mathbf{q}^{12}\mathbf{t}^9+\mathbf{q}^{14}\mathbf{t}^9+\mathbf{q}^{14}\mathbf{t}^{11}+ 2\mathbf{q}^{16}\mathbf{t}^{11}+\mathbf{q}^{18}\mathbf{t}^{11}+\mathbf{q}^{20}\mathbf{t}^{13}+\mathbf{q}^{22}\mathbf{t}^{13})\mathbf{a}^2+(\mathbf{q}^6\mathbf{t}^8+\mathbf{q}^{10}\mathbf{t}^{10}+\mathbf{q}^{12}\mathbf{t}^{12}+\mathbf{q}^{14}\mathbf{t}^{12}+\mathbf{q}^{18}\mathbf{t}^{14})\mathbf{a}^4$

$\mathcal{P}(K_{4,7}) = 1+\mathbf{q}^{22}\mathbf{t}^{14}+\mathbf{q}^4\mathbf{t}^2+\mathbf{q}^6\mathbf{t}^4+\mathbf{q}^8\mathbf{t}^4+\mathbf{q}^8\mathbf{t}^6+\mathbf{q}^{10}\mathbf{t}^6+\mathbf{q}^{12}\mathbf{t}^6+\mathbf{q}^{30}\mathbf{t}^{16}+ \mathbf{q}^{28}\mathbf{t}^{14}+\mathbf{q}^{22}\mathbf{t}^{12}+2\mathbf{q}^{16}\mathbf{t}^{10}+2\mathbf{q}^{12}\mathbf{t}^8+\mathbf{q}^{36}\mathbf{t}^{18}+\mathbf{q}^{32}\mathbf{t}^{16}+\mathbf{q}^{14}\mathbf{t}^{10}+\mathbf{q}^{18}\mathbf{t}^{12}+2\mathbf{q}^{20}\mathbf{t}^{12}+ \mathbf{q}^{14}\mathbf{t}^8+\mathbf{q}^{28}\mathbf{t}^{16}+\mathbf{q}^{16}\mathbf{t}^8+\mathbf{q}^{18}\mathbf{t}^{10}+\mathbf{q}^{20}\mathbf{t}^{10}+\mathbf{q}^{26}\mathbf{t}^{14}+2\mathbf{q}^{24}\mathbf{t}^{14}+\mathbf{q}^{24}\mathbf{t}^{12}+(\mathbf{q}^2\mathbf{t}^3+\mathbf{q}^4\mathbf{t}^5+$

$$q^{6}t^{5}+q^{6}t^{7}+2q^{8}t^{7}+q^{10}t^{7}+3q^{10}t^{9}+2q^{12}t^{9}+2q^{12}t^{11}+q^{14}t^{9}+4q^{14}t^{11}+2q^{16}t^{11}+3q^{16}t^{13}+q^{18}t^{11}+4q^{18}t^{13}+q^{18}t^{15}+2q^{20}t^{13}+3q^{20}t^{15}+q^{22}t^{13}+4q^{22}t^{15}+2q^{24}t^{15}+2q^{24}t^{17}+q^{26}t^{15}+3q^{26}t^{17}+2q^{28}t^{17}+q^{30}t^{17}+q^{30}t^{19}+q^{32}t^{19}+q^{34}t^{19})a^{2}+(q^{6}t^{8}+q^{8}t^{10}+q^{10}t^{10}+q^{10}t^{12}+2q^{12}t^{12}+q^{14}t^{12}+3q^{14}t^{14}+2q^{16}t^{14}+q^{16}t^{16}+q^{18}t^{14}+3q^{18}t^{16}+2q^{20}t^{16}+q^{20}t^{18}+q^{22}t^{16}+3q^{22}t^{18}+2q^{24}t^{18}+q^{26}t^{18}+q^{26}t^{20}+q^{28}t^{20}+q^{30}t^{20})a^{4}+(q^{12}t^{15}+q^{16}t^{17}+q^{18}t^{19}+q^{20}t^{19}+q^{24}t^{21})a^{6}$$

$$\mathcal{P}(K_{5,8}) = 1+2q^{40}t^{22}+2q^{32}t^{18}+q^{32}t^{16}+3q^{38}t^{22}+3q^{34}t^{20}+2q^{42}t^{24}+q^{40}t^{20}+q^{38}t^{20}+3q^{36}t^{22}+2q^{36}t^{20}+q^{34}t^{18}+q^{36}t^{18}+q^{4}t^{2}+q^{6}t^{4}+q^{8}t^{4}+q^{34}t^{22}+q^{8}t^{6}+q^{10}t^{6}+q^{12}t^{6}+q^{10}t^{8}+q^{22}t^{12}+2q^{12}t^{8}+2q^{16}t^{12}+4q^{24}t^{16}+q^{22}t^{16}+3q^{22}t^{14}+4q^{32}t^{20}+q^{24}t^{12}+3q^{20}t^{14}+2q^{20}t^{12}+3q^{18}t^{12}+2q^{14}t^{10}+2q^{24}t^{14}+3q^{26}t^{16}+2q^{26}t^{18}+2q^{40}t^{24}+q^{14}t^{8}+q^{42}t^{22}+2q^{30}t^{20}+3q^{30}t^{18}+q^{26}t^{14}+q^{28}t^{14}+4q^{28}t^{18}+2q^{28}t^{16}+q^{30}t^{16}+q^{16}t^{8}+q^{18}t^{10}+q^{20}t^{10}+2q^{16}t^{10}+q^{46}t^{24}+q^{44}t^{22}+2q^{44}t^{24}+q^{46}t^{26}+q^{48}t^{24}+q^{48}t^{26}+q^{50}t^{26}+q^{52}t^{26}+q^{56}t^{28}+(q^{2}t^{3}+q^{4}t^{5}+q^{6}t^{5}+q^{6}t^{7}+2q^{8}t^{7}+q^{8}t^{9}+q^{10}t^{7}+3q^{10}t^{9}+2q^{12}t^{9}+4q^{12}t^{11}+q^{14}t^{9}+4q^{14}t^{11}+3q^{14}t^{13}+2q^{16}t^{11}+6q^{16}t^{13}+q^{18}t^{11}+q^{16}t^{15}+4q^{18}t^{13}+7q^{18}t^{15}+2q^{20}t^{13}+7q^{20}t^{15}+q^{22}t^{13}+4q^{20}t^{17}+4q^{22}t^{15}+9q^{22}t^{17}+2q^{24}t^{15}+q^{22}t^{19}+7q^{24}t^{17}+q^{26}t^{15}+7q^{24}t^{19}+4q^{26}t^{17}+10q^{26}t^{19}+2q^{28}t^{17}+2q^{26}t^{21}+7q^{28}t^{19}+q^{30}t^{17}+8q^{28}t^{21}+4q^{30}t^{19}+10q^{30}t^{21}+2q^{32}t^{19}+2q^{30}t^{23}+7q^{32}t^{21}+q^{34}t^{19}+7q^{32}t^{23}+4q^{34}t^{21}+9q^{34}t^{23}+2q^{36}t^{21}+q^{34}t^{25}+7q^{36}t^{23}+q^{38}t^{21}+4q^{36}t^{25}+4q^{38}t^{23}+7q^{38}t^{25}+2q^{40}t^{23}+6q^{40}t^{25}+q^{42}t^{23}+q^{40}t^{27}+4q^{42}t^{25}+3q^{42}t^{27}+2q^{44}t^{25}+4q^{44}t^{27}+q^{46}t^{25}+3q^{46}t^{27}+2q^{48}t^{27}+q^{48}t^{29}+q^{50}t^{27}+q^{50}t^{29}+q^{52}t^{29}+q^{54}t^{29})a^{2}+(q^{6}t^{8}+q^{8}t^{10}+q^{10}t^{10}+2q^{10}t^{12}+2q^{12}t^{12}+q^{12}t^{14}+q^{14}t^{12}+4q^{14}t^{14}+q^{14}t^{16}+2q^{16}t^{14}+5q^{16}t^{16}+q^{18}t^{14}+5q^{18}t^{16}+4q^{18}t^{18}+2q^{20}t^{16}+7q^{20}t^{18}+q^{22}t^{16}+2q^{20}t^{20}+5q^{22}t^{18}+8q^{22}t^{20}+2q^{24}t^{18}+8q^{24}t^{20}+q^{26}t^{18}+4q^{24}t^{22}+5q^{26}t^{20}+10q^{26}t^{22}+2q^{28}t^{20}+q^{26}t^{24}+8q^{28}t^{22}+q^{30}t^{20}+5q^{28}t^{24}+5q^{30}t^{22}+10q^{30}t^{24}+2q^{32}t^{22}+q^{30}t^{26}+8q^{32}t^{24}+q^{34}t^{22}+4q^{32}t^{26}+5q^{34}t^{24}+8q^{34}t^{26}+2q^{36}t^{24}+7q^{36}t^{26}+q^{38}t^{24}+2q^{36}t^{28}+5q^{38}t^{26}+4q^{38}t^{28}+2q^{40}t^{26}+5q^{40}t^{28}+q^{42}t^{26}+4q^{42}t^{28}+q^{42}t^{30}+2q^{44}t^{28}+q^{44}t^{30}+q^{46}t^{28}+2q^{46}t^{30}+q^{48}t^{30}+q^{50}t^{30})a^{4}+(q^{12}t^{15}+q^{14}t^{17}+q^{16}t^{17}+q^{16}t^{19}+2q^{18}t^{19}+q^{18}t^{21}+q^{20}t^{19}+3q^{20}t^{21}+2q^{22}t^{21}+3q^{22}t^{23}+q^{24}t^{21}+4q^{24}t^{23}+q^{24}t^{25}+2q^{26}t^{23}+4q^{26}t^{25}+q^{28}t^{23}+4q^{28}t^{25}+2q^{28}t^{27}+2q^{30}t^{25}+4q^{30}t^{27}+q^{32}t^{25}+4q^{32}t^{27}+q^{32}t^{29}+2q^{34}t^{27}+3q^{34}t^{29}+q^{36}t^{27}+3q^{36}t^{29}+2q^{38}t^{29}+q^{38}t^{31}+q^{40}t^{29}+q^{40}t^{31}+q^{42}t^{31}+q^{44}t^{31})a^{6}+(q^{20}t^{24}+q^{24}t^{26}+q^{26}t^{28}+q^{28}t^{28}+q^{30}t^{30}+q^{32}t^{30}+q^{36}t^{32})a^{8}$$

$$\mathcal{P}(K_{7,10}) = 1+2q^{40}t^{22}+11q^{38}t^{28}+2q^{32}t^{18}+q^{32}t^{16}+3q^{38}t^{22}+3q^{34}t^{20}+3q^{42}t^{24}+15q^{44}t^{30}+q^{40}t^{20}+q^{38}t^{20}+10q^{44}t^{28}+5q^{36}t^{22}+2q^{36}t^{20}+11q^{44}t^{32}+12q^{42}t^{28}+10q^{36}t^{24}+q^{34}t^{18}+q^{36}t^{18}+7q^{40}t^{30}+q^{4}t^{2}+q^{6}t^{4}+q^{8}t^{4}+10q^{34}t^{24}+7q^{34}t^{22}+14q^{42}t^{30}+10q^{40}t^{26}+7q^{42}t^{26}+3q^{36}t^{28}+q^{8}t^{6}+q^{10}t^{6}+q^{12}t^{6}+q^{10}t^{8}+q^{22}t^{12}+2q^{12}t^{8}+2q^{42}t^{32}+q^{12}t^{10}+3q^{18}t^{14}+3q^{16}t^{12}+5q^{24}t^{16}+5q^{22}t^{16}+3q^{22}t^{14}+5q^{26}t^{20}+5q^{32}t^{20}+4q^{28}t^{22}+q^{24}t^{12}+4q^{20}t^{14}+3q^{20}t^{16}+2q^{20}t^{12}+q^{24}t^{20}+6q^{24}t^{18}+q^{22}t^{18}+3q^{18}t^{12}+8q^{28}t^{20}+2q^{14}t^{10}+2q^{24}t^{14}+3q^{26}t^{16}+6q^{26}t^{18}+5q^{40}t^{24}+7q^{38}t^{24}+11q^{38}t^{26}+4q^{46}t^{34}+q^{14}t^{8}+14q^{40}t^{28}+12q^{36}t^{26}+q^{42}t^{22}+6q^{34}t^{26}+7q^{30}t^{20}+3q^{30}t^{18}+8q^{30}t^{22}+2q^{30}t^{24}+q^{26}t^{14}+q^{28}t^{14}+9q^{32}t^{22}+9q^{32}t^{24}+5q^{28}t^{18}+2q^{28}t^{16}+q^{30}t^{16}+q^{16}t^{8}+q^{18}t^{10}+q^{20}t^{10}+2q^{16}t^{10}+q^{14}t^{12}+q^{56}t^{40}+q^{46}t^{24}+q^{98}t^{52}+q^{84}t^{50}+q^{70}t^{36}+q^{72}t^{36}+q^{68}t^{34}+q^{82}t^{42}+q^{96}t^{52}+q^{84}t^{42}+q^{94}t^{52}+q^{92}t^{46}+q^{94}t^{48}+q^{78}t^{40}+q^{90}t^{46}+q^{98}t^{50}+q^{102}t^{52}+q^{96}t^{48}+q^{104}t^{52}+q^{100}t^{50}+q^{88}t^{44}+q^{100}t^{52}+q^{86}t^{44}+q^{108}t^{54}+q^{86}t^{50}+9q^{76}t^{46}+14q^{66}t^{42}+8q^{78}t^{46}+5q^{86}t^{48}+7q^{68}t^{44}+11q^{64}t^{42}+2q^{66}t^{44}+q^{44}t^{22}+5q^{44}t^{26}+7q^{46}t^{28}+12q^{46}t^{30}+q^{48}t^{36}+2q^{44}t^{24}+16q^{46}t^{32}+3q^{46}t^{26}+q^{74}t^{38}+q^{80}t^{40}+q^{76}t^{38}+q^{62}t^{32}+q^{66}t^{34}+2q^{72}t^{38}+10q^{56}t^{34}+7q^{58}t^{34}+5q^{64}t^{36}+3q^{62}t^{34}+2q^{64}t^{34}+3q^{66}t^{36}+5q^{56}t^{32}+5q^{60}t^{34}+2q^{68}t^{36}+$$

$$q^{64}t^{32} + 7q^{50}t^{30} + 16q^{60}t^{38} + 10q^{64}t^{38} + 4q^{80}t^{48} + 9q^{76}t^{44} + 12q^{72}t^{44} + 10q^{72}t^{42} +$$
$$7q^{78}t^{44} + 10q^{68}t^{40} + 7q^{74}t^{42} + 11q^{70}t^{42} + 5q^{72}t^{40} + 5q^{68}t^{38} + 3q^{70}t^{38} + 6q^{82}t^{46} +$$
$$2q^{78}t^{48} + 16q^{62}t^{40} + 3q^{72}t^{46} + 11q^{70}t^{44} + 12q^{66}t^{40} + 3q^{92}t^{50} + 5q^{82}t^{48} + 15q^{64}t^{40} +$$
$$10q^{74}t^{44} + 3q^{86}t^{46} + 3q^{78}t^{42} + 2q^{76}t^{40} + 4q^{88}t^{48} + 6q^{84}t^{48} + 3q^{74}t^{40} + 6q^{74}t^{46} +$$
$$5q^{76}t^{42} + 5q^{80}t^{44} + 5q^{84}t^{46} + 2q^{88}t^{46} + 2q^{96}t^{50} + 2q^{92}t^{48} + 8q^{80}t^{46} + 2q^{80}t^{42} +$$
$$3q^{82}t^{44} + 2q^{84}t^{44} + 3q^{90}t^{50} + 3q^{90}t^{48} + 2q^{94}t^{50} + 12q^{62}t^{38} + 7q^{70}t^{40} + 2q^{60}t^{32} +$$
$$7q^{62}t^{36} + 3q^{58}t^{32} + 10q^{60}t^{36} + 7q^{66}t^{38} + 12q^{58}t^{36} + 3q^{88}t^{50} + 14q^{68}t^{42} + 5q^{48}t^{28} +$$
$$10q^{48}t^{30} + 12q^{50}t^{32} + q^{48}t^{24} + 16q^{48}t^{32} + 2q^{48}t^{26} + 17q^{50}t^{34} + 14q^{48}t^{34} + q^{50}t^{26} +$$
$$6q^{50}t^{36} + q^{52}t^{38} + 17q^{58}t^{38} + 6q^{58}t^{40} + 3q^{50}t^{28} + 12q^{54}t^{34} + q^{52}t^{26} + 3q^{54}t^{30} +$$
$$5q^{52}t^{30} + q^{54}t^{28} + q^{56}t^{28} + 16q^{56}t^{36} + q^{60}t^{42} + q^{58}t^{30} + 2q^{56}t^{30} + 7q^{54}t^{32} + q^{60}t^{30} +$$
$$14q^{60}t^{40} + 4q^{62}t^{42} + 7q^{54}t^{38} + 15q^{52}t^{36} + 18q^{54}t^{36} + 16q^{52}t^{34} + 15q^{56}t^{38} + 10q^{52}t^{32} +$$
$$2q^{52}t^{28} + (q^2t^3 + q^4t^5 + q^6t^5 + q^6t^7 + 2q^8t^7 + q^8t^9 + q^{10}t^7 + 3q^{10}t^9 + q^{10}t^{11} +$$
$$2q^{12}t^9 + 4q^{12}t^{11} + q^{14}t^9 + q^{12}t^{13} + 4q^{14}t^{11} + 5q^{14}t^{13} + 2q^{16}t^{11} + 6q^{16}t^{13} + q^{18}t^{11} +$$
$$6q^{16}t^{15} + 4q^{18}t^{13} + 9q^{18}t^{15} + 2q^{20}t^{13} + 5q^{18}t^{17} + 7q^{20}t^{15} + q^{22}t^{13} + 12q^{20}t^{17} +$$
$$4q^{22}t^{15} + 3q^{20}t^{19} + 11q^{22}t^{17} + 2q^{24}t^{15} + 14q^{22}t^{19} + 7q^{24}t^{17} + q^{26}t^{15} + 2q^{22}t^{21} +$$
$$16q^{24}t^{19} + 4q^{26}t^{17} + 13q^{24}t^{21} + 12q^{26}t^{19} + 2q^{28}t^{17} + q^{24}t^{23} + 21q^{26}t^{21} + 7q^{28}t^{19} +$$
$$q^{30}t^{17} + 11q^{26}t^{23} + 18q^{28}t^{21} + 4q^{30}t^{19} + 24q^{28}t^{23} + 12q^{30}t^{21} + 2q^{32}t^{19} + 7q^{28}t^{25} +$$
$$25q^{30}t^{23} + 7q^{32}t^{21} + q^{34}t^{19} + 25q^{30}t^{25} + 19q^{32}t^{23} + 4q^{34}t^{21} + 3q^{30}t^{27} + 31q^{32}t^{25} +$$
$$12q^{34}t^{23} + 2q^{36}t^{21} + 21q^{32}t^{27} + 27q^{34}t^{25} + 7q^{36}t^{23} + q^{38}t^{21} + q^{32}t^{29} + 37q^{34}t^{27} +$$
$$19q^{36}t^{25} + 4q^{38}t^{23} + 13q^{34}t^{29} + 35q^{36}t^{27} + 12q^{38}t^{25} + 2q^{40}t^{23} + 37q^{36}t^{29} + 28q^{38}t^{27} +$$
$$7q^{40}t^{25} + q^{42}t^{23} + 6q^{36}t^{31} + 44q^{38}t^{29} + 19q^{40}t^{27} + 4q^{42}t^{25} + 29q^{38}t^{31} + 37q^{40}t^{29} +$$
$$12q^{42}t^{27} + 2q^{44}t^{25} + q^{38}t^{33} + 49q^{40}t^{31} + 28q^{42}t^{29} + 7q^{44}t^{27} + q^{46}t^{25} + 16q^{40}t^{33} +$$
$$48q^{42}t^{31} + 19q^{44}t^{29} + 4q^{46}t^{27} + 44q^{42}t^{33} + 38q^{44}t^{31} + 12q^{46}t^{29} + 2q^{48}t^{27} + 6q^{42}t^{35} +$$
$$56q^{44}t^{33} + 28q^{46}t^{31} + 7q^{48}t^{29} + q^{50}t^{27} + 29q^{44}t^{35} + 50q^{46}t^{33} + 19q^{48}t^{31} + 4q^{50}t^{29} +$$
$$q^{44}t^{37} + 55q^{46}t^{35} + 38q^{48}t^{33} + 12q^{50}t^{31} + 2q^{52}t^{29} + 12q^{46}t^{37} + 60q^{48}t^{35} + 28q^{50}t^{33} +$$
$$7q^{52}t^{31} + q^{54}t^{29} + 39q^{48}t^{37} + 51q^{50}t^{35} + 19q^{52}t^{33} + 4q^{54}t^{31} + 3q^{48}t^{39} + 61q^{50}t^{37} +$$
$$38q^{52}t^{35} + 12q^{54}t^{33} + 2q^{56}t^{31} + 18q^{50}t^{39} + 62q^{52}t^{37} + 28q^{54}t^{35} + 7q^{56}t^{33} + q^{58}t^{31} +$$
$$45q^{52}t^{39} + 51q^{54}t^{37} + 19q^{56}t^{35} + 4q^{58}t^{33} + 4q^{52}t^{41} + 63q^{54}t^{39} + 38q^{56}t^{37} + 12q^{58}t^{35} +$$
$$2q^{60}t^{33} + 19q^{54}t^{41} + 62q^{56}t^{39} + 28q^{58}t^{37} + 7q^{60}t^{35} + q^{62}t^{33} + q^{54}t^{43} + 45q^{56}t^{41} +$$
$$51q^{58}t^{39} + 19q^{60}t^{37} + 4q^{62}t^{35} + 4q^{56}t^{43} + 61q^{58}t^{41} + 38q^{60}t^{39} + 12q^{62}t^{37} + 2q^{64}t^{35} +$$
$$18q^{58}t^{43} + 60q^{60}t^{41} + 28q^{62}t^{39} + 7q^{64}t^{37} + q^{66}t^{35} + 39q^{60}t^{43} + 50q^{62}t^{41} + 19q^{64}t^{39} +$$
$$4q^{66}t^{37} + 3q^{60}t^{45} + 55q^{62}t^{43} + 38q^{64}t^{41} + 12q^{66}t^{39} + 2q^{68}t^{37} + 12q^{62}t^{45} + 56q^{64}t^{43} +$$
$$28q^{66}t^{41} + 7q^{68}t^{39} + q^{70}t^{37} + 29q^{64}t^{45} + 48q^{66}t^{43} + 19q^{68}t^{41} + 4q^{70}t^{39} + q^{64}t^{47} +$$
$$44q^{66}t^{45} + 37q^{68}t^{43} + 12q^{70}t^{41} + 2q^{72}t^{39} + 6q^{66}t^{47} + 49q^{68}t^{45} + 28q^{70}t^{43} + 7q^{72}t^{41} +$$
$$q^{74}t^{39} + 16q^{68}t^{47} + 44q^{70}t^{45} + 19q^{72}t^{43} + 4q^{74}t^{41} + 29q^{70}t^{47} + 35q^{72}t^{45} + 12q^{74}t^{43} +$$
$$2q^{76}t^{41} + q^{70}t^{49} + 37q^{72}t^{47} + 27q^{74}t^{45} + 7q^{76}t^{43} + q^{78}t^{41} + 6q^{72}t^{49} + 37q^{74}t^{47} +$$
$$19q^{76}t^{45} + 4q^{78}t^{43} + 13q^{74}t^{49} + 31q^{76}t^{47} + 12q^{78}t^{45} + 2q^{80}t^{43} + 21q^{76}t^{49} + 25q^{78}t^{47} +$$
$$7q^{80}t^{45} + q^{82}t^{43} + q^{76}t^{51} + 25q^{78}t^{49} + 18q^{80}t^{47} + 4q^{82}t^{45} + 3q^{78}t^{51} + 24q^{80}t^{49} +$$
$$12q^{82}t^{47} + 2q^{84}t^{45} + 7q^{80}t^{51} + 21q^{82}t^{49} + 7q^{84}t^{47} + q^{86}t^{45} + 11q^{82}t^{51} + 16q^{84}t^{49} +$$
$$4q^{86}t^{47} + 13q^{84}t^{51} + 11q^{86}t^{49} + 2q^{88}t^{47} + q^{84}t^{53} + 14q^{86}t^{51} + 7q^{88}t^{49} + q^{90}t^{47} +$$
$$2q^{86}t^{53} + 12q^{88}t^{51} + 4q^{90}t^{49} + 3q^{88}t^{53} + 9q^{90}t^{51} + 2q^{92}t^{49} + 5q^{90}t^{53} + 6q^{92}t^{51} +$$
$$q^{94}t^{49} + 6q^{92}t^{53} + 4q^{94}t^{51} + 5q^{94}t^{53} + 2q^{96}t^{51} + 4q^{96}t^{53} + q^{98}t^{51} + q^{96}t^{55} + 3q^{98}t^{53} +$$
$$q^{98}t^{55} + 2q^{100}t^{53} + q^{100}t^{55} + q^{102}t^{53} + q^{102}t^{55} + q^{104}t^{55} + q^{106}t^{55})a^2 + (q^6t^8 +$$
$$q^8t^{10} + q^{10}t^{10} + 2q^{10}t^{12} + 2q^{12}t^{12} + 2q^{12}t^{14} + q^{14}t^{12} + 4q^{14}t^{14} + 3q^{14}t^{16} + 2q^{16}t^{14} +$$
$$6q^{16}t^{16} + q^{18}t^{14} + 2q^{16}t^{18} + 5q^{18}t^{16} + 9q^{18}t^{18} + 2q^{20}t^{16} + 2q^{18}t^{20} + 8q^{20}t^{18} + q^{22}t^{16} +$$
$$11q^{20}t^{20} + 5q^{22}t^{18} + q^{20}t^{22} + 14q^{22}t^{20} + 2q^{24}t^{18} + 11q^{22}t^{22} + 9q^{24}t^{20} + q^{26}t^{18} +$$
$$q^{22}t^{24} + 19q^{24}t^{22} + 5q^{26}t^{20} + 10q^{24}t^{24} + 16q^{26}t^{22} + 2q^{28}t^{20} + 25q^{26}t^{24} + 9q^{28}t^{22} +$$
$$q^{30}t^{20} + 8q^{26}t^{26} + 24q^{28}t^{24} + 5q^{30}t^{22} + 26q^{28}t^{26} + 17q^{30}t^{24} + 2q^{32}t^{22} + 5q^{28}t^{28} +$$

$$
\begin{aligned}
&34q^{30}t^{26}+9q^{32}t^{24}+q^{34}t^{22}+26q^{30}t^{28}+26q^{32}t^{26}+5q^{34}t^{24}+2q^{30}t^{30}+42q^{32}t^{28}+\\
&17q^{34}t^{26}+2q^{36}t^{24}+19q^{32}t^{30}+39q^{34}t^{28}+9q^{36}t^{26}+q^{38}t^{24}+q^{32}t^{32}+47q^{34}t^{30}+\\
&27q^{36}t^{28}+5q^{38}t^{26}+13q^{34}t^{32}+51q^{36}t^{30}+17q^{38}t^{28}+2q^{40}t^{26}+43q^{36}t^{32}+41q^{38}t^{30}+\\
&9q^{40}t^{28}+q^{42}t^{26}+5q^{36}t^{34}+64q^{38}t^{32}+27q^{40}t^{30}+5q^{42}t^{28}+32q^{38}t^{34}+56q^{40}t^{32}+\\
&17q^{42}t^{30}+2q^{44}t^{28}+2q^{38}t^{36}+65q^{40}t^{34}+42q^{42}t^{32}+9q^{44}t^{30}+q^{46}t^{28}+18q^{40}t^{36}+\\
&73q^{42}t^{34}+27q^{44}t^{32}+5q^{46}t^{30}+56q^{42}t^{36}+58q^{44}t^{34}+17q^{46}t^{32}+2q^{48}t^{30}+7q^{42}t^{38}+\\
&81q^{44}t^{36}+42q^{46}t^{34}+9q^{48}t^{32}+q^{50}t^{30}+34q^{44}t^{38}+78q^{46}t^{36}+27q^{48}t^{34}+5q^{50}t^{32}+\\
&2q^{44}t^{40}+75q^{46}t^{38}+59q^{48}t^{36}+17q^{50}t^{34}+2q^{52}t^{32}+16q^{46}t^{40}+90q^{48}t^{38}+42q^{50}t^{36}+\\
&9q^{52}t^{34}+q^{54}t^{32}+50q^{48}t^{40}+80q^{50}t^{38}+27q^{52}t^{36}+5q^{54}t^{34}+4q^{48}t^{42}+87q^{50}t^{40}+\\
&59q^{52}t^{38}+17q^{54}t^{36}+2q^{56}t^{34}+23q^{50}t^{42}+94q^{52}t^{40}+42q^{54}t^{38}+9q^{56}t^{36}+q^{58}t^{34}+\\
&q^{50}t^{44}+58q^{52}t^{42}+81q^{54}t^{40}+27q^{56}t^{38}+5q^{58}t^{36}+7q^{52}t^{44}+91q^{54}t^{42}+59q^{56}t^{40}+\\
&17q^{58}t^{38}+2q^{60}t^{36}+26q^{54}t^{44}+94q^{56}t^{42}+42q^{58}t^{40}+9q^{60}t^{38}+q^{62}t^{36}+q^{54}t^{46}+\\
&58q^{56}t^{44}+80q^{58}t^{42}+27q^{60}t^{40}+5q^{62}t^{38}+7q^{56}t^{46}+87q^{58}t^{44}+59q^{60}t^{42}+17q^{62}t^{40}+\\
&2q^{64}t^{38}+23q^{58}t^{46}+90q^{60}t^{44}+42q^{62}t^{42}+9q^{64}t^{40}+q^{66}t^{38}+q^{58}t^{48}+50q^{60}t^{46}+\\
&78q^{62}t^{44}+27q^{64}t^{42}+5q^{66}t^{40}+4q^{60}t^{48}+75q^{62}t^{46}+58q^{64}t^{44}+17q^{66}t^{42}+2q^{68}t^{40}+\\
&16q^{62}t^{48}+81q^{64}t^{46}+42q^{66}t^{44}+9q^{68}t^{42}+q^{70}t^{40}+34q^{64}t^{48}+73q^{66}t^{46}+27q^{68}t^{44}+\\
&5q^{70}t^{42}+2q^{64}t^{50}+56q^{66}t^{48}+56q^{68}t^{46}+17q^{70}t^{44}+2q^{72}t^{42}+7q^{66}t^{50}+65q^{68}t^{48}+\\
&41q^{70}t^{46}+9q^{72}t^{44}+q^{74}t^{42}+18q^{68}t^{50}+64q^{70}t^{48}+27q^{72}t^{46}+5q^{74}t^{44}+32q^{70}t^{50}+\\
&51q^{72}t^{48}+17q^{74}t^{46}+2q^{76}t^{44}+2q^{70}t^{52}+43q^{72}t^{50}+39q^{74}t^{48}+9q^{76}t^{46}+q^{78}t^{44}+\\
&5q^{72}t^{52}+47q^{74}t^{50}+26q^{76}t^{48}+5q^{78}t^{46}+13q^{74}t^{52}+42q^{76}t^{50}+17q^{78}t^{48}+2q^{80}t^{46}+\\
&19q^{76}t^{52}+34q^{78}t^{50}+9q^{80}t^{48}+q^{82}t^{46}+q^{76}t^{54}+26q^{78}t^{52}+24q^{80}t^{50}+5q^{82}t^{48}+\\
&2q^{78}t^{54}+26q^{80}t^{52}+16q^{82}t^{50}+2q^{84}t^{48}+5q^{80}t^{54}+25q^{82}t^{52}+9q^{84}t^{50}+q^{86}t^{48}+\\
&8q^{82}t^{54}+19q^{84}t^{52}+5q^{86}t^{50}+10q^{84}t^{54}+14q^{86}t^{52}+2q^{88}t^{50}+11q^{86}t^{54}+8q^{88}t^{52}+\\
&q^{90}t^{50}+q^{86}t^{56}+11q^{88}t^{54}+5q^{90}t^{52}+q^{88}t^{56}+9q^{90}t^{54}+2q^{92}t^{52}+2q^{90}t^{56}+6q^{92}t^{54}+\\
&q^{94}t^{52}+2q^{92}t^{56}+4q^{94}t^{54}+3q^{94}t^{56}+2q^{96}t^{54}+2q^{96}t^{56}+q^{98}t^{54}+2q^{98}t^{56}+q^{100}t^{56}+\\
&q^{102}t^{56})a^4+(q^{12}t^{15}+q^{14}t^{17}+q^{16}t^{17}+2q^{16}t^{19}+2q^{18}t^{19}+3q^{18}t^{21}+q^{20}t^{19}+4q^{20}t^{21}+\\
&3q^{20}t^{23}+2q^{22}t^{21}+7q^{22}t^{23}+q^{24}t^{21}+3q^{22}t^{25}+5q^{24}t^{23}+10q^{24}t^{25}+2q^{26}t^{23}+\\
&3q^{24}t^{27}+9q^{26}t^{25}+q^{28}t^{23}+12q^{26}t^{27}+5q^{28}t^{25}+2q^{26}t^{29}+15q^{28}t^{27}+2q^{30}t^{25}+\\
&13q^{28}t^{29}+10q^{30}t^{27}+q^{32}t^{25}+q^{28}t^{31}+21q^{30}t^{29}+5q^{32}t^{27}+12q^{30}t^{31}+17q^{32}t^{29}+\\
&2q^{34}t^{27}+q^{30}t^{33}+27q^{32}t^{31}+10q^{34}t^{29}+q^{36}t^{27}+9q^{32}t^{33}+26q^{34}t^{31}+5q^{36}t^{29}+\\
&28q^{34}t^{33}+18q^{36}t^{31}+2q^{38}t^{29}+6q^{34}t^{35}+37q^{36}t^{33}+10q^{38}t^{31}+q^{40}t^{29}+26q^{36}t^{35}+\\
&28q^{38}t^{33}+5q^{40}t^{31}+3q^{36}t^{37}+44q^{38}t^{35}+18q^{40}t^{33}+2q^{42}t^{31}+19q^{38}t^{37}+42q^{40}t^{35}+\\
&10q^{42}t^{33}+q^{44}t^{31}+q^{38}t^{39}+45q^{40}t^{37}+29q^{42}t^{35}+5q^{44}t^{33}+11q^{40}t^{39}+54q^{42}t^{37}+\\
&18q^{44}t^{35}+2q^{46}t^{33}+38q^{42}t^{39}+44q^{44}t^{37}+10q^{46}t^{35}+q^{48}t^{33}+5q^{42}t^{41}+61q^{44}t^{39}+\\
&29q^{46}t^{37}+5q^{48}t^{35}+24q^{44}t^{41}+59q^{46}t^{39}+18q^{48}t^{37}+2q^{50}t^{35}+q^{44}t^{43}+54q^{46}t^{41}+\\
&45q^{48}t^{39}+10q^{50}t^{37}+q^{52}t^{35}+12q^{46}t^{43}+70q^{48}t^{41}+29q^{50}t^{39}+5q^{52}t^{37}+37q^{48}t^{43}+\\
&61q^{50}t^{41}+18q^{52}t^{39}+2q^{54}t^{37}+4q^{48}t^{45}+66q^{50}t^{43}+45q^{52}t^{41}+10q^{54}t^{39}+q^{56}t^{37}+\\
&17q^{50}t^{45}+74q^{52}t^{43}+29q^{54}t^{41}+5q^{56}t^{39}+q^{50}t^{47}+44q^{52}t^{45}+62q^{54}t^{43}+18q^{56}t^{41}+\\
&2q^{58}t^{39}+6q^{52}t^{47}+69q^{54}t^{45}+45q^{56}t^{43}+10q^{58}t^{41}+q^{60}t^{39}+21q^{54}t^{47}+74q^{56}t^{45}+\\
&29q^{58}t^{43}+5q^{60}t^{41}+q^{54}t^{49}+44q^{56}t^{47}+61q^{58}t^{45}+18q^{60}t^{43}+2q^{62}t^{41}+6q^{56}t^{49}+\\
&66q^{58}t^{47}+45q^{60}t^{45}+10q^{62}t^{43}+q^{64}t^{41}+17q^{58}t^{49}+70q^{60}t^{47}+29q^{62}t^{45}+5q^{64}t^{43}+\\
&q^{58}t^{51}+37q^{60}t^{49}+59q^{62}t^{47}+18q^{64}t^{45}+2q^{66}t^{43}+4q^{60}t^{51}+54q^{62}t^{49}+44q^{64}t^{47}+\\
&10q^{66}t^{45}+q^{68}t^{43}+12q^{62}t^{51}+61q^{64}t^{49}+29q^{66}t^{47}+5q^{68}t^{45}+24q^{64}t^{51}+54q^{66}t^{49}+\\
&18q^{68}t^{47}+2q^{70}t^{45}+q^{64}t^{53}+38q^{66}t^{51}+42q^{68}t^{49}+10q^{70}t^{47}+q^{72}t^{45}+5q^{66}t^{53}+\\
&45q^{68}t^{51}+28q^{70}t^{49}+5q^{72}t^{47}+11q^{68}t^{53}+44q^{70}t^{51}+18q^{72}t^{49}+2q^{74}t^{47}+19q^{70}t^{53}+\\
&37q^{72}t^{51}+10q^{74}t^{49}+q^{76}t^{47}+q^{70}t^{55}+26q^{72}t^{53}+26q^{74}t^{51}+5q^{76}t^{49}+3q^{72}t^{55}+\\
&28q^{74}t^{53}+17q^{76}t^{51}+2q^{78}t^{49}+6q^{74}t^{55}+27q^{76}t^{53}+10q^{78}t^{51}+q^{80}t^{49}+9q^{76}t^{55}+\\
&21q^{78}t^{53}+5q^{80}t^{51}+12q^{78}t^{55}+15q^{80}t^{53}+2q^{82}t^{51}+q^{78}t^{57}+13q^{80}t^{55}+9q^{82}t^{53}+
\end{aligned}
$$

$$\mathbf{q^{84}t^{51} + q^{80}t^{57} + 12q^{82}t^{55} + 5q^{84}t^{53} + 2q^{82}t^{57} + 10q^{84}t^{55} + 2q^{86}t^{53} + 3q^{84}t^{57} + 7q^{86}t^{55} + q^{88}t^{53} + 3q^{86}t^{57} + 4q^{88}t^{55} + 3q^{88}t^{57} + 2q^{90}t^{55} + 3q^{90}t^{57} + q^{92}t^{55} + 2q^{92}t^{57} + q^{94}t^{57} + q^{96}t^{57})a^6 + (q^{20}t^{24} + q^{22}t^{26} + q^{24}t^{26} + 2q^{24}t^{28} + 2q^{26}t^{28} + 2q^{26}t^{30} + q^{28}t^{28} + 4q^{28}t^{30} + 3q^{28}t^{32} + 2q^{30}t^{30} + 6q^{30}t^{32} + q^{32}t^{30} + 2q^{30}t^{34} + 5q^{32}t^{32} + 8q^{32}t^{34} + 2q^{34}t^{32} + 2q^{32}t^{36} + 8q^{34}t^{34} + q^{36}t^{32} + 9q^{34}t^{36} + 5q^{36}t^{34} + q^{34}t^{38} + 13q^{36}t^{36} + 2q^{38}t^{34} + 8q^{36}t^{38} + 9q^{38}t^{36} + q^{40}t^{34} + q^{36}t^{40} + 16q^{38}t^{38} + 5q^{40}t^{36} + 6q^{38}t^{40} + 15q^{40}t^{38} + 2q^{42}t^{36} + 18q^{40}t^{40} + 9q^{42}t^{38} + q^{44}t^{36} + 4q^{40}t^{42} + 21q^{42}t^{40} + 5q^{44}t^{38} + 14q^{42}t^{42} + 16q^{44}t^{40} + 2q^{46}t^{38} + 2q^{42}t^{44} + 25q^{44}t^{42} + 9q^{46}t^{40} + q^{48}t^{38} + 11q^{44}t^{44} + 23q^{46}t^{42} + 5q^{48}t^{40} + 23q^{46}t^{44} + 16q^{48}t^{42} + 2q^{50}t^{40} + 5q^{46}t^{46} + 30q^{48}t^{44} + 9q^{50}t^{42} + q^{52}t^{40} + 17q^{48}t^{46} + 24q^{50}t^{44} + 5q^{52}t^{42} + 2q^{48}t^{48} + 28q^{50}t^{46} + 16q^{52}t^{44} + 2q^{54}t^{42} + 9q^{50}t^{48} + 32q^{52}t^{46} + 9q^{54}t^{44} + q^{56}t^{42} + 21q^{52}t^{48} + 24q^{54}t^{46} + 5q^{56}t^{44} + 3q^{52}t^{50} + 30q^{54}t^{48} + 16q^{56}t^{46} + 2q^{58}t^{44} + 10q^{54}t^{50} + 32q^{56}t^{48} + 9q^{58}t^{46} + q^{60}t^{44} + q^{54}t^{52} + 21q^{56}t^{50} + 24q^{58}t^{48} + 5q^{60}t^{46} + 3q^{56}t^{52} + 28q^{58}t^{50} + 16q^{60}t^{48} + 2q^{62}t^{46} + 9q^{58}t^{52} + 30q^{60}t^{50} + 9q^{62}t^{48} + q^{64}t^{46} + 17q^{60}t^{52} + 23q^{62}t^{50} + 5q^{64}t^{48} + 2q^{60}t^{54} + 23q^{62}t^{52} + 16q^{64}t^{50} + 2q^{66}t^{48} + 5q^{62}t^{54} + 25q^{64}t^{52} + 9q^{66}t^{50} + q^{68}t^{48} + 11q^{64}t^{54} + 21q^{66}t^{52} + 5q^{68}t^{50} + 14q^{66}t^{54} + 15q^{68}t^{52} + 2q^{70}t^{50} + 2q^{66}t^{56} + 18q^{68}t^{54} + 9q^{70}t^{52} + q^{72}t^{50} + 4q^{68}t^{56} + 16q^{70}t^{54} + 5q^{72}t^{52} + 6q^{70}t^{56} + 13q^{72}t^{54} + 2q^{74}t^{52} + 8q^{72}t^{56} + 8q^{74}t^{54} + q^{76}t^{52} + q^{72}t^{58} + 9q^{74}t^{56} + 5q^{76}t^{54} + q^{74}t^{58} + 8q^{76}t^{56} + 2q^{78}t^{54} + 2q^{76}t^{58} + 6q^{78}t^{56} + q^{80}t^{54} + 2q^{78}t^{58} + 4q^{80}t^{56} + 3q^{80}t^{58} + 2q^{82}t^{56} + 2q^{82}t^{58} + q^{84}t^{56} + 2q^{84}t^{58} + q^{86}t^{58} + q^{88}t^{58})a^8 + (q^{30}t^{35} + q^{32}t^{37} + q^{34}t^{37} + q^{34}t^{39} + 2q^{36}t^{39} + q^{36}t^{41} + q^{38}t^{39} + 3q^{38}t^{41} + q^{38}t^{43} + 2q^{40}t^{41} + 3q^{40}t^{43} + q^{42}t^{41} + q^{40}t^{45} + 4q^{42}t^{43} + 3q^{42}t^{45} + 2q^{44}t^{43} + 5q^{44}t^{45} + q^{46}t^{43} + 3q^{44}t^{47} + 4q^{46}t^{45} + 6q^{46}t^{47} + 2q^{48}t^{45} + q^{46}t^{49} + 6q^{48}t^{47} + q^{50}t^{45} + 5q^{48}t^{49} + 4q^{50}t^{47} + 7q^{50}t^{49} + 2q^{52}t^{47} + 3q^{50}t^{51} + 6q^{52}t^{49} + q^{54}t^{47} + 6q^{52}t^{51} + 4q^{54}t^{49} + q^{52}t^{53} + 8q^{54}t^{51} + 2q^{56}t^{49} + 3q^{54}t^{53} + 6q^{56}t^{51} + q^{58}t^{49} + 6q^{56}t^{53} + 4q^{58}t^{51} + q^{56}t^{55} + 7q^{58}t^{53} + 2q^{60}t^{51} + 3q^{58}t^{55} + 6q^{60}t^{53} + q^{62}t^{51} + 5q^{60}t^{55} + 4q^{62}t^{53} + 6q^{62}t^{55} + 2q^{64}t^{53} + q^{62}t^{57} + 5q^{64}t^{55} + q^{66}t^{53} + 3q^{64}t^{57} + 4q^{66}t^{55} + 3q^{66}t^{57} + 2q^{68}t^{55} + 3q^{68}t^{57} + q^{70}t^{55} + q^{68}t^{59} + 3q^{70}t^{57} + q^{70}t^{59} + 2q^{72}t^{57} + q^{72}t^{59} + q^{74}t^{57} + q^{74}t^{59} + q^{76}t^{59} + q^{78}t^{59})a^{10} + (q^{42}t^{48} + q^{46}t^{50} + q^{48}t^{52} + q^{50}t^{52} + q^{52}t^{54} + q^{54}t^{54} + q^{54}t^{56} + q^{56}t^{56} + q^{58}t^{56} + q^{60}t^{58} + q^{62}t^{58} + q^{66}t^{60})a^{12}}$$

Appendix A. Acknowledgements

This research is supported in part by Berkeley Center for Theoretical Physics, by the National Science Foundation (award number 0855653), by the Institute for the Physics and Mathematics of the Universe, by the US Department of Energy under Contract DE-AC02-05CH11231. The work of S.S. is also partly supported by Ministry of Education and Science of the Russian Federation under contract 14.740.11.5194, by RFBR grant 10-01-00536 and by joint grants 09-02-93105-CNRSL, 09-02-91005-ANF.

References

[1] E. Witten, "Quantum Field Theory and the Jones Polynomial," Commun. Math. Phys. **121**, 351 (1989).

[2] V.F.R. Jones "Index for subfactors", Invent. Math. 72 (1983) 1-25; V.F.R. Jones "A polynomial invariant for knots via von Neumann algebras", Bull. Amer. Math. Soc. 12 (1985) 103-112; V.F.R. Jones, "Hecke algebra representations of braid groups and link polynomials". Ann. of Math. (2) 126 (1987), no. 2, 335-388.

[3] P. Freyd, D. Yetter, J. Hoste, W.B.R. Lickorish, K. Millett, and A. Ocneanu, "A New Polynomial Invariant of Knots and Links". Bulletin of the American Mathematical Society 12 (2) (1985) 239-246. doi:10.1090/S0273-0979-1985-15361-3.

[4] M. Khovanov, "A Categorification Of The Jones Polynomial," Duke. Math. J. 101 (2000) 359-426.

[5] M. Khovanov and L. Rozansky, "Matrix factorizations and link homology I", arXiv:math/0401268.
[6] M. Khovanov and L. Rozansky, "Matrix factorizations and link homology II", arXiv:math/0505056.
[7] N. M. Dunfield, S. Gukov, and J. Rasmussen, "The Superpotential For Knot Homologies," Experiment. Math. 15 (2006) 129, math/0505662.
[8] E. Witten, "Chern-Simons gauge theory as a string theory," Prog. Math. **133**, 637-678 (1995). [hep-th/9207094].
[9] S. Gukov, A. S. Schwarz, and C. Vafa, "Khovanov-Rozansky Homology And Topological Strings," Lett. Math. Phys. 74 (2005) 53-74, hep-th/0412243.
[10] M. Aganagic and S. Shakirov, "Knot Homology from Refined Chern-Simons Theory," [arXiv:1105.5117 [hep-th]].
[11] N. A. Nekrasov, "Seiberg-Witten prepotential from instanton counting," Adv. Theor. Math. Phys. 7(2003), 831-864, arXiv:hep-th/0206161.
[12] E. Witten, "Fivebranes and Knots," Quant. topol. 3 (2012), 1-137, arXiv:1101.3216.
[13] I.G. Macdonald, "A new class of symmetric functions", Publ. I.R.M.A. Strasbourg, 372/S-20, Actes 20 Seminaire Lotharingien (1988), 131-171.
[14] I.G. Macdonald, "Orthogonal polynomials associated with root systems", preprint (1988).
[15] I. Cherednik, "Macdonald's Evaluation Conjectures and Difference Fourier Transform", Invent. Math. 125 (1995), 119-145, arXiv:q-alg/9412016
[16] A. Kirillov, Jr., "On an inner product in modular tensor categories", J. Amer. Math Soc. 9(1996), 1135-1169, arXiv:q-alg/9508017
[17] I. Cherednik, "Double Affine Hecke Algebras and Macdonald's Conjectures", The Annals of Mathematics, Second Series, Vol. 141, No. 1 (Jan., 1995), pp. 191-216
[18] E. P. Verlinde, "Fusion Rules and Modular Transformations in 2D Conformal Field Theory," Nucl. Phys. **B300**, 360 (1988).
[19] G. W. Moore, N. Seiberg, "Lectures On Rcft,"
[20] I. Cherednik and V. Ostrik, "From Double Affine Hecke Algebra to Fourier Transform", Selecta Math. (N.S.) 9, no. 2, 161-249, (2003).
[21] C. Beasley, E. Witten, "Non-Abelian localization for Chern-Simons theory," J. Diff. Geom. **70**, 183-323 (2005). [hep-th/0503126].
[22] S. K. Hansen, "Reshetikhin-Turaev Invariants of Seifert 3-Manifolds and a Rational Surgery Formula," Algebr. Geom. Topol. 1 (2001) 627-686, math.GT/0111057.
[23] R. Lawrence and L. Rozansky, "Witten-Reshetikhin-Turaev Invariants of Seifert Manifolds," Commun. Math. Phys. 205 (1999) 287-314
[24] L. Rozansky, "Residue Formulas for the Large k Asymptotics of Witten's Invariants of Seifert Manifolds: The Case of SU(2)," Commun. Math. Phys. 178 (1996) 27-60, hep-th/9412075
[25] D. Freed and R. Gompf, "Computer Calculation of Witten's 3-Manifold Invariant," Commun. Math. Phys. 141 (1991) 79-117
[26] L. Jeffrey, "On Some Aspects of Chern-Simons Gauge Theory," D.Phil. thesis, University of Oxford, 1991
[27] L. Jeffrey, "Chern-Simons-Witten Invariants of Lens Spaces and Torus Bundles," and the Semiclassical Approximation, Commun. Math. Phys. 147 (1992) 563-604.
[28] S. Garoufalidis, "Relations Among 3-Manifold Invariants," Ph.D. thesis, University of Chicago, 1992.
[29] J.R. Neil, "Combinatorial Calculation of the Various Normalizations of the Witten Invariants for 3-Manifolds," J. Knot Theory Ramifications 1 (1992) 407-449.
[30] L. Rozansky, "A Large k Asymptotics of Witten's Invariant of Seifert Manifolds," Comm. Math. Phys. 171 (1995) 279-322, hep-th/9303099.
[31] J.Rasmussen, "Khovanov-Rozansky homology of two-bridge knots and links", Duke Math J. 136 (2007), 551-583, arXiv:math.GT/0508510.
[32] J.Rasmussen, "Some differentials on Khovanov-Rozansky homology", arXiv:math/0607544.
[33] Gorsky, Oblomkov, Rasmussen, private communication
[34] H. Ooguri and C. Vafa, "Knot invariants and topological strings," Nucl. Phys. B **577**, 419 (2000) [arXiv:hep-th/9912123].
[35] C. Taubes, "Lagrangians for the Gopakumar-Vafa conjecture," Adv. Theor. Math. Phys. 5 (2001), 139-163, math.DG/0201219.

[36] R. Dijkgraaf, C. Vafa, E. Verlinde, "M-theory and a topological string duality," [hep-th/0602087].

[37] M. Aganagic and M. Yamazaki, "Open BPS Wall Crossing and M-theory," Nucl. Phys. B834 (2010) 258-272, 0911.5342.

[38] S. Cecotti, A. Neitzke, C. Vafa, "R-Twisting and 4d/2d Correspondences," [arXiv:1006.3435 [hep-th]].

[39] M. C. N. Cheng, R. Dijkgraaf, C. Vafa, "Non-Perturbative Topological Strings And Conformal Blocks," [arXiv:1010.4573 [hep-th]].

[40] T. J. Hollowood, A. Iqbal, C. Vafa, "Matrix models, geometric engineering and elliptic genera," JHEP **0803**, 069 (2008). [hep-th/0310272].

[41] C. Beasley, "Localization for Wilson Loops in Chern-Simons Theory," [arXiv:0911.2687 [hep-th]].

[42] C. Beasley, "Remarks on Wilson Loops and Seifert Loops in Chern-Simons Theory," AMS/IP Atud. Adv. Math. 50, AMS (2011), 1-17, arXiv:1012.5064.

[43] L. F. Alday, D. Gaiotto, S. Gukov, Y. Tachikawa, H. Verlinde, "Loop and surface operators in N=2 gauge theory and Liouville modular geometry," JHEP **1001**, 113 (2010). [arXiv:0909.0945 [hep-th]].

[44] O. Aharony, A. Hanany, K. A. Intriligator, N. Seiberg and M. J. Strassler, "Aspects of N=2 supersymmetric gauge theories in three-dimensions," Nucl. Phys. B **499**, 67 (1997). [hep-th/9703110].

[45] M. Aganagic, J. Park, C. Popescu and J. H. Schwarz, "Dual D-brane actions," Nucl. Phys. B **496**, 215 (1997). [hep-th/9702133].

[46] M. Aganagic, J. Park, C. Popescu and J. H. Schwarz, "World volume action of the M theory five-brane," Nucl. Phys. B **496**, 191 (1997). [hep-th/9701166].

[47] J. H. Schwarz, "Coupling a selfdual tensor to gravity in six-dimensions," Phys. Lett. B **395**, 191 (1997). [hep-th/9701008].

[48] D. Gaiotto and E. Witten, "Knot Invariants from Four-Dimensional Gauge Theory," arXiv:1106.4789

[49] E. Witten, "Khovanov Homology And Gauge Theory," arXiv:1108.3103

[50] S.Gukov and E.Witten, "Rigid Surface Operators", Advanced Theor. Math. Phys. 14 (2010) 87-177, arXiv:0804.1561

[51] D. Gaiotto, "Surface Operators in N = 2 4d Gauge Theories," [arXiv:0911.1316 [hep-th]].

[52] C. Kozcaz, S. Pasquetti and N. Wyllard, JHEP **1008**, 042 (2010). [arXiv:1004.2025 [hep-th]].

[53] L. F. Alday and Y. Tachikawa, Lett. Math. Phys. **94**, 87 (2010). [arXiv:1005.4469 [hep-th]].

[54] T. Dimofte, S. Gukov and L. Hollands, "Vortex Counting and Lagrangian 3-manifolds," Lett. Math. Phys. **98**, 225 (2011). [arXiv:1006.0977 [hep-th]].

[55] M.Taki, "Surface Operator, Bubbling Calabi-Yau and AGT Relation", JHEP 1107 (2011) 047, arXiv:1007.2524

[56] H. Awata, H. Fuji, H. Kanno, M. Manabe, Y. Yamada, "Localization with a Surface Operator, Irregular Conformal Blocks and Open Topological String",arXiv:1008.0574

[57] L. F. Alday, D. Gaiotto, S. Gukov, Y. Tachikawa, H. Verlinde, "Loop and surface operators in N=2 gauge theory and Liouville modular geometry," JHEP **1001**, 113 (2010). [arXiv:0909.0945 [hep-th]].

[58] S. Elitzur, G. W. Moore, A. Schwimmer, N. Seiberg, "Remarks on the Canonical Quantization of the Chern-Simons-Witten Theory," Nucl. Phys. **B326**, 108 (1989).

[59] G. t'Hooft, "Planar Diagram Theory for Strong Interactions", Nucl. Phys. 72 (1974) 461.

[60] R. Gopakumar and C. Vafa, "On the gauge theory/geometry correspondence," Adv. Theor. Math. Phys. **3**, 1415 (1999) [arXiv:hep-th/9811131].

[61] M. Aganagic and C. Vafa, "Mirror symmetry, D-branes and counting holomorphic discs," [hep-th/0012041].

[62] M. Aganagic, A. Klemm and C. Vafa, "Disk Instantons, Mirror Symmetry and the Duality Web," Z.Naturforsch. A57 (2002) 1-28, [hep-th/0105045].

[63] M. Aganagic, M. Mariño and C. Vafa, "All loop topological string amplitudes from Chern-Simons theory," Comm. Math. Phys. 247 (2004), 467-512, hep-th/0206164.

[64] A. Brini, B. Eynard and M. Marino, "Torus knots and mirror symmetry," [arXiv:1105.2012 [hep-th]].

[65] R. Dijkgraaf, C. Vafa, "Toda Theories, Matrix Models, Topological Strings, and N=2 Gauge Systems," [arXiv:0909.2453 [hep-th]].

[66] M. Aganagic, M. C. N. Cheng, R. Dijkgraaf, D. Krefl, C. Vafa, "Quantum Geometry of Refined Topological Strings," [arXiv:1105.0630 [hep-th]].
[67] M. Aganagic, K. Schaeffer, "Orientifolds and the Refined Topological String", to appear
[68] M. Aganagic, A. Klemm, M. Marino, C. Vafa, "The Topological vertex," Commun. Math. Phys. **254**, 425-478 (2005). [hep-th/0305132].
[69] A.Iqbal and C.Kozcaz, Refined Hopf Link Revisited, arXiv:1111.0525
[70] A. Iqbal, C. Kozcaz, C. Vafa, "The Refined Topological Vertex", JHEP 2009, no.10, 069, hep-th/0701156.
[71] S. Gukov, A. Iqbal, C. Kozcaz, and C. Vafa, "Link Homologies and the Refined Topo- logical Vertex," Comm. Math. Phys. 298 (2010), 757-785, arXiv:0705.1368.
[72] H. Awata and Y. Yamada, Five-dimensional AGT Relation and the Deformed beta-ensemble, Prog.Theor.Phys.124 (2010) 227-262
[73] H.Awata and Y. Yamada, Five-dimensional AGT Conjecture and the Deformed Virasoro Algebra, JHEP 1001 (2010) 125

Department of Mathematics and Center for Theoretical Physics, University of California, Berkeley, USA

Department of Mathematics, University of California, Berkeley, USA and Institute for Theoretical and Experimental Physics, Moscow, Russia

Proceedings of Symposia in Pure Mathematics
Volume **85**, 2012

Elusive Worldsheet Instantons in Heterotic String Compactifications

Paul S. Aspinwall and M. Ronen Plesser

ABSTRACT. We compute the spectrum of massless gauge singlets in some heterotic string compactifications using Landau–Ginzburg, orbifold and non-linear σ-model methods. This probes the worldsheet instanton corrections to the quadratic terms in the spacetime superpotential. Previous results predict that some of these states remain massless when instanton effects are included. We find vanishing masses in many cases not covered by these predictions. However, we discover that in the case of the Z-manifold the corrections do not vanish. Despite this, in all the examples studied, we find that the massless spectrum in the orbifold limit agrees with the nonlinear σ-model computation.

1. Introduction

The oldest approach to phenomenology in string theory is to compactify the heterotic string on a Calabi–Yau manifold X together with a choice of vector bundle $E \to X$. Among compactifications with $\mathcal{N} = 1$ spacetime supersymmetry, such models are unique in having a relatively straightforward worldsheet formulation in terms of a $(0, 2)$ superconformal field theory, and can be studied beyond the supergravity approximation. The effects of worldsheet instantons in these models present some formidable technical challenges and are far from completely understood.

In this paper we consider the moduli space of $(0, 2)$-theories. Unlike $(2, 2)$-theories we may have obstructions to first order deformations and, correspondingly, we have a notion of a superpotential in would-be moduli fields. The superpotential is subject to instanton corrections at any degree. A linear correction destabilizes the vacuum completely while a quadratic correction removes a first-order deformation. Higher terms in the superpotential affect the obstructions to the first-order deformations.

One may approach the computation of worldsheet instantons in two ways. The direct way is to explicitly attack the geometry of the rational curves in the target space. This has been pursued in works such as [**1**–**6**]. In this method one needs to compute Pfaffians associated to each curve and then sum over all curves while correctly taking into account multiple covers, etc. In principle this allows a complete computation of the full superpotential; in practice even the quadratic terms present a daunting challenge.

2010 *Mathematics Subject Classification.* Primary 81T30, Secondary 14D15 14E16 14J33.

The second approach, which avoids such formidable computations, is to compare a perturbative non-linear σ-model count of massless states with an exact conformal field theory method such as Landau–Ginzburg theories. The difference will yield precisely the instanton corrections to the quadratic terms in the superpotential. Surprisingly, this difference is frequently zero.

The history of worldsheet instantons destabilizing the vacuum is interesting. One knows that $(2,2)$ vacua are stable and so the "standard embedding" where E is the tangent bundle $T(X)$ yields a zero superpotential. Beyond that it was expected that a generic CY 3-fold would have rational curves giving generic instanton contributions leading to a nonzero superpotential, indicating that the compactification specified by E is not in fact a supersymmetric vacuum [**1**]. Thus, anything other than a very specially chosen vector bundle on a particular Calabi–Yau manifold would fail to give a good heterotic string compactification.

It was then realized [**2**, **3**] that the contribution to worldsheet instanton corrections from a single rational curve would be zero if the bundle E split nontrivially over the curve. While trivial splitting is generic, one could imagine finding special examples where every rational curve had a nontrivial splitting and thus the total instanton correction would be zero.

It was further realized that, in many cases which were understood as an exact conformal field theory, even if single instantons contributed nontrivially to the superpotential, there must a cancellation to produce a zero net result [**7**]. This was further explored for the quintic threefold in [**4**]. This apparently miraculous cancellation was explained in work by Beasley and Witten [**8**]. This latter paper also indicated that such a cancellation must happen in general when X is a complete intersection in a toric variety V and either

(1) E is a pullback of a vector bundle on V or
(2) E is in the form of a monad naturally realized by the gauged linear σ-model.

This accounts for a very wide range of possibilities and naturally one might ask the question: *Can we ever measure nonzero instanton contributions to the superpotential from the exact conformal field theory?*

We will address this question in the easiest of settings, namely where E is the tangent bundle $T(X)$. The conformal field theory exhibits $(2,2)$ worldsheet supersymmetry. The model has an E_6 spacetime gauge symmetry, and the spectrum includes massless E_6 singlet chiral multiplets corresponding to first-order deformations of T. In the supergravity approximation, these are classically counted by $h^1(\mathrm{End}(T))$, and perturbative corrections in α' to the spacetime superpotential are prohibited by supersymmetry. We further restrict our attention to terms in the superpotential quadratic in these fields, which represent instanton-induced mass terms.

If worldsheet instantons contribute to these terms in the superpotential then the number of these massless gauge singlet chiral superfields, computed in the conformal field theory, will be smaller than $h^1(\mathrm{End}(T))$. Thus we may look for worldsheet instantons by comparing classical (i.e., perturbative nonlinear σ-model) $h^1(\mathrm{End}(T))$ computations with exact results from the worldsheet. In this paper we will use mirror symmetry to compute the exact result. In particular, we will compute the spectrum from a Landau–Ginzburg realization of the mirror.

Any deformation of E that can be associated to a simple deformation of the gauged linear σ-model Lagrangian is a truly marginal deformation of the conformal field theory, protected from worldsheet instanton-induced obstruction at any degree by the Beasley–Witten result. Thus we need to focus on more complicated deformations.

We will see that the cancellation miracle persists in some cases beyond Beasley-Witten (as had previously been observed in [**9**]). We will also find an example where the number of massless singlets is drastically reduced from $h^1(\mathrm{End}(T))$ and thus we do indeed have instanton corrections. This example is given by the Z-manifold, i.e., a crepant resolution of $T^6/\mathbb{Z}_3$.

Even though the Z-manifold suffers from instanton correction we will see that it surprisingly obeys the "heterotic McKay correspondence". That is, we may correctly compute $h^1(\mathrm{End}(T))$ by counting untwisted singlets from a 6-torus and then adding in 27 copies of the twisted singlets from a $\mathbb{Z}_3$-quotient singularity as explained in [**10**]. This is unexpected from the point of view of string theory. Unlike the McKay correspondence which, from the point of view of string theory, relates two computations of an invariant quantity valid in different regions of the moduli space, the quantity we are computing here – the number of massless E_6 singlet fields – is not invariant and changes as we move about the moduli space. The calculation of [**10**] is valid on the orbifold locus, a nine-dimensional subspace describing the space with quotient singularities (unresolved). As we shall see, upon resolving the singularities some of these modes acquire a mass. $h^1(\mathrm{End}(T))$, on the other hand, computes the number of fields whose masses vanish exponentially as the volume of the space increases to infinity (in such a way that the sizes of all holomorphic curves grow). It is interesting that these numbers agree.

2. Deformations of the Tangent Bundle

Let V be a toric variety and let $X \subset V$ be the desired Calabi–Yau threefold given as a complete intersection given by s equations $f^a = 0$, for $a = 0, \ldots, s-1$.

We review the construction of V to fix notation. Let $x_0, \ldots, x_{N-1}$ be the homogeneous coordinates on V. This is the homogeneous coordinate ring in the sense of Cox [**11**]

$$R = \mathbb{C}[x_0, \ldots, x_{N-1}]. \tag{1}$$

We have a short exact sequence

$$0 \longrightarrow M \longrightarrow \mathbb{Z}^{\oplus N} \xrightarrow{\ \Phi\ } D \longrightarrow 0, \tag{2}$$

where D is a lattice[1] of rank r. Each column of the matrix Φ can be thought of as a $\mathrm{U}(1)^r$ charge vector of the coordinates x_i. That is, R has the structure of an r-multigraded ring.

The toric variety is given as

$$V = \frac{\mathrm{Spec}(R) - Z(B)}{(\mathbb{C}^*)^r}, \tag{3}$$

where B is the "irrelevant ideal" in R and $Z(B)$ is the associated subvariety of $\mathbb{C}^N$. B is determined combinatorially from the fan describing V.

[1] Assumed to be torsion-free.

Let $\mathbf{v}$ denote an element of the lattice D, i.e., an r-vector. If M is a multi-graded R-module then we may shift multi-gradings to form $M(\mathbf{v})$ in the usual way. Correspondingly, if $\mathscr{O}_V$ is the structure sheaf of V, then we may denote by $\mathscr{O}_V(\mathbf{v})$ the twisted sheaf associated to the module $R(\mathbf{v})$. Line bundles on V correspond to $\mathscr{O}_V(\mathbf{v})$ for various $\mathbf{v} \in D$. If V is smooth then every element of D defines a line bundle.

Let $\mathbf{q}_i$ denote the row vectors of the transpose of Φ. That is, $\mathbf{q}_i$ represents the multi-grading of the homogeneous coordinate x_i. Let T_V be the tangent sheaf of V. Assuming V is smooth, we have the generalization of the Euler exact sequence for a toric variety [**12**]

$$0 \longrightarrow \mathscr{O}_V^{\oplus r} \xrightarrow{x_i \mathbf{q}_i} \bigoplus_{i=0}^{N-1} \mathscr{O}_V(\mathbf{q}_i) \longrightarrow T_V \longrightarrow 0. \tag{4}$$

Let

$$\mathbf{Q} = \sum_{i=0}^{N-1} \mathbf{q}_i. \tag{5}$$

For the complete intersection X we have the adjunction exact sequence:

$$0 \longrightarrow T_X \longrightarrow T_{V|X} \longrightarrow \bigoplus_a \mathscr{O}_X(\mathbf{Q}_a) \longrightarrow 0, \tag{6}$$

where $\mathbf{Q}_a$ is the multi-degree of the equation f^a. The Calabi–Yau condition is $\sum_a \mathbf{Q}_a = \mathbf{Q}$.

Since all the sheaves in (4) are locally free, we may restrict to X and the sequence will remain exact. Combining this with the sequence (6) yields the following fact. The tangent sheaf is given by the cohomology at the middle term of

$$0 \longrightarrow \mathscr{O}_X^{\oplus r} \xrightarrow{x_i \mathbf{q}_i} \bigoplus_i \mathscr{O}_X(\mathbf{q}_i) \xrightarrow{\partial_i W_a} \bigoplus_a \mathscr{O}_X(\mathbf{Q}_a) \longrightarrow 0. \tag{7}$$

Obviously we may deform this complex to

$$0 \longrightarrow \mathscr{O}_X^{\oplus r} \xrightarrow{E} \bigoplus_i \mathscr{O}_X(\mathbf{q}_i) \xrightarrow{J} \bigoplus_a \mathscr{O}_X(\mathbf{Q}_a) \longrightarrow 0, \tag{8}$$

for generic matrices E and J of the correct multi-degree. By varying E and J we produce a family of sheaves containing the tangent sheaf. This family of sheaves can be understood in terms of the gauged linear σ-model [**13, 14**]. As such these are deformations of the $(0,2)$-model protected by Beasley–Witten. In general this is a subspace of the moduli space of $(0,2)$ deformations. Correspondingly, there are deformations of T_X that this description does not capture. These deformations are not protected and worldsheet instanton effects can be nontrivial.

We thus need a more complete description of the space of deformations of T_X. The first order deformations of T_X are given by the vector space

$$\operatorname{Ext}^1(T_X, T_X) = H^1(X, \operatorname{End}(T)). \tag{9}$$

We may follow [**9**] and compute this as follows. The sheaf $\operatorname{End}(T)$ can be written as the cohomology of (7) tensored with its dual. That is,

$$\bigoplus_a \mathscr{O}_X(-\mathbf{Q}_a)^{\oplus r} \longrightarrow \begin{matrix} \bigoplus_{i,a} \mathscr{O}_X(\mathbf{q}_i - \mathbf{Q}_a) \\ \oplus \\ \bigoplus_i \mathscr{O}_X(-\mathbf{q}_i)^{\oplus r} \end{matrix} \longrightarrow \underline{\begin{matrix} \mathscr{O}_X^{\oplus r^2} \\ \oplus \\ \bigoplus_{i,j} \mathscr{O}_X(\mathbf{q}_i - \mathbf{q}_j) \\ \oplus \\ \mathscr{O}_X \end{matrix}} \longrightarrow \begin{matrix} \bigoplus_i \mathscr{O}_X(\mathbf{q}_i)^{\oplus r} \\ \oplus \\ \bigoplus_{i,a} \mathscr{O}_X(\mathbf{Q}_a - \mathbf{q}_i) \end{matrix} \longrightarrow \bigoplus_a \mathscr{O}_X(\mathbf{Q}_a)^{\oplus r} \tag{10}$$

The cohomology is at the term we have underlined. We will consider the underlined term position "zero" in the complex. If a sheaf is presented as the cohomology (at position zero) of a complex

$$\cdots \longrightarrow \mathscr{E}^{-1} \longrightarrow \mathscr{E}^{0} \longrightarrow \mathscr{E}^{1} \longrightarrow \cdots, \tag{11}$$

then there is a spectral sequence [**9**] converging to the cohomology of the sheaf whose E_1 term is given by

$$E_1^{p,q} = H^q(X, \mathscr{E}^p). \tag{12}$$

As explained in [**9**], there is a strong resemblance between row 0 (i.e., $q = 0$) and the first order deformations of the linear σ-model (8). The space $H^0(V, \mathscr{O}_V(\mathbf{q}))$ is the space of global sections of $\mathscr{O}_V(\mathbf{q})$ and its dimension is counted by the number of monomials in R with multi-degree $\mathbf{q}$. We can count the deformations of (8) by considering all possible matrices of polynomials E and J such that $J.E = 0$ and then subtracting the number of reparametrizations induced by changes of homogeneous coordinates. This amounts to computing the cohomology in the middle term of the sequence:

$$\begin{matrix} H^0(\mathscr{O}_V^{\oplus r^2}) \\ \oplus \\ \bigoplus_{i,j} H^0(\mathscr{O}_V(\mathbf{q}_i - \mathbf{q}_j)) \\ \oplus \\ H^0(\mathscr{O}_V) \end{matrix} \longrightarrow \begin{matrix} \bigoplus_i H^0(\mathscr{O}_V(\mathbf{q}_i)^{\oplus r}) \\ \oplus \\ \bigoplus_i H^0(\mathscr{O}_V(\mathbf{Q} - \mathbf{q}_i)) \end{matrix} \longrightarrow H^0(\mathscr{O}_V(\mathbf{Q})^{\oplus r}). \tag{13}$$

Comparing this to the zeroth cohomology of (10) we see two obvious differences:

(1) The computation is on V rather than X and
(2) The left two terms of (10) are missing.

In simple cases these differences have no effect and the zeroth row of the spectral sequence accurately represents the deformations as seen by the linear σ-model.

For the quintic threefold in $\mathbb{P}^4$ this is the complete story. There is no contribution to $h^1(\mathrm{End}(T))$ from any rows other than the zeroth row. Thus all deformations of T are understood in terms of the linear σ-model. They are unobstructed and cannot be spoiled by worldsheet instantons.

3. The Octic

Let us summarize the results of [**9**] where X is the resolved octic hypersurface in $\mathbb{P}^4_{\{2,2,2,1,1\}}$.

- There are 179 deformations of T that are seen by the gauged linear σ-model and are given by the bottom row of the spectral sequence.
- The total count of $h^1(\mathrm{End}(T))$ depends on the complex structure. Generically it is 188 while for the Fermat hypersurface it is 200. All even values between these extremes can be obtained by choosing suitable octic defining equations.
- For generic values of the map E the number of deformations is 188. Thus some, if not all, of the extra 12 deformations associated to the Fermat complex structure are obstructed.

This jumping of the value of $h^1(\mathrm{End}(T))$ shows that we have a nontrivial superpotential for the singlets.

At the Gepner point we have 206 singlets associated with the bundle data. (That is, we have 206 singlets aside from deformations of complex structure, deformations of Kähler form and partners of the $\mathrm{U}(1)^4$ gauge symmetry.) By deforming the superpotential of the *mirror* Landau–Ginzburg theory we lose 6 of these 206. Thus 6 singlets are an artifact of being at the Gepner radius.

On deforming the superpotential of the Landau–Ginzburg theory we can lose up to 12 more singlets. In fact we get *perfect* agreement between the Landau–Ginzburg theory and the geometrical result. For any defining equation, i.e., superpotential, the number of computed singlets associated to $h^1(\mathrm{End}(T))$ is between 188 and 200 and the Landau–Ginzburg matches the geometry.[2]

One should be able to, in principle, compute the precise obstruction theory for the bundle and compute all correlation functions between the singlets in the Landau–Ginzburg model. This would allow a precise comparison of the superpotential between geometry and the exact result. This, in turn, would show if there were any instanton corrections to the superpotential. We have not done this but the agreement above does show that there are no corrections that would affect the *masses* of the singlets.

This shows that there is some "miracle" that kills instanton corrections to these $(0,2)$-models that goes beyond [**7, 8**]. Indeed we have checked many examples of Calabi–Yau *hypersurfaces* in toric varieties and we always find agreement of singlet counting between classical and exact methods.

It is natural to ask, therefore, whether this unnatural agreement persists for all $(0,2)$-models with the standard embedding. We will see that it is not the case.

4. The Z-Manifold

4.1. Geometry. Let T be the 2-torus given by $z \in \mathbb{C}$ under the identification $z \mapsto z+1$ and $z \mapsto z+\omega$, where $\omega = \exp(2\pi i/3)$. Let Z_0 be the orbifold $(T \times T \times T)/\mathbb{Z}_3$ where the $\mathbb{Z}_3$ is generated by the action

$$g : (z_0, z_1, z_2) \mapsto (\omega z_0, \omega z_1, \omega z_2). \tag{14}$$

[2]Note that this Landau–Ginzburg computation is stuck at small radius while the classical computation is stuck at large radius. It is just conceivable that this perfect agreement at large and small radius is spoiled at intermediate radii. We will assume this is not the case.

This is the well-known Z-orbifold introduced in [**10**] in which the associated conformal field theory was written in terms of free fields and the spectrum computed exactly.

The $\mathbb{Z}_3$ action has 27 fixed points yielding 27 singularities in Z_0. These can be resolved by blowing up each point with a $\mathbb{P}^2$ exceptional set to yield the Z-manifold.

It is very easy to compute the Hodge numbers of Z by using homology. Each blow-up introduces a 4-cycle in terms of the exceptional $\mathbb{P}^2$. Adding this 27 to the 9 invariant cycles from the covering 6-torus yields $b_2 = h^{1,1} = 36$. Similarly one can argue that $h^{2,1} = 0$ and the Z-manifold is rigid.

It is considerably harder to compute $h^1(\mathrm{End}(T))$. String theory implies there is a way to add local contributions of the blow-ups to some global contribution of the 6-torus but we do not know how to *rigorously* formulate this geometrically.[3] Instead we use another construction of the Z-manifold from which we do know how to extract $h^1(\mathrm{End}(T))$.

Since toric varieties offer a tractable path, we embed Z into a toric variety. Unfortunately, it cannot be embedded as a hypersurface since all such hypersurfaces have a mirror [**15**]. We can, however write it as the complete intersection of 3 equations.

The computation of $h^1(\mathrm{End}(T))$ for the Z-manifold is very lengthy and we relegate the details to an appendix. The result is that

$$h^1(\mathrm{End}(T)) = 208. \tag{15}$$

It turns out that, of these 208, only 6 come from the bottom row of the spectral sequence and are thus protected by Beasley–Witten from instanton corrections.

4.2. The Landau–Ginzburg Picture. At a point in the $(2,2)$ moduli space, the superconformal nonlinear sigma model on the Z-manifold is equivalent to a Gepner model [**16**] . This is most directly seen by recalling that the sigma model on an elliptic curve, at precisely the complex structure exhibiting a $\mathbb{Z}_3$ symmetry mentioned in the previous section, is equivalent at one point in its Kähler moduli space to a Gepner model, a $\mathbb{Z}_3$ quotient of the product $A_1^{\otimes 3}$. The construction of the Z as an orbifold then shows that at a point in the nine-dimensional moduli space of the orbifold theory (before blowing up) the model is equivalent to a $(\mathbb{Z}_3)^4$ quotient of the product of superconformal minimal models $A_1^{\otimes 9}$.

In this form, it is straightforward to count the 270 massless E_6 singlet fields at this point in the moduli space [**17**]. Of these, of course, 36 are the $(2,2)$ moduli, but this still leaves 234 singlets, more than the 208 found at large radius above. One part of the discrepancy is clear. At the Gepner point the theory exhibits a $\mathrm{U}(1)^6 \times \mathrm{SU}(3)$ gauge symmetry, which is broken by generic (Kähler) deformations. This leads to D-term masses (via the Higgs mechanism) for 14 of the singlet fields. Masses for any of the remaining 220 fields are generated by the spacetime superpotential.

In the Gepner model mentioned (and the associated Landau–Ginzburg model) the $(2,2)$ moduli are all twisted fields under the orbifold projection, making it difficult to study the model away from this one point. To get around this we study instead the mirror of the Z-manifold. At a point in its moduli space, the superconformal theory on this is equivalent to a quotient of the same Gepner model [**18**]. The construction leads to a $\mathbb{Z}_3^5$ quotient of the product of minimal models.

[3] An ALE space must be deformed before it is glued in and the torus metric must be deformed away from being flat. Such deformations may *a priori* affect $h^1(\mathrm{End}(T))$.

It is simpler, in this case, to note that since an elliptic curve is its own mirror, up to a relabeling of the fields this is equivalent to a $\mathbb{Z}_3^2$ quotient. As expected, the mirror model has a 36-dimensional space of $(2,2)$ deformations. Of these, 30 are untwisted under the quotient and can be represented in the associated Landau–Ginzburg model as deformations of the worldsheet superpotential. The methods of [**19**] then enable a computation of the number of massless singlet fields at any point in this 30-dimensional subspace of the full moduli space. The orbifold locus intersects this subspace along a three-dimensional subspace.

The Landau–Ginzburg model contains nine $(2,2)$ chiral superfields which we denote X^i, Y^i, Z^i, $i = 1\ldots 3$, interacting via a cubic superpotential. The model has a U(1) R-symmetry under which the fields all have charge 1/3. In the infrared this model flows to a $(2,2)$ superconformal field theory with a $\mathrm{U}(1)\times\mathrm{U}(1)$ R-symmetry. The Gepner model (after GSO projection) is a $\mathbb{Z}_6$ orbifold of this, and our mirror model is a further quotient by a $\mathbb{Z}_3$ generated by $(X^i, Y^j, Z^k) \mapsto (\omega X^i, \omega^2 Y^j, Z^k)$. The most general invariant superpotential can be written as

$$W = A_{ijk}X^iX^jX^k + B_{ijk}Y^iY^jY^k + C_{ijk}Z^iZ^jZ^k + D_{ijk}X^iY^jZ^k\ . \tag{16}$$

The 57 parameters in this superpotential are subject to an action of $\mathrm{GL}(3)^3$, which allows us to bring it to the form

$$\begin{aligned} W = \sum (X^i)^3 - 3A_x X^1X^2X^3 + \sum (Y^i)^3 - 3A_y Y^1Y^2Y^3 \\ + \sum (Z^i)^3 - 3A_z Z^1Z^2Z^3 + D_{ijk}X^iY^jZ^k\ . \end{aligned} \tag{17}$$

This is our explicit representation of the 30-dimensional family. The orbifold locus corresponds to $D_{ijk} = 0$.

We can consider $(0,2)$ deformations of this as a Landau–Ginzburg model; this is a degenerate version ($E = 0$) of the gauged linear sigma model counting (8) and we find 82 such deformations, providing a lower bound on the number of massless singlets at any point in the moduli space.

Supersymmetric ground states are found using the left-moving $N = 2$ superconformal algebra in the cohomology of the right-moving supercharge $\overline{Q}$ and classified by their charges $(q,\bar{q})$ under the U(1) symmetry contained in this algebra and under the right-moving U(1) symmetry inherited from the $(2,2)$ structure. States are described as excitations by the lowest oscillator modes of free bosonic fields ϕ_a^i (labeled by $a = x, y, z$ and $i = 1, 2, 3$) with charges $(1/3, 1/3)$ and left-moving fermionic fields γ_a^i with charges $(-2/3, 1/3)$.

The generator of the Gepner quotient acts as $e^{\pi i q}$ while the additional $\mathbb{Z}_3$ quotient acts as $\phi_a^i \to \omega^{w_a}\phi_a^i$; $\gamma_a^i \to \omega^{w_a}\gamma_a^i$ with $w = (1, -1, 0)$. Twisted vacua are labeled by $(k; l)$ with $k = 0, \ldots, 5$ and $l = 0, 1, 2$. It should be emphasized that this orbifold construction of the mirror of the Z-manifold is distinct from the orbifold construction of the Z-orbifold itself as in [**10**]. In particular twisted states in one orbifold need not correspond to twisted states in the other. In the $(k; l)$ sector the fields have a twisted moding

$$\begin{aligned} \phi_a^i(z) &= \sum_{s\in\mathbb{Z}-\nu_a} x_{a(s)}^i z^{-s-1/6}, & \gamma_a^i(z) &= \sum_{s\in\mathbb{Z}-\widetilde{\nu}_a} \gamma_{a(s)}^i z^{-s-2/3} \\ 2\partial\overline{\phi}_{ai}(z) &= \sum_{s\in\mathbb{Z}+\nu_a} \rho_{ai(s)} z^{-s-5/6}, & \overline{\gamma}_{ai}(z) &= \sum_{s\in\mathbb{Z}+\widetilde{\nu}_a} \overline{\gamma}_{ai(s)} z^{-s-5/6}, \end{aligned} \tag{18}$$

where

$$\nu_a(k;l) = \frac{k}{6} + \frac{lw_a}{3} \pmod 1 \qquad 0 \leq \nu_a < 1$$

$$\widetilde{\nu}_a(k;l) = \frac{-k}{3} + \frac{lw_a}{3} \pmod 1 \qquad -1 < \tilde{\nu}_a \leq 0 \tag{19}$$

The ground state energy and charges of the twisted vacua are

$$\begin{aligned} E(k;l) &= -\frac{5}{8} + \frac{3}{2}\sum_a \Big(\nu_a(1-\nu_a) + \widetilde{\nu}_a(1+\widetilde{\nu}_a)\Big) \quad k \text{ odd} \\ q(k;l) &= -\sum_a \left(2\widetilde{\nu}_a + \nu_a + \tfrac{1}{2}\right) \\ \overline{q}(k;l) &= \sum_a \left(\widetilde{\nu}_a + 2\nu_a - \tfrac{1}{2}\right) . \end{aligned} \tag{20}$$

For k even $E(k;l) = 0$.

Massless fermions arise in R (odd k) sectors and correspond to excitations with $E = 0$. E_6 singlets are characterized by $q = 0$. The right-moving charge $\overline{q}$ of a state determines the spacetime multiplet to which the fermion belongs: $\overline{q} = \pm\frac{3}{2}$ are fermions in vector multiplets while $\overline{q} = \pm\frac{1}{2}$ are fermions in chiral multiplets. We construct states with $E = q = 0$ by acting on $|k;l\rangle$ with the lowest excited modes, which we denote

$$x^i_a \equiv x^i_{a(-\nu_a)}, \quad \rho_{ai} \equiv \rho_{ai(\nu_a-1)}, \quad \gamma^i_a \equiv \gamma^i_{a(-1-\widetilde{\nu}_a)}, \quad \overline{\gamma}_{ai} \equiv \overline{\gamma}_{ai(\widetilde{\nu}_a)}. \tag{21}$$

In describing the $\overline{Q}$ cohomology we will also need the conjugate modes

$$x^\dagger{}_{ai} \equiv \rho_{ai(\nu_a)}, \quad \rho^{\dagger i}{}_a \equiv x^i_{a(1-\nu_a)}, \quad \gamma^\dagger{}_{ai} \equiv \overline{\gamma}_{ai(1+\widetilde{\nu}_a)}, \quad \overline{\gamma}^{\dagger i}{}_a \equiv \gamma^i_{a(-\widetilde{\nu}_a)}. \tag{22}$$

Acting on these states, the $\overline{Q}$ operator takes the general form

$$\overline{Q} = \sum_{a,i} \left\{\gamma^i_a \partial_{ai} W_{1+\widetilde{\nu}_i} + \overline{\gamma}^\dagger{}_{ai}\partial_{ai}W_{\widetilde{\nu}_i}\right\}. \tag{23}$$

The quotient breaks the S_9 permutation symmetry of the product to a subgroup. Of interest to us is an unbroken S_3 subgroup permuting the a indices of the fields. This relabels the generators of the quotient group and correspondingly permutes the twisted sectors.

4.2.1. *Untwisted States.* In the untwisted R sector $(k;l) = (1;0)$ we have $\nu_a = \frac{1}{6}$, $\widetilde{\nu}_a = -\frac{1}{3}$ so that the ground state has $E(1;0) = -1$, $q(1;0) = 0$, $\overline{q}(1;0) = -\frac{3}{2}$. The complex of $E = q = 0$ states upon which $\overline{Q}$ acts is

$$\begin{array}{ccc} \bar{q} = -\frac{3}{2} & & \bar{q} = -\frac{1}{2} \\ & & \\ \phi^i_a \rho_{aj}|1;0\rangle_{27} & & \phi^i_a\phi^j_a\gamma^k_a|1;0\rangle_{54} \\ \oplus & \xrightarrow{\overline{Q}} & \oplus \\ \overline{\gamma}_{ai}\gamma^j_a|1;0\rangle_{27} & & \phi^i_a\phi^j_b\gamma^k_c|1;0\rangle_{81} \qquad (a \neq b \neq c) \end{array} \tag{24}$$

where subscripts on kets indicate the dimension of the space of states of a given form; repeated indices on fields are not summed. We write $\overline{Q}$ as a sum of three terms

$$\overline{Q} = \overline{Q}_G + \overline{Q}_O + \overline{Q}_D . \tag{25}$$

$\overline{Q}_G$, the supercharge at the Gepner point, is in this sector given by

$$\overline{Q}_G = 6\sum_{a,i} \gamma^i_a x^i_a \rho^{\dagger i}_{\ a} + 3\overline{\gamma}^{\dagger i}_{\ a}\left(x^i_a\right)^2 . \tag{26}$$

This has a nine-dimensional kernel spanned by

$$J^i_a = \tfrac{1}{3}\left(x^i_a\rho_{ai} - 2\overline{\gamma}_{ai}\gamma^i_a\right)|1;0\rangle, \tag{27}$$

indicating that the enhanced gauge group at the Gepner point has rank eight (one generator is the $\mathrm{U}(1) \subset E_6$), and that the untwisted sector gives rise to 90 chiral singlets. $\overline{Q}_O$, the additional charge at a generic point on the orbifold locus $D = 0$ is in this sector given by

$$\overline{Q}_O = -3\sum_a A_a \sum_{i\neq j\neq k}\left[\gamma^i_a(x^j_a\rho^{\dagger k}_{\ a} + x^k_a\rho^{\dagger j}_{\ a}) + \overline{\gamma}^{\dagger i}_{\ a}x^j_a x^k_a\right] . \tag{28}$$

Adding this to $\overline{Q}_G$ with generic A_a reduces the dimension of the kernel to three, spanned by $J_a = \sum_i J^i_a$, indicating an enhanced gauge group of rank two, and a total of 84 chiral singlets. Adding

$$\begin{aligned}\overline{Q}_D = \sum_{ijk} D_{ijk}\big[&\gamma^i_x(y^j\rho^{\dagger k}_{\ z} + z^k\rho^{\dagger j}_{\ y}) + \overline{\gamma}^{\dagger i}_{\ x}y^j z^k \\ &+ \gamma^j_y(x^i\rho^{\dagger k}_{\ z} + z^k\rho^{\dagger i}_{\ x}) + \overline{\gamma}^{\dagger j}_{\ y}x^i z^k + \gamma^k_z(y^j\rho^{\dagger i}_{\ x} + x^i\rho^{\dagger j}_{\ y}) + \overline{\gamma}^{\dagger k}_{\ z}x^i y^j\big]\end{aligned} \tag{29}$$

leaves a one-dimensional kernel (guaranteed by the quasihomogeneity of W) generated by $J = \sum_a J_a$, indicating the gauge symmetry is reduced to E_6 and leaving 82 neutral chiral multiplets.

4.2.2. *Twisted States.* Since the S_3 symmetry permutes the twisted sectors, a calculation in any one is sufficient to produce the entire spectrum. In the $(k;l) = (1;1)$ sector we have $\nu(1;1) = (\frac{1}{2},\frac{5}{6},\frac{1}{6})$, $\widetilde{\nu}(1;1) = (0,-\frac{2}{3},-\frac{1}{3})$ and hence $E(1;1) = -\frac{1}{2}$, $q(1;1) = -1$, $\overline{q}(1;1) = \frac{1}{2}$. The complex of $E = q = 0$ states is

$$\begin{array}{ccccccc} \bar{q} = -\frac{3}{2} & & \bar{q} = -\frac{1}{2} & & \bar{q} = \frac{1}{2} & & \bar{q} = \frac{3}{2} \\ & & \rho_x\overline{\gamma}^2_x|1;1\rangle_9 & & x\overline{\gamma}_x|1;1\rangle_9 & & \\ & & \oplus & & \oplus & & \\ & & z\rho^2_y\overline{\gamma}^2_x|1;1\rangle_{54} & & z^2\rho_y\overline{\gamma}_x|1;1\rangle_{54} & & \\ \rho^3_y\overline{\gamma}^3_x|1;1\rangle_{10} & \xrightarrow{\overline{Q}} & \oplus & \xrightarrow{\overline{Q}} & \oplus & \xrightarrow{\overline{Q}} & z^3|1;1\rangle_{10} \\ & & \rho_y\overline{\gamma}_x\overline{\gamma}_z|1;1\rangle_{27} & & z\overline{\gamma}_z|1;1\rangle_9 & & \\ & & \oplus & & \oplus & & \\ & & \rho_y\overline{\gamma}^3_x\gamma_y|1;1\rangle_9 & & z\overline{\gamma}^2_x\gamma_y|1;1\rangle_{27} & & \end{array} \tag{30}$$

In this sector we have

$$\begin{aligned}\overline{Q}_G = 3\sum_i \Big[&\gamma^i_x\left(\rho^\dagger_{\ xi}\right)^2 + 2\overline{\gamma}^{\dagger i}_{\ x}x^i\rho^{\dagger i}_{\ x} + \gamma^i_y\left(\rho^\dagger_{\ yi}\right)^2 + 2\overline{\gamma}^{\dagger i}_{\ y}y^i\rho^{\dagger i}_{\ y} \\ &+2\gamma^i_z z^i\rho^\dagger_{\ zi} + \overline{\gamma}^{\dagger i}_{\ y}\left(z^i\right)^2\Big] .\end{aligned} \tag{31}$$

Acting on the $\overline{q} = \frac{3}{2}$ space this has a one-dimensional kernel, which is generated by $\rho_{y1}\rho_{y2}\rho_{y3}\overline{\gamma}^3_x|1;1\rangle$; the six vector multiplets arising from twisted sectors fill out the enhanced gauge group $\mathrm{U}(1)^6 \times \mathrm{SU}(3)$ with the Cartan subgroup generated by the charges associated to $J_a - \frac{1}{3}J$.

Acting on the $\overline{q} = \frac{1}{2}$ space we find a 39-dimensional kernel, so that each twisted sector contributes 30 massless E_6 singlet chiral multiplets; when added to the 90 we found in the untwisted sector this reproduces the expected 270 singlets at the Gepner point.

The analysis is made easier by organizing the zero-energy states into SU(3) multiplets and recalling that $\overline{Q}_G$ as well as $\overline{Q}_O$ are SU(3) invariant and that $\overline{Q}$ acts within a given twist sector. We find

(32)
$$\begin{array}{ccccccc}
\bar{q} = -\frac{3}{2} & & \bar{q} = -\frac{1}{2} & & \bar{q} = \frac{1}{2} & & \bar{q} = \frac{3}{2} \\[2ex]
\mathbf{1}^{34} & & \mathbf{1}^{36} \oplus \mathbf{3}^{27} & & \mathbf{1}^{36} \oplus \overline{\mathbf{3}}^{27} & & \mathbf{1}^{34} \\
\oplus & & \oplus & & \oplus & & \oplus \\
\mathbf{8}^{10} & \xrightarrow{\overline{Q}} & \mathbf{8}^{9} & \xrightarrow{\overline{Q}} & \mathbf{8}^{9} & \xrightarrow{\overline{Q}} & \mathbf{8}^{10} \\
 & & \oplus & & \oplus & & \\
 & & \mathbf{1}^{54} \oplus \mathbf{3}^{108} \oplus \overline{\mathbf{3}}^{54} & & \mathbf{1}^{54} \oplus \mathbf{3}^{54} \oplus \overline{\mathbf{3}}^{108} & &
\end{array}$$

where the first row represents the contribution of the untwisted sectors and the last the contributions of the twisted sectors. Adjoints, in the second row, are counted separately as they are assembled from twisted and untwisted states. In the untwisted sector we computed above that $\overline{Q}_G$ has rank 27 when acting on the space of SU(3) singlets. Acting on the adjoints it has the maximum rank possible. Since untwisted states with $\overline{q} > 0$ arise in the conjugate $(5;0)$ sector the map $\overline{q} = -\frac{1}{2} \to \overline{q} = \frac{1}{2}$ vanishes for untwisted states. In the twisted sector, acting on the SU(3) singlet states $\overline{Q}_G$ has rank 36. In the $(1;1)$ sector the three-dimensional kernel at $\overline{q} = -\frac{1}{2}$ is spanned by $\rho_{xi}\overline{\gamma}_{xj}\overline{\gamma}_{xk}|1;1\rangle$ with $i \neq j \neq k$. The rank on the SU(3) charged fields is maximal and we find that the physical states are vector multiplets in the representation

(33)
$$\mathbf{1}^7 \oplus \mathbf{8},$$

of which one of the singlets is a Cartan generator of E_6, and chiral multiplets in the representation

(34)
$$\underbrace{\mathbf{1}^9 \oplus \mathbf{3}^{27}}_{\text{untwisted}} \oplus \underbrace{\mathbf{1}^{18} \oplus \mathbf{3}^{54}}_{\text{twisted}}.$$

The (2,2) moduli are $\mathbf{1}^3 \oplus \mathbf{3}^9 \oplus \mathbf{1}^6$ as expected.

Now we move away from the Gepner point but remain mirror to the *orbifold* by adding $\overline{Q}_O$. In the untwisted sector, as noted above, the rank of the map between SU(3) singlet states increases from 27 to 33. This is the Higgs mechanism breaking the enhanced gauge symmetry $\mathrm{SU}(3) \times \mathrm{U}(1)^6$ to SU(3) by removing a vector multiplet and a chiral multiplet for each of the six broken generators of the gauge group. In the twisted sector we find that the rank of the map on singlets increases from 36 to 48. In the $(1;1)$ sector the one-dimensional kernel at $\overline{q} = \frac{1}{2}$ is now spanned by $\sum_{ijk} \epsilon^{ijk} \rho_{xi}\overline{\gamma}_{xj}\overline{\gamma}_{xk}|1;1\rangle$. The cohomology thus contains 9 vector multiplets in the representation $\mathbf{1}\oplus\mathbf{8}$ and 252 chiral multiplets in the representation

(35)
$$\underbrace{\mathbf{1}^3 \oplus \mathbf{3}^{27}}_{\text{untwisted}} \oplus \underbrace{\mathbf{1}^{6} \oplus \mathbf{3}^{54}}_{\text{twisted}},$$

as was found in [**10**].

Finally we blow-up the orbifold in the mirror by adding $\overline{Q}_D$. Now the enhanced gauge symmetry is completely broken. In the untwisted sector we saw that the rank of $\overline{Q}$ grows by two. This is again the Higgs mechanism for the Cartan elements of the SU(3) gauge symmetry, and it is accompanied by a corresponding change in rank by one in the map $\overline{q} = -\frac{3}{2} \to \overline{q} = -\frac{1}{2}$ in each twisted sector. In addition, we find that the rank of the map $\overline{q} = -\frac{1}{2} \to \overline{q} = \frac{1}{2}$ increases from 62 to 80, indicating that 108 chiral multiplets are lifted by spacetime superpotential interactions (F-terms). This leaves a total of 136 E_6 neutral chiral multiplets at generic points in the moduli space.

Comparing these numbers to the results in the previous section we find that a worldsheet instanton generated spacetime superpotential leads to mass terms for 108 of the perturbatively massless scalars found by the large-radius analysis.

5. Summary and Discussion

It is surprising that $h^1(\mathrm{End}(T))$ is so resilient to instanton corrections. It would seem that there are mechanisms beyond those of [**7**, **8**] that protect these states from acquiring masses away from the large radius limit. From the examples we have considered, it is tempting to conjecture that a good candidate for a class of models in which there are no corrections is that of hypersurfaces in toric varieties. Typically these have contributions to $h^1(\mathrm{End}(T))$ above the zeroth row of the spectral sequence and, in all cases we have considered, their masses are not affected by instantons.

The fact that there seems to be a large class of models without corrections should lead to some interesting mathematics. The agreement between the cohomology of the complexes of the Landau–Ginzburg picture and the spectral sequence of the σ-model is not yet understood. One may also be able to make curious statements such as $h^1(\mathrm{End}(T))$ being equal for mirror pairs of Calabi–Yau threefolds.

On the other hand, we now have an example where instantons *do* give mass to $H^1(\mathrm{End}(T))$ modes. An interesting next step will be to enumerate these instantons more precisely from the conformal field theory computation.

In the case of the Z-manifold, we find that of the 252 massless E_6 singlet fields in the Z-orbifold spectrum, 8 acquire D-term masses via the Higgs mechanism and 108 acquire F-term masses via a spacetime superpotential as we blow up the quotient singularities. Removing the 36 $(2,2)$ moduli, this leaves precisely 100 massless singlets corresponding to bundle deformations at a generic point in the (30-dimensional) moduli space. Comparing this to the geometric computation in the large-radius limit we find again that of the 208 singlet fields counted by $h^1(\mathrm{End}(T))$ 108 acquire masses via instanton-induced superpotential terms. It is natural to conjecture that one can identify the full 208-dimensional space of massless states at the orbifold locus with $H^1(\mathrm{End}(T))$ and that the two sets of F-terms are associated to the same superpotential.

This counting argument suggests that even in this case where instanton contributions are nonzero there is a McKay-like correspondence for local contributions to $h^1(\mathrm{End}(T))$ from an orbifold resolution. This would imply that there is some local picture for noncompact Calabi–Yau threefolds where instantons can be shown to cancel. Clearly further investigation of this would be interesting.

Acknowledgments

We thank I. Melnikov and E. Miller for useful discussions. This work was partially supported by NSF grants DMS–0606578 and DMS–0905923. Any opinions, findings, and conclusions or recommendations expressed in this material are those of the authors and do not necessarily reflect the views of the National Science Foundation.

Appendix A. $h^1(\mathrm{End}(T))$ for the Z-manifold

Z is the resolution of $T^6/\mathbb{Z}_3$ at the 27 fixed points of the orbifold action. We may write the 2-torus as a cubic in $\mathbb{P}^2$. Hence Z can be written as a complete intersection of three "cubics" in V where V is the crepant desingularization of

$$V' = \frac{\mathbb{P}^2 \times \mathbb{P}^2 \times \mathbb{P}^2}{\mathbb{Z}_3}. \tag{36}$$

V is a toric variety as follows. The 1-dimensional rays of the associated fan are given by the rows of the matrix

$$\begin{pmatrix} 1 & 1 & -2 & 1 & 1 & 0 \\ -1 & -1 & 2 & -1 & -1 & -1 \\ 0 & 0 & 0 & 0 & 0 & 1 \\ 0 & 0 & 0 & -1 & -1 & 0 \\ 0 & 0 & 0 & 1 & 0 & 0 \\ 0 & 0 & 0 & 0 & 1 & 0 \\ -1 & -1 & -1 & 0 & 0 & 0 \\ 1 & 0 & 1 & 0 & 0 & 0 \\ 0 & 1 & 0 & 0 & 0 & 0 \\ 0 & 0 & -1 & 0 & 0 & 0 \end{pmatrix} \tag{37}$$

The homogeneous coordinates $(x_0, \dots, x_9)$ are associated to these rays. The Stanley-Reisner ideal is given by

$$I = \langle x_0x_1x_2, x_3x_4x_5, x_0x_3x_6, x_6x_7x_8, x_1x_2x_9, x_4x_5x_9, x_7x_8x_9 \rangle. \tag{38}$$

Z is now the complete intersection in V associated to the ideal $\langle f^0, f^1, f^2 \rangle$ where

$$\begin{aligned} f^0 &= x_1^3 + x_2^3 + x_0^3 x_9 \\ f^1 &= x_4^3 + x_5^3 + x_3^3 x_9 \\ f^2 &= x_7^3 + x_8^3 + x_6^3 x_9. \end{aligned} \tag{39}$$

We have a representation of the sheaf $\mathrm{End}(T_Z)$ in terms of a complex of sums of line bundles on Z given by (10). It is more convenient for computational purposes if everything is written in terms of line bundles over the ambient toric variety V.

To see how to do this let us consider the much simpler case of the tangent bundle of the quintic 3-fold. This is given as the complex

$$\mathcal{O}_X \xrightarrow{x_i} \mathcal{O}_X(1)^{\oplus 5} \xrightarrow{\partial_i f} \mathcal{O}_X(5). \tag{40}$$

Let $\mathcal{O}$ denote the structure sheaf of the ambient $\mathbb{P}^4$. $\mathcal{O}_X$ is then equivalent, in the derived category, to

$$\mathcal{O}(-5) \xrightarrow{f} \mathcal{O}. \tag{41}$$

Thus, by mapping cones, the complex given by the last two terms of (40) is given by

$$\mathcal{O}(-4)^{\oplus 5} \xrightarrow{\begin{pmatrix} -fI \\ \partial_i f \end{pmatrix}} \begin{matrix} \mathcal{O}(1)^{\oplus 5} \\ \oplus \\ \mathcal{O} \end{matrix} \xrightarrow{(\partial_i f \;\; f)} \mathcal{O}(5). \tag{42}$$

The complete complex (40) is then given by the mapping cone of a map from the complex representing $\mathscr{O}_X$ to the complex (42). This morphism is given by the chain map

$$
\begin{array}{ccccc}
\mathscr{O}(-5) & \xrightarrow{\quad f \quad} & \mathscr{O} & & \\
\Big\downarrow{\scriptstyle x_i} & & \Big\downarrow{\scriptstyle \left(\begin{smallmatrix} -x_i \\ 5 \end{smallmatrix}\right)} & & \\
\mathscr{O}(-4)^{\oplus 5} & \xrightarrow{\left(\begin{smallmatrix} -fI \\ \partial_i f \end{smallmatrix}\right)} & \begin{matrix}\mathscr{O}(1)^{\oplus 5} \\ \oplus \\ \mathscr{O}\end{matrix} & \xrightarrow{(\,\partial_i f \;\; f\,)} & \mathscr{O}(5).
\end{array} \tag{43}
$$

Note, in particular, the need for the "5" to produce a chain map. The tangent sheaf of the quintic can thus be represented by the chain complex

$$
\mathscr{O}(-5) \xrightarrow{\left(\begin{smallmatrix} -f \\ x_i \end{smallmatrix}\right)} \begin{matrix}\mathscr{O} \\ \oplus \\ \mathscr{O}(-4)^{\oplus 5}\end{matrix} \xrightarrow{\left(\begin{smallmatrix} -x_i & -fI \\ 5 & \partial_i f \end{smallmatrix}\right)} \begin{matrix}\mathscr{O}(1)^{\oplus 5} \\ \oplus \\ \mathscr{O}\end{matrix} \xrightarrow{(\,\partial_i f \;\; f\,)} \mathscr{O}(5). \tag{44}
$$

We need to extend this construction to complete intersections. To this end we may prove the following theorem analogously to the above.

THEOREM 1. *Consider a commutative diagram of sheaves on V of the following form:*

$$
\begin{array}{ccccccccc}
\cdots & \longrightarrow & A^{n-1} & \xrightarrow{a_{n-1}} & A^n & \xrightarrow{a_n} & A^{n+1} & \longrightarrow & \cdots \\
 & & \Big\downarrow{\scriptstyle h_{n-1}} & & \Big\downarrow{\scriptstyle h_n} & & \Big\downarrow{\scriptstyle h_{n+1}} & & \\
\cdots & \longrightarrow & B^{n-1} & \xrightarrow{b_{n-1}} & B^n & \xrightarrow{b_n} & B^{n+1} & \longrightarrow & \cdots \\
 & & \Big\downarrow{\scriptstyle g_{n-1}} & & \Big\downarrow{\scriptstyle g_n} & & \Big\downarrow{\scriptstyle g_{n+1}} & & \\
\cdots & \longrightarrow & C^{n-1} & \xrightarrow{c_{n-1}} & C^n & \xrightarrow{c_n} & C^{n+1} & \longrightarrow & \cdots \\
 & & \Big\downarrow{\scriptstyle f_{n-1}} & & \Big\downarrow{\scriptstyle f_n} & & \Big\downarrow{\scriptstyle f_{n+1}} & & \\
\cdots & \longrightarrow & D^{n-1} & \xrightarrow{d_{n-1}} & D_n & \xrightarrow{d_n} & D^{n+1} & \longrightarrow & \cdots
\end{array} \tag{45}
$$

satisfying

(1) *Each column is exact except it has cohomology L^n supported on Z in the last position.*

(2) *the composition of two horizontal maps is zero when restricted to Z.*

The complex $L^\bullet$ is quasi-isomorphic to the complex with terms $A^{n+3} \oplus B^{n+2} \oplus C^{n+1} \oplus D^n$ and differentials

$$
\begin{pmatrix}
a_{n+3} & (-1)^n \alpha_{n+2} & \beta_{n+1} & (-1)^n \gamma_n \\
(-1)^n h_{n+3} & b_{n+2} & (-1)^n \delta_{n+1} & \epsilon_n \\
0 & (-1)^n g_{n+2} & c_{n+1} & (-1)^n \zeta_n \\
0 & 0 & (-1)^n f_{n+1} & d_n
\end{pmatrix} : \begin{pmatrix} A^{n+3} \\ B^{n+2} \\ C^{n+1} \\ D^n \end{pmatrix} \longrightarrow \begin{pmatrix} A^{n+4} \\ B^{n+3} \\ C^{n+2} \\ D^{n+1} \end{pmatrix} \tag{46}
$$

where

$$
\begin{aligned}
\zeta_n &= f_{n+2}^{-1} d_{n+1} d_n \\
\delta_n &= g_{n+2}^{-1}(c_{n+1}c_n - \zeta_n f_n) \\
\epsilon_n &= g_{n+3}^{-1}(c_{n+2}\zeta_n - \zeta_{n+1} d_n) \\
\alpha_n &= h_{n+2}^{-1}(b_{n+1}b_n - \delta_n g_n) \\
\beta_n &= h_{n+3}^{-1}(b_{n+2}\delta_n - \delta_{n+1}c_n + \epsilon_n f_n) \\
\gamma_n &= h_{n+4}^{-1}(b_{n+3}\epsilon_n - \delta_{n+2}\zeta_n + \epsilon_{n+1} d_n)
\end{aligned} \tag{47}
$$

Let the multi-degree of the equation f_i be denoted by $\mathbf{Q}_i$. The sheaf $\mathscr{O}_Z$ can be resolved in terms of the Koszul complex:

$$
\mathscr{O}(-\textstyle\sum_i \mathbf{Q}_i) \xrightarrow{h} \bigoplus_i \mathscr{O}(-\sum_{j\neq i}\mathbf{Q}_j) \xrightarrow{g} \bigoplus_i \mathscr{O}(-\mathbf{Q}_i) \xrightarrow{f} \mathscr{O} \longrightarrow \mathscr{O}_Z, \tag{48}
$$

where $\mathscr{O}$ is the structure sheaf of V.

Applying the theorem to this and the complex (10) yields a rather messy complex of line bundles on V representing $\operatorname{End}(T)$ for the Z-manifold. The explicit form is too large to give here and we will write it more concisely as

$$
\begin{aligned}
\mathscr{F}^{-5}_{(12)} \longrightarrow \mathscr{F}^{-4}_{(106)} \longrightarrow \mathscr{F}^{-3}_{(371)} \longrightarrow \mathscr{F}^{-2}_{(667)} \longrightarrow \mathscr{F}^{-1}_{(667)} \longrightarrow \\
\mathscr{F}^{0}_{(371)} \longrightarrow \mathscr{F}^{1}_{(106)} \longrightarrow \mathscr{F}^{2}_{(12)}
\end{aligned} \tag{49}
$$

where each term is a sum of line bundles on V where the rank is denoted by the subscript.

Now, to compute the cohomology we have a spectral sequence $E_1^{p,q} = H^q(\mathscr{F}^p)$. This yields

(50)

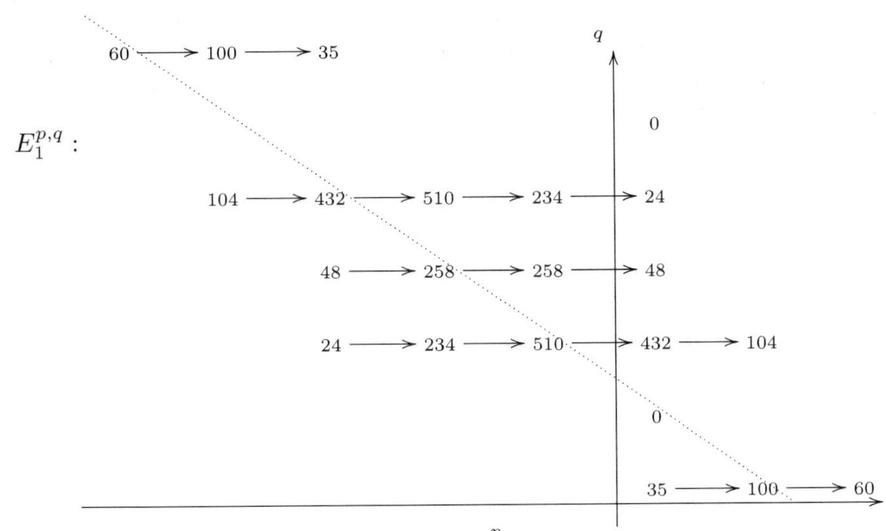

where each number represents the dimension of the space at each position. The dotted line shows the terms that will ultimately contribute to $h^1(\operatorname{End}(T))$. Now we need to compute the maps above and take cohomology to proceed to $E_2^{p,q}$.

To do this we follow the method given in [**9**]. We may regard the spectral sequence as having arisen from a double complex with the given complex maps in the horizontal direction and Čech cohomology in the vertical direction.

Let $B = (m_1, m_2, \ldots, m_l)$, where m_i are monomials, be the irrelevant ideal which is the Alexander dual [**20**] of the Stanley–Reisner ideal (38). An open cover of V that can be used to compute Čech cohomology is then given by the set of $U_i = V - Z(m_i)$. This is equivalent to a local cohomology computation as explained in [**21**]. This local cohomology approach to computing the cohomology of line bundles has also been explored in the context of string theory in [**22–24**]. We refer to the appendix of [**9**] for a review of the ideas required here.

An element of the underlying double complex is given by a Čech cochain which is a collection of Laurent monomials $\{c_{j_1,j_2,\ldots}\}$ each of which takes values in the localization $R_{m_{j_1},m_{j_2},\ldots}$. This structure simplifies a little once we go to the E_1 stage by taking vertical cohomology. Each cochain must be coclosed and so we require exact cancellations on certain multiple overlaps of open patches U_i. Suppose we fix one of the $c_{j_1,j_2,\ldots}$'s to be a fixed monomial of homogeneous coordinates. In straightforward cases, such a cancellation tends to require that the other $c_{j_1,j_2,\ldots}$'s are given (perhaps up to some fixed constant) by the exact same monomial. Thus we specify a given dimension of the space of cohomology by specifying a particular monomial. In other words, computing the dimension of the cohomology amounts to counting Laurent monomials of a certain form.

Whether this simple counting method works depends on the finely-graded Betti numbers $\operatorname{Tor}_i^R(I, \mathbb{C})_\alpha$ as in Corollary 3.1 of [**25**]. They are required to all be 0 or 1. While there are certainly counterexamples where these Betti numbers are bigger than one, such examples are combinatorially quite complicated and, fortunately, the Z-manifold does not fall into this class.

The upshot of all this is that we can represent entries in $E_1^{p,q}$ simply by Laurent monomials (rather than collections of monomials associated to intersections of patches). The combinatorics of counting such monomials has been discussed in [**21, 24**]. What is even nicer is that it easily follows that the horizontal maps in the $E_1^{p,q}$ stage of the spectral sequence are given by the obvious maps between monomials inherited from the original complex (49). An exercise in Macaulay 2 programming then yields:

(51) $E_2^{p,q}$:

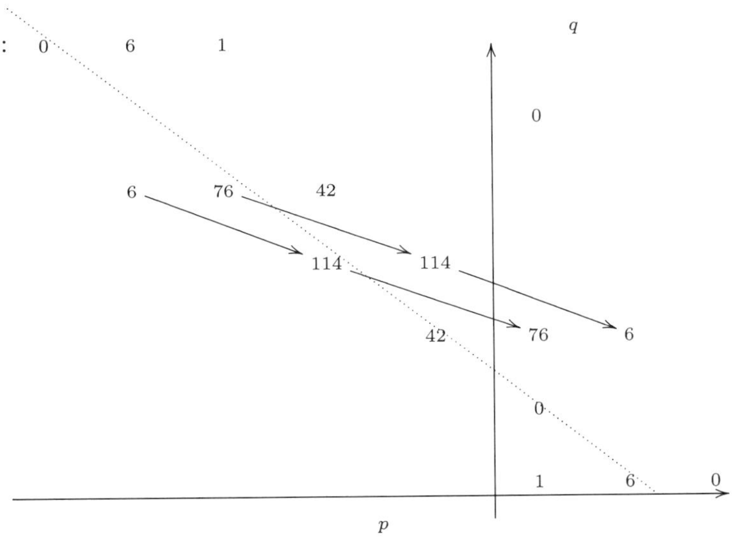

We now need to compute the d_2 maps shown in the above diagram. These are computed following the staircases of maps as described in [**26**]. We need to go right-down-right. An example of this was described completely explicitly in [**9**] and so we will be brief here. The composition of two right maps obviously gives zero in the underlying complex (49). However, the downwards map serves to scramble some of the signs in the map because of the sign choices in Čech cohomology. Thus, the d_2 maps need not be zero. Fortunately most of the entries in the matrices representing d_2 can be shown to be zero because of the large number of zeroes in the horizontal maps. The result is:

(52) $E_3^{p,q}$:

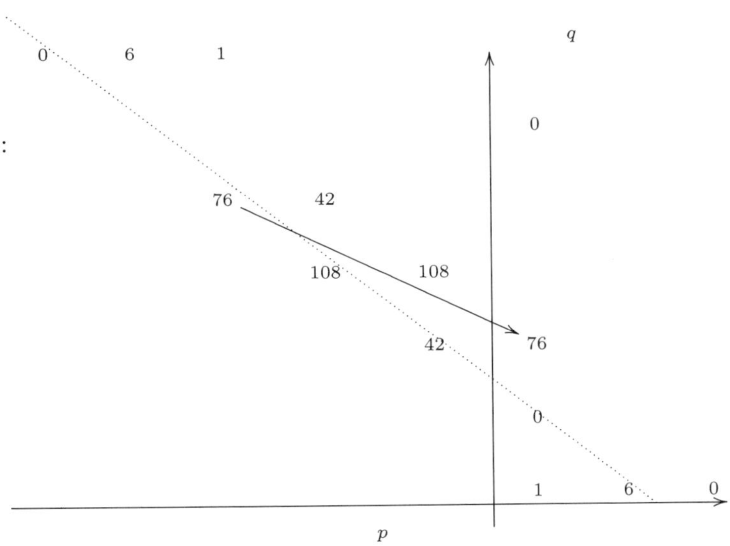

Mercilessly, the spectral sequence is not done with us yet and we need to compute one d_3 map. This can be tackled using the same method as we used for d_2. The result is:

(53)

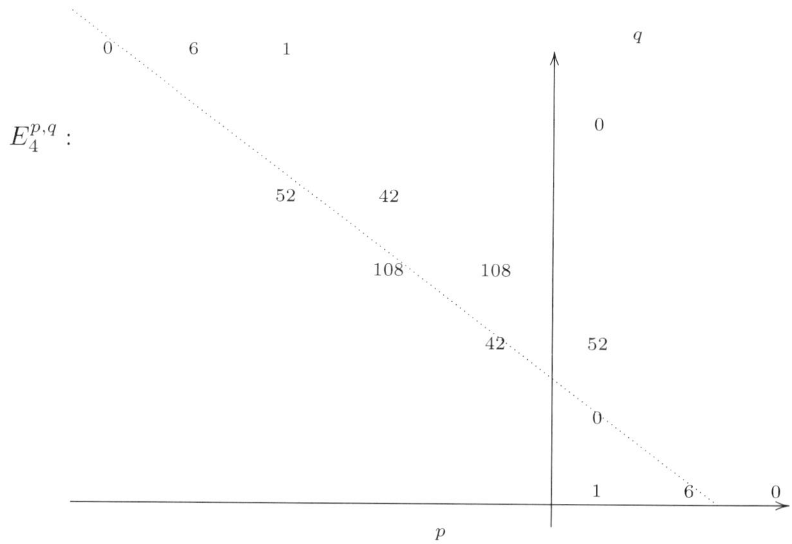

Finally the spectral sequence degenerates and so

$$h^1(\mathrm{End}(T)) = 208. \tag{54}$$

References

[1] M. Dine, N. Seiberg, X. G. Wen, and E. Witten, *Nonperturbative Effects on the String World-Sheet*, Nucl. Phys. **B278** (1986) 769–789, and Nucl. Phys. **B289** (1987) 319–363.

[2] J. Distler, *Resurrecting (2,0) Compactifications*, Phys. Lett. **188B** (1987) 431–436.

[3] J. Distler and B. R. Greene, *Aspects of (2,0) String Compactifications*, Nucl. Phys. **B304** (1988) 1–62.

[4] P. Berglund et al., *On the Instanton Contributions to the Masses and Couplings of E_6 Singlets*, Nucl. Phys. **B454** (1995) 127–163, arXiv:hep-th/9505164.

[5] E. Witten, *World Sheet Corrections via D Instantons*, JHEP **0002** (2000) 030, arXiv:hep-th/9907041.

[6] E. I. Buchbinder, R. Donagi, and B. A. Ovrut, *Superpotentials for Vector Bundle Moduli*, Nucl. Phys. **B653** (2003) 400–420, arXiv:hep-th/0205190.

[7] E. Silverstein and E. Witten, *Criteria for conformal invariance of (0,2) models*, Nucl. Phys. **B444** (1995) 161–190, arXiv:hep-th/9503212.

[8] C. Beasley and E. Witten, *Residues and world-sheet instantons*, JHEP **10** (2003) 065, arXiv:hep-th/0304115.

[9] P. S. Aspinwall, I. V. Melnikov, and M. R. Plesser, *(0,2) Elephants*, arXiv:1008.2156.

[10] L. Dixon, J. A. Harvey, C. Vafa, and E. Witten, *Strings on Orbifolds*, Nucl. Phys. **B261** (1985) 678–686, and **B274** (1986) 285–314.

[11] D. A. Cox, *The Homogeneous Coordinate Ring of a Toric Variety*, J. Algebraic Geom. **4** (1995) 17–50, alg-geom/9210008.

[12] V. V. Batyrev and D. A. Cox, *On the Hodge Structure of Projective Hypersurfaces in Toric Varieties*, Duke Math. J **75** (1994) 293–338, arXiv:alg-geom/9306011.

[13] E. Witten, *Phases of $N = 2$ Theories in Two Dimensions*, Nucl. Phys. **B403** (1993) 159–222, hep-th/9301042.

[14] M. Kreuzer, J. McOrist, I. V. Melnikov, and M. R. Plesser, *(0,2) Deformations of Linear Sigma Models*, arXiv:1001.2104.

[15] V. V. Batyrev, *Dual Polyhedra and Mirror Symmetry for Calabi–Yau Hypersurfaces in Toric Varieties*, J. Alg. Geom. **3** (1994) 493–535.

[16] D. Gepner, *Space-Time Supersymmetry in Compactified String Theory and Superconformal Models*, Nucl. Phys. **B296** (1988) 757.

[17] C. A. Lütken and G. G. Ross, *Taxonomy of Heterotic Superconformal Field Theories*, Phys. Lett. **B213** (1988) 152.

[18] B. R. Greene and M. R. Plesser, *Duality in Calabi–Yau Moduli Space*, Nucl. Phys. **B338** (1990) 15–37.

[19] S. Kachru and E. Witten, *Computing the complete massless spectrum of a Landau- Ginzburg orbifold*, Nucl. Phys. **B407** (1993) 637–666, arXiv:hep-th/9307038.

[20] E. Miller and B. Sturmfels, *Combinatorial Commutative Algebra*, Springer, 2005.

[21] D. Eisenbud, M. Mustaţă, and M. Stillman, *Cohomology on Toric Varieties and Local Cohomology with Monomial Supports*, J. Symbolic Comput. **29** (2000) 583–600, arXiv:math/0001159.

[22] C. P. Herzog and R. L. Karp, *Exceptional Collections and D-Branes Probing Toric Singularities*, JHEP **02** (2006) 061, arXiv:hep-th/0507175.

[23] P. S. Aspinwall, *D-Branes on Toric Calabi–Yau Varieties*, arXiv:0806.2612.

[24] R. Blumenhagen, B. Jurke, T. Rahn, and H. Roschy, *Cohomology of Line Bundles: A Computational Algorithm*, J. Math. Phys. **51** (2010) 103525, arXiv:1003.5217.

[25] M. Mustaţă, *Local Cohomology at Monomial Ideals*, J. Symbolic Comput. **29** (2000) 709–720, arXiv:math/0001153.

[26] R. Bott and L. W. Tu, *Differential Forms in Algebraic Topology*, Springer-Verlag, New York, 1982.

CENTER FOR GEOMETRY AND THEORETICAL PHYSICS, BOX 90318, DUKE UNIVERSITY, DURHAM, NC 27708-0318

Proceedings of Symposia in Pure Mathematics
Volume **85**, 2012

The Largest Mathieu Group and (Mock) Automorphic Forms

Miranda C. N. Cheng and John F. R. Duncan

Abstract. We review the relationship between the largest Mathieu group and various modular objects, including recent progress on the relation to mock modular forms. We also review the connections between these mathematical structures and string theory on $K3$ surfaces.

Contents

1. Cusp Forms and M_{24}

We start by recalling some facts about the largest Mathieu group, M_{24}. The group M_{24} may be characterised as the automorphism group of the unique doubly even self-dual binary code of length 24 with no words of weight 4, also known as the *(extended) binary Golay code.* In other words, there is a unique (up to permutation) set, $\mathcal{G}$ say, of length 24 binary codewords (sequences of 0's and 1's) such that any other length 24 codeword has even overlap with all the codewords of $\mathcal{G}$ if and only if this word itself is in $\mathcal{G}$, and the number of 1's in each codeword of $\mathcal{G}$ is divisible by but not equal to 4. The group of permutations of the 24 coordinates that preserves the set $\mathcal{G}$ is the sporadic group M_{24}. See, for instance, [**1**].

As such, M_{24} naturally admits a permutation representation of degree 24, which we denote R and call the *defining representation* of M_{24}, and via R we may assign a *cycle shape* to each of its elements. For example, to the identity element we associate the cycle shape 1^{24}, to an element of M_{24} that is a product of 12 mutually

2010 *Mathematics Subject Classification.* Primary 11F22; Secondary 11F11, 11F37, 11F46, 11F50, 20C34, 20C35, 58J26, 81T30, 81T40, 14J28.

$[g]$	cycle shape	$\eta_g(\tau)$	k_g	n_g	N_g	h_g
$1A$	1^{24}	$\eta^{24}(\tau)$	12	1	1	1
$2A$	$1^8 2^8$	$\eta^8(\tau)\eta^8(2\tau)$	8	2	2	1
$2B$	2^{12}	$\eta^{12}(2\tau)$	6	2	4	2
$3A$	$1^6 3^6$	$\eta^6(\tau)\eta^6(3\tau)$	6	3	3	1
$3B$	3^8	$\eta^8(3\tau)$	4	3	9	3
$4A$	$2^4 4^4$	$\eta^4(2\tau)\eta^4(4\tau)$	4	4	8	2
$4B$	$1^4 2^2 4^4$	$\eta^4(\tau)\eta^2(2\tau)\eta^4(4\tau)$	5	4	4	1
$4C$	4^6	$\eta^6(4\tau)$	3	4	16	4
$5A$	$1^4 5^4$	$\eta^4(\tau)\eta^4(5\tau)$	4	5	5	1
$6A$	$1^2 2^2 3^2 6^2$	$\eta^2(\tau)\eta^2(2\tau)\eta^2(3\tau)\eta^2(6\tau)$	4	6	6	1
$6B$	6^4	$\eta^4(6\tau)$	2	6	36	6
$7AB$	$1^3 7^3$	$\eta^3(\tau)\eta^3(7\tau)$	3	7	7	1
$8A$	$1^2 2^1 4^1 8^2$	$\eta^2(\tau)\eta(2\tau)\eta(4\tau)\eta^2(8\tau)$	3	8	8	1
$10A$	$2^2 10^2$	$\eta^2(2\tau)\eta^2(10\tau)$	2	10	20	2
$11A$	$1^2 11^2$	$\eta^2(\tau)\eta^2(11\tau)$	2	11	11	1
$12A$	$2^1 4^1 6^1 12^1$	$\eta(2\tau)\eta(4\tau)\eta(6\tau)\eta(12\tau)$	2	12	24	2
$12B$	12^2	$\eta^2(12\tau)$	1	12	144	12
$14AB$	$1^1 2^1 7^1 14^1$	$\eta(\tau)\eta(2\tau)\eta(7\tau)\eta(14\tau)$	2	14	14	1
$15AB$	$1^1 3^1 5^1 15^1$	$\eta(\tau)\eta(3\tau)\eta(5\tau)\eta(15\tau)$	2	15	15	1
$21AB$	$3^1 21^1$	$\eta(3\tau)\eta(21\tau)$	1	21	63	3
$23AB$	$1^1 23^1$	$\eta(\tau)\eta(23\tau)$	1	23	23	1

TABLE 1. The cycle shapes, weights (k_g), levels (N_g) and orders (n_g) of the 26 conjugacy classes of the sporadic group M_{24}. The length of the shortest cycle is $h_g = N_g/n_g$. The naming of the conjugacy classes follows the ATLAS convention [**2**]. We write $7AB$, for example, to indicate that the entries of the incident row are valid for both the conjugacy classes $7A$ and $7B$.

commuting transpositions we associate the cycle shape 2^{12}, and so on. More generally, any cycle shape arising from an element of M_{24} (or S_{24}, for that matter) is of the form

$$i_1{}^{\ell_1} i_2{}^{\ell_2} \cdots i_r{}^{\ell_r}, \quad \sum_{s=1}^{r} \ell_s \, i_s = 24 \,,$$

for some $\ell_s \in \mathbb{Z}^+$ and $1 \le i_1 < \cdots < i_r \le 23$ with $r \ge 1$. Clearly, the cycle shape of an element $g \in M_{24}$ depends only on its conjugacy class, denoted by $[g]$, although different conjugacy classes can share the same cycle shape. We write

$$\chi(g) = \mathrm{tr}_R \, g$$

for the character attached to this 24-dimensional representation of M_{24}. The decomposition of R into irreducible representations of M_{24} is given by

$$R = \rho_1 \oplus \rho_{23}$$

(cf. Table 3). The value of the character $\chi(g)$ equals the number of fixed points of g in the action on the set of 24 points. In particular, note that $\chi(g) = \ell_1$ in case $i_1 = 1$ and $\chi(g) = 0$ otherwise.

It turns out that the cycle shapes of M_{24} have many special properties that will be important for the understanding of the modular properties of the associated

McKay–Thompson series which we will discuss shortly. First, the M_{24} cycle shapes are privileged in that they are all of the so-called *balanced type* [**3**], meaning that for each $g \in M_{24}$ there exists a positive integer N_g such that if $\prod i_s^{\ell_s}$ is the cycle shape associated to g then

$$\prod_s i_s^{\ell_s} = \prod_s \left(\frac{N_g}{i_s}\right)^{\ell_s} .$$

We will refer to the number N_g as the "*level* of g".

If g has cycle shape $i_1^{\ell_1}\cdots i_r^{\ell_r}$ then the order of g is the least common multiple of the i_s's. A second special property of $[g] \subset M_{24}$ is that the order of g coincides with the length i_r of the longest cycle in the cycle shape. Henceforth we will denote $n_g = i_r$.

Finally, observe that for all $g \in M_{24}$ the level N_g defined above equals the product of the shortest and the longest cycle. Hence we have $n_g h_g = N_g$ where h_g denotes the length of the shortest cycle in the cycle shape. Moreover, we also have the property $h_g|n_g$ and $h_g|12$. This is reminiscent of the fact, significant in monstrous moonshine [**3**], that the normaliser of $\Gamma_0(N)/\langle\pm \mathrm{Id}\rangle$ in $PSL_2(\mathbb{R})$ (cf. (1.2)) is a group containing a conjugate of $\Gamma_0(n/h)/\langle\pm\mathrm{Id}\rangle$ where $n = N/h$ and h is the largest divisor of 24 such that $h^2|N$. We set $k_g = \sum_{s=1}^r \ell_s/2$ to be half of the total number of cycles and call it the *weight* of g. Of course, N_g, n_g, h_g and k_g depend only on the conjugacy class $[g]$ containing g. All the values of these parameters can be found in Table 1.

To each element $g \in M_{24}$ we can attach an *eta-product*, to be denoted η_g, which is the function on the upper half-plane given by

$$\eta_g(\tau) = \prod_s \eta(i_s\tau)^{\ell_s} \tag{1.1}$$

where $\prod_s i_s^{\ell_s}$ is the cycle shape attached to g, and $\eta(\tau)$ is the Dedekind eta function satisfying $\eta(\tau) = q^{1/24}\prod_{n\in\mathbb{Z}^+}(1-q^n)$ for $q = e(\tau)$, where here and everywhere else in the paper we use the shorthand notation

$$e(x) = e^{2\pi i x} .$$

As was observed in [**4, 5**], the eta-product η_g associated to an element $g \in M_{24}$ (or rather, to its conjugacy class $[g]$) is a cusp form of weight k_g for the group $\Gamma_0(N_g)$, with a *Dirichlet character* ς_g that is trivial if the weight k_g is even and is otherwise defined, in terms of the Jacobi symbol $(\frac{n}{m})$, by

$$\varsigma_g(\gamma) = \begin{cases} \left(\frac{N_g}{d}\right)(-1)^{\frac{d-1}{2}}, & d \text{ odd},\\ \left(\frac{N_g}{d}\right), & d \text{ even},\end{cases}$$

in case d is the lower right entry of $\gamma \in \Gamma_0(N_g)$. Let's recall that

$$\Gamma_0(N) = \left\{\gamma \,\middle|\, \gamma = \begin{pmatrix} a & b\\ c & d\end{pmatrix} \in SL_2(\mathbb{Z})\,,\ c \equiv 0 \mod N\right\} , \tag{1.2}$$

and a cusp form is a modular form which vanishes at every cusp. Namely it vanishes whenever $\tau \to \infty$ or $\alpha \in \mathbb{Q}$.

Interestingly, these cusp forms point to the existence of an infinite-dimensional $\mathbb{Z}$-graded module for M_{24}; a fact we will now explain. It is by now a routine task to construct a module for M_{24} whose graded dimension is the reciprocal of the cusp form $\eta^{24}(\tau)$. To see this, let us consider the defining 24-dimensional representation R. To obtain a q-series we also need an extra $\mathbb{Z}$-grading. Hence we will take a copy of this representation, denoted by R_k, for each positive integer k and consider the direct sum

$$V = R_1 \oplus R_2 \oplus \cdots = \bigoplus_{k=1}^{\infty} R_k \, .$$

Furthermore we would like to consider the graded Fock space $\mathcal{H}$ built from the above infinite-dimensional vector space V

$$\mathcal{H} = \bigoplus_{n=0}^{\infty} S^n V = \bigoplus_{k=0}^{\infty} \mathcal{H}_k \quad , \quad S^n V = V^{\otimes n}/S_n \, ,$$

where the $\mathbb{Z}$-grading is naturally inherited from that of V. The associated partition function is then given by

$$Z(\tau) = \sum_{k=0}^{\infty} q^{k-1} \mathrm{dim} \mathcal{H}_k = \frac{1}{\eta^{24}(\tau)} \, .$$

In the language of vertex operator algebras, this is simply the partition function of the conformal field theory (CFT) with 24 free chiral bosons. Namely, we consider 24 pairs of creation and annihilation operators $a^{(i)}_{-k}, a^{(i)}_{k}, i = 1, \ldots, 24$ for each positive integer k, which furnish the 24-dimensional representation R_k and satisfy the commutation relation

$$[a^{(i)}_k, a^{(j)}_\ell] = \delta_{ij}\delta_{k+\ell,0} \, .$$

The quantum states of this theory are built from the unique vacuum state $|0\rangle$, characterised by the condition $a^{(i)}_n|0\rangle = 0$ for all $n \geq 0, i = 1, \ldots, 24$, and take the form

$$a^{(i_1)}_{-n_1} \cdots a^{(i_N)}_{-n_N}|0\rangle \quad , \quad n_k > 0, \;\; 1 \leq i_k \leq 24 \, ,$$

while the Hamiltonian of the theory is given by the operator

$$\hat{H} = \sum_{i=1}^{24} \sum_{n=1}^{\infty} n \, a^{(i)}_{-n} \, a^{(i)}_{n} - 1 \, .$$

Hence the partition function of this theory is indeed given by the modular form

$$Z(\tau) = \mathrm{tr}_{\mathcal{H}} \, q^{\hat{H}} = \frac{1}{\eta^{24}(\tau)} \, .$$

Notice that, physically the (-1)-shift in the Hamiltonian operator corresponds to the ground state energy in the quantum theory with $\mathcal{H}$ as the space of quantum states (the *Hilbert space*).

From the above construction, in particular the fact that the M_{24}-action commutes with the $\mathbb{Z}$-grading, we conclude that not only $\mathcal{H}$ but also each of its components $\mathcal{H}_k$ are M_{24}-modules. Hence, for each element $g \in M_{24}$ we can consider the following *twisted* partition function, or *McKay-Thompson series*

$$Z_g(\tau) = \mathrm{tr}_{\mathcal{H}} \, g q^{\hat{H}} = \sum_{k=0}^{\infty} q^{k-1} \big(\mathrm{tr}_{\mathcal{H}_k} \, g \big) \, . \tag{1.3}$$

To be more precise, in the above formula g really stands for the corresponding linear map acting on the representation in question. Note that taking the identity element $g = \mathbf{1}$, we recover the usual partition function $\mathrm{tr}_{\mathcal{H}} \, q^{\hat{H}} = Z(\tau)$ encoding the graded dimension of the representation. Moreover, it is clear from their definition that these McKay-Thompson series depend only on the conjugacy class $[g]$ of the group element. Namely, we have

$$Z_g(\tau) = Z_{hgh^{-1}}(\tau)\,.$$

From the description of $\mathcal{H}$, it is now easy to prove that they are given by the eta-products as

(1.4)

$$Z_g(\tau) = q^{-1} \exp\left(\sum_{k=1}^{\infty} \mathrm{tr}_V \frac{(q^{\hat{H}} g)^k}{k}\right) = q^{-1} \exp\left(\sum_{n=1}^{\infty}\sum_{k=1}^{\infty} \frac{q^{nk}\ \mathrm{tr}_R(g^k)}{k}\right) = \frac{1}{\eta_g(\tau)}\,.$$

In the path integral language, the summing over k in the expression for $\log Z_k$ can be thought of as a sum over how many times the time circle of the world-volume wraps around the time circle in the target space.

In other words, we can equate the McKay-Thompson series of the infinite-dimensional M_{24}-module $\mathcal{H}$, given by the twisted partition function of the physical theory $Z_g(\tau)$, to a weight $-k_g$, level N_g modular form $1/\eta_g(\tau)$. The first few Fourier coefficients of the twisted partition functions and the corresponding M_{24}-modules $\mathcal{H}_k$ in terms of its decomposition into irreducible representations can be found in Table 4.

A few comments are in order here. First, historically the relation between the $\eta_g(\tau)$ and M_{24} was phrased in terms of the eta-products themselves as opposed to their reciprocals as we described above. In that case, underlying the Fourier coefficients is an infinite-dimensional, $\mathbb{Z}$-graded *super* (or *virtual*) module of M_{24}, corresponding to considering instead the following $\mathbb{Z}/2$-graded Hilbert space

$$\tilde{\mathcal{H}} = \tilde{\mathcal{H}}_{\bar{0}} \oplus \tilde{\mathcal{H}}_{\bar{1}}, \quad \tilde{\mathcal{H}}_{\bar{k}} = \bigoplus_{n=0}^{\infty} \Lambda^{2n+k} V$$

which is the exterior (super)algebra of the infinite-dimensional vector space V, where V is regarded as having odd parity [**4, 6**]. A super G-module is just a G-module W with a G-invariant $\mathbb{Z}/2$-grading $W = W_{\bar{0}} \oplus W_{\bar{1}}$. Given such a module we may consider the operator $F : W \to W$ for which the subspaces $W_{\bar{k}}$ are eigenspaces with eigenvalue k for F. Then the supertrace of an operator A on W is defined by setting

$$\mathrm{str}_W A = \mathrm{tr}_W (-1)^F A.$$

In physical terms such a $\mathbb{Z}/2$-grading often indicates the presence of fermions, and the operator F is called the *fermion number operator.*

Second, as the alert reader may have noticed, starting from $\eta^{24}(\tau)$ we can build not only an infinite-dimensional module $\mathcal{H}$ for M_{24} but this same space (or vertex operator algebra) admits an action by the full symmetric group S_{24}, or even $O_{24}(\mathbb{C})$[1]. In this sense, the relation between these 21 eta-products and M_{24} is not as distinguished as one might have hoped from the point of view of moonshine. Nevertheless, as we will see later in §4, together with various weak Jacobi forms,

[1]Although in general the corresponding twisted partition function won't be multiplicative if $g \in S_{24}$ but $g \notin M_{24}$. A series $\sum_{n=0}^{\infty} a_n q^n$ with $q = e^{2\pi i \tau}$ and $\tau \in \mathbb{H}$ is said to be multiplicative if $a_n a_m = a_{mn}$ whenever $(n,m) = 1$. Of the 1575 possible partitions of 24 only 30 correspond to multiplicative eta-products [**5**]. All of the 30 multiplicative eta-products are recovered from a signed permutation action of a group of the form $2^{24} : M_{24}$ and 21 of them arise from the action of M_{24}; these are the 21 eta-products listed in Table 1.

mock modular forms and Siegel modular forms, they do tie together and form a closely-knit net of modular objects attached to the sporadic group M_{24}.

Third, while we have seen that the eta-products $\eta_g(\tau)$ are cusp forms of level N_g, from the following CFT arguments we expect a further, more refined modular property. To explain this, let us first discuss a certain natural generalisation of the above discussion: Given a vertex operator algebra $\mathcal{H}$ and two commuting automorphisms g and h we expect there to exist a so-called *h-twisted module* $\mathcal{H}^{(h)}$ for $\mathcal{H}$ and one can consider the twisted twisted sector partition function

$$Z\left(\tau; g\underset{h}{\Box}\right) \tag{1.5}$$

defined as the (rationally) graded trace of g on $\mathcal{H}^{(h)}$, thereby generalizing the twisted partition function $Z_g(\tau)$ which is recovered by taking h to be the identity. The significance of the twisted partition functions (1.5) in the context of monstrous moonshine was first observed by S. Norton in [**7**] and dubbed *generalised moonshine.* It is conjectured in loc. cit. that any such function is either constant or a generator for the field of functions on some discrete subgroup of $PSL_2(\mathbb{R})$ in case $\mathcal{H}$ is the moonshine module vertex operator algebra and g and h are a commuting pair of elements of the Monster simple group. An analogue of generalised moonshine for commuting pairs of elements of M_{24} is discussed in [**8**].

Recall that the equivalence between the Hamiltonian and Lagrangian formulations of quantum field theory implies that the insertion of the elements g and h as in (1.5) has an interpretation as altering the boundary condition of the path integral over the maps from an elliptic curve world-sheet into the target space. Identifying the modular group $SL_2(\mathbb{Z})$ as the mapping class group acting on the elliptic curve world-sheet, regarding Z as being associated with an appropriate 'chiral half' of a conformal field theory, we conclude that the following hold:

$$Z\left(\tau; g\underset{h}{\Box}\right) = \varepsilon(\gamma, g, h)\, Z\left(\frac{a\tau+b}{c\tau+d}; g^ah^b\underset{g^ch^d}{\Box}\right), \quad \gamma = \begin{pmatrix} a & b \\ c & d \end{pmatrix} \in SL_2(\mathbb{Z}),$$

where $\varepsilon(\gamma, g, h)$ is a phase. Specializing the above equation to the case that h is the identity we conclude that the McKay-Thompson series Z_g must be invariant up to a phase under the group $\Gamma_0(n_g)$ where n_g is the order of the group.

Indeed, the eta-product η_g also defines a cusp form of weight k_g on the larger (or equal) group $\Gamma_0(n_g) \supseteq \Gamma_0(N_g)$ if we allow for a slightly more subtle multiplier system. Note that this change in multiplier system only comes into play when g is in one of the fixed-point-free classes of M_{24} with $\chi(g) = 0$, for otherwise $\Gamma_0(n_g) = \Gamma_0(N_g)$. Our convention is to say that a function $\xi : \Gamma \to \mathbb{C}^*$ is a *multiplier system for* Γ *of weight* w in case the identity

$$\xi(\gamma\sigma)\,\mathrm{jac}(\gamma\sigma,\tau)^{w/2} = \xi(\gamma)\xi(\sigma)\,\mathrm{jac}(\gamma,\sigma\tau)^{w/2}\,\mathrm{jac}(\sigma,\tau)^{w/2} \tag{1.6}$$

holds for all $\gamma, \sigma \in \Gamma$ and $\tau \in \mathbb{H}$. Writing $\gamma = \left(\begin{smallmatrix} a & b \\ c & d \end{smallmatrix}\right)$, we have in the above formula $\gamma\tau = \frac{a\tau+b}{c\tau+d}$, and the Jacobian

$$\mathrm{jac}(\gamma,\tau) = (c\tau+d)^{-2}\ .$$

In detail, for $\gamma \in \Gamma$, we define the *slash operator* $|_{\xi,w}$ *of weight* w *associated with the multiplier* ξ by setting

$$(f|_{\xi,w}\gamma)(\tau) = \xi(\gamma)f(\gamma\tau)\,\mathrm{jac}(\gamma,\tau)^{w/2} \tag{1.7}$$

so that a holomorphic function $f : \mathbb{H} \to \mathbb{C}$ is a modular form of weight w and multiplier ξ for Γ if and only if $(f|_{\xi,w}\gamma)(\tau) = f(\tau)$ for all $\gamma \in \Gamma$. In the present case we have

$$\left(\frac{1}{\eta_g}\Big|_{\xi_g,k_g}\gamma\right)(\tau) = \frac{1}{\eta_g(\tau)}, \quad \text{for all} \quad \gamma \in \Gamma_0(n_g), \tag{1.8}$$

where $\xi_g(\gamma) = \rho_{n_g|h_g}(\gamma)\varsigma_g(\gamma)$ with

$$\rho_{n|h}(\gamma) = e(-\tfrac{1}{(\gamma\infty-\gamma 0)nh}) = e(-cd/nh)\ . \tag{1.9}$$

Note that $\rho_{n_g|h_g}$ is actually a character on $\Gamma_0(n_g)$ since we have that $xy \equiv 1 \pmod{h_g}$ implies $x \equiv y \pmod{h_g}$ by virtue of the fact that $h_g = N_g/n_g$ is a divisor of 24 for every $g \in M_{24}$ [**3**, §3].

2. Mock Modular Forms and M_{24}

In the previous section we have seen the relation between M_{24} and eta-quotients. Recently, there has been an unexpected observation relating M_{24} and a weight 1/2 mock modular form. Namely, consider the q-series

$$\begin{aligned} H(\tau) = \frac{-2E_2(\tau) + 48F_2^{(2)}(\tau)}{\eta(\tau)^3} &= 2q^{-\frac{1}{8}}\left(-1 + 45q + 231q^2 + \cdot\right) \\ &= q^{-\frac{1}{8}}\left(-2 + \sum_{n=1}^{\infty} t_n q^n\right). \end{aligned}$$

where

$$F_2^{(2)}(\tau) = \sum_{\substack{r>s>0 \\ r-s\equiv 1 \bmod 2}} (-1)^r\, s\, q^{rs/2} = q + q^2 - q^3 + q^4 + \dots\ ,$$

then the observation made in [**9**] is that the first few t_n's read

$$2 \times 45,\ 2 \times 231,\ 2 \times 770,\ 2 \times 2277,\ 2 \times 5796,\ \dots$$

and the integers 45, 231, 770, 2277 and 5796 are dimensions of irreducible representations of M_{24}.

This function $H(\tau)$, defined in (2.4), enjoys a special relationship with the group $SL_2(\mathbb{Z})$; namely, it is a *weakly holomorphic mock modular form of weight* 1/2 on $SL_2(\mathbb{Z})$ with *shadow* $24\,\eta(\tau)^3$ [**10, 11**]. This means that if we define the *completion* $\hat{H}(\tau)$ of the holomorphic function $H(\tau)$ by setting

$$\hat{H}(\tau) = H(\tau) + 24\,(4i)^{-1/2} \int_{-\bar{\tau}}^{\infty} (z+\tau)^{-1/2}\overline{\eta(-\bar{z})^3}\mathrm{d}z, \tag{2.1}$$

then $\hat{H}(\tau)$ transforms as a modular form of weight 1/2 on $SL_2(\mathbb{Z})$ with multiplier system conjugate to that of $\eta(\tau)^3$. In other words, we have

$$\big(\hat{H}(\tau)|_{\epsilon^{-3},1/2}\gamma\big)(\tau) = \epsilon(\gamma)^{-3}\hat{H}(\gamma\tau)\,\mathrm{jac}(\gamma,\tau)^{1/4} = \hat{H}(\tau)$$

for $\gamma \in SL_2(\mathbb{Z})$, where $\epsilon : SL_2(\mathbb{Z}) \to \mathbb{C}^*$ is the multiplier system for $\eta(\tau)$ satisfying

$$\big(\eta|_{\epsilon,1/2}\gamma\big)(\tau) = \eta(\tau)\ .$$

See §A for an explicit description of ϵ.

More generally, a holomorphic function $h(\tau)$ on $\mathbb{H}$ is called a *(weakly holomorphic) mock modular form of weight* w for a discrete group Γ (e.g. a congruence subgroup of $SL_2(\mathbb{R})$) if it has at most exponential growth as $\tau \to \alpha$ for any $\alpha \in \mathbb{Q}$, and if there exists a holomorphic modular form $f(\tau)$ of weight $2-w$ on Γ such that $\hat{h}(\tau)$, given by

$$\hat{h}(\tau) = h(\tau) + (4i)^{w-1} \int_{-\bar{\tau}}^{\infty} (z+\tau)^{-w} \overline{f(-\bar{z})} \mathrm{d}z, \tag{2.2}$$

is a (non-holomorphic) modular form of weight w for Γ for some multiplier system ψ say. In this case the function f is called the *shadow* of the mock modular form h and ψ is called the multiplier system of h. Evidently ψ is the conjugate of the multiplier system of f. The completion $\hat{h}(\tau)$ satisfies interesting differential equations. For instance, completions of mock modular forms were identified as Maass forms in [**12**] and this led to a solution of the longstanding Andrews–Dragonette conjecture (cf. loc. cit.). As was observed in [**11**] we have the identity

$$2^{1-w}\pi\Im(\tau)^w \frac{\partial \hat{h}(\tau)}{\partial \bar{\tau}} = -2\pi i \overline{f(\tau)}$$

when f is the shadow of h.

Given the observation regarding the first few Fourier coefficients of $H(\tau)$ indicated above one would like to conjecture that the entire set of values t_n for $n \in \mathbb{Z}^+$ encode the graded dimension of a naturally defined $\mathbb{Z}$-graded M_{24}-module $K = \bigoplus_{n=1}^{\infty} K_n$ with $\dim K_n = t_n$. Of course, this conjecture by itself is an empty statement, since all positive integers can be expressed as dimensions of representations of any group, since we may always consider trivial representations. However, the fact that the first few t_n can be written so nicely in terms of irreducible representations suggests that the K_n should generally be non-trivial, and given any particular guess for a M_{24}-module structure on the K_n we can test its merit by considering the twists of this mock modular form H obtained by replacing the identity element in $\dim K_n = \mathrm{tr}_{K_n} \mathbf{1}$ with an element g of the group M_{24}. We would then call the resulting q-series

$$q^{-\frac{1}{8}} \left(-2 + \sum_{n=1}^{\infty} \mathrm{tr}_{K_n} g q^n \right) \tag{2.3}$$

the *McKay–Thompson series* attached to g. A non-trivial connection between mock modular forms and M_{24} arises if all such McKay–Thompson series of the M_{24}-module K display interesting (mock) modular properties. In fact, since a function with good modular properties is generally determined by the first few of its Fourier coefficients, it is easier in practice to guess the McKay–Thompson series than it is to guess the representations K_n. Not long after the original observation was announced in [**9**] candidates for the McKay–Thompson series had been proposed for all conjugacy classes $[g] \subset M_{24}$ in [**13, 14, 15, 16**], and with functions $\tilde{T}_g(\tau)$ defined as in Table 2 the following result was established.

PROPOSITION 2.1. *Let $H : \mathbb{H} \to \mathbb{C}$ be given by*

$$H(\tau) = \frac{-2E_2(\tau) + 48F_2^{(2)}(\tau)}{\eta(\tau)^3} = 2q^{-\frac{1}{8}}\left(-1 + 45q + 231q^2 + \cdot\right) \tag{2.4}$$

where

$$F_2^{(2)}(\tau) = \sum_{\substack{r>s>0 \\ r-s=1 \bmod 2}} (-1)^r \, s \, q^{rs/2} = q + q^2 - q^3 + q^4 + \cdots\cdots .$$

$[g]$	$\chi(g)$	$\tilde{T}_g(\tau)$
$1A$	24	0
$2A$	8	$16\Lambda_2$
$2B$	0	$-24\Lambda_2+8\Lambda_4=2\eta(\tau)^8/\eta(2\tau)^4$
$3A$	6	$6\Lambda_3$
$3B$	0	$2\eta(\tau)^6/\eta(3\tau)^2$
$4A$	0	$4\Lambda_2-6\Lambda_4+2\Lambda_8=2\eta(2\tau)^8/\eta(4\tau)^4$
$4B$	4	$4(-\Lambda_2+\Lambda_4)$
$4C$	0	$2\eta(\tau)^4\eta(2\tau)^2/\eta(4\tau)^2$
$5A$	4	$2\Lambda_5$
$6A$	2	$2(-\Lambda_2-\Lambda_3+\Lambda_6)$
$6B$	0	$2\eta(\tau)^2\eta(2\tau)^2\eta(3\tau)^2/\eta(6\tau)^2$
$7AB$	3	Λ_7
$8A$	2	$-\Lambda_4+\Lambda_8$
$10A$	0	$2\eta(\tau)^3\eta(2\tau)\eta(5\tau)/\eta(10\tau)$
$11A$	2	$2(\Lambda_{11}-11\eta(\tau)^2\eta(11\tau)^2)/5$
$12A$	0	$2\eta(\tau)^3\eta(4\tau)^2\eta(6\tau)^3/\eta(2\tau)\eta(3\tau)\eta(12\tau)^2$
$12B$	0	$2\eta(\tau)^4\eta(4\tau)\eta(6\tau)/\eta(2\tau)\eta(12\tau)$
$14AB$	1	$(-\Lambda_2-\Lambda_7+\Lambda_{14}-14\eta(\tau)\eta(2\tau)\eta(7\tau)\eta(14\tau))/3$
$15AB$	1	$(-\Lambda_3-\Lambda_5+\Lambda_{15}-15\eta(\tau)\eta(3\tau)\eta(5\tau)\eta(15\tau))/4$
$21AB$	0	$(7\eta(\tau)^3\eta(7\tau)^3/\eta(3\tau)\eta(21\tau)-\eta(\tau)^6/\eta(3\tau)^2)/3$
$23AB$	1	$(\Lambda_{23}-23f_{23,1}+23f_{23,2})/11$

TABLE 2. In this table we collect the data that via equation (2.5) and (3.8) define the weight 1/2 (mock) modular forms $H_g(\tau)$ and the weak Jacobi forms $\mathcal{Z}_g(\tau,z)$. For $N\in\mathbb{Z}_+$, we denote by Λ_N the weight 2 modular form on $\Gamma_0(N)$ given by $\Lambda_N = Nq\frac{d}{dq}(\log\frac{\eta(N\tau)}{\eta(\tau)})$ (cf. (A.2)). For $N=23$, there are two newforms and we choose the basis $f_{23,1}=\eta_{23AB}^2$ and $f_{23,2}$ given in (A.6).

Then for all $g\in M_{24}$, the function

$$H_g(\tau)=\frac{\chi(g)}{24}H(\tau)-\frac{\tilde{T}_g(\tau)}{\eta(\tau)^3}\,, \tag{2.5}$$

is a (mock) modular form for $\Gamma_0(N_g)$ of weight 1/2 with shadow $\chi(g)\eta(\tau)^3$. Moreover, we have

$$\hat{H}_g(\tau)=\psi(\gamma)\,\mathrm{jac}(\gamma,\tau)^{1/4}\,\hat{H}_g(\gamma\tau),$$

for $\gamma\in\Gamma_0(n_g)$ where

$$\hat{H}_g(\tau)=H_g(\tau)+\chi(g)\,(4i)^{-1/2}\int_{-\bar{\tau}}^{\infty}(z+\tau)^{-1/2}\overline{\eta(-\bar{z})^3}\mathrm{d}z.$$

and the multiplier system is given by $\psi(\gamma)=\epsilon(\gamma)^{-3}\rho_{n_g|h_g}(\gamma)$.

(See [**11**] for more information regarding the particular expression for $H(\tau)$ given in (2.4).)

Notice that the extra multiplier $\rho_{n|h}$ that appears when $h_g\neq 1$ is the same as that of the inverse eta-product $1/\eta_g(\tau)$ (cf. (1.9)) which are also related to M_{24}, a fact that is in accordance with the 1/2- and 1/4-BPS spectrum of the $\mathcal{N}=4$, $d=4$

theory obtained by $K3 \times T^2$ compactification of the type II string theory, as will be discussed in more detail in §5.

Our discussion above leads to the following conjecture.

CONJECTURE 2.2. *The weight 1/2 (mock) modular forms H_g defined in (2.5) satisfy*

$$H_g(\tau) = q^{-\frac{1}{8}}\left(-2 + \sum_{n=1}^{\infty} q^n \left(\mathrm{tr}_{K_n} g\right)\right) \tag{2.6}$$

for a certain $\mathbb{Z}$-graded, infinite-dimensional M_{24}-module $K = \bigoplus_{n=1}^{\infty} K_n$.

Moreover, the representations K_n are even in the sense that they can all be written in the form $K_n = k_n \oplus k_n^*$ for some M_{24}-modules k_n where k_n^* denotes the module dual to k_n.

The first few Fourier coefficients of the q-series $H_g(\tau)$ and the corresponding M_{24}-representations are given in Table 7.

A proof of for the first part of the above conjecture, namely the existence of an M_{24}-module $K = \bigoplus_{n=1}^{\infty} K_n$ such that (2.6) holds, has been attained very recently [**17**]. However, it is important to stress that, unlike the eta-product moonshine described in §1, the nature and origin of the M_{24}-module K remains mysterious.

The compelling modular properties of the functions H_g constitute strong evidence for the existence of the M_{24}-module K. Yet stronger evidence would be furnished by a uniform construction of the (mock) modular forms H_g. Such a construction was established recently in [**18**]. The construction presented there, in terms of Rademacher sums, is of further interest in that it points to an extension of the connection between M_{24} and $K3$ surfaces, expressed in CFT terms via the H_g (cf. §3), to a connection in the context of quantum gravity. Also, by comparison with the Rademacher sum expressions for the functions of monstrous moonshine [**19**], the Rademacher sum construction of the H_g suggests a reformulation of the crucial genus zero property of monstrous moonshine that incorporates, and emphasises the distinguished nature of, the functions H_g attached to M_{24}.

THEOREM 2.3. [**18**] *Let $g \in M_{24}$. Let $\mathrm{reg}(\gamma,\tau)$ be the regularisation factor given by $\mathrm{reg}(\gamma,\tau) = 1$ in case γ is upper triangular (i.e. $\gamma \cdot \infty = \infty$) and*

$$\mathrm{reg}(\gamma,\tau) = e\left(\tfrac{\gamma\tau-\gamma\infty}{8}\right) e\left(\tfrac{\gamma\infty-\gamma\tau}{8}, \tfrac{1}{2}\right) \tag{2.7}$$

otherwise, where $e(x,s)$ is the following generalisation of the exponential function:

$$e(x,s) = \sum_{m\geq 0} \frac{(2\pi i x)^{m+s}}{\Gamma(m+s+1)}\,. \tag{2.8}$$

Given $\Gamma < SL_2(\mathbb{Z})$ containing the group Γ_∞ of upper triangular matrices in $SL_2(\mathbb{Z})$, and given a character ρ on Γ define the Rademacher sum

$$R_{\Gamma,\rho}(\tau) = \lim_{K\to\infty} \sum_{\gamma\in(\Gamma_\infty\backslash\Gamma)_{<K}} \psi(\gamma) e\left(-\tfrac{\gamma\tau}{8}\right) \mathrm{reg}(\gamma,\tau)\,\mathrm{jac}(\gamma,\tau)^{1/4}$$

where $\psi(\gamma) = \rho_{n_g|h_g}(\gamma)\epsilon(\gamma)^{-3}$ and $(\Gamma_\infty\backslash\Gamma)_{<K}$ denotes the set of cosets of Γ_∞ in Γ having a representative with lower row (c,d) satisfying the bounds $0 \leq c < K$ and $|d| < K^2$. Then $H_g(\tau) = -2R_{\Gamma,\rho}(\tau)$ when $\Gamma = \Gamma_0(n_g)$ and $\rho = \rho_{n_g|h_g}$.

3. Weak Jacobi Forms and M_{24}

In the previous section we have discussed the relation between M_{24} and a set of (mock) modular forms. The origin and the structure of the corresponding M_{24}-module K remains to be explored, in contrast to the situation of the cusp forms of §1, whose relation to M_{24} can be fully understood in terms of the 24-dimensional permutation representation R. But so far, the observation of the relation between M_{24} and mock modular forms was in fact mostly made via a weak Jacobi form [**9**]. This weak Jacobi form has the property that it incorporates both the infinite-dimensional M_{24}-module K and the permutation representation R, a feature that will become even more significant in our later discussion in §4 about Siegel modular forms. Perhaps the best way to motivate this observation is to put it in the context of the elliptic genus of $K3$ surfaces.

For a compact complex manifold M with $\dim_{\mathbb{C}} M = d_0$, the elliptic genus is defined as the character-valued Euler characteristic of the formal vector bundle [**20, 21, 22, 23, 24**]

$$\mathbf{E}_{q,y} = y^{d/2} \bigwedge\nolimits_{-y^{-1}} T_M^* \bigotimes\nolimits_{n\geq 1} \bigwedge\nolimits_{-y^{-1}q^n} T_M^* \bigotimes\nolimits_{n\geq 1} \bigwedge\nolimits_{-yq^n} T_M \bigotimes\nolimits_{n\geq 0} S_{q^n}(T_M \oplus T_M^*),$$

where T_M and T_M^* are the holomorphic tangent bundle and its dual, and we adopt the notation

$$\bigwedge\nolimits_q V = 1 + qV + q^2 \bigwedge\nolimits^2 V + \ldots, \qquad S_q V = 1 + qV + q^2 S^2 V + \cdots\cdots,$$

with $S^k V$ denoting the k-th symmetric power of V. In other words, we have

$$\mathbf{EG}(\tau, z; M) = \int_M ch(\mathbf{E}_{q,y}) \mathrm{Td}(M). \tag{3.1}$$

From the above definition we see that this topological quantity reduces to the familiar ones: the Euler number, the signature, and the $\hat{A}$-genus of M, when we specialise z to $z = 0, \tau/2, (\tau+1)/2$, respectively, when z and τ are such that $y = e(z)$ and $q = e(\tau)$.

When M has vanishing first Chern class—in particular when M is a Calabi–Yau manifold—its elliptic genus $\mathbf{EG}(\tau, z; M)$ transforms nicely under the group $SL_2(\mathbb{Z}) \ltimes \mathbb{Z}^2$ and is in fact a weak Jacobi form of weight zero and index $d_0/2$ [**24**]. We say a function $\phi : \mathbb{H} \times \mathbb{C} \to \mathbb{C}$ is a weak Jacobi form of weight w and index t if it transforms under $SL_2(\mathbb{Z}) \ltimes \mathbb{Z}^2$ as follows:

$$\begin{aligned} \phi(\tau, z) &= \mathrm{jac}(\gamma, \tau)^{w/2} e(-t\tfrac{cz^2}{c\tau+d})\, \phi(\gamma(\tau, z)), \quad \gamma = \begin{pmatrix} a & b \\ c & d \end{pmatrix} \in SL_2(\mathbb{Z}) \\ \phi(\tau, z) &= e(t(\lambda^2\tau + 2\lambda z))\, \phi(\tau, z + \lambda\tau + \mu), \quad \lambda, \mu \in \mathbb{Z} \end{aligned} \tag{3.2}$$

where $\gamma(\tau, z) = (\frac{a\tau+b}{c\tau+d}, \frac{z}{c\tau+d})$ and if the Fourier expansion of ϕ in $q = e(\tau)$ and $y = e(z)$ involves no negative powers of q.

The fact that the elliptic genus of a Calabi–Yau manifold M is a Jacobi form can be proven from the definition (3.1) [**25**], but is especially clear from the point of view of conformal field theories. When the target space manifold M is Calabi–Yau the corresponding supersymmetric sigma model with $\mathcal{N} = 2$ supersymmetry is also conformal. The spectrum of the supersymmetric states is encoded in the function

$$\mathbf{EG}(\tau, z; M) = \mathrm{tr}_{\mathcal{H}_{\mathrm{RR}}} \left((-1)^{J_0 + \bar{J}_0} y^{J_0} q^{L_0 - c/24} \bar{q}^{\bar{L}_0 - c/24} \right) \tag{3.3}$$

where $c = 3d_0$ is the central charge of the conformal field theory, and $J_0, L_0, \bar{J}_0, \bar{L}_0$ are the zero modes of the left- and right-moving copies of the $U(1)$ and Virasoro parts of the $\mathcal{N} = 2$ superconformal algebra, respectively, and $\mathcal{H}_{RR}$ denotes the space of quantum states in the Ramond–Ramond sector. The action of the $\mathcal{N} = (2,2)$ superconformal algebra then predicts that the right hand side of (3.3) transforms as in (3.2) with $w = 0$ and $t = d_0/2$. As our notation suggests, the Jacobi forms attached to M by (3.1) and (3.3) coincide. To understand this we should view the vector bundle $\mathbf{E}_{q,y}$ as the space of the Ramond–Ramond quantum states of the CFT built by acting on the Ramond sector ground states with the left-moving bosonic operators α_{-n} and the fermionic operators ψ^i_{-n}, $\psi^{\bar{\imath}}_{-n}$.

When written in the form (3.3) the elliptic genus may be interpreted as computing the graded dimension of the cohomology of the right-moving supercharge [**23**], graded by the left-moving quantum numbers. At the special value $z = 0$ the above definition reduces to that of a Witten index enumerating Ramond–Ramond ground states.

The Jacobi form property greatly constrains the possibilities for the elliptic genus of a Calabi–Yau manifold M. For example, when M is a $K3$ surface the Euler number $\mathbf{EG}(\tau, z = 0; K3) = 24$ forces the elliptic genus to be

$$\mathcal{Z}(\tau, z) = \mathbf{EG}(\tau, z; K3) = 8 \sum_{i=2,3,4} \left(\frac{\theta_i(\tau, z)}{\theta_i(\tau, 0)} \right)^2 , \tag{3.4}$$

as computed in [**26**] (cf. §A.2 for the definition of $\theta_i(\tau, z)$). In particular this function is independent of the choice of $K3$ surface and so there is no ambiguity in writing $K3$ in place of M in (3.4).

It is to be expected that some further special properties should hold for the elliptic genus when M is a $K3$ surface because two (complex) dimensional Calabi–Yau manifolds are not only Kähler but also hyper-Kähler. As a result, the $U(1)_R$ symmetry can be extended to $SU(2)_R$ and the sigma model has enhanced $\mathcal{N} = 4$ superconformal symmetry for both the left and the right movers. This leads to a specific decomposition of the elliptic genus of an $\mathcal{N} = 4$ SCFT that we will now explain.

Since the underlying CFT admits an action by the $\mathcal{N} = 4$ superconformal algebra the Hilbert space decomposes into a direct sum of (unitary) irreducible representations of this algebra. Hence the elliptic genus (3.3) can be written as a sum of characters of representations of this algebra with some multiplicities (cf. (3.7)). A natural embedding of the $U(1)$ current algebra of the $\mathcal{N} = 2$ superconformal algebra into the $SU(2)$ current algebra of the $\mathcal{N} = 4$ superconformal algebra is obtained by choosing $J_0^3 = J_0$. As a result, the $\mathcal{N} = 4$ highest weight representations are again labeled by two quantum numbers h, ℓ, corresponding to the operators L_0, J_0^3 respectively, and the character of an irreducible representation $V_{h,\ell}$ say is defined as

$$ch_{h,\ell}(\tau, z) = \operatorname{tr}_{V_{h,\ell}} \left((-1)^{J_0} y^{J_0} q^{L_0 - c/24} \right) .$$

For the central charge $c = 6$, there are two supersymmetric (also called ‘BPS’ or ‘massless’) representations in the Ramond sector and they have the quantum numbers

$$h = \tfrac{1}{4} \, , \; \ell = 0, \tfrac{1}{2} \, .$$

Their characters are given by [**27, 28, 29**]

$$\begin{aligned} ch_{\frac{1}{4},0}(\tau,z) &= \frac{\theta_1^2(\tau,z)}{\eta^3(\tau)}\mu(\tau,z), \\ ch_{\frac{1}{4},\frac{1}{2}}(\tau,z) &= q^{-\frac{1}{8}}\frac{\theta_1(\tau,z)}{\eta^3(\tau)} - 2\frac{\theta_1(\tau,z)^2}{\eta^3(\tau)}\mu(\tau,z), \end{aligned} \tag{3.5}$$

where $\mu(\tau, z)$ denotes the so-called Appell-Lerch sum

$$\mu(\tau,z) = \frac{-iy^{1/2}}{\theta_1(\tau,z)}\sum_{\ell=-\infty}^{\infty}\frac{(-1)^\ell y^\ell q^{\ell(\ell+1)/2}}{1-yq^\ell}.$$

Notice that the supersymmetric representations have non-vanishing Witten index

$$ch_{\frac{1}{4},0}(\tau,z=0)=1, \quad ch_{\frac{1}{4},0}(\tau,z=0)=-2.$$

On the other hand the massive (or 'non-BPS' or 'non-supersymmetric') representations with

$$h=\tfrac{1}{4}+n, \quad \ell=\tfrac{1}{2}, \qquad n=1,2,\ldots$$

have the character given by

$$ch_{\frac{1}{4}+n,\frac{1}{2}}(\tau,z) = q^{-\frac{1}{8}+n}\frac{\theta_1^2(\tau,z)}{\eta^3(\tau)}$$

which has, by definition, vanishing Witten index $ch_{\frac{1}{4}+n,\frac{1}{2}}(\tau,z=0)=0$.

Rewriting the $K3$ elliptic genus in terms of these characters, we arrive at the following specific expression for the weak Jacobi form $\mathcal{Z}(\tau,z)$ [**26, 10, 30**]

$$\mathcal{Z}(\tau,z) = \frac{\theta_1^2(\tau,z)}{\eta^3(\tau)}\left(a\,\mu(\tau,z) + q^{-1/8}\big(b+\sum_{n=1}^{\infty}t_nq^n\big)\right) \tag{3.6}$$

where $a, b \in \mathbb{Z}$ and $t_n \in \mathbb{Z}$ for all positive integers n, and the t_n count the number of non-supersymmetric representations with $h = n + 1/4$ which contribute to the elliptic genus. As the notation suggests, from (3.4) we can compute the above integers to be $a = 24, b = -2$, and the first few t_n's are indeed as we have seen in the last section

$$2\times 45,\ 2\times 231,\ 2\times 770,\ 2\times 2277,\ 2\times 5796,\ldots$$

In other words, the mock modular form $H(\tau)$ (cf. (2.4)) can be interpreted as the generating function of multiplicities of massive irreducible representations of the $\mathcal{N} = 4$ superconformal algebra in the $K3$ elliptic genus.

In this context, the origin of the mock modularity of $H(\tau)$ can be understood in the following way. From the fact that the elliptic genus (3.6) transforms nicely under $SL_2(\mathbb{Z})$, so must the combination $24\mu(\tau,z) + H(\tau)$. Now, the Appell-Lerch sum itself does not transform nicely, rather its non-holomorphic completion

$$\hat{\mu}(\tau;z) = \mu(\tau;z) - \frac{1}{2}\int_{-\bar{\tau}}^{\infty} dz\,\frac{\eta^3(-z)}{\sqrt{i(z+\tau)}}$$

transforms like a weight 1/2 theta function. This has been demonstrated in [**31**] as part of a systematic treatment of mock theta-functions. Therefore, the mock modularity of the q-series $H(\tau)$ is directly related to the mock modularity of the massless (BPS) $\mathcal{N} = 4$ characters (3.5).

In (3.6) we have seen the appearance of the infinite-dimensional M_{24}-module K in the elliptic genus $\mathcal{Z}(\tau,z)$. One might wonder whether M_{24} acts on the other part of the decomposition as well. A simple observation is that $a = 24$ is the dimension of the defining permutation representation R of M_{24}, and hence a naive guess will be that M_{24} acts on the massless $\mathcal{N} = 4$ multiplets as the direct sum of R and a two-dimensional (odd) trivial representation and this suggests that we write $24 = \mathrm{tr}_R\,\mathbf{1}$. Together with this assumption, the conjecture (2.2) of the previous section implies that M_{24} acts on the states of the $K3$ sigma model contributing to the elliptic genus and moreover commutes with the superconformal algebra. Therefore, it is natural to consider also the twisted (or equivariant) elliptic genus which is expected to have the decomposition

$$\begin{aligned}\mathcal{Z}_g(\tau,z) &= \frac{\theta_1^2(\tau,z)}{\eta^3(\tau)}\left(\chi(g)\,\mu(\tau,z) + H_g(\tau)\right)\\ &= \sum_{n\geq 0,\ell\in\mathbb{Z}} c_g(4n-\ell^2)q^n y^\ell \quad \text{with } \chi(g) = \mathrm{tr}_R\, g.\end{aligned} \tag{3.7}$$

Again, a non-trivial connection between weak Jacobi forms and M_{24} arises if all such $\mathcal{Z}_g(\tau,z)$ display interesting modular properties. Moreover, as we mentioned in §1, consistency with the CFT interpretation requires this to be true. More specifically it requires that $\mathcal{Z}_g(\tau,z)$ transform nicely under the action of $\Gamma_0(n_g)$. Indeed, from the (mock) modularity of $H_g(\tau)$ (cf. (2.1)) it is now easy to show

PROPOSITION 3.1. *For all $g \in M_{24}$, the function*

$$\mathcal{Z}_g(\tau,z) = \frac{\chi(g)}{12}\,\varphi_{0,1}(\tau,z) + \tilde{T}_g(\tau)\,\varphi_{-2,1}(\tau,z) \tag{3.8}$$

is a weak Jacobi form of weight 0 and index 1 for the group $\Gamma_0(N_g)$. Moreover, we have

$$\mathcal{Z}_g(\tau,z) = \rho_{n_g|h_g}(\gamma)e\left(-\tfrac{cz^2}{c\tau+d}\right)\mathcal{Z}_g(\gamma(\tau,z))$$

for $\gamma \in \Gamma_0(n_g)$.

In the above $\varphi_{0,1}(\tau,z)$ and $\varphi_{-2,1}(\tau,z)$ are weight 0 and weight 2, index 1 weak Jacobi forms whose explicit expressions can be found in Appendix A. The weight two modular forms $\tilde{T}_g(\tau)$ are listed in Table 2.

As for M_{24}-modules, the conjecture (2.2) on the relationship between $H_g(\tau)$ and M_{24} now implies that the Fourier coefficients $c_g(4n-\ell^2)$ of the twisted $K3$ elliptic genus encode the supercharacter of a $\mathbb{Z}$-graded super (or virtual) representation $\hat{K} = \bigoplus \hat{K}_{4n-\ell^2}$ for M_{24}. More specifically, we have the following conjecture.

CONJECTURE 3.2. *The Fourier coefficients $c_g(k)$ of*

$$\mathcal{Z}_g(\tau,z) = \sum_{n\geq 0,\ell\in\mathbb{Z}} c_g(4n-\ell^2)q^n y^\ell$$

satisfy $c_g(k) = \mathrm{str}_{\hat{K}_k} g$ for $k \equiv 0,3 \pmod 4$ and $k \geq -1$ where $\hat{K}_k$ is the degree k component of an infinite-dimensional, $\mathbb{Z}$-graded super (or virtual) module $\hat{K}$ for M_{24}.

$$\hat{K} = \left(\bigoplus_{\ell=0}^{\infty} \hat{K}_{4\ell-1}\right) \bigoplus \left(\bigoplus_{\ell=0}^{\infty} \hat{K}_{4\ell}\right).$$

Moreover, when $k \neq 0$ the representation $\hat{K}_k$ is even in the sense that is has the form $\hat{k}_k \oplus \hat{k}_k^*$ where $\hat{k}_k$ is a certain super representation of M_{24} and $\hat{k}_k^*$ denotes its dual.

The first few coefficients $c_g(k)$ and the corresponding M_{24}-supermodule $\hat{K}_k$ can be found in Tables 5 and 6. Notice that, from the expression (3.7) one can see that the existence of the $\mathbb{Z}$-graded, infinite-dimensional M_{24}-module $K = \bigoplus_{n=1}^{\infty} K_n$ underlying the mock modular forms $H_g(\tau)$ as conjectured in Conjecture 2.2 implies the existence of the $\mathbb{Z}$-graded super (or virtual) module $\hat{K}$ in the above conjecture.

We conclude this section with a related conjecture which seeks to identify a mechanism that will explain why the largest Mathieu group should be related to the modular objects described in this section. In this sense, this is a stronger conjecture than Conjecture 3.2.

CONJECTURE 3.3. *The elliptic cohomology of $K3$ surfaces furnishes an infinite-dimensional, $\mathbb{Z}$-graded M_{24}-supermodule. More precisely, the cohomology of the right-moving supercharge as computed by the elliptic genus (3.3) is an infinite-dimensional $\mathbb{Z}$-graded M_{24}-supermodule whose characters (the twisted elliptic genera) are given by $\mathcal{Z}_g(\tau, z)$.*

In particular, if there exist special points in the CFT moduli space where the Abelian subgroup $\langle g \rangle \subset M_{24}$ generated by an element $g \in M_{24}$ is a symmetry of the full Hilbert space (and not just the cohomology), at such a point we have

$$\mathcal{Z}_g(\tau, z) = \mathrm{tr}_{\mathcal{H}_{RR}} \left(g \, (-1)^{J_0 + \bar{J}_0} y^{J_0} q^{L_0 - c/24} \bar{q}^{\bar{L}_0 - c/24} \right) . \tag{3.9}$$

As evidence for this conjecture we observe that (3.9) is known to be true in the case of conformal field theories corresponding to certain special points in the $K3$ moduli space where the group $\langle g \rangle$ is realised as hyper-Kähler automorphisms of the $K3$ surface. This is related to the result of Mukai [**32**] (see also [**33**]) that any finite group of hyper-Kähler automorphisms of a $K3$ surface can be embedded in M_{23}—the maximal subgroup of M_{24} fixing a point in the defining permutation representation R. Mukai's result can be obtained by considering the relationship between the action of M_{24} on a certain Niemeier lattice and the cohomology lattice of the $K3$ surface in question. Explicit computations along these lines can be found in [**34**]. An interesting stringy analogue of Mukai's result has been obtained in [**35**]. For $g \in 2A, 3A, 4B, 5A, 6A, 7A$, the twisted $K3$ elliptic genus has been computed in this way in [**36, 37, 38**]. On the other hand, the elliptic cohomology can remain invariant even when the $K3$ surface is not invariant. In the generic situation where g is a symmetry of the cohomology but not the full Hilbert space, one should adopt the above formula and first compute the cohomology by taking the trace on the right-moving sector before inserting the operator g.

4. Siegel Modular Forms and M_{24}

In §3 we have discussed the relation between M_{24} and a set of weak Jacobi forms. It turns out that there is a set of Siegel modular forms that are closely related to these weak Jacobi forms via the so-called Borcherds, or exponential lift. As a result, we expect the relation described in §3 to be translated into a relation between these Siegel modular forms and M_{24}. In particular, both the permutation representation R and the infinite-dimensional representation K play an evident role in defining these Siegel forms. It is the goal of this section to describe and discuss these connections.

First we will consider the relation between the weak Jacobi forms appearing in §3 and Siegel modular forms. Recall that it is natural to consider the weak Jacobi forms $\mathcal{Z}_g(\tau,z)$ as twisted elliptic genera of $K3$ surfaces as this interpretation makes the modularity as well as the connection to M_{24} and the mock modular forms in §2 manifest. Similarly, we can arrive at Siegel modular forms by considering the elliptic genera of the symmetric powers $S^N M = M^N/S_N$ of a Calabi–Yau manifold M. In [**39, 40**], an elegant expression for the generating function of the $\mathbf{EG}(\tau,z;S^N M)$ for a d_0 (complex) dimensional Calabi–Yau manifold M in terms of $\mathbf{EG}(\tau,z;M)$ has been obtained by identifying such a generating function with the partition function of second-quantised strings on $S^1 \times M$. This identification gives

$$\begin{aligned}\sum_{N=0}^{\infty} p^N\,\mathbf{EG}(\tau,z;S^N M) &= \exp\left(\sum_{N=0}^{\infty} p^N (T_N\,\mathbf{EG})(\tau,z;M)\right)\\ &= \prod_{\substack{n>0,\\ m\geq 0,\\ \ell\in\mathbb{Z}}}\left(\frac{1}{1-p^n q^m y^\ell}\right)^{c(2d_0 nm-\ell^2)}\end{aligned} \tag{4.1}$$

where

$$\mathbf{EG}(\tau,z;M) = \sum_{n,\ell} c(2d_0 nm-\ell^2) q^n y^\ell$$

is the elliptic genus of M itself and T_N is the Hecke operator mapping a Jacobi form $\phi(\tau,z)$ of weight zero and index t to a form of weight zero and index Nt. Explicitly, we have

$$(T_N\phi)(\tau,z) = \frac{1}{N}\sum_{\substack{ad=N\\ b \bmod d}} \phi\left(\frac{a\tau+b}{d}, az\right). \tag{4.2}$$

The right hand side of (4.2) has a natural interpretation as a sum over degree N maps between elliptic curves $E \to E'$, mapping the two 1-cycles of E according to $\left(\begin{smallmatrix} a & b\\ 0 & d\end{smallmatrix}\right)$, and summing over N gives the instanton contributions to the free energy of the free string theory. This leads to a path-integral derivation of the expression (4.1) for the second-quantised elliptic genus. As the elliptic genus of a single copy of M is a weak Jacobi form, one might guess that this second-quantised elliptic genus, encoding elliptic genera of all the symmetric powers of M through the parameter p, also has interesting automorphy. It was observed in [**39, 25**] that, up to a factor which has the effect of symmetrising p and q, the second-quantised elliptic genus is in fact a Siegel modular form when M is Calabi–Yau.

To explain this recall that the *genus two Siegel upper half space* $\mathbb{H}_2$ may be realised concretely as the set of symmetric 2×2 matrices with complex entries whose imaginary parts are positive definite.

$$\mathbb{H}_2 = \{\Omega \in M_2(\mathbb{C}) \mid \Omega = \Omega^t,\ \det\Im(\Omega) > 0,\ \operatorname{tr}\Im(\Omega) > 0\} \tag{4.3}$$

We may introduce (complex) coordinates σ, τ and z on $\mathbb{H}_2$ by requiring that

$$\Omega = \begin{pmatrix} \sigma & z\\ z & \tau\end{pmatrix} \tag{4.4}$$

and then $\sigma,\tau \in \mathbb{H}$, where $\mathbb{H}$ is the upper half plane, and $z \in \mathbb{C}$ satisfies $\Im(z)^2 < \Im(\sigma)\Im(\tau)$. A function $\Phi_k(\Omega) : \mathbb{H}_2 \to \mathbb{C}$ is called a *Siegel modular form of weight* k for a group $\Gamma^{(2)} \subset Sp_2(\mathbb{Q})$ if

$$\Phi_k((A\Omega+B)(C\Omega+D)^{-1}) = (C\Omega+D)^k\Phi_k(\Omega)$$

for all

$$\begin{pmatrix} A & B \\ C & D \end{pmatrix} \in \Gamma^{(2)} .$$

As alluded to above, after introducing a factor $A(\Omega)$, sometimes referred to as the Hodge anomaly, which depends only on the Hodge numbers of M, the second-quantised elliptic genus of an even-dimensional Calabi–Yau manifold M is a Siegel modular form of some weight for the full Siegel modular group $Sp_2(\mathbb{Z})$. More precisely, upon setting $p = e(\sigma)$ one obtains that

$$\frac{1}{\Phi_k(\Omega)} = A(\Omega) \sum_{N=0}^{\infty} p^N \, \mathbf{EG}(\tau, z; S^N M)$$

is a Siegel modular form of weight $-k = -\frac{1}{2}\chi'_{d_0}(M)$ for $\dim_{\mathbb{C}} M = 2d_0$ where

$$\chi'_n(M) = \sum_{m=1}^{2d_0} (-1)^{n+m} h^{n,m}(M).$$

See [**25**] for a complete description of the modular property.

In the case of interest to us, where M is a $K3$ surface, we have [**52**]

$$\begin{aligned} \Phi_{10}(\Omega) &= pqy \prod_{\substack{m,n,l\in\mathbb{Z} \\ (m,n,l)>0}} (1 - p^m q^n y^l)^{c(4mn-l^2)} \\ &= pqy \prod_{(n,m,\ell)>0} \exp\left(- \sum_{k=1}^{\infty} \frac{c(4nm-\ell^2)}{k} (p^m q^n y^\ell)^k \right) \end{aligned} \tag{4.5}$$

for the reciprocal of the $A(\Omega)$-corrected second-quantised elliptic genus, where the condition denoted $(m,n,l)>0$ is that $m, n \geq 0$ and $l < 0$ when $m = n = 0$ and the values $c(k)$ are the Fourier coefficients of the $K3$ elliptic genus (3.7). This function $\Phi_{10}(\Omega)$ is a cusp form of weight 10 for $Sp_2(\mathbb{Z})$ and is typically taken as one of four standard generators of the full ring of Siegel modular forms for $Sp_2(\mathbb{Z})$ [**41, 42**].

The elliptic genus plays an evident role in defining the automorphic form $\Phi_{10}(\Omega)$. Mathematically, this is a special case of a general method, known as the *Borcherds* or *exponential lift*, for transforming the data of a weak Jacobi form into a Siegel modular form [**51, 52, 53**]. In §3 we have seen that, apart from $\mathcal{Z}(\tau, z)$, we have in fact a weak Jacobi form $\mathcal{Z}_g(\tau, z)$ for each conjugacy class $[g]$ of M_{24}, and a natural question to ask at this point is whether there are also Siegel modular forms $\Phi_g(\Omega)$ obtained from these by a similar lifting. It turns out that, both from the algebraic and geometric points of view, there are natural candidates for such functions on $\mathbb{H}_2$.

From the interpretation of the Hecke operator as summing over degree N maps between elliptic curves and the interpretation of twisting as changing the boundary condition in the path integral, we see that we should replace the Hecke operator with the *equivariant* (or *twisted*) Hecke operator:

$$(T_N \mathcal{Z}_g)(\tau, z) = \frac{1}{N} \sum_{\substack{ad=N \\ b \bmod d}} \mathcal{Z}_{g^a}\left(\frac{a\tau + b}{d}, az \right). \tag{4.6}$$

(See [**54, 55**] for a more thorough geometric description of the equivariant Hecke operator.) From the above argument we see that the corresponding twisted object is [13]

$$\Phi_g(\Omega) = pqy \prod_{(n,m,\ell)>0} \exp\left(-\sum_{k=1}^{\infty} \frac{c_{g^k}(4nm-\ell^2)}{k}(p^m q^n y^\ell)^k\right). \tag{4.7}$$

Note that the above expression can again be rewritten in an infinite-product form. The functions $\Phi_g(\Omega)$ for $g \in 2A, 3A, 5A, 7A$ were proposed previously as partition functions for 1/4-BPS states twisted by a $\mathbb{Z}/p$ symmetry of the theory for $p = 2, 3, 5, 7$ in [**56, 57, 58, 38, 59**]. See also [**60**] for a closely related discussion.

The zeros of a Siegel modular form constitute its *rational quadratic divisor.* For instance, $\Phi_{10}(\Omega)$ has a double zero at $z = 0$

$$\lim_{z\to 0} (2\pi i z)^{-2}\, \Phi_{10}(\Omega) = \eta(\tau)^{24}\eta(\sigma)^{24}. \tag{4.8}$$

Moreover, the zeros of $\Phi_{10}(\Omega)$ are given by the images of $z = 0$ under the Siegel modular group $Sp_2(\mathbb{Z})$.

Interestingly, this limit recovers (two copies of) the cusp form $\eta_g(\tau)$ discussed in §1 for the case that g is the identity element. In fact, this property holds more generally. In §3 we have seen how the weak Jacobi forms $\mathcal{Z}_g(\tau, z)$ incorporate both the permutation representation R, which is responsible for the connection between cusp forms $\eta_g(\tau)$ and M_{24}, and the infinite-dimensional M_{24} representation K underlying the connection to the mock modular forms $H_g(\tau)$. The data of both these M_{24}-modules is thus contained in $\Phi_g(\Omega)$ via the lifting procedure described above. Through direct computation we see that taking the limit $z \to 0$ amounts to "forgetting" the data of the massive module K. And more generally, using (1.4) it can be shown [**13**] that

$$\lim_{z\to 0} (2\pi i z)^{-2}\, \Phi_g(\Omega) = \eta_g(\tau)\, \eta_g(\sigma) \tag{4.9}$$

for all $g \in M_{24}$ with $\Phi_g(\Omega)$ defined as in (4.7). (See also [**61**] for a slightly different approach.)

It is natural to ask about the automorphic properties of the $\Phi_g(\Omega)$. As $\Phi_{10}(\Omega)$—corresponding to the identity element of M_{24}—is a Siegel modular form for the full Siegel modular group, it is reasonable to guess that each $\Phi_g(\Omega)$ for $g \in M_{24}$ is a Siegel modular form for some congruence subgroup of $Sp_2(\mathbb{Z})$. Inspired by physical considerations to be discussed in the next section we are led to the following.

CONJECTURE 4.1. *For $g \in M_{24}$ the function $\Phi_g(\Omega)$ is a Siegel modular form, with a possibly non-trivial multiplier, for a subgroup of $Sp_2(\mathbb{Z})$ containing $\Gamma_0^{(2)}(N_g)$, where*

$$\Gamma_0^{(2)}(N) = \left\{ \begin{pmatrix} A & B \\ C & D \end{pmatrix} \in Sp_2(\mathbb{Z}) \middle| \, C \equiv 0 \bmod N \right\}.$$

Note that $\Gamma_0^{(2)}(N)$ is the genus two analogue of $\Gamma_0(N)$ defined in (1.2). As evidence in support of Conjecture 4.1 we remark that the functions $\Phi_g(\Omega)$ for $g \in 2A, 3A, 4B$ have been discussed in [**62**] and have been verified to transform in the required way under $\Gamma_0^{(2)}(N_g)$ with $N_g = 2, 3, 4$ respectively. The automorphic property of $\Phi_g(\Omega)$ has also been discussed for $g \in 5A, 7A$ in [**36**].

We conlcude this section with some discussion of open questions. First, apart from the infinite products we have seen in (4.5) and (4.7) another way to construct a Siegel modular form of a given weight is to apply the *Maass* or *additive lift* to a Jacobi form of the same weight. See for instance [**53, 62**]. It seems that the cusp form $\eta_g(\tau)$ can be used to perform an additive lift construction of $\Phi_g(\Omega)$ for some conjugacy classes of M_{24} as well [**63, 13, 61**]; however, so far there is no known formulation that applies to all elements of M_{24}.

Second, recall that (4.8) may be obtained as the *pinching limit* ($\epsilon \to 0$) of the genus two bosonic string amplitude associated to gluing pairs of tori [**39, 64**]. It would be interesting to have an analogous interpretation for general $\Phi_g(\Omega)$. We refer to [**65**] for a related discussion relevant to the special case that $g \in 2A$.

Last but not least, the weight five Siegel modular form $\Delta_5 = \Phi_{10}^{1/2}$ coincides with the denominator of a generalised Kac–Moody superalgebra [**52**]. The relationship between $\Phi_{10}(\Omega)$, the Jacobi form $\mathcal{Z}(\tau, z)$ and representations of M_{24} (cf. §3) suggests the possibility of an action of M_{24} on the roots of this or some closely related superalgebra. Since the Siegel modular form $\Phi_{10}(\Omega)$, together with its twists $\Phi_g(\Omega)$, unifies all the automorphic objects discussed above it would be very interesting to have a concrete M_{24}-module that realises these functions $\Phi_g(\Omega)$. The existence of the generalised Kac–Moody superalgebra attached to Δ_5 suggests that such a construction might be possible. See [**66**] for some related discussions. We mention here that early results relating Siegel modular forms to infinite-dimensional Kac–Moody algebras appeared in [**67**]. In particular, evidence for a family of generalisations of the Maass lift (mentioned above) is provided in loc. cit. According to [**68**] the subalgebra of the generalised Kac–Moody superalgebra attached to Δ_5 generated by its real roots appears as a subalgebra of the hyperbolic Kac–Moody algebra considered in [**67**]. Interesting connections between Δ_5, moduli of $K3$ surfaces and (an "automorphic correction" of) the hyperbolic Kac–Moody algebra of [**67**] are established in [**69**].

5. String Theory and M_{24}

So far we have phrased our discussions in terms of automorphic forms and the relatively familiar language of conformal field theories and vertex operator algebras. However, much useful intuition can be been derived from the physical system of quantum black holes in the $K3 \times T^2$-compactified string theory—another indication that the largest Mathieu group is closely connected to $K3$ surfaces in a way that is yet to be fully understood. In this final section we will describe how the various mathematical objects and relations discussed above can be understood in the context of this physical setup. We will assume familiarity with the basics of string theory.

First we start with the string theoretic interpretation of the cusp forms of §1. Apart from being the partition function of the theory with 24 free chiral bosons, the function $1/\eta(\tau)^{24}$, which is $1/\eta_g(\tau)$ for g the identity of M_{24}, encodes the graded numbers of the supersymmetric (perturbative) states in heterotic string theory. As a result, the duality between type II and heterotic string theories dictates that this function is also the generating function for the index enumerating BPS states preserving 8 of the 16 supercharges of type II string theory compactified on $K3 \times T^2$. In the event that a particular $K3$ surface admits a hyper-Kähler automorphism that can be identified with the cyclic subgroup $\langle g \rangle$ of M_{24} [**32, 33**] we can also compute the 1/2-BPS partition function twisted by the symmetry g of the Hilbert space. As can be expected from the discussion in §1 these turn out to be none other than inverses $Z_g(\tau) = 1/\eta_g(\tau)$ of cusp forms attached to M_{24} in the cases that have been

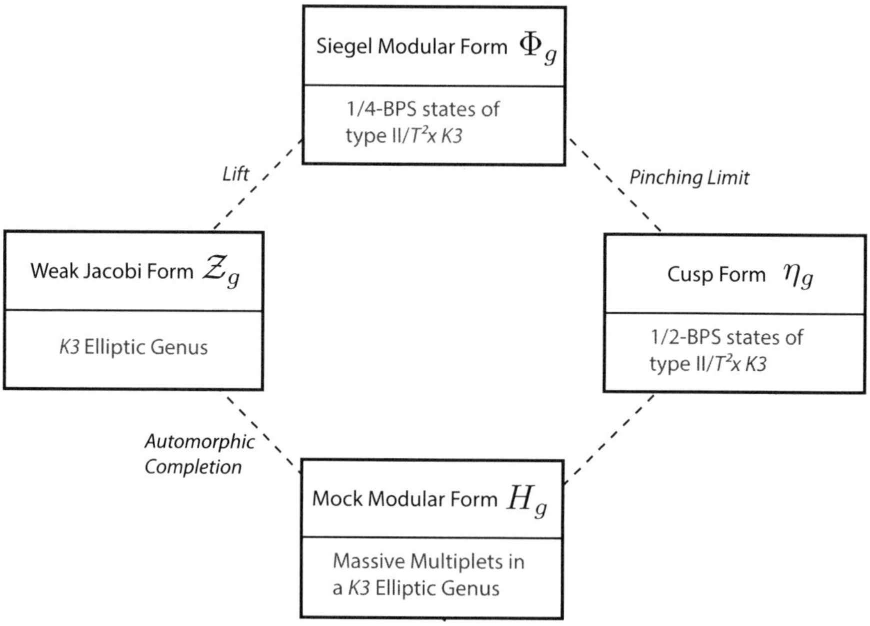

FIGURE 1. The relations between the automorphic objects discussed in this review and the physical objects associated to them.

explicitly verified; viz. $g \in 2A, 3A, 4B, 5A$ [**45, 59**]. More generally, if the 1/2-BPS spectrum of the $K3 \times T^2$-compactified type II string theory has a symmetry generated by $g \in M_{24}$ we can expect the corresponding twisted partition function to be given by $Z_g(\tau) = 1/\eta_g(\tau)$.

Next we turn to the physical interpretation of the connection between representations of M_{24} and the Siegel modular form $\Phi_{10}(\Omega)$ arising from the fact that the latter can be identified (cf. (4.5))with the Borcherds lift of $\mathcal{Z}(\tau, z)$. Recall that $\Phi_{10}(\Omega)$ is almost equal to the generating function of the elliptic genera of the symmetric powers $S^N K3$ and only differs from this by the Hodge anomaly $A(\Omega)$. Physically this extra factor $A(\Omega)$ accounts for the further compactification on $K3 \times T^2$ [**43, 44**]. As a result, the function $1/\Phi_{10}$ can be interpreted as counting 1/4-BPS states of type II superstring theory compactified on $K3 \times T^2$ [**39**]. (See [**45, 47, 46**] for a review of this topic and for further references.) There are interesting subtleties regarding the poles of the function $1/\Phi_{10}$ for which we refer also to [**48, 49, 50, 11**]. With this interpretation, Conjecture 3.2 suggests that these 1/4-BPS states of $K3 \times T^2$-compactified superstring theory furnish representations of M_{24}, which then suggests that we can compute the twisted 1/4-BPS partition function for the theory and arrive at $\Phi_g(\Omega)$ as given in (4.7).

It is interesting to observe that (4.9) must hold in order for the above string theoretic interpretation of $1/\Phi_g(\Omega)$ and $1/\eta_g(\tau)$ to be consistent [**13**]. This is because the number of states preserving 1/2 and 1/4 of the supersymmetries are related through the presence of the so-called two-centred bound states, which are bound states preserving 1/4 of the supersymmetries of two particles that preserve 1/2 of the supersymmetries individually. As alluded to earlier, the zeros of the Siegel modular form $\Phi_{10}(\Omega)$ play a special role in the physical counting of BPS states. They correspond to the poles of the partition function $1/\Phi_{10}$ and their physical significance is different depending on whether or not they intersect the (two-sheeted) hyperboloid $\langle\Im(\Omega),\Im(\Omega)\rangle = -R$ when R is taken to be very large (and real). On the other hand, the poles which do intersect the hyperboloid, such as those at $z=0$, control the absence or presence of the aforementioned two-centred bound states [**48, 49, 50, 11**]. Recall that such 1/2-BPS states can be thought of as quantum states in a theory with 24 free bosons and hence admit an action by M_{24} according to the discussion of §1; this dictates (4.9). We summarise the relations between the cusp forms, weak Jacobi forms, Siegel modular forms and their physical interpretations in Figure 1.

The above discussion constitutes strong evidence that the largest Mathieu group acts as symmetries of the supersymmetric states of $K3\times T^2$-compactified superstring theory at any point in the moduli space of the theory. An understanding of this is yet to be achieved but some related results have been obtained recently in [**34, 35**]. This problem is extremely interesting since a positive solution will provide us with a way to understand the relation between M_{24} and the automorphic objects discussed in the present paper through string theory.

Appendix A. Modular Forms

A.1. Dedekind Eta Function. The *Dedekind eta function*, denoted $\eta(\tau)$, is a holomorphic function on the upper half-plane defined by the infinite product

$$\eta(\tau)=q^{1/24}\prod_{n\geq 1}(1-q^n)$$

where $q=e(\tau)=e^{2\pi i\tau}$. It is a modular form of weight 1/2 for the modular group $SL_2(\mathbb{Z})$ with multiplier $\epsilon: SL_2(\mathbb{Z})\to\mathbb{C}^*$, which means that

$$\epsilon(\gamma)\eta(\gamma\tau)\,\mathrm{jac}(\gamma,\tau)^{1/4}=\eta(\tau)$$

for all $\gamma=\left(\begin{smallmatrix}a&b\\c&d\end{smallmatrix}\right)\in SL_2(\mathbb{Z})$, where $\mathrm{jac}(\gamma,\tau)=(c\tau+d)^{-2}$. The *multiplier system* ϵ may be described explicitly as

$$\epsilon\begin{pmatrix}a&b\\c&d\end{pmatrix}=\begin{cases}e(-b/24), & c=0,\ d=1\\ e(-(a+d)/24c+s(d,c)/2+1/8), & c>0\end{cases}\tag{A.1}$$

where $s(d,c)=\sum_{m=1}^{c-1}(d/c)((md/c))$ and $((x))$ is 0 for $x\in\mathbb{Z}$ and $x-\lfloor x\rfloor-1/2$ otherwise. We can deduce the values $\epsilon(a,b,c,d)$ for $c<0$, or for $c=0$ and $d=-1$, by observing that $\epsilon(-\gamma)=\epsilon(\gamma)e(1/4)$ for $\gamma\in SL_2(\mathbb{Z})$.

Let T denote the element of $SL_2(\mathbb{Z})$ such that $\mathrm{tr}(T)=2$ and $T\tau=\tau+1$ for $\tau\in\mathbb{H}$. Observe that

$$\epsilon(T^m\gamma)=\epsilon(\gamma T^m)=e(-m/24)\epsilon(\gamma)$$

for $m\in\mathbb{Z}$.

A.2. Theta Functions. The *Jacboi theta functions* $\theta_i(\tau,z)$ are defined as follows for $q=e(\tau)$ and $y=e(z)$.

$$\theta_1(\tau,z)=-iq^{1/8}y^{1/2}\prod_{n=1}^{\infty}(1-q^n)(1-yq^n)(1-y^{-1}q^{n-1})$$

$$\theta_2(\tau,z) = q^{1/8}y^{1/2}\prod_{n=1}^{\infty}(1-q^n)(1+yq^n)(1+y^{-1}q^{n-1})$$

$$\theta_3(\tau,z) = \prod_{n=1}^{\infty}(1-q^n)(1+y\,q^{n-1/2})(1+y^{-1}q^{n-1/2})$$

$$\theta_4(\tau,z) = \prod_{n=1}^{\infty}(1-q^n)(1-y\,q^{n-1/2})(1-y^{-1}q^{n-1/2})$$

A.3. Weak Jacobi Forms. According to [**70**, Thm. 9.3] (see also [**67**, §7]) the ring of weak Jacobi forms of even weight is generated by the functions

$$\begin{aligned}\varphi_{0,1}(\tau,z) &= 4\left(\left(\frac{\theta_2(\tau,z)}{\theta_2(\tau,0)}\right)^2+\left(\frac{\theta_3(\tau,z)}{\theta_3(\tau,0)}\right)^2+\left(\frac{\theta_4(\tau,z)}{\theta_4(\tau,0)}\right)^2\right),\\ \varphi_{-2,1}(\tau,z) &= -\frac{\theta_1(\tau,z)^2}{\eta(\tau)^6},\end{aligned}$$

over the ring of modular forms, where $\varphi_{0,1}$ has weight 0 and index 1 and $\varphi_{-2,1}$ has weight -2 and index 1.

A.4. Forms of Higher Level. The congruence subgroups of the modular group $SL_2(\mathbb{Z})$ that are most relevant for this paper are

$$\Gamma_0(N) = \left\{\begin{bmatrix} a & b\\ c & d\end{bmatrix}\in SL_2(\mathbb{Z}), c\equiv 0 \bmod N\right\}.$$

For $N>1$ a (non-zero) modular form of weight two for $\Gamma_0(N)$ is given by

$$\begin{aligned}\Lambda_N(\tau) &= N\,q\partial_q \log\left(\frac{\eta(N\tau)}{\eta(\tau)}\right)\\ &= \frac{N(N-1)}{24}\left(1+\frac{24}{N-1}\sum_{k>0}\sigma(k)(q^k-Nq^{Nk})\right)\end{aligned}\tag{A.2}$$

where $\sigma(k)$ is the divisor function $\sigma(k)=\sum_{d|k} d$.

A modular form on $\Gamma_0(N)$ is a modular form on $\Gamma_0(M)$ whenever $N|M$ and for some small N the space of forms of weight 2 is spanned by the $\Lambda_d(\tau)$ for d a divisor of N. In the case that $N=11$ we have the *newform*

$$f_{11}(\tau)=\eta^2(\tau)\eta^2(11\tau)\tag{A.3}$$

which is cusp form of weight 2 for $\Gamma_0(11)$ that is not a multiple of $\Lambda_{11}(\tau)$. We meet the newforms

$$f_{14}(\tau) = \eta(\tau)\eta(2\tau)\eta(7\tau)\eta(14\tau),\tag{A.4}$$

$$f_{15}(\tau) = \eta(\tau)\eta(3\tau)\eta(5\tau)\eta(15\tau),\tag{A.5}$$

at $N=14$ and $N=15$, respectively, and together with f_N the functions $\Lambda_d(\tau)$ for $d|N$ span the space of weight 2 forms on $\Gamma_0(N)$ for $N=11,14,15$.

For $N=23$ there is a two-dimensional space of newforms. We may use the basis

$$\begin{aligned}f_{23,1}(\tau) &= \eta_{23AB}(\tau)^2=\eta(\tau)^2\eta(23\tau)^2\\ \text{(A.6)}\,f_{23,2}(\tau) &= \frac{\eta(\tau)^3\eta(23\tau)^3}{\eta(2\tau)\eta(46\tau)}+4\eta(\tau)\eta(2\tau)\eta(23\tau)\eta(46\tau)+4\eta(2\tau)^2\eta(46\tau)^2\end{aligned}$$

(but note that these are not Hecke eigenforms). See [**71**] and Chapter 4.D of [**72**] for more details.

A discussion of the ring of weak Jacobi forms of higher level can be found in [**73**]. From its Proposition 6.1 we conclude that the space of weak Jacobi forms of weight 0 and index 1 for $\Gamma_0(N)$ is spanned by $\varphi_{0,1}(\tau, z)$ and the functions $f(\tau)\varphi_{-2,1}(\tau, z)$ where $f(\tau)$ is a modular form of weight 2 for $\Gamma_0(N)$.

Appendix B. Character Table

classes	$1A$	$2A$	$2B$	$3A$	$3B$	$4A$	$4B$	$4C$	$5A$	$6A$	$6B$	$7A$	$\overline{7A}$	$8A$	$10A$	$11A$	$12A$	$12B$	$14A$	$\overline{14A}$	$15A$	$\overline{15A}$	$21A$	$\overline{21A}$	$23A$	$\overline{23A}$
	1	1	1	1	1	1	1	1	1	1	1	1	1	1	1	1	1	1	1	1	1	1	1	1	1	1
	23	7	-1	5	-1	-1	3	-1	3	1	-1	2	2	1	-1	1	-1	-1	0	0	0	0	-1	-1	0	0
	45	-3	5	0	3	-3	1	1	0	0	-1	e_7	$\bar{e}_7$	-1	0	1	0	1	$-e_7$	$-\bar{e}_7$	0	0	e_7	$\bar{e}_7$	-1	-1
	$\overline{45}$	-3	5	0	3	-3	1	1	0	0	-1	$\bar{e}_7$	e_7	-1	0	1	0	1	$-\bar{e}_7$	$-e_7$	0	0	$\bar{e}_7$	e_7	-1	-1
	231	7	-9	-3	0	-1	-1	3	1	1	0	0	0	-1	1	0	-1	0	0	0	e_{15}	$\bar{e}_{15}$	0	0	1	1
	$\overline{231}$	7	-9	-3	0	-1	-1	3	1	1	0	0	0	-1	1	0	-1	0	0	0	$\bar{e}_{15}$	e_{15}	0	0	1	1
	252	28	12	9	0	4	4	0	2	1	0	0	0	0	2	-1	1	0	0	0	-1	-1	0	0	-1	-1
	253	13	-11	10	1	-3	1	1	3	-2	1	1	1	-1	-1	0	0	1	-1	-1	0	0	1	1	0	0
	483	35	3	6	0	3	3	3	-2	2	0	0	0	-1	-2	-1	0	0	0	0	1	1	0	0	0	0
	770	-14	10	5	-7	2	-2	-2	0	1	1	0	0	0	0	0	-1	1	0	0	0	0	0	0	e_{23}	$\bar{e}_{23}$
	$\overline{770}$	-14	10	5	-7	2	-2	-2	0	1	1	0	0	0	0	0	-1	1	0	0	0	0	0	0	$\bar{e}_{23}$	e_{23}
	990	-18	-10	0	3	6	2	-2	0	0	-1	e_7	$\bar{e}_7$	0	0	0	0	1	e_7	$\bar{e}_7$	0	0	e_7	$\bar{e}_7$	1	1
	$\overline{990}$	-18	-10	0	3	6	2	-2	0	0	-1	$\bar{e}_7$	e_7	0	0	0	0	1	$\bar{e}_7$	e_7	0	0	$\bar{e}_7$	e_7	1	1
	1035	27	35	0	6	3	-1	3	0	0	2	-1	-1	1	0	1	0	0	-1	-1	0	0	-1	-1	0	0
	1035	-21	-5	0	-3	3	3	-1	0	0	1	$2e_7$	$2\bar{e}_7$	-1	0	1	0	-1	0	0	0	0	$-e_7$	$-\bar{e}_7$	0	0
	$\overline{1035}$	-21	-5	0	-3	3	3	-1	0	0	1	$2\bar{e}_7$	$2e_7$	-1	0	1	0	-1	0	0	0	0	$-\bar{e}_7$	$-e_7$	0	0
	1265	49	-15	5	8	-7	1	-3	0	1	0	-2	-2	1	0	0	-1	0	0	0	0	0	1	1	0	0
	1771	-21	11	16	7	3	-5	-1	1	0	-1	0	0	-1	1	0	0	-1	0	0	1	1	0	0	0	0
	2024	8	24	-1	8	8	0	0	-1	-1	0	1	1	0	-1	0	-1	0	1	1	-1	-1	1	1	0	0
	2277	21	-19	0	6	-3	1	-3	-3	0	2	2	2	-1	1	0	0	0	0	0	0	0	-1	-1	0	0
	3312	48	16	0	-6	0	0	0	-3	0	-2	1	1	0	1	1	0	0	-1	-1	0	0	1	1	0	0
	3520	64	0	10	-8	0	0	0	0	-2	0	-1	-1	0	0	0	0	0	1	1	0	0	-1	-1	1	1
	5313	49	9	-15	0	1	-3	-3	3	1	0	0	0	-1	-1	0	1	0	0	0	0	0	0	0	0	0
	5544	-56	24	9	0	-8	0	0	-1	1	0	0	0	0	-1	0	1	0	0	0	-1	-1	0	0	1	1
	5796	-28	36	-9	0	-4	4	0	1	-1	0	0	0	0	1	-1	-1	0	0	0	1	1	0	0	0	0
	10395	-21	-45	0	0	3	-1	3	0	0	0	0	0	1	0	0	0	0	0	0	0	0	0	0	-1	-1

TABLE 3. Character table of M_{24}. See [**2**]. We adopt the naming system of [**2**] and use the notation $e_n = \frac{1}{2}(-1 + i\sqrt{n})$.

Appendix C. Tables of Decompositions

$1/\eta_g(\tau)$	$1A$	$2A$	$2B$	$3A$	$3B$	$4A$	$4B$	$4C$	$5A$	$6A$	$6B$	$7A$	$\overline{7A}$	$8A$	$10A$	$11A$	$12A$	$12B$	$14A$	$\overline{14A}$	$15A$	$\overline{15A}$	$21A$	$\overline{21A}$	$23A$	$\overline{23A}$
q^0	24	8	0	6	0	0	4	0	4	2	0	3	3	2	0	2	0	0	1	1	1	1	0	0	1	1
q^1	324	52	12	27	0	4	16	0	14	7	0	9	9	6	2	5	1	0	3	3	2	2	0	0	2	2
q^2	3200	256	0	104	8	0	48	0	40	16	0	22	22	12	0	10	0	0	4	4	4	4	1	1	3	3
q^3	25650	1122	90	351	0	18	142	6	105	39	0	51	51	28	5	20	3	0	9	9	6	6	0	0	5	5
q^4	176256	4352	0	1080	0	0	368	0	256	80	0	108	108	52	0	36	0	0	12	12	10	10	0	0	7	7
q^5	1073720	15640	520	3107	44	56	928	0	590	175	4	221	221	104	10	65	5	0	23	23	17	17	2	2	11	11
q^6	5930496	52224	0	8424	0	0	2176	0	1296	336	0	432	432	184	0	110	0	0	32	32	24	24	0	0	15	15
q^7	30178575	165087	2535	21762	0	175	4979	27	2740	666	0	819	819	341	20	185	10	0	55	55	37	37	0	0	22	22
q^8	143184000	495872	0	53976	192	0	10864	0	5600	1232	0	1506	1506	580	0	300	0	0	76	76	56	56	3	3	30	30
q^9	639249300	1428612	10908	129141	0	468	23184	0	11130	2289	0	2706	2706	1010	38	481	15	0	122	122	81	81	0	0	42	42

	ρ_1	ρ_{23}	ρ_{45}	$\rho_{\overline{45}}$	ρ_{231}	$\rho_{\overline{231}}$	ρ_{252}	ρ_{253}	ρ_{483}	ρ_{770}	$\rho_{\overline{770}}$	ρ_{990}	$\rho_{\overline{990}}$	$\rho_{1035'}$	ρ_{1035}	$\rho_{\overline{1035}}$	ρ_{1265}	ρ_{1771}	ρ_{2024}	ρ_{2277}	ρ_{3312}	ρ_{3520}	ρ_{5313}	ρ_{5544}	ρ_{5796}	ρ_{10395}
$\mathcal{H}_1$	1	1	0	0	0	0	0	0	0	0	0	0	0	0	0	0	0	0	0	0	0	0	0	0	0	0
$\mathcal{H}_2$	3	3	0	0	0	0	1	0	0	0	0	0	0	0	0	0	0	0	0	0	0	0	0	0	0	0
$\mathcal{H}_3$	6	8	0	0	0	0	3	2	1	0	0	0	0	0	0	0	1	0	0	0	0	0	0	0	0	0
$\mathcal{H}_4$	14	20	0	0	0	0	12	6	4	0	0	0	0	1	0	0	3	0	0	0	1	3	0	0	0	0
$\mathcal{H}_5$	27	48	0	0	0	0	33	22	15	0	0	0	0	3	0	0	15	1	0	3	6	16	3	0	0	3
$\mathcal{H}_6$	59	110	0	0	0	0	97	61	51	0	0	0	0	19	0	0	54	10	9	17	34	69	25	6	9	26
$\mathcal{H}_7$	114	249	0	0	6	6	255	174	161	3	3	3	3	70	3	3	190	45	47	88	158	276	147	59	71	194
$\mathcal{H}_8$	235	552	0	0	32	32	687	457	498	40	40	39	39	301	39	39	633	220	269	393	694	1042	758	418	490	1088
$\mathcal{H}_9$	460	1217	1	1	169	169	1783	1235	1504	255	255	296	296	1126	294	294	2152	994	1252	1730	2850	3870	3528	2354	2656	5544
$\mathcal{H}_{10}$	924	2677	24	24	731	731	4754	3294	4575	1425	1425	1675	1675	4329	1699	1699	7207	4391	5592	7131	11460	14340	15393	11758	13026	25565

TABLE 4. i) The first ten Fourier coefficients of the McKay-Thompson series $Z_g(\tau) = \sum_{k=0}^{\infty} q^{k-1}(\mathrm{tr}_{\mathcal{H}_k}\, g) = 1/\eta_g(\tau)$. ii) The decomposition of the first ten M_{24}-modules $\mathcal{H}_k$ into irreducible representations.

ℓ \ $[g]$	$1A$	$2A$	$2B$	$3A$	$3B$	$4A$	$4B$	$4C$	$5A$	$6A$	$6B$	$7A$	$\overline{7A}$	$8A$	$10A$	$11A$	$12A$	$12B$	$14A$	$\overline{14A}$	$15A$	$\overline{15A}$	$21A$	$\overline{21A}$	$23A$	$\overline{23A}$
0	2	2	2	2	2	2	2	2	2	2	2	2	2	2	2	2	2	2	2	2	2	2	2	2	2	2
1	-128	0	0	-2	4	16	0	8	2	6	12	5	5	8	10	4	10	8	7	7	8	8	11	11	10	10
2	-1026	-2	-2	0	0	62	-2	14	4	16	40	17	17	30	28	8	32	20	19	19	25	25	35	35	32	32
3	-5504	0	0	4	4	176	0	24	6	36	108	47	47	88	70	18	92	48	49	49	69	69	95	95	85	85
4	-23550	2	2	-6	12	450	2	50	10	74	260	110	110	226	162	34	234	104	114	114	169	169	236	236	209	209
5	-86400	0	0	0	0	1072	0	88	20	144	576	239	239	536	340	60	544	208	245	245	380	380	539	539	471	471

ℓ \ ρ	ρ_1	ρ_{23}	ρ_{45}	$\rho_{\overline{45}}$	ρ_{231}	$\rho_{\overline{231}}$	ρ_{252}	ρ_{253}	ρ_{483}	ρ_{770}	$\rho_{\overline{770}}$	ρ_{990}	$\rho_{\overline{990}}$	$\rho_{1035'}$	ρ_{1035}	$\rho_{\overline{1035}}$	ρ_{1265}	ρ_{1771}	ρ_{2024}	ρ_{2277}	ρ_{3312}	ρ_{3520}	ρ_{5313}	ρ_{5544}	ρ_{5796}	ρ_{10395}
0	2	0	0	0	0	0	0	0	0	0	0	0	0	0	0	0	0	0	0	0	0	0	0	0	0	0
1	8	-2	-1	-1	0	0	0	0	0	0	0	0	0	0	0	0	0	0	0	0	0	0	0	0	0	0
2	24	-6	-5	-5	-1	-1	0	0	0	0	0	0	0	0	0	0	0	0	0	0	0	0	0	0	0	0
3	64	-16	-15	-15	-5	-5	0	0	0	-1	-1	0	0	0	0	0	0	0	0	0	0	0	0	0	0	0
4	154	-40	-40	-40	-15	-15	0	0	0	-5	-5	0	0	0	0	0	0	0	0	-2	0	0	0	0	0	0
5	342	-90	-97	-97	-40	-40	0	0	0	-15	-15	0	0	0	0	0	0	0	0	-10	0	0	0	0	-2	0

TABLE 5. The first few Fourier coefficients $c_g(4\ell-1)$ and the corresponding super representation $\hat{K}_{4\ell-1}$ of M_{24}.

ℓ \ $[g]$	$1A$	$2A$	$2B$	$3A$	$3B$	$4A$	$4B$	$4C$	$5A$	$6A$	$6B$	$7A$	$\overline{7A}$	$8A$	$10A$	$11A$	$12A$	$12B$	$14A$	$\overline{14A}$	$15A$	$\overline{15A}$	$21A$	$\overline{21A}$	$23A$	$\overline{23A}$
0	20	4	-4	2	-4	-4	0	-4	0	-2	-4	-1	-1	-2	-4	-2	-4	-4	-3	-3	-3	-3	-4	-4	-3	-3
1	216	-8	8	0	0	-24	0	-8	-4	-8	-16	-8	-8	-12	-12	-4	-12	-8	-8	-8	-10	-10	-14	-14	-14	-14
2	1616	16	-16	-4	-4	-80	0	-16	-4	-20	-52	-22	-22	-40	-36	-12	-44	-28	-26	-26	-34	-34	-46	-46	-40	-40
3	8032	-32	32	4	-8	-224	0	-32	-8	-44	-136	-60	-60	-112	-88	-20	-116	-56	-60	-60	-86	-86	-120	-120	-110	-110
4	33048	56	-56	0	0	-568	0	-56	-12	-88	-320	-132	-132	-284	-196	-40	-292	-128	-140	-140	-210	-210	-294	-294	-256	-256
5	117280	-96	96	-8	-8	-1312	0	-96	-20	-168	-696	-292	-292	-656	-404	-68	-664	-240	-292	-292	-458	-458	-652	-652	-572	-572

ℓ \ ρ	ρ_1	ρ_{23}	ρ_{45}	$\rho_{\overline{45}}$	ρ_{231}	$\rho_{\overline{231}}$	ρ_{252}	ρ_{253}	ρ_{483}	ρ_{770}	$\rho_{\overline{770}}$	ρ_{990}	$\rho_{\overline{990}}$	$\rho_{1035'}$	ρ_{1035}	$\rho_{\overline{1035}}$	ρ_{1265}	ρ_{1771}	ρ_{2024}	ρ_{2277}	ρ_{3312}	ρ_{3520}	ρ_{5313}	ρ_{5544}	ρ_{5796}	ρ_{10395}
0	-3	1	0	0	0	0	0	0	0	0	0	0	0	0	0	0	0	0	0	0	0	0	0	0	0	0
1	-10	2	2	2	0	0	0	0	0	0	0	0	0	0	0	0	0	0	0	0	0	0	0	0	0	0
2	-32	8	6	6	2	2	0	0	0	0	0	0	0	0	0	0	0	0	0	0	0	0	0	0	0	0
3	-80	20	20	20	6	6	0	0	0	2	2	0	0	0	0	0	0	0	0	0	0	0	0	0	0	0
4	-190	50	50	50	20	20	0	0	0	6	6	0	0	0	0	0	0	0	0	4	0	0	0	0	0	0
5	-412	108	120	120	50	50	0	0	0	20	20	0	0	0	0	0	0	0	0	12	0	0	0	0	4	0

TABLE 6. The first few Fourier coefficients $c_g(4\ell)$ and the corresponding super representation $\hat{K}_{4\ell}$ of M_{24}.

n \ $[g]$	$1A$	$2A$	$2B$	$3A$	$3B$	$4A$	$4B$	$4C$	$5A$	$6A$	$6B$	$7A$	$\overline{7A}$	$8A$	$10A$	$11A$	$12A$	$12B$	$14A$	$\overline{14A}$	$15A$	$\overline{15A}$	$21A$	$\overline{21A}$	$23A$	$\overline{23A}$
1	90	-6	10	0	6	-6	2	2	0	0	-2	-1	-1	-2	0	2	0	2	1	1	0	0	-1	-1	-2	-2
2	462	14	-18	-6	0	-2	-2	6	2	2	0	0	0	-2	2	0	-2	0	0	0	-1	-1	0	0	2	2
3	1540	-28	20	10	-14	4	-4	-4	0	2	2	0	0	0	0	0	-2	2	0	0	0	0	0	0	-1	-1
4	4554	42	-38	0	12	-6	2	-6	-6	0	4	4	4	-2	2	0	0	0	0	0	0	0	-2	-2	0	0
5	11592	-56	72	-18	0	-8	8	0	2	-2	0	0	0	0	2	-2	-2	0	0	0	2	2	0	0	0	0
6	27830	86	-90	20	-16	6	-2	6	0	-4	0	-2	-2	2	0	0	0	0	2	2	0	0	-2	-2	0	0
7	61686	-138	118	0	30	6	-10	-2	6	0	-2	2	2	-2	-2	-2	0	-2	2	2	0	0	2	2	0	0
8	131100	188	-180	-30	0	-4	4	-12	0	2	0	-3	-3	0	0	2	2	0	-1	-1	0	0	0	0	0	0
9	265650	-238	258	42	-42	-14	10	10	-10	2	6	0	0	-2	-2	0	-2	-2	0	0	2	2	0	0	0	0

n \ ρ	ρ_1	ρ_{23}	ρ_{45}	$\rho_{\overline{45}}$	ρ_{231}	$\rho_{\overline{231}}$	ρ_{252}	ρ_{253}	ρ_{483}	ρ_{770}	$\rho_{\overline{770}}$	ρ_{990}	$\rho_{\overline{990}}$	$\rho_{1035'}$	ρ_{1035}	$\rho_{\overline{1035}}$	ρ_{1265}	ρ_{1771}	ρ_{2024}	ρ_{2277}	ρ_{3312}	ρ_{3520}	ρ_{5313}	ρ_{5544}	ρ_{5796}	ρ_{10395}
1	0	0	1	1	0	0	0	0	0	0	0	0	0	0	0	0	0	0	0	0	0	0	0	0	0	0
2	0	0	0	0	1	1	0	0	0	0	0	0	0	0	0	0	0	0	0	0	0	0	0	0	0	0
3	0	0	0	0	0	0	0	0	0	1	1	0	0	0	0	0	0	0	0	0	0	0	0	0	0	0
4	0	0	0	0	0	0	0	0	0	0	0	0	0	0	0	0	0	0	0	2	0	0	0	0	0	0
5	0	0	0	0	0	0	0	0	0	0	0	0	0	0	0	0	0	0	0	0	0	0	0	0	2	0
6	0	0	0	0	0	0	0	0	0	0	0	0	0	0	0	0	0	0	0	0	0	2	0	0	0	2
7	0	0	0	0	0	0	0	0	0	0	0	0	0	0	0	0	0	2	2	0	0	0	2	2	2	2
8	0	0	0	0	0	0	0	0	0	0	0	1	1	0	1	1	2	0	0	2	2	2	4	2	2	6
9	0	0	0	0	0	0	0	0	2	2	2	0	0	2	2	2	0	2	2	2	4	4	4	8	8	10

TABLE 7. The first few Fourier coefficients of the terms $q^{-\frac{1}{8}+n}$ in the q-series $H_g(\tau)$ and the corresponding representation K_n. Notice that the representations always come in conjugate pairs.

References

1. J. H. Conway and N. J. A. Sloane, *Sphere packings, lattices and groups.* Springer-Verlag, 1999.
2. J. Conway, R. Curtis, S. Norton, R. Parker, and R. Wilson, *Atlas of finite groups. Maximal subgroups and ordinary characters for simple groups. With comput. assist. from J. G. Thackray.* Oxford: Clarendon Press, 1985.
3. J. H. Conway and S. P. Norton, "Monstrous Moonshine," *Bull. London Math. Soc.* **11** (1979) 308 339.
4. G. Mason, "M_{24} and Certain Automorphic Forms," *Contemporary Mathematics* **45** (1985) 223–244.
5. D. Dummit, H. Kisilevsky, and J. McKay, "Multiplicative products of eta-functions.," *Contemp. Math. (Finite groups - coming of age, Proc. CMS Conf., Montreal/Que. 1982)* **45** (1985) .
6. D. Ford and J. McKay, "Ramifications of Ramanunjan's Work on η-products," *Proc. Indian Acad. Sci.* **99, No.3** (1989) 221–229.
7. S. Norton, "Generalized Moonshine," *Proc. Symp. Pure Math* **47** (1987) 208–209.
8. G. Mason, "G-elliptic systems and the genus zero problem for M_{24}," *Bull. Am. Math. Soc., New Ser. 25* **1** (1991) 45–53.
9. T. Eguchi, H. Ooguri and Y. Tachikawa, "Notes on the K3 Surface and the Mathieu group M_{24}," *Exper. Math* **20** (2011) 91 [arXiv:1004.0956 [hep-th]].
10. T. Eguchi and K. Hikami, "Superconformal algebras and mock theta functions," *J.Phys.A* **42:304010,2009** (Dec., 2008) , `0812.1151`.
11. A. Dabholkar, S. Murthy, and D. Zagier, "Quantum Black Hole, Wall-Crossing, and Mock Modular Forms," *to appear* .
12. K. Bringmann and K. Ono, "The $f(q)$ mock theta function conjecture and partition ranks," *Invent. Math.* **165** (2006) no. 2, 243–266. `http://dx.doi.org/10.1007/s00222-005-0493-5`.
13. M. C. N. Cheng, "$K3$ Surfaces, $N = 4$ Dyons, and the Mathieu Group M_{24}," *Commun. Num. Theor. Phys* **4** (2010) 623, `1005.5415`.
14. M. R. Gaberdiel, S. Hohenegger, and R. Volpato, "Mathieu twining characters for $K3$," *JHEP* **1009** (2010) 058, `1006.0221`.
15. M. R. Gaberdiel, S. Hohenegger, and R. Volpato, "Mathieu Moonshine in the elliptic genus of $K3$," *JHEP* **1010** (2010) 062, `1008.3778`.
16. T. Eguchi and K. Hikami, "Note on Twisted Elliptic Genus of $K3$ Surface," *Phys. Lett. B* **694** (2011) 446, `1008.4924`.
17. T. Gannon, 2011. Personal communication.
18. M. C. N. Cheng and J. F. R. Duncan, "On Rademacher Sums, the Largest Mathieu Group, and the Holographic Modularity of Moonshine," `1110.3859`.
19. J. F. R. Duncan and I. B. Frenkel, "Rademacher sums, moonshine and gravity," Commun. Num. Theor. Phys. **5** (2011) 849, `0907.4529`.
20. S. Ochanine, "Sur les genres multiplicatifs définis par des intégrales elliptiques. (french) [on multiplicative genera defined by elliptic integrals]," *Topology. An International Journal of Mathematics* **26** (1987) 143.
21. E. Witten, "Elliptic Genera and Quantum Field Theory," *Commun. Math. Phys.* **109** (1987) 525.
22. P. S. Landweber, ed., *Elliptic curves and modular forms in algebraic topology.* Springer-Verlag, Berlin, 1988.
23. E. Witten, "On the Landau-Ginzburg description of $N = 2$ minimal models," *Int.J.Mod.Phys.A* **9:4783-4800,1994** (Int.J.Mod.Phys.A9:4783-4800,1994) , `hep-th/9304026`.
24. T. Kawai, Y. Yamada, and S.-K. Yang, "Elliptic genera and n=2 superconformal field theory," *Nucl.Phys.B* **414:191-212,1994**, `hep-th/9306096`.
25. V. Gritsenko, "Elliptic genus of Calabi-Yau manifolds and Jacobi and Siegel modular forms," *St. Petersbg. Math. J.* **11**, No. 5, 781-804 (2000), `math/9906190`.
26. T. Eguchi, H. Ooguri, A. Taormina, and S.-K. Yang, "Superconformal Algebras and String Compactification on Manifolds with SU(N) Holonomy," *Nucl. Phys.* **B315** (1989) 193.
27. T. Eguchi and A. Taormina, "Unitary Representations of $N = 4$ Superconformal Algebra," *Phys. Lett.* **B196** (1987) 75.

28. T. Eguchi and A. Taormina, "Character Formulas for the $N = 4$ Superconformal Algebra," *Phys. Lett.* **B200** (1988) 315.
29. T. Eguchi and A. Taormina, "On the Unitary Representations of $N = 2$ and $N = 4$ Superconformal Algebras," *Phys. Lett.* **B210** (1988) 125.
30. T. Eguchi and K. Hikami, "Superconformal Algebras and Mock Theta Functions 2. Rademacher Expansion for $K3$ Surface," *Communications in Number Theory and Physics* **3,** (Apr., 2009) 531–554, 0904.0911.
31. S. Zwegers, *Mock Theta Functions.* PhD thesis, Utrecht University, 2002.
32. S. Mukai, "Finite groups of automorphisms of $K3$ surfaces and the Mathieu group," *Invent. Math.* **94, No.1** (1988) 183–221.
33. S. Kondo, Shigeyuki (Mukai, "Niemeier lattices, Mathieu groups, and finite groups of symplectic automorphisms of $K3$ surfaces. With an appendix by Shigeru Mukai.," *Duke Math. J* **92, No.3** (1998) 593–603.
34. A. Taormina and K. Wendland, "The symmetries of the tetrahedral Kummer surface in the Mathieu group M_{24}," 1008.0954.
35. M. R. Gaberdiel, S. Hohenegger, and R. Volpato, "Symmetries of $K3$ sigma models," 1106.4315.
36. J. R. David, D. P. Jatkar, and A. Sen, "Product Representation of Dyon Partition Function in CHL Models," *JHEP* **0606:064,2006** (JHEP 0606:064,2006) , hep-th/0602254.
37. S. Govindarajan and K. G. Krishna, "Generalized Kac-Moody Algebras from CHL dyons," *JHEP* **0904:032,2009** (July, 2008) , 0807.4451.
38. A. Sen, "Discrete Information from CHL Black Holes," JHEP **1011** (2010) 138, 1002.3857.
39. R. Dijkgraaf, E. Verlinde, and H. Verlinde, "Counting Dyons in N=4 String Theory," *Nucl.Phys.B* **484:543-561,1997** , hep-th/9607026.
40. R. Dijkgraaf, G. W. Moore, E. P. Verlinde, and H. L. Verlinde, "Elliptic genera of symmetric products and second quantized strings," *Commun. Math. Phys.* **185** (1997) 197–209, arXiv:hep-th/9608096.
41. J.-I. Igusa, "On Siegel modular forms of genus two (I)," *Am. J. Math.* **84** (1962) 175–200.
42. J.-I. Igusa, "On Siegel modular forms of genus two (II)," *Am. J. Math.* **86** (1964) 392–412.
43. J. R. David and A. Sen, "CHL Dyons and Statistical Entropy Function from D1-D5 System," *JHEP* **0611:072,2006** (JHEP 0611:072,2006) , hep-th/0605210.
44. D. Shih, A. Strominger, and X. Yin, "Recounting Dyons in N=4 String Theory," *JHEP* **0610:087,2006** , hep-th/0505094.
45. A. Sen, "Black hole entropy function, attractors and precision counting of microstates," *Gen.Rel.Grav.* **40:2249-2431,2008** (Aug., 2007) , 0708.1270.
46. M. C. N. Cheng, "The spectra of supersymmetric states in string theory," 0807.3099.
47. A. Dabholkar, "Cargese lectures on black holes, dyons, and modular forms," Nucl. Phys. Proc. Suppl. **171** (2007) 2.
48. A. Sen, "Walls of Marginal Stability and Dyon Spectrum in N=4 Supersymmetric String Theories," *JHEP* **0705:039,2007**, hep-th/0702141.
49. M. C. N. Cheng and E. Verlinde, "Dying Dyons Don't Count," *JHEP* **09** (2007) 070, arXiv:0706.2363 [hep-th].
50. A. Dabholkar, D. Gaiotto, and S. Nampuri, "Comments on the spectrum of CHL dyons," *JHEP* **01** (2008) 023, arXiv:hep-th/0702150.
51. R. E. Borcherds, "Automorphic forms with singularities on Grassmannians," *Invent. Math.* **132**, No.3, 491-562 (1998).
52. V. A. Gritsenko and V. V. Nikulin, "Siegel automorphic form corrections of some Lorentzian Kac-Moody algebras," *Amer.J.Math.* **119** (1991) 181.
53. V. A. Gritsenko and V. V. Nikulin, "Automorphic forms and Lorentzian Kac-Moody algebras. part ii," *Int. J. Math.* **9**, No.2, 201-275 (1998), alg-geom/9611028.
54. S. Carnahan, "Generalized moonshine i: Genus zero functions," *Algebra and Number Theory* **4:6** (Dec., 2008) 649–679, 0812.3440.
55. N. Ganter, "Hecke operators in equivariant elliptic cohomology and generalized moonshine," *[Harnad, John (ed.) et al., Groups and symmetries. From Neolithic Scots to John McKay. Selected papers of the conference, Montreal, Canada, April 27–29, 2007. Providence, RI: American Mathematical Society (AMS). CRM Proceedings and Lecture Notes 47, 173-209 (2009; Zbl 1206.55006)]* .

56. J. R. David, D. P. Jatkar, and A. Sen, "Dyon Spectrum in Generic N=4 Supersymmetric Z_N Orbifolds," *JHEP* **0701:016,2007** , hep-th/0609109.
57. A. Dabholkar and S. Nampuri, "Spectrum of Dyons and Black Holes in CHL orbifolds using Borcherds Lift," *JHEP* **0711:077,2007** , hep-th/0603066.
58. A. Sen, "A Twist in the Dyon Partition Function," 0911.1563.
59. S. Govindarajan and K. G. Krishna, "BKM Lie superalgebras from dyon spectra in Z_N-CHL orbifolds for composite N," *JHEP* **1005** (2010) 014, 0907.1410.
60. S. Govindarajan, "Unravelling Mathieu moonshine," 1106.5715.
61. T. Eguchi and K. Hikami, "Twisted elliptic genus for $K3$ and Borcherds product," 1112.5928.
62. V. Gritsenko and F. Clery, "The Siegel modular forms of genus 2 with the simplest divisor," *Proc. Lond. Math. Soc.* **(3) 102**, No. 6, 1024-1052 (2011), 0812.3962.
63. D. P. Jatkar and A. Sen, "Dyon Spectrum in CHL Models," *JHEP* **0604:018,2006**, hep-th/0510147.
64. M. P. Tuite, "Genus two meromorphic conformal field theory," *CRM Proceedings & Lecture Notes* **30,** (2001) 231–251., math/9910136.
65. A. Dabholkar and D. Gaiotto, "Spectrum of CHL Dyons from Genus-Two Partition Function," *JHEP* **0712:087,2007**, hep-th/0612011.
66. S. Govindarajan, "Brewing moonshine for Mathieu," 1012.5732.
67. A. J. Feingold and I. B. Frenkel, "A hyperbolic Kac-Moody algebra and the theory of Siegel modular forms of genus 2," *J. Math. Ann.* **263** (1983) 87–144.
68. A. J. Feingold and H. Nicolai, "Subalgebras of hyperbolic Kac-Moody algebras," in *Kac-Moody Lie algebras and related topics*, vol. 343 of *Contemp. Math.*, pp. 97–114. Amer. Math. Soc., Providence, RI, 2004.
69. V. A. Gritsenko and V. V. Nikulin, "Igusa modular forms and "the simplest" Lorentzian Kac-Moody algebras," *Mat. Sb.* **187** (1996) no. 11, 27–66. http://dx.doi.org/10.1070/SM1996v187n11ABEH000171.
70. M. Eichler and D. Zagier, *The theory of Jacobi forms*. Birkhäuser, 1985.
71. http://modi.countnumber.de/index.php?chap=ell.newforms/ell.newforms.html.
72. C. Itzykson, ed., *From Number Theory to Physics*. Springer, 1992.
73. H. Aoki and T. Ibukiyama, "Simple graded rings of Siegel modular forms, differential operators and Borcherds products," *Int. J. Math.* **16, No. 3** (2005) 249–279.

DEPARTMENT OF MATHEMATICS AND DEPARTMENT OF PHYSICS, HARVARD UNIVERSITY, CAMBRIDGE, MA 02138, USA

Current address: Université Paris 7, UMR CNRS 7586, Paris, France

E-mail address: chengm@math.jussieu.fr

DEPARTMENT OF MATHEMATICS, CASE WESTERN RESERVE UNIVERSITY, CLEVELAND, OH 44106, USA

E-mail address: john.duncan@case.edu

Proceedings of Symposia in Pure Mathematics
Volume **85**, 2012

(0,2) Quantum Cohomology

Ron Donagi, Joshua Guffin, Sheldon Katz, and Eric Sharpe

ABSTRACT. We review progress on a heterotic analogue of quantum cohomology, known as 'quantum sheaf cohomology'. Whereas ordinary quantum cohomology is computed by intersection theory on a moduli space of curves, quantum sheaf cohomology is computed via sheaf cohomology on a moduli space of curves.

1. Introduction

Quantum sheaf cohomology is a relatively new theory which is a simultaneous generalization of quantum cohomology and sheaf cohomology. Ordinary quantum cohomology can be defined for any complex manifold X, and computes intersection theory on a moduli space of curves in X. Quantum sheaf cohomology instead computes sheaf cohomology over a moduli space of curves in X. It is defined by a pair $(X, \mathcal{E})$, where X is a complex manifold and $\mathcal{E} \to X$ is a holomorphic vector bundle, satisfying the constraints

$$\wedge^{\rm top}\mathcal{E}^* \cong K_X, \quad {\rm ch}_2(\mathcal{E}) = {\rm ch}_2(TX)$$

A pair $(X, \mathcal{E})$ satisfying these properties is sometimes known as "omalous." Given an omalous pair, quantum sheaf cohomology is a family of ring structures on

$$\oplus_{p,q} H^p(X, \wedge^q \mathcal{E}^*)$$

Physically, quantum sheaf cohomology arises in the study of heterotic strings, where it computes nonperturbative corrections to charged-matter couplings. For example, if we compactify a ten-dimensional $E_8 \times E_8$ string on a Calabi-Yau three-fold with a rank three SU(3) gauge bundle, then one E_8 is broken to E_6, and the massless spectrum of the theory contains matter transforming under $\mathbf{27}$ and $\overline{\mathbf{27}}$ representations of E_6. In the special case that the gauge bundle matches the tangent bundle, the $\overline{\mathbf{27}}^3$ couplings are computed by the A model topological field theory, and quantum cohomology. More generally, when the gauge bundle is not the same as the tangent bundle, neither the A model nor quantum cohomology are relevant. It is in these more general cases that quantum sheaf cohomology is relevant, and computes nonperturbative corrections to *e.g.* $\overline{\mathbf{27}}^3$ couplings.

2000 *Mathematics Subject Classification.* Primary 14J81, Secondary 14N35, 14M25.
RD was partially supported by NSF grant DMS-0908487 and NSF-RTG 0636606.
SK was partially supported by NSF grant DMS-05-55678.
ES was partially supported by NSF grants DMS-0705381, PHY-0755614.

Quantum sheaf cohomology was first conjectured to exist by [**1**], where certain (0,2) Landau-Ginzburg model mirrors were also worked out, particularly for $\mathbb{P}^1 \times \mathbb{P}^1$. The details were initially worked out in [**14**], which has been followed up by a number of papers [**2, 8, 9, 11, 12, 15–20, 22, 25–28**]. The work [**2**] gave a general set of circumstances in which the correlators appearing in quantum sheaf cohomology would form a ring. The paper [**17**] derived quantum sheaf cohomology relations from physical aspects of linear sigma models, for toric cases and what they called 'linear' deformations of the tangent bundle. The recent papers [**8, 9**] extended the results of [**17**] to all deformations of toric tangent bundles, most of which are nonlinear deformations.

This paper surveys the recent progress [**8, 9**] and puts it in some context. In section 2 we explain the sense in which quantum sheaf cohomology generalizes quantum cohomology, and in section 3 we discuss it in the context of (0,2) mirror symmetry. The physics of quantum sheaf cohomology is reviewed in section 4, while the highlights of the mathematical arguments in [**8, 9**] are discussed in section 5. The special case of $\mathbb{P}^1 \times \mathbb{P}^1$, whose study initiated this project, is described in detail in section 6, followed by a list of open problems in the last section.

2. Quantum cohomology

The idea behind quantum cohomology is that the ordinary multiplication of cohomology classes on the variety X can be perturbed into a power series whose coefficients encode Gromov-Witten invariants, or intersections in moduli spaces of maps to X.

Fix a class $\beta \in H_2(X, \mathbb{Z})$, and let $M_\beta := M_{g,k}(X, \beta)$ be the moduli space of maps to X from a curve of genus g with k marked points. The complex structure of the curve, as well as the location of the marked points, are free to vary. We will usually restrict attention to $g = 0$.

There are several natural maps: the evaluation map $e : M_\beta \times \mathbb{P}^1 \to X$, the projection $\pi : M_\beta \times \mathbb{P}^1 \to M_\beta$, and k sections $s_i : M_\beta \to M_\beta \times \mathbb{P}^1$ for $i = 1, \ldots, k$ corresponding to the marked points.

The correlation function $\langle a_1, ..., a_k \rangle_\beta$ of cohomology classes $a_1, ..., a_k$ on X, or the degree β piece of the quantum product of the a_i's, is by definition the ordinary product on M_β of the induced classes $s_i^* e^* a_i$. Making sense of this requires several technicalities. In particular, we need to replace M_β by a good compactification X_β, and we need a well behaved notion of a virtual fundamental class.

For the formal definition, introduce the Novikov ring of X, which is the ring of formal power series $\Lambda := \mathbb{Z}[[H_2^+(X)]]$ on the semigroup $H_2^+(X)$ of effective classes in H_2. Elements of Λ are written

$$\lambda = \Sigma_{\beta \in H_2^+(X)} \lambda_\beta q^\beta, \tag{2.1}$$

where the q^β are formal symbols satisfying $q^{\beta_1+\beta_2} = q^{\beta_1} q^{\beta_2}$. Other variants of Λ are possible, and are frequently encountered. In this work we will not need power series, so we may as well replace Λ by the semigroup ring $\Lambda := \mathbb{Z}[H_2^+(X)]$.

The quantum cohomology ring is a Λ-algebra $QH^*(X)$. The additive structure, i.e. the underlying Λ-module, is $H^*(X) \otimes \Lambda$. By Poincaré duality, the quantum product $a_1 * a_2$ of elements $a_1, a_2 \in H^*(X)$ can be defined by specifying its Poincaré pairing with an arbitrary $a_3 \in H^*(X)$:

$$\langle a_1 * a_2, a_3 \rangle = \sum_\beta \langle a_1, a_2, a_3 \rangle_\beta q^\beta.$$

The multiplication is then extended to all of $QH^*(X)$ by Λ-linearity. If we set the quantum variables q to 0, i.e. take the constant coefficients in (2.1), we recover the classical multiplication on X.

The specific theory of quantum cohomology depends on the choice of a compactification. Several compactifications X_β of M_β are known. We will focus on two of these: the stable map, or NLSM (non-linear sigma model) compactification $\overline{M}_{0,3}(X, \beta)$, and the toric, or GLSM (=gauged linear sigma model) compactification, which we will denote by $\mathcal{M}_\beta$.

The stable map compactification is due to Kontsevich, and is defined for any smooth projective variety X, including Calabi-Yaus and Fanos. A point of the Kontsevich compactification explicitly depends on the location of the marked points. It is essentially geometric: the objects being parametrized are honest maps to X from various degenerations of the original curve Σ. This implies the existence of a universal curve over it, and makes possible the definition of Gromov-Witten invariants, or intersection numbers on all of X_β. We will denote the resulting quantum cohomology by $QH^*_{NLSM}(X)$.

On the other hand we have the toric, or GLSM compactification. This is defined for toric varieties X, and we will see in section 4 below that it is a toric variety itself. The moduli of the marked points do not enter into the definition of the GLSM compactification. It can be described by a global quotient construction, or constructed from a fan which is obtained from the fan for X by replicating each edge an appropriate number of times. The data it parametrizes is algebraic rather than geometric, so there is no universal object over it, making the definition of correlation functions harder. (Many of the technical difficulties addressed in the present work are due to the absence of a universal curve.) Still, we get a quantum cohomology ring, denoted $QH^*_{GLSM}(X)$. The additive structure is that of a free Λ-module, obtained as a quotient of the symmetric algebra $S^*H^2(X) \otimes \Lambda$ by certain relations. It is not clear whether this is isomorphic (as a Λ-module) to $H^*(X) \otimes \Lambda$.

In most of this article we focus exclusively on the GLSM. Here we make a few comments on the NLSM. It is obtained by using Kontsevich's stable map compactification $\overline{M}_\beta := \overline{M}_{g,k}(X, \beta)$ of $M_\beta := M_{g,k}(X, \beta)$. In this case, the quantum product is called the small quantum cohomology ring. The big quantum cohomology ring is defined using the complete set of all Gromov-Witten invariants of X, which is equivalent to the Frobenius manifold structure of $H^*(X)$. The small quantum cohomology, on the other hand, only involves the three-point functions. Physically, the big quantum cohomology ring includes gravity, while the small quantum cohomology ring does not. In general, gravity refers to variation of the metric. The relevant parameter for us is the complex structure of the source curve, which is determined by the metric. Since we restrict to genus zero, this amounts simply to the location of the k punctures. The big quantum cohomology ring encodes all Gromov-Witten invariants, *i.e.* intersection numbers on the Kontsevich compactification $\overline{M}_\beta$ of M_β, while the small quantum cohomology ring encodes only Gromov invariants, i.e. intersections on a fiber of $p : \overline{M}_\beta \to \overline{M}_{g,k}$. When $g = 0, k = 3$, the moduli space $\overline{M}_{g,k}$ is just a point, so the fiber of p is all of $\overline{M}_\beta$ and there is no difference between the two notions. However, as soon as k becomes bigger than 3, $\overline{M}_{0,k}$ becomes positive dimensional and the two notions diverge. For example, consider $X = \mathbb{P}^2$. Its ordinary cohomology is $H^*(X) = \mathbb{Z}[H]/(H^3)$, where H is the class of a line and H^2 the class of a point. The fact that there is a unique line through two distinct points says that $\langle H^2, H^2\rangle_1^{gravity} = 1$. Similarly, the fact that there is a unique conic through five general points says that $\langle H^2, H^2, H^2, H^2, H^2\rangle_2^{gravity} = 1$. If we fix only 4 points, we get a pencil of conics. As the conic varies in this pencil, the cross ratio of the 4 marked points on it varies linearly, so there is a unique conic in the pencil on which the cross ratio takes a specified value. This translates into $\langle H^2, H^2, H^2, H^2\rangle_2^{nogravity} = 1$.

The GLSM is a variant of the small quantum cohomology. The two objects $QH^*_{NLSM}(X)$ and $QH^*_{GLSM}(X)$ are distinct rings, but are related by the following results of Givental [**10**]:

- There is a "mirror map", a (typically nonlinear) change in the coordinate variables q^β, which identifies the ring $QH^*_{NLSM}(X)$ with $QH^*_{GLSM}(X)$.
- The mirror map is completely determined by the (gravitational) Gromov-Witten theory of X.
- When X is Fano, the mirror map is the identity, so the two theories are completely equivalent.

The triviality of the mirror map in the Fano case is explained in [**7**].

In the next section we discuss what is known about the (0,2) version of mirror symmetry. In sections 4 and 5 we study quantum sheaf cohomology proper. This is a generalization $QH^*_{GLSM}(X, \mathcal{E})$ of $QH^*_{GLSM}(X)$, or by the above remarks also of the small quantum cohomology . It is based on the same GLSM compactification of M_β as for $QH^*_{GLSM}(X)$. The new feature is that it depends on a sheaf $\mathcal{E}$ on X. The case in which this sheaf is the tangent bundle TX is essentially that of $QH^*_{GLSM}(X)$.

3. (0,2) mirror symmetry

Work on quantum sheaf cohomology is motivated in part by a generalization of mirror symmetry, known as (0,2) mirror symmetry. Schematically, the simplest incarnation of ordinary mirror symmetry is a symmetry in which pairs of Calabi-Yau manifolds (of the same dimension) are described by the same conformal field theory. At the level of SCFT's, ordinary mirror symmetry appears as a sign flip in a global U(1) symmetry on the worldsheet, which has the effect of exchanging massless states and corresponding cohomology groups. For example, if X_1 and X_2 are mirror Calabi-Yau manifolds, then

$$H^{p,q}(X_1) \cong H^{n-p,q}(X_2)$$

where n is the complex dimension of X_1 and X_2.

Similarly, the simplest incarnation of (0,2) mirror symmetry is a symmetry in which pairs of omalous pairs $(X_1, \mathcal{E}_1)$, $(X_2, \mathcal{E}_2)$, for X_1, X_2 Calabi-Yau of the same dimension (n) and $\mathcal{E}_1$, $\mathcal{E}_2$ of the same rank (r), are described by the same conformal field theory. Just as in ordinary mirror symmetry, massless states are exchanged, but here massless states are counted by sheaf cohomology, hence one has relations of the form

$$H^q(X_1, \wedge^p \mathcal{E}_1) \cong H^q(X_2, \wedge^p \mathcal{E}_2^*).$$

In the special case that $\mathcal{E} = TX$, (0,2) mirror symmetry reduces to ordinary mirror symmetry. (Indeed, note that the sheaf cohomology exchange above reduces to an exchange of Hodge numbers in this special case.) In more general (0,2) mirror pairs, X_1 and X_2 need not be ordinary-mirror-symmetric to one another.

(0,2) mirror symmetry is not nearly as well understood as ordinary mirror symmetry, but a number of basics are known. For example:

- Numerical evidence. Some groups such as [**5**] have written computer programs to compute dimensions of sheaf cohomology groups in omalous pairs in large numbers of examples, and have verified numerically that 'most' of the time, omalous pairs with mirror symmetric sheaf cohomology groups do exist.
- An analogue of the Greene-Plesser orbifold construction is known for (0,2) mirrors [**6**].
- A (0,2) analogue of Hori-Vafa-Morrison-Plesser's construction [**13, 23**] of ordinary mirrors is now known [**1**].
- For deformations of the tangent bundle, in many cases there now exists a (0,2) version of the monomial-divisor mirror map [**21**].
- Finally, this proceedings contribution is concerned with the (0,2) analogue of quantum cohomology.

4. Generalities on quantum sheaf cohomology

Ordinary quantum cohomology is derived from correlation functions in a topological field theory known as the A model [**29**]. Those correlation functions amount to intersection theory computations on a moduli space $\mathcal{M}$ of curves. Quantum sheaf cohomology is derived from correlation functions in a pseudo-topological field theory known as the A/2 model. Here, correlation functions amount to sheaf cohomology computations in a sheaf $\mathcal{F}$ over a moduli space $\mathcal{M}$ of curves.

4.1. Definition of A/2 model. Let us briefly review the A and A/2 models.

The A model topological field theory is a twist of the $(2,2)$ supersymmetric nonlinear sigma model

$$\frac{1}{\alpha'}\int_{\Sigma} d^2 z \left((g_{\mu\nu} + iB_{\mu\nu})\,\partial\phi^{\mu}\overline{\partial}\phi^{\nu} + \frac{i}{2}g_{\mu\nu}\psi_+^{\mu}D_{\overline{z}}\psi_+^{\nu} + \frac{i}{2}g_{\mu\nu}\psi_-^{\mu}D_z\psi_-^{\nu} \right.$$
$$\left. + R_{i\overline{\jmath}k\overline{l}}\psi_+^i\psi_+^{\overline{\jmath}}\psi_-^k\psi_-^{\overline{l}} \right),$$

where $\phi : \Sigma \to X$ is a map from the worldsheet Σ into the space X in which the string propagates, and the $\psi_{\pm}^{\mu}$ are fermionic superpartners of the coordinates ϕ^{μ} on X. Specifically, the A model is defined by twisting worldsheet fermions into worldsheet scalars and vectors as follows [**29**]:

$$\psi_+^i\ (\equiv \psi_z^i) \in \Gamma_{C^\infty}\left(K \otimes \phi^* T^{1,0}X\right), \qquad \psi_-^i\ (\equiv \chi^i) \in \Gamma_{C^\infty}\left(\left(\phi^* T^{0,1}X\right)^*\right),$$
$$\psi_+^{\overline{\imath}}\ (\equiv \chi^{\overline{\imath}}) \in \Gamma_{C^\infty}\left(\left(\phi^* T^{1,0}X\right)^*\right), \qquad \psi_-^{\overline{\imath}}\ (\equiv \psi_{\overline{z}}^{\overline{\imath}}) \in \Gamma_{C^\infty}\left(\overline{K} \otimes \phi^* T^{0,1}X\right).$$

After the twist, there is a nilpotent scalar operator Q, known as the BRST operator, whose action on the fields above is schematically as follows:

$$\delta\phi^i \propto \chi^i, \quad \delta\phi^{\overline{\imath}} \propto \chi^{\overline{\imath}}, \quad \delta\chi^i = 0 = \delta\chi^{\overline{\imath}}, \quad \delta\psi_{\overline{z}}^{\overline{\imath}} \neq 0, \quad \delta\psi_z^i \neq 0.$$

The states of the theory are BRST-closed dimension zero operators (modulo BRST-exact operators), which constrains them to be of the form

$$b_{i_1\cdots i_p \bar{\imath}_1 \cdots \bar{\imath}_q}(\phi)\chi^{i_1}\cdots\chi^{i_p}\chi^{\bar{\imath}_1}\cdots\chi^{\bar{\imath}_q}.$$

Witten [**29**] observed that the states above are in one-to-one correspondence with differential forms

$$b_{i_1\cdots i_p \bar{\imath}_1 \cdots \bar{\imath}_q}(\phi) dz^{i_1}\wedge\cdots\wedge dz^{i_p}\wedge d\bar{z}^{\bar{\imath}_1}\wedge\cdots\wedge d\bar{z}^{\bar{\imath}_q}$$

and the BRST operator Q with the exterior derivative d, hence the states are in one-to-one correspondence with elements of $H^{p,q}(X)$.

The A/2 model is a twist of the $(0,2)$ nonlinear sigma model

$$\frac{1}{\alpha'}\int_\Sigma d^2 z\left((g_{\mu\nu}+iB_{\mu\nu})\,\partial\phi^\mu\overline{\partial}\phi^\nu \;+\; \frac{i}{2}g_{\mu\nu}\psi_+^\mu D_{\bar{z}}\psi_+^\nu \;+\; \frac{i}{2}h_{\alpha\beta}\lambda_-^\alpha D_z\lambda_-^\beta \right.$$
$$\left. + \, F_{i\bar{\jmath}a\bar{b}}\psi_+^i\psi_+^{\bar{\jmath}}\lambda_-^a\lambda_-^{\bar{b}}\right),$$

in which the fermions couple to bundles as follows:

$$\begin{aligned} \psi_+^i &\in \Gamma_{C^\infty}\left(K\otimes\phi^*T^{1,0}X\right), & \lambda_-^a &\in \Gamma_{C^\infty}\left(\phi^*\overline{\mathcal{E}}^*\right), \\ \psi_+^{\bar{\imath}} &\in \Gamma_{C^\infty}\left(\left(\phi^*T^{1,0}X\right)^*\right), & \lambda_-^{\bar{a}} &\in \Gamma_{C^\infty}\left(\overline{K}\otimes\phi^*\overline{\mathcal{E}}\right), \end{aligned}$$

In the special case that $\mathcal{E} = TX$, the A/2 model becomes the A model.

Anomaly cancellation in the A/2 model requires

$$\wedge^{\rm top}\mathcal{E}^* \;\cong\; K_X, \;\; {\rm ch}_2(\mathcal{E}) \;=\; {\rm ch}_2(TX).$$

The second statement is the Green-Schwarz anomaly cancellation condition generic to all heterotic theories; the first is a condition specific to the A/2 twist, an analogue of the condition that the closed string B model can only propagate on spaces X such that $K_X^{\otimes 2}$ is trivial [**26, 29**]. Note that in the case $\mathcal{E} = TX$, the constraints above are satisfied trivially.

In fact, a specific choice of isomorphism $\wedge^{\rm top}\mathcal{E}^* \cong K_X$ is part of the data needed to define the path integral. Although both left- and right-movers have been twisted, the theory defined by the twisting above is not a topological field theory, since the worldsheet does not have supersymmetry on left-movers. Nevertheless it is sufficiently close to a true topological field theory to enable mathematical computations.

After the twist, there is a BRST operator Q as before, now acting as

$$\delta\phi^i \;=\; 0,\;\; \delta\phi^{\bar{\imath}} \,\propto\, \psi_+^{\bar{\imath}},\;\; \delta\psi_+^{\bar{\imath}} \;=\; 0 \;=\; \delta\lambda_-^a,\;\; \delta\psi_+^i \,\neq\, 0,\;\; \delta\lambda_-^{\bar{a}} \,\neq\, 0.$$

The states of the A/2 model generalizing the A model states are of the form

$$b_{\bar{\imath}_1\cdots\bar{\imath}_q a_1\cdots a_p}\psi_+^{\bar{\imath}_1}\cdots\psi_+^{\bar{\imath}_q}\lambda_-^{a_1}\cdots\lambda_-^{a_p}.$$

Proceeding in an analogous fashion, the BRST operator Q is identified with $\overline{\partial}$, and the states above are identified with bundle-valued differential forms, counted by sheaf cohomology $H^q(X,\wedge^p\mathcal{E}^\vee)$.

4.2. Formal computations. Schematically, we compute correlation functions in the A/2 model as follows. A correlation function will have the form

$$\langle \mathcal{O}_1 \cdots \mathcal{O}_n \rangle = \sum_\beta \int_{\mathcal{M}_\beta} \omega_1 \wedge \cdots \wedge \omega_n$$

(if there are no $\psi_+^{\bar{\imath}}$, λ_-^a zero modes). $\mathcal{M}_\beta$ is a moduli space of curves of degree β, ω_i is an element of $H^q(\mathcal{M}_\beta, \wedge^p \mathcal{F}^*)$ induced by the element of $H^q(X, \wedge^p \mathcal{E}^*)$ corresponding to $\mathcal{O}_i$. The sheaf $\mathcal{F}$ is induced from $\mathcal{E}$. For example, if the moduli space $\mathcal{M}$ admits a universal instanton $\alpha : \mathcal{M}_\beta \times \mathbb{P}^1 \to X$, then $\mathcal{F} = R^0 \pi_* \alpha^* \mathcal{E}$.

For classical contributions ($\beta = 0$), the integrand above is an element of

$$H^{\sum q_i}\left(X, \wedge^{\sum p_i} \mathcal{E}^*\right)$$

and so will vanish unless

$$\sum q_i = \dim X, \quad \sum p_i = \operatorname{rank} \mathcal{E}.$$

Furthermore, the constraint $\wedge^{\text{top}} \mathcal{E}^* \cong K_X$ now plays an important role, as it guarantees that the sheaf cohomology group above is isomorphic to the complex line $\mathbb{C}$.

More generally ($\beta \neq 0$), the integrand will be an element of

$$H^{\sum q_i}\left(\mathcal{M}_\beta, \wedge^{\sum p_i} \mathcal{F}^*\right)$$

and so will vanish unless

$$\sum q_i = \dim \mathcal{M}_\beta, \quad \sum p_i = \operatorname{rank} \mathcal{F}.$$

Grothendieck-Riemann-Roch tells us that the conditions for $(X, \mathcal{E})$ to be omalous, namely

$$\wedge^{\text{top}} \mathcal{E}^* \cong K_X, \quad \text{ch}_2(\mathcal{E}) = \text{ch}_2(TX)$$

imply that, at least formally, $\wedge^{\text{top}} \mathcal{F}^* \cong K_{\mathcal{M}_\beta}$, which is what we need for the sheaf cohomology group above to be isomorphic to the complex line $\mathbb{C}$.

If there are $\psi_+^{\bar{\imath}}$, λ_-^a zero modes, we have to work harder. In the ordinary A model, this would involve adding an Euler class factor, corresponding physically to using four-fermi terms to soak up the extra zero modes. In the A/2 model, one also brings down four-fermi terms, for the same reason. Their mathematical interpretation seems to be as follows. If there exists a universal instanton α over the moduli space $\mathcal{M}_\beta$, define

$$\mathcal{F}_1 \equiv R^1 \pi_* \alpha^* \mathcal{E}, \quad \text{Obs} \equiv R^1 \pi_* \alpha^* TX,$$

then one can formally identify the contribution of each four-fermi term with an insertion of an element of

$$H^1\left(\mathcal{M}_\beta, \mathcal{F}^* \otimes \mathcal{F}_1 \otimes (\text{Obs})^*\right).$$

Altogether the integrand is then an element of

$$H^{\text{top}}\left(\mathcal{M}_\beta, \wedge^{\text{top}} \mathcal{F}^* \otimes \wedge^{\text{top}} \mathcal{F}_1 \otimes \wedge^{\text{top}} \text{Obs}^*\right)$$

and Grothendieck-Riemann-Roch formally implies, in the same fashion as before, that

$$\det \mathcal{F}^* \otimes \det \mathcal{F}_1 \otimes \det \text{Obs}^* \cong K_{\mathcal{M}_\beta},$$

hence we get a number.

4.3. Linear sigma model compactifications. To actually carry out these computations, we need to pick compactifications of moduli spaces, as well as a procedure for extending induced sheaves over compactification divisors.

We will work with toric varieties and linear sigma model compactifications of moduli spaces. Briefly, a linear sigma model compactification of a moduli space of curves in a toric variety is built by expressing the toric variety as a $\mathbb{C}^\times$ quotient, expanding the homogeneous coordinates in zero modes, and then taking those zero modes to be homogeneous coordinates on the moduli space (with the same $\mathbb{C}^\times$ quotient and weightings as the original homogeneous coordinates).

An example will clarify. Consider the toric variety $\mathbb{P}^{N-1}$. Let $x_1, \ldots, x_N$ be homogeneous coordinates. The projective space $\mathbb{P}^{N-1}$ is expressed as a $\mathbb{C}^\times$ quotient in which each homogeneous coordinate has weight 1. Now, for a moduli space of degree d maps $\phi : \mathbb{P}^1 \to \mathbb{P}^{N-1}$, expand each homogeneous coordinate in a basis of sections of $\phi^*\mathcal{O}(1) = \mathcal{O}(d)$:

$$x_i \;=\; x_{i0}u^d \;+\; x_{i1}u^{d-1}v \;+\; \cdots \;+\; x_{id}v^d,$$

where u, v are homogeneous coordinates on $\mathbb{P}^1$. Then, we take the $\{x_{ij}\}$ to be homogeneous coordinates on the moduli space $\mathcal{M}$, each of weight 1 under a single $\mathbb{C}^\times$ quotient:

$$\mathcal{M}_{\mathrm{LSM}} \;=\; \mathbb{P}^{N(d+1)-1}.$$

In particular, a linear sigma model moduli space of a toric variety is another toric variety. (Furthermore, if the original toric variety is smooth, so too will be the linear sigma model moduli space.)

4.4. Induced sheaves. In the present case, we need not only a linear sigma model moduli space, but also induced sheaves $\mathcal{F}$, $\mathcal{F}_1$ over that moduli space. This is also straightforward for bundles appearing in (0,2) GLSM's. In such theories, all bundles are built from monads, i.e. they are kernels, cokernels or cohomologies of short exact sequences in which the other terms are sums of line bundles.

In physics terms, briefly, the prescription for the sheaf induced by any one line bundle is to expand the left-moving worldsheet fermions in a basis of zero modes, and identify each basis element with a line bundle of the same $\mathbb{C}^\times$ weights as the original line bundle.

In mathematical language, the prescription is to lift each line bundle to a natural line bundle on $\mathbb{P}^1 \times \mathcal{M}$, then push forward to $\mathcal{M}$. Specifically, lift each line bundle L on a toric variety X to a line bundle $\mathcal{L}$ on $\mathbb{P}^1 \times \mathcal{M}$, and use additivity and multiplicativity to lift all sums and powers. In other words,

$$\oplus_i {L_i}^{\otimes n_i} \;\longrightarrow\; X$$

lifts to

$$\oplus_i \mathcal{L}_i^{\otimes n_i} \;\longrightarrow\; \mathbb{P}^1 \times \mathcal{M}.$$

After lifting, these bundles are then pushed forward to $\mathcal{M}$.

Let us consider an example in which the rank of the Picard group is 1, so there is a universal subbundle S such that all the L_i are its powers. Consider the completely reducible bundle

$$\mathcal{E} \;=\; \oplus_a \mathcal{O}(n_a)$$

on $\mathbb{P}^{N-1}$. Corresponding to the universal subbundle

$$S := \mathcal{O}(-1) \;\longrightarrow\; \mathbb{P}^{N-1}$$

is

$$\mathcal{S} = \pi_1^* \mathcal{O}_{\mathbb{P}^1}(-d) \otimes \pi_2^* \mathcal{O}_{\mathcal{M}}(-1) \longrightarrow \mathbb{P}^1 \times \mathcal{M}$$

The lift of $\mathcal{E}$ is

$$\oplus_a \mathcal{S}^{\otimes - n_a} \longrightarrow \mathbb{P}^1 \times \mathcal{M}$$

which pushes forward to

$$\begin{aligned}\mathcal{F} &= \oplus_a H^0\left(\mathbb{P}^1, \mathcal{O}(n_a d)\right) \otimes_{\mathbb{C}} \mathcal{O}(n_a) \\ \mathcal{F}_1 &= \oplus_a H^1\left(\mathbb{P}^1, \mathcal{O}(n_a d)\right) \otimes_{\mathbb{C}} \mathcal{O}(n_a)\end{aligned}$$

This generalizes trivially to other toric varieties, and even to Grassmannians (where the linear sigma model moduli spaces are Quot schemes). Physically, the effect is equivalent to expanding worldsheet fermions in a basis of zero modes, and identifying each basis element with a line bundle of the same $\mathbb{C}^\times$ weights as the original line bundle.

Typically, we do not have a single reducible bundle, but rather a kernel or cokernel of some map between sums of line bundles. The induced map is constructed by expanding homogeneous coordinates in zero modes, and picking out coefficients of factors of powers of u, v (homogeneous coordinates on $\mathbb{P}^1$). Let us consider an example, to make this more clear. Let X be the Hirzebruch surface $\mathbb{F}_n$, expressed as a $(\mathbb{C}^\times)^2$ quotient with homogeneous coordinates a, b, s, t of weights

a	b	s	t
1	1	0	n
0	0	1	1.

We describe a deformation $\mathcal{E}$ of the tangent bundle of $\mathbb{F}_n$ as the cokernel

$$0 \longrightarrow \mathcal{O}^{\oplus 2} \xrightarrow{E} \mathcal{O}(1,0)^{\oplus 2} \oplus \mathcal{O}(0,1) \oplus \mathcal{O}(n,1) \longrightarrow \mathcal{E} \longrightarrow 0,$$

where

$$E = \left[\begin{array}{cc} Ax & Bx \\ \gamma_1 s & \gamma_2 s \\ \alpha_1 t + s f_1(a,b) & \alpha_2 t + s f_2(a,b) \end{array} \right],$$

with

$$x \equiv \left[\begin{array}{c} a \\ b \end{array} \right],$$

A, B constant 2×2 matrices, γ_1, γ_2, α_1, α_2 constants, and $f_{1,2}(a,b)$ homogeneous polynomials of degree n. The map E induces a map E' by expanding fields in zero modes and picking out terms with the same product of homogeneous coordinates on $\mathbb{P}^1$. In the degree $(1,0)$ sector, we expand

$$\begin{aligned} a &= a_0 u + a_1 v, \\ b &= b_0 u + b_1 v, \\ s &= s_0, \\ t &= t_0 u^n + t_1 u^{n-1} v + \cdots + t_n v^n, \end{aligned}$$

where u, v are homogeneous coordinates on $\mathbf{P}^1$. Then, in the original map E, we replace each field a, b, s, t by its expansion in zero modes above, and pick out terms with the same homogeneous coordinates. In this fashion, we find E' is given by

$$\left[\begin{array}{cc} \begin{bmatrix} A & 0 \\ 0 & A \end{bmatrix} x' & \begin{bmatrix} B & 0 \\ 0 & B \end{bmatrix} x' \\ \gamma_1 s_0 & \gamma_2 s_0 \\ \begin{array}{c} \alpha_1 t_0 + s_0 f_{10} a_0^n + s_0 f_{11} a_0^{n-1} b_0 \\ + \cdots + s_0 f_{1n} b_0^n \end{array} & \begin{array}{c} \alpha_2 t_0 + s_0 f_{20} a_0^n + s_0 f_{21} a_0^{n-1} b_0 \\ + \cdots + s_0 f_{2n} b_0^n \end{array} \\ \begin{array}{c} \alpha_1 t_1 + s_0 f_{10}(n a_0^{n-1} a_1) \\ + s_0 f_{11} a_0^{n-1} b_1 \\ + (n-1) s_0 f_{11} a_0^{n-2} a_1 b_0 + \cdots \end{array} & \begin{array}{c} \alpha_2 t_1 + s_0 f_{20}(n a_0^{n-1} a_1) \\ + s_0 f_{21} a_0^{n-1} b_1 \\ + (n-1) s_0 f_{21} a_0^{n-2} a_1 b_0 + \cdots \end{array} \\ \cdots & \cdots \end{array}\right],$$

where

$$x' = [a_0, b_0, a_1, b_1]^T$$

and

$$f_i(a,b) = f_{i0} a^n + f_{i1} a^{n-1} b + \cdots + f_{in} b^n.$$

In E', the lines with t_0, for example, correspond to coefficients of u^n, the lines with t^1 correspond to coefficients of $u^{n-1}v$, and so forth.

For bundles built as kernels and cokernels, the pushforward generates a long exact sequence. For example, consider the bundle $\mathcal{E}$ built as a cokernel below:

$$0 \longrightarrow \mathcal{O}^{\oplus k} \longrightarrow \oplus_i \mathcal{O}(\vec{q}_i) \longrightarrow \mathcal{E} \longrightarrow 0$$

over some toric variety X. Applying the procedure outlined above, we get the long exact sequence

$$\begin{array}{lllll} 0 & \longrightarrow \oplus_k H^0(\mathcal{O}) \otimes \mathcal{O} & \longrightarrow \oplus_i H^0(\mathcal{O}(\vec{q}_i \cdot \vec{d})) \otimes \mathcal{O}(\vec{q}_i) & \longrightarrow \mathcal{F} & \\ & \longrightarrow \oplus_k H^1(\mathcal{O}) \otimes \mathcal{O} & \longrightarrow \oplus_i H^1(\mathcal{O}(\vec{q}_i \cdot \vec{d})) \otimes \mathcal{O}(\vec{q}_i) & \longrightarrow \mathcal{F}_1 & \longrightarrow 0, \end{array}$$

which simplifies to the statements

$$0 \longrightarrow \mathcal{O}^{\oplus k} \longrightarrow \oplus_i H^0(\mathcal{O}(\vec{q}_i \cdot \vec{d})) \otimes \mathcal{O}(\vec{q}_i) \longrightarrow \mathcal{F} \longrightarrow 0,$$

$$\mathcal{F}_1 \cong \oplus_i H^1(\mathcal{O}(\vec{q}_i \cdot \vec{d})) \otimes \mathcal{O}(\vec{q}_i).$$

It can be shown that if $\mathcal{E}$ is locally-free, then $\mathcal{F}$ will also be locally-free.

As a consistency check, let us examine the special case that $\mathcal{E} = TX$. The tangent bundle of a (compact, smooth) toric variety can be expressed as a cokernel

$$0 \longrightarrow \mathcal{O}^{\oplus k} \longrightarrow \oplus_i \mathcal{O}(\vec{q}_i) \longrightarrow TX.$$

Applying the previous ansatz, we have

$$0 \longrightarrow \mathcal{O}^{\oplus k} \longrightarrow \oplus_i H^0(\mathcal{O}(\vec{q}_i \cdot \vec{d})) \otimes \mathcal{O}(\vec{q}_i) \longrightarrow \mathcal{F} \longrightarrow 0,$$

$$\mathcal{F}_1 \cong \oplus_i H^1(\mathcal{O}(\vec{q}_i \cdot \vec{d})) \otimes \mathcal{O}(\vec{q}_i).$$

In this case, for $\mathcal{E} = TX$, we expect $\mathcal{F} = T\mathcal{M}$ and $\mathcal{F}_1$ to be the obstruction sheaf in the sense of [**3**, **24**]. The sequences above have precisely the right form, and it can be shown that the induced maps are also correct.

5. Mathematical derivation and proof via toric geometry

In this section, we describe our results for deformations of the tangent bundle of a toric variety in the language of algebraic geometry. For more details, see [**8**].

First, we establish some terminology. Let $X = X_\Sigma$ be a smooth projective toric variety of dimension n associated to a fan Σ. We will denote by $\Sigma(1)$ the set of one-dimensional cones of the fan, and we will write $\bigoplus_\rho$ and $\sum_\rho$ in place of $\bigoplus_{\rho\in\Sigma(1)}$ and $\sum_{\rho\in\Sigma(1)}$, respectively. To each $\rho \in \Sigma(1)$ is associated a torus-invariant Weil divisor denoted D_ρ, a unique minimal generator v_ρ of ρ, and a canonical section $x_\rho \in H^0(X, \mathcal{O}_X(D_\rho))$. These canonical sections freely generate the homogeneous coordinate ring of X:

$$S := \mathbb{C}[x_\rho \mid \rho \in \Sigma(1)]. \tag{5.1}$$

The homogeneous coordinate ring S has a natural grading by $\mathrm{Pic}(X)$, which assigns to x_ρ the degree $[D_\rho] \in \mathrm{Pic}(X)$.

Let $W = H^2(X, \mathbb{C})$. Then the tangent bundle of X fits into the *toric Euler sequence*:

$$0 \longrightarrow \mathcal{O}_X \otimes_{\mathbb{C}} W^\vee \xrightarrow{E_0} \bigoplus_\rho \mathcal{O}_X(D_\rho) \longrightarrow TX \longrightarrow 0. \tag{5.2}$$

Thinking of E_0 as an element of $\bigoplus_\rho S_{[D_\rho]} \otimes W$, the ρth component of E_0 is $x_\rho \otimes [D_\rho]$.

To define a half-twisted (0,2) GLSM, we need a presentation of a vector bundle $\mathcal{E}$ obtained from (5.2) by simply changing the map E_0:

$$0 \longrightarrow \mathcal{O}_X \otimes_{\mathbb{C}} W^\vee \xrightarrow{E} \bigoplus_\rho \mathcal{O}_X(D_\rho) \longrightarrow \mathcal{E} \longrightarrow 0. \tag{5.3}$$

Our results can be expressed concisely and explicitly in terms of E. The components of E are a collection of W-valued sections E_ρ of $\mathcal{O}_X(D_\rho)$. We write

$$E_\rho = \sum_{\rho'} a_{\rho\rho'} x_{\rho'} + \cdots, \tag{5.4}$$

where $a_{\rho\rho'} \in W$, and any nonlinear terms in the x's have been omitted. Note that $a_{\rho\rho'}$ vanishes unless D_ρ is linearly equivalent to $D_{\rho'}$.

We form matrices out of certain components of E. Let c denote a linear equivalence class within the set of divisors D_ρ. Let A_c be the W-valued $|c| \times |c|$ matrix whose entries are $a_{\rho\rho'}$, where ρ and ρ' vary within the linear equivalence class c. Finally, put

$$Q_c = \det A_c. \tag{5.5}$$

Our results can be described simply in terms of these Q_c.

We now turn to the development.

Dualizing (5.2) gives

$$0 \longrightarrow \mathcal{E}^\vee \longrightarrow \bigoplus_\rho \mathcal{O}_X(-D_\rho) \xrightarrow{E^\vee} \mathcal{O}_X \otimes_{\mathbb{C}} W \longrightarrow 0 \tag{5.6}$$

We will compute the classical sheaf cohomology of $\mathcal{E}^\vee$ and its exterior powers by using (5.6) and an associated generalized Koszul complex. Then we will repeat the calculation in nontrivial instanton sectors, from which we can prove the quantum sheaf cohomology relations.

In order to simplify notation, put $Z = \oplus_\rho \mathcal{O}(-D_\rho)$. Let $\mathcal{E}$ be a deformation of the tangent bundle as in (5.3). The dual of the exact sequence (5.3) induces an injection $\wedge^k \mathcal{E}^\vee \to \wedge^k Z$ and maps $\alpha_j : \wedge^j Z \otimes \mathrm{Sym}^{k-j} W \to \wedge^{j-1} Z \otimes \mathrm{Sym}^{k-j+1} W$ defined as

$$(5.7)\ \alpha_j : (z_1 \wedge \cdots \wedge z_j) \otimes s \mapsto \sum_{\ell=1}^{j} (-1)^{\ell-1} (z_1 \wedge \cdots \wedge \hat{z}_\ell \wedge \cdots \wedge z_j) \otimes \left[E^\vee(z_\ell) \odot s\right],$$

where E is the injection in (5.3) and $\odot$ is multiplication in $\mathrm{Sym}^* W$. These maps may be arranged into an exact sequence

$$(5.8)\quad 0 \to \wedge^k \mathcal{E}^\vee \to \wedge^k Z \to \wedge^{k-1} Z \otimes W \to \cdots \to Z \otimes \mathrm{Sym}^{k-1} W \to \mathrm{Sym}^k W \otimes \mathcal{O} \to 0.$$

Recall that a collection of edges $K \subset \Sigma(1)$ is a *primitive collection* if K does not span any cone in Σ, but every proper subcollection of K does. Equivalently, the intersection of the divisors D_ρ with $\rho \in K$ is empty, but the intersection of any proper subset of these divisors is nonempty.

If K is a primitive collection, consider the element $v = \sum_{\rho \in K} v_\rho$ of the toric lattice. Then v lies in the relative interior of a unique cone $\sigma \in \Sigma$. Letting K^- denote the set of edges of σ, then we have an identity

$$(5.9)\qquad v = \sum_{\{\rho \in K^-\}} c_\rho v_\rho,$$

where each $c_\rho > 0$. Combining the definition of v with (5.9), we have an identity

$$\sum_{\rho \in K} v_\rho = \sum_{\rho \in K^-} c_\rho v_\rho$$

which we rewrite in the form

$$(5.10)\qquad \sum a_\rho v_\rho = 0$$

By dualizing the standard toric exact sequence

$$(5.11)\qquad 0 \to M \to \mathbb{Z}^{\Sigma(1)} \to \mathrm{Pic}(X) \to 0,$$

we see that relations among the v's such as (5.10) can be induced by intersection with elements of $H_2(X,\mathbb{Z})$. This implies that there is a class $\beta_K \in H_2(X,\mathbb{Z})$ such that $D_\rho \cdot \beta_K = a_\rho$ for each $\rho \in \Sigma(1)$. See [**4**] for more details, and [**8**] where this notation was used previously.

The main result needed for the toric computation is

PROPOSITION 5.1. *Let Σ be a simplicial fan and $K \subset \Sigma(1)$. Setting $k = |K|$ and $D_K = \sum_{\rho \in K} D_\rho$ as before, we have that*

(1) *For all $\ell \geq k$, $H^\ell(X, \mathcal{O}_X(-D_K)) = 0$*
(2) *If $\cap_{\rho \in K} D_\rho \neq \emptyset$, then for all $\ell \in \mathbb{Z}$, $H^\ell(X, \mathcal{O}_X(-D_K)) = 0$.*
(3) *If K is a primitive collection,*

$$H^\ell(X, \mathcal{O}_X(-D_K)) \simeq \begin{cases} \mathbb{C} & \ell = k-1 \\ 0 & \textit{otherwise} \end{cases}$$

(4) *If K is not a primitive collection, then $H^{k-1}(X, \mathcal{O}_X(-D_K)) = 0$.*

In particular, for $K = \{\rho\}$, Proposition 5.1 implies that

$$H^0(\mathcal{O}(-D_\rho) = H^1(\mathcal{O}(-D_\rho)) = 0.$$

Then the long exact cohomology sequence associated to (5.6) simplifies to an isomorphism

$$W \simeq H^1(X, \mathcal{E}^\vee).$$

It is not hard to show that $H^q(X, \wedge^p \mathcal{E}^\vee)$ vanishes unless $p = q$ and the algebra $\oplus_p H^p(X, \wedge^p \mathcal{E}^\vee)$ is generated multiplicatively by $H^1(\mathcal{E}^\vee)$. It follows that this algebra is a quotient of Sym$W$. See [**8**] for more details.

To describe this quotient, we break up (5.8) into short exact sequences and compute cohomology using Proposition 5.1. This procedure will be illustrated for $\mathbb{P}^1 \times \mathbb{P}^1$ in Section 6.1. The computation works because the multiplication map

$$m_k : \mathrm{Sym}^k W \to H^k(X, \wedge^k \mathcal{E}^\vee). \tag{5.12}$$

can be identified with the extension class in

$$\mathrm{Ext}^k(\mathrm{Sym}^k W \otimes \mathcal{O}_X, \wedge^k \mathcal{E}^\vee) \simeq \mathrm{Hom}(\mathrm{Sym}^k W, H^k(X, \wedge^k \mathcal{E}^\vee)$$

associated to the generalized Koszul complex (5.8).

Let K be a primitive collection and let Q_K be the product of the Q_c over all linear equivalence classes c. This makes sense, as it is proven in [**8**] that any primitive collection K is a union of linear equivalence classes.

DEFINITION 5.2. The *Stanley-Reisner ideal of* $\mathcal{E}$ is

$$SR(X, \mathcal{E}) = (Q_K \mid K \text{ a primitive collection of } \Sigma) \subset \mathrm{Sym}^* W. \tag{5.13}$$

REMARK 5.3. If $\mathcal{E} = TX$, then $Q_K = \prod_{\rho \in K} [D_\rho]$, and $SR(X, TX)$ is the image of the usual Stanley-Reisner ideal $SR(X)$ in $\mathrm{Sym}^* W$ under the natural quotient $S \to \mathrm{Sym}^* W$ which takes x_ρ to $[D_\rho]$. See [**8**] for more details.

THEOREM 5.4. *Let X be a smooth projective toric variety, $W = \mathrm{Pic}(X) \otimes_{\mathbb{Z}} \mathbb{C}$, and $\mathcal{E} \to X$ a toric deformation of the tangent bundle of X. Then as a graded algebra we have*

$$\oplus_p H^p(X, \wedge^p \mathcal{E}^\vee) \simeq \mathrm{Sym}^* W / SR(X, \mathcal{E}) \tag{5.14}$$

The (unnormalized) classical correlation functions are obtained by multiplication in $\oplus_p H^p(X, \wedge^p \mathcal{E}^\vee)$ followed by projection to the 1-dimensional vector space $H^n(X, \wedge^n \mathcal{E}^\vee) \simeq H^n(X, K_X) \simeq \mathbb{C}$.

Turning to the correlation functions in an instanton sector indexed by $\beta \in H_2(X, \mathbb{Z})$, recall that the GLSM moduli space $\mathcal{M}_\beta$ has nothing to do with the fermions in the theory: it is the same as the GLSM moduli space $\mathcal{M}_\beta$ in the (2,2) theory. In particular, it is itself a toric variety. We have already illustrated this for $X = \mathbb{P}^{N-1}$ in Section 4.3.

Furthermore, the sheaf $\mathcal{E}$ induces a sheaf $\mathcal{F}_\beta$ on $\mathcal{M}_\beta$ which is a deformation of the tangent bundle of $\mathcal{M}_\beta$. We have already explained how this works for the Hirzebruch surface $\mathbb{F}_n$ in Section 4.4. Rather than explain this in general, we will content ourselves with the example already given and by further illustration in the case $X = \mathbb{P}^1 \times \mathbb{P}^1$ later. The general case is explained in [**8**].

It follows from the toric construction that $\oplus_p H^p(\mathcal{M}_\beta, \wedge^p \mathcal{F}_\beta)$ is also a quotient of Sym$W$. The convenient fact is that the quotient only depends on the Q_c that we have already defined. To state the result, we define the integer valued function

$$h^0(x) = h^0(\mathbb{P}^1, \mathcal{O}_{\mathbb{P}^1}(x)) = \max(0, x+1). \tag{5.15}$$

Direct computation shows that

$$\oplus_p H^p(\mathcal{M}_\beta, \mathcal{F}_\beta) = \mathrm{Sym}^* W / \widehat{SR}(\mathcal{M}_\beta, \mathcal{F}_\beta),$$

where

$$\widehat{SR}(\mathcal{M}_\beta, \mathcal{F}_\beta) = \big(Q_{K_\beta} \mid K \text{ a primitive collection of } \Sigma\big) \subset \mathrm{Sym}^* W$$

and

$$Q_{K_\beta} = \prod_{c \in [K]} Q_c^{h^0(D_c \cdot \beta)}. \tag{5.16}$$

Here, by $D_c \cdot \beta$ we mean $D_\rho \cdot \beta$ for any ρ in the linear equivalence class c. An example explaining the genesis of the exponent in (5.16) is given in Section 6.2.

The dimension n_β of $\mathcal{M}_\beta$ can be expressed in terms of the function h^0 as

$$n_\beta := \dim \mathcal{M}_\beta = \sum_\rho \left(h^0(d_\rho^\beta)\right) - h^2(X). \tag{5.17}$$

We now turn to the correlation functions in an instanton sector β. The correlation functions in sector β can be nonzero only if the operators have degree $c_1(X) \cdot \beta + \dim(X)$. Since $c_1(X) = \sum_\rho D_\rho$, it follows that

$$c_1(X) \cdot \beta = \sum_\rho d_\rho^\beta,$$

which implies that

$$c_1(X) \cdot \beta + \dim(X) = \sum_\rho \left(d_\rho^\beta + 1\right) - h^2(X), \tag{5.18}$$

where $d_\rho^\beta = D_\rho \cdot \beta$. The formula (5.18) plays the role of the virtual or expected dimension of Gromov-Witten theory. Note that (5.17) and (5.18) differ only in that $h^0(d_\rho^\beta)$ in(5.17) is replaced by $d_\rho^\beta + 1$ in (5.18). These are in fact equal, unless $d_\rho^\beta \le -2$ for some ρ. Such a situation is the analogue of excess dimension in Gromov-Witten theory. In our situation, we have both the excess dimension of $\mathcal{M}_\beta$ and the excess rank of $\mathcal{F}_\beta$.

In this case, to compensate, we need something to play the role of the obstruction classes of Gromov-Witten theory. These were called four-fermi terms in [**14**] because of how they arose in the path integral.

Define $h^1(x) = h^1(\mathcal{O}_{\mathbb{P}^1}(x))$ in analogy with Definition 5.15. Then our formula for the four-fermi terms is

$$F_\beta = \prod_c Q_c^{h^1(d_c^\beta)}, \tag{5.19}$$

where the product is over all linear equivalence classes c of the divisors D_ρ.

At last, we can define the correlation functions in sector β. Let

$$p \in \mathrm{Sym}^{c_1(X) \cdot \beta + \dim X} W.$$

A simple computation of degree shows that $pF_\beta \in \mathrm{Sym}^{n_\beta} W$. For any $P \in \mathrm{Sym}^* W$, let $[P]$ denote the equivalence class of P in the quotient $\mathrm{Sym}^* W/\widehat{SR}(\mathcal{M}_\beta, \mathcal{F}_\beta)$. Then the unnormalized correlation function is defined as

$$\langle p\rangle_\beta = [pF_\beta], \tag{5.20}$$

identified with an element of the 1 dimensional vector space $H^{n_\beta}(\mathcal{M}_\beta, \wedge^{n_\beta}\mathcal{F}_\beta^\vee)$.

In lieu of normalizing the correlation functions, we can instead compare correlation functions in different instanton sectors. It suffices to compare "enough" different instanton sectors.

DEFINITION 5.5. For classes $\beta, \beta' \in H_2(X, \mathbb{Z})$, we say that β' *dominates* β if $\beta' - \beta$ is effective and $h^0(d_c^{\beta'}) \geq h^0(d_c^\beta)$ for all linear equivalence classes c of the irreducible toric divisors D_ρ.

If β' dominates β, we define the expression

$$R_{\beta'\beta} = \prod_c Q_c^{h^0(d_c^{\beta'}) - h^0(d_c^\beta)} \in \mathrm{Sym}^* W. \tag{5.21}$$

Then the comparison is simply induced by multiplication by $R_{\beta\beta'}$. This works because of the following lemma.

LEMMA 5.6. *Suppose that β' dominates β. Then*

$$R_{\beta'\beta}\left(SR(\mathcal{M}_\beta, \mathcal{F}_\beta)\right) \subset SR(X_{\beta'}, \mathcal{F}_{\beta'}).$$

If $p \in \mathrm{Sym}^d W$ with $d \neq c_1(X)\cdot\beta + \dim X$, we define $\langle p\rangle_\beta = 0$. Using the $R_{\beta'\beta}$, we treat the correlation functions as living in the same one-dimensional vector space, so they can be added over β.

For any $p \in \mathrm{Sym}^* W$ we define the correlation function

$$\langle p\rangle = \sum_\beta \langle p\rangle_\beta q^\beta.$$

For each primitive collection K, let $[K]$ be the set of linear equivalence classes of the D_ρ with $\rho \in K$, and let $[K^-]$ be the set of linear equivalence classes of the D_ρ with $\rho \in K^-$.

PROPOSITION 5.7. *For any primitive collection K, we have a quantum cohomology relation*

$$\prod_{c\in[K]} Q_c = q^{\beta_K} \prod_{c\in[K^-]} Q_c^{-d_c^{\beta_K}}. \tag{5.22}$$

Unwinding the definitions, this means that for any $p \in \mathrm{Sym}\, W$ and $\beta \in H_2(X, \mathbb{Z})$ we have

$$\langle p \prod_{c\in[K]} Q_c\rangle_{\beta+\beta_K} = \langle p \prod_{c\in[K^-]} Q_c^{-d_c^{\beta_K}}\rangle_\beta.$$

Proof: Choosing β' dominating both β and $\beta + \beta_K$, we have to show the equality

$$R_{\beta',\beta+\beta_K} F_{\beta+\beta_K} Y \prod_{c\in K} Q_c = R_{\beta'\beta} F_\beta Y \prod_{c\in[K^-]} Q_c^{-D_c\cdot\beta_K} \tag{5.23}$$

as elements of the quotient ring

$$\mathrm{Sym}^* W / SR(X_{\beta'}, \mathcal{F}_{\beta'}).$$

In fact, we will see that (5.23) holds in $\mathrm{Sym}^* W$. For this it suffices to show

$$R_{\beta',\beta+\beta_K} F_{\beta+\beta_K} \prod_{c \in K} Q_c = R_{\beta'\beta} F_\beta \prod_{c \in [K^-]} Q_c^{-D_c \cdot \beta_K} \tag{5.24}$$

Both sides are products of powers of the Q_c, so it suffices to check the exponents. This is done by direct computation. See [**8**] for more details.

6. Example on $\mathbb{P}^1 \times \mathbb{P}^1$

Let us work through a specific example. Consider $X = \mathbb{P}^1 \times \mathbb{P}^1$ and let $\psi, \tilde{\psi} \in H^2(X, \mathbb{Z})$ be the generators of the respective $H^2(\mathbb{P}^1, \mathbb{Z})$ summands. Consider a vector bundle $\mathcal{E}$ given as the cokernel

$$0 \longrightarrow \mathcal{O} \oplus \mathcal{O} \stackrel{*}{\longrightarrow} \mathcal{O}(1,0)^2 \oplus \mathcal{O}(0,1)^2 \longrightarrow \mathcal{E} \longrightarrow 0 \tag{6.1}$$

where

$$* = \left[\begin{array}{cc} Ax & Bx \\ C\tilde{x} & D\tilde{x} \end{array} \right]$$

where A, B, C, D are 2×2 matrices, and

$$x = \left[\begin{array}{c} x_1 \\ x_2 \end{array} \right], \quad \tilde{x} = \left[\begin{array}{c} \tilde{x}_1 \\ \tilde{x}_2 \end{array} \right]$$

are arrays of homogeneous coordinates on the two $\mathbb{P}^1$ factors. Here we have used the basis $\psi, \tilde{\psi}$ of $W \simeq H^2(X, \mathbb{C})$ to identify

$$W^\vee \otimes \mathcal{O} \simeq \mathcal{O} \oplus \mathcal{O}.$$

The bundle $\mathcal{E}$ is a deformation of the tangent bundle. It is straightforward to check that $(X, \mathcal{E})$ is an omalous pair. From this point forward, ψ and $\tilde{\psi}$ will be identified with elements of $H^1(X, \mathcal{E}^\vee)$. We will show:

THEOREM 6.1. *The quantum sheaf cohomology ring for $X = \mathbb{P}^1 \times \mathbb{P}^1$ is $\mathbf{C}[\psi, \tilde{\psi}]$ modulo the relations*

$$\det\left(A\psi + B\tilde{\psi}\right) = q, \quad \det\left(C\psi + D\tilde{\psi}\right) = \tilde{q}.$$

Note as a consistency check that in the special case that

$$A = D = I_{2\times 2}, \quad B = C = 0$$

the bundle $\mathcal{E} = TX$, and the quantum sheaf cohomology relations become

$$\psi^2 = q, \quad \tilde{\psi}^2 = \tilde{q},$$

which is exactly right for the ordinary quantum cohomology ring of $\mathbb{P}^1 \times \mathbb{P}^1$.

6.1. Classical ring relations. Now, let us compute the quantum sheaf cohomology relations above. As in Section 5, we dualize and write $\mathcal{E}^*$ as a kernel

$$0 \longrightarrow \mathcal{E}^* \longrightarrow Z \xrightarrow{*} W \otimes \mathcal{O} \longrightarrow 0 \tag{6.2}$$

where $W = H^2(\mathbb{P}^1 \times \mathbb{P}^1) \simeq \mathbb{C}^2$ and

$$Z = \mathcal{O}(-1,0)^2 \oplus \mathcal{O}(0,-1)^2.$$

Recall for this example the Koszul resolution (5.8)

$$0 \longrightarrow \wedge^2\mathcal{E}^* \longrightarrow \wedge^2 Z \longrightarrow Z \otimes W \longrightarrow \mathrm{Sym}^2 W \otimes \mathcal{O} \longrightarrow 0.$$

Now, let us break this long exact sequence into a pair of short exact sequences:

$$0 \longrightarrow \wedge^2\mathcal{E}^* \longrightarrow \wedge^2 Z \longrightarrow Q \longrightarrow 0, \tag{6.3}$$

$$0 \longrightarrow Q \longrightarrow Z \otimes W \longrightarrow \mathrm{Sym}^2 W \otimes \mathcal{O} \longrightarrow 0. \tag{6.4}$$

The second short exact sequence (6.4) gives a map

$$\mathrm{Sym}^2 W = H^0(\mathrm{Sym}^2 W \otimes \mathcal{O}) \longrightarrow H^1(Q),$$

and the first (6.3) gives

$$H^1(Q) \longrightarrow H^2(\wedge^2\mathcal{E}^*).$$

The composition of these two maps, is precisely the classical contribution to correlation functions.

Let us study these in more detail. Starting from the second short exact sequence (6.4) above, the associated long exact sequence is

$$H^0(Z \otimes W) \longrightarrow H^0(\mathrm{Sym}^2 W \otimes \mathcal{O}) \longrightarrow H^1(Q) \longrightarrow H^1(Z \otimes W).$$

However, since Z is a sum of $\mathcal{O}(-1,0)$, $\mathcal{O}(0,-1)$'s, we see

$$H^0(Z \otimes W) = 0 = H^1(Z \otimes W),$$

so we have

$$\mathrm{Sym}^2 W = H^0(\mathrm{Sym}^2 W \otimes \mathcal{O}) \xrightarrow{\sim} H^1(Q).$$

From the first short exact sequence (6.3), the associated long exact sequence is

$$H^1(\wedge^2 Z) \longrightarrow H^1(Q) \longrightarrow H^2(\wedge^2\mathcal{E}^*) \longrightarrow H^2(\wedge^2 Z).$$

Here,

$$H^2(\wedge^2 Z) = 0,$$

but

$$\begin{aligned} H^1(\wedge^2 Z) &= H^1\left(\mathbf{P}^1, \mathcal{O}(-2,0) \oplus \mathcal{O}(0,-2)\right), \\ &= \mathbb{C} \oplus \mathbb{C}, \end{aligned}$$

and so the map

$$H^1(Q) \longrightarrow H^2(\wedge^2\mathcal{E}^*)$$

has a two-dimensional kernel. Since $\mathrm{Sym}^2 W \simeq H^1(Q)$, chasing through the diagrams will allow us to compute two elements $Q, \tilde{Q} \in \mathrm{Sym}^2 W$ which span the kernel.

Thus we have computed the classical ring relations

$$Q := \det\left(\psi A + \tilde{\psi} B\right) = 0 = \tilde{Q} := \det\left(\psi C + \tilde{\psi} D\right).$$

6.2. Quantum corrections. Now, let us turn to quantum corrections. Let $\mathcal{M}_\beta$ denote the (GLSM) moduli space of instantons of fixed degree β, and $\mathcal{F}_\beta$ the induced sheaf on $\mathcal{M}$. Since $\mathcal{E}$ is a deformation of TX, $\mathcal{F}_\beta$ is a deformation of $T\mathcal{M}_\beta$. Thus, in principle, we can treat quantum corrections in the same fashion as classical ring relations.

Work in degree (d,e), so that

$$\mathcal{M}_{(d,e)} \;=\; \mathbb{P}^{2d+1} \times \mathbb{P}^{2e+1}.$$

The induced sheaf $\mathcal{F}_{(d,e)}$ is given by

$$0 \longrightarrow \mathcal{O}^2 \longrightarrow \oplus_1^{2d+2}\mathcal{O}(1,0) \oplus \oplus_1^{2e+2}\mathcal{O}(0,1) \longrightarrow \mathcal{F}_{(d,e)} \longrightarrow 0,$$

which we shall write as

$$0 \longrightarrow W^* \otimes \mathcal{O} \longrightarrow Z^* \longrightarrow \mathcal{F}_{(d,e)} \longrightarrow 0$$

(defining W, Z appropriately).

Here, correlation functions are linear maps

$$\mathrm{Sym}^{2e+2e+2}\left(H^1(\mathcal{F}^*_{(d,e)})\right)\left(= \mathrm{Sym}^{2d+2e+2}W\right) \longrightarrow H^{2e+2e+2}\left(\wedge^{\mathrm{top}}\mathcal{F}^*_{(d,e)}\right) = \mathbb{C}.$$

As before, we compute the (kernel of the) correlation functions using a Koszul resolution of $\wedge^{\mathrm{top}}\mathcal{F}^*_{(d,e)}$:

$$0 \longrightarrow \wedge^{\mathrm{top}}\mathcal{F}^*_{(d,e)} \longrightarrow \wedge^{2d+2e+2}Z \longrightarrow \wedge^{2d+2e+1}Z\otimes W \longrightarrow \wedge^{2d+2e}Z\otimes \mathrm{Sym}^2 W$$
$$\longrightarrow \cdots \longrightarrow Z\otimes\mathrm{Sym}^{2d+2e+1}W \longrightarrow \mathrm{Sym}^{2d+2e+2}W\otimes\mathcal{O} \longrightarrow 0.$$

Briefly, the (long exact) Koszul resolution factors into a sequence of short exact sequences of the form

$$0 \longrightarrow S_i \longrightarrow \wedge^i Z\otimes \mathrm{Sym}^{2d+2e+2-i}W \longrightarrow S_{i-1} \longrightarrow 0 \tag{6.5}$$

and the coboundary maps

$$\delta:\; H^i(S_i) \longrightarrow H^{i+1}(S_{i+1})$$

factor the map determining the correlation functions:

$$H^0\left(\mathrm{Sym}^{2d+2e+2}W\otimes\mathcal{O}\right) \longrightarrow H^1(S_1) \stackrel{\delta}{\longrightarrow} H^2(S_2) \stackrel{\delta}{\longrightarrow} \cdots$$
$$\cdots \stackrel{\delta}{\longrightarrow} H^{2d+2e+1}(S_{2d+2e+1}) \stackrel{\delta}{\longrightarrow} H^{2d+2e+2}\left(\wedge^{\mathrm{top}}\mathcal{F}^*_{(d,e)}\right).$$

So, to evaluate the correlation function, we compute the coboundary maps.

Referring to the short exact sequence (6.5), it can be shown that the $\wedge^i Z$ only have nonzero cohomology in degrees $2d+2$, $2e+2$. Thus, the coboundary maps $\delta: H^i(S_i) \to H^{i+1}(S_{i+1})$ are mostly isomorphisms. The remaining maps have computable kernels.

The result is as follows. For fixed instanton degree (d,e), the sheaf cohomology lives in

$$\mathrm{Sym}^{\cdot}W/(Q^{d+1},\tilde{Q}^{e+1})$$

where

$$Q \;=\; \det\left(\psi A + \tilde{\psi}B\right),\quad \tilde{Q} \;=\; \det\left(\psi C + \tilde{\psi}D\right).$$

as before.

For example, for degree $(1,0)$ maps, $\mathcal{M} = \mathbb{P}^3\times\mathbb{P}^1$, and $\mathcal{F}_{(1,0)}$ is defined by

$$0 \longrightarrow W^*\otimes\mathcal{O} \stackrel{*'}{\longrightarrow} \mathcal{O}(1,0)^4 \oplus \mathcal{O}(0,1)^2 \longrightarrow \mathcal{F}_{(1,0)} \longrightarrow 0,$$

where

$$*' = \left[\begin{array}{cc} \left[\begin{array}{cc} A & 0 \\ 0 & A \end{array} \right] y & \left[\begin{array}{cc} B & 0 \\ 0 & B \end{array} \right] y \\ C\tilde{x} & D\tilde{x} \end{array} \right]$$

(where $\tilde{x}$ is as before, a 2-vector of homogeneous coordinates on $\mathbb{P}^1$, and y is a 4-vector of homogeneous coordinates on $\mathbb{P}^3$). The kernel is generated by

$$\det \left(\psi \left[\begin{array}{cc} A & 0 \\ 0 & A \end{array} \right] + \tilde{\psi} \left[\begin{array}{cc} B & 0 \\ 0 & B \end{array} \right] \right) = \det(\psi A + \tilde{\psi} B)^2, \quad \det(\psi C + \tilde{\psi} D).$$

6.3. Quantum sheaf cohomology. So far we have discussed correlation functions in sectors of fixed instanton degree as maps

$$\mathrm{Sym}^n W \longrightarrow H^n(\wedge^n \mathcal{F}^*_\beta) = \mathbb{C}$$

with computable kernels. However, we are interested in ring relations. In physics language, we are interested in nonperturbative corrections to entire correlation functions, which relate different instanton sectors, not correlation functions in fixed instanton degree.

Let us turn again to the example on $\mathbb{P}^1 \times \mathbb{P}^1$. We have outlined how

$$\langle \mathcal{O}_1 \cdots \mathcal{O}_n \rangle_{(a,b)} \in \mathrm{Sym}^n W / (Q^{a+1}, \tilde{Q}^{b+1}),$$

where

$$Q = \det \left(\psi A + \tilde{\psi} B \right), \quad \tilde{Q} = \det \left(\psi C + \tilde{\psi} D \right).$$

We are looking for ring relations that relate correlation functions in different instanton degrees. To be consistent, these relations should map ideals to ideals.

Existence of such ring relations (physically, existence of OPE's) implies relations of the form

$$\langle \mathcal{O} \rangle_{(a,b)} \propto \langle \mathcal{O} R_{a,b,a',b'} \rangle_{(a',b')}$$

for some $R_{a,b,a',b'}$, which should map ideals to ideals. The $R_{a,b,a',b'}$'s determine quantum sheaf cohomology ring relations (OPE's).

In the present $\mathbb{P}^1 \times \mathbb{P}^1$ example, in order for any relation of the form above to be compatible with ideals, we need

$$\langle \mathcal{O} \rangle_{(a,b)} \propto \langle \mathcal{O} Q^{a'-a} \tilde{Q}^{b'-b} \rangle_{(a',b')}. \tag{6.6}$$

Define q, $\tilde{q}$ to encode the proportionality constant, as

$$\langle \mathcal{O} \rangle_{(a,b)} = q^{a'-a} \tilde{q}^{b'-b} \langle \mathcal{O} Q^{a'-a} \tilde{Q}^{b'-b} \rangle_{(a',b')},$$

then the quantum sheaf cohomology ring relations (OPE's) are

$$Q = q, \quad \tilde{Q} = \tilde{q}.$$

In fact, we have done more than this: the toric analysis done earlier proves that the proportionality constant in (6.6) is exactly 1.

Thus we have established Theorem 6.1.

7. Conclusions and open problems

We have described the recent progress on the construction and calculation of $(0,2)$ quantum cohomology. Little is known, and much remains to be explored:

- Give a cleaner definition of quantum sheaf cohomology for the GLSM from which the $R_{\beta\beta'}$ can be derived from first principles.
- Give a mathematically rigorous definition of quantum sheaf cohomology for the NLSM.
- Compute the quantum sheaf cohomology for hypersurfaces and complete intersections.
- Compute the quantum sheaf cohomology for omalous bundles which are not deformations of the tangent bundle, e.g. $\mathcal{O}_X^n \oplus TX$.
- Formulate $(0,2)$ mirror symmetry and prove it.
- What do the $(0,2)$ correlation functions "count"? Presumably these are virtual invariants of the pullback of the defining bundle via the map when that makes sense.
- Define quantum sheaf cohomology for orbifolds.
- Do the quantum sheaf cohomology relations that we have described generate all possible relations? It seems that even in the $(2,2)$ where this is believed to be true, a proof has never been written down.

References

[1] Allan Adams, Anirban Basu, and Savdeep Sethi, *(0,2) duality*, Adv.Theor.Math.Phys. **7** (2004), 865–950, available at `hep-th/0309226`.

[2] Allan Adams, Jacques Distler, and Morten Ernebjerg, *Topological heterotic rings*, Adv.Theor.Math.Phys. **10** (2006), 657–682, available at `hep-th/0506263`.

[3] Paul S. Aspinwall and David R. Morrison, *Topological field theory and rational curves*, Progr. Math. **160** (1998), 141–175, available at `alg-geom/9701016`.

[4] V. Batyrev, *Quantum cohomology rings of toric manifolds*, Journés de géométrie algébrique d'orsay (orsay, 1992), 1993, pp. 9–34.

[5] Ralph Blumenhagen, Rolf Schimmrigk, and Andreas Wisskirchen, *(0,2) mirror symmetry*, Nucl.Phys. **B486** (1997), 598–628, available at `hep-th/9609167`.

[6] Ralph Blumenhagen and Savdeep Sethi, *On orbifolds of (0,2) models*, Nucl.Phys. **B491** (1997), 263–278, available at `hep-th/9611172`.

[7] David A. Cox and Sheldon Katz, *Mirror symmetry and algebraic geometry*, Mathematical Surveys and Monographs, vol. 68, American Mathematical Society, Providence, RI, 1999. MR1677117 (2000d:14048)

[8] Ron Donagi, Josh Guffin, Sheldon Katz, and Eric Sharpe, *A mathematical theory of quantum sheaf cohomology*, 2011. arXiv: 1110.3751.

[9] ———, *Physical aspects of quantum sheaf cohomology for deformations of tangent bundles of toric varieties*, 2011. arXiv: 1110.3752.

[10] A. Givental, *A mirror theorem for toric complete intersections*, Topological field theory, primitive forms and related topics (kyoto, 1996), 1998, pp. 141–175.

[11] Josh Guffin, *Quantum sheaf cohomology, a precis*, 2011. arXiv: 1101.1305.

[12] Josh Guffin and Sheldon Katz, *Deformed quantum cohomology and (0,2) mirror symmetry*, JHEP **1008** (2010), 109, available at `0710.2354`.

[13] Kentaro Hori and Cumrun Vafa, *Mirror symmetry*, 2000. hep-th/0002222.

[14] Sheldon H. Katz and Eric Sharpe, *Notes on certain (0,2) correlation functions*, Commun.Math.Phys. **262** (2006), 611–644, available at `hep-th/0406226`.

[15] Maximilian Kreuzer, Jock McOrist, Ilarion V. Melnikov, and M.Ronen Plesser, *(0,2) deformations of linear sigma models*, JHEP **1107** (2011), 044, available at `1001.2104`.

[16] Jock McOrist, *The revival of (0,2) linear sigma models*, Int.J.Mod.Phys. **A26** (2011), 1–41, available at `1010.4667`.

[17] Jock McOrist and Ilarion V. Melnikov, *Half-twisted correlators from the Coulomb branch*, JHEP **0804** (2008), 071, available at 0712.3272.
[18] ———, *Summing the instantons in half-twisted linear sigma models*, JHEP **0902** (2009), 026, available at 0810.0012.
[19] ———, *Old issues and linear sigma models*, 2011. arXiv: 1103.1322.
[20] Ilarion V. Melnikov, *(0,2) Landau-Ginzburg models and residues*, JHEP **0909** (2009), 118, available at 0902.3908.
[21] Ilarion V. Melnikov and M.Ronen Plesser, *A (0,2) Mirror Map*, JHEP **1102** (2011), 001, available at 1003.1303.
[22] Ilarion V. Melnikov and Savdeep Sethi, *Half-twisted (0,2) Landau-Ginzburg models*, JHEP **0803** (2008), 040, available at 0712.1058.
[23] David R. Morrison and M. Ronen Plesser, *Towards mirror symmetry as duality for two-dimensional abelian gauge theories*, Strings '95 (Los Angeles, CA, 1995), 1996, pp. 374–387. MR1660724 (99m:81224)
[24] David R. Morrison and M.Ronen Plesser, *Summing the instantons: Quantum cohomology and mirror symmetry in toric varieties*, Nucl.Phys. **B440** (1995), 279–354, available at hep-th/9412236.
[25] Eric Sharpe, *Notes on correlation functions in* $(0,2)$ *theories*, Snowbird lectures on string geometry, 2006, pp. 93–104. MR2222532 (2007e:81100)
[26] ———, *Notes on certain other (0,2) correlation functions*, Adv.Theor.Math.Phys. **13** (2009), 33–70, available at hep-th/0605005.
[27] Meng-Chwan Tan, *Two-dimensional twisted sigma models and the theory of chiral differential operators*, Adv.Theor.Math.Phys. **10** (2006), 759–851, available at hep-th/0604179.
[28] ———, *Two-dimensional twisted sigma models, the mirror chiral de Rham complex, and twisted generalised mirror symmetry*, JHEP **0707** (2007), 013, available at 0705.0790. Companion paper to hep-th/0604179.
[29] Edward Witten, *Mirror manifolds and topological field theory*, Essays on mirror manifolds, 1992, pp. 120–158. MR1191422 (94c:81194)

Department of Mathematics, David Rittenhouse Lab., 209 S. 33rd St., Philadelphia, PA 19104
E-mail address: donagi@math.upenn.edu

Department of Mathematics, David Rittenhouse Lab., 209 S. 33rd St., Philadelphia, PA 19104
E-mail address: josh.guffin@gmail.com

Department of Mathematics, University of Illinois, 1409 W. Green Street, Urbana, IL 61801
E-mail address: katz@math.uiuc.edu

Department of Physics, MC 0435, Virginia Tech, 910 Drillfield Drive, Blacksburg, VA 24061
E-mail address: ersharpe@vt.edu

Proceedings of Symposia in Pure Mathematics
Volume **85**, 2012

Foundations of Quantum Field Theory

Michael R. Douglas

ABSTRACT. A short introduction, from a physicist's point of view, to nonperturbative and potentially rigorous definitions of quantum field theory, and the questions we would like to study with them.

1. Introduction

Quantum field theory and string theory have had a significant impact on mathematics, and mathematics is essential in studying quantum field theory and string theory, as we have seen in every talk at this conference. This leads to a natural question, which many have asked:

How can we make mathematically precise definitions of
quantum field theory and string theory ?

This question has become all the more important given the many mathematical applications of QFT which have been found in recent years, such as mirror symmetry and Seiberg-Witten theory just to name two. At present the standard mathematical approach to these applications is to use QFT to motivate interesting conjectures, but then to base all the subsequent work on reformulations of the conjectures which have rigorous foundations. It would be interesting, and perhaps simpler and more direct, to try to prove the QFT formulations of the conjectures. Of course, this requires rigorous foundations for QFT. This is a clear mathematical motivation to pursue the subject, but there are other mathematical and even physical motivations, which we will explain shortly.

There is a long history of mathematically precise approaches to quantum field theory, including

- axiomatic quantum field theory [**49, 53**]
- constructive quantum field theory [**23, 42**]
- algebraic quantum field theory [**24, 57**]
- functional integration and approximating expansions [**11, 37, 43**]
- vertex algebras and conformal field theory [**17, 18, 20**]
- probabilistic approaches, such as Schramm-Loewner evolution [**7, 34, 51**]
- chiral algebras and factorization algebras [**5, 12**]
- topological field theory and higher category theory [**10, 36**]

2000 *Mathematics Subject Classification.* Primary 81T08; secondary 60J67, 81T05.

One sees that the subject is very diverse, and the first goal of this overview is just to give the reader a basic idea of the different assumptions, techniques and goals of these various approaches.

A second goal is to explain why the question has not been settled, after more than fifty years of work. Probably the main reason is that for much of this period, the only general approach we had to QFT was the renormalized perturbative expansion around free field theory, *i.e.* Feynman diagrams. While very powerful, as is well known (and as we briefly review in §4), this series does not converge; it is at best asymptotic. While there are methods to sum asymptotic series such as Borel resummation, this approach has had limited results. Indeed, resummation of divergent series expansions is generally not a preferred approach in other areas of mathematics or physics; the standard treatments of quantum mechanics or classical PDE don't do this. On the other hand, perturbative renormalization theory is very well developed and has a certain beauty to it. Even if it is not the full story, it seems to be an essential part of the story.

The direct approach to solving this problem is constructive quantum field theory, in which one splits up and rearranges the perturbative series to make it convergent, as we discuss in §6. This has been successful, but at the cost of greatly complicating the expansion, and this is perhaps the main reason this approach has not found more applications. It is not clear at this point whether this complication is essential, or could be overcome by further developments. Recent work bringing ideas from homotopy theory into QFT, such as [**11**], might help in simplifying these powerful techniques.

The other response to the difficulties of perturbation theory is to use a non-perturbative starting point, with lattice QFT and solvable conformal field theory (CFT) being the most successful so far. Both of these have been made rigorous in examples, but again at a price. For example, the work [**3**] of Balaban establishing a continuum limit of 4d lattice Yang-Mills theory, in the words of [**42**] "reaches the limits of human communicability" and has not yet had the impact that such a fundamental result deserves.

It seems fair to say that the questions which attracted the most attention in this area, such as the construction of four-dimensional Yang-Mills theory, still appear difficult from all points of view – see the Clay Millennium Prize write-ups [**8**]. Thus progress in the near term probably depends as much or more on identifying easy questions with significant consequences (the proverbial "low-hanging fruit") as on any technical advances.

In what follows, I am going to try to motivate the goal of proving rigorous results about QFT in ways that I think should interest physicists as well as mathematicians. In addition to some general arguments in §2, this interest will follow from three main points:

(1) Some questions are very fundamental and not so difficult to answer rigorously. Well known examples are the CPT and spin-statistics theorems.
(2) Rigor enables one to work in more generality and more abstractly.
(3) Rigor shows that one's basic definitions are consistent and complete, and that no-go theorems are meaningful.

As point (1) is well accepted, let me illustrate points (2) and (3) with some possible applications of the ideas we are about to review. As our first example, let us consider the many physical approaches to nonperturbative computations in

QFT, including lattice field theory, series resummation techniques, the conformal bootstrap, supersymmetry and localization, and so on. Practical computational techniques often lead to approximate results and thus natural questions arise as to how to define the convergence of an approximation scheme, and how to compare approximate results coming out of the different techniques to get well motivated error bounds. While this is a very common element of the physics work, so far as we know these questions have not been systematically studied. This is probably because they are mathematical at heart, and require a certain measure of rigor to obtain convincing answers.

Another problem whose solution requires at least some measure of rigor is to prove that a QFT with particular properties does *not* exist. In physics, there is a saying that "no-go theorems," to the effect that a particular structure or object does not exist, would better be renamed "go-go theorems," as there are so many examples in which loopholes in the arguments, or changes to the assumptions, inspired a way to construct the object in question. While this is a fine attitude for the exploratory phase of a subject, of course its value diminishes with time, and eventually the focus must inevitably change to mapping out the possibilities and excluding what is not possible in a more conclusive way.

As a very basic example where we could use a rigorous no-go theorem, we would like an upper bound on the space-time dimension of an interacting QFT. Note that there *are* interacting statistical field theories in arbitrary space-time dimension, defined by functional integrals with higher derivatives in the quadratic part of the action. These do not count because they have indefinite norm Hilbert spaces, but this shows that a convincing argument must use the positive definiteness of the norm (or "unitarity"), while simple arguments based on perturbation theory and stability do not.[1]

Until about fifteen years ago, the universal belief was that interacting unitary QFTs did not exist in space-time dimensions $D > 4$. But it is now believed, based on arguments from string theory [**60**], that there are interacting supersymmetric theories in $D \leq 6$. Is this really the limit, or is our current intuition based on string theory and supersymmetry flawed as well? This question seems fundamental enough to belong under point (1) of the list above.

Conversely, once we have the ability to prove that certain QFTs do not exist, serious efforts can begin to map out the entire space of QFTs which do exist. While there are very few classes of QFT for which we can confidently say that we know all the possibilities, at least in some cases one can make conjectures which are at least not manifestly incorrect. For example, there is a very well motivated conjecture, which we discuss in §9, describing the families of $D = 2$ superconformal field theories which in the weak coupling limit correspond to sigma models with Calabi-Yau target spaces. A variety of ideas on how to describe the entire "space of theories" in $D = 2$ are surveyed in [**14**].

These are just a few of the questions for which rigorous foundations would enable us to find better and more convincing answers. Adding to these the original mathematical motivations, it seems to me that the rigorous study of QFT deserves more attention from our math-string theory community.

---

[1] A similar point can be made for quantum gravity in $D = 4$ and this is one of the main reasons why the nonexistence of a QFT reducing to Einstein gravity at low energies is not yet proven; see [**58**] for a recent discussion.

2. The role of rigor in the math-physics interface

> I am acutely aware of the fact that the marriage between mathematics and physics, which was so enormously fruitful in past centuries, has recently ended in divorce.
> Freeman Dyson, *Missed Opportunities*, 1972 [**16**].

Up through the early 19th century, mathematics and theoretical physics were more or less aspects of a single subject, and many of the greats – Newton, Laplace, Lagrange, Gauss, and so on, worked on both. This changed during the 19th century, primarily because the subject became too large for any one person to cover, but just as importantly for natural reasons coming from the intellectual evolution and needs of the two fields. Especially, in the late 19th century, with the growing use of abstraction and nonconstructive arguments, paradoxes and contradictions emerged in various mathematical fields – not just logic and set theory, but also algebraic geometry, functional analysis, and other areas. These problems were eventually solved by the development of rigorous foundations grounded in set theory, which reached their current form in the early 20th century. Part II of [**39**] gives a good overview of this history; see [**32**] for a very brief recent introduction.

It took many years for the bulk of mathematics to be tied to these foundations. In many areas, such as algebraic geometry and PDE, this was not fully accomplished until the 1950's and 60's. By now, providing foundations is an expected part of the development of a new field. In doing this, rarely must one start from scratch with logic and set theory. No matter how novel the ideas, their roots will lie within established fields of mathematics, and one can build upon existing foundations. In the case of QFT, this starting point will include functional analysis, and perhaps other mathematical subfields (as we discuss below). A classic textbook on functional analysis with applications to mathematical physics is [**41**].

Despite the advantages of having rigorous foundations, one can do a great deal of theoretical science without knowing them, and not many theoretical physicists (these days) take the time to master or even learn them. Of course, this diversity of levels of rigor and points of view is by no means unique to mathematical physics. Even within pure mathematics, one might adopt an even more rigorous and explicit style than is usual, requiring that the core of the discussion consist only of statements which can be explicitly translated into the terms of logic and set theory – this is sometimes called "formal rigor." This brings its own advantages, for example one can verify such statements on a computer [**25**]. Nevertheless, many mathematicians feel that they do not need this level of rigor in their own work, or might even feel that it obscures the essential points.

There are also many nonrigorous levels of discussion. One might argue that some of these distinctions are essentially sociological – a theoretical physicist might look down his nose at a quantum chemist, and so on. But most working theoretical physicists agree on the distinction between papers which are solidly grounded in the theoretical framework of their time, versus more speculative or more loosely reasoned work. Within a large community which agrees on such a standard, which has furthermore been tested by unexpected discoveries which were then accomodated within their framework, it is very natural to feel that one has attained truths, even if they may not have found their final and most precise expression. In the words of Richard Feynman, "we can know more than we can prove."

Indeed, to this day there is still debate over the value or necessity for rigorous results in theoretical physics, and the subject of quantum field theory has long been at the front lines of this debate. There are many points of view about this (see [**54, 29, 1, 55, 9**]), but I will simply state my own, which is that mathematical rigor is often valuable, even leaving aside the role it plays in defining and regulating the field of mathematics (as a physicist I will not get into this aspect). It is valuable because of the clarity it brings to many discussions, and as part of the necessary price for vastly increased generality.

This second point is easier to explain, being less subjective, so let me do so in an example drawn from classical field theories such as the Maxwell, Yang-Mills, or Einstein equations. In both mathematics and physics, a point of prime importance is to describe the boundary conditions required to uniquely determine a solution. All physics textbooks discuss this, and generally are satisfied with the principle that one needs to specify the fields and their time derivatives on a space-like hypersurface, perhaps adding some details about subtleties such as those caused by gauge invariance. This principle is then realized in various ways in the discussions of specific classes of solutions or techniques for computing approximate solutions.

Suppose one wants to understand this point for a large class of solutions. For example, this comes up in understanding singularities in general relativity. As shown by Hawking and Ellis [**26**], very generally, time evolution in general relativity leads to singularities, such as black hole formation or a "big crunch." To make this claim convincing, one must show that this happens for "physically reasonable" initial data, as opposed to initial data which is artificially constructed to generate singularities, say by focusing a finite amount of energy into a single point. Thus one must start by reformulating the vague term "physically reasonable" as some precise condition on the initial data, which for simpler physical theories would rule out singularity formation, before going on to study general relativity.[2] Textbooks such as [**56**] thus include mathematical discussions of the initial value problem, even if they are not directly concerned with rigor.

All of this physics might only be the "heuristic" introduction to the topic in a mathematical textbook on hyperbolic PDE, which might focus on necessary and sufficient conditions for a solution to exist and be unique, which can be proven and applied to a wide variety of equations. Although by this point one is asking very different questions, even this more abstract discussion can be useful in physics. For example, in string theory one knows that Einstein's equations get corrections due to the finite size of the string (the "α' corrections"). Are these corrections relevant for the singularity theorems? Can they lead to superluminal propagation and a breakdown of causality? Since it is not known how to explicitly compute all of the corrections, results which apply to a variety of equations in this class could be very helpful to have. But it is no coincidence that these successively more general levels of discussion are "more mathematical" and place a greater emphasis on rigor.

While the first point about clarity of exposition is more subjective, it is also real, as was brought home to me during my time as an editor for Communications in Mathematical Physics. Let us begin with an observation which will be familiar to many in our subfield, namely the divergence in style between mathematics and

[2]For nonlinear PDE, a typical condition of this type is that the initial data should have a finite Sobolev norm, meaning roughly that the integrals of certain powers of the function and its derivatives are finite.

physics papers. Modern mathematics papers are immediately recognizable by their deductive or "definition-theorem-proof" structure, which is enshrined in journal style sheets and TeX macro packages. Physics papers tend to be more discursive, and often tell a story, beginning by explaining why the subject is important, then recapitulating its development, and only then coming to the questions at hand. These might be addressed by making a claim and then supporting it with examples, or perhaps working through long computations. One might also prove theorems – of course physics papers contain many theorems, often with proofs which could be made rigorous without essential difficulty. But proof is rarely the central point, rather it is one of many tools for exploration, explanation and understanding.[3]

This view of proof as just one among many approaches towards the larger goal of understanding and explanation is held by some mathematicians as well (see for example [**55**]), and frees us to look for the advantages of each style and approach. To condense what could be a long discussion, a discursive style is more flexible and, one might argue, is better suited to situations in which the basic concepts are still unclear and essentials might be lost by formalization. One can freely bring in facts and observations from many different sources, postponing the considerable effort of reconciling and understanding them more deeply until the value of doing so becomes clear. The discursive style has advantages in complicated situations as well, as one can simply proceed to do a calculation or make one's point without being forced to consider all of the potential complications of other similar problems.

But, speaking for myself, there are many situations where the "definition-theorem-proof" style is clearer than a discursive style. This is particularly true in formalizing a set of ideas, but also in trying to explain a subject to a broader audience. Many mathematicians say they find physics discussions frustrating because "they never define what they are talking about." Even if the definition is only one part of a good explanation, it is a very important part, and doing it right justifies a good deal of work. Rigor, from this point of view, is the acid test that one has "done it right." As with earlier physical theories, someday this test will be passed for QFT as well.

This is most of what we will say to justify the project of developing QFT on rigorous foundations. Once we accept this goal, the first choice one must make is "which foundations." One's first thought might be to introduce new axioms to define QFT, phrased solely in terms of the generally accepted foundations of functional analysis. Such axioms, described in §3, are a necessary starting point. But unlike high school geometry (and many more sophisticated topics), so far there has been only limited success in deducing interesting general facts about QFT from the axioms. In itself this is not a criticism – after all the axiomatic definition of a manifold is also far from the interesting theorems or applications – but shows that we need further tools and techniques to construct and discuss the examples. People have taken these from several existing fields of mathematics, such as

- Probability theory
- Operator algebras
- Abstract algebra (category theory, sheaf theory, homotopy theory)

and the subsequent discussion will branch off in these diverse directions.

[3]Conversely, the mathematics literature contains many conjectures, and it is clearly not true that rigor disallows the discussion of what cannot be proven. The difference is that a mathematical conjecture is supposed to be stated precisely so that it could in principle be proven or disproven.

3. Axiomatic quantum field theory

This answers the question "what is a QFT" by adding to the well-accepted axioms for quantum mechanics (Dirac, von Neumann, etc.) further axioms that express the properties of locality, causality, and Lorentz invariance. The three most commonly used sets of axioms are

- The Wightman axioms [**53**]. These describe a QFT in terms of fields which are functions on Minkowski space-time, whose values are operators acting on a Hilbert space.
- The Haag-Kastler axioms [**24**]. These also describe a QFT on a Minkowski space-time, but instead of postulating explicit field operators, one instead postulates a C^*-algebra of fields.
- The Osterwalder-Schrader (O-S) axioms ([**49**], §II and [**23**], §6), which govern correlation functions in the Euclidean space-time obtained by Wick rotation.

Other important axiom systems include the Nelson axioms (see [**49**]), and Segal's modular functor definition of 2d conformal field theory [**47**], which we discuss briefly below and in §5. In addition there are more specialized axiom systems (not obviously applying to all QFTs) which we discuss in §8, §9. As we are merely giving a taste of the subject, we will not list the axioms in detail, and only try to give some sense of what they are and what one can do with them.

The central results of axiomatic QFT are the "reconstruction theorems," according to which a QFT defined within one set of axioms determines a unique QFT in another set. For some of the axioms, this is straightforward. For example, all of these axiom systems include invariance under the symmetry group of space-time. If we can justify the analytic continuation which relates Euclidean and Minkowski space-time, then clearly the symmetry groups will be related.

What is less obvious is the translation of statements which cannot be made directly in a given axiom system. For example, how does one state the condition that field operators act on a Hilbert space, in a formulation such as O-S which does not refer to operators? The well known answer is that this becomes a condition called "reflection positivity," which states that for all sets of k disjoint points (with $x_1 > 0$),

$$\langle\prod_{i=1}^{k} \phi_i(\vec{x}_i)\phi_i(R\,\vec{x}_i)\rangle > 0, \tag{3.1}$$

where $R\vec{x} = R(x_1, x_2, \ldots, x_D) = (-x_1, x_2, \ldots, x_D)$ is the reflection in the hypersurface $x_1 = 0$. It is clear that this is necessary, so that states

$$\prod_{i=1}^{k} \phi_i(R\,\vec{x}_i)|0\rangle \tag{3.2}$$

in a Hilbert space of states defined on the hypersurface $x_1 = 0$ will have positive norm. But how does one show that this is sufficient?

To do this, one proceeds by analogy to a universal construction which, given an operator algebra $\mathcal{A}$, an involution $*$ and a positive linear functional, produces a representation on a Hilbert space $\mathcal{H}$, the Gel′fand-Naimark-Segal (GNS) construction. Starting with $\mathcal{A}$, we take as a provisional definition of the Hilbert space $\mathcal{H}_0$ all products of operators with $x_1 < 0$. While this is compatible with a physics definition in which the Hilbert space is generated by all states of the form Eq. (3.2), logically it is different as we are regarding Eq. (3.2) as the *definition* of the Hilbert space. A Hilbert space is of course a linear space with an involution and a norm derived from an inner product, complete in this norm. To get this additional structure, we define the involution to be reflection R combined with Hermitian conjugation, while the norm squared of a state Eq. (3.2) is given by Eq. (3.1). Completeness follows from the completeness of $\mathcal{A}$.

While so far this may seem tautological, in fact this does not yet define a Hilbert space, because $\mathcal{H}_0$ contains null vectors – for example, two combination of operators related by time evolution will create the same physical state, so their difference is a null vector. The Hilbert space $\mathcal{H}$ of the GNS construction is the quotient of $\mathcal{A}$ by the null vectors, and is clearly a representation of $\mathcal{A}$.

While the validity of this construction would be a theorem for a bounded C^*-algebra $\mathcal{A}$, of course the operator algebras of QFT are not bounded, and one must place further conditions on Eq. (3.1) for this to work. These are the "regularity" axioms of O-S, which state that the partition function in the presence of a source (as in Eq. (5.2)) is analytic and does not grow too fast as the source becomes large. In this sense, reflection positivity is not sufficient by itself. Without it, one expects to run into trouble, because there is no O-S axiom which directly corresponds to locality (say, commutativity of space-like separated operators). The relation between regularity and locality is not entirely obvious (to me) however.

Perhaps the most well known results of axiomatic QFT are the CPT and spin-statistics theorems, and results (such as bounds on operator dimensions) which follow from the classification of unitary representations of space-time symmetry algebras. They are both fundamental and experimentally testable. To oversimplify a bit, they follow from mathematical properties of two-point correlation functions; in other words they identify certain properties of free field theory which are in fact universal to QFT.

There is clearly a lot of scope for new axiom systems to take advantage of additional structure, for example supersymmetry, and to cover more general cases, for example QFT in curved space-time [**30, 28**]. In addition there are many known relations between different presentations of QFT which should be turned into reconstruction theorems.

As an example, according to the "bootstrap" approach to conformal field theory, a CFT is determined by the complete set of two- and three-point correlation functions (of all operators), or equivalently the operator product coefficients. Clearly it would be useful to prove that this data determines a QFT in the other presentations; of course one needs convergence (or regularity) conditions for this to be so. This has been done for $D = 2$ rational CFT in many works (we return to this in §9) but not (as far as we know) for non-rational CFT or $D > 2$. Some steps in this direction are in [**27**].

Similarly, one might try to relate Segal's axioms to QFT in curved space. This would probably require adding regularity conditions to Segal's axioms, a point also made in [**33**]. One of the key points in Segal's axioms is the gluing condition, which describes the gluing of space-time manifolds along their boundaries. This is related to the Markov property, which we discuss in §5.

4. Structure and problems of perturbation theory

> In the thirties, under the demoralizing influence of quantum-theoretic perturbation theory, the mathematics required of a theoretical physicist was reduced to a rudimentary knowledge of the Latin and Greek alphabets.
>
> Res Jost, as quoted in *PCT, Spin and Statistics, and All That*, Streater and Wightman, 1964.

All textbooks on quantum field theory begin by defining the field theories of free bosons and fermions. They then develop a perturbation theory which expresses observables as a series expansion in 'coupling constants' which control the interactions. The main result of this theory, from a physics point of view, is the proof of perturbative renormalizability and unitarity of Yang-Mills theory in up to four dimensions, coupled to a fairly general theory of scalars and fermions.

Now, as long as one only tries to define the coefficients of this expansion, and does not try to sum it, there is no essential difficulty in making these definitions rigorous, as has been done in [**2, 11, 23, 42, 49**] and many other works. This has found various mathematical applications – for example, the expansion of Chern-Simons theory leads to knot and 3-manifold invariants, while the numerical values of loop diagrams have intriguing algebraic geometric and number theoretic properties.

The physical development of perturbation theory, and consideration of issues such as gauge invariance and constraints, led to many simultaneous or rediscoveries of mathematical techniques for dealing with analogous issues. The culmination of this development was the Batalin-Vilkovisky formalism, which turned out to reproduce a sizable fraction of homological and homotopical algebra [**52**]. A recent discussion from a mathematical point of view is [**11**].

For many physicists, BV quantization is a very technical subject which one calls upon when one needs to treat technically challenging problems in quantization, such as the case of gauge symmetries with field-dependent structure constants, or which only close after using the equations of motion. In mathematics the corresponding framework is far more general and conceptual. One of its slogans is that "rigid algebraic structures become flexible in the homotopy category," meaning that even if one cannot deform a structure, one can often generalize it by introducing a cohomology and introducing higher products which are zero in cohomology, so that the original defining properties hold up to homotopy. An example relevant for string theory is the homotopy category of loops in a manifold, which starts from the observation that concatenation of parameterized loops defines a non-associative multiplication, which is associative up to homotopy (reparameterizing the loop). This can be formalized in an A_∞-structure, and this lies behind the BV structure of closed string field theory [**31**]. Similariy, the general problem that regularization and renormalization breaks the symmetries of quantum field theory (gauge symmetry, supersymmetry, and so on), can be addressed by defining larger structures in which these symmetries hold up to homotopy. This is the "hidden geometry" behind BV which may eventually be recognized as one of the key foundational principles of QFT.

Even given a deep understanding of perturbation theory, it does not provide a rigorous definition of QFT. As explained in the textbooks, perturbation theory around the free boson leads to an asymptotic series expansion, with zero radius of convergence. This follows from simple combinatorics (the number of graphs with ℓ loops grows as $\ell!$, or from physical arguments (a theory with coupling $g < 0$ tends to be unstable, so physical observables should not be analytic around $g = 0$).

The essential phenomenon is already present in zero dimensions, in which the functional integral reduces to finite-dimensional integrals such as

$$Z = \int_{-\infty}^{\infty} dx\ e^{-x^2 - gx^4}. \tag{4.1}$$

Expanding in g and integrating, this is

$$Z = \sum_{n \geq 0} \frac{(-g)^n}{n!} \Gamma[2n + \frac{1}{2}] \tag{4.2}$$

and the coefficients grow in magnitude as $n!$. On the other hand, for Re $g > 0$ the integral is clearly more convergent than a Gaussian. Mathematically, the large ϕ behavior of the integrand is different for $g = 0$ and $g \neq 0$, so that $g = 0$ is an essential singularity of Z. We will see that it arises from large ϕ in §6.

Since the large order behavior is a property of the actual coefficients of the coupling constant expansion, it will not be solved by changing the technique one uses to compute them. For example, in recent years there has been remarkable progress in developing simpler descriptions of gauge theory amplitudes than the original Feynman diagram expansion, using twistors, new recursion relations, relations to hyperbolic volumes and periods on Grassmannians, and so on. While these techniques look very different from Feynman diagrams (and each other), since they still compute terms in the perturbative expansion, any attempt to use them to define finite N Yang-Mills theory will still face this problem.[4]

Rather, one must change the whole approach, either somehow avoiding series expansions altogether, changing the nature of the expansion, or else by perturbing around other starting points. Thus, given a QFT T_0, we could define

$$Z[g] = \langle e^{g \int d^D x\ \mathcal{O}} \rangle_{T_0} \tag{4.3}$$

and hope to get a convergent expansion. In fact this is already the case for the free fermion. Because of sign cancellations, determinants are much better behaved than bosonic integrals.

Renormalized perturbation theory around a nontrivial UV fixed point is a standard physics technique; see [**21**] for a recent discussion. It is believed that perturbation theory by a marginal operator around a conformal fixed point with a discrete spectrum of operator dimensions should have finite radius of convergence. It is certainly true that exact results (say for anomalous dimensions and OPE coefficients) are analytic and thus this could be true; however there do not seem to be any results in the literature for such a radius of convergence for an interacting CFT, even in 2d and even non-rigorous. Such results would be very valuable to have; we discuss one application in §9.

---

[4]One might imagine that superconformal theories are different, and at this point nobody really knows the large order behavior for $N = 4$ super Yang-Mills theory. Note also that if one first takes the large N limit, the loop expansion has different (exponential) asymptotics, and finite radius of convergence. This has been used in constructive QFT, for example the "wrong sign planar Φ^4 theory in $D = 4$" is discussed in [**42**], §II.5.

5. The language of probability theory

There is nothing in the definition of QFT that requires us to use perturbation theory, and one might decide to go back to one of the physical starting points, say the Euclidean functional integral, and try to define it rigorously, perhaps in a nonconstructive way. For definiteness, consider a scalar field $\Phi : M \to \mathbb{R}$, and the integral

$$Z = \int [d\Phi] \, e^{-\int_M d^D x \, |\partial\Phi(x)|^2 + g\Phi(x)^n} \equiv \int [d\mu_g], \tag{5.1}$$

where M is a D-dimensional manifold. Although one often hears and reads frightening statements such as "there is no analog of Lebesgue measure $[d\Phi]$ on an infinite dimensional space" [**59**], this is of course a red herring as we only need to define the integral with an exponential factor as in Eq. (5.1), which is highly convergent.

In fact, many analogous but simpler problems of integration over infinite dimensional spaces are mathematically well understood, such as a wide class of integrals over paths (the case $D = 1$), and Gaussian integrals in any dimension. Given a non-negative measure such as Eq. (5.1), one can think of it as a probability density function (by dividing by Z), and thus much of this work falls under the rubric of probability theory and stochastic analysis. Thus one can find the rigorous theory of the Gaussian integral, already adapted for QFT, in many sources [**23, 49, 48**]. Since the additional term in Eq. (5.1) (for real $g > 0$) only makes the integral more convergent, it should be easy to show that this integral makes sense.

In fact, given the expectation values of general products of operators, or equivalently the partition function with a general source

$$Z[J] = \int [d\mu_g] \, e^{i \int_M d^D x \, J(x)\Phi(x)}, \tag{5.2}$$

(this is the analog of the "characteristic function" in probability theory), one can reconstruct the measure $[d\mu_g]$. One can show (by Minlos' theorem; see chapter 6 of [**23**]) that the O-S axioms *guarantee* that this measure exists and thus, if the QFT in question exists at all, it *must* have a Euclidean functional integral description. The question is simply how to get it.

One key property of many probability distributions is the "Markov property." For a time series $\phi(t_i)$ (or "stochastic process"), this states that the joint probability distribution of a set of $\phi(t_i)$ with $t_i \geq T$ for any given T depends only on the probability distribution for $\phi(T)$, call this $P[\phi(T)]$, and not on any $\phi(t_j)$ with $t_j < T$. This is what one usually means by "locality" in the special case $D = 1$. It implies that the dynamics can be described by a first order linear PDE satisfied by $P[\phi(T)]$, the Fokker-Planck equation. This is the Euclidean time analog of the Schrödinger equation, and standard techniques in functional analysis can be used to define all of the structures referred to in the axioms of §3.

The analog of the Markov property for Euclidean QFT is that, given the probability distribution for fields $\phi(x_i)$ with all x_i on the boundary of some (bounded) region R, the probability distribution for fields $\phi(y_i)$ with y_i in R should be determined. In fact this was one of Nelson's axioms ([**49**], §IV.1). These axioms made precise the idea that the measure in Eq. (5.1) is a local probability measure, where locality is the Markov property.

Note that the Markov property is closely related to the gluing property of Segal's axioms. Given a manifold Σ and a hypersurface Γ which divides it into two parts Σ_1 and Σ_2, the gluing property states that

$$U_\Sigma(V_1, V_2) = \mathrm{Tr}_V\, U_{\Sigma_1}(V_1, V) U_{\Sigma_2}(V, V_2), \tag{5.3}$$

where U_Σ is the multilinear functional associated to Σ and V is a state in the Hilbert space associated with the common boundary Γ. By the Markov property, we can obtain the right hand side (with a particular V) from the left by specifying the fields on Γ.

Interestingly, Nelson's axioms are *stronger* than the O-S axioms ([**49**], §IV.4). For example, a quantum mechanics with the Hamiltonian $H = p^4 + p^2 + q^2$ satisfies the O-S axioms but not the Nelson axioms, since one can show that Euclidean time evolution of an everywhere positive wave function need not lead to an everywhere positive wave function. On the other hand, this quantum mechanics is still governed by a local (in time) Schrödinger equation, so perhaps this difference has to do with the probability interpretation rather than locality.

6. Constructive quantum field theory

Again, we can only summarize the basic ideas of this; some excellent introductions to this topic are [**23, 42, 43**], while [**37**] explains the exact RG in this framework.

Let us return to the $D = 0$ case Eq. (4.1). Expanding in g and applying the saddle point method to the nth term, one finds the integral is dominated by $x \sim \pm\sqrt{2n}$, so the problem with large orders is coming from large $|x|$. Thus, a simple way to get a convergent series expansion is to break up the region of integration,

$$Z = \left(\int_{-C}^{C} dx \; + \int_{C}^{\infty} dx \; + \int_{-\infty}^{-C} dx \right) e^{-x^2 - gx^4}. \tag{6.1}$$

Expanding the first integral in g, the terms with order $n > C$ now have exponential growth,

$$\int_{-C}^{C} dx\; e^{-x^2} x^{4n} \sim e^{-C^2} C^{4n}. \tag{6.2}$$

On the other hand, for $\mathrm{Re}\, g > 0$, the large field regions produce very small corrections $\sim \exp -gC^4$. Thus, at least from an analytic standpoint in which one is trying to get controlled estimates as opposed to exact results, the fact that the interactions suppress the measure makes the large order problem easy to deal with.

For Eq. (5.1) with $D \geq 1$, this idea is more challenging to apply, because the interaction term which suppresses the measure is simple in position space, while the kinetic term and everything else about perturbation theory is simpler in momentum space. Following [**22**], the standard CQFT solution to this problem is the cluster expansion.

We start by partitioning space-time into small cells R_α, say non-overlapping hypercubes. We then do the analog of Eq. (6.1) for each cell, schematically

$$\begin{aligned} \int [d\Phi] \;&=\; \prod_\alpha \int [d\Phi|_{R_\alpha}] \\ &\Rightarrow\; \prod_\alpha \left(\int_{-C}^{C} [d\Phi|_{R_\alpha}] \;+ \int_{C}^{\infty} [d\Phi|_{R_\alpha}] \;+ \int_{-\infty}^{-C} [d\Phi|_{R_\alpha}] \right). \end{aligned} \tag{6.3}$$

This can be rewritten as a "polymer expansion," with terms which correspond to subsets of the R_α which sit in the large field regime.

We then combine this expansion with the standard weak coupling expansion in the regions in which Φ is small. By analogy to Eq. (6.2), this leads to a convergent weak coupling expansion. The remaining step is to bound the contribution of the large field regions. For a superrenormalizable or even asymptotically free theory, this can be done, because for sufficiently small regions R_α the field is approximately Gaussian.

As discussed in [**23, 42**] and references there, this suffices to construct super-renormalizable theories such as the "Landau-Ginzburg theory" (scalar fields with polynomial interactions) in $D = 2$. This case is a bit easier as the only renormalization required is to normal order the interaction, so one can replace the partitioning of space-time as in Eq. (6.3) with a modification of the coupling expansion. But the essential improvement is still to split off and bound the large field region. One can see that this is necessary by considering a normal ordered polynomial interaction at different scales, say

$$:V(\Phi):|_\mu = g\Phi^4 + (m^2 + g\log\frac{\mu}{\Lambda})\Phi^2 + \text{const}\cdot(m^2, g, \Lambda). \tag{6.4}$$

As $\mu/\Lambda \to 0$, this becomes unbounded below, and one must show that this happens in such a small region of field space that its contribution is negligible.

For more complicated theories, such as scalars in $D \geq 3$, the cluster expansion by itself does not really suffice. A next step is to do a "multiscale" expansion, in which the propagators are decomposed into momentum slices such as

$$\frac{1}{p^2+m^2} = \int_0^\infty d\tau\; e^{-\tau(p^2+m^2)} \tag{6.5}$$

$$= \sum_{n\in\mathbb{Z}^+} \int_n^{n+1} d\tau\; e^{-\tau(p^2+m^2)}. \tag{6.6}$$

This allows separating the contributions of different scales to the renormalization group, and thus reorganizing the large RG contributions to effective couplings into a controlled expansion. These techniques enabled the construction of Φ^4 theory in $D = 3$, and even in $D = 4$ with a UV cutoff (required since this is not asymptotically free). These ideas were also used to rigorously define the RG near a Gaussian fixed point [**6**] and are perhaps easiest to learn in the context of this problem [**37**].

Fermionic fields do not suffer from the large field problem, and many rigorous results were obtained, for example the construction of the Gross-Neveu model (fermions with a quartic interaction) in $D = 2$ and even in $D = 3$, where there is a nontrivial fixed point for large but finite N. Various fermion models of many-body physics have also been discussed [**44, 46**].

Attempts were made to define $D = 4$ Yang-Mills theory in a small box (so that the coupling would remain weak), but without complete success [**42**]. Besides the complexity of the multiple series expansions, another major difficulty was that a successful analysis tends to require some sort of positivity in the intermediate results; in other words summing the expansions should not lead to large cancellations, which are almost impossible to control. This is of course very dependent on the specifics of the expansion – for example, a naive computation of the determinant of a matrix involves drastic cancellations, whereas its computation as the product of the eigenvalues involves no cancellations.

To summarize, a variety of interesting QFTs in $D = 2$ and even $D = 3$ have been constructed using these methods, and systematic application of the existing methods might well suffice to construct the "general" superrenormalizable QFT in $D = 2$, meaning a gauged Laudau-Ginzburg model with fermions.

7. Other probabilistic approaches

This rubric covers various approaches which use the mathematical theory of probability to make arguments which need not have simple translations into the language of functional integrals and perturbative expansions. One of the most interesting examples is the concept of Schramm-Loewner evolution, which defines a family of random distributions of one-dimensional curves or interfaces (a "random curve" for short). We refer to [**7, 40, 34**] for detailed discussions and only quote the basic definition,

$$\partial_t g_t(z) = \frac{2/\kappa}{g_t(z) + B_t}. \tag{7.1}$$

Here κ is a parameter (determining the distribution SLE_κ) and B_t is Brownian motion, in other words a random walk on the real line depending on a 'time' t. This equation determines $g_t(z)$, a family of conformal maps associated to a random curve $\gamma(t)$ in the upper half-plane. The point is that, starting with the simple and linear random curve B_t, and applying the highly nonlinear transformation Eq. (7.1), one gets a natural *conformally invariant* family of random curves. These satisfy uniqueness properties which make them the more or less inevitable description of any family of curves with some local definition in terms of the degrees of freedom of a two-dimensional conformal field theory.

As an example, consider the Ising model on a disk with the following boundary conditions: choose two points a and b on the boundary; then the spins to the left of a and the right of b are up, while the spins to the right of a and the left of b are down. For any particular spin configuration, one might try to associate a curve through the disk from a to b which separates the region controlled by the up boundary condition from that controlled by the down boundary condition; in some sense having up spins on one side and down on the other side. This defines a family of random curves, whose properties would be related to correlation functions of the spin operator. By arguments in [**45**], this family must be distributed according to SLE with $\kappa = 3$.

In a 2d lattice theory, one can make a precise definition of an interface as a boundary between clusters of up and down spins. One can then hope that, in the continuum limit, this will go over to the family of curves described by SLE. Indeed, using similar ideas (a bit before SLE was defined), Smirnov showed that this type of observable is conformally invariant in both 2d percolation and the critical 2d Ising model [**50**].

A somewhat simpler argument [**51**], not directly using SLE, proves conformal invariance of a local Ising model observable, essentially the expectation value of a fermion in the presence of a pair of spin fields (points at which the fermion has a -1 monodromy). By making careful definitions in the lattice model, one can get a version of this which is a "discrete holomorphic function," satisfying a discrete analog of the Cauchy-Riemann equations [**35**]. For example, one can define a discrete holomorphic function f on vertices of a square lattice, $z \in \mathbb{Z} + i\mathbb{Z}$, as one satisfying[5]

$$f(z+1)-f(z+i) \over 1-i} = \frac{f(z+1+i)-f(z)}{1+i}. \tag{7.2}$$

Given suitable regularity conditions, in the continuum limit these will become analytic functions.

Discrete holomorphic functions satisfy some of the properties of continuum analytic functions, but not all. For example, differentiation and contour integration lead to analytic functions. On the other hand, the product of two discrete holomorphic functions *need not* be discrete holomorphic. This leads to numerous difficulties in copying the continuum CFT results, and at present there are only partial analogs of this construction for other CFTs such as $\mathbb{Z}_N$ parafermions [**40**].

Stepping back, a general lesson one can draw is that rigorous definitions can sometimes make it possible to formulate and work with highly nonlinear relations such as Eq. (7.1). Another illuminating example is [**15**], in which analogs of the operators $: \exp \alpha\phi :$ used in the DDK approach to 2d Liouville theory are defined by applying the nonlinear expression $: \exp \alpha\phi :$ to a free field ϕ, and used to compute quantum gravity-like observables.

8. Algebraic approaches and locality

One of the central defining properties of QFT is locality, and there have been many mathematical attempts to specify what this means. The original physics definition was for field operators in Minkowski space-time and states that these commute at space-like separation. This is one of the Wightman axioms and is clearly important, but it is hard to use deductively, and by itself it does not seem to capture all of what is intuitively meant by locality.

A generalization is to consider the entire algebra of operators $\mathcal{A}_R$ in a subregion R of space-time, and to require that the algebras associated with two causally disjoint regions (*i.e.* such that the light-cone of one is disjoint from the closure of the other) commute. This is one of the Haag-Kastler axioms, and is the starting point for algebraic QFT [**24**]. Rather than consider a particular list of subsets, it is better to consider *all* open subsets R_i of space-time. The inclusion relation $R_i \subset R_j$ then implies an embedding of algebras $\mathcal{A}_{R_i} \to \mathcal{A}_{R_j}$, satisfying the commutativity conditions. This structure is called a net of operator algebras.

The dual (in the mathematical sense) to a net is a presheaf. This is also a system of subsets, inclusion relations, and associated linear spaces $\mathcal{F}_{R_i}$, but now $R_i \subset R_j$ implies the existence of a linear map $\mathcal{F}_{R_j} \to \mathcal{F}_{R_i}$. If a choice of compatible elements $f_i \in \mathcal{F}_{R_i}$ determines a unique $\mathcal{F}_R$ in $R = \bigcup R_i$, we have a sheaf.[6] The prototypical example is the sheaf of functions over a manifold, since functions can be restricted to smaller regions, and patched back together.

Given a net of operator algebras, the structure of the correlation functions of operators in all the regions R_i is a presheaf, or perhaps a sheaf. Thus these structures provide a natural mathematical framework for expressing locality. The commutativity of causally disjoint regions is natural for Lorentzian causal structures. But what replaces it in Euclidean or Riemannian space-times?

[5]This is a bit different from the definition used in [**51**].

[6]This is a bit simplified.

From many points of view, the most natural definition of locality is that one can decompose space-time into bounded regions R_i, such that all observables in a given region R_i (the partition function, correlation functions of local operators, and so on) depend on the other regions $R_j \neq R_i$ only through boundary conditions, in other words a choice from a set determined in some way by the boundary ∂R_i. This is the content of the Markov property of §5, and the functorial definition of Segal. In these frameworks, one must explicitly define the space of boundary conditions. As has been found in physics, it is difficult to do this within the framework of renormalization theory, since cutoff theories are not local.

An alternate approach is to use a universal construction as in Eq. (3.2) to implicitly define this space in terms of the space of possible operators on the disk or ball which could be used to fill in the boundary. Physically, this is done in the "state-operator correspondence" of conformal field theory, combined with the assertion that the operator product expansion converges, so that products of operators can be replaced with sums over single operators. A related mathematical definition is the factorization algebra of Beilinson and Drinfeld [**5, 18**], in which correlation functions are sections of sheaves over the product of several copies of space-time (the locations of operators), satisfying additional consistency conditions corresponding to associativity of the OPE.

Given these fairly simple abstract descriptions of locality, and the intuition that locality is a central property of QFT, one would hope that fairly simple abstract arguments could be made to derive the physical consequences of locality. So far these questions seem largely open. As an example, it is unio.p.versally believed that, whenever one has a family of QFTs depending on a parameter g, the family can be realized by perturbing by a local operator around any of the members, as in Eq. (4.3) (this claim takes its simplest form for CFT and marginal deformations). So far as we know, this has not been proven in any rigorous framework, in part because one needs nontrivial information to show that Eq. (4.3) makes sense.

9. 2d conformal field theory

This is a particularly interesting special case, much more familiar to mathematicians, and a variety of interacting QFTs of this type have been rigorously constructed though algebraic techniques. Besides citing some of this literature, we will try to explain why, from a physics point of view, half of the subject (the "non-rational CFTs") still awaits a comparable mathematical understanding.

One can take two broad points of view to 2d CFT. One is as a special case of the general framework we have been discussing. The other is to start with "chiral CFT," in which correlation functions are meromorphic functions on a one complex dimensional Euclidean space-time, satisfying braiding relations. One can then argue that a more general CFT can be thought of in the same terms, by a double analytic continuation from 2d Minkowski space-time to two complex dimensions. The resulting correlation functions are meromorphic in the two complex variables and satisfy braiding relations, structure which makes the mathematics far more tractable. Most axiom systems for CFT, such as [**20**], take this starting point.

Perhaps the simplest interacting quantum field theories with a rigorous definition are the 2d CFTs with global symmetry $G \times G$, in other words the Wess-Zumino-Witten models. One can construct the Hilbert space in terms of the highest weight representations of the affine Kac-Moody algebra $\hat{\mathfrak{g}}$. One then needs to show that the highest weight representations can be created from the vacuum by primary fields which obey the CFT axioms including the constraints of conformal invariance. This has been done in various ways – within the theory of vertex operator algebras [**17**], and in an operator algebraic approach in [**19, 57**].

The result of these constructions is a chiral CFT. While these have physical realizations, for example as edge states in Chern-Simons theories, the CFTs defined as functional integrals (such as the Wess-Zumino-Witten models) are "non-chiral," with single-valued fields whose correlation functions are not meromorphic. These are the theories which satisfy the axioms of §3 and make direct contact with the rest of our discussion.

Although it is an additional step, many non-chiral CFTs, those which are rational and quasi-rational, can be rigorously defined using the same techniques. The WZW models and many other solved CFTs such as the BPZ minimal models are rational, essentially meaning that the Hilbert space and correlation functions are finite sums of products derived from two chiral CFTs. The other solved models are "quasi-rational," in which one uses infinite sums but of polynomial growth. This covers the sigma models with torus target space, and orbifolds of these.

While these two classes are very important, there are strong physical arguments that they do not cover all of the CFTs. On general grounds, one expects perturbation theory by a marginal operator (as in Eq. (4.3)) to spoil rationality. Thus, if there are families of CFTs other than the torus sigma models and orbifolds, then *a priori* one can expect that non-rational (neither rational nor quasi-rational) CFTs exist. While elementary examples of families of CFTs are not known, the superconformal theories with $(2,2)$ and $(4,4)$ supersymmetry do come in families. We briefly summarize the arguments for this, referring to [**13, 38**] for more details. In the weak coupling limit, these can be defined as supersymmetric sigma models with Calabi-Yau and hyper-Kähler target spaces respectively. The effective coupling constant is the inverse curvature radius and, although one expects that perturbation theory is asymptotic, it is very plausible that these models exist at weak coupling. If so, then one can argue physically that these theories come in moduli spaces (the marginal perturbations are exactly marginal) with known geometries (this follows from arguments using topological twisting and so-called "special geometry."). To confirm this picture, one has other functional integral constructions, such as the gauged linear sigma model, and also arguments, starting with Gepner, that various isolated points in these moduli spaces are rational CFTs, which could be rigorously defined using existing techniques. All this gives strong reasons to believe that these CFTs exist and come in families, which for the $(2,2)$ theories cannot be equivalent to tori or orbifolds.

The existence and nature of these non-rational CFTs is arguably *the* outstanding open question in 2d conformal field theory. In some cases, such as the sigma models with K3 target space, it remains possible that all of the theories are quasi-rational, but this is not at all clear. For target space a three complex dimensional Calabi-Yau, this seems very unlikely, but one can still hypothesize that a dense subset of theories are rational or quasi-rational, in which case the existing techniques would probably suffice to define the others. A fairly direct way to decide this would be to get control over the shifts in operator dimensions in perturbation theory around a rational theory (as in §4), as this hypothesis implies very strong constraints on these shifts.

My own suspicion is that the rational and quasi-rational theories are not dense, and that the existing algebraic CFT methods will not suffice to construct non-rational theories. Thus it would be very interesting to construct examples of these theories, even at weak but finite coupling, in order to explore these questions. If one starts from the gauged linear sigma model, which is superrenormalizable, this would seem to be well within the scope of the existing methods in CQFT.[7]

10. Where to go next?

While we were not able to do more than survey the rigorous approaches to QFT, we have made the point that there are many results in this area and that for at least some QFTs we have several such definitions. We also raised some conceptual questions for which the existing physics techniques do not seem adequate and for which these approaches might be useful.

While we largely focused on approaches rather than applications, coming back to the original mathematical motivations of the introduction, a good test of any approach would be to rigorously define a supersymmetric QFT with a corresponding topologically twisted model, and show that the corresponding observables actually agree. For example, there are many models such as Donaldson theory or the 2d A-twisted $(2,2)$ sigma models in which certain correlation functions are believed to be given exactly by integrals over appropriate moduli spaces of instantons, with a measure derived by a one-loop computation (this is usually called "supersymmetric localization"). While this has been checked in the examples which admit rigorous definition, it has never been proven from first principles. The $(2,2)$ linear sigma models can probably be defined using the existing techniques of constructive QFT and this leads to a precise conjecture, which we will discuss elsewhere.

Finally, it seems to me that a central theme of the most interesting recent developments is to take constructions which were motivated by QFT and the physical and mathematical questions to which it gives rise, and place them within a broader context in mathematics. Among the many examples, let us come back to the constructive QFT techniques of §6. A central part of this is the use of series expansions to isolate and control different contributions to the functional integral, and it seems to me that much could be done to systematize this and integrate it with other parts of mathematics which deal with similar problems. The general problem of decomposing finite dimensional manifolds into regions to facilitate defining and computing integrals is of course very important and was the original motivation for homology theory, leading to much of topology. While the problems of constructive QFT are far more intricate, involving infinite-dimensional integrals, perturbative expansions and renormalization, one would still hope that the intuitions based on finite dimensions can be adapted to this case. This is the intuition which made the functional integral so powerful in physics and we would certainly hope that someday it will be a larger part of mathematics.

Acknowledgements

I would like to thank Alain Connes, Kevin Costello, Arthur Jaffe, Vincent Rivasseau, Albert Schwarz, Dennis Sullivan, A. J. Tolland, and Steve Zelditch for discussions on these topics, and Steve Shenker for emphasizing the question about field theories in $D > 6$.

[7]I am indebted to A. J. Tolland for many discussions about this idea.

References

1. M. Atiyah *et al*, Responses to "Theoretical Mathematics: Toward a cultural synthesis of mathematics and theoretical physics" by A. Jaffe and F. Quinn, Bull.Am.Math.Soc.30:178-207, 1994, [arXiv:math/9404229].
2. S. Axelrod and I. M. Singer, "Chern-Simons perturbation theory I," [arXiv:hep-th/9110056]; "Chern-Simons perturbation theory II," J. Diff. Geom. 39, 1 (1994), 173-213.
3. T. Balaban, "Renormalization group approach to lattice gauge field theories. I," Comm. Math. Phys. **109**, 249-301, (1987).
4. A. A. Belavin, A. M. Polyakov, A. B. Zamolodchikov, "Infinite Conformal Symmetry in Two-Dimensional Quantum Field Theory," Nucl. Phys. **B241**, 333-380 (1984).
5. A. Beilinson and V. G. Drinfeld, *Chiral algebras*, AMS, 2004.
6. D. C. Brydges, P. K. Mitter, B. Scoppola, "Critical (Phi**4)(3, epsilon)," Commun. Math. Phys. **240**, 281-327 (2003). [hep-th/0206040].
7. J. Cardy, "SLE for theoretical physicists," Annals Phys. 318 (2005) 81-118, [arXiv:cond-mat/0503313v2].
8. The Clay Millenium Prize for Yang-Mills and Mass Gap, `http://www.claymath.org/millennium/Yang-Mills_Theory/`
9. Communications in Mathematical Physics: Mission and Standards of Presentation; Comm. Math. Phys. **244** 1 (2004), [http://dx.doi.org/10.1007/s00220-003-0990-6].
10. Costello, "Topological conformal field theories and Calabi-Yau categories," Adv. Math. 210 (1), 2007.
11. K. Costello, *Renormalization and Effective Field Theory*, AMS, 2011.
12. K. Costello and O. Gwilliam, "Factorization algebras in perturbative quantum field theory," `http://math.northwestern.edu/~costello/factorization_public.html`
13. *Dirichlet branes and mirror symmetry*, Volume 4 of Clay mathematics monographs, eds. M. R. Douglas and M. Gross, AMS, 2009.
14. M. R. Douglas, "Spaces of Quantum Field Theories," [arXiv:1005.2779 [hep-th]].
15. B. Duplantier and S. Sheffield, "Liouville Quantum Gravity and KPZ," Invent. Math. **185** (2011) 333-393, [arXiv:0808.1560 [math.PR]].
16. F. J. Dyson, "Missed Opportunities," Bull. of the AMS **78** 5 (1972), 635–652.
17. I. Frenkel, Yi-Zhi Huang, and J. Lepowsky, *On axiomatic approaches to vertex operator algebras and modules*, Memoirs of the AMS Series 494, AMS 1993.
18. E. Frenkel and D. Ben-Zvi, *Vertex algebras and algebraic curves*, AMS, 2001.
19. F. Gabbiani and J. Frohlich, "Operator algebras and conformal field theory," Commun. Math. Phys. **155**, 569 (1993).
20. M. R. Gaberdiel and P. Goddard, "Axiomatic conformal field theory," Commun. Math. Phys. **209**, 549-594 (2000). [hep-th/9810019].
21. M. R. Gaberdiel, A. Konechny and C. Schmidt-Colinet, "Conformal perturbation theory beyond the leading order," J. Phys. AA **42**, 105402 (2009) [arXiv:0811.3149 [hep-th]].
22. J. Glimm, A. Jaffe, T. Spencer, "The particle structure of the weakly coupled p(phi)(2) model and other applications of high temperature expansions. 2. The cluster expansion. (talk)," In *Erice 1973, Constructuve Quantum Field Theory*, Berlin 1973, 199-242.
23. J. Glimm and A. Jaffe, *Quantum Physics: A Functional Integral Point of View*, Springer, New York, 1981.
24. R. Haag, *Local quantum physics: fields, particles, algebras*, 2nd ed, Springer, 1996.
25. J. Harrison, "Formal ProofTheory and Practice," Notices of the AMS, Dec 2008, 1395–1406. See also the articles by Gonthier, Hales, and Wiedijk in this issue.
26. S. W. Hawking, G. F. R. Ellis, "The Large scale structure of space-time," Cambridge University Press, Cambridge, 1973.
27. S. Hollands and H. Olbermann, "Perturbative quantum field theory via vertex algebras," J. Math. Phys. 50, Issue 11, pp. 112304-112304-42 (2009) [arXiv:0906.5313].
28. S. Hollands and R. M. Wald, "Axiomatic quantum field theory in curved spacetime," Commun. Math. Phys. **293**, 85 (2010) [arXiv:0803.2003 [gr-qc]].
29. A. Jaffe and F. Quinn, "Theoretical mathematics": Toward a cultural synthesis of mathematics and theoretical physics, Bull.Am.Math.Soc. 29 (1993) 1-13, [arXiv:math/9307227].
30. A. Jaffe and G. Ritter, J. Math. Phys. 49, 052301, 2008.

31. H. Kajiura and J. Stasheff, "Open-closed homotopy algebra in mathematical physics," J. Math. Phys. **47**, 023506 (2006) [hep-th/0510118].
32. S. G. Krantz, "The History and Concept of Mathematical Proof," `http://citeseerx.ist.psu.edu/viewdoc/download?doi=10.1.1.100.8702& rep=rep1& type=pdf`.
33. I. Kriz, Mathematical Reviews MR2079383 (2005h:81334).
34. G. F. Lawler, "Schramm-Loewner Evolution," Lecture notes from course given at Park City/IAS Institute in 2007, [arXiv:math.PR/0712.3256[.
35. L . Lovász, "Discrete analytic functions: an exposition," in *Surveys in differential geometry* Vol. IX, 241273. Int. Press, Somerville, MA, 2004.
36. J. Lurie, `http://math.mit.edu/~lurie/papers/cobordism.pdf`
37. P. K. Mitter, "The Exact renormalization group," [math-ph/0505008].
38. W. Nahm and K. Wendland, "A Hiker's guide to K3: Aspects of N=(4,4) superconformal field theory with central charge c = 6," Commun. Math. Phys. **216**, 85 (2001) [hep-th/9912067].
39. *The Princeton Companion to Mathematics*, eds. T. Gowers, J. Barrow-Green and I. Leader, Princeton University Press, 2008.
40. M. A. Rajabpour and J. Cardy, "Discretely Holomorphic Parafermions in Lattice Z(N) Models," J.Phys.A40:14703-14713, 2008.
41. M. Reed and B. Simon, *Methods of Modern Mathematical Physics*, 4 volumes, Academic Press, 1980.
42. V. Rivasseau, *From Perturbative to Constructive Renormalization*, Princeton University Press, 1991.
43. V. Rivasseau, "Constructive Field Theory in Zero Dimension," Adv. Math. Phys. **2010**, 180159 (2010). [arXiv:0906.3524 [math-ph]].
44. V. Rivasseau, "Introduction to the Renormalization Group with Applications to Non-Relativistic Quantum Electron Gases," [arXiv:1102.5117 [math-ph]].
45. S. Rohde and O. Schramm, "Basic properties of SLE," Annals Math. 161 (2005) 879-920, [arXiv:math/0106036].
46. M. Salmhofer, *Renormalization: an introduction*, Springer, 1999.
47. G. Segal, "The definition of conformal field theory," in *Topology, geometry and quantum field theory*, 421577, LMS Lecture Note Ser. 308, Cambridge Univ. Press, 2004.
48. S. Sheffield, "Gaussian free fields for mathematicians," [arXiv:math/0312099].
49. B. Simon, *The* $P(\Phi)_2$ *Euclidean Quantum Field Theory*, Princeton University Press, 1974.
50. S. Smirnov, "Critical percolation in the plane," C. R. Acad. Sci. Paris, t. 333, Série I, p. 239244, 2001.
51. S. Smirnov, "Discrete Complex Analysis and Probability," in *Proceedings of the International Congress of Mathematicians (ICM), Hyderabad, India, 2010*, [arXiv:1009.6077].
52. J. Stasheff, "The (Secret?) homological algebra of the Batalin-Vilkovisky approach," in volume 219 of *Contemporary mathematics*, eds. M. Henneaux *et al*, AMS 1998, [hep-th/9712157].
53. R. F. Streater and A. S. Wightman, *PCT, spin and statistics, and all that*, Princeton University Press, 2000.
54. W. E. Thirring, *A course in mathematical physics*, 4 volumes, Springer, 1992.
55. W. P. Thurston, "On proof and progress in mathematics," [arXiv:math/9404236].
56. R. M. Wald, *General Relativity*, Univ. of Chicago Press, 1984.
57. A. Wassermann, "Operator Algebras and Conformal Field Theory III. Fusion of positive energy representations of LSU(N) using bounded operators," [arXiv:math/9806031v1 [math.OA]].
58. S. Weinberg, "Effective Field Theory, Past and Future," PoS C **D09**, 001 (2009) [arXiv:0908.1964 [hep-th]].
59. `http://en.wikipedia.org/wiki/Infinite-dimensional_Lebesgue_measure`
60. E. Witten, "Some comments on string dynamics," [hep-th/9507121].

Simons Center for Geometry and Physics, Stony Brook, NY 11794 USA *and* I.H.E.S., Le Bois-Marie, Bures-sur-Yvette, 91440 France

E-mail address: `mdouglas@scgp.stonybrook.edu`

Proceedings of Symposia in Pure Mathematics
Volume **85**, 2012

Homological algebra of knots and BPS states

Sergei Gukov and Marko Stošić

ABSTRACT. It is known that knot homologies admit a physical description as spaces of open BPS states. We study operators and algebras acting on these spaces. This leads to a very rich story, which involves wall crossing phenomena, algebras of closed BPS states acting on spaces of open BPS states, and deformations of Landau-Ginzburg models.

One important application to knot homologies is the existence of "colored differentials" that relate homological invariants of knots colored by different representations. Based on this structure, we formulate a list of properties of the colored HOMFLY homology that categorifies the colored HOMFLY polynomial. By calculating the colored HOMFLY homology for symmetric and anti-symmetric representations, we find a remarkable "mirror symmetry" between these triply-graded theories.

CONTENTS

1. Setting the stage

Quantum knot invariants were introduced in 1980's [**1, 2**]: for every representation R of a Lie algebra $\mathfrak{g}$, one can define a polynomial invariant $\overline{P}^{\mathfrak{g},R}(K)$ of a knot K. Its *reduced* version is

$$P^{\mathfrak{g},R}(K) = \frac{\overline{P}^{\mathfrak{g},R}(K)}{\overline{P}^{\mathfrak{g},R}(\bigcirc)}, \tag{1.1}$$

where $\bigcirc$ denotes the unknot.

A categorification of the polynomial $P^{\mathfrak{g},R}(K)$ (or its *unreduced* version $\overline{P}^{\mathfrak{g},R}(K)$) is a doubly-graded homology theory[1] $\mathcal{H}^{\mathfrak{g},R}(K)$ whose graded Euler characteristic is equal to $P^{\mathfrak{g},R}(K)$. In other words, if $\mathcal{P}^{\mathfrak{g},R}(K)(q,t)$ denotes the Poincaré polynomial of $\mathcal{H}^{\mathfrak{g},R}(K)$, then we have

$$P^{\mathfrak{g},R}(K)(q) \;=\; \mathcal{P}^{\mathfrak{g},R}(K)(q,t=-1)\,.$$

Unlike $P^{\mathfrak{g},R}(K)$, the explicit combinatorial definition of $\mathcal{H}^{\mathfrak{g},R}(K)$ exists for very few choices of $\mathfrak{g}$ and R. However, physics insights based on BPS state counting and Landau-Ginzburg theories predict various properties and a very rigid structure of these homology theories.

One of the first results was obtained in [**3**] for $\mathfrak{g} = sl(N)$ and its fundamental representation $R = \square$. This work builds on a physical realization of knot homologies as spaces of BPS states [**4, 5**]:

$$\mathcal{H}_{\mathrm{knot}} \;=\; \mathcal{H}_{\mathrm{BPS}}\,. \tag{1.2}$$

Among other things, this relation predicts the existence of a polynomial knot invariant $\mathcal{P}^{\square}(K)(\boldsymbol{a},q,t)$, sometimes called the *superpolynomial*, such that for all sufficiently large N one has

$$\mathcal{P}^{sl(N),\square}(K)(q,t) \;=\; \mathcal{P}^{\square}(K)(\boldsymbol{a}=q^N,q,t)\,. \tag{1.3}$$

Moreover, the polynomial $\mathcal{P}^{\square}(K)(\boldsymbol{a},q,t)$ has nonnegative coefficients and is equal to the Poincaré polynomial of a triply graded homology theory $\mathcal{H}^{\square}(K)$ that categorifies the reduced two-variable HOMFLY polynomial $P^{\square}(K)(\boldsymbol{a},q)$, and similarly for the unreduced invariants. This triply graded theory comes equipped with a collection of differentials $\{d_N\}$, such that the homology of $\mathcal{H}^{\square}(K)$ with respect to d_N is isomorphic to $\mathcal{H}^{sl(N),\square}(K)$.

There are only two triply-graded knot homologies that have been studied in the literature up to now. Besides the above-mentioned HOMFLY homology, the second triply-graded theory, proposed in [**6**], similarly unifies homological knot invariants for the N-dimensional vector representation $R = V$ of $\mathfrak{g} = so(N)$ and $\mathfrak{g} = sp(N)$. This triply-graded theory $\mathcal{H}^{\mathrm{Kauff}}(K)$ comes with a collection of differentials $\{d_N\}$, such that the homology with respect to d_N for $N > 1$ is isomorphic to $\mathcal{H}^{so(N),V}(K)$, while the homology with respect to d_N for even $N < 0$ is isomorphic to $\mathcal{H}^{sp(-N),V}(K)$. Since the graded Euler characteristic of $\mathcal{H}^{\mathrm{Kauff}}(K)$ is equal to the (reduced) Kauffman polynomial of K, $\mathcal{H}^{\mathrm{Kauff}}(K)$ is called the Kauffman homology of a knot K.

One way to discover differentials acting on all of these knot homology theories is via studying deformations of the potentials and matrix factorizations in the corresponding Landau-Ginzburg theories (see section 3 for details). In particular, in the case of the Kauffman homology one finds a peculiar deformation that leads to a "universal" differential $d_{\rightarrow}$ and its conjugate $d_{\leftarrow}$, such that the homology with respect to these differentials is, in both cases, isomorphic to the triply-graded HOMFLY homology $\mathcal{H}^{\square}(K)$.

A careful reader may notice that most of the existent results reviewed here deal with the fundamental or vector representations of classical Lie algebras (of Cartan type A, B, C, or D). In this paper, we do roughly the opposite: we focus mainly on $\mathfrak{g} = sl(N)$ but vary the representation R. In particular, we propose infinitely

[1] All homologies in this paper are defined over $\mathbb{Q}$.

many triply-graded homology theories associated with arbitrary symmetric (S^r) and anti-symmetric (Λ^r) representation of $sl(N)$. Moreover, these colored HOMFLY homology theories come equipped with differentials, such that the homology, say, with respect to $d_N^{S^r}$ is isomorphic to $\mathcal{H}^{sl(N),S^r}(K)$, and similarly for $R = \Lambda^r$.

Remarkably, in addition to the differentials labeled by N (for a given r) we also find *colored* differentials that allow to pass from one triply graded theory to another, thus relating homological knot invariants associated with different representations!

Specifically, for each pair of positive integers (r, m) with $r > m$, we find a differential $d_{r\to m}$, such that the homology of $\mathcal{H}^{S^r}(K)$ with respect to $d_{r\to m}$ is isomorphic to $\mathcal{H}^{S^m}(K)$. Similarly, in the case of anti-symmetric representations, we find an infinite sequence of triply-graded knot homologies $\mathcal{H}^{\Lambda^r}(K)$, one for every positive integer r, equipped with colored differentials that allow to pass between two triply-graded theories with different values of r.

The colored differentials are a part of a larger algebraic structure that becomes manifest in a physical realization of knot homologies as spaces of BPS states. As it often happens in physics, the same physical system may admit several mathematical descriptions; a prominent example is the relation between Donaldson-Witten and Seiberg-Witten invariants of 4-manifolds that follows from physics of supersymmetric gauge theories in four dimensions [**7**]. Similarly, the space of BPS states in (1.2) admits several (equivalent) descriptions depending on how one looks at the system of five-branes in eleven-dimensional M-theory [**4**] relevant to this problem.

Specifically, for knots in a 3-sphere $\mathbf{S}^3$ the relevant system is a certain configuration of five-branes in M-theory on $\mathbb{R} \times M_4 \times X$, where $M_4 \cong \mathbb{R}^4$ is a 4-manifold with isometry group $U(1)_P \times U(1)_F$ and X is a non-compact toric Calabi-Yau 3-fold (both of which will be discussed below in more detail). And, if one looks at this M-theory setup from the vantage point of the Calabi-Yau space X, one finds a description of BPS states via enumerative geometry of X. Furthermore, for simple knots and links that preserve toric symmetry of the Calabi-Yau 3-fold X the study of enumerative invariants reduces to a combinatorial problem of counting certain 3d partitions (= fixed points of the 3-torus action [**8**]), hence, providing a combinatorial formulation of knot homologies in terms of 3d partitions [**9, 10**].

On the other hand, if one looks at this M-theory setup from the vantage point of the 4-manifold M_4, one can express the counting of BPS invariants in terms of equivariant instanton counting on M_4. In this approach (see *e.g.* [**11**]), the "quantum" q-grading and the homological t-grading on the space (1.2) originate from the equivariant action of $U(1)_P \times U(1)_F$ on M_4.

A closely related viewpoint, that will be very useful to us in what follows, is based on the five-brane world-volume theory [**12**]. Let us briefly review the basic ingredients of this approach that will make the relation to the setup of [**4**] more apparent. In both cases, knot homology is realized as the space of BPS states and, as we shall see momentarily, the physical realization of the triply-graded knot homology proposed in [**4**] is essentially the large-N dual of the system realizing the doubly-graded knot homology in [**12**]. This is very typical for systems with $SU(N)$ gauge symmetry[2] which often admit a dual "holographic" description that comprises all N in the same package and leads to useful computational techniques [**13**].

[2]The same is true for other classical groups.

In the case of $sl(N)$ homological knot invariants, the five-brane configuration described in [**12**, sec. 6] is the following:

$$\begin{array}{rcl} \text{space-time} & : & \mathbb{R} \times T^*W \times M_4 \\ N \text{ M5-branes} & : & \mathbb{R} \times W \times D \\ \text{M5-brane} & : & \mathbb{R} \times L_K \times D \end{array} \tag{1.4}$$

Here, W is a 3-manifold and $D \cong \mathbb{R}^2$ is the "cigar" in the Taub-NUT space $M_4 \cong \mathbb{R}^4$. The Lagrangian submanifold $L_K \subset T^*W$ is the conormal bundle to the knot $K \subset W$; in particular,

$$L_K \cap W = K\,. \tag{1.5}$$

In all our applications, we consider $W = \mathbf{S}^3$ (or, a closely related case of $W = \mathbb{R}^3$). Similarly, the setup of [**4**] can be summarized as

$$\begin{array}{rcl} \text{space-time} & : & \mathbb{R} \times X \times M_4 \\ \text{M5-brane} & : & \mathbb{R} \times L_K \times D \end{array} \tag{1.6}$$

where X is the resolved conifold, *i.e.* the total space of the $\mathcal{O}(-1) \oplus \mathcal{O}(-1)$ bundle over $\mathbb{C}\mathbf{P}^1$. From the way we summarized (1.4) and (1.6), it is clear that they have many identical elements. The only difference is that (1.4) has extra M5-branes supported on $\mathbb{R} \times W \times D$, whereas (1.6) has a different space-time (with a 2-cycle in the Calabi-Yau 3-fold X), which is exactly what one expects from a holographic duality or large-N transition [**14, 15**].

Indeed, what is important for the purpose of studying the space of BPS states, $\mathcal{H}_{\text{BPS}}$, is that both (1.4) and (1.6) preserve the same amount of supersymmetry and have the same symmetries:

- **time translations:** both systems have a translation symmetry along the time direction (denoted by the $\mathbb{R}$ factor in (1.4) and (1.6)). Therefore, in both cases, one can ask for a space of BPS-states — on multiple M5-branes in (1.4), and on a single M5-brane in (1.6) — which is precisely what was proposed as a candidate for the $sl(N)$ knot homology (resp. HOMFLY homology).
- **rotation symmetries:**

$$U(1)_P \times U(1)_F \tag{1.7}$$

Here, the two $U(1)$ factors correspond, respectively, to the rotation symmetry of the tangent and normal bundle of $D \cong \mathbb{R}^2$ in a 4-manifold $M_4 \cong \mathbb{R}^4$. In particular, in both frameworks (1.4) and (1.6), the former is responsible for the q-grading of $\mathcal{H}_{BPS}$, which corresponds to the conserved angular momentum derived from the rotation symmetry of D.

A well-known feature of the large-N duality is that the rank of the gauge group turns into a geometric parameter of the dual system (*cf.* [**13**] or [**14**]). In the present case, it is the Kähler modulus of the Calabi-Yau 3-fold $X = \mathcal{O}_{\mathbf{CP}^1}(-1) \oplus \mathcal{O}_{\mathbf{CP}^1}(-1)$:

$$N \;\sim\; \log(\boldsymbol{a}) \;=\; \mathrm{Vol}(\mathbb{C}\mathbf{P}^1)\,. \tag{1.8}$$

The reason we denote the Kähler parameter by $\log(\boldsymbol{a})$ rather than $\boldsymbol{a}$ is that with this convention $\boldsymbol{a} = q^N$ is the standard variable of the HOMFLY polynomial / knot homology.

Another feature familiar to the practitioners of the refined / motivic Donaldson-Thomas theory is that $\mathcal{H}_{\text{BPS}}$ can jump as one varies stability conditions [**16, 17,**

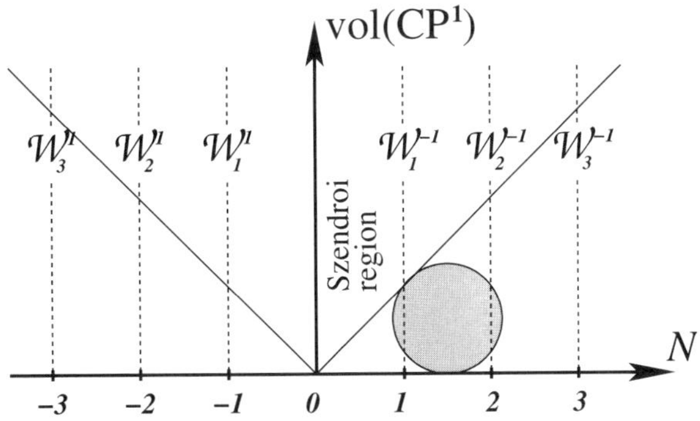

FIGURE 1. For the conifold X, the space of stability conditions is one-dimensional. It is parametrized by the Kähler parameter $N \sim \log(a) = \mathrm{Vol}(\mathbb{C}\mathbf{P}^1)$. This space is divided by walls of marginal stability into a set of chambers, which can be identified with the set of integers $\mathbb{Z}$.

18, 19, 20, 21]. Thus, in a closely related type IIA superstring compactification on a Calabi-Yau 3-fold X, the stability parameters are the Kähler moduli of X, and in the present case there is only one Kähler modulus (1.8) given by the volume of the $\mathbb{C}\mathbf{P}^1$ cycle in X. Therefore, we conclude that the space (1.2) can jump as one changes the stability parameter $N \sim \mathrm{Vol}(\mathbb{C}\mathbf{P}^1)$.

Luckily, in the case where X is the total space of the $\mathcal{O}(-1) \oplus \mathcal{O}(-1)$ bundle over $\mathbb{C}\mathbf{P}^1$ relevant to our applications, the wall-crossing behavior of the refined BPS invariants has been studied in the literature [**22, 23, 18**]. The one-dimensional space of stability conditions is divided into a set of chambers illustrated in Figure 1. In each chamber, $\mathcal{H}_{\mathrm{BPS}}$ is constant and the jumps of closed BPS states occur at the walls $\mathcal{W}_n^{\pm 1}$ characterized by different types of "fragments":

$$\begin{array}{lcl} \mathcal{W}_n^1 & : & D2/D0 \text{ fragments} \\ \mathcal{W}_n^{-1} & : & \overline{D2}/D0 \text{ fragments} \\ \mathcal{W}_n^0 & : & D0 \text{ fragments} \end{array} \tag{1.9}$$

Notice, the set of chambers in this model can be identified with $\mathbb{Z}$, the set of integer numbers. As we explain in the next section, this is not a coincidence. Namely, as we shall see, every fragment corresponds to a differential acting on the space (1.2), so that in the present example one finds a set of differentials $\{d_N\}$ labeled by $N \in \mathbb{Z}$.

The differentials $\{d_N\}$ are part of the homological algebra of knots / BPS states, depending on whether one prefers to focus on the left or right side of the relation (1.2). For larger representations, in addition to the differentials $\{d_N\}$ one finds colored differentials that allow to pass between homology theories associated with different R. Even though a combinatorial definition of the majority of such theories, with all the differentials, is still missing, their structure (deduced from physics) is so rigid that enables computation of the homology groups for many knots and passes a large number of consistency checks.

In particular, by computing the triply-graded homologies $\mathcal{H}^{S^r}(K)$ and $\mathcal{H}^{\Lambda^r}(K)$ for various knots, we find the following surprising symmetry between the two theories:

$$\mathcal{H}^{\Lambda^r}_{i,j,*}(K) \cong \mathcal{H}^{S^r}_{i,-j,*}(K)\,. \tag{1.10}$$

One of the implication is that $\mathcal{H}^{S^r}(K)$ and $\mathcal{H}^{\Lambda^r}(K)$ can be combined into a single homology theory!

CONJECTURE 1.1. For every positive integer r, there exists a triply-graded theory $\mathcal{H}^r(K)$ together with a collection of differentials $\{d^r_N\}$, with $N \in \mathbb{Z}$, such that the homology of $\mathcal{H}^r(K)$ with respect to d^r_N, for $N > 0$, is isomorphic to $\mathcal{H}^{sl(N),S^r}(K)$, while the homology of $\mathcal{H}^r(K)$ with respect to d^r_N, for $N < 0$, is isomorphic (up to a simple regrading) to $\mathcal{H}^{sl(-N),\Lambda^r}(K)$.

Moreover, it is tempting to speculate that the symmetry (1.10) extends to all representations:

$$\text{“mirror symmetry”} \ : \quad \mathcal{H}^{\lambda}(K) \cong \mathcal{H}^{\lambda^t}(K)\,, \tag{1.11}$$

where λ and λ^t are a pair of Young tableaux related by transposition (mirror reflection across the diagonal), *e.g.*

$\lambda =$ 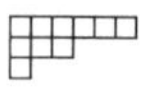 $\longleftrightarrow$ $\lambda^t =$ 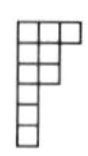

The symmetry (1.11) has not been discussed in physical or mathematical literature before.

While we offer its interpretation in section 5.3, we believe the mirror symmetry for colored knot homology (1.11) deserves a more careful study, both in physics as well as in mathematics. In particular, its deeper understanding should lead to the “categorification of level-rank duality” in Chern-Simons theory, which is the origin of the simpler, decategorified version of (1.11):

$$P^{\lambda}(K)(\boldsymbol{a},q) = P^{\lambda^t}(K)(\boldsymbol{a},q^{-1}) \tag{1.12}$$

for colored HOMFLY polynomials [**24, 25, 26, 27**], and extends the familiar symmetry $q \leftrightarrow q^{-1}$ of the ordinary HOMFLY polynomial. We plan to pursue the categorification of level-rank duality and to study the new, homological symmetry (1.11) in the future work.

Organization of the paper. We start by explaining in section 2 that, in general, the space of open BPS states forms a representation of the algebra of closed BPS states. Then, in section 3 we review elements of the connection between string realizations (1.4)–(1.6) of knot homologies and Landau-Ginzburg models that play an important role in mathematical formulations of certain knot homologies based on Lie algebra $\mathfrak{g}$ and its representation R. In particular, we illustrate in simple examples how the corresponding potentials $W_{\mathfrak{g},R}$ can be derived from the physical setup (1.4)–(1.6) and how deformations of these potentials lead to various differentials acting on $\mathcal{H}^{\mathfrak{g},R}(K)$. This gives another way to look at the algebra acting on (1.2). Based on these predictions, in section 4 we summarize the mathematical structure of the triply-graded homology $\mathcal{H}^{S^r}(K)$, together with its computation for small knots. Section 5 lists the analogous properties of the homology associated with anti-symmetric representations, and explains the explicit form of the “mirror symmetry" (1.10) between symmetric and anti-symmetric triply-graded theories. Unreduced triply-graded theory for symmetric and anti-symmetric representations is briefly discussed in section 6. In appendix A we collect the list of our notations, whereas in appendix B we present the computations of the S^2, Λ^2 and Kauffman triply-graded homology for knots 8_{19} and 9_{42}. These particular examples of "thick" knots provide highly non-trivial tests of all the properties of the homologies presented in the paper. Appendix C contains the computation of the S^3 and Λ^3 homology of the figure-eight knot 4_1. Finally, appendix D collects some notations and calculations relevant to the unreduced colored HOMFLY polynomial of the unknot discussed in section 6.

2. Algebra of BPS states and its representations

Differentials in knot homology form a part of a larger algebraic structure that has an elegant interpretation in the geometric / physical framework. Because this algebraic structure has analogs in more general string / M-theory compactifications, in this section we shall consider aspects of such structure for an arbitrary Calabi-Yau 3-fold X with extra branes supported on a general Lagrangian submanifold $L \subset X$, *e.g.*

$$\begin{array}{rcl} \text{space-time} & : & \mathbb{R} \times X \times M_4 \\ \text{M5-brane} & : & \mathbb{R} \times L \times D \end{array} \tag{2.1}$$

For applications to knot homologies, one should take X to be the total space of the $\mathcal{O}(-1) \oplus \mathcal{O}(-1)$ bundle over $\mathbb{C}\mathbf{P}^1$ and L_K to be the Lagrangian submanifold determined by a knot K [**15, 28, 29**]. Then, (2.1) becomes precisely the setup (1.6), in which homological knot invariants are realized as spaces of refined BPS states, *cf.* (1.2).

In fact, there are *two* spaces of BPS states relevant to this particular problem and its variants based on a more general 3-fold X. One is the space of refined *closed* BPS states, denoted as $\mathcal{H}_{\text{BPS}}^{\text{closed}}$, and the other is called the space of refined *open* BPS states, $\mathcal{H}_{\text{BPS}}^{\text{open}}$. The difference is that, while the latter contains BPS particles in the presence of defects[3], the former comprises only those BPS states which are still present in a theory when all defects are removed. In the description [**11**] via equivariant instanton counting on a 4-manifold M_4, the defect (M5-brane) corresponds to a particular ramification along the divisor $D \subset M_4$, the so-called surface operator.

On the other hand, if one looks at the general setup (2.1) from the vantage point of the Calabi-Yau space X, then $\mathcal{H}_{\text{BPS}}^{\text{closed}}$ and $\mathcal{H}_{\text{BPS}}^{\text{open}}$ can be formulated in terms of enumerative invariants of X and (X, L) that "count", respectively, closed holomorphic curves embedded in X and bordered holomorphic Riemann surfaces $(\Sigma, \partial\Sigma) \hookrightarrow (X, L)$ with boundary on the Lagrangian submanifold L. As a way to remember this, it is convenient to keep in mind that

- $\mathcal{H}_{\text{BPS}}^{\text{closed}}$ depends only on the Calabi-Yau space X
- $\mathcal{H}_{\text{BPS}}^{\text{open}}$ depends on both the Calabi-Yau space X and the Lagrangian submanifold $L \subset X$

---

[3]M5-branes in the M-theory setup (2.1)

In applications to knots, open (resp. closed) BPS states are represented by open (resp. closed) membranes in the M-theory setup (1.6) or by bound states of D0 and D2 branes in its reduction to type IIA string theory. It is the space of open BPS states that depends on the choice of the knot K and, therefore, provides a candidate for homological knot invariant in (1.2).

In general, the space of BPS states is $\Gamma \oplus \mathbb{Z}$-graded, where Γ is the "charge lattice" and the extra $\mathbb{Z}$-grading comes from the (half-integer) spin of BPS states, such that $2j_3 \in \mathbb{Z}$. For example, in the case of closed BPS states, the charge lattice is usually just the cohomology lattice of the corresponding Calabi-Yau 3-fold X,

$$\Gamma = H^{\text{even}}(X;\mathbb{Z}) \,. \tag{2.2}$$

In the case of open BPS states, Γ also depends on the choice of the Lagrangian submanifold $L \subset X$.

When X is the total space of the $\mathcal{O}(-1)\oplus\mathcal{O}(-1)$ bundle over $\mathbb{C}\mathbf{P}^1$ and $L = L_K$, as in application to knot homologies, the lattice Γ is two-dimensional for both open and closed BPS states. As a result, both $\mathcal{H}_{\text{BPS}}^{\text{closed}}$ and $\mathcal{H}_{\text{BPS}}^{\text{open}}$ are $\mathbb{Z}\oplus\mathbb{Z}\oplus\mathbb{Z}$-graded. In particular, the space of open BPS states is graded by spin $2j_3 \in \mathbb{Z}$ and by charge $\gamma = (n,\beta) \in \Gamma$, where the degree $\beta \in H_2(X, L_K) \cong \mathbb{Z}$ is sometimes called the "D2-brane charge" and $n \in \mathbb{Z}$ is the "D0-brane charge." In relation to knot homologies (1.2), these become the three gradings of the theory categorifying the colored HOMFLY polynomial:

$$\begin{aligned} \text{“}\boldsymbol{a}-\text{grading”} &= \beta \in H_2(X, L_K) \cong \mathbb{Z} \\ \text{“}q-\text{grading”} &= n \in \mathbb{Z} \\ \text{“}t-\text{grading”} &= 2j_3 \in \mathbb{Z} \end{aligned} \tag{2.3}$$

Now, let us discuss the algebraic structure that will help us understand the origin of differentials acting on the triply-graded vector space $\mathcal{H}_{\text{knot}} = \mathcal{H}_{\text{BPS}}^{\text{open}}$. The fact that $\mathcal{H}_{\text{BPS}}^{\text{closed}}$ forms an algebra is well appreciated in physics [**30**] as well as in math literature [**31**]. Less appreciated, however, is the fact that $\mathcal{H}_{\text{BPS}}^{\text{open}}$ forms a representation of the algebra $\mathcal{H}_{\text{BPS}}^{\text{closed}}$:

$$\boxed{\begin{array}{lc} \text{refined open BPS states :} & \mathcal{H}_{\text{BPS}}^{\text{open}} \\ & \circlearrowleft \\ \text{refined closed BPS states :} & \mathcal{H}_{\text{BPS}}^{\text{closed}} \end{array}} \tag{2.4}$$

Indeed, two closed BPS states, $\mathcal{B}_1$ and $\mathcal{B}_2$, of charge $\gamma_1, \gamma_2 \in \Gamma$ can form a bound state, $\mathcal{B}_{12}$ of charge $\gamma_1 + \gamma_2$, as a sort of "extension" of $\mathcal{B}_1$ and $\mathcal{B}_2$,

$$0 \to \mathcal{B}_2 \to \mathcal{B}_{12} \to \mathcal{B}_1 \to 0 \,, \tag{2.5}$$

thereby defining a product on $\mathcal{H}_{\text{BPS}}^{\text{closed}}$:

$$\begin{array}{ccc} \mathcal{H}_{\text{BPS}}^{\text{closed}} \otimes \mathcal{H}_{\text{BPS}}^{\text{closed}} & \longrightarrow & \mathcal{H}_{\text{BPS}}^{\text{closed}} \\ (\,\mathcal{B}_1\,,\ \mathcal{B}_2\,) & \mapsto & \mathcal{B}_{12} \end{array} \tag{2.6}$$

Similarly, a bound state of a closed BPS state $\mathcal{B}_1^{\text{closed}} \in \mathcal{H}_{\text{BPS}}^{\text{closed}}$ with an open BPS state $\mathcal{B}_2^{\text{open}} \in \mathcal{H}_{\text{BPS}}^{\text{open}}$ is another open BPS state $\mathcal{B}_{12}^{\text{open}} \in \mathcal{H}_{\text{BPS}}^{\text{open}}$:

$$(\mathcal{B}_1^{\text{closed}}, \mathcal{B}_2^{\text{open}}) \mapsto \mathcal{B}_{12}^{\text{open}} \tag{2.7}$$

This defines an action of the algebra of closed BPS states on the space of open BPS states.

The process of formation or fragmentation of a bound state in (2.6) and (2.7) takes place when the binding energy vanishes. Since the energy of a BPS state is given by the absolute value of the central charge[4] function $\mathcal{Z} : \Gamma \to \mathbb{C}$ this condition can be written as

$$\left|\mathcal{Z}(\gamma_1 + \gamma_2)\right| - \left|\mathcal{Z}(\gamma_1)\right| - \left|\mathcal{Z}(\gamma_2)\right| = 0 \tag{2.8}$$

for a process that involves either $\mathcal{B}_{12} \to \mathcal{B}_1 + \mathcal{B}_2$ or its inverse $\mathcal{B}_1 + \mathcal{B}_2 \to \mathcal{B}_{12}$. A particular instance of the relation (2.8) is when the central charge of the fragment vanishes:

$$\mathcal{Z}(\gamma_{\text{fragment}}) = 0 \tag{2.9}$$

Then, a fragment becomes massless and potentially can bind to any other BPS state of charge γ. When combined with (2.4), it implies that closed BPS states of zero mass correspond to operators acting on the space of open BPS states $\mathcal{H}_{\text{BPS}}^{\text{open}}$. The degree of the operator is determined by the spin and charge of the corresponding BPS state, as in (2.3).

For example, when X is the total space of the $\mathcal{O}(-1) \oplus \mathcal{O}(-1)$ bundle over $\mathbb{CP}^1$, as in application to knot homologies, we have

$$\exp(\mathcal{Z}) = \boldsymbol{a}^{\beta} q^n , \tag{2.10}$$

where we used the relation (1.8) between $\boldsymbol{a}$ and $\text{Vol}(\mathbb{CP}^1)$. Therefore, for special values of $\boldsymbol{a}$ and q we have the following massless fragments, *cf.* (1.9):

$$\begin{array}{lcl} \boldsymbol{a} = q^{-N} & : & D2/D0 \text{ fragments} \\ \boldsymbol{a} = q^{N} & : & \overline{D2}/D0 \text{ fragments} \\ q = 1 & : & D0 \text{ fragments} \end{array} \tag{2.11}$$

Moreover, the D2/D0 fragments obey the Fermi-Dirac statistics (see *e.g.* [**18, 23**]) and, therefore, lead to anti-commuting operators (*i.e.* differentials) on $\mathcal{H}_{\text{BPS}}^{\text{open}}$.

To summarize, we conclude that various specializations of the parameters (stability conditions) are accompanied by the action of commuting and anti-commuting operators on $\mathcal{H}_{\text{BPS}}^{\text{open}}$. The algebra of these operators is precisely the algebra of closed BPS states $\mathcal{H}_{\text{BPS}}^{\text{closed}}$. Mathematical candidates for the algebra of closed BPS states include variants of the Hall algebra [**32**], which by definition encodes the structure of the space of extensions (2.5):

$$[\mathcal{B}_1] \cdot [\mathcal{B}_2] = \sum_{\mathcal{B}_{12}} |0 \to \mathcal{B}_2 \to \mathcal{B}_{12} \to \mathcal{B}_1 \to 0| \, [\mathcal{B}_{12}] \tag{2.12}$$

In the present case, the relevant algebras include the motivic Hall algebra [**17**], the cohomological Hall algebra [**31**], and its various ramifications, *e.g.* cluster algebras.

[4]The central charge function is a linear function in the sense that $\mathcal{Z}(\gamma_1+\gamma_2) = \mathcal{Z}(\gamma_1)+\mathcal{Z}(\gamma_2)$, *i.e.* it defines a homomorphism $\mathcal{Z} \in \text{Hom}(\Gamma, \mathbb{C})$.

Therefore, the problem can be approached by studying representations of these algebras, as will be described elsewhere.

3. B-model and matrix factorizations

Let us denote by $\mathcal{H}^{\mathfrak{g},R}$ a homology theory of knots and links colored by a representation R of the Lie algebra $\mathfrak{g}$. Many such homology theories can be constructed using categories of matrix factorizations [**33, 34, 35, 36, 37, 38, 39, 40**]. In this approach, one of the main ingredients is a polynomial function $W_{\mathfrak{g},R}$ called the *potential*, associated to every segment of a link (or, more generally, of a tangle) away from crossings. For example, for the fundamental representation of $\mathfrak{g} = sl(N)$ the potential is a function of a single variable,

$$W_{sl(N),\square}(x) \quad = \quad x^{N+1} \,. \tag{3.1}$$

In physics, matrix factorizations are known [**41, 42, 43, 44, 45, 46**] to describe D-branes and topological defects in Landau-Ginzburg models which, in the present context, are realized on the two-dimensional part of the five-brane world-volume in (1.4) or (1.6). More precisely, it was advocated in [**6**] that reduction of the M-theory configuration (1.4) on one of the directions in D and a T-duality along the time direction gives a configuration of intersecting D3-branes in type IIB string theory, such that the effective two-dimensional theory on their common world-volume provides a physical realization of the Landau-Ginzburg model that appears in the mathematical constructions.

In particular, this interpretation was used to deduce potentials $W_{\mathfrak{g},R}$ associated to many Lie algebras and representations. Indeed, since away from crossings every segment of the knot K is supposed to be described by a Landau-Ginzburg theory with potential $W_{\mathfrak{g},R}$, we can approximate this local problem by taking $W = \mathbb{R}^3$ and $K = \mathbb{R}$ in (1.4). Then, we also have $L_K = \mathbb{R}^3$ and the reduction (plus T-duality) of (1.4) gives type IIB theory in flat space-time with two sets of D3-branes supported on 4-dimensional hyperplanes in $\mathbb{R}^{10}$: one set supported on $\mathbb{R} \times W$, and another supported on $\mathbb{R} \times L_K$. The space of open strings between these two groups of D3-branes contains information about the potential $W_{\mathfrak{g},R}$.

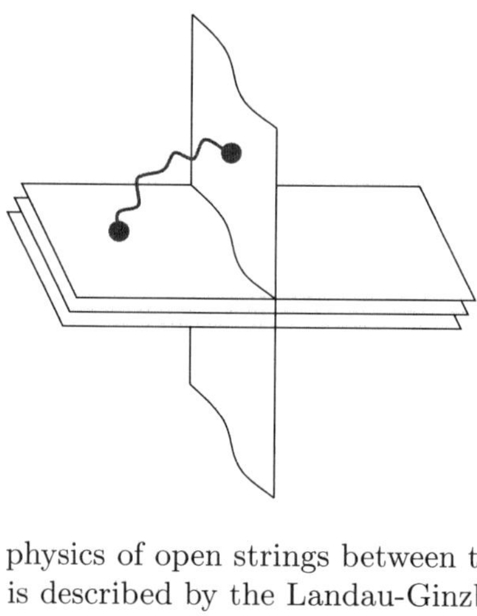

FIGURE 2. The physics of open strings between two stacks of Lagrangian branes is described by the Landau-Ginzburg model with potential $W_{\mathfrak{g},R}$.

For example, in the case of the fundamental representation of $sl(N)$, the first stack consists of N D3-branes and the second only contains a single D3-brane. The open strings between these two stacks of D3-branes transform in the bifundamental representation $(\mathbf{N}, +1)$ under the gauge symmetry $U(N) \times U(1)$ on the D3-branes. The Higgs branch of this two-dimensional theory is the Kähler quotient of the vector space $\mathbb{C}^N$ parametrized by the bifundamental chiral multiplets, modulo $U(1)$ gauge symmetry of a single D3-brane supported on $\mathbb{R} \times L_K$:

$$\mathbb{C}^N // U(1) \cong \mathbf{CP}^{N-1}. \tag{3.2}$$

The chiral ring of this theory on the intersection of D3-branes is precisely the Jacobi ring of the potential (3.1).

Following similar arguments one can find potentials associated to many other Lie algebras and representations [**6**], such that

$$\mathcal{H}^{\mathfrak{g},R}(\bigcirc) = \mathcal{J}(W_{\mathfrak{g},R}). \tag{3.3}$$

For example, the arguments that lead to (3.1) can be easily generalized to $R = \Lambda^r$, the r-th anti-symmetric representation of $sl(N)$. The only difference is that, in this case, the corresponding brane systems (1.4) and (1.6) contain r coincident M5-branes supported on $\mathbb{R} \times L_K \times D$. Following the same arguments as in the case of the fundamental representation ($r = 1$) and zooming in closely on the local geometry of the brane intersection, after all the dualities we end up with a system of intersecting D3-branes in flat ten-dimensional space-time,

$$\begin{array}{rcl} N \text{ D3-branes} & : & \mathbb{R} \times W \\ r \text{ D3}'\text{-branes} & : & \mathbb{R} \times L_K \end{array} \tag{3.4}$$

where, as in the previous discussion, for the purpose of deriving $W_{\mathfrak{g},R}$ we can approximate $W \simeq \mathbb{R}^3$ and $L_K \simeq \mathbb{R}^3$, so that $W \cap L_K = \mathbb{R}$. Now, the open strings between two sets of D3-branes in (3.4) transform in the bifundamental representation $(\mathbf{N}, \mathbf{r})$ under the gauge symmetry $U(N) \times U(r)$ on the D3-branes. Here, if we want to "integrate out" open strings ending on the D3$'$-branes, only the second gauge factor should be considered dynamical, while $U(N)$ should be treated as a global symmetry of the two-dimensional $U(r)$ gauge theory on the brane intersection. In the infrared this theory flows to a sigma-model based on the Grassmannian manifold:

$$\text{Gr}(r, N) = \frac{U(N)}{U(r) \times U(N-r)}. \tag{3.5}$$

The potential of the corresponding Landau-Ginzburg model [**47**] is a homogeneous polynomial of degree $N+1$,

$$W_{sl(N),\Lambda^r}(z_1, \ldots, z_r) = x_1^{N+1} + \ldots + x_r^{N+1}, \tag{3.6}$$

where the right-hand side should be viewed as a function of the variables z_i of degree $\deg(z_i) = i$, $i = 1, \ldots, r$, which are the elementary symmetric polynomials in the x_j,

$$z_i = \sum_{j_1 < j_2 < \cdots < j_i} x_{j_1} x_{j_2} \cdots x_{j_i}.$$

We shall return to the discussion of the potential $W_{sl(N),\Lambda^r}$ later in this section. In the case of more general representations, one needs to consider various sectors of the $U(r)$ gauge theory on $\mathbb{R} \times L_K$ labeled by non-trivial flat connections (Wilson lines) around the codimension-2 locus where D3$'$-branes meet D3-branes, *cf.* [**15, 48**].

In this paper we are mostly interested in knots colored by symmetric and anti-symmetric representations of $\mathfrak{g} = sl(N)$, even though much of the present discussion can be easily generalized to other Lie algebras and representations. Thus, for a symmetric representation $R = S^r$ of $\mathfrak{g} = sl(N)$ one finds that the corresponding potential $W_{sl(N),S^r}(z_1, \ldots, z_r)$ is a homogeneous polynomial of degree $N + r$ in variables z_i of degree $i = 1, \ldots, r$, much like (3.6). Moreover, the explicit form of such potentials can be conveniently expressed through a generating function [**6**]:

$$\sum_N (-1)^N t^{N+r} W_{sl(N),S^r}(z_1, \ldots, z_r) \;=\; (1 + \sum_{i=1}^r t^i z_i) \log(1 + \sum_{i=1}^r t^i z_i)\,, \tag{3.7}$$

which in the basic case $N = r = 2$ gives

$$W_{sl(2),\square\square} \;=\; z_1^4 - 6z_1^2 z_2 + 6z_2^2\,. \tag{3.8}$$

Instead of going through the derivation of this formula we can use a simple trick based on the well known isomorphism $sl(2) \cong so(3)$ under which a vector representation of $so(3)$ is identified with the adjoint representation of $sl(2)$. Indeed, it implies that (3.8) should be identical to the well known potential

$$W_{so(3),V} \;=\; x^2 + xy^2 \tag{3.9}$$

in the $so(3)$ homology theory, *cf.* [**6, 37**]. It is easy to verify that the potentials (3.8) and (3.9) are indeed related by a simple change of variables.[5]

Moreover, the fact that the adjoint representation of $sl(2)$ is identical to the vector representation of $so(3)$ implies that

$$\boxed{\mathcal{H}^{sl(2),\square\square}(K) \;\cong\; \mathcal{H}^{so(3),V}(K)} \tag{3.10}$$

should hold for every knot K. In particular, it should hold for the unknot. And, since $\mathcal{H}^{so(3),V}(\bigcirc)$ is 3-dimensional, it follows

$$\dim \mathcal{H}^{sl(2),\square\square}(\bigcirc) \;=\; \dim \mathcal{H}^{so(3),V}(\bigcirc) \;=\; 3\,. \tag{3.11}$$

This is indeed what one finds in physical realizations of knot homologies reviewed in section 1. In the framework of [**4**] the colored homology $\mathcal{H}^{sl(2),\square\square}$ of the unknot was computed in [**9**] using localization with respect to the toric symmetry of the Calabi-Yau space X. Similarly, in the gauge theory framework [**12**] the moduli space of solutions on $\mathbb{R}^2$ with a single defect operator in the adjoint representation of the gauge group $G = SU(2)$ is the weighted projective space $\mathbb{WCP}^2_{(1,1,2)}$ (= the space of Hecke modifications [**49**], see also [**50**]). In this approach, the colored homology $\mathcal{H}^{sl(2),\square\square}(\bigcirc)$ is given by the L^2 cohomology of the moduli space $\mathbb{WCP}^2_{(1,1,2)}$ which is 3-dimensional, in agreement with (3.11).

[5]Explicitly, the change of variables that relates (3.8) and (3.9) is given by:

$$\begin{aligned} x &= -\frac{1}{2}(\sqrt{2}+\sqrt{6})\, z_1^2 + \sqrt{6}\, z_2 \\ y &= 2^{1/4}\, z_1 \end{aligned}$$

3.1. Colored differentials. One of the reasons why we carefully reviewed the properties of the potentials $W_{\mathfrak{g},R}$ is that they hold a key to understanding the colored differentials. Namely, in doubly-graded knot homologies constructed from matrix factorizations differentials that relate different theories are in one-to-one correspondence with deformations of the potentials [**51, 52, 53, 54, 55, 6**]:

$$\boxed{\text{differentials on } \mathcal{H}^{\mathfrak{g},R} \quad \Longleftrightarrow \quad \text{deformations of } W_{\mathfrak{g},R}} \tag{3.12}$$

For example, deformations of the potential (3.1) of the form $\Delta W = \beta x^{M+1}$ with $M < N$, correspond to differentials d_M that relate $sl(N)$ and $sl(M)$ knot homologies (with $R = \square$).

More generally, one can consider deformations ΔW of the potential $W_{\mathfrak{g},R}$ such that $\deg \Delta W < \deg W_{\mathfrak{g},R}$ and

$$W_{\mathfrak{g},R} + \Delta W \;\simeq\; W_{\mathfrak{g}',R'} \tag{3.13}$$

for some Lie algebra $\mathfrak{g}'$ and its representation R'. Here, the symbol "$\simeq$" means that the critical point(s) of the deformed potential is locally described by the new potential $W_{\mathfrak{g}',R'}$. A deformation of this form leads to a spectral sequence that relates knot homologies $\mathcal{H}^{\mathfrak{g},R}$ and $\mathcal{H}^{\mathfrak{g}',R'}$. With the additional assumption that the spectral sequence converges after the first page one arrives at (3.12). Moreover, the difference

$$\deg W_{\mathfrak{g},R} - \deg \Delta W \tag{3.14}$$

gives the q-grading of the corresponding differential. Notice, the condition $\deg \Delta W < \deg W_{\mathfrak{g},R}$ implies that this q-grading is positive.

For example, among deformations of the degree-4 potential (3.8) one finds $\Delta W = z_1^3$, which leads to a differential of q-degree 1 that relates $\mathcal{H}^{sl(2),\square\!\square}$ and $\mathcal{H}^{sl(2),\square}$. This deformation has an obvious analog for higher rank S^2-colored homology; it deforms the homogeneous polynomial $W_{sl(N),\square\!\square}(z_1, z_2)$ of degree $N+2$ in such a way that the deformed potential has a critical point described by the potential $W_{\mathfrak{g}',R'} = W_{sl(N),\square}(z_1)$ of degree $N+1$. Therefore, it leads to a colored differential of q-degree 1, such that

$$\left(\mathcal{H}^{sl(N),\square\!\square},\, d_{\text{colored}}\right) \;\cong\; \mathcal{H}^{sl(N),\square}\,. \tag{3.15}$$

In section 4 we present further evidence for the existence of a differential with such properties not only in the doubly-graded $sl(N)$ theory but also in the triply-graded knot homology that categorifies the colored HOMFLY polynomial.

Similar colored differentials exist in other knot homologies associated with more general Lie algebras and representations. Basically, a knot homology associated to a representation R of the Lie algebra $\mathfrak{g}$ comes equipped with a set of colored differentials that, when acting on $\mathcal{H}^{\mathfrak{g},R}$, lead to homological invariants associated with smaller representations (and, possibly, Lie algebras),

$$\dim R' \;<\; \dim R\,. \tag{3.16}$$

While it would be interesting to perform a systematic classification of such colored differentials using the general principle (3.12), in this paper we limit ourselves only to symmetric and anti-symmetric representations of $\mathfrak{g} = sl(N)$.

As we already discussed earlier, when $R = \Lambda^r$ is the r-th anti-symmetric representation of $sl(N)$ the corresponding Landau-Ginzburg potential (3.6) is a homogeneous polynomial of degree $N+1$. Equivalently, the potentials with a fixed value of r can be organized into a generating function, analogous to (3.7):

$$\sum_N (-1)^N t^{N+1} W_{sl(N),\Lambda^r}(z_1,\ldots,z_r) \; = \; \log(1 + \sum_{i=1}^{r} t^i z_i)\,. \tag{3.17}$$

For example, in the first non-trivial case of $r = 2$ there are only two variables, $z_1 = x_1 + x_2$ and $z_2 = x_1 x_2$. For $N = 2$ one finds a "trivial" potential $W_{sl(2),\begin{smallmatrix}\square\\\square\end{smallmatrix}}$ of degree 3, which corresponds to the fact that the anti-symmetric representation $R = \Lambda^2$ (also denoted $R = \begin{smallmatrix}\square\\\square\end{smallmatrix}$) is trivial in $sl(2)$. For $N = 3$, the existence of the anti-symmetric tensor ϵ^{ijk} identifies the second anti-symmetric representation $R = \begin{smallmatrix}\square\\\square\end{smallmatrix}$ with the fundamental representation of $sl(3)$. The next case in this sequence, $N = 4$, is the first example where the second anti-symmetric representation is not related to any other representation of $sl(4)$. According to (3.6) and (3.17), the corresponding potential is a homogeneous polynomial of degree 5,

$$W_{sl(4),\begin{smallmatrix}\square\\\square\end{smallmatrix}} \; = \; \frac{z_1^5}{5} - z_1^3 z_2 + z_1 z_2^2\,. \tag{3.18}$$

Before studying deformations of this potential, we note that by a simple change of variables it is related to the potential

$$W_{so(6),V} \; = \; x^5 + xy^2 \tag{3.19}$$

associated to a vector representation of $so(6)$. This is a manifestation of the well known isomorphism $sl(4) \cong so(6)$ under which the six-dimensional anti-symmetric representation $R = \begin{smallmatrix}\square\\\square\end{smallmatrix}$ of $sl(4)$ is identified with the vector representation of $so(6)$. This isomorphism can help us understand deformations of the potential $W_{sl(4),\begin{smallmatrix}\square\\\square\end{smallmatrix}} = W_{so(6),V}$. Indeed, the deformations of $W_{so(N),V}$ were already studied in [**6**]; they include several deformations which lead to *canceling* differentials and a deformation by $\Delta W = y^2$ that leads to a *universal* differential $\mathcal{H}^{so(N),V} \rightsquigarrow \mathcal{H}^{sl(N-2),\square}$.

In view of the relation $W_{sl(4),\begin{smallmatrix}\square\\\square\end{smallmatrix}}(z_1,z_2) = W_{so(6),V}(x,y)$, these deformations (and the corresponding differentials) should be present in the $sl(4)$ theory as well. In particular, there are deformations of $W_{sl(4),\begin{smallmatrix}\square\\\square\end{smallmatrix}}$ that lead to canceling differentials and, more importantly, there is a deformation by $\Delta W = y^2$ that leads to the universal differential which relates $\mathcal{H}^{so(6),V} \cong \mathcal{H}^{sl(4),\begin{smallmatrix}\square\\\square\end{smallmatrix}}$ and $\mathcal{H}^{sl(4),\square}$. Note, from the viewpoint of the $sl(4)$ knot homology, this is exactly the colored differential d_{colored} that does not change the rank of the Lie algebra, but changes the representation. Making use of (3.17) it is easy to verify that, for all values of N, the potential $W_{sl(N),\begin{smallmatrix}\square\\\square\end{smallmatrix}}$ admits a deformation by terms of degree N that leads to $W_{sl(N),\square}$ and, therefore, to the analog of (3.15):

$$\left(\mathcal{H}^{sl(N),\begin{smallmatrix}\square\\\square\end{smallmatrix}},\, d_{\mathrm{colored}}\right) \; \cong \; \mathcal{H}^{sl(N),\square}\,. \tag{3.20}$$

Much like in the case of the symmetric representations, this colored differential as well as canceling differentials come from the triply-graded theory that categorifies the $\begin{smallmatrix}\square\\\square\end{smallmatrix}$-colored HOMFLY polynomial (see section 5 for details).

4. Colored HOMFLY homology

In this section we propose structural properties of the triply-graded theory categorifying the colored version of the reduced HOMFLY polynomial. The central role in this intricate network of structural properties belongs to the colored differentials, whose existence we already motivated in the previous sections.

4.1. Structural properties. Let N and r be positive integers, and let $\mathcal{H}^{sl(N),S^r}(K)$ denote a reduced doubly-graded homology theory categorifying $P_N^{S^r}(K)$, the polynomial invariant of a knot K labeled with the r-th symmetric representation of $sl(N)$. $\mathcal{P}_N^{S^r}$ denotes the Poincaré polynomial of $\mathcal{H}^{sl(N),S^r}(K)$. Motivated by physics, we expect that such theories with a given value of r have a lot in common.

CONJECTURE 4.1. For a knot K and a positive integer r, there exists a finite polynomial $\mathcal{P}^{S^r}(K) \in \mathbb{Z}_+[\boldsymbol{a}^{\pm 1}, q^{\pm 1}, t^{\pm 1}]$ such that

$$\mathcal{P}_N^{S^r}(K)(q,t) = \mathcal{P}^{S^r}(K)(\boldsymbol{a} = q^N, q, t), \tag{4.1}$$

for all sufficiently large N.

Since the left-hand side of (4.1) is a Poincaré polynomial of a homology theory, all coefficients of $\mathcal{P}^{S^r}(K)(\boldsymbol{a}, q, t)$ must be nonnegative. This suggests that there exists a triply-graded homology theory whose Poincaré polynomial is equal to $\mathcal{P}^{S^r}(K)(\boldsymbol{a}, q, t)$, and whose Euler characteristic is equal to the normalized S^r-colored two-variable HOMFLY polynomial.

As in the case of ordinary HOMFLY homology [**3**] (that, in fact, corresponds to $r = 1$) and in the case of Kauffman homology [**6**], this triply-graded theory comes with the additional structure of differentials, that will imply Conjecture 4.1. In particular, for each positive integer r we have a triply-graded homology theory of a knot K. Moreover, these theories come with additional structure of differentials that, as in (3.16), allow us to pass from the homology theory with $R = S^r$ to theories with $R' = S^m$ and $m < r$.

Thus, we arrive to our main conjecture that describes the structure of the triply-graded homology categorifying the S^r-colored HOMFLY polynomials:

CONJECTURE 4.2. For every positive integer r there exists a triply-graded homology theory $\mathcal{H}_*^{S^r} = \mathcal{H}_{i,j,k}^{S^r}(K)$ that categorifies the reduced two-variable S^r-colored HOMFLY polynomial of K. It comes with a family of differentials $\{d_N^{S^r}\}$, with $N \in \mathbb{Z}$, and also with an additional collection of *universal colored* differentials $d_{r \to m}$, for every $1 \leq m < r$, satisfying the following properties:

- **Categorification:** $\mathcal{H}_*^{S^r}$ categorifies P^{S^r} :
$$\chi(\mathcal{H}_*^{S^r}(K)) = P^{S^r}(K).$$
- **Anticommutativity:** The differentials $\{d_N^{S^r}\}$ anticommute[6]:
$$d_N^{S^r} d_M^{S^r} = -d_M^{S^r} d_N^{S^r}.$$
- **Finite support:**
$$\dim(\mathcal{H}_*^{S^r}) < +\infty.$$

[6] *cf.* comments following (2.11)

• **Specializations:** For $N > 1$, the homology of $\mathcal{H}_*^{S^r}(K)$ with respect to $d_N^{S^r}$ is isomorphic to $\mathcal{H}^{sl(N),S^r}(K)$:

$$\left(\mathcal{H}_*^{S^r}(K), d_N^{S^r}\right) \cong \mathcal{H}^{sl(N),S^r}(K).$$

• **Canceling differentials:** The differentials $d_1^{S^r}$ and $d_{-r}^{S^r}$ are *canceling*: the homology of $\mathcal{H}_*^{S^r}(K)$ with respect to the differentials $d_1^{S^r}$ and $d_{-r}^{S^r}$ is one-dimensional, with the gradings of the remaining generators being simple invariants of the knot K.

• **Vertical Colored differentials:** The differentials $d_{1-k}^{S^r}$, for $1 \leq k \leq r-1$, have $\boldsymbol{a}$-degree -1, and the homology of $\mathcal{H}_*^{S^r}(K)$ with respect to the differential $d_{1-k}^{S^r}$ is isomorphic, after simple regrading that preserves $\boldsymbol{a}$- and t-gradings, to the k-colored homology $\mathcal{H}_*^{S^k}(K)$.

• **Universal Colored differentials:** For any positive integer m, with $m < r$, the differentials $d_{r\to m}$ have $\boldsymbol{a}$-degree zero, and the homology of $\mathcal{H}_*^{S^r}(K)$ with respect to the colored differential $d_{r\to m}$ is isomorphic (after regrading) to the m-colored homology $\mathcal{H}_*^{S^m}(K)$:

$$\left(\mathcal{H}_*^{S^r}(K), d_{r\to m}\right) \cong \mathcal{H}_*^{S^m}(K).$$

A combinatorial definition of a triply-graded theory with the structure given in Conjecture 4.2, as well as of the homologies $\mathcal{H}^{sl(N),S^r}(K)$ for $r > 1$ and $N > 2$, still does not exist in the literature.

Even though there is no such combinatorial definition, one can use any combination of the above axioms as a definition, and the remaining properties as consistency checks. In particular, one can obtain various consequence of the Conjecture 4.2 and properties of the triply-graded homology $\mathcal{H}^{S^r}$, along with the predictions for the triply-graded homology of simple knots.

In the rest of this section we give a summary of these properties, including some non-trivial checks.

4.2. A word on grading conventions. So far we summarized the general structural properties of the colored knot homology. Now we are about to make it concrete and derive explicit predictions for colored homology groups of simple knots. This requires committing to specific grading conventions, as well as other choices that may affect the form of the answer. It is important to realize, however, that none of these affect the very existence of the structural properties, which are present with any choices and merely may look different. While some of these choices will be discussed in section 6.2, here we focus on

- choices that associate various formulae to a Young tableaux λ versus its transpose λ^t;
- choices of grading, *e.g.* grading conventions used in this paper (that we sometimes refer to as "old") and grading conventions used in most of the existent literature [**3, 58, 59**] (that we sometimes call "new" in view of the forthcoming work [**60**] based on this choice).

The first choice here breaks the symmetry ("mirror symmetry") between representations S^r and Λ^r. Indeed, since in view of the Conjecture 1.1 the triply-graded homologies associated with these representations are essentially identical and can be packaged in a single theory $\mathcal{H}^r$, one has a choice whether S^r homologies arise for $N > 0$ or $N < 0$.

The second choice listed here starts with different grading assignments, but turns out to be exactly the same as the first choice. In other words, the "old" gradings and "new" gradings are related by "mirror symmetry." Another way to describe this is to note, that in grading conventions of this paper the S^r-colored superpolynomials are related (by a simple change of variables) to the Λ^r-colored invariants that one would find by following the same steps in grading conventions of *e.g.* [**3, 58, 59**]:

$$\mathcal{P}^{S^r}_{\text{here}} = \mathcal{P}^{\Lambda^r}_{\text{elsewhere}} \tag{4.2}$$

Note, that the S^r-colored invariant is related to the Λ^r-colored invariant, and vice versa. The explicit change of variables in this transformation is sensitive to even more elementary redefinitions, such as $\boldsymbol{a} \to \boldsymbol{a}^2$ and $q \to q^2$ which is ubiquitous in knot theory literature. For example, with one of the most popular choices of $\boldsymbol{a}$- and q-grading, the transformation of variables / gradings looks like:

$$\begin{aligned} A &\mapsto \boldsymbol{a}t^3, \\ q &\mapsto \frac{1}{qt^2}, \\ t &\mapsto \frac{1}{q}. \end{aligned} \tag{4.3}$$

The moral of the story is that, besides the grading conventions used in the earlier literature, the present paper offers yet another choice of grading conventions consistent with all the structural properties. And the relation between the two grading conventions can be viewed as a manifestation of mirror symmetry (1.10). Keeping these words of caution in mind, now let us take a closer look at the structure of the colored knot homology.

4.3. Consequences of Conjecture 4.2. First of all, our main Conjecture 4.2 implies the Conjecture 4.1. Indeed, in order to be consistent with the specialization $\boldsymbol{a} = q^N$ from (4.1), the q-degree of the differential $d_N^{S^r}$ must be proportional to N. Since $\mathcal{H}_*^{S^r}$ has finite support, this leads to the Conjecture 4.1, with $\mathcal{P}^{S^r}(K)$ being the Poincaré polynomial of $\mathcal{H}_*^{S^r}(K)$.

More precisely, the differentials $d_N^{S^r}$, $N \geq 1$, are expected to have the following degrees:

$$\deg(d_N^{S^r}) = (-1, N, -1), \quad N > 0,$$

which is consistent with the specialization $\boldsymbol{a} = q^N$ and the formula (3.14) that determines the q-grading of the corresponding differential in the doubly-graded theory. In fact, the differential $d_N^{S^r}$ acts on the the following bi-graded chain complex:

$$\mathcal{C}_{p,k}^{sl(N),S^r} = \bigoplus_{iN+j=p} \mathcal{H}_{i,j,k}^{S^r},$$

and has q-degree 0, and t-degree -1. The homology of $\mathcal{C}^{sl(N),S^r}$ with respect to $d_N^{S^r}$ is isomorphic to $\mathcal{H}^{sl(N),S^r}$.

In general, the degrees of the differentials $d_N^{S^r}$, for $N \in \mathbb{Z}$ are given by:

$$\begin{aligned} \deg(d_N^{S^r}) &= (-1, N, -1), \quad N \geq 1-r, \\ \deg(d_N^{S^r}) &= (-1, N, -3), \quad N \leq -r. \end{aligned}$$

We note that for every $r \geq 1$, and every $N \in \mathbb{Z}$, the degree of the differential $d_N^{S^r}$ has the form $\deg(d_N^{S^r}) = (-1, N, *)$.

4.3.1. *Canceling differentials.* Canceling differentials appear in all conjectural triply-graded theories, including the ordinary HOMFLY homology and the Kauffman homology. The defining property of a canceling differential is that the homology of the triply-graded theory with respect to this differential is "trivial", *i.e.* isomorphic to the homology of the unknot. In reduced theory, this means that the resulting homology is one-dimensional. Furthermore, the degree of the remaining generator depends in a particularly simple way on the knot.

In the case of the colored HOMFLY homology $\mathcal{H}^{S^r}$, the canceling differentials are $d_1^{S^r}$ and $d_{-r}^{S^r}$. Their degrees are:

$$\begin{aligned} \deg(d_1^{S^r}) &= (-1, 1, -1), \\ \deg(d_{-r}^{S^r}) &= (-1, -r, -3). \end{aligned}$$

Note that for $r = 1$ this agrees with the gradings of the canceling differentials in the ordinary triply-graded HOMFLY homology. (Keep in mind, though, the conventions we are using in this paper, see Remark A.1.) For either of the two canceling differentials, the degree of the surviving generator depends only on the S-invariant[7] of a knot K, introduced in [**3**]. In particular, the surviving generators have the following $(\boldsymbol{a}, q, t)$-degrees:

$$\begin{aligned} \deg\left(\mathcal{H}_*^{S^r}(K), d_1^{S^r}\right) &= (rS, -rS, 0), \\ \deg\left(\mathcal{H}_*^{S^r}(K), d_{-r}^{S^r}\right) &= (rS, r^2S, 2rS). \end{aligned} \tag{4.4}$$

Note, that the remaining generator with respect to $d_1^{S^r}$ has t-degree equal to zero.

4.3.2. *Vertical Colored differentials.* Arguably, the most interesting feature of the colored triply-graded theory is the existence of colored differentials. They allow to pass from the homology theory for a representation $R = S^r$ to the homology theory for another representation $R' = S^m$, with $m < r$.

The first group of colored differentials are "vertical" colored differentials $d_{1-k}^{S^r}$, for $1 \leq k \leq r-1$. As said before, the degrees of these differentials are

$$\deg(d_{1-k}^{S^r}) = (-1, 1-k, -1), \quad 1 \leq k \leq r-1. \tag{4.5}$$

The homology of $\mathcal{H}^{S^r}$ with respect to the differential $d_{1-k}^{S^r}$, for any $k = 1, \ldots, r-1$, is (up to a simple regrading) isomorphic to $\mathcal{H}^{S^k}$.

[7] Again, the value that we are using here is half of the value defined in [**3**].

In particular, up to an overall shift of the $\boldsymbol{a}$-grading, the $\boldsymbol{a}$- and t-gradings of the homologies match. More precisely, the Poincare polynomial of the homology $(\mathcal{H}^{S^r}, d_{1-k}^{S^r})$ satisfies:

$$(\mathcal{H}^{S^r}, d_{1-k}^{S^r})(\boldsymbol{a}, q=1, t) = \boldsymbol{a}^{(r-k)S}\, \mathcal{P}^{S^k}(\boldsymbol{a}, q=1, t), \tag{4.6}$$

where the S-invariant is defined by (4.4). The q-grading is controlled in the following way: the canceling differentials $d_1^{S^k}$ and $d_{-k}^{S^k}$ of $\mathcal{H}^{S^k}$ correspond to two canceling differentials of $(\mathcal{H}^{S^r}, d_{1-k}^{S^r})$ of degrees $(-1, 1+r-k, -1)$ and $(-1, -r, -3)$, respectively.

In the particular case of the differential $d_0^{S^r}$ that allows passage from S^r-colored homology to the uncolored S^1-homology, the explicit regrading is particularly simple:

$$(\mathcal{H}^{S^r}, d_0^{S^r})(\boldsymbol{a}, q, t) = \boldsymbol{a}^{(r-1)S}\, \mathcal{P}^{\square}(\boldsymbol{a}, q^r, t). \tag{4.7}$$

4.3.3. *Universal Colored differentials.* The universal colored differentials in the triply-graded theory are "universal" analogs of the colored differentials in the doubly-graded theory (3.15):

$$\mathcal{H}^{sl(N),S^r} \rightsquigarrow \mathcal{H}^{sl(N),S^{r-1}}.$$

As explained in section 3, these colored differentials come from the deformations of the Landau-Ginzburg potentials.

Since colored differentials are universal in the triply-graded theory — *i.e.* they relate colored triply-graded homologies without even specializing to doubly-graded $sl(N)$ theories — they should have zero $\boldsymbol{a}$-degree. This property distinguishes them clearly from the vertical colored differentials.

Furthermore, as explained in section 3, the basic colored differential $d_{r\to(r-1)}$ should have q-degree equal to 1. Then, it is easy to see that consistency of the theory also requires this differential to have zero t-grading. Combining all of these facts, we conclude that the $(\boldsymbol{a}, q, t)$-degree of the differentials $d_{r\to(r-1)}$ is equal to $(0,1,0)$.

More generally, we expect that the degree of the differential $d_{r\to m}$ depends only on the difference $r-m$ and the homology of $\mathcal{H}_*^{S^r}(K)$ with respect to the colored differential $d_{r\to m}$ is isomorphic (up to regrading) to $\mathcal{H}_*^{S^m}(K)$. The explicit form of the regrading for the colored differential $d_{2\to1}$ is as follows: the Poincaré polynomial of the homology $(\mathcal{H}^{\square\square}, d_{2\to1})$ is equal to $\mathcal{P}^{\square}(\boldsymbol{a}^2, q^2, t^2 q)$. Put differently,

$$\left(\mathcal{H}^{\square\square}(K), d_{2\to1}\right)_{2i,k+2j,2k} \cong \mathcal{H}^{\square}_{i,j,k}(K). \tag{4.8}$$

In general, the explicit regrading is very subtle (unlike the case of the vertical colored differentials).

We point out that, for $m < m_1 < r$, the homology $((\mathcal{H}^{S^r}, d_{r\to m_1}), d_{m_1\to m})$ does not need to coincide with the homology $(\mathcal{H}^{S^r}, d_{r\to m})$.

4.4. Second symmetric representation. In this subsection we focus on the case $r=2$, *i.e.* on the homology $\mathcal{H}^{\square\square}(K)$. Specialization of the above mentioned properties to $r=2$ gives:

- Two canceling differentials $d_1^{\square\square}$ and $d_{-2}^{\square\square}$ that have degrees:

$$\begin{aligned}\deg(d_1^{\square\square}) &= (-1,1,-1),\\ \deg(d_{-2}^{\square\square}) &= (-1,-2,-3).\end{aligned}$$

- The generator that survives the differential $d_1^{\square\square}$ has degree $(2S,-2S,0)$.
- The generator that survives the differential $d_{-2}^{\square\square}$ has degree $(2S,4S,4S)$.
- The vertical colored differential $d_0^{\square\square}$ has degree $(-1,0,-1)$.
- The Poincaré polynomial of the homology $(\mathcal{H}^{\square\square}(K),d_0^{\square\square})$ is equal to $\boldsymbol{a}^S\mathcal{P}^{\square}(K)(\boldsymbol{a},q^2,t)$.
- The colored differential $d_{2\to1}$ has degree $(0,1,0)$.
- The Poincaré polynomial of the homology $(\mathcal{H}^{\square\square}(K),d_{2\to1})$ is equal to $\mathcal{P}^{\square}(K)(\boldsymbol{a}^2,q^2,t^2q)$.

In addition, the homology of $\mathcal{H}^{\square\square}(K)$ with respect to the differential $d_2^{\square\square}$ should be isomorphic (after specialization $\boldsymbol{a}=q^2$) to $\mathcal{H}^{sl(2),\square\square}(K)$. To find the latter homology one can use the isomorphism (3.10) with $\mathcal{H}^{so(3),V}(K)$ which, in turn, can be obtained from the triply-graded Kauffman homology $\mathcal{H}^{\text{Kauff}}(K)$ studied in [**6**]. Indeed, the doubly-graded homology $\mathcal{H}^{so(3),V}(K)$ is isomorphic to the homology of $\mathcal{H}^{\text{Kauff}}(K)$ with respect to the corresponding differential d_3 from [**6**], after the specialization $\lambda=q^2$. Usually, that differential d_3 acts trivially; in particular, this is the case for all knots that we analyze below.

The structure of the homology $\mathcal{H}^{\square\square}(K)$ with the above differentials allows us to compute it for various small knots, as we shall illustrate next.

4.4.1. *$\mathcal{H}^{S^2}$ and $\mathcal{P}^{S^2}$ for small knots.* The homology $\mathcal{H}^{\square\square}(K)$ and the superpolynomial $\mathcal{P}^{\square\square}(K)(\boldsymbol{a},q,t)$ (= the Poincaré polynomial of $\mathcal{H}^{\square\square}(K)$) must satisfy the following properties:

- specialization to $t=-1$ gives the reduced $\square\square$-colored HOMFLY polynomial $P^{\square\square}(K)$.[8]
- specialization to $\boldsymbol{a}=q^2$ gives the Poincaré polynomial $\mathcal{P}_2^{\square\square}(K)$ of the homology $\mathcal{H}^{sl(2),\square\square}(K)$. This homology is isomorphic to $\mathcal{H}^{so(3),V}(K)$. To find the latter one, we use the results from Table 3 of [**6**], if available.[9]

[8]In order to find the colored HOMFLY polynomial $P^{\square\square}(K)$ one can use *e.g.* equation (3.25) in [**26**] and the values for the BPS invariants $\widehat{N}_{\square\square,g,Q}$ tabulated in section 6 of that paper. The result gives the unreduced two-variable colored HOMFLY polynomial. In order to find the reduced polynomial, one should divide the unreduced polynomial by:

$$\frac{(\boldsymbol{a}-\boldsymbol{a}^{-1})(\boldsymbol{a}q-\boldsymbol{a}^{-1}q^{-1})}{(q-q^{-1})(q^2-q^{-2})}.$$

The results from [**26**] enable us to compute the reduced $\square\square$-colored HOMFLY polynomial for the knots 3_1, 4_1, 5_1 and 6_1. Another useful source of the colored HOMFLY polynomials and their specializations to $\boldsymbol{a}=q^2$ and $\boldsymbol{a}=q^3$ is the KnotAtlas [**56**], which the reader may want to consult for many other knots.

[9]In all examples we have computed, the values $i+j$ for all nontrivial Kauffman homology groups $\mathcal{H}^{\text{Kauff}}_{i,j,k}(K)$ have the same parity. Thus the differential d_3 on the Kauffman homology, which is of degree $(-1,2,-1)$, is trivial. Consequently, the Poincaré polynomial $\mathcal{P}^{so(3),V}(K)(q,t)$ is equal to the $\lambda=q^2$ specialization of the Poincaré superpolynomial of the Kauffman homology $\mathcal{P}^{\text{Kauff}}(K)(\lambda=q^2,q,t)$ in all our examples.

• $\mathcal{H}^{\square\square}(K)$ comes equipped with the differentials described in section 4.4.

These requirements are more than sufficient to determine the colored superpolynomial for many small knots. As the first example, we consider the trefoil knot:

Example. *The trefoil knot* 3_1

The reduced $\square\square$-colored HOMFLY polynomial of the trefoil knot is equal to (see *e.g.* [**26, 57**]):

$$P^{\square\square}(3_1) = \boldsymbol{a}^2 q^{-2} + \boldsymbol{a}^2 q + \boldsymbol{a}^2 q^2 + \boldsymbol{a}^2 q^4 - \boldsymbol{a}^3 - \boldsymbol{a}^3 q - \boldsymbol{a}^3 q^3 - \boldsymbol{a}^3 q^4 + \boldsymbol{a}^4 q^3 \,.$$

The homology $\mathcal{H}^{so(3),V}(3_1)$ is computed in [**6**] (see eq. (6.14) and Table 3). Hence we have:

$$\mathcal{P}_2^{\square\square}(3_1) = q^2 + q^5 t^2 + q^6 t^2 + q^6 t^3 + q^7 t^3 + q^8 t^4 + q^9 t^5 + q^{10} t^5 + q^{11} t^6$$

From these two expressions we immediately deduce the colored superpolynomial of the trefoil[10]:

$$\mathcal{P}^{\square\square}(3_1) = \boldsymbol{a}^2(q^{-2} + qt^2 + q^2t^2 + q^4t^4) + \boldsymbol{a}^3(t^3 + qt^3 + q^3t^5 + q^4t^5) + \boldsymbol{a}^4 q^3 t^6 . \tag{4.9}$$

Note that its specializations to $t = -1$ and $\boldsymbol{a} = q^2$ give $P^{\square\square}(3_1)$ and $\mathcal{P}_2^{\square\square}(3_1)$, respectively. Moreover, the corresponding homology $\mathcal{H}^{\square\square}(3_1)$ also enjoys the action of two canceling differentials and one colored differential. In order to visualize this homology, we represent each generator by a dot in the $(\boldsymbol{a}, q)$-plane, with a label denoting its t-grading. In Figure 3, the canceling differential $d_1^{\square\square}$ is represented by a blue arrow, the canceling differential $d_{-2}^{\square\square}$ is represented by a red arrow, the colored differential $d_{2\to 1}$ is represented by a magenta arrow, while the vertical colored differential $d_0^{\square\square}$ is represented by dashed light blue arrow.

The generator that survives $d_1^{\square\square}$ has degree $(2,-2,0)$, while the one that survives $d_{-2}^{\square\square}$ has degree $(2,4,4)$. Both are consistent with the S-invariant of the trefoil $S(3_1) = 1$ and the general discussion in section 4.4. The Poincaré polynomial of the homology with respect to the colored differential $d_{2\to 1}$ is equal to:

$$\boldsymbol{a}^2 q^{-2} + \boldsymbol{a}^2 q^4 t^4 + \boldsymbol{a}^4 q^3 t^6 \,,$$

while the Poincaré polynomial of the homology with respect to the vertical colored differential $d_0^{\square\square}$ is equal to:

$$\boldsymbol{a}^2 q^{-2} + \boldsymbol{a}^2 q^2 t^4 + \boldsymbol{a}^3 t^3 \,.$$

A careful reader will notice that the last two expressions are equal to $\mathcal{P}^{\square}(3_1)(\boldsymbol{a}^2, q^2, t^2 q)$ and $\boldsymbol{a}\mathcal{P}^{\square}(3_1)(\boldsymbol{a}, q^2, t)$, respectively, where $\mathcal{P}^{\square}(3_1)(\boldsymbol{a}, q, t)$ is the ordinary superpolynomial, whose explicit form is written in (A.1).

---

[10] As discussed in section 4.2, there are two different possibilities for grading conventions. Besides the grading conventions discussed in most of this paper, there are also "new" grading conventions where the $\boldsymbol{a}$ and q degrees are both twice the value of the corresponding degrees in the conventions that we are using in this paper, while the t-degree change is more subtle. The value of the colored superpolynomial of the trefoil in the "new" gradings is given by

$$\mathcal{P}^{\square\square}_{\text{new grad.}}(3_1) = \boldsymbol{a}^4(q^{-4} + q^2t^4 + q^4t^6 + q^8t^8) + \boldsymbol{a}^6(t^5 + q^2t^7 + q^6t^9 + q^8t^{11}) + \boldsymbol{a}^8 q^6 t^{12}.$$

We note that in these gradings, the answer coincides with [**58, 59**].

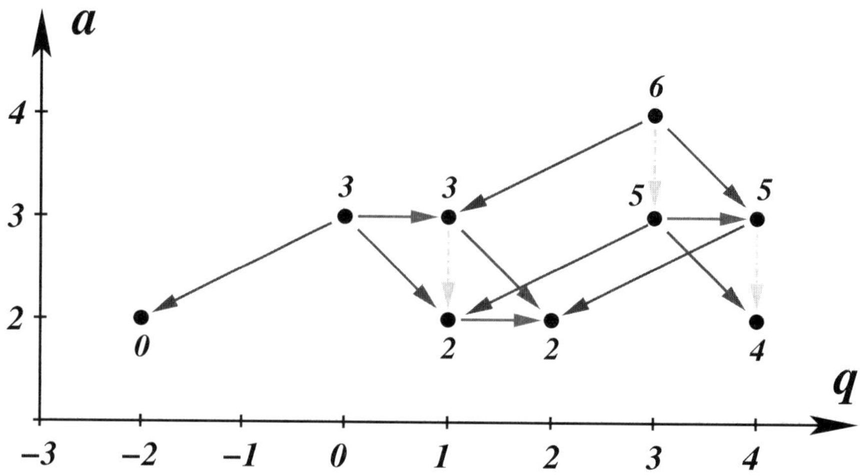

FIGURE 3. The reduced S^2-colored homology of the trefoil knot.

This computation can be easily extended to many other small knots. We list the results for all prime knots with up to 6 crossings in Tables 1 and 2. The fact that the structure described in this section works beautifully for all knots with up to 6 crossings is already an impressive test of our main Conjecture 4.2. However, to convince even hard-boiled skeptics, in appendix B we carry out a much more challenging computation of the colored HOMFLY homology for "thick" knots 8_{19} and 9_{42}.

We notice that all computations of colored homologies in this paper are done by hand, only by using the existence and properties of the differentials described in this section. Moreover, in majority of the cases only a few of the differentials are used to obtain the result, which than matched perfectly all the remaining properties.

smaller[3]

4.5. Differentials for higher symmetric representations. Now let us consider the triply graded homology $\mathcal{H}^{S^r}$ of knots and links colored by the representation $R = S^r$ with more general $r \geq 1$. Much as in the case $r = 2$ considered in the previous subsection, we expect that $\mathcal{H}^{S^r}$ comes equipped with the following differentials:

- canceling differential $d_1^{S^r}$ of degree $(-1, 1, -1)$, whose homology is one-dimensional and consists of a degree $(rS, -rS, 0)$ generator;
- canceling differential $d_{-r}^{S^r}$ of degree $(-1, -r, -3)$, which leaves behind a one-dimensional homology with a generator of degree $(rS, r^2S, 2rS)$;
- for every $1 \leq k < r$, there exists a vertical colored differential $d_{1-k}^{S^r}$ of degree $(-1, 1-k, -1)$, such that the homology of $\mathcal{H}^{S^r}$ with respect to $d_{1-k}^{S^r}$ is isomorphic to $\mathcal{H}^{S^k}$;
- for every $1 \leq m < r$, there exists a universal colored differential $d_{r \to m}$ which, when acting on $\mathcal{H}^{S^r}$, leaves behind the homology $\mathcal{H}^{S^m}$. In particular, the colored differential $d_{r \to (r-1)}$ has degree $(0, 1, 0)$.

Knot	$\mathcal{P}^{\square\square}$
3_1	$\boldsymbol{a}^2(q^2t^2+q^{-2}+qt^2+q^4t^4)+\boldsymbol{a}^3(q^4t^5+q^3t^5+t^3+qt^3)+\boldsymbol{a}^4q^3t^6$
4_1	$\boldsymbol{a}^{-2}q^{-2}t^{-4}+(\boldsymbol{a}^{-1}q^{-3}+\boldsymbol{a}^{-1}q^{-2})t^{-3}+(q^{-3}+\boldsymbol{a}^{-1}q^{-1}+\boldsymbol{a}^{-1})t^{-2}+$ $+(q^{-2}+q^{-1}+\boldsymbol{a}^{-1}+\boldsymbol{a}^{-1}q)t^{-1}+(q^{-1}+3+q)+(q^2+q+\boldsymbol{a}+\boldsymbol{a}q^{-1})t+$ $+(q^3+\boldsymbol{a}q+\boldsymbol{a})t^2+(\boldsymbol{a}q^3+\boldsymbol{a}q^2)t^3+\boldsymbol{a}^2q^2t^4$
5_1	$\boldsymbol{a}^4q^{-4}+(\boldsymbol{a}^4q^{-1}+\boldsymbol{a}^4)t^2+(\boldsymbol{a}^5q^{-2}+\boldsymbol{a}^5q^{-1})t^3+$ $+(\boldsymbol{a}^4q^2+\boldsymbol{a}^4q^3+\boldsymbol{a}^4q^4)t^4+(\boldsymbol{a}^5q+2\boldsymbol{a}^5q^2+\boldsymbol{a}^5q^3)t^5+$ $+(\boldsymbol{a}^4q^5+\boldsymbol{a}^4q^6+\boldsymbol{a}^6q)t^6+(\boldsymbol{a}^5q^4+2\boldsymbol{a}^5q^5+\boldsymbol{a}^5q^6)t^7+$ $+(\boldsymbol{a}^4q^8+\boldsymbol{a}^6q^4+\boldsymbol{a}^6q^5)t^8+(\boldsymbol{a}^5q^7+\boldsymbol{a}^5q^8)t^9+\boldsymbol{a}^6q^7t^{10}.$
5_2	$\boldsymbol{a}^2q^{-2}+(\boldsymbol{a}^2q^{-1}+\boldsymbol{a}^2)t+(2\boldsymbol{a}^2q+\boldsymbol{a}^2q^2+\boldsymbol{a}^3q^{-2}+\boldsymbol{a}^3q^{-1})t^2+$ $+(\boldsymbol{a}^2q^2+\boldsymbol{a}^2q^3+2\boldsymbol{a}^3+2\boldsymbol{a}^3q)t^3+(\boldsymbol{a}^2q^4+2\boldsymbol{a}^3q+3\boldsymbol{a}^3q^2+\boldsymbol{a}^3q^3+\boldsymbol{a}^4)t^4+$ $+(2\boldsymbol{a}^3q^3+2\boldsymbol{a}^3q^4+\boldsymbol{a}^4+2\boldsymbol{a}^4q+\boldsymbol{a}^4q^2)t^5+$ $+(\boldsymbol{a}^3q^4+\boldsymbol{a}^3q^5+\boldsymbol{a}^4q^2+3\boldsymbol{a}^4q^3+\boldsymbol{a}^4q^4)t^6+$ $+(\boldsymbol{a}^4q^3+2\boldsymbol{a}^4q^4+\boldsymbol{a}^4q^5+\boldsymbol{a}^5q^2+\boldsymbol{a}^5q^3)t^7+$ $+(\boldsymbol{a}^4q^6+\boldsymbol{a}^5q^3+\boldsymbol{a}^5q^4)t^8+(\boldsymbol{a}^5q^5+\boldsymbol{a}^5q^6)t^9+\boldsymbol{a}^6q^5t^{10}.$

TABLE 1. Colored superpolynomial for prime knots with up to 5 crossings.

4.5.1. *Colored superpolynomial $\mathcal{P}^{S^3}$ for the trefoil.* It can be computed by requiring that its specialization to $t=-1$ equals the reduced S^3-colored HOMFLY polynomial and that it enjoys the action of the canceling and the first colored differentials of appropriate degrees. In particular, according to the general rules (4.4), we require that the remaining generator with respect to the $d_1^{\square\square\square}$ action has degree $(3,-3,0)$, while the remaining generator with respect to the action of $d_{-3}^{\square\square\square}$ has degree $(3,9,6)$. For the colored differential $d_{3\to 2}$ we require that the remaining homology should have rank 9, just like $\mathcal{H}^{\square\square}(3_1)$. From these, we obtain the following result:

$$\begin{aligned}\mathcal{P}^{S^3}(3_1) \;=\;& \boldsymbol{a}^3q^{-3}+(\boldsymbol{a}^3q+\boldsymbol{a}^3q^2+\boldsymbol{a}^3q^3)t^2+(\boldsymbol{a}^4+\boldsymbol{a}^4q+\boldsymbol{a}^4q^2)t^3+\\ &+(\boldsymbol{a}^3q^5+\boldsymbol{a}^3q^6+\boldsymbol{a}^3q^7)t^4+(\boldsymbol{a}^4q^4+2\boldsymbol{a}^4q^5+2\boldsymbol{a}^4q^6+\boldsymbol{a}^4q^7)t^5+\\ &+(\boldsymbol{a}^5q^4+\boldsymbol{a}^5q^5+\boldsymbol{a}^5q^6+\boldsymbol{a}^3q^9)t^6+(\boldsymbol{a}^4q^8+\boldsymbol{a}^4q^9+\boldsymbol{a}^4q^{10})t^7+\\ &+(\boldsymbol{a}^5q^8+\boldsymbol{a}^5q^9+\boldsymbol{a}^5q^{10})t^8+\boldsymbol{a}^6q^9t^9.\end{aligned}$$

Note, there exists a differential $d_{3\to 1}$ on $\mathcal{H}^{S^3}(3_1)$ of degree $(0,4,2)$, such that the homology with respect to this differential is of rank 3, as $\mathcal{H}^{\square}(3_1)$.

Knot	$\mathcal{P}^{\square\square}$
6_1	$\boldsymbol{a}^{-2}q^{-2}t^{-4}+(\boldsymbol{a}^{-1}q^{-3}+\boldsymbol{a}^{-1}q^{-2})t^{-3}+(\boldsymbol{a}^{-1}q^{-2}+2\boldsymbol{a}^{-1}q^{-1}+q^{-3}+\boldsymbol{a}^{-1})t^{-2}+$ $+(q^{-3}+\boldsymbol{a}^{-1}+3q^{-2}+2q^{-1}+\boldsymbol{a}^{-1}q)t^{-1}+(\boldsymbol{a}q^{-3}+2q^{-1}+\boldsymbol{a}q^{-2}+5+q)+$ $+(\boldsymbol{a}q^{-2}+1+3q+3\boldsymbol{a}q^{-1}+2q^{2}+2\boldsymbol{a})t+$ $+(\boldsymbol{a}q^{-1}+4\boldsymbol{a}+4\boldsymbol{a}q+q^{3}+\boldsymbol{a}^{2}q^{-1}+\boldsymbol{a}q^{2})t^{2}+$ $+(\boldsymbol{a}^{2}q^{-1}+\boldsymbol{a}q+2\boldsymbol{a}^{2}+3\boldsymbol{a}q^{2}+2\boldsymbol{a}q^{3}+\boldsymbol{a}^{2}q)t^{3}+$ $+(\boldsymbol{a}^{2}+2\boldsymbol{a}^{2}q+\boldsymbol{a}q^{3}+3\boldsymbol{a}^{2}q^{2}+\boldsymbol{a}q^{4}+\boldsymbol{a}^{2}q^{3})t^{4}+$ $+(\boldsymbol{a}^{2}q^{2}+2\boldsymbol{a}^{2}q^{3}+\boldsymbol{a}^{3}q+\boldsymbol{a}^{3}q^{2}+\boldsymbol{a}^{2}q^{4})t^{5}+$ $+(\boldsymbol{a}^{3}q^{2}+\boldsymbol{a}^{3}q^{3}+\boldsymbol{a}^{2}q^{5})t^{6}+(\boldsymbol{a}^{3}q^{4}+\boldsymbol{a}^{3}q^{5})t^{7}+\boldsymbol{a}^{4}q^{4}t^{8}.$
6_2	$\boldsymbol{a}^{4}q^{6}t^{8}+(\boldsymbol{a}^{3}q^{6}+\boldsymbol{a}^{3}q^{7})t^{7}+(\boldsymbol{a}^{4}q^{3}+\boldsymbol{a}^{2}q^{7})t^{6}+(2\boldsymbol{a}^{3}q^{3}+2\boldsymbol{a}^{3}q^{4})t^{5}+$ $+(\boldsymbol{a}^{4}+\boldsymbol{a}^{2}q^{3}+3\boldsymbol{a}^{2}q^{4}+\boldsymbol{a}^{2}q^{5})t^{4}+(2\boldsymbol{a}^{3}+2\boldsymbol{a}^{3}q+\boldsymbol{a}q^{4}+\boldsymbol{a}q^{5})t^{3}+(3\boldsymbol{a}^{2}q+\boldsymbol{a}^{2}q^{2})t^{2}+$ $+(\boldsymbol{a}^{3}q^{-3}+\boldsymbol{a}^{3}q^{-2}+\boldsymbol{a}q+\boldsymbol{a}q^{2})t+(\boldsymbol{a}^{2}q^{-3}+3\boldsymbol{a}^{2}q^{-2}+\boldsymbol{a}^{2}q^{-1}+q^{2})+$ $+(\boldsymbol{a}q^{-2}+\boldsymbol{a}q^{-1})t^{-1}+\boldsymbol{a}^{2}q^{-5}t^{-2}+(\boldsymbol{a}q^{-5}+\boldsymbol{a}q^{-4})t^{-3}+q^{-4}t^{-4}+$ $+(1+q)(1+\boldsymbol{a}^{-1}qt^{-1})(1+\boldsymbol{a}^{-1}q^{-2}t^{-3})\times[\boldsymbol{a}^{4}q^{4}t^{7}+(\boldsymbol{a}^{4}q^{3}+\boldsymbol{a}^{3}q^{4})t^{6}+$ $+(\boldsymbol{a}^{4}q+\boldsymbol{a}^{3}q^{3})t^{5}+(\boldsymbol{a}^{3}q^{2}+\boldsymbol{a}^{3}q)t^{4}+\boldsymbol{a}^{3}t^{3}+(\boldsymbol{a}^{3}q^{-2}+\boldsymbol{a}^{2})t^{2}].$
6_3	$\boldsymbol{a}^{2}q^{5}t^{6}+(\boldsymbol{a}q^{5}+\boldsymbol{a}q^{6})t^{5}+(\boldsymbol{a}^{2}q^{2}+q^{6})t^{4}+(2\boldsymbol{a}q^{2}+2\boldsymbol{a}q^{3})t^{3}+$ $+(\boldsymbol{a}^{2}q^{-1}+q^{2}+3q^{3}+q^{4})t^{2}+(2\boldsymbol{a}q^{-1}+2\boldsymbol{a}+\boldsymbol{a}^{-1}q^{3}+\boldsymbol{a}^{-1}q^{4})t$ $+(q^{-1}+5+q)+(\boldsymbol{a}q^{-4}+\boldsymbol{a}q^{-3}+2\boldsymbol{a}^{-1}+2\boldsymbol{a}^{-1}q)t^{-1}+(q^{-4}+3q^{-3}+q^{-2}+\boldsymbol{a}^{-2}q)t^{-2}+$ $+(2\boldsymbol{a}^{-1}q^{-3}+2\boldsymbol{a}^{-1}q^{-2})t^{-3}+(q^{-6}+\boldsymbol{a}^{-2}q^{-2})t^{-4}+(\boldsymbol{a}^{-1}q^{-6}+\boldsymbol{a}^{-1}q^{-5})t^{-5}+\boldsymbol{a}^{-2}q^{-5}t^{-6}+$ $+(1+q)(1+\boldsymbol{a}^{-1}qt^{-1})(1+\boldsymbol{a}^{-1}q^{-2}t^{-3})\times[\boldsymbol{a}^{2}q^{3}t^{5}+(\boldsymbol{a}^{2}q+\boldsymbol{a}q^{3})t^{4}+$ $+(\boldsymbol{a}^{2}+\boldsymbol{a}q^{2}+\boldsymbol{a}q)t^{3}+3\boldsymbol{a}t^{2}+(\boldsymbol{a}q^{-1}+\boldsymbol{a}q^{-2}+1)t+(\boldsymbol{a}q^{-3}+q^{-1})+q^{-3}t^{-1}].$

TABLE 2. Colored superpolynomial for prime knots with 6 crossings.

Also, there exists a differential $d_0^{\square\square\square}$ of degree $(-1,0,-1)$ such that the Poincare polynomial of $(\mathcal{H}^{\square\square\square}(3_1), d_0^{\square\square\square})$ is equal to $\boldsymbol{a}^2\ \mathcal{P}^{\square}(3_1)(\boldsymbol{a},q^3,t)$.

Finally, there exists a differential $d_{-1}^{\square\square\square}$ of degree $(-1,-1,-1)$ such that the homology$(\mathcal{H}^{\square\square\square}(3_1), d_{-1}^{\square\square\square})$ is isomorphic to $\mathcal{H}^{\square\square}(3_1)$.

In Appendix C we compute also the S^3-colored homology of the figure-eight knot 4_1.

4.5.2. *Size of the homology.* Computations show that for a knot K, the rank of the homology $\mathcal{H}^{S^r}$ grows exponentially with r. In particular, this makes the computation of the homologies $\mathcal{H}^{S^r}(K)$ difficult for large r. (In fact, even for $r>2$ the size of the homology is too big to make computations practical.) To be more precise, for all thin and torus knots studied here we find:

$$\operatorname{rank}\mathcal{H}^{S^r}(K) \;=\; \left(\operatorname{rank}\mathcal{H}^{S}(K)\right)^r . \tag{4.10}$$

5. Mirror symmetry for knots

In this section, we observe a remarkable "mirror symmetry" relation (1.11) between two completely different triply-graded homology theories associated with symmetric and anti-symmetric representations of $sl(N)$, which will allow us to formulate even a bigger theory that will contain both. As a first step, however, we need to extend the discussion in section 4 to the HOMFLY homology colored by anti-symmetric representations of $sl(N)$.

5.1. Anti-symmetric representations. Much as for the symmetric representation S^r, we can repeat the analysis for the anti-symmetric representations Λ^r of $sl(N)$.

In particular, for every positive integer r there exists a triply-graded homology theory $\mathcal{H}^{\Lambda^r}(K)$, together with the collection of differentials $\{d_N^{\Lambda^r}\}$, $N \in \mathbb{Z}$, such that the homology with respect to $d_N^{\Lambda^r}$ is isomorphic to $\mathcal{H}^{sl(N),\Lambda^r}(K)$. Moreover, it comes equipped with the collection of "universal" colored differentials, like in the case of the symmetric representations. The homologies $\mathcal{H}^{\Lambda^r}(K)$, together with all the differentials, satisfy the same properties as $\mathcal{H}^{S^r}(K)$ from Conjecture 4.2.

Again, we have two canceling differentials, this time $d_{-1}^{\Lambda^r}$ and $d_r^{\Lambda^r}$ of $(\boldsymbol{a}, q, t)$-degrees $(-1, -1, -3)$ and $(-1, r, -1)$, respectively. The origin of these differentials is clear, and can be inferred either from deformations of B-model potentials, as in section 3, or from basic representation theory. For example, the fact the representation Λ^r of $sl(r)$ is trivial gives rise to a canceling differential $d_r^{\Lambda^r}$.

Another basic fact is $\Lambda^r \cong \Lambda$ for $\mathfrak{g} = sl(r+1)$, which leads to the relation

$$\mathcal{H}^{sl(r+1),\Lambda^r}(K) \;\cong\; \mathcal{H}^{sl(r+1),\Box}(K)\,. \tag{5.1}$$

For the triply-graded theory $\mathcal{H}^{\Lambda^r}(K)$, this relation implies that the $\boldsymbol{a} = q^{r+1}$ specializations of the homologies $(\mathcal{H}^{\Lambda^r}(K), d_{r+1}^{\Lambda^r})$ and $(\mathcal{H}^{\Box}(K), d_{r+1}^{\Box})$ should be isomorphic.

Like in the case of symmetric representations, all the required properties allow computation of the anti-symmetric homology for various small knots. Below, we provide the details for the trefoil knot.

Using the isomorphism $so(6) \cong sl(4)$ under which the vector representation of $so(6)$ is identified with the anti-symmetric representation of $sl(4)$, we conclude

$$\mathcal{H}^{sl(4),\boxminus}(K) \;\cong\; \mathcal{H}^{so(6),V}(K) \tag{5.2}$$

From this relation[11] we immediately find

$$\mathcal{H}^{sl(4),\boxminus}(3_1) = q^4 + q^6t^2 + q^7t^2 + q^8t^3 + q^9t^3 + q^{10}t^4 + q^{11}t^5 + q^{12}t^5 + q^{13}t^6 \tag{5.3}$$

Also, for $\mathfrak{g} = sl(3)$ we have $\Lambda^2 \cong \Lambda$, which implies another useful relation

$$\mathcal{H}^{sl(3),\boxminus}(K) \;\cong\; \mathcal{H}^{sl(3),\Box}(K) \tag{5.4}$$

For the trefoil, this gives:

$$\mathcal{H}^{sl(3),\boxminus}(3_1) = q^2 + q^4t^2 + q^6t^3 \tag{5.5}$$

---

[11] Here, in comparing the two homology theories we take into account that q-gradings differ by a factor of 2.

Combining this data with the colored HOMFLY polynomial

$$P^{\boxminus}(\boldsymbol{a},q) = \boldsymbol{a}^2(q^{-4}+q^{-2}+q^{-1}+q^2) - \boldsymbol{a}^3(q^{-4}+q^{-3}+q^{-1}+1) + \boldsymbol{a}^4 q^{-3} \tag{5.6}$$

we easily find the anti-symmetric version of the superpolynomial for the trefoil knot:

(5.7)

$$\mathcal{P}^{\boxminus}(3_1) = \boldsymbol{a}^2(q^{-4}+q^{-2}t^2+q^{-1}t^2+q^2t^4)+\boldsymbol{a}^3(q^{-4}t^3+q^{-3}t^3+q^{-1}t^5+t^5)+\boldsymbol{a}^4q^{-3}t^6$$

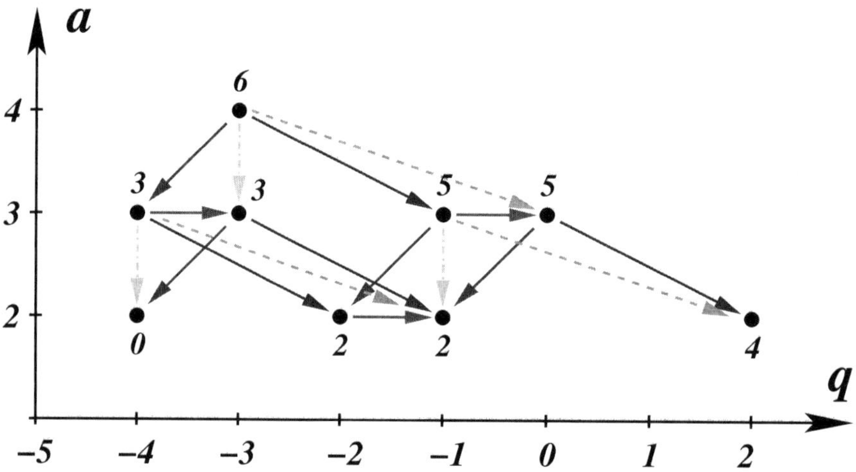

FIGURE 4. The reduced Λ^2-colored homology of the trefoil

The homology $\mathcal{H}^{\boxminus}(3_1)$ is shown on the Figure 4 below. It has the following differentials:

- canceling differential $d_{-1}^{\boxminus}$ of $(\boldsymbol{a},q,t)$-degree $(-1,-1,-3)$
- canceling differential $d_2^{\boxminus}$ of $(\boldsymbol{a},q,t)$-degree $(-1,2,-1)$ reflects the fact that Λ^2 is trivial in $sl(2)$ theory
- differential $d_3^{\boxminus}$ of $(\boldsymbol{a},q,t)$-degree $(-1,3,-1)$ reflects the fact that $\Lambda^2 \cong \Lambda$ in $sl(3)$ theory and gives, *cf.* (5.5):

$$\boldsymbol{a}^2q^{-4}t^0 + \boldsymbol{a}^2q^{-2}t^2 + \boldsymbol{a}^3q^{-3}t^3 \overset{a=q^3}{=} \mathcal{H}^{sl(3),\boxminus}(3_1) \tag{5.8}$$

- differential $d_0^{\boxminus}$ of $(\boldsymbol{a},q,t)$-degree $(-1,0,-3)$ gives:

$$\boldsymbol{a}^2q^{-2}t^2 + \boldsymbol{a}^2q^2t^4 + \boldsymbol{a}^3q^0t^5 = \boldsymbol{a}t^2\mathcal{P}^{\square}(3_1)(\boldsymbol{a},q^2,t). \tag{5.9}$$

- universal differential $d_{2\to1}^{\boxminus}$ of $(\boldsymbol{a},q,t)$-degree $(0,1,0)$ gives:

$$\boldsymbol{a}^2q^{-4}t^0 + \boldsymbol{a}^2q^2t^4 + \boldsymbol{a}^4q^{-3}t^6 = \mathcal{P}^{\square}(\boldsymbol{a}^2,q^4,q^{-1}t^2) \tag{5.10}$$

5.2. Mirror symmetry for knot homology. By computing the triply-graded homologies $\mathcal{H}^{S^r}(K)$ and $\mathcal{H}^{\Lambda^r}(K)$ for various small knots, we discover the following remarkable symmetry between these two classes of theories, labeled by $R = S^r$ and $R = \Lambda^r$:

$$\mathcal{H}^{\Lambda^r}_{i,j,*}(K) \cong \mathcal{H}^{S^r}_{i,-j,*}(K). \tag{5.11}$$

Furthermore, this symmetry extends to the differentials as well. More precisely, let

$$\phi : \mathcal{H}^{S^r}(K) \to \mathcal{H}^{\Lambda^r}(K), \tag{5.12}$$

be the isomorphisms from (5.11). Then

$$\phi\, d_N^{S^r} = d_{-N}^{\Lambda^r}\, \phi, \quad N \in \mathbb{Z}. \tag{5.13}$$

As the first illustration of the mirror symmetry, let us compare the second symmetric and anti-symmetric homology for the trefoil knot. From Figures 3 and 4 it is clear that "mirror symmetry" is manifest both for the homology (5.11) and for the differentials (5.13). The explicit t-grading change in (5.11) in this case is given by:

$$\mathcal{H}^{\square\!\square\,\text{(column)}}_{i,j,k}(3_1) \cong \mathcal{H}^{\square\!\square}_{i,-j,4i-k-4}(3_1) \tag{5.14}$$

for the trefoil knot.

The first implication of the mirror symmetry is that one can combine $\mathcal{H}^{S^r}(K)$ and $\mathcal{H}^{\Lambda^r}(K)$ into a single homology theory. By setting $\mathcal{H}^r(K)$ to be $\mathcal{H}^{S^r}(K)$, we obtain the Conjecture 1.1. More precisely, we conjecture the following:

CONJECTURE 5.1. For every positive integer r there exists a triply-graded homology theory $\mathcal{H}^r_*(K) = \mathcal{H}^{S^r}_{i,j,k}(K)$, that comes with a family of differentials $\{d^r_N\}$, with $N \in \mathbb{Z}$, and also with an additional collection of *universal colored* differentials $d_{r\to m}$, for every $1 \le m < r$, satisfying the following properties:

- **Mirror Symmetry**

$$\mathcal{H}^r_{i,j,*}(K) \cong \mathcal{H}^{S^r}_{i,j,*}(K) \cong \mathcal{H}^{\Lambda^r}_{i,-j,*}(K).$$

- **Categorification:** $\mathcal{H}^r_*$ categorifies P^{S^r} and P^{Λ^r} :

$$\chi(\mathcal{H}^r_*(K)) = P^{S^r}(K)(a,q) = P^{\Lambda^r}(K)(a,q^{-1}).$$

- **Anticommutativity:** The differentials $\{d^{S^r}_N\}$ anticommute[12]:

$$d^r_N d^r_M = -d^r_M d^r_N.$$

- **Finite support:**

$$\dim(\mathcal{H}^r_*) < +\infty.$$

- **Specializations:** For $N > 1$, the homology of $\mathcal{H}^r_*(K)$ with respect to d^r_N is isomorphic to $\mathcal{H}^{sl(N),S^r}(K)$:

$$(\mathcal{H}^r_*(K), d^r_N) \cong \mathcal{H}^{sl(N),S^r}(K).$$

For $N \le -2r$, the homology of $\mathcal{H}^r_*(K)$ with respect to d^r_N is isomorphic to $\mathcal{H}^{sl(-N),\Lambda^r}(K)$:

$$(\mathcal{H}^r_*(K), d^r_N) \cong \mathcal{H}^{sl(-N),\Lambda^r}(K).$$

- **Canceling differentials:** The differentials d^r_1 and d^r_{-r} are *canceling*: the homology of $\mathcal{H}^r_*(K)$ with respect to the differentials d^r_1 and d^r_{-r} is one-dimensional. This reflects the fact that S^r representation of $sl(1)$ and Λ^r representation of $sl(r)$ are trivial.

[12] *cf.* comments following (2.11)

• $sl(N)$ **Colored differentials:** For every $1 \le k \le r-1$, the homology of $\mathcal{H}^r_*(K)$ with respect to the differential d^r_{-r-k} is isomorphic to $\mathcal{H}^k(K)$:

$$(\mathcal{H}^r_*(K), d^r_{-r-k}) \cong \mathcal{H}^k(K), \quad 1 \le k \le r-1,$$

reflecting the fact that $\Lambda^r \cong \Lambda^k$ for $sl(r+k)$, where $1 \le k \le r-1$.

• **Vertical Colored differentials:** The differentials d^r_{1-k}, for $1 \le k \le r-1$, have $\boldsymbol{a}$-degree -1, and the homology of $\mathcal{H}^r_*(K)$ with respect to the differential d^r_{1-k} is isomorphic (after simple regrading that preserves $\boldsymbol{a}$- and t-gradings), to $\mathcal{H}^k_*(K)$:

$$(\mathcal{H}^r_*(K), d^r_{1-k}) \cong \mathcal{H}^k(K), \quad 1 \le k \le r-1.$$

• **Universal Colored differentials:** For any positive integer m, with $m < r$, the differentials $d_{r \to m}$have $\boldsymbol{a}$-degree zero, and the homology of $\mathcal{H}^r_*(K)$ with respect to the colored differential $d_{r\to m}$ is isomorphic (after regrading) to $\mathcal{H}^m_*(K)$:

$$(\mathcal{H}^r_*(K), d_{r\to m}) \cong \mathcal{H}^m_*(K).$$

Example. *$\mathcal{H}^3$ for the trefoil knot*

As another example of the symmetry (5.11), and this time for degree higher than 2, let us consider $\mathcal{H}^3(3_1)$. Besides the two canceling differentials d^3_1 and d^3_{-3}, we also have the differential d^3_{-4} of degree $(-1,-4,-3)$. The homology with respect to this differential is equal to

$$(\mathcal{H}^3(3_1), d^3_{-4}) = \boldsymbol{a}^3 q^7 t^4 + \boldsymbol{a}^3 q^9 t^6 + \boldsymbol{a}^4 q^8 t^7. \tag{5.15}$$

The explicit t-grading change in (5.11) for the trefoil and $r=3$ is given by:

$$\mathcal{H}^{\yng(1,1,1)}_{i,j,k}(3_1) \cong \mathcal{H}^{\yng(3)}_{i,-j,4i-k-6}(3_1). \tag{5.16}$$

With this change of gradings, the "mirror image" of (5.15) is equal to $\boldsymbol{a}^3q^{-9}t^0 + \boldsymbol{a}^3q^{-7}t^2 + \boldsymbol{a}^4q^{-8}t^3$ which, according to the Conjecture 1.1, should be equal to the homology $(\mathcal{H}^{\yng(1,1,1)}(3_1), d_4^{\yng(1,1,1)})$. In particular, for $\boldsymbol{a} = q^4$ it implies that the Poincaré polynomial of $(\mathcal{H}^{\yng(1,1,1)}(3_1), d_4^{\yng(1,1,1)})$ is equal to $q^3 + q^5t^2 + q^8t^3$. The latter, in turn, is equal to $\mathcal{P}^{\yng(1)}(3_1)(\boldsymbol{a} = q^4, q, t)$ (see (A.1)), in agreement with the isomorphism $\Lambda^3 \cong \Lambda$ for $sl(4)$.

In Appendix C we compute $\mathcal{H}^3$ of the figure-eight knot and show that it satisfies all of the properties from conjecture 5.1.

It is straightforward to check that the triply-graded colored HOMFLY homologies of other knots that we computed in Section 4 satisfy all the required properties of $\mathcal{H}^2$.

As for the t-grading change in (5.11) for an arbitrary knot, ϕ sends a generator x of (a,q,t)-degree (i,j,k) to a generator of degree $(i,-j,4i-k+2r\delta'(x))$. Here $\delta'(x)$ is a certain grading of the generator x, generalizing the δ-grading of the ordinary HOMFLY homology $\mathcal{H}^{\yng(1)}$. In the case of thin knots, δ'-grading of all generators is equal to the S-invariant of knots.

Now, since the t-gradings of x and $\phi(x)$ have the same parity, by decategorifying (5.11) we get the following simple and beautiful relation between the colored HOMFLY polynomials:

$$P^{S^r}(K)(\boldsymbol{a},q) = P^{\Lambda^r}(K)(\boldsymbol{a},q^{-1}). \tag{5.17}$$

To the best of our knowledge, this relation has not been observed before. It generalizes the symmetry $q \leftrightarrow q^{-1}$ of the (ordinary) HOMFLY polynomial.

Based on the above observations about the mirror symmetry for symmetric and anti-symmetric representations, we speculate that this symmetry extends to arbitrary representations: for a representation R of $sl(N)$ that corresponds to a partition λ, we conjecture:

$$\mathcal{H}^{\lambda}(K) \cong \mathcal{H}^{\lambda^t}(K)\,, \tag{5.18}$$

where λ^t is the dual (transpose) partition of λ. Here, we also tacitly assume that for every partition λ there exists a triply-graded homology theory $\mathcal{H}^{\lambda}(K)$ categorifying the λ-colored HOMFLY polynomial.

Furthermore, decategorifying the isomorphism (5.18) we obtain the following symmetry of the colored HOMFLY polynomials (see appendix A for conventions):

$$P^{\lambda}(K)(\boldsymbol{a},q) \;=\; P^{\lambda^t}(K)(\boldsymbol{a},q^{-1})\,. \tag{5.19}$$

5.3. Physical interpretation. The only evidence for the mirror symmetry (5.18) and for its polynomial version (5.19) comes from the physical interpretation of knot homologies / polynomials in terms of BPS invariants. Indeed, the symmetry (5.19) of the reduced colored HOMFLY polynomial is a direct consequence of the corresponding symmetry for the unreduced colored HOMFLY polynomial:

$$\overline{P}^{\lambda^t}(K)(\boldsymbol{a},q) \;=\; (-1)^{|\lambda|}\,\overline{P}^{\lambda}(K)(\boldsymbol{a},q^{-1})\,, \tag{5.20}$$

where $|\lambda|$ is the total number of boxes in the partition λ. This symmetry, in turn, follows[13] from the explicit form of the colored HOMFLY polynomial [**15**]:

$$\overline{P}^{\lambda}(K)(\boldsymbol{a},q) \;=\; \sum_{i,j} \frac{\boldsymbol{a}^i q^j}{q-q^{-1}} N^{\lambda}_{i,j}(K) \tag{5.21}$$

written in terms of the ordinary (that is, unrefined) BPS invariants $N^{\lambda}_{i,j}(K)$ and from the property of the integer BPS invariants [**26**, eq. (2.17)]:

$$N^{\lambda^t}_{i,j}(K) \;=\; (-1)^{|\lambda|+1}\,N^{\lambda}_{i,-j}(K)\,. \tag{5.22}$$

Indeed, when combined, (5.21) and (5.22) imply that under $q \leftrightarrow q^{-1}$ the unreduced colored HOMFLY polynomial of every knot has parity $(-1)^{|\lambda|}$. In particular, this is true for the unknot. Hence, the normalized colored HOMFLY polynomial, defined as the ratio of $\overline{P}^{\lambda}(K)$ and $\overline{P}^{\lambda}(\bigcirc)$, enjoys the symmetry (5.19).

There is a refined / homological version of (5.20) and (5.22) that leads to (5.18). Much as the colored HOMFLY polynomial can be written in terms of the unrefined

[13] Another way to derive (5.20) is to use the properties of the Clebsch-Gordon coefficients for the symmetric group $C_{\lambda\mu^t\nu^t} = C_{\lambda\mu\nu}$ and the characters $S_{\lambda^t}(q) = (-1)^{|\lambda|-1}S_{\lambda}(q^{-1})$ in eq.(2.6) of [**61**], which describes the geometric origin of the q-dependence in the colored HOMFLY polynomail.

BPS invariants $N^{\lambda}_{i,j}(K)$, the unnormalized superpolynomial of every knot K can be expressed in terms of the refined integer BPS invariants [4]:

$$\overline{\mathcal{P}}^{\lambda}(K)(\boldsymbol{a},q,t) \;=\; \sum_{i,j,k} \frac{\boldsymbol{a}^i q^j t^k}{q-q^{-1}} D^{\lambda}_{i,j,k}(K)\,. \tag{5.23}$$

The refined BPS invariants $D^{\lambda}_{i,j,k}(K)$ enjoy a symmetry that generalizes (5.22),

$$D^{\lambda^t}_{i,j,*}(K) \;=\; D^{\lambda}_{i,-j,*}(K)\,, \tag{5.24}$$

and follows from the CPT symmetry of the five-brane theory in (1.4) or (1.6). The normalized / reduced version of the symmetry (5.24) is precisely (5.18).

Further details and interpretation of the mirror symmetry (5.18) for the triply-graded knot homology will appear elsewhere.

6. Unreduced colored HOMFLY homology

Here we compute the unreduced colored superpolynomial and the colored HOMFLY polynomial of the unknot and the Hopf link by using the refined topological vertex approach from [9]. The formulas obtained there are partition functions, presented in the form of the quotient of two infinite series. Below we find the explicit closed form expressions for the unreduced S^r-colored HOMFLY homology of the unknot and the Hopf link. More precisely, we evaluate eq. (67) of [9], according to which the unreduced superpolynomial (= the Poincaré polynomial of the unreduced triply-graded colored homology) of the Hopf link with components colored by partitions λ and μ is given by:

$$\overline{\mathcal{P}}^{\lambda\mu}(\text{Hopf}) = (-1)^{|\lambda|+|\mu|}\left(\frac{q_1}{q_2}\right)^{|\lambda||\mu|}\left(Q^{-1}\sqrt{\frac{q_1}{q_2}}\right)^{\frac{|\lambda|+|\mu|}{2}} \times \frac{Z_{\lambda\mu}}{Z_{\emptyset\emptyset}}\,, \tag{6.1}$$

where

$$Z_{\lambda\mu} = \sum_{\nu}(-Q)^{|\nu|} q_2^{\frac{||\nu||}{2}} q_1^{\frac{||\nu^t||}{2}} \widetilde{Z}_{\nu}(q_1,q_2)\widetilde{Z}_{\nu^t}(q_2,q_1) s_{\lambda}(q_2^{-\rho}q_1^{-\nu^t}) s_{\mu}(q_2^{-\rho}q_1^{-\nu^t})\,.$$

The unreduced superpolynomial of the unknot colored by λ is obtained by setting $\mu = \emptyset$ in (6.1).

The change of variables from topological strings variables (Q,q_1,q_2) to knot theory variables $(\boldsymbol{a},q,t)$ used in this paper is given by:

$$\begin{aligned} \sqrt{q_2} &= q, \\ \sqrt{q_1} &= -tq, \\ Q &= -t\boldsymbol{a}^{-2}. \end{aligned} \tag{6.2}$$

In particular, the specialization $q_1 = q_2$ corresponds to the specialization $t = -1$ in the homological knot invariants.

By expanding the product of the Schur functions as

$$s_{\lambda}s_{\mu} = \sum_{\varphi} c^{\varphi}_{\lambda,\mu} s_{\varphi}\,,$$

where $c^{\varphi}_{\lambda,\mu}$ is the Littlewood-Richardson coefficient, we obtain:

$$\begin{aligned} Z_{\lambda\mu} &= \sum_{\varphi} c^{\varphi}_{\lambda,\mu} \sum_{\nu} (-Q)^{|\nu|} q_2^{\frac{||\nu||}{2}} q_1^{\frac{||\nu^t||}{2}} \widetilde{Z}_{\nu}(q_1,q_2) \widetilde{Z}_{\nu^t}(q_2,q_1) s_{\varphi}(q_2^{-\rho} q_1^{-\nu^t}) \\ &= \sum_{\varphi} c^{\varphi}_{\lambda,\mu} Z_{\varphi} \,. \end{aligned}$$

Replacing this in (6.1) gives

$$\begin{aligned} \overline{\mathcal{P}}^{\lambda\mu}(\mathrm{Hopf}) &= (-1)^{|\lambda|+|\mu|} \left(\frac{q_1}{q_2}\right)^{|\lambda||\mu|} \left(Q^{-1}\sqrt{\frac{q_1}{q_2}}\right)^{\frac{|\lambda|+|\mu|}{2}} \times \frac{Z_{\lambda\mu}}{Z_{\emptyset\emptyset}} = \\ &= (-1)^{|\lambda|+|\mu|} \left(\frac{q_1}{q_2}\right)^{|\lambda||\mu|} \left(Q^{-1}\sqrt{\frac{q_1}{q_2}}\right)^{\frac{|\lambda|+|\mu|}{2}} \sum_{\varphi} c^{\varphi}_{\lambda,\mu} \frac{Z_{\varphi}}{Z_{\emptyset\emptyset}} = \\ &= \left(\frac{q_1}{q_2}\right)^{|\lambda||\mu|} \sum_{\varphi} c^{\varphi}_{\lambda,\mu} (-1)^{|\varphi|} \left(Q^{-1}\sqrt{\frac{q_1}{q_2}}\right)^{\frac{|\varphi|}{2}} \frac{Z_{\varphi}}{Z_{\emptyset\emptyset}} = \\ &= \left(\frac{q_1}{q_2}\right)^{|\lambda||\mu|} \sum_{\varphi} c^{\varphi}_{\lambda,\mu} \overline{\mathcal{P}}^{\varphi}(\bigcirc) \,. \end{aligned}$$

Equivalently, in the knot theory variables $(\boldsymbol{a}, q, t)$ we found the following simple formula for the superpolynomial of the Hopf link expressed in terms of that of the unknot:

$$\overline{\mathcal{P}}^{\lambda\mu}(\mathrm{Hopf}) = t^{2|\lambda||\mu|} \sum_{\varphi} c^{\varphi}_{\lambda,\mu} \overline{\mathcal{P}}^{\varphi}(\bigcirc) \,. \tag{6.3}$$

Thus, in order to compute the unreduced superpolynomial of the Hopf link, it suffices to compute the superpolynomial of the unknot from (6.1).

6.1. Unreduced colored HOMFLYPT polynomial and homology of the unknot. Below we give the results for the unknot derived from (6.1). The notations and computations are summarized in Appendix D.

The quantum $sl(N)$ invariant (that is, $\boldsymbol{a} = q^N$ specialization of the colored HOMFLY polynomial) is given by:

$$\begin{aligned} \overline{P}^{\lambda}(\bigcirc)(\boldsymbol{a} = q^N, q) &= q^{-2\sum_{x\in\lambda} c(x)} \begin{bmatrix} N \\ \lambda^t \end{bmatrix} = \\ &= q^{-2(n(\lambda^t)-n(\lambda))} \begin{bmatrix} N \\ \lambda^t \end{bmatrix} = q^{-\kappa(\lambda)} \begin{bmatrix} N \\ \lambda^t \end{bmatrix} . \end{aligned}$$

In particular, for the r-th symmetric representation S^r we find

$$\overline{P}^{S^r}(\bigcirc)(\boldsymbol{a} = q^N, q) = q^{-r(r-1)} \begin{bmatrix} N+r-1 \\ r \end{bmatrix} , \tag{6.4}$$

whereas for the anti-symmetric representation $R = \Lambda^r$ we have

$$\overline{P}^{\Lambda^r}(\bigcirc)(\boldsymbol{a} = q^N, q) = q^{r(r-1)} \begin{bmatrix} N \\ r \end{bmatrix} . \tag{6.5}$$

The two-variable polynomial $\overline{P}^{\lambda}(\bigcirc)(\boldsymbol{a}, q)$ can be obtained from the above expressions by replacing q^N with $\boldsymbol{a}$ and the q-binomial coefficients by two-variable polynomials in the following way:

$$\begin{bmatrix} N \\ r-j \end{bmatrix} \leftrightarrow \frac{(-1)^{r-j}\boldsymbol{a}^{r-j}q^{r-j}}{(1-q^2)(1-q^4)\dots(1-q^{2r})}\sum_{l=0}^{r}(-1)^l\boldsymbol{a}^{-2l}q^{l(r-j-1)} \times \begin{bmatrix} r-j \\ l \end{bmatrix}(1-q^{2(r-j+1)})\cdots(1-q^{2r}), \tag{6.6}$$

where the left hand side is the $\boldsymbol{a}=q^N$ specialization of the right hand side. This last formula follows from

$$\prod_{i=0}^{n-1}(1+q^{2i}z) = \sum_{j=0}^{n} q^{j(n-1)}\begin{bmatrix} n \\ j \end{bmatrix} z^j .$$

In particular, for the symmetric representation we have:

$$\overline{P}^{S^r}(\bigcirc)(\boldsymbol{a},q) = \frac{(-1)^r\boldsymbol{a}^r q^r}{(1-q^2)(1-q^4)\dots(1-q^{2r})}q^{-2r(r-1)}\sum_{l=0}^{r}(-1)^l\boldsymbol{a}^{-2l}q^{l(r-1)} \times \begin{bmatrix} r \\ l \end{bmatrix} q^{2(r-l)(r-1)}. \tag{6.7}$$

Now, the formula for the $\boldsymbol{a}=q^N$ specialization of the S^r-colored superpolynomial for the unknot is obtained by using the following quantum binomial coefficients formula:

$$\begin{bmatrix} N+r-1 \\ r \end{bmatrix} = q^{r(r-1)}\sum_{j=0}^{r-1} q^{-j(N+r-1)}\begin{bmatrix} r-1 \\ j \end{bmatrix}\begin{bmatrix} N \\ r-j \end{bmatrix}. \tag{6.8}$$

Then, the Poincaré polynomial of the S^r-colored $sl(N)$ homology of the unknot is obtained by adding a factor t^{-2j} in every summand in the above expression for the quantum binomial coefficient in (6.4):

$$\begin{aligned} \overline{\mathcal{P}}^{sl(N),S^r}(\bigcirc)(q,t) &= \overline{\mathcal{P}}^{S^r}(\bigcirc)(\boldsymbol{a}=q^N,q,t) = \\ &= \sum_{j=0}^{r-1} q^{-j(N+r-1)}\begin{bmatrix} r-1 \\ j \end{bmatrix}\begin{bmatrix} N \\ r-j \end{bmatrix} t^{-2j}. \end{aligned} \tag{6.9}$$

Note that the corresponding homology $\overline{\mathcal{H}}^{sl(N),S^r}(\bigcirc)$ is finite-dimensional.

We list some particular instances of (6.9) for small r:

$$\overline{\mathcal{P}}^{sl(N),S^1}(\bigcirc) = [N], \tag{6.10}$$

$$\overline{\mathcal{P}}^{sl(N),S^2}(\bigcirc) = \begin{bmatrix} N \\ 2 \end{bmatrix} + q^{-(N+1)}[N]t^{-2}, \tag{6.11}$$

$$\overline{\mathcal{P}}^{sl(N),S^3}(\bigcirc) = \begin{bmatrix} N \\ 3 \end{bmatrix} + q^{-(N+2)}[2]\begin{bmatrix} N \\ 2 \end{bmatrix} t^{-2} + q^{-2(N+2)}[N]t^{-4}. \tag{6.12}$$

Specifying further the value of N, one finds the following expressions:

$$\begin{aligned}
\overline{\mathcal{P}}^{sl(2),S^2}(\bigcirc) &= 1+q^{-2}t^{-2}+q^{-4}t^{-2}, \\
\overline{\mathcal{P}}^{sl(3),S^2}(\bigcirc) &= q^2+1+q^{-2}+q^{-2}t^{-2}+q^{-4}t^{-2}+q^{-6}t^{-2}, \\
\overline{\mathcal{P}}^{sl(3),S^3}(\bigcirc) &= 1+q^{-2}t^{-2}+2q^{-4}t^{-2}+2q^{-6}t^{-2}+q^{-8}t^{-2}+q^{-8}t^{-4}+q^{-10}t^{-4} \\
&\quad +q^{-12}t^{-4}, \\
\overline{\mathcal{P}}^{sl(4),S^2}(\bigcirc) &= q^4+q^2+2+q^{-2}+q^{-4}+q^{-2}t^{-2}+q^{-4}t^{-2}+q^{-6}t^{-2}+q^{-8}t^{-2}.
\end{aligned}$$

The expression for the whole triply-graded superpolynomial is obtained from (6.9) by using (6.6).

In the case of anti-symmetric representations, the entire homology of the unknot is concentrated in the homological degree zero, and thus the Λ^r superpolynomial of the unknot coincides with its Λ^r-colored HOMFLY polynomial:

$$\overline{\mathcal{P}}^{sl(N),\Lambda^r}(\bigcirc) = q^{r(r-1)} \begin{bmatrix} N \\ r \end{bmatrix}. \tag{6.13}$$

6.2. Comparison with other approaches. Much of the present paper is devoted to exploring the structure — motivated from physics — of the colored knot homology, namely its reduced version. A combinatorial or group theoretic definition of such theory is still waiting to be discovered. However, in the case of the unreduced theory, which we sketched in this section, there have been several attempts to define the colored knot homology, especially in low rank. Therefore, we conclude this section with a brief comparison to other approaches.

Unfortunately, the structure of the colored differentials becomes more obscure (alternatively, more interesting!) in the unreduced version of the colored knot homology.[14] This, in part, is the reason why we kept our discussion here very brief, relegating a more thorough analysis to future work. Another reason, which will become clear in a moment, is that even a quick look at the unknot exposes a number of questions that need to be understood in order to relate and unify different formulations:

- singularities in moduli spaces (of BPS configurations): $\dim \mathcal{H}_{\text{BPS}} < \infty$ versus $\dim \mathcal{H}_{\text{BPS}} = \infty$
- framing dependence in the colored knot homology
- colored homological invariants versus cabling
- analog of wall crossing phenomena in mathematical formulations of colored knot homologies
- the role of the "preferred direction" in the combinatorial formulation based on 3d partitions
- proper interpretation of formal expressions, or

$$\frac{1-q^2}{1-q^2} = (1-q^2)(1+q^2+q^4+q^6+\ldots) \qquad \text{versus} \qquad \frac{1-q^2}{1-q^2} = 1 \tag{6.14}$$

[14]This is familiar from the ordinary, non-colored knot homology [**3, 6**].

In addition, each formulation typically involves individual choices and subtleties, which may also affect the form of the answer. In fact, even the total dimension of the colored homology may depend on some (or, perhaps, all) of these choices.[15]

While good understanding of these aspects is still lacking, many approaches to colored knot homology seem to agree on one general feature: the unreduced $sl(N)$ homology has finite support only for certain sufficiently small representations. For example, in [**9**, eq. (67)] this corresponds to the fact that for general representations there is no way to clear the denominators. This should be compared with the fundamental representation of $sl(N)$, where every existent approach leads to a homology with finite support. The simplest example that belongs to the "grey territory" is the second symmetric representation $R = S^2$ of $sl(N)$. For $N = 2$, this corresponds to the adjoint representation of $sl(2)$ and, as we saw in (3.11), physics realizations [**4, 12**] lead to a 3-dimensional knot homology $\overline{\mathcal{H}}^{sl(2),\square\square}(\bigcirc)$ categorifying the colored Jones polynomial of the unknot,

$$\overline{P}_2^{\square\square}(\bigcirc) \;=\; [3] \;=\; q^{-2} + 1 + q^2\,. \tag{6.15}$$

On the other hand, some mathematical formulations lead to a theory with infinite support (which can be attributed to several gaps in the present understanding and the above-listed questions). For example, fixing[16] a typo in [**62**, Proposition 3.4], one finds the following candidate for the Poincaré polynomial of the colored unknot homology:

$$\overline{\mathcal{P}}_2^{\square\square}(\bigcirc) \;=\; q^{-2}t^2 + 1 + q^2t^{-2} + \frac{q^{-2}+q^{-2}t}{1-t^2q^{-4}}\,. \tag{6.16}$$

The structure of the corresponding homology theory is clear: the first three terms reproduce (upon specializing to $t=-1$) the colored Jones polynomial (6.15) and the quotient in the last term corresponds to the infinite-dimensional contribution to the homology, all of which disappears upon taking the Euler characteristic.

Similar structure emerges in other frameworks, in particular in approaches based on categorification of the Jones-Wenzl projectors. The Jones-Wenzl projectors appear in decomposing the finite dimensional representations of the quantum group $U_q(\mathfrak{sl}_2)$ and, as such, play a key role in the definition and computation of quantum group invariants of knots and 3-manifolds. Several ways to categorify the Jones-Wenzl projectors have been proposed in the literature, *e.g.* the topological categorification [**63**] and the Lie theoretic categorification [**64**] which agree (up to Koszul duality). In particular, the latter approach leads to a theory that categorifies (6.15) by replacing the middle term with infinite-dimensional homology whose Poincaré polynomial equals

$$\overline{\mathcal{P}}_2^{\square\square}(\bigcirc) \;=\; q^{-2} + \frac{1}{[2][2]}(q+q^{-1})^2 + q^2\,, \tag{6.17}$$

where $[2] = q+q^{-1}$ and the authors of [**64**] instruct us to interpret $\frac{1}{[2]}$ as a power series $q-q^3+q^5-q^7+\ldots$. This power series is familiar to physicists as a trace ("partition function") over the infinite-dimensional Hilbert space $\mathcal{H}_{\text{Bose}} = H^*(\mathbb{C}\mathbf{P}^\infty) =$

---

[15] We hope that at least some of these delicate aspects are washed away when one passes to the reduced theory, as it happens in the non-colored case. This is one of the reasons why in the present paper we mainly consider the reduced homology.

[16] We thank E. Gorsky for pointing this out.

$\mathbb{C}[x]$ of a harmonic oscillator / single boson,

$$P_{\text{Bose}} = \frac{1}{1-q^2} = 1 + q^2 + q^4 + q^6 + \dots . \tag{6.18}$$

Partition function of a single fermion has a similar form, except that fermions contribute to the numerator instead of the denominator. Indeed, the trace over a two-dimensional Hilbert space of a single fermion looks like

$$P_{\text{Fermi}} = 1 - q^2 , \tag{6.19}$$

in agreement with a well-known fact that contributions of bosons and fermions cancel each other, *cf.* (6.14). Therefore, instead of canceling the ratio in the middle term of (6.17), the authors of [**64**] instruct us to interpret it as a Hilbert space of two bosons and two fermions. Note, due to the presence of bosonic states this Hilbert space is infinite-dimensional, as opposed to a much smaller, finite-dimensional space that one might infer by simplifying the ratio. Similarly, (6.16) contains one boson (due to the factor $\frac{1}{1-t^2q^{-4}}$ in the last term), *etc.*

If this, however, is the proper interpretation of (6.17), then one immediately runs into a general question of how to interpret formal expressions like (6.14) and when to clear denominators. The answer to this question will certainly affect many calculations of Poincaré polynomials, in particular calculations based on [**9**, eq. (67)] that has non-trivial numerators and denominators, as well as similar calculations in other frameworks.

A novel physical framework that appears to be closely related to knot homology is the so-called "refined Chern-Simons theory." Although Lagrangian definition of this theory is not known at present, its partition function was conjectured [**58**] to compute topological invariants of knots and 3-manifolds that preserve an extra rotation symmetry. This includes torus knots and Seifert 3-manifolds. The rotation symmetry gives rise to an extra quantum number, so that for torus knots and Seifert 3-manifolds the refined Chern-Simons theory leads to a striking prediction: the space (1.2) is quadruply-graded rather than triply-graded in these cases.

In simple examples, the fourth grading (coming from the extra rotation symmetry of a 3-manifold) is determined by the other three gradings (2.3). It would be interesting to study under which conditions this happens; when it does, the partition function of the $SU(N)$ refined Chern-Simons theory computes the specialization of the superpolynomial to $\boldsymbol{a} = q^N$. Assuming this is the case for the unknot colored by the second symmetric representation, the $SU(2)$ refined Chern-Simons theory gives:

$$\overline{\mathcal{P}}_2^{\square\square}(\bigcirc) = - \frac{(q^2t + q^{-2})}{(q - q^{-1})} \frac{(q^4t^2 + q^{-2}t^{-1})}{(q^3t^2 - q^{-1})} . \tag{6.20}$$

The corresponding Hilbert space contains at least two "bosons" (due to two factors in the denominator of (6.20)) and, therefore, leads to a version of colored homology with infinite support.

In our quick tour through different ways of categorifying the colored Jones polynomial of the unknot (6.15) we saw theories with finite support as well as theories with infinite support, in fact, of different kind (with different number of "bosons" / factors in the denominator). One would hope that all these theories correspond to different choices (of framing, chamber, regularization, ...) and with a proper understanding of the above-mentioned issues could be unified in a single framework. One piece of evidence that it might be possible comes from the fact that all physical and geometrical approaches agree when the corresponding moduli spaces are non-singular, as *e.g.* for minuscule representations. Therefore, we hope to see a much bigger story, only small elements of which have been revealed so far.

Acknowledgments We would like to thank S. Cautis, M. Marino, K. Schaeffer, Y. Soibelman, C. Stroppel, C. Vafa, J. Walcher, and E. Witten for valuable discussions. The work of SG is supported in part by DOE Grant DE-FG03-92-ER40701 and in part by NSF Grant PHY-0757647. Opinions and conclusions expressed here are those of the authors and do not necessarily reflect the views of funding agencies. MS was partially supported by the Portuguese Fundação para a Ciência e a Tecnologia through ISR/IST plurianual funding and through the project number PTDC/MAT/101503/2008, New Geometry and Topology. MS was also partially supported by the Ministry of Science of Serbia, project no. 174012.

Appendix A. Notations

K denotes a knot. $\bigcirc$ denotes the unknot.

For every positive integer r we have:

• $\overline{P}_N^{S^r}(K)(q)$ or $\overline{P}_N^{\overbrace{\square\square\cdots\square}^{r}}(K)(q)$ denotes the unreduced one-variable polynomial quantum invariant of K, labelled with the r-th symmetric representation of $sl(N)$.

• $P_N^{S^r}(K)(q)$ or $P_N^{\overbrace{\square\square\cdots\square}^{r}}(K)(q)$ denotes the reduced (a.k.a. normalized) one-variable polynomial quantum invariant of K, labeled by the r-th symmetric representation of $sl(N)$. It is obtained from the unnormalized polynomial $\overline{P}_N^{S^r}(K)(q)$ by

$$P_N^{S^r}(K) = \frac{\overline{P}_N^{S^r}(K)}{\overline{P}_N^{S^r}(\bigcirc)},$$

so that $P_N^{S^r}(\bigcirc) = 1$.

• $P^{S^r}(K)(a,q)$ or $P^{\overbrace{\square\square\cdots\square}^{r}}(K)(\boldsymbol{a},q)$ denotes the reduced two-variable colored HOMFLY polynomial of K. The normalization is $P^{S^r}(\bigcirc) = 1$. In particular

$$P^{S^r}(K)(\boldsymbol{a} = q^N, q) = P_N^{S^r}(K)(q)$$

• $\mathcal{H}^{S^r}(K)$ or $\mathcal{H}^{\overbrace{\square\square\cdots\square}^{r}}(K)$ denotes a reduced triply-graded homology theory $\mathcal{H}^{S^r}_{i,j,k}(K)$ that categorifies the two-variable colored HOMFLY polynomial $P^{S^r}(K)$:

$$\chi(\mathcal{H}^{S^r}(K)) = P^{S^r}(K).$$

The homological grading of $\mathcal{H}^{S^r}_{i,j,k}(K)$ is its third grading, and its (doubly-graded) Euler characteristic is given by:

$$\chi(\mathcal{H}^{S^r}(K)) = \sum_{i,j,k} (-1)^k q^i t^j (\dim \mathcal{H}^{S^r}_{i,j,k}(K)).$$

• $\mathcal{P}^{S^r}(K)$ or $\mathcal{P}^{\overbrace{\square\square\cdots\square}^{r}}(K)$ denotes the Poincaré polynomial of $\mathcal{H}^{S^r}(K)$, *i.e.*

$$\mathcal{P}^{S^r}(K)(\boldsymbol{a},q,t) \;=\; \sum_{i,j,k} \boldsymbol{a}^i q^j t^k (\dim \mathcal{H}^{S^r}_{i,j,k}(K)).$$

Specialization to $t=-1$ gives $P^{S^r}(K)$:

$$\mathcal{P}^{S^r}(K)(\boldsymbol{a},q,t=-1) \;=\; P^{S^r}(K)(\boldsymbol{a},q)\,.$$

We also call $\mathcal{P}^{S^r}(K)$ the S^r*-colored superpolynomial.*

• $\mathcal{H}^{sl(N),S^r}(K)$ denotes the reduced doubly-graded homology theory categorifying $P_N^{S^r}(K)$.

• $\mathcal{P}_N^{S^r}(K)$ denotes the Poincaré polynomial of $\mathcal{H}^{sl(N),S^r}(K)$. In particular

$$\mathcal{P}_N^{S^r}(K)(q,t) \;=\; \sum_{j,k} q^j t^k (\dim \mathcal{H}^{sl(N),S^r}_{j,k}(K))\,.$$

$$\mathcal{P}_N^{S^r}(K)(q,t=-1) \;=\; P_N^{S^r}(K)(q)\,.$$

• The corresponding polynomials and homologies for the anti-symmetric representations are denoted in the same way with S^r replaced by Λ^r.

• Unreduced versions of the homology and polynomials are denoted by putting a bar.

All (tri-)degrees are $(\boldsymbol{a},q,t)$-degrees.

REMARK A.1. **Conventions for the superpolynomial in the vector representation:** The case $r=1$ corresponds to the (ordinary) HOMFLY polynomial and the Khovanov-Rozansky homology. The corresponding superpolynomial, together with the structure of the triply-graded homology was studied in [**3**].

Here we shall use slightly different conventions: in the superpolynomial, we replace $\boldsymbol{a}$ and q from [**3**] with $\boldsymbol{a}^{1/2}$ and $q^{1/2}$, respectively. For example, the superpolynomial of the trefoil knot becomes:

$$\mathcal{P}^{\square}(3_1) = \boldsymbol{a}q^{-1} + \boldsymbol{a}qt^2 + \boldsymbol{a}^2t^3. \tag{A.1}$$

Also, by $S(K)$, or just S, we mean half of the value of $S(K)$ from [**3**].

This way, the degrees of canceling differentials d_1 and d_{-1} from [**3**] become $(-1,1,-1)$ and $(-1,-1,-3)$, respectively. Also, from now on we denote these two differentials by $d_1^{\square}$ and $d_{-1}^{\square}$, respectively.

Appendix B. Kauffman and S^2 homologies of the knots 8_{19} and 9_{42}

In this appendix we compute the S^2 and Kauffman homologies of 8_{19} and 9_{42}. Since both knots are homologically thick knots, these computations give highly non-trivial examples of our main Conjecture 4.2 (as well as Conjecture 2 of [**6**] for the Kauffman homology), especially because the size of both homologies is rather large and because they have to obey a large list of structural properties described in sections 4 and 5.

Before we start our computations, we point out that knot 9_{42} here is the mirror image of 9_{42} from [**56**]. The superpolynomial of 9_{42} is given by[17] [**3**]:

(B.1) $\mathcal{P}^{\square}(9_{42})(\boldsymbol{a},q,t) = \boldsymbol{a}(q^{-1}t^2+qt^4)+(q^{-2}t^{-1}+1+2t+q^2t^3)+\boldsymbol{a}^{-1}(q^{-1}t^{-2}+q).$

In particular, the reduced $sl(2)$ Khovanov homology of 9_{42} is:

$$\begin{aligned} Kh(9_{42})(q,t) &= \mathcal{P}^{\square}(9_{42})(\boldsymbol{a}=q^2,q,t) = \\ &= q^{-6}t^{-2}+q^{-4}t^{-1}+q^{-2}+1+2t+q^2t^2+q^4t^3+q^6t^4\,. \end{aligned}$$

The S invariant is $S(9_{42}) = 0$. Moreover, the δ-grading of a generator x of the homology $\mathcal{H}^{\square}(K)$ in our conventions is given by:

$$\delta(x) = t(x) - 2\boldsymbol{a}(x) - q(x).$$

All generators of the homology of thin knots have the same value of the δ-grading. However, for 9_{42}, the generator $1(=\boldsymbol{a}^0q^0t^0)$ has δ-grading 0, while the remaining 8 generators have δ-grading equal to -1.

As for 8_{19}, here it is the mirror image of 8_{19} from the Knot Atlas [**56**]. This knot is also known as the positive (3,4)-torus knot. Its superpolynomial is given by:

$$\begin{aligned} \mathcal{P}^{\square}(8_{19})(\boldsymbol{a},q,t) &= \boldsymbol{a}^3q^{-3}+\boldsymbol{a}^4q^{-2}t^3+\boldsymbol{a}^3q^{-1}t^2+\boldsymbol{a}^4t^5+\boldsymbol{a}^3qt^4+\boldsymbol{a}^4q^2t^7+\boldsymbol{a}^4q^3t^6 \\ &\quad +\boldsymbol{a}^5t^8+\boldsymbol{a}^4q^{-1}t^5+\boldsymbol{a}^4q^1t^7+\boldsymbol{a}^3t^4. \end{aligned}$$

The first seven generators have δ-degree equal to -3, while the remaining four have δ-grading equal to -2.

Before explaining the result for the S^2-homology, we first consider the Kauffman homologies.

B.0.1. *Kauffman homology of* 8_{19} *and* 9_{42}. The Kauffman polynomial of 9_{42} can be written as:

(B.2)

$$F(9_{42})(\boldsymbol{a},q) = 1+(1-\boldsymbol{a}^{-1}q)(1+\boldsymbol{a}^{-1}q^{-1})(1-\boldsymbol{a}^{-2})\cdot\boldsymbol{a}^2(q^{-6}-q^{-4}+q^{-2}+q^2-q^4+q^6).$$

The Kauffman homology of 9_{42} that we have computed has 209 generators. We present its Poincaré polynomial in a structured form:

$$\begin{aligned} \mathcal{P}^{\text{Kauff}}(9_{42})(\boldsymbol{a},q,t) &= 1+(1+\boldsymbol{a}^{-1}qt^{-1})(1+\boldsymbol{a}^{-1}q^{-1}t^{-2})(1+\boldsymbol{a}^{-2}t^{-3})\times \\ &\quad\times\{\boldsymbol{a}^2(q^{-6}t^2+q^{-4}t^3+q^{-2}t^4+q^2t^6+q^4t^7+q^6t^8)+ \\ &\quad\quad+(1+t)\left[(\boldsymbol{a}^3+\boldsymbol{a}t^{-3})(q^{-3}t^4+q^{-1}t^5+qt^6+q^3t^7)\right. \\ &\quad\quad\left.+2\boldsymbol{a}^2t^4\right]\} \end{aligned}$$

The Kauffman polynomial of 8_{19} can be written as:

$$F(8_{19})(\boldsymbol{a},q) = (\boldsymbol{a}^6-\boldsymbol{a}^8)(q^{-6}+q^{-2}+1+q^2+q^6)+(\boldsymbol{a}^7-\boldsymbol{a}^9)(q^{-5}-q^5)+\boldsymbol{a}^{10}.$$

---

[17]Note that the $\boldsymbol{a}$ and q gradings that we are using in this paper are half of those from [**3**], see Remark A.1

The Kauffman homology of 8_{19} that we have computed has 89 generators. Its Poincaré polynomial is given by:

$$\begin{aligned}\mathcal{P}^{\text{Kauff}}(8_{19})(\boldsymbol{a},q,t) \;=\;& (\boldsymbol{a}^6+\boldsymbol{a}^8t^3)(q^{-6}+q^{-2}t^2+t^4+q^2t^4+q^6t^6)+\\ &+(\boldsymbol{a}^7+\boldsymbol{a}^9t^3)(q^{-5}t^2+q^5t^7)+\boldsymbol{a}^{10}t^{10}+\\ &+(1+t^{-1})(\boldsymbol{a}^7q^{-1}t^5+\boldsymbol{a}^7q^1t^6+\boldsymbol{a}^8t^7+\boldsymbol{a}^9q^{-1}t^8+\boldsymbol{a}^9q^1t^9)+\\ &+(1+t^{-1})(1+\boldsymbol{a}^{-1}qt^{-1})(1+\boldsymbol{a}^{-2}t^{-3})(1+\boldsymbol{a}^{-1}q^{-1}t^{-2})\times\\ &\quad\times(\boldsymbol{a}^{13}q^{-1}t^{14}+\boldsymbol{a}^{13}q^1t^{15}+\boldsymbol{a}^{11}q^{-3}t^{10}+\boldsymbol{a}^{11}q^3t^{13}).\end{aligned}$$

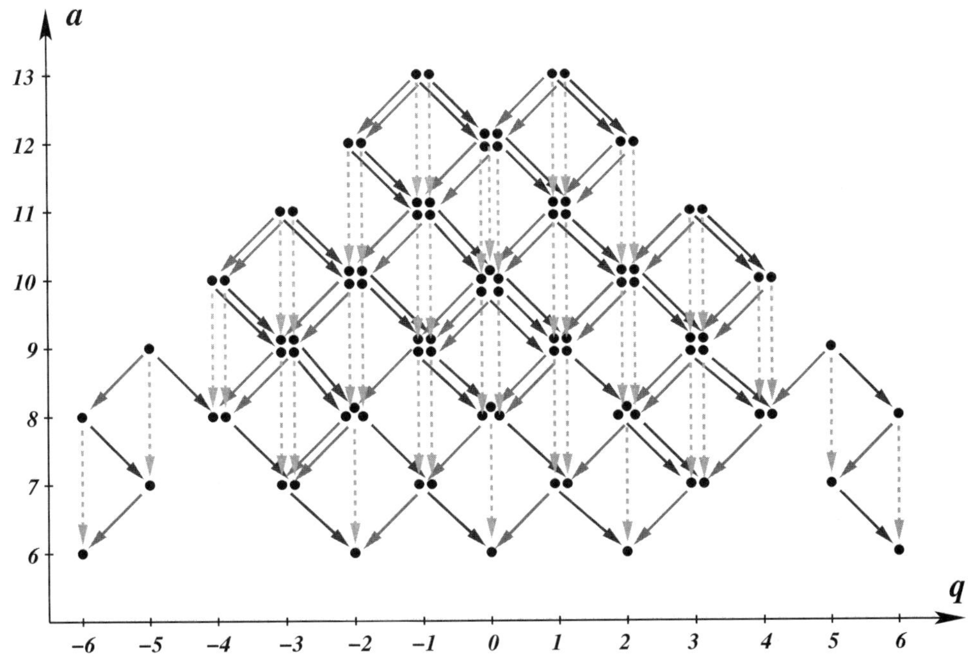

FIGURE 5. The reduced Kauffman homology of the knot $T_{3,4} = 8_{19}$. To avoid clutter, we show only canceling differentials d_2, d_1 and d_0 (represented by red, green, and blue arrows, respectively).

Both results meet the desired properties of the Kauffman homology (see Section 6 of [**6**]):

- Specialization to $t=-1$ gives the Kauffman polynomial:
$$\mathcal{P}^{\text{Kauff}}(\boldsymbol{a},q,t=-1)=F(\boldsymbol{a},q).$$
- There exist three canceling differentials d_2, d_1 and d_0 of degrees $(-1,1,-1)$, $(-2,0,-3)$ and $(-1,-1,-2)$, respectively. Indeed, from the form we write $\mathcal{P}^{\text{Kauff}}(9_{42})$, it is obvious that only $1=\boldsymbol{a}^0q^0t^0$ survives in the homology with respect to any of these differentials. As for $\mathcal{H}^{\text{Kauff}}(8_{19})$, the surviving generators for d_2, d_1 and d_0 respectively have degrees $(6,-6,0)$, $(12,0,12)$ and $(6,6,6)$, as expected since the S-invariant for 8_{19} is $S(8_{19})=6$.
- There exist two universal differentials $d_{\rightarrow}$ and $d_{\leftarrow}$ of degrees $(0,2,1)$ and $(0,-2,-1)$, such that the homology with respect to these differentials is

isomorphic (up to regrading) to the triply-graded HOMFLY homology. More precisely, they satisfy:

$$(\mathcal{H}^{\text{Kauff}}, d_{\rightarrow}) = \mathcal{P}^{\square}(\boldsymbol{a}^2q^{-2}, q^2, t),$$
$$(\mathcal{H}^{\text{Kauff}}, d_{\leftarrow}) = \mathcal{P}^{\square}(\boldsymbol{a}^2q^2t^2, q^2, t).$$

Again, it is straightforward to check that such differentials exist in both homologies $\mathcal{H}^{\text{Kauff}}(9_{42})$ and $\mathcal{H}^{\text{Kauff}}(8_{19})$.

- There exists a differential d_{-2} of degree $(-1,-3,-3)$, such that the $\boldsymbol{a}=q^{-3}$ specialization of the homology $(\mathcal{H}^{\text{Kauff}}, d_{-2})$ is isomorphic to the homology $\mathcal{H}^{sp(2),V}$.

  Moreover, the triply-graded version holds: the Poincaré polynomial of the triply-graded homology $(\mathcal{H}^{\text{Kauff}}, d_{-2})$ is equal to $t^S R(a^{1/2}q^{-1/2}, t)$, where $R(q,t) = \mathcal{P}^{\square}(\boldsymbol{a}=q^2, q, t)$. This is true for 8_{19}, 9_{42} and for all prime knots with up to 6 crossings.

  Note that this generalizes and corrects the value predicted in [**6**]. We also note that $(\mathcal{H}^{\text{Kauff}}, d_{-2})$ is significantly smaller than $\mathcal{H}^{\text{Kauff}}$: for 8_{19} it has only 11 generators, and for 9_{42} it has only 9 generators.

- Finally, although not explicitly stated in [**6**], the Kauffman homology enjoys the symmetry $q \leftrightarrow q^{-1}$:

$$\mathcal{H}^{\text{Kauff}}_{i,j,*} \cong \mathcal{H}^{\text{Kauff}}_{i,-j,*}\,.$$

B.0.2. *S^2-colored homologies of 8_{19} and 9_{42}.* In order to present the Poincaré polynomial of the S^2-colored homology of 9_{42} in a nice form and to show that all the expected properties are satisfied, we write it in the following structured form:

$$\begin{aligned}
&\mathcal{P}^{\square\square}(9_{42})(\boldsymbol{a},q,t) = \\
&= \{1 + (1+\boldsymbol{a}^{-1}qt^{-1})(1+\boldsymbol{a}^{-1}q^{-2}t^{-3})(1+\boldsymbol{a}^{-1}q^{-3}t^{-3})(1+\boldsymbol{a}^{-1}t^{-1})(\boldsymbol{a}^2q^6t^8+\boldsymbol{a}^2t^4)\} + \\
&\quad +(1+q)(1+\boldsymbol{a}^{-1}qt^{-1})(1+\boldsymbol{a}^{-1}q^{-2}t^{-3})(1+\boldsymbol{a}^{-1}q^{-3}t^{-3})(\boldsymbol{a}^2q^2t^6+\boldsymbol{a}q^3t^5) + \\
&\quad +(1+q)(1+\boldsymbol{a}^{-1}qt^{-1})(1+\boldsymbol{a}^{-1}q^{-2}t^{-3})(\boldsymbol{a}q^{-1}t^2+\boldsymbol{a}qt^4) + \\
&\quad +(1+q)(1+\boldsymbol{a}^{-1}qt^{-1})(1+\boldsymbol{a}^{-1}q^{-2}t^{-3})(1+\boldsymbol{a}^{-1}q^{-3}t^{-3}) \times \\
&\qquad\quad \times(\boldsymbol{a}^2q^2t^6+\boldsymbol{a}q^3t^6+\boldsymbol{a}q^3t^5+2\boldsymbol{a}qt^5+\boldsymbol{a}t^4+qt^4+t^3) + \\
&\quad +(1+q)(1+\boldsymbol{a}^{-1}qt^{-1})(1+\boldsymbol{a}^{-1}q^{-2}t^{-3})(1+\boldsymbol{a}^{-1}q^{-3}t^{-3})(1+t^{-1}) \times \\
&\qquad\quad \times(\boldsymbol{a}^2q^4t^8+\boldsymbol{a}q^4t^7+\boldsymbol{a}^2qt^6+\boldsymbol{a}q^2t^5+q^3t^5+\boldsymbol{a}t^4+t^3).
\end{aligned}$$

Similarly, for 8_{19} we have

$$\begin{aligned}
&\mathcal{P}^{\square\square}(8_{19})(\boldsymbol{a},q,t) = \\
&= \{(1+\boldsymbol{a}^{-1}qt^{-1})(1+\boldsymbol{a}^{-1}q^{-2}t^{-3})(1+\boldsymbol{a}^{-1}q^{-3}t^{-3})(\boldsymbol{a}^{10}q^8t^{16}+\boldsymbol{a}^9q^8t^{15}) + \\
&\quad +(1+\boldsymbol{a}^{-1}qt^{-1})(1+\boldsymbol{a}^{-1}q^{-2}t^{-3})(\boldsymbol{a}^8q^{-1}t^6+\boldsymbol{a}^8q^5t^{10}+\boldsymbol{a}^8q^{11}t^{14}) + \boldsymbol{a}^6q^{-6} + \\
&\quad +(1+\boldsymbol{a}^{-1}qt^{-1})(\boldsymbol{a}^7q^{-4}t^3+\boldsymbol{a}^7q^{-1}t^5+\boldsymbol{a}^7q^2t^7+\boldsymbol{a}^7q^5t^9+\boldsymbol{a}^7q^8t^{11}+\boldsymbol{a}^7q^{11}t^{13})\} + \\
&\quad +(1+q)(1+\boldsymbol{a}^{-1}qt^{-1})(1+\boldsymbol{a}^{-1}q^{-2}t^{-3})(\boldsymbol{a}^8q^2t^8+\boldsymbol{a}^8q^6t^{10}+\boldsymbol{a}^8q^8t^{12}+\boldsymbol{a}^8q^9t^{14}) + \\
&\quad +(1+q)(1+\boldsymbol{a}^{-1}qt^{-1})(1+\boldsymbol{a}^{-1}q^{-2}t^{-3})(1+\boldsymbol{a}^{-1}q^{-3}t^{-3})(\boldsymbol{a}^9q^3t^{11}+\boldsymbol{a}^9q^6t^{13}+\boldsymbol{a}^9q^9t^{15}).
\end{aligned}$$

The two homologies from above and their mirror images satisfy a large part of the properties of the S^2-colored homology from Conjecture 4.2 and of the Λ^2 homology from Section 5:

- There exist canceling differentials $d_1^{\yng(2)}$ and $d_{-2}^{\yng(2)}$ of degrees $(-1,1,-1)$ and $(-1,-2,-3)$, respectively. The remaining generator for both differentials is $\boldsymbol{a}^0 q^0 t^0 = 1$ in the case of 9_{42}, whereas for 8_{19} the remaining generators have degrees $(6,-6,0)$ and $(6,12,12)$, respectively.
- There exists colored differential $d_{2\to 1}$ of degree $(0,1,0)$, such that the homology with respect to it is equal to $\mathcal{P}^{\square}(\boldsymbol{a}^2, q^2, t^2 q)$.
- $\mathcal{P}^{\yng(2)}(\boldsymbol{a}, q, t=-1)$ is equal to the S^2-colored HOMFLY polynomial.
- There exists a differential $d_{-3}^{\yng(2)}$ of degree $(-1,-3,-3)$, such that the homology with respect to it is very small. In the case of 9_{42} it has only 9 generators:

$$(\mathcal{H}^{\yng(2)}(9_{42}), d_{-3}^{\yng(2)}) = \boldsymbol{a}(q^{-1}t^2+qt^4)+(q^{-2}t^{-1}+1+2t+q^2t^3)+\boldsymbol{a}^{-1}(q^{-1}t^{-2}+q), \tag{B.3}$$

while in the case of 8_{19} we have

$$\begin{aligned}(\mathcal{H}^{\yng(2)}(8_{19}), d_{-3}^{\yng(2)}) \;=\;& (\boldsymbol{a}^6q^{12}t^{12}+\boldsymbol{a}^7q^{11}t^{13}+\boldsymbol{a}^6q^{10}t^{10}+\boldsymbol{a}^7q^9t^{11}+\boldsymbol{a}^6q^8t^8+\boldsymbol{a}^7q^7t^9+\\ &+\boldsymbol{a}^6q^6t^6)+(\boldsymbol{a}^8q^{10}t^{14}+\boldsymbol{a}^7q^9t^{11}+\boldsymbol{a}^7q^{11}t^{13}+\boldsymbol{a}^6q^{10}t^{10}).\end{aligned} \tag{B.4}$$

Note that from the formulas for $\mathcal{P}^{\yng(2)}$, for both knots, the last two lines have a factor $(1+\boldsymbol{a}^{-1}q^{-3}t^{-3})$ and so the corresponding homology gets canceled automatically by $d_{-3}^{\yng(2)}$. Thus, it is enough to check the above formulas only for the remaining part, which is a straightforward computation.

- There exists a differential $d_0^{\yng(2)}$ of degree $(-1,0,-1)$ such that the homology with respect to it is equal to $\boldsymbol{a}^S\mathcal{P}^{\square}(\boldsymbol{a}, q^2, t)$.

- There exists a differential $d_2^{\yng(2)}$ of degree $(-1,2,-1)$ such that the homology with respect to it, after specializing $\boldsymbol{a}=q^2$, is isomorphic to $\mathcal{H}^{sl(2),\yng(2)}$. The latter one is isomorphic to $\mathcal{H}^{so(3),V}$, where V denotes the vector representation of $so(3)$, and to obtain its Poincare polynomial we use the result for the Kauffman homology we computed in section B.0.1. In particular, for both knots we have that

$$\mathcal{P}_2^{\yng(2)}(q,t) = \mathcal{P}^{so(3),V}(q,t) = \mathcal{P}^{\mathrm{Kauff}}(\boldsymbol{a}=q^2,q,t).$$

Now, the "mirror image" of $\mathcal{H}^{\yng(2)}$ is also behaving quite well. To that end, let $\mathcal{H}^{\yng(1,1)}$ be a homology obtained from $H^{\yng(2)}$ as in Section 5:

$$\mathcal{H}^{\yng(1,1)}_{i,j,k} \cong \mathcal{H}^{\yng(2)}_{i,-j,k'}.$$

The transformation $k \mapsto k'$ depends also on δ'-grading.

Since for $\mathcal{H}^{\yng(2)}$ of 9_{42} and 8_{19} the properties of the S^2-colored homology listed above are satisfied, it can be easily seen (by "mirroring" the differentials) that the mirror homology $\mathcal{H}^{\yng(1,1)}$ obtained in this way satisfies the properties of the anti-symmetric Λ^2-colored homology:

- There exist canceling differentials $d_{-1}^{\yng(1,1)}$ and $d_2^{\yng(1,1)}$ of degrees $(-1,-1,-3)$ and $(-1,2,-1)$, respectively.
- There exists colored differential $d_{2\to 1}^{\yng(1,1)}$ of degree $(0,1,0)$, such that the homology with respect to it has Poincaré polynomial equal to $\mathcal{P}^{\square}(\boldsymbol{a}^2, q^4, t^2q^{-1})$.
- $\mathcal{P}^{\yng(1,1)}(\boldsymbol{a}, q, t=-1)$ is equal to the Λ^2-colored HOMFLY polynomial.

- There exists a differential $d_3^{\begin{smallmatrix}\square\\\square\end{smallmatrix}}$ of degree $(-1,3,-1)$ such that the homology of $\mathcal{H}^{\begin{smallmatrix}\square\\\square\end{smallmatrix}}$ with respect to it is isomorphic to $\mathcal{H}^{\square}$, both specialized to $\boldsymbol{a}=q^3$.

The last property in fact holds even on the level of triply-graded homologies (without specialization $a=q^3$), as can be seen from (B.3) and (B.4). We also note that the isomorphism of $(\mathcal{H}^{\begin{smallmatrix}\square\\\square\end{smallmatrix}}, d_3^{\begin{smallmatrix}\square\\\square\end{smallmatrix}})$ and $\mathcal{H}^{\square}$ as triply-graded theories, also holds for all prime knots with up to 6 crossings.

Appendix C. $\mathcal{H}^3$ homology of the figure-eight knot

The Poincare polynomial of the $\mathcal{H}^3$ homology of the figure-eight knot 4_1 is given by:

$$\begin{aligned}
\mathcal{P}^3(4_1)=\mathcal{P}^{\square\!\square\!\square}(4_1) \;=\; & 1+(1+\boldsymbol{a}^{-1}qt^{-1})(1+\boldsymbol{a}^{-1}t^{-1})(1+\boldsymbol{a}^{-1}q^{-1}t^{-1}) \\
& \times(1+\boldsymbol{a}^{-1}q^{-3}t^{-3})(1+\boldsymbol{a}^{-1}q^{-4}t^{-3})(1+\boldsymbol{a}^{-1}q^{-5}t^{-3}) \\
& \times\boldsymbol{a}^3q^6t^6 \\
& +(1+q+q^2)(1+\boldsymbol{a}^{-1}qt^{-1})(1+\boldsymbol{a}^{-1}q^{-3}t^{-3})\boldsymbol{a}t^2 \\
& +(1+q+q^2)(1+\boldsymbol{a}^{-1}qt^{-1})(1+\boldsymbol{a}^{-1}t^{-1})(1+\boldsymbol{a}^{-1}q^{-3}t^{-3}) \\
& \times(1+\boldsymbol{a}^{-1}q^{-4}t^{-3})\boldsymbol{a}^2q^2t^4.
\end{aligned}$$

This homology categorifies the $\square\!\square\!\square$-colored HOMFLY polynomial of 4_1;

$$\mathcal{P}^3(4_1)(\boldsymbol{a},q,t=-1)=P^{\square\!\square\!\square}(4_1)(\boldsymbol{a},q).$$

Furthermore, this homology has all of the wanted properties - namely, there exist following differential ons $\mathcal{H}^3(4_1)$:

- canceling differential d_1^3 of degree $(-1,1,-1)$, leaving $1=\boldsymbol{a}^0q^0t^0$ as remaining generator.
- canceling differential d_{-3}^3 of degree $(-1,-3,-3)$, also leaving $1=\boldsymbol{a}^0q^0t^0$ as remaining generator.
- differential d_{-4}^3 of degree $(-1,-4,-3)$, such that

$$(\mathcal{H}^3(4_1),d_{-4}^3)\cong\mathcal{H}^1(4_1).$$

- differential d_{-5}^3 of degree $(-1,-5,-3)$, such that

$$(\mathcal{H}^3(4_1),d_{-5}^3)\cong\mathcal{H}^2(4_1).$$

- vertical colored differential d_0^3 of degree $(-1,0,-1)$, such that

$$(\mathcal{H}^3(4_1),d_0^3)\cong\mathcal{H}^1(4_1).$$

In particular

$$(\mathcal{H}^3(4_1),d_0^3)(\boldsymbol{a},q,t)=\mathcal{P}^1(4_1)(\boldsymbol{a},q^3,t).$$

- vertical colored differential d_{-1}^3 of degree $(-1,-1,-1)$, such that

$$(\mathcal{H}^3(4_1),d_{-1}^3)\cong\mathcal{H}^2(4_1).$$

In particular

$$(\mathcal{H}^3(4_1),d_{-1}^3)(\boldsymbol{a},q=1,t)=\mathcal{P}^2(4_1)(\boldsymbol{a},q=1,t).$$

Appendix D. Computation of the unreduced homology of the unknot

For a nonnegative integer N we define the quantum dimension $[N]$ to be

$$[N] = \frac{q^N - q^{-N}}{q - q^{-1}}.$$

Also

$$[N]! = [N][N-1]\dots[1],$$

and

$$\begin{bmatrix} N \\ k \end{bmatrix} = \frac{[N][N-1]\dots[N-k+1]}{[k]!}.$$

For a partition $\lambda = (\lambda_1, \lambda_2, \dots)$, we set:

$$\begin{aligned} |\lambda| &= \sum_i \lambda_i, \\ n(\lambda) &= \sum_i (i-1)\lambda_i, \\ m(\lambda) &= \sum_i i\lambda_i, \\ ||\lambda||^2 &= \sum_i \lambda_i^2. \end{aligned}$$

By λ^t, we denote the dual (conjugate) partition of λ. We have

$$\begin{aligned} m(\lambda) &= n(\lambda) + |\lambda|, \\ 2m(\lambda) &= ||\lambda^t||^2 + |\lambda|. \end{aligned}$$

We also define

$$\kappa(\lambda) := 2(||\lambda||^2 - ||\lambda^t||^2) = 2(n(\lambda^t) - n(\lambda)).$$

We identify a partition and its corresponding Young diagram.
For a box $x = (i,j) \in \lambda$, we define its content and the hook length by:

$$\begin{aligned} c(x) &= j - i, \\ h(x) &= \lambda_i + \lambda_j^t - i - j + 1. \end{aligned}$$

For a partition λ and positive integer N, we set:

$$\begin{bmatrix} N \\ \lambda \end{bmatrix} := \prod_{x \in \lambda} \frac{[N - c(x)]}{[h(x)]}.$$

Then we have

$$s_\lambda(q^{1-N}, q^{3-N}, \dots, q^{N-3}, q^{N-1}) = \begin{bmatrix} N \\ \lambda^t \end{bmatrix}. \tag{D.1}$$

Of course, the following holds

$$s_\lambda(1, q^2, q^4, \dots, q^{2(N-2)}, q^{2(N-1)}) = q^{(N-1)|\lambda|} s_\lambda(q^{1-N}, q^{3-N}, \dots, q^{N-3}, q^{N-1}).$$

D.1. Colored HOMFLY polynomial of the unknot. We compute the specialization of the superpolynomial of the unknot at $q_1 = q_2$ (*i.e.* at $t = -1$) from the equation (6.1). According to (6.3), this gives the value of the polynomial of the Hopf link as well. We denote this specialization by $\overline{P}^{\lambda}(\bigcirc)$, *i.e.*

$$\overline{P}^{\lambda}(\bigcirc) := \overline{\mathcal{P}}^{\lambda}(\bigcirc)|_{q_1=q_2} = \overline{\mathcal{P}}^{\lambda}(\bigcirc)|_{t=-1}\,.$$

We also denote $Z'_\lambda := Z_\lambda|_{q_1=q_2}$, and so

$$\overline{P}^{\lambda}(\bigcirc) = (-1)^{|\lambda|}(Q^{-1})^{\frac{|\lambda|}{2}} \times \frac{Z'_\lambda}{Z'_\emptyset},$$

with

$$Z'_\lambda = \sum_\nu (-Q)^{|\nu|} s_\nu(q_2^{-\rho}) s_{\nu^t}(q_2^{-\rho}) s_\lambda(q_2^{-\rho-\nu^t}).$$

By using the following identity for the Schur functions

$$s_\lambda(q_2^{-\rho-\nu^t}) s_{\nu^t}(q_2^{-\rho}) = q_2^{-\frac{\kappa(\lambda)}{2}} \sum_\eta s_{\nu^t/\eta}(q_2^{-\rho}) s_{\lambda^t/\eta}(q_2^{-\rho}) \tag{D.2}$$

in the expression for Z'_λ, we get

$$\begin{aligned} Z'_\lambda &= \sum_\nu (-Q)^{|\nu|} s_\nu(q_2^{-\rho}) q_2^{-\frac{\kappa(\lambda)}{2}} \sum_\eta s_{\nu^t/\eta}(q_2^{-\rho}) s_{\lambda^t/\eta}(q_2^{-\rho}) = \\ &= q_2^{-\frac{\kappa(\lambda)}{2}} \sum_\eta s_{\lambda^t/\eta}(q_2^{-\rho}) \sum_\nu (-Q)^{|\nu|} s_\nu(q_2^{-\rho}) s_{\nu^t/\eta}(q_2^{-\rho}) = \\ &= q_2^{-\frac{\kappa(\lambda)}{2}} \sum_\eta s_{\lambda^t/\eta}(q_2^{-\rho}) \sum_\nu (-Q)^{|\nu|} s_\nu(q_2^{-\rho}) \sum_\varphi c^{\nu^t}_{\eta,\varphi} s_\varphi(q_2^{-\rho}) = \\ &= q_2^{-\frac{\kappa(\lambda)}{2}} \sum_\eta s_{\lambda^t/\eta}(q_2^{-\rho}) \sum_\varphi s_\varphi(q_2^{-\rho}) \sum_\nu (-Q)^{|\eta|+|\varphi|} c^{\nu^t}_{\eta,\varphi} s_\nu(q_2^{-\rho}) = \\ &= q_2^{-\frac{\kappa(\lambda)}{2}} \sum_\eta s_{\lambda^t/\eta}(q_2^{-\rho}) (-Q)^{|\eta|} \sum_\varphi (-Q)^{|\varphi|} s_\varphi(q_2^{-\rho}) s_{\varphi^t}(q_2^{-\rho}) s_{\eta^t}(q_2^{-\rho}) = \\ &= q_2^{-\frac{\kappa(\lambda)}{2}} \sum_\eta (-Q)^{|\eta|} s_{\lambda^t/\eta}(q_2^{-\rho}) s_{\eta^t}(q_2^{-\rho}) \times Z'_\emptyset. \end{aligned}$$

Hence, the value of the colored HOMFLY polynomial invariant of the unknot labeled by λ is given by:

$$\overline{P}^{\lambda}(\bigcirc) = q_2^{-\frac{\kappa(\lambda)}{2}} \cdot (-1)^{|\lambda|}(Q^{-1})^{\frac{|\lambda|}{2}} \sum_\eta (-Q)^{|\eta|} s_{\lambda^t/\eta}(q_2^{-\rho}) s_{\eta^t}(q_2^{-\rho})\,. \tag{D.3}$$

In particular, the last formula tells us that (up to an overall factor) P_λ is a polynomial in Q of degree $|\lambda|$. Also, at $Q = 1$ one finds that $\overline{P}_\lambda|_{Q=1} = 0$ for $\lambda \neq 0$.

Written in terms of the knot-theoretical variables $(\boldsymbol{a}, q)$ the polynomial becomes:

$$\overline{P}^{\lambda}(\bigcirc)(\boldsymbol{a}, q) = q^{-\kappa(\lambda)} \cdot (-1)^{|\lambda|} \boldsymbol{a}^{\lambda} \sum_\eta (-1)^{|\eta|} \boldsymbol{a}^{-2|\eta|} s_{\lambda^t/\eta}(q^{-2\rho}) s_{\eta^t}(q^{-2\rho})\,. \tag{D.4}$$

Finally, when $\boldsymbol{a} = q^N$ one can show that the above expression without the factor $q^{-\kappa(\lambda)}$ is equal to $\begin{bmatrix} N \\ \lambda^t \end{bmatrix}$.

References

[1] N. Reshetikhin, V. Turaev, *Invariants of 3-manifolds via link polynomials and quantum groups*, Invent. Math. **103 (1)** (1991), 547–597.

[2] E. Witten, *Quantum field theory and Jones polynomial*, Commun. Math. Phys. **121** (1989), 351–399.

[3] N. M. Dunfield, S. Gukov, J. Rasmussen, *The Superpolynomial for Knot Homologies*, Experiment. Math. **15** (2006) 129, math.gt/0505662, 44pp.

[4] S. Gukov, A. S. Schwarz, C. Vafa, *Khovanov-Rozansky homology and topological strings*, Lett. Math. Phys. **74** (2005) 53, hep-th/0412243.

[5] S. Gukov, *Surface Operators and Knot Homologies*, proceedings of the International Congress on Mathematical Physics 2006, Fortsch. Phys. **55** (2007) 473-490, arXiv:0706.2369.

[6] S. Gukov, J. Walcher, *Matrix factorizations and Kauffman homology*, hep-th/0512298, 45pp.

[7] E. Witten, *Monopoles and four manifolds*, Math. Res. Lett. **1** (1994) 769, hep-th/9411102.

[8] A. Okounkov, N. Reshetikhin, C. Vafa, *Quantum Calabi-Yau and classical crystals*, hep-th/0309208.

[9] S. Gukov, A. Iqbal, C. Kozcaz, C. Vafa, *Link homologies and the refined topological vertex*, arXiv:0705.1368.

[10] A. Iqbal, C. Kozcaz, *Refined Hopf Link Revisited*, arXiv:1111.0525.

[11] T. Dimofte, S. Gukov, L. Hollands, *Vortex Counting and Lagrangian 3-manifolds*, arXiv:1006.0977.

[12] E. Witten, *Fivebranes and Knots*, arXiv:1101.3216.

[13] O. Aharony, S. S. Gubser, J. M. Maldacena, H. Ooguri, Y. Oz, *Large N field theories, string theory and gravity*, Phys. Rept. **323** (2000) 183-386, hep-th/9905111.

[14] R. Gopakumar, C. Vafa, *On the gauge theory/geometry correspondence*, Adv. Theor. Math. Phys. **3** (1999) 1415-1443, hep-th/9811131.

[15] H. Ooguri, C. Vafa, *Knot invariants and topological strings*, Nucl. Phys. **B577** (2000) 419-438, hep-th/9912123.

[16] F. Denef, G. Moore, *Split states, entropy enigmas, holes and halos*, hep-th/0702146.

[17] M. Kontsevich, Y. Soibelman, *Stability structures, motivic Donaldson-Thomas invariants and cluster transformations*, arXiv:0811.2435.

[18] T. Dimofte, S. Gukov, *Refined, Motivic, and Quantum*, Lett. Math. Phys. **91** (2010) 1, arXiv:0904.1420, 31pp.

[19] K. Behrend, J. Bryan, B. Szendroi, *Motivic degree zero Donaldson-Thomas invariants*, arXiv:0909.5088.

[20] S. Cecotti, C. Vafa, *BPS Wall Crossing and Topological Strings*, arXiv:0910.2615.

[21] T. Dimofte, S. Gukov, Y. Soibelman, *Quantum Wall Crossing in N=2 Gauge Theories*, Lett. Math. Phys. **95** (2011) 1-25, arXiv:0912.1346.

[22] K. Nagao, H. Nakajima, *Counting invariant of perverse coherent sheaves and its wall-crossing*, arXiv:0809.2992.

[23] D. Jafferis, G. Moore, *Wall crossing in local Calabi Yau manifolds*, arXiv:0810.4909.

[24] S.G. Naculich, H.J. Schnitzer, *Duality between $SU(N)_k$ and $SU(k)_N$ WZW models*, Nucl. Phys. **B347** (1990) 687.

[25] S.G. Naculich, H.A. Riggs, H.J. Schnitzer, *Group level duality in WZW models and Chern-Simons theory*, Phys. Lett. **B246** (1990) 417; E.J. Mlawer, S.G. Naculich, H.A. Riggs, H.J. Schnitzer, *Group-level duality of WZW coefficients and Chern-Simons link observable*, Nucl. Phys. **B352** (1991) 863.

[26] J. M. F. Labastida, M. Marino, C. Vafa, *Knots, links and branes at large N*, JHEP **0011** (2000) 007, hep-th/0010102.

[27] K. Liu, P. Peng, *New Structure of Knot Invariants*, arXiv:1012.2636.

[28] C. Taubes, *Lagrangians for the Gopakumar-Vafa conjecture*, Geom. Topol. Monogr. **8** (2006) 73-95, arXiv:math/0201219.

[29] S. Koshkin, *Conormal bundles to knots and the Gopakumar-Vafa conjecture*, arXiv:math.DG/0503248.

[30] J. Harvey, G. Moore, *On the algebras of BPS states*, Commun. Math. Phys. **197** (1998) 489-519, hep-th/9609017.

[31] M. Kontsevich, Y. Soibelman, *Cohomological Hall algebra, exponential Hodge structures and motivic Donaldson-Thomas invariants*, arXiv:1006.2706.

[32] O. Schiffmann, *Lectures on Hall algebras,* arXiv:math/0611617.
[33] M. Khovanov, L. Rozansky, *Matrix factorizations and link homology,* Fund. Math. **199** (2008) 1-91, math/0401268.
[34] M. Khovanov, L. Rozansky, *Topological Landau-Ginzburg models on a world-sheet foam,* Adv. Theor. Math. Phys. **11** (2007) 233-260, hep-th/0404189.
[35] M. Khovanov, L. Rozansky, *Matrix factorizations and link homology II,* Geometry & Topology **12** (2008) 1387-1425, math.QA/0505056.
[36] M. Khovanov, *Triply-graded link homology and Hochschild homology of Soergel bimodules,* Int. Journal of Math., vol. **18**, no. 8 (2007) 869-885, arXiv:math/0510265.
[37] M. Khovanov, L. Rozansky, *Virtual crossings, convolutions and a categorification of the $SO(2N)$ Kauffman polynomial,* arXiv:math/0701333.
[38] M. Mackaay, M. Stošić, P. Vaz, *The 1,2-coloured HOMFLY-PT link homology,* Trans. Amer. Math. Soc. **363** (2011), 2091-2124, arXiv:0809.0193.
[39] Y. Yonezawa, *Quantum $(sl_n, \wedge V_n)$ link invariant and matrix factorizations,* arXiv:0906.0220.
[40] H. Wu, *A colored $sl(N)$-homology for links in S^3,* arXiv:0907.0695.
[41] M. Kontsevich, unpublished.
[42] A. Kapustin, Y. Li, *D-branes in Landau-Ginzburg models and algebraic geometry,* JHEP **0312** (2003) 005, hep-th/0210296.
[43] D. Orlov, *Triangulated Categories of Singularities and D-Branes in Landau-Ginzburg Orbifold,* math.AG/0302304.
[44] I. Brunner, M. Herbst, W. Lerche, B. Scheuner, *Landau-Ginzburg realization of open string TFT,* JHEP **0611** (2006) 043, hep-th/0305133.
[45] K. Hori, J. Walcher, *F-term equations near Gepner points,* JHEP **0501** (2005) 008, hep-th/0404196.
[46] I. Brunner, D. Roggenkamp, *B-type defects in Landau-Ginzburg models,* JHEP **0708** (2007) 093, arXiv:0707.0922.
[47] E. Witten, *The Verlinde algebra and the cohomology of the Grassmannian,* in "Cambridge 1993, Geometry, topology, and physics" 357-422, hep-th/9312104.
[48] S. Gukov, E. Witten, *Rigid Surface Operators,* Adv. Theor. Math. Phys. **14** (2010) 87-177, arXiv:0804.1561.
[49] A. Kapustin and E. Witten, *Electric-magnetic Duality And The Geometric Langlands Program,* Communications in Number Theory and Physics **1** (2007) 1-236, arXiv:hep-th/0604151.
[50] D. Gaiotto, E. Witten, *Knot Invariants from Four-Dimensional Gauge Theory,* arXiv:1106.4789.
[51] E.S. Lee, *Khovanov's invariants for alternating links,* math.GT/0210213.
[52] D. Bar-Natan, *Khovanov's Homology for Tangles and Cobordisms,* math.GT/0410495.
[53] P. Turner, *Calculating Bar-Natan's characteristic-two Khovanov homology,* math.GT/0411225.
[54] M. Khovanov, *Link homology and Frobenius extensions,* math.QA/0411447.
[55] B. Gornik, *Note on Khovanov link cohomology,* math.QA/0402266.
[56] The Knot Atlas, `http://katlas.org`.
[57] X.-S. Lin, H. Zheng, *On the Hecke algebras and the colored HOMFLY polynomial,* Trans. Amer. Math. Soc. **362** (2010), 1-18, arXiv:math/0601267.
[58] M. Aganagic, S. Shakirov, *Knot Homology from Refined Chern-Simons Theory,* arXiv:1105.5117.
[59] P. Dunin-Barkowski, A. Mironov, A. Morozov, A. Sleptsov and A. Smirnov, *Superpolynomials for toric knots from evolution induced by cut-and-join operators,* arXiv:1106.4305.
[60] work in progress.
[61] M. Marino, C. Vafa, *Framed Knots at Large N,* hep-th/0108064.
[62] B. Webster, *Knot invariants and higher representation theory II: the categorification of quantum knot invariants,* arXiv:1005.4559.
[63] B. Cooper, V. Krushkal, *Categorification of the Jones-Wenzl Projectors,* arXiv:1005.5117.
[64] I. Frenkel, C. Stroppel, J. Sussan, *Categorifying fractional Euler characteristics, Jones-Wenzl projector and 3j-symbols,* arXiv:1007.4680.
[65] B. Cooper, M. Hogancamp, V. Krushkal, *$SO(3)$ Homology of Graphs and Links,* Algebr. Geom. Topol. **11** (2011) 2137-2166, arXiv:1012.3672.

CALIFORNIA INSTITUTE OF TECHNOLOGY, PASADENA, CA 91125, USA,
MAX-PLANCK-INSTITUT FÜR MATHEMATIK, VIVATSGASSE 7, D-53111 BONN, GERMANY.
E-mail address: gukov@theory.caltech.edu

INSTITUTO DE SISTEMAS E ROBOTICA AND CAMGSD, INSTITUTO SUPERIOR TECNICO, TORRE NORTE, PISO 7, AV. ROVISCO PAIS, 1049-001 LISBON, PORTUGAL,
MATHEMATICAL INSTITUTE SANU, KNEZ MIHAILOVA 36, 11000 BEOGRAD, SERBIA.
E-mail address: mstosic@math.ist.utl.pt

Proceedings of Symposia in Pure Mathematics
Volume 85, 2012

Motivic structures in quantum field theory

Matilde Marcolli

ABSTRACT. This is a writeup of the lecture given by the author at the String–Math 2011 conference in Philadelphia. It gives an overview of recent work of the author, in collaboration with Aluffi and with Ceyhan, on some aspects of the occurrence of motivic structures in perturbative quantum field theory.

1. Motives and quantum fields

The theory of motives originated with Grothendieck's idea of a "universal cohomology" theory for algebraic varieties, underlying the different realizations (de Rham, Betti, étale cohomologies, Hodge structures). The first detailed account of Grothendieck's theory of motives was given by Manin in [**36**].

In this sense, categories of motives "interpolate" between categories of varieties and their cohomologies. In particular, since categories of motives are constructed by replacing the stricter notion of morphisms of algebraic varieties with a more general notion of *correspondences*, they tend to be better behaved than the category of varieties itself. In the best possible case, one obtains an *abelian* category. This is indeed the case when one works with the category of *pure motives*, which means motives associated to smooth projective varieties, with the correspondences given by algebraic cycles modulo the *numerical* equivalence relation. In this case, a result of Jannsen [**29**] shows that the category one obtains is abelian.

1.1. Pure motives. More precisely, the objects of the category of pure motives are triples (X, p, m) of a smooth projective variety X, an endomorphism $p \in \mathrm{End}(X)$ with $p^2 = p$, and an integer $m \in \mathbb{Z}$. Morphisms are given by

$$\mathrm{Hom}((X,p,m),(Y,q,n)) = q\mathrm{Corr}^{m-n}_{/\sim,\mathbb{Q}}(X,Y)\,p,$$

where $\mathrm{Corr}^{m-n}_{/\sim,\mathbb{Q}}(X,Y)$ denotes the $\mathbb{Q}$-vector space of equivalence classes of algebraic cycles in $X \times Y$ of codimension $m - n$. The algebraic cycles are considered modulo the numerical equivalence relation, which means that a cycle Z is numerically trivial ($Z \sim 0$) if its intersection pairing with arbitrary other cycles is zero. The composition of morphisms

$$\mathrm{Corr}(X,Y) \times \mathrm{Corr}(Y,Z) \to \mathrm{Corr}(X,Z)$$

is induced by the intersection product in $X \times Y \times Z$,

$$(\pi_{X,Z})_*(\pi^*_{X,Y}(\alpha) \bullet \pi^*_{Y,Z}(\beta)).$$

Here $\pi_{X,Z}$, $\pi_{X,Y}$, $\pi_{Y,Z}$ are the projection maps from $X \times Y \times Z$ to $X \times Z$, $X \times Y$, and $Y \times Z$, respectively, $\bullet$ denotes the intersection product, and α and β are (linear combinations of) cycles in $X \times Y$ and $Y \times Z$, respectively, as illustrated in the following diagram:

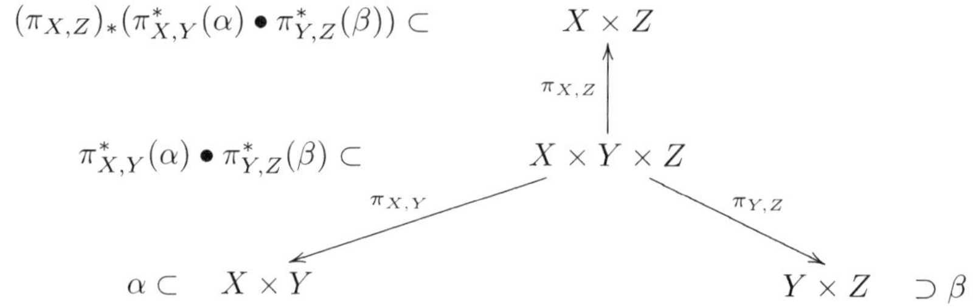

The category $\mathcal{M}_{num,\mathbb{Q}}(k)$ of numerical pure motives obtained in this way is a semi-simple abelian category [**29**]. The Tate objects in the category of pure motives are defined as the objects of the form $\mathbb{Q}(m) = \mathbb{Q}(1)^{\otimes m}$. The Tate motive $\mathbb{Q}(1)$ is defined as the formal inverse of the Lefschetz motive $\mathbb{L}$, which is characterized by the property that the motive of the projective line $\mathbb{P}^1$ is of the form $1 + \mathbb{L}$.

1.2. Mixed motives. In view of the applications of motives to quantum field theory, however, one cannot restrict oneself to the setting of smooth projective varieties. In fact, the varieties that occur in that context are typically *singular* hypersurfaces. The theory of motives of varieties that are not smooth projective is much more complicated than the case of pure motives and it goes under the name of *mixed motives*.

In this case, one obtains only a *triangulated category* $\mathcal{DM}$ of mixed motives, for which there are several different (but equivalent) constructions, by Voevodsky [**42**], Hanamura [**27**], and Levine [**32**]. The triangulated structure reflects the long exact sequences in cohomology associated to embeddings of subvarieties $Y \subset X$,

$$\mathfrak{m}(Y) \to \mathfrak{m}(X) \to \mathfrak{m}(X \smallsetminus Y) \to \mathfrak{m}(Y)[1],$$

where $\mathfrak{m}(X)$ denotes the object in the category $\mathcal{DM}$ defined by a variety X, and the homotopy invariance property

$$\mathfrak{m}(X \times \mathbb{A}^1) = \mathfrak{m}(X)(-1)[2],$$

where (-1) denotes the Tate twist, obtained by tensoring with the Tate motive.

Inside the triangulated category $\mathcal{DM}$ of mixed motives one can identify a subcategory $\mathcal{DMT} \subset \mathcal{DM}$, which is generated by the Tate objects $\mathbb{Q}(m)$. This is called the (triangulated) category of mixed Tate motives. In the case of varieties defined over a number field, a crucial vanishing result [**33**] allows the construction of an abelian category of mixed Tate motives, obtained as the heart of a t-structure in this triangulated category. For the reader interested in a gentle introduction to the subject, a brief account of motives for physicists can be found in [**38**], while a general overview on motives and quantum field theory is given in the author's monograph [**37**].

1.3. Motives and periods. An aspect of the theory of motives which is of direct relevance to quantum field theory is its relation to periods of varieties. Periods are a special class of numbers that can be obtained by integrating an algebraic differential form over a cycle defined by algebraic equations in an algebraic variety, see [**30**]. Interestingly, the motivic nature of the variety determines what kind of numbers can arise as periods. In particular, it was conjectured (and recently proved by Francis Brown [**16**]) that the periods of mixed Tate motives over $\mathbb{Z}$ are $\mathbb{Q}[(2\pi i)^{-1}]$-linear combinations of multiple zeta values. The latter are numbers of the form

$$\zeta(n_1,\dots,n_r) = \sum_{0<k_1<\dots<k_r} \frac{1}{k_1^{n_1}\cdots k_r^{n_r}}$$

with $n_j \geq 1$ and $n_r \geq 2$.

This very special class of numbers is precisely the original source of the connection between motives and quantum field theory. An extensive investigation by Broadhurst and Kreimer [**15**] revealed the pervasive occurrence of multiple zeta values in computations of residues of Feynman integrals. A way to begin to understand how that may relate to some underlying motivic properties (see [**12**]) is to express the Feynman integral computations in terms of the well known Feynman parametric form.

1.4. Feynman amplitudes. For a massless (Euclidean) perturbative scalar field theory, the Feynman rules prescribe propagators associated to the internal edges e_i of each Feynman graph Γ that are given by quadratic forms q_i in the momentum variables and momentum conservation laws at all vertices. The resulting Feynman amplitude is then expressed as an integral of the form ([**10**], [**28**], [**37**])

$$U(\Gamma) = \int \frac{\delta(\sum_{i=1}^n \epsilon_{v,i}k_i + \sum_{j=1}^N \epsilon_{v,j}p_j)}{q_1\cdots q_n}\, d^Dk_1\cdots d^Dk_n \tag{1.1}$$

where $n = \#E_{int}(\Gamma)$, $N = \#E_{ext}(\Gamma)$ are the numbers of internal and external edges of the graph, the q_i are quadratic forms in the momentum variables k_i, the p_j are assigned external momenta, and δ is the Dirac delta function imposing the momentum conservation law; D is the spacetime dimension, and

$$\epsilon_{e,v} = \begin{cases} +1 & t(e)=v \\ -1 & s(e)=v \\ 0 & \text{otherwise,} \end{cases}$$

with $s(e)$ and $t(e)$ the source and target vertices of the oriented egde e. In the following, we will use the notation $U(\Gamma)$ for the Feynman amplitude, as in (1.1), or $U(\Gamma,p)$ when we want to explicitly stress the dependence on the datum of the external momenta p.

These Feynman amplitudes satisfy some formal properties that make it possible to reduce the combinatorics of graphs involved in the perturbative expansion. In fact, one can consider only connected graphs, since

$$U(\Gamma_1 \amalg \Gamma_2, p) = U(\Gamma_1,p_1)U(\Gamma_2,p_2)$$

and one can further reduce to 1PI graphs, namely those graphs that cannot be disconnected by the removal of a single internal edge. Indeed, for a connected graph described as $\Gamma = \cup_{v\in T}\Gamma_v$, a tree with vertices replaced by 1PI graphs Γ_v, one has

$$U(\Gamma,p) = \prod_{v\in T} U(\Gamma_v,p_v)\frac{\delta((p_v)_e - (p_{v'})_e)}{q_e((p_v)_e)}.$$

These formal properties can be abstracted to define formal "algebro-geometric Feynman rules", which can be obtained, for instance, though invariants of singular varieties based Chern classes, or from classes in the Grothendieck ring of varieties, and which behave, in certain respects, like physical Feynman rules, see [**3**], [**4**], [**1**].

1.5. Parametric form of Feynman integrals. The Feynman parametric form of the amplitude is obtained from (1.1) by passing to Schwinger parameters through the identity

$$q_1^{-k_1}\cdots q_n^{-k_n} = \frac{1}{\Gamma(k_1)\cdots\Gamma(k_n)}\int_{\mathbb{R}^n} e^{-(s_1q_1+\cdots+s_nq_n)}\, s_1^{k_1-1}\cdots s_n^{k_n-1}\, ds_1\cdots ds_n.$$

The so called Feynman trick then expresses

$$\frac{1}{q_1\cdots q_n} = (n-1)!\int \frac{\delta(1-\sum_{i=1}^n t_i)}{(t_1q_1+\cdots+t_nq_n)^n}\, dt_1\cdots dt_n.$$

Let $\{\ell_r\}_{r=1,\ldots,b_1(\Gamma)}$ denote a choice of a basis of $H_1(\Gamma)$, namely a collection of loops ℓ_r of edges in the graph Γ that generate the first homology. With a change of variables $k_i = u_i + \sum_{r=1}^{\ell}\eta_{ir}x_r$, where

$$\eta_{ir} = \begin{cases} \pm 1 & \text{edge } \pm e_i \in \text{ loop } \ell_r \\ 0 & \text{otherwise} \end{cases}$$

one then obtains the Feynman amplitude in the form

$$U(\Gamma) = \frac{\Gamma(n-D\ell/2)}{(4\pi)^{\ell D/2}}\int_{\sigma_n}\frac{\omega_n}{\Psi_\Gamma(t)^{D/2}V_\Gamma(t,p)^{n-D\ell/2}}$$

where the integration is over the simplex $\sigma_n = \{t\in\mathbb{R}^n_+ \,|\, \sum_i t_i = 1\}$, with ω_n the volume form. The integrand is given by the *graph polynomials*: the Kirchhoff polynomial. This can be written as a determinant involving the matrix η_{ir}, and also explicitly as a sum over spanning trees Tcof the graph Γ,

$$\Psi_\Gamma(t) = \det M_\Gamma(t) = \sum_T\prod_{e\notin T} t_e \quad \text{with} \quad (M_\Gamma)_{kr}(t) = \sum_{i=0}^n t_i\eta_{ik}\eta_{ir}.$$

One also has

$$V_\Gamma(t,p) = \frac{P_\Gamma(t,p)}{\Psi_\Gamma(t)} \quad \text{with} \quad P_\Gamma(p,t) = \sum_{C\subset\Gamma} s_C\prod_{e\in C} t_e.$$

Here the sum is over cut-sets C (complements of spanning trees plus one edge), and $s_C = (\sum_{v\in V(\Gamma_1)} P_v)^2$ are functions of the external momenta, with $P_v = \sum_{e\in E_{ext}(\Gamma), t(e)=v} p_e$, with the conservation law $\sum_{e\in E_{ext}(\Gamma)} p_e = 0$. One has $\deg\Psi_\Gamma = b_1(\Gamma) = \deg P_\Gamma - 1$. In the stable range where $-n + D\ell/2 \geq 0$, the Feynman amplitude is an integral of a ratio of polynomials of the form

$$U(\Gamma) = \frac{\Gamma(n-D\ell/2)}{(4\pi)^{\ell D/2}}\int_{\sigma_n}\frac{P_\Gamma(t,p)^{-n+D\ell/2}\omega_n}{\Psi_\Gamma(t)^{-n+D(\ell+1)/2}},$$

while in the simpler log divergent case with $n = D\ell/2$ one has

$$\int_{\sigma_n}\frac{\omega_n}{\Psi_\Gamma(t)^{D/2}}.$$

1.6. Divergences and periods. In this form, one sees the ultraviolet divergence in the form of a pole of a Gamma function. Up to that divergent factor, the "residue" is given by the integral

$$\int_{\sigma_n} \frac{P_\Gamma(t,p)^{-n+D\ell/2}\omega_n}{\Psi_\Gamma(t)^{-n+D(\ell+1)/2}}. \tag{1.2}$$

Unfortunately, this integral itself is still in general divergent. These infrared divergences come from the intersections of the domain of integration with the locus defined by the vanishing of the polynomial in the denominator. Thus, these integrals need to be further regularized and renormalized, before one can properly interpret them as periods. This issue is discussed at length in the work of Bloch, Esnault, Kreimer [**12**], and Bloch and Kreimer [**13**], [**14**], where algebro-geometric techniques based on blowups, monodromies, and limiting mixed Hodge structures are used to address the renormalization problem for these divergences.

For simplicity of exposition, I am going to ignore here these important divergence issues and refer the readers to [**12**], [**13**], [**14**] for an appropriate discussion. If the integrals (1.2) were convergent, they would indeed be periods. In fact, one can define the graph hypersurfaces as

$$\hat{X}_\Gamma = \{t \in \mathbb{A}^n \mid \Psi_\Gamma(t) = 0\},$$

in the affine case. Since the polynomial Ψ_Γ is homogenous with deg $= b_1(\Gamma)$, one can also consider the projective hypersurface

$$X_\Gamma = \{t \in \mathbb{P}^{n-1} \mid \Psi_\Gamma(t) = 0\}.$$

In terms of explicit calculations of classes in the Grothendieck ring, it is often convenient to go back and forth between the affine formulation in terms of $\hat{X}_\Gamma$ and the projective one in terms of X_Γ.

The domain of integration in (1.2) is the simplex σ_n, which is a chain with non-trivial boundary. Thus, the integral can be regarded as a period for a *relative* cohomology, namely

$$H^{n-1}(\mathbb{P}^{n-1} \smallsetminus X_\Gamma, \Sigma_n \smallsetminus (\Sigma_n \cap X_\Gamma)),$$

where $\Sigma_n = \{\prod_i t_i = 0\}$ is a divisor that contains the boundary $\partial\sigma_n$ of the domain of integration. (We are assuming we work in the stable range $-n+D\ell/2 \geq 0$, where the other graph polynomial $P_\Gamma(t,p)$ does not appear in the denominator.) When one takes the divergences into account, as in [**12**], [**13**], this relative cohomology is in fact replaced by a similar relative cohomology involving a toric variety obtained as an iterated blowup of $\mathbb{P}^{n-1}$. More precisely, the divergences happen where $X_\Gamma \cap \sigma_n \neq \emptyset$. This locus is clearly contained inside the divisor $\Sigma_n \supset \partial\sigma_n$ of coordinate hyperplanes. In fact, since the graph polynomial has positive coefficients, it has no real solutions in the interior of the positive quadrant, so the only place where the hypersurface can meet the simplex is along the boundary $\partial\sigma_n$. One proceeds as in [**13**] to an iterated series of blowups of $\mathbb{P}^{n-1}$ along the coordinate linear spaces defined by edges of 1PI subgraphs. The result is a toric variety $P(\Gamma)$. The iterated blowups that define $P(\Gamma)$ have the effect of separating the strict transform of X_Γ from the non-negative real points. Deforming the integration chain leads to a monodromy problem and the subtraction of divergences is then achieved in terms of Poincaré residues and limiting mixed Hodge structures (see [**13**]).

1.7. The polynomial countability question. Thus, from this point of view, the question of why one finds multiple zeta values when evaluating residues of Feynman integrals can be reformulated as a question on whether a relative cohomology group of a hypersurface complement is a realization of a mixed Tate motive. Very early into the development of the subject, Kontsevich conjectured that the graph hypersurfaces X_Γ themselves may be always mixed Tate. More precisely, the conjecture was formulated in terms of a closely related property, *polynomial countability.* A variety X that is defined over $\mathbb{Z}$ admits reductions X_p modulo various primes p. These give varieties defined over finite fields $\mathbb{F}_p$ and one can count the number of their algebraic points $X_p(\mathbb{F}_q)$ for field extensions $\mathbb{F}_q$ with $q = p^m$. The variety X is polynomially countable if the counting function $\#X_p(\mathbb{F}_q)$ is a polynomial in q with $\mathbb{Z}$ coefficients. (One can also consider a weaker conditions where this holds for all but finitely many primes.) This polynomial countability property is related to the mixed Tate nature of the motive of X through the relation between the counting of points over finite fields and the classes in the Grothendieck ring of varieties that we discuss more precisely below. The conjecture was, at first, verified for all graphs with up to twelve edges in [**40**], but was later disproved by a remarkable general result of Belkale and Brosnan [**8**], which showed that, on the contrary, the graph hypersurfaces can be "arbitrarily complicated" as motives, in a sense that we'll discuss more precisely below. Recently, explicit counterexamples were found by Doryn [**22**] and by Brown and Schnetz [**19**]. The counterexample identified in [**22**] fails the polynomially countable condition, but may still satisfy a weaker version for all but finitely many primes, while the example of [**19**] exhibits a more substantial failure of polynomial countability.

1.8. Virtual motives as universal Euler characteristics. While the category of mixed motives has a very complicated construction, and it is generally hard to work directly with the mixed motives themselves, there is a very good invariant of motives that can be more easily computed and which is very useful in order to detect properties of motives. This is given by the class in the Grothendieck ring of varieties $K_0(\mathcal{V})$.

The generators $[X]$ of $K_0(\mathcal{V})$ are the isomorphism classes of smooth quasi-projective varieties and the relations are the inclusion-exclusion (or scissor congruence) relation

$$[X] = [X \smallsetminus Y] + [Y],$$

for a closed embedding of a smooth closed subvariety $Y \subset X$ and the product relation that gives the ring structure,

$$[X] \cdot [Y] = [X \times Y].$$

Classes in the Grothendieck ring $K_0(\mathcal{V})$ are also called *virtual motives.*

The Grothendieck ring has an alternative presentation, as shown in [**9**], where the generators are isomorphism classes of *smooth projective* varieties and the relations are given by the blowup formula $[Bl_Y(X)] = [X] - [Y] + [E]$, with E the exceptional divisor of the blowup, together with the trivial relation $[\emptyset] = 0$ and the ring structure given again by the product.

In the Grothendieck ring of varieties the (virtual) Tate motives are the elements of the subring $\mathbb{Z}[\mathbb{L}]$, where $\mathbb{L} = [\mathbb{A}^1]$ is the class of the affine line, the Lefschetz motive.

Notice that, if X is a variety that is neither smooth nor projective, but which has a stratification by strata that are smooth (quasi)projective, then one has a corresponding well defined class $[X]$ in $K_0(\mathcal{V})$, obtained in terms of the classes of the strata. Thus, if X is a variety whose motive $\mathfrak{m}(X)$ in the category of mixed Tate motives is mixed Tate, one knows that the corresponding class $[X]$ will also be mixed Tate as a virtual motive, that is, an element of the subring $\mathbb{Z}[\mathbb{L}]$. This is essentially because a mixed Tate motive has a filtration whose graded pieces are pure Tate motives. Subject to some conjectures (such as the Tate conjecture, see the discussion of this issue in [**7**], for instance), a converse statement would also hold, namely if the class in the Grothendieck ring is a virtual Tate motive, the the variety is a (mixed) Tate motive. The conjectural statement needed here is essentially the fact that the numbers N_p of points over finite fields suffice to determine the motive. Since the counting of points over finite fields is an additive invariant that factors through the Grothendieck ring, the class $[X]$ would then also suffice.

The class $[X]$ of a variety in the Grothendieck ring $K_0(\mathcal{V})$ is regarded as a "universal Euler characteristic" for the variety, [**9**]. The reason lies in the following fact. An *additive invariant* of varieties is a function on algebraic varieties with values in a commutative ring $\mathcal{R}$ which satisfies $\chi(X) = \chi(Y)$, for isomorphic varieties $X \cong Y$, and

$$\chi(X) = \chi(Y) + \chi(X \smallsetminus Y),$$

for a closed embedding of a subvariety $Y \subset X$, and

$$\chi(X \times Y) = \chi(X)\chi(Y).$$

Any additive invariant of varieties factors through the Grothendieck ring, as a ring homomorphism

$$\chi : K_0(\mathcal{V}) \to \mathcal{R}.$$

One readily recognizes that the properties of additive invariants are exactly those that are satisfied by the Euler characteristic: inclusion-exclusion and multiplicativity on products. Indeed, examples of additive invariants are: the topological Euler characteristic or the virtual Hodge polynomial in the case of complex varieties; the couting of algebraic points in the case of varieties over finite fields; the Gillet–Soulé motivic Euler characteristic $\chi_{mot} : K_0(\mathcal{V})[\mathbb{L}^{-1}] \to K_0(\mathcal{M})$, given by $\chi_{mot}(X) = [(X, id, 0)]$ when X is smooth projective and by a complex $\chi_{mot}(X) = W^{\cdot}(X)$ in more general case, as defined in [**25**].

2. Virtual motives of graph hypersurfaces

In this section we recall some results on the virtual motives of the graph hypersurfaces. We focus on the recent work of Aluffi and the author ([**2**]–[**5**]).

2.1. Graph hypersurfaces and the Grothendieck ring. Suppose X is a variety defined over $\mathbb{Z}$. The conditions that a variety X is a mixed Tate motive, that its class $[X]$ in the Grothendieck ring $K_0(\mathcal{V})$ is a virtual mixed Tate motive (it lies in the $\mathbb{Z}[\mathbb{L}]$ subring), and that X is polynomially countable are closely related. If X is a mixed Tate motive then its virtual motive is mixed Tate and the counting of algebraic points for the reductions modulo the various primes is a polynomial function for all but finitely many primes (to account for primes of bad reduction). In fact, if one assumes certain conjectures (Tate conjecture), then knowledge of the

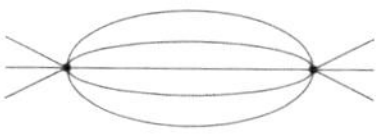

counting function $\#X_p(\mathbb{F}_q)$ determines the motives, so that the converse would also be true.

The result of Belkale and Brosnan [**8**] mentioned above, which disproved the Kontsevich conjecture on polynomial countability of the graph hypersurfaces can then be stated in the following way: the graph hypersurfaces X_Γ generate the Grothendieck ring $K_0(\mathcal{V})$, *localized* at $\mathbb{L}^n - \mathbb{L}$, for all $n > 1$.

This means that, in an appropriate sense, the graph hypersurfaces can be arbitrarily complicated as motives: any class in the (localized) Grothendieck ring can be obtained from graph hypersurfaces, including those that are not mixed Tate. Thus, as the graphs become more and more combinatorially complicated, one expects to see more and more non-mixed-Tate examples appear among them, even though the results of [**40**] show that the classes remain mixed Tate for all graphs with up to twelve edges. The occurrence of the first explicit counterexamples of [**22**] and of [**19**] shortly past the previously explored range confirms this understanding. The result of Belkale and Brosnan [**8**] is a very deep and elaborate result, which depends on a universality theorem for matroids.

It is worth pointing out that there is an interesting dichotomy in the Grothendieck ring between what happens when one *inverts* $\mathbb{L}$ and when one instead *sets* $\mathbb{L}$ *to zero*. As shown by Larsen and Lunts [**31**] in the context of motivic integration, setting $\mathbb{L}$ to zero in the Grothendieck ring $K_0(\mathcal{V}_{\mathbb{C}})$ gives the ring of *stable birational equivalence classes* of varieties

$$\mathbb{Z}[SB] = K_0(\mathcal{V})|_{\mathbb{L}=0}.$$

It was shown by Aluffi and the author in [**5**], by a very simple and direct argument, that the graph hypersurfaces span the subring $\mathbb{Z}$ inside $\mathbb{Z}[SB]$. Thus, while the graph hypersurfaces are "as general as possible" when one inverts $\mathbb{L}^n - \mathbb{L}$, they are just equivalent to points when one sets $\mathbb{L} = 0$. This property depends upon a deletion–contraction type formula, that we discuss in more detail below.

It is also interesting to point out that the localization of $K_0(\mathcal{V})$ at all the elements $\mathbb{L}^n - \mathbb{L}$ is isomorphic to the Grothendieck ring of special Artin stacks, as proved by Toën in [**41**], though it is not yet clear what this identification may be saying in this quantum field theory context.

2.2. Computing in the Grothendieck ring. While computing virtual motives is generally easier than working in the category of mixed motives, it still happens rarely that one can give a completely explicit calculation of the class $[X_\Gamma]$ of a graph hypersurface X_Γ. We recall here a sufficiently simple example from [**2**], where one can carry out a calculation completely explicitly.

The *banana graphs* are graphs with two vertices and n parallel edges between then, as in the figure.

The graph polynomial of the n-th banana graph Γ_n is

$$\Psi_{\Gamma_n}(t) = t_1 \cdots t_n \left(\frac{1}{t_1} + \cdots + \frac{1}{t_n}\right)$$

and the class in the Grothendieck ring turns out to be explicitly given by the formula

$$[X_{\Gamma_n}] = \frac{\mathbb{L}^n - 1}{\mathbb{L} - 1} - \frac{(\mathbb{L}-1)^n - (-1)^n}{\mathbb{L}} - n\,(\mathbb{L}-1)^{n-2}, \tag{2.1}$$

with $\mathbb{L} = [\mathbb{A}^1]$ the Lefschetz motive, as above.

The reason why, in this case, one can carry out the computation explicitly is that one can compare the hypersurface of a planar graph with that of the dual graph via the Cremona transformation.

The Cremona transformation is the map

$$\mathcal{C} : (t_1 : \cdots : t_n) \mapsto (\frac{1}{t_1} : \cdots : \frac{1}{t_n})$$

defined outside the singularities locus $\mathcal{S}_n$ of the divisor of coordinate hyperplanes $\Sigma_n = \{\prod_i t_i = 0\}$. This is the locus defined by the ideal

$$I_{\mathcal{S}_n} = (t_1 \cdots t_{n-1}, t_1 \cdots t_{n-2} t_n, \cdots, t_1 t_3 \cdots t_n).$$

The relation between the graph polynomial of a planar graph and that of the dual graph is then given through the Cremona transformation as

$$\Psi_\Gamma(t_1, \ldots, t_n) = (\prod_e t_e) \Psi_{\Gamma^\vee}(t_1^{-1}, \ldots, t_n^{-1})$$

so that we have

$$\mathcal{C}(X_\Gamma \cap (\mathbb{P}^{n-1} \smallsetminus \Sigma_n)) = X_{\Gamma^\vee} \cap (\mathbb{P}^{n-1} \smallsetminus \Sigma_n).$$

This gives an isomorphism of X_Γ and $X_{\Gamma^\vee}$ outside of Σ_n, see [**2**] for more details.

For a banana graph Γ_n the dual graph $\Gamma_n^\vee$ is just a polygon with n sides, for which the graph hypersurface $X_{\Gamma_n^\vee} = \mathcal{L}$ is just a hyperplane in $\mathbb{P}^{n-1}$. One then computes separately the contribution to the virtual motive coming from the intersection with the locus Σ_n,

$$[X_{\Gamma_n}] = [X_{\Gamma_n} \cap \Sigma_n] + [X_{\Gamma_n} \smallsetminus \Sigma_n]$$

and from the complement. One finds, in terms of the class $\mathbb{T} = [\mathbb{G}_m] = [\mathbb{A}^1] - [\mathbb{A}^0]$ of the multiplicative group,

$$[\mathcal{L} \smallsetminus \Sigma_n] = [\mathcal{L}] - [\mathcal{L} \cap \Sigma_n] = \frac{\mathbb{T}^{n-1} - (-1)^{n-1}}{\mathbb{T} + 1}$$

$$X_{\Gamma_n} \cap \Sigma_n = \mathcal{S}_n \quad \text{with} \quad [\mathcal{S}_n] = [\Sigma_n] - n\mathbb{T}^{n-2},$$

so that $[X_{\Gamma_n}] = [\mathcal{S}_n] + [\mathcal{L} \smallsetminus \Sigma_n]$ gives the formula (2.1).

2.3. Sums over graphs. When one works with graphs that are not necessarily planar, one can still look at the image of the graph hypersurfaces under the Cremona transformation. The dual hypersurface obtained in that way is then not necessarily a graph hypersurface, but it can still give useful information on the original graph hypersurface. This approach was used by Bloch in [**11**] to prove a very interesting result about virtual motives of certain sums of graphs.

If one considers only graphs that have no looping edges or parallel edges, then such graphs can always be realized as subgraphs of the complete graph on the same number of vertices. By analyzing the graph hypersurface of the complete graph, and its relation to those of subgraphs, through the Cremona transformation method, Bloch showed in [**11**] that the sum

$$S_N = \sum_{\#V(\Gamma)=N} [X_\Gamma] \frac{N!}{\#\mathrm{Aut}(\Gamma)} \tag{2.2}$$

is always in the subring $\mathbb{Z}[\mathbb{L}]$ of virtual Tate motives, even though the individual terms $[X_\Gamma]$ will, in general, not be in $\mathbb{Z}[\mathbb{L}]$.

This result suggests that, although individual graphs may give non-mixed-Tate contributions, the sum over graphs in Feynman amplitudes may still be mixed Tate. There are two subtle problems in this, though: the first is that the typical sum over graphs in Feynman amplitudes is a sum over graphs with a fixed number of loops, not of vertices, and it is not clear whether a result like (2.2) may be expected in that case. The second problem is with the fact that periods do not factor through the Grothendieck ring, so the (renormalized) Feynman amplitudes $U(\Gamma)$ also will not decompose well according to a decomposition of the variety X_Γ as a class $[X_\Gamma]$ in the Grothendieck ring. Thus, even if a result analogous to (2.2) could be obtained for the more physical sum over loops, the result for classes in the Grothendieck ring would not directly imply a result of the same type for the Feynman amplitudes.

2.4. Deletion–contraction relations. The graph polynomial satisfies a deletion–contraction relation. Namely, for a graph Γ with $n \geq 2$ edges, with $\deg \Psi_\Gamma = \ell > 0$, one has

$$\Psi_\Gamma = t_e \Psi_{\Gamma \smallsetminus e} + \Psi_{\Gamma/e}$$

where the polynomials of the deletion and the contraction are, respectively, given by

$$\Psi_{\Gamma \smallsetminus e} = \frac{\partial \Psi_\Gamma}{\partial t_n} \quad \text{and} \quad \Psi_{\Gamma/e} = \Psi_\Gamma|_{t_n=0}.$$

It is a well known result about the graph polynomials (Proposition 5.2 of [**39**]) that the irreducible components of a graph hypersurface are themselves graph hypersurfaces of subgraphs that cannot be disconnected by removing a single vertex. Thus, one can assume that the graph hypersurface is irreducible, or work separately on each irreducible component. One can then apply the following general fact, proved in [**4**]. For $X = \{\psi = 0\} \subset \mathbb{P}^{n-1}$, and $Y = \{F = 0\} \subset \mathbb{P}^{n-2}$ with a polynomial ψ satisfying a decomposition

$$\psi(t_1, \ldots, t_n) = t_n F(t_1, \ldots, t_{n-1}) + G(t_1, \ldots, t_{n-1}),$$

one obtains that the projection from $(0 : \cdots : 0 : 1)$ gives an isomorphism

$$X \smallsetminus (X \cap \overline{Y}) \xrightarrow{\sim} \mathbb{P}^{n-2} \smallsetminus Y,$$

where $\overline{Y}$ is the cone of Y in $\mathbb{P}^{n-1}$.

This leads directly to a form of deletion–contraction relation for the virtual motives of the affine graph hypersurfaces $\widehat{X}_\Gamma \subset \mathbb{A}^n$, as proved by Aluffi and the author in [**4**]:

$$[\mathbb{A}^n \smallsetminus \widehat{X}_\Gamma] = \mathbb{L} \cdot [\mathbb{A}^{n-1} \smallsetminus (\widehat{X}_{\Gamma \smallsetminus e} \cap \widehat{X}_{\Gamma/e})] - [\mathbb{A}^{n-1} \smallsetminus \widehat{X}_{\Gamma \smallsetminus e}]$$

if e not a bridge (an edge whose removal would increase the number of connected components of the graph) or a looping edge (an edge with source and target at the same vertex);

$$[\mathbb{A}^n \smallsetminus \widehat{X}_\Gamma] = \mathbb{L} \cdot [\mathbb{A}^{n-1} \smallsetminus \widehat{X}_{\Gamma \smallsetminus e}] = \mathbb{L} \cdot [\mathbb{A}^{n-1} \smallsetminus \widehat{X}_{\Gamma/e}]$$

if e bridge;

$$[\mathbb{A}^n \smallsetminus \widehat{X}_\Gamma] = (\mathbb{L} - 1) \cdot [\mathbb{A}^{n-1} \smallsetminus \widehat{X}_{\Gamma \smallsetminus e}] = (\mathbb{L} - 1) \cdot [\mathbb{A}^{n-1} \smallsetminus \widehat{X}_{\Gamma/e}]$$

if e looping edge.

One certainly does not expect an actual deletion–contraction relation for the virtual motives, expressing the class $[\mathbb{A}^n \smallsetminus \widehat{X}_\Gamma]$ solely as a function of the classes $[\mathbb{A}^{n-1} \smallsetminus \widehat{X}_{\Gamma \smallsetminus e}]$ and $[\mathbb{A}^{n-1} \smallsetminus \widehat{X}_{\Gamma/e}]$ of the deletion and the contraction. In fact, if one has such a relation, it would imply that the virtual motives always remain mixed Tate, since we know they are mixed Tate for small graphs. Indeed, the deletion–contraction formula described above is more subtle, and it involves the more complicated term

$$[\mathbb{A}^{n-1} \smallsetminus (\widehat{X}_{\Gamma\smallsetminus e} \cap \widehat{X}_{\Gamma/e})]$$

with the intersection of the hypersurfaces of the deletion and the contraction.

It is precisely this intersection $\widehat{X}_{\Gamma\smallsetminus e} \cap \widehat{X}_{\Gamma/e}$ that is difficult to control motivically. In fact, two varieties can be mixed Tate but intersect along a non-mixed-Tate locus. So, if one tries to use this deletion–contraction formula to compute the class of $\widehat{X}_\Gamma$, for increasingly complicated graphs, in terms of the less complicated $\Gamma \smallsetminus e$ and Γ/e, one will at some point run into non-mixed Tate examples where, although both $\widehat{X}_{\Gamma\smallsetminus e}$ and $\widehat{X}_{\Gamma/e}$ are mixed Tate, the intersection $\widehat{X}_{\Gamma\smallsetminus e} \cap \widehat{X}_{\Gamma/e}$ is not.

2.5. Recursive operations on graphs. Although the presence of the term

$$[\mathbb{A}^{n-1} \smallsetminus (\widehat{X}_{\Gamma\smallsetminus e} \cap \widehat{X}_{\Gamma/e})]$$

in the deletion-contraction formula can make it difficult to explicitly compute the class of $\widehat{X}_\Gamma$ using this method, there are some operations on graphs for which one can carry out the computation explicitly and reduce the formula above to a real deletion-contraction formula. This happens because of a useful cancellation in the intersection term, [**4**].

Here we use the notation $\mathbb{U}(\Gamma) = [\mathbb{A}^n \smallsetminus \widehat{X}_\Gamma]$, which is suggested by the abstract "Feynman rules" type properties of this virtual motive (see [**3**]).

The first such operation is the replacement of an edge of the graph by a number of parallel edges between the same pair of vertices, namely Γ_{me} is obtained from Γ by replacing an edge e by m parallel edges, with $\Gamma_{0e} = \Gamma \smallsetminus e$ and $\Gamma_e = \Gamma$. Then one obtains as in [**4**] an explicit generating function

$$\begin{aligned} \sum_{m\geq 0} \mathbb{U}(\Gamma_{me}) \frac{s^m}{m!} &= \frac{e^{\mathbb{T}s} - e^{-s}}{\mathbb{T}+1} \mathbb{U}(\Gamma) \\ &+ \frac{e^{\mathbb{T}s} + \mathbb{T}e^{-s}}{\mathbb{T}+1} \mathbb{U}(\Gamma \smallsetminus e) \\ &+ \left(s\, e^{\mathbb{T}s} - \frac{e^{\mathbb{T}s} - e^{-s}}{\mathbb{T}+1} \right) \mathbb{U}(\Gamma/e), \end{aligned}$$

in the case where e not bridge nor looping edge, with $\mathbb{T} = [\mathbb{G}_m] \in K_0(\mathcal{V})$. The case of bridges or looping edges gives rise to a similar formula.

This formula is obtained in [**4**] by first considering the case of the doubling of an edge, for which the inclusion-exclusion formula for the virtual motive gives

$$\mathbb{U}(\Gamma_{2e}) = \mathbb{L} \cdot [\mathbb{A}^n \smallsetminus (\hat{X}_\Gamma \cap \hat{X}_{\Gamma_o})] - \mathbb{U}(\Gamma)$$

$$[\hat{X}_\Gamma \cap \hat{X}_{\Gamma_o}] = [\hat{X}_{\Gamma/e}] + (\mathbb{L}-1) \cdot [\hat{X}_{\Gamma\smallsetminus e} \cap \hat{X}_{\Gamma/e}],$$

where Γ_o denotes the graph obtained by attaching a looping edge to Γ/e. As one can see, in this case a cancellation occurs that eliminates the term with the intersection of the hypersurfaces of the deletion and the contraction and one obtains the recursion formula

$$\mathbb{U}(\Gamma_{2e}) = (\mathbb{L}-2) \cdot \mathbb{U}(\Gamma) + (\mathbb{L}-1) \cdot \mathbb{U}(\Gamma \smallsetminus e) + \mathbb{L} \cdot \mathbb{U}(\Gamma/e),$$

which only uses the data $\mathbb{U}(\Gamma)$, $\mathbb{U}(\Gamma \smallsetminus e)$, and $\mathbb{U}(\Gamma/e)$.

A similar recursive formula exists for the operation of splitting an edge by inserting valence two vertices. An example of an application of these recursive formulae is the computation of the virtual motives for "lemon graphs" and chains of polygons, [**4**]. If Λ_m is the lemon graph m wedges and Γ_m^Λ is the graph obtained by replacing an edge e of Γ with a "lemon wedge" Λ_m, one finds that the virtual motives are given by the generating function:

$$\sum_{m\geq 0} \mathbb{U}(\Gamma_m^\Lambda) s^m = \frac{(1-(\mathbb{T}+1)s)\,\mathbb{U}(\Gamma) + (\mathbb{T}+1)\mathbb{T}s\,\mathbb{U}(\Gamma\smallsetminus e) + (\mathbb{T}+1)^2 s\,\mathbb{U}(\Gamma/e)}{1-\mathbb{T}(\mathbb{T}+1)s - \mathbb{T}(\mathbb{T}+1)^2 s^2},$$

in the case where e is not bridge nor looping edge, and by a similar formula otherwise. The recursive relation is in this case of the form

$$\mathbb{U}(\Lambda_{m+1}) = \mathbb{T}(\mathbb{T}+1)\mathbb{U}(\Lambda_m) + \mathbb{T}(\mathbb{T}+1)^2\mathbb{U}(\Lambda_{m-1})$$

and the terms $a_m = \mathbb{U}(\Lambda_m)$ form a *divisibility sequence*, namely $\mathbb{U}(\Lambda_{m-1})$ divides $\mathbb{U}(\Lambda_{n-1})$ if m divides n.

3. Other approaches

We discuss in this section some other closely related approaches to the question of the motivic nature of the residues of Feynman integrals.

3.1. Determinant hypersurfaces and Schubert cells. Another way of looking at the problem of the nature of the periods and motives involved in the Feynman integral computations is to use the description of the Kirchhoff polynomial Ψ_Γ as a determinant (the matrix-tree theorem) to map the relevant period computation from the complement $\mathbb{A}^n \smallsetminus \hat{X}_\Gamma$, whose motive is hard to compute and generally not mixed Tate, to the complement of a determinant hypersurface, which is always mixed Tate. The cost of doing this is to trade the simpler divisor Σ_n that contains the boundary $\delta\sigma_n$ of the domain of integration, for a more complicated locus whose motivic nature may be complicated to compute explicitly. This is the point of view proposed by Aluffi and the author in [**6**], which we recall briefly here.

More precisely, one has a map

$$\Upsilon : \mathbb{A}^n \to \mathbb{A}^{\ell^2}, \quad \Upsilon(t)_{kr} = \sum_i t_i \eta_{ik}\eta_{ir},$$

which identifies the affine graph hypersurface with the preimage

$$\hat{X}_\Gamma = \Upsilon^{-1}(\hat{\mathcal{D}}_\ell)$$

of the determinant hypersurface $\hat{\mathcal{D}}_\ell = \{\det(x_{ij}) = 0\}$.

The virtual motive of the determinant hypersurface complement is much simpler than that of the graph hypersurfaces. It is a mixed Tate motive with class in the Grothendieck ring given by the explicit formula

$$[\mathbb{A}^{\ell^2} \smallsetminus \hat{\mathcal{D}}_\ell] = \mathbb{L}^{\binom{\ell}{2}} \prod_{i=1}^{\ell} (\mathbb{L}^i - 1).$$

Thus, when Υ is an embedding, one can compute the Feynman amplitude as

$$U(\Gamma) = \int_{\Upsilon(\sigma_n)} \frac{\mathcal{P}_\Gamma(x,p)^{-n+D\ell/2}\omega_\Gamma(x)}{\det(x)^{-n+(\ell+1)D/2}}.$$

If $\hat{\Sigma}_\Gamma$ is a normal crossings divisor in $\mathbb{A}^{\ell^2}$ with $\Upsilon(\partial\sigma_n) \subset \hat{\Sigma}_\Gamma$, then the main question becomes whether the motive

$$\mathfrak{m}(\mathbb{A}^{\ell^2} \smallsetminus \hat{\mathcal{D}}_\ell, \hat{\Sigma}_\Gamma \smallsetminus (\hat{\Sigma}_\Gamma \cap \hat{\mathcal{D}}_\ell))$$

is a mixed Tate motive. Here the notation $\mathfrak{m}(X,Y)$ is used to denote the object in the category of mixed motives underlying the relative cohomology $H^*(X,Y)$.

It is possible to identify some explicit combinatorial conditions on the graph Γ that ensure that the map

$$\Upsilon : \mathbb{A}^n \smallsetminus \hat{X}_\Gamma \hookrightarrow \mathbb{A}^{\ell^2} \smallsetminus \hat{\mathcal{D}}_\ell$$

is an embedding. One has the following conditions (see [**6**]): Γ is a closed 2-cell embedded graph $\iota : \Gamma \hookrightarrow S_g$ in a Riemann surface of genus g, with $S_g \smallsetminus \Gamma$ union of open disks (faces). The closure of each is a disk. Two faces have at most one edge in common. Every edge in the boundary of two faces.

Then a sufficient condition that ensures that the map to the determinant hypersurface complement is an embedding is that Γ is a 3-edge-connected graph with closed 2-cell embedding of face width ≥ 3. The face width is the largest $k \in \mathbb{N}$ such that every non-contractible simple closed curve in S_g intersects Γ at least k times. (It is set to be ∞ for planar graphs.) Notice that, in the usual physics terminology, a 2-edge-connected graph is called a 1PI graph, while the 2-vertex-connected condition conjecturally implies face width ≥ 2.

One can then look again at the question of identifying the motive $\mathfrak{m}(X,Y)$.

One first observes that the divisor $\hat{\Sigma}_\Gamma$ is always contained, $\hat{\Sigma}_\Gamma \subset \hat{\Sigma}_{\ell,g}$, in a divisor $\hat{\Sigma}_{\ell,g}$, for $f = \ell - 2g + 1$, where

$$\hat{\Sigma}_{\ell,g} = L_1 \cup \cdots \cup L_{\binom{f}{2}}$$

is a union of linear spaces L_i, so that $\hat{\Sigma}_{\ell,g}$ can be described by equations of the form

$$\begin{cases} x_{ij} & = 0 \quad 1 \leq i < j \leq f-1 \\ x_{i1} + \cdots + x_{i,f-1} & = 0 \quad 1 \leq i \leq f-1 \end{cases}$$

and

$$\mathfrak{m}(\mathbb{A}^{\ell^2} \smallsetminus \hat{\mathcal{D}}_\ell, \hat{\Sigma}_{\ell,g} \smallsetminus (\hat{\Sigma}_{\ell,g} \cap \hat{\mathcal{D}}_\ell)), \tag{3.1}$$

where $\hat{\Sigma}_{\ell,g}$ is a normal crossings divisor containing $\Upsilon_\Gamma(\partial\sigma_n) \subset \hat{\Sigma}_{\ell,g}$, which depends only on $\ell = b_1(\Gamma)$ and on g, the minimum genus of a Riemann surface S_g in which the graph can be embedded.

As observed in [**6**] a sufficient condition for the motive $\mathfrak{m}(X,Y)$ of the form (3.1) to be mixed Tate is that the *varieties of frames* are mixed Tate motives, where for a collection $V_1, \ldots, V_\ell$ of linear subspaces in a fixed ℓ-dimensional vector space, the variety of frames $\mathbb{F}(V_1, \ldots, V_\ell)$ is defined as

$$\mathbb{F}(V_1, \ldots, V_\ell) := \{(v_1, \ldots, v_\ell) \in \mathbb{A}^{\ell^2} \smallsetminus \hat{\mathcal{D}}_\ell \,|\, v_k \in V_k\}.$$

In the case of two or three subspaces this can be verified directly (see [**6**]): for two subspaces with $d_{12} = \dim(V_1 \cap V_2)$ one has

$$[\mathbb{F}(V_1, V_2)] = \mathbb{L}^{d_1+d_2} - \mathbb{L}^{d_1} - \mathbb{L}^{d_2} - \mathbb{L}^{d_{12}+1} + \mathbb{L}^{d_{12}} + \mathbb{L}.$$

In the case of three subspaces with $D = \dim(V_1 + V_2 + V_3)$, one has

$$[\mathbb{F}(V_1, V_2, V_3)] = (\mathbb{L}^{d_1} - 1)(\mathbb{L}^{d_2} - 1)(\mathbb{L}^{d_3} - 1)$$

$$-(\mathbb{L}-1)((\mathbb{L}^{d_1}-\mathbb{L})(\mathbb{L}^{d_{23}}-1)+(\mathbb{L}^{d_2}-\mathbb{L})(\mathbb{L}^{d_{13}}-1)+(\mathbb{L}^{d_3}-\mathbb{L})(\mathbb{L}^{d_{12}}-1)$$
$$+(\mathbb{L}-1)^2(\mathbb{L}^{d_1+d_2+d_3-D}-\mathbb{L}^{d_{123}+1})+(\mathbb{L}-1)^3$$

However, one cannot use these cases to start an induction argument.

One can see the difficulty by another formulation given in [**6**]: consider the locus $Flag_{\ell,\{d_i,e_i\}}(\{V_i\})$ of complete flags $0 \subset E_1 \subset E_2 \subset \cdots \subset E_\ell = E$, with $\dim E_i \cap V_i = d_i$ and $\dim E_i \cap V_{i+1} = e_i$. One can ask when these are mixed Tate, for all choices of d_i, e_i.

The variety of frames $\mathbb{F}(V_1, \ldots, V_\ell)$ is a fibration over this $Flag_{\ell,\{d_i,e_i\}}(\{V_i\})$ and the virtual motive satsifies

$$[\mathbb{F}(V_1,\ldots,V_\ell)] = [Flag_{\ell,\{d_i,e_i\}}(\{V_i\})](\mathbb{L}^{d_1}-1)(\mathbb{L}^{d_2}-\mathbb{L}^{e_1})(\mathbb{L}^{d_3}-\mathbb{L}^{e_2})\cdots(\mathbb{L}^{d_r}-\mathbb{L}^{e_{r-1}})$$

where $Flag_{\ell,\{d_i,e_i\}}(\{V_i\})$ is an intersection of unions of Schubert cells in flag varieties. The motivic nature of such loci is known to be a very delicate question related to the Kazhdan–Lusztig conjecture. So one sees where the difficulties resurface under this point of view.

3.2. Motives and Feynman amplitudes in configuration spaces. In what we reviewed so far, the point of view is always on the approach to perturbative quantum field theory where the Feynman integrals of individual Feynman graphs are computed in momentum space. One can also take the dual, configuration space point of view and analyze the question of the occurrence of motives in that setting. I will recall here some recent work of Ceyhan and the author [**20**] that explores the motivic setup for Feynman integrals in configuration space.

In this setting, the singularities of the Feynman amplitudes occur along a hypersurface Z_Γ in X^{V_Γ}, for X the spacetime manifold, whose real locus is along the diagonals

$$\Delta_e = \{(x_v)_{v\in V_\Gamma} \,|\, x_{v_1} = x_{v_2} \text{ for } \partial_\Gamma(e) = \{v_1, v_2\}\}.$$

The configuration space itself, for a given Feynman graph Γ is given by

$$Conf_\Gamma(X) = X^{V_\Gamma} \smallsetminus \bigcup_{e\in E_\Gamma} \Delta_e = X^{V_\Gamma} \smallsetminus \cup_{\gamma\subset \mathcal{G}_\Gamma} \Delta_\gamma,$$

with $\mathcal{G}_\Gamma$ the set of subgraphs that are *induced* (they contain all edges of Γ between the given subset of vertices) and 2-vertex-connected.

To treat the divergences along the diagonals, one can embed the configuration space

$$Conf_\Gamma(X) \hookrightarrow \prod_{\gamma\in\mathcal{G}_\Gamma} \mathrm{Bl}_{\Delta_\gamma} X^{V_\Gamma}$$

inside an iterated blowup along a family of diagonals associated to suitable subgraphs. This gives rise to a particular case of what is more generally known as the De Concini–Procesi "wonderful compactifications" [**21**], which in turn generalize the Fulton–MacPherson compactification [**24**]. The compactification can be explicitly described in terms of strata

$$\overline{Conf}_\Gamma(X) = Conf_\Gamma(X) \cup \bigcup_{\mathcal{N}\in\ \mathcal{G}-\text{nests}} X^\circ_{\mathcal{N}}$$

of a stratification by $\mathcal{G}$-nests of subgraphs, see the work of Li Li, [**34**], [**35**], for more details on the more general construction of wonderful compactifications for arrangements of subvarieties.

For these wonderful compactifications, which have the purpose of blowing up the locus where the singularities of the Feynman amplitudes occur, one can explicitly compute the motive, both in the category of Voevodsky motives (in the case where X is quasi-projective) and as a virtual motive. In the smooth projective case the Chow motive computation follows from the general result of Li Li for the wonderful compactifications [**35**].

In the Voevodsky category of mixed motives, we find (see [**20**]) an explicit expression for the motive $\mathfrak{m}(\overline{Conf}_\Gamma(X))$.

One uses the notation $M_{\mathcal{N}} := \{(\mu_\gamma)_{\Delta_\gamma \in \mathcal{G}_\Gamma} : 1 \leq \mu_\gamma \leq r_\gamma - 1,\ \mu_\gamma \in \mathbb{Z}\}$ with $r_\gamma = r_{\gamma,\mathcal{N}} := \dim(\cap_{\gamma' \in \mathcal{N}:\gamma' \subset \gamma} \Delta_{\gamma'}) - \dim \Delta_\gamma$ and $\|\mu\| := \sum_{\Delta_\gamma \in \mathcal{G}_\Gamma} \mu_\gamma$, as in [**35**]. Then the motives of the compactification is given by

$$\mathfrak{m}(\overline{Conf}_\Gamma(X)) = \mathfrak{m}(X^{V_\Gamma}) \oplus \bigoplus_{\mathcal{N} \in \mathcal{G}_\Gamma\text{-nests}, \mu \in M_{\mathcal{N}}} \mathfrak{m}(X^{V_{\Gamma/\delta_{\mathcal{N}}(\Gamma)}})(\|\mu\|)[2\|\mu\|]$$

where the notation $\Gamma/\delta_{\mathcal{N}}(\Gamma)$ means the quotient $\Gamma//(\gamma_1 \cup \cdots \cup \gamma_r)$, where $\mathcal{N} = \{\gamma_1, \ldots, \gamma_r\}$ is the $\mathcal{G}$-nest and the quotient double bar $\Gamma//\gamma$ means the graph obtained from Γ by shrinking each connected component of the subgraph γ to a (different) vertex.

One can compute in a similar way the virtual motive in the Grothendieck ring of varieties, which is given by (see [**20**])

$$[\overline{Conf}_\Gamma(X)] = [X]^{|V_\Gamma|} + \sum_{\mathcal{N} \in \mathcal{G}_\Gamma\text{-nests}} [X]^{|V_{\Gamma/\delta_{\mathcal{N}}(\Gamma)}|} \sum_{\mu \in M_{\mathcal{N}}} \mathbb{L}^{\|\mu\|}.$$

In both cases, the key to obtaining these explicit formulae is the presence of blowup formulae: for mixed motives ([**42**])

$$\mathfrak{m}(\mathrm{Bl}_V(Y)) \cong \mathfrak{m}(Y) \oplus \bigoplus_{k=1}^{\mathrm{codim}_Y(V)-1} \mathfrak{m}(V)(k)[2k],$$

and for the virtual motives the Bittner relations ([**9**])

$$[\mathrm{Bl}_V(Y)] = [Y] - [V] + [E] = [Y] + [V]([\mathbb{P}^{\mathrm{codim}_Y(V)-1}] - 1),$$

where E is the exceptional divisor. One then obtains the formulae for the wonderful compactifications of the configuration spaces of Feynman graphs using their description as iterated blowups.

In particular, one sees from these explicit formulae that $\overline{Conf}_\Gamma(X)$ is a mixed Tate motives whenever X is. Moreover, to regularize the Feynman integrals one can lift the amplitude computation to the blowup $\overline{Conf}_\Gamma(X)$. This generates some ambiguities related to monodromies along exceptional divisors of the iterated blowups (the scenario is reminiscent of the situation one encounters in momentum space in [**13**]). The residues of Feynman integrals can then be expressed in terms of periods on hypersurface complement in $\overline{Conf}_\Gamma(X)$ and some of these residues can be computed using Poincaré residues, in terms of periods on intersections of divisors of the stratification, see [**20**].

4. Further comments, questions and perspectives

All the original cases computed by Broadhurst and Kreimer in [**15**] have now been rigorously proved to be periods of mixed Tate motives in recent work of Francis Brown, [**17**], by mapping the period computation to moduli space $\bar{\mathcal{M}}_{0,n}$ and using results on multiple zeta values as periods of $\bar{\mathcal{M}}_{0,n}$, of Goncharov and Manin [**26**] and of Brown [**18**].

Heuristically, the fact that the graph hypersurfaces remain mixed Tate for a surprisingly large size of the graph (the first counterexample found by Doryn has 14 edges), is related to the fact that these hypersurfaces are very singular, with singularities in low codimension, which allows for the motive to be much "simpler" than that of less singular hypersurfaces of the same degree. It is in any case an interesting question to study the singularities of the graph hypersurfaces and have good invariant that estimate how singular they are. The Chern–Schwartz–MacPherson classes of singular varieties are a very refined invariant that measures how singular a variety is. These Chern classes of singular varieties can be conveniently assembled into an invariant of the graph hypersurface complement that behaves like an algebro-geometric Feynman rule, in the sense of [**3**]. It is shown in the recent work of Aluffi [**1**] that these invariants also satisfy a deletion-contraction relation and recursions formulae, similar to those we recalled above for the virtual motive.

While the general result of Belkale and Brosnan [**8**] and the recent explicit examples of Doryn [**22**] and Brown and Schnetz [**19**] show that graph hypersurfaces become non-mixed-Tate as the graphs become sufficiently large, the question remains of whether the specific period given by the Feynman amplitude may still be itself a mixed Tate period. In fact, in principle, it may happen that the specific piece of the cohomology that is involved in that period computation remains mixed Tate even if the variety itself is no longer a mixed Tate motive. Some computation of the middle cohomology carried out by Doryn in [**23**] are compatible with this possibility. However, there are good reasons at this point to believe that this guess itself may be too strong and that Feynman amplitudes that are non-mixed-Tate periods will also appear for sufficiently large graphs. A good source of evidence in this direction is in the recent paper [**19**], where Brown and Schnetz show that, for physically significant graphs one can write the virtual motive of the graph hypersurface in the form $[X_\Gamma] \equiv c_2(\Gamma)\mathbb{L}^2 \bmod \mathbb{L}^3$, where what they denote $c_2(\Gamma)$ is a new invariant, a class in the Grothendieck ring of varieties, which they expect will be closely related to the "framing of the motive", which would be the smallest piece of the motive carrying the information on the period. Thus, the fact that these $c_2(\Gamma)$ will eventually be non-mixed-Tate might be used to prove that the Feynman amplitudes themselves will cease to be mixed Tate for appropriate graphs.

References

[1] P. Aluffi, *Chern classes of graph hypersurfaces and deletion-contraction*, arXiv:1106.1447, to appear in Moscow Mathematical Journal.

[2] P. Aluffi, M. Marcolli, *Feynman motives of banana graphs*, Communications in Number Theory and Physics, Vol.3 (2009) N.1, 1–57.

[3] P. Aluffi, M. Marcolli, *Algebro-geometric Feynman rules*, International Journal of Geometric Methods in Modern Physics, Vol.8 (2011) N.1, 203–237.

[4] P. Aluffi, M. Marcolli, *Feynman motives and deletion-contraction relations*, in "Topology of Algebraic Varieties and Singularities", Contemporary Mathematics, Vol.538 (2011) 21–64.

[5] P. Aluffi, M. Marcolli *Graph hypersurfaces and a dichotomy in the Grothendieck ring*, Letters in Mathematical Physics, (2011) Vol.95, 223–232.

[6] P. Aluffi, M. Marcolli, *Parametric Feynman integrals and determinant hypersurfaces*, Advances in Theoretical and Mathematical Physics, Vol.14 (2010) 911–963.

[7] Y. André, *An introduction to motivic zeta functions of motives*, in "Motives, quantum field theory, and pseudodifferential operators", pp.3–17, Clay Math. Proc., 12, Amer. Math. Soc., Providence, RI, 2010.
[8] P. Belkale, P. Brosnan, *Matroids, motives, and a conjecture of Kontsevich*, Duke Math. J. 116 (2003), no. 1, 147–188.
[9] F. Bittner, *The universal Euler characteristic for varieties of characteristic zero*, Compos. Math. 140 (2004), no. 4, 1011–1032.
[10] J. Bjorken, S. Drell, Relativistic Quantum Mechanics, McGraw-Hill, 1964, and Relativistic Quantum Fields, McGraw-Hill, 1965.
[11] S. Bloch, *Motives associated to sums of graphs*, in "The geometry of algebraic cycles", 137–143, Clay Math. Proc., 9, Amer. Math. Soc., Providence, RI, 2010.
[12] S. Bloch, H. Esnault, D. Kreimer, *On motives associated to graph polynomials*, Comm. Math. Phys. 267 (2006), no. 1, 181–225.
[13] S. Bloch, D. Kreimer, *Mixed Hodge structures and renormalization in physics*, Commun. Number Theory Phys. 2 (2008), no. 4, 637–718.
[14] S. Bloch, D. Kreimer, *Feynman amplitudes and Landau singularities for one-loop graphs*, Commun. Number Theory Phys. 4 (2010), no. 4, 709–753.
[15] D. Broadhurst, D. Kreimer, *Association of multiple zeta values with positive knots via Feynman diagrams up to 9 loops*, Phys. Lett. B 393 (1997), no. 3-4, 403–412.
[16] F. Brown, *Mixed Tate motives over* $\mathbb{Z}$, Annals of Math. Vol.175 (2012) N.2, 949–976.
[17] F. Brown, *On the periods of some Feynman integrals*, arXiv:0910.0114.
[18] F. Brown, *Multiple zeta values and periods of moduli spaces* $\mathfrak{M}_{0,n}$, Ann. Sci. Éc. Norm. Supér. (4) 42 (2009), no. 3, 371–489.
[19] F. Brown, O. Schnetz, *A* $K3$ *in* ϕ^4, arXiv:1006.4064.
[20] Ö. Ceyhan, M. Marcolli, *Feynman integrals and motives of configuration spaces*, arXiv:1012.5485, to appear in Communications in Mathematical Physics.
[21] C. De Concini, C. Procesi, *Wonderful models of subspace arrangements*, Selecta Math. (N.S.) Vol.1 (1995) N.3, 459–494.
[22] D. Doryn, *On one example and one counterexample in counting rational points on graph hypersurfaces*, Lett. Math. Phys. 97 (2011), no. 3, 303–315.
[23] D. Doryn, *Cohomology of graph hypersurfaces associated to certain Feynman graphs*, Commun. Number Theory Phys. 4 (2010), no. 2, 365–415.
[24] W. Fulton, R. MacPherson, *A compactification of configuration spaces*, Ann. of Math. (2) 139 (1994), no. 1, 183–225.
[25] H. Gillet, C. Soulé, *Descent, motives and K-theory*, J. Reine Angew. Math. 478 (1996), 127–176.
[26] A. Goncharov, Yu.I. Manin, *Multiple* ζ*-motives and moduli spaces* $\bar{M}_{0,n}$, Compos. Math. 140 (2004), no. 1, 1–14.
[27] M. Hanamura, *Mixed motives and algebraic cycles. I*, Math. Res. Lett. 2 (1995), no. 6, 811–821.
[28] C. Itzykson, J.B. Zuber, Quantum Field Theory, Dover Publications, 2006.
[29] U. Jannsen, *Motives, numerical equivalence, and semi-simplicity*, Invent. Math. 107 (1992), no. 3, 447–452.
[30] M. Kontsevich, D. Zagier, *Periods*, in "Mathematics unlimited—-2001 and beyond", 771–808, Springer, Berlin, 2001.
[31] M. Larsen, V. Lunts, *Motivic measures and stable birational geometry*, Mosc. Math. J. 3 (2003), no. 1, 85–95.
[32] M. Levine, *Mixed motives*, Mathematical Surveys and Monographs, 57. American Mathematical Society, Providence, RI, 1998.
[33] M. Levine, *Tate motives and the vanishing conjectures for algebraic K-theory*, in "Algebraic K-theory and algebraic topology" 167–188, NATO Adv. Sci. Inst. Ser. C Math. Phys. Sci., 407, Kluwer, 1993.
[34] L. Li, *Wonderful compactification of an arrangement of subvarieties*, Michigan Math. J. 58 (2009), no. 2, 535–563.
[35] L. Li, *Chow Motive of Fulton–MacPherson configuration spaces and wonderful compactifications*, Michigan Math. J. 58 (2009), no. 2, 565–598.
[36] Yu.I. Manin, *Correspondences, motifs and monoidal transformations*, Mat. Sb. (N.S.) 77 (119) 1968 475–507.

[37] M. Marcolli, *Feynman motives*, World Scientific, 2010.

[38] A. Rej, *Motives: an introductory survey for physicists. With an appendix by Matilde Marcolli*, Contemp. Math., 539, "Combinatorics and physics", 377–415, Amer. Math. Soc., Providence, RI, 2011.

[39] M. Sato, T. Miwa, M. Jimbo, T. Oshima, *Holonomy Structure of Landau Singularities and Feynman Integrals*, Publ. RIMS, Kyoto Univ. 12 Suppl. (1977), 387–438.

[40] J. Stembridge, *Counting points on varieties over finite fields related to a conjecture of Kontsevich*, Ann. Comb. 2 (1998), no. 4, 365–385.

[41] B. Toën, *Anneaux de Grothendieck des n-champs d'Artin*, arXiv:math/0509098.

[42] V. Voevodsky, *Triangulated categories of motives over a field*, in "Cycles, transfer and motivic homology theories", pp. 188–238, Annals of Mathematical Studies, Vol. 143, Princeton, 2000.

Mathematics Department, Mail Code 253-37, Caltech, 1200 E. California Blvd. Pasadena, CA 91125, USA

E-mail address: matilde@caltech.edu

Proceedings of Symposia in Pure Mathematics
Volume **85**, 2012

On 2d TQFTs whose values are holomorphic symplectic varieties

Gregory W. Moore and Yuji Tachikawa

ABSTRACT. For a simple and simply-connected complex algebraic group $G_{\mathbb{C}}$, we conjecture the existence of a functor $\eta_{G_{\mathbb{C}}}$ from the category of 2-bordisms to the category of holomorphic symplectic varieties with Hamiltonian action, such that gluing of boundaries corresponds to the holomorphic symplectic quotient with respect to the diagonal action of $G_{\mathbb{C}}$. We describe various properties of $\eta_{G_{\mathbb{C}}}$ obtained via string-theoretic analysis. Mathematicians are urged to construct $\eta_{G_{\mathbb{C}}}$ rigorously.

1. Introduction

Physicists have been gathering evidence for the existence of a class of interacting superconformal quantum field theories in six dimensions with $\mathcal{N} = (2,0)$ supersymmetry. These theories are believed to have an ADE classification, i.e. they are classified by a connected, simply connected and simply laced compact Lie group G, and appear to lie at the heart of a large number of results and constructions in physical mathematics. We will denote these theories by $S[G]$. For a discussion of 6d $\mathcal{N} = (2,0)$ theory aimed at mathematicians, see [**45, 44**].

Let C be a punctured Riemannian surface and consider the theories $S[G]$ on $\mathbb{R}^{1,3} \times C$. Combining the conformality of $S[G]$ with a partial topological twist one can argue that the resulting theory depends only on the conformal structure of C. Consequently, these partially twisted theories should satisfy factorization properties reminiscent of those of two-dimensional topological and conformal field theories. In Section 2

2010 *Mathematics Subject Classification.* Primary: 57R56, Secondary: 18D10, 53C26, 53D20, 81T45.

below we provide a slightly more precise description of these factorization properties, although we hasten to add that this description will not be completely satisfactory to mathematicians. This state of affairs can, however, be ameliorated. Several geometric objects, such as certain branches of moduli spaces of vacua, can be associated to the theories $S[G]$ on C, and these geometric objects inherit the factorization properties of the parent theory. The assignment of such geometric objects to the data of C *is* something which is susceptible to rigorous mathematical discussion. In this note we provide a simple example. Namely, we show how the "maximal-dimension Higgs branch of the theory $S[G]$ on C" provides an example of a two-dimensional topological field theory valued in a symmetric monoidal category which is not simply a category of vector spaces. This formulation captures some results originally found and described in physics language in the literature. See, for examples, [**3, 4, 24, 23, 22, 20, 19, 6, 37, 7, 8, 11, 41, 25**].

The existence of this two-dimensional field theory relies on the existence of certain holomorphic symplectic manifolds satisfying properties listed in Section 3. We hope mathematicians find the formulation of this topological field theory sufficiently precise and interesting to provide a rigorous construction of the required manifolds.

2. Physical Background

In this section we sketch in a little more detail the sense in which the theories $S[G]$ define a generalization of two-dimensional conformal field theory. Readers who are only interested in the two-dimensional topological field theory or those who prefer rigorously formulated mathematics can safely skip this section and proceed to Section 3.

Quantum field theories can have "defects," or "defect operators." These are operators or degrees of freedom which can be placed on positive codimension subspaces of spacetime. Thus, for example, a local operator is a point defect, a Wilson line operator is a line defect, and so on. The six-dimensional theories $S[G]$ have certain supersymmetric codimension-two defects. For our purposes all we need to know about these defects is that they are specified by a homomorphism $\rho : \mathfrak{su}(2) \to \mathfrak{g}$. Let us now consider an oriented surface C of genus g with n punctures with a Riemannian metric of finite volume. We study the theory $S[G]$ on $\mathbb{R}^{1,3} \times C$ with the codimension-two defects located at the punctures. We denote the punctures p_a assigned with the homomorphisms ρ_a collectively as D, and refer to the pair (C, D) as a decorated surface.

The theory $S[G]$ on $\mathbb{R}^{1,3} \times C$ admits a partial topological twist such that, when one takes the long distance limit, (i.e. the "Kaluza-Klein reduction") the result is a four-dimensional theory with $d = 4, \mathcal{N} = 2$ supersymmetry, which we will denote $S_G(C, D)$. This construction was introduced in [**30, 43, 21, 19**]. These theories, known as *theories of class S*, have many properties. One is that the defects D_a associated to punctures $p_a \in C$ have a global symmetry group [1] $G_a \subset G$, the commutant of the image of ρ. The global symmetry group of the theory $S_G(C, D)$ includes $\prod_a G_a$. Moreover, the theory $S_G(C, D)$ only depends on the *conformal structure* on C (but up to some subtleties). This property results from the topological twisting. Furthermore, the space of coupling constants is the moduli space $\mathcal{M}_{g,n}$ of complex structures on C, and the Deligne-Mumford boundaries correspond to weak coupling limits of the theory.

In [**19**], D. Gaiotto argued that at the boundaries of $\mathcal{M}_{g,n}$ the theories $S_G(C, D)$ should exhibit factorization properties analogous to those enjoyed by two-dimensional conformal field theories and two-dimensional topological field theories. Let us consider two decorated surfaces (C_L, D_L) and (C_R, D_R) and let us pick punctures $p_L \in C_L$ and $p_R \in C_R$ both with $\rho_{p_L} = \rho_{p_R} = 0$ so that $G_{p_L} = G_{p_R} = G$. Now, on the one hand, we can consider the product theory, and gauge the diagonal subgroup of its global $G \times G$ symmetry. In $d = 4, \mathcal{N} = 2$ gauge theory the only essential parameter introduced in this operation is a coupling constant $q = e^{2\pi i \tau}$. Call the resulting theory $S_L \times_{G,q} S_R$. On the other hand, choosing local coordinates z_L, z_R at p_L, p_R such that $p_{L,R}$ is at $z_{L,R} = 0$, we can form the punctured surface $C_L \times_q C_R$ by setting $z_L z_R = q$ with data D_{LR} at the remaining punctures. Gaiotto's crucial statement is that the two constructions lead to the same theory:

$$S_L \times_{G,q} S_R = S_G(C_L \times_q C_R, D_{LR}). \tag{2.1}$$

We caution the reader that for some pairs (C, D) (e.g. when C is a sphere with two punctures) the theory $S_G(C, D)$ does not exist as a genuine four-dimensional theory, and hence in some situations (2.1) must be interpreted with care. Nevertheless, this result points to the existence of a generalization of two-dimensional conformal field theory where we have a two-dimensional field theory whose target category is something like the "symmetric monoidal category of four-dimensional $\mathcal{N} = 2$ theories."

[1]The physics literature is rather imprecise about which compact form of the group one should choose, and the issue is nontrivial for reasons discussed in [**45**]. We will gloss over that point in this Section, and it will not affect the more precise mathematical statements of Sections 3 and 4.

The idea of a topological field theory whose target category is a general symmetric monoidal category has appeared quite often before in the formal study of topological field theories. One recent discussion may be found in [**29**]. If $\mathcal{F}$ is an n-dimensional topological field theory then for any fixed compact k-dimensional manifold K the functor $\mathcal{F}_K$ whose domain is the the $(n-k)$-dimensional bordism category and which is defined by $\mathcal{F}_K : M \mapsto \mathcal{F}(M \times K)$ is an $(n-k)$ dimensional topological field theory. This is a TFT version of Kaluza-Klein reduction along K. Therefore, $K \mapsto \mathcal{F}_K$ is an example of a k-dimensional field theory whose target category is the category of $(n-k)$-dimensional field theories. In an analogous way, we would like to regard S_G as a "functor" from decorated surfaces: $S_G : (C, D) \to S_G(C, D)$, where C is endowed with a conformal structure, as an example of a "two-dimensional conformal field theory whose target category is the category of four-dimensional $\mathcal{N} = 2$ theories." This statement can surely be made more precise, and it would be worthwhile doing so. However, in the absence of a mathematically rigorous formulation of an $\mathcal{N} = 2$, $d = 4$ quantum field theory it cannot be made fully rigorous.

In any case, we will view S_G as something analogous to a functor. Now, physicists know that many different mathematical objects can be associated to a 4d $\mathcal{N} = 2$ theory "functorially." Composing them with S_G, we expect to have operations which associate to (C, D) simpler mathematical objects, which might be rigorously formulated. Let us mention a few examples:

(1) Take the maximal-dimension Higgs branch $\mathcal{H}(\mathcal{T})$ of a 4d $\mathcal{N} = 2$ theory $\mathcal{T}$. Denoting this operation by $\mathcal{H}$, we consider the composed operation $\eta_G = \mathcal{H} \circ S_G$. This operation η_G associates a hyper-Kähler manifold to a punctured 2d surface C. The construction preserves SU(2) R-symmetry, which is manifested as the SO(3) isometry of the hyper-Kähler manifold rotating the three complex structures. Supersymmetry implies that the Higgs branch should be independent of coupling constant, and hence it is believed that $\eta_G(C)$ only depends on the topology of C. Since S_G is a 'functor' and $\mathcal{H}$ is a natural operation, we may expect that η_G is susceptible of a precise mathematical definition as a well-defined functor from the category of 2-bordisms to the category of hyper-Kähler manifolds. Due to a subtlety which we come back to in Section 5, we need to regard the image of η_G as a holomorphic symplectic variety to define a genuine TQFT. The SO(3) isometry of the hyper-Kähler manifold is now manifested as the $\mathbb{C}^\times$ action ψ_t on the variety

under which the symplectic form ω is rescaled, i.e. $\psi_t^*(\omega) = t^{-2}\omega$. In the next section we will show that η_G determines a 2d TQFT whose values are holomorphic symplectic varieties. In other words, $\eta_G(C)$ is the "amplitude" associated to the surface C.

(2) We could also talk about the composition $\mathcal{C} \circ S_G$, where $\mathcal{C}(\mathcal{T})$ of a 4d $\mathcal{N} = 2$ theory $\mathcal{T}$ is the "Coulomb branch of $\mathcal{T}$ on $\mathbb{R}^3 \times S^1$." As a holomorphic symplectic variety this is the algebraic integrable system canonically associated to $\mathcal{T}$. (See, for examples [**27, 13**] for reviews of early work and [**21**] for a recent discussion.) For a punctured Riemann surface C, $\mathcal{C} \circ S_G(C)$ is the moduli space of a Hitchin system on C. It is to be emphasized that $\eta_G(C)$ is *not* to be confused with $\mathcal{C} \circ S_G(C)$. Rather, $\eta_G(C)$ is the "3d mirror" to $\mathcal{C} \circ S_G(C)$ [**26, 23**]. Understanding the factorization properties of $\mathcal{C} \circ S_G$ expected from physics appears to be challenging.

(3) It should also be emphasized that there are many other objects which can be extracted from a "4d $\mathcal{N} = 2$ theory." Another class of objects is defined for each four-manifold X, and is called the partition function on X and denoted by Z_X. For compact X this will be a complex number. (In practice, X is endowed with extra data such as parameters of equivariant cohomology, when it has symmetries, or the data of external gauge fields coupling to global symmetries, in which case it is a function of these parameters.) Z_X has factorization properties analogous to the correlation functions of a conformal field theory or of a topological field theory on C. There are some notable examples of this in the literature. For example $Z_{S^4} \circ S_G(C)$ for $G = \mathrm{SU}(2)$ turns out to be related to the correlation functions of local operators on C of the Liouville conformal field theory [**1**]. This statement has been generalized to the 2d Toda field theory of type G with W-symmetry in [**46**]. The local operators are inserted at the punctures of C, and determined by the defect data. On the other hand, $Z_{S^3\times S^1} \circ S_G(C)$ is independent of the complex structure. It is in general a two-parameter deformation of the 2d q-deformed two-dimensional Yang-Mills theory in the zero-area limit [**16, 17, 18**].

The rest of the note is devoted to the formulation of $\eta_G = \mathcal{H} \circ S_G$. Conjecturally, the functor η_G also exists for non-simply-laced G. There is no known 6d $\mathcal{N} = (2,0)$ theory corresponding to non-simply-laced

Lie algebras. Therefore, when G is not simply-laced, one first puts a suitable 6d $\mathcal{N} = (2,0)$ theory on S^1 with an automorphism twist to produce 5D super Yang-Mills theories with gauge group $G^\vee$ [**42, 2**]. Compactifying such a theory on the punctured surface C produces a 3d theory, whose moduli space of vacua contains one branch of the form $\eta_G[C]$ and another branch which is the moduli space of a $G^\vee$-Hitchin system.

3. Axioms

3.1. The source and the target categories. Here we describe the basic properties of our functor. First we specify the source and the target categories. We take the source category to be the bordism category Bo_2, i.e. the objects are closed oriented one-dimensional manifolds (i.e. disjoint unions of multiple S^1s) and a morphism from B_1 to B_2 is a two-dimensional oriented manifold C whose boundary is $B_1 \sqcup (-B_2)$. Bo_2 is a symmetric monoidal category with duality under the standard operations.

The target category HS is a category of holomorphic symplectic varieties with Hamiltonian action defined as follows. Let us start with the category structure:

- The elements of Obj(HS) are complex algebraic semi-simple groups (including the trivial group 1).
- For $G_\mathbb{C}, G_\mathbb{C}' \in \mathrm{Obj(HS)}$, an element of $\mathrm{Hom}(G_\mathbb{C}, G_\mathbb{C}')$ is a triple $([X], G_\mathbb{C}, G_\mathbb{C}')$. Here X is a holomorphic symplectic variety with a $\mathbb{C}^\times$ action ψ_t such that $\psi_t^*(\omega) = t^{-2}\omega$ where $t \in \mathbb{C}^\times$, together with a holomorphic Hamiltonian action of $G_\mathbb{C} \times G_\mathbb{C}'$ commuting with ψ_t. We identify an X with symplectic form ω with an X' with symplectic form ω' if there is a holomorphic isomorphism $f : X \to X'$ commuting with the $G_\mathbb{C} \times G_\mathbb{C}'$ action and the $\mathbb{C}^\times$ action such that $f^*(\omega') = \omega$. $[X]$ denotes the resulting equivalence class. To lighten the notation we write informally $X \in \mathrm{Hom}(G_\mathbb{C}, G_\mathbb{C}')$ but one must bear in mind that a morphism has an ordered pair of groups $G_\mathbb{C}$ and $G_\mathbb{C}'$.
- For $X \in \mathrm{Hom}(G_\mathbb{C}', G_\mathbb{C})$ and $Y \in \mathrm{Hom}(G_\mathbb{C}, G_\mathbb{C}'')$, their composition $Y \circ X \in \mathrm{Hom}(G_\mathbb{C}', G_\mathbb{C}'')$ is defined as the holomorphic symplectic quotient

$$Y \circ X := X \times Y // G_\mathbb{C} = \{(x,y) \in X \times Y \mid \mu_X(x) + \mu_Y(y) = 0\}/G_\mathbb{C} \tag{3.1}$$

where $\mu_X : X \to \mathfrak{g}_\mathbb{C}^*$ and $\mu_Y : Y \to \mathfrak{g}_\mathbb{C}^*$ are moment maps of the $G_\mathbb{C}$ action on X and Y, respectively. Here $\mathfrak{g}_\mathbb{C}$ is the Lie algebra of $G_\mathbb{C}$.

- The identity $\mathrm{id}_{G_{\mathbb{C}}} \in \mathrm{Hom}(G_{\mathbb{C}}, G_{\mathbb{C}})$ is $T^*G_{\mathbb{C}}$ which has Hamiltonian $G_{\mathbb{C}} \times G_{\mathbb{C}}$ action.

We must check that these definitions define a category. The associativity of the composition follows readily from the definition. To see that $T^*G_{\mathbb{C}}$ acts as the identity consider, for example $T^*G_{\mathbb{C}} \circ X$. We can identify $T^*G_{\mathbb{C}} \cong G_{\mathbb{C}} \times \mathfrak{g}_{\mathbb{C}}$ using left- or right-invariant forms. Denoting an element of $T^*G_{\mathbb{C}}$ by (g, a) the moment map condition is $a + \mu(x) = 0$, which eliminates a. The induced two-form on the solution space is $G_{\mathbb{C}}$-invariant and basic. The quotient by $G_{\mathbb{C}}$ allows us to gauge g to 1, thus giving a holomorphic isomorphism with the original space X with its symplectic form.

The category HS is a symmetric monoidal category. The monoidal structure is given by the following operations, which are obviously symmetric:

- For $G_{\mathbb{C}}, G_{\mathbb{C}}' \in \mathrm{Obj(HS)}$, $G_{\mathbb{C}} \times G_{\mathbb{C}}' \in \mathrm{Obj(HS)}$ is the Cartesian product of groups.
- For $X \in \mathrm{Hom}(G_{\mathbb{C}}, G_{\mathbb{C}}')$ and $Y \in \mathrm{Hom}(H_{\mathbb{C}}, H_{\mathbb{C}}')$, $X \times Y \in \mathrm{Hom}(G_{\mathbb{C}} \times H_{\mathbb{C}}, G_{\mathbb{C}}' \times H_{\mathbb{C}}')$ is also the Cartesian product of X and Y.

Note that the trivial group 1 is the unit under this operation.

The category HS also comes with duality. Here by a symmetric monoidal category with duality we mean one such that

- For any object A, there is a dual object A^*.
- There are basic morphisms $p_A \in \mathrm{Hom}(A \times A^*, 1)$ and $q_A \in \mathrm{Hom}(1, A^* \times A)$ so that the following identities (sometimes called the "zig-zag identities" or "S diagram") hold:

$$(p_A \times \mathrm{id}_A) \circ (\mathrm{id}_A \times q_A) = \mathrm{id}_A \quad (\mathrm{id}_{A^*} \times p_A) \circ (q_A \times \mathrm{id}_{A^*}) = \mathrm{id}_{A^*}, . \tag{3.2}$$

The duality structure for Bo_2 is well known: It is simply the statement that the standard S-cobordism is equal to the tube. The duality structure for HS is given by:

- For an object $G_{\mathbb{C}} \in \mathrm{Obj(HS)}$, we define its dual to be $G_{\mathbb{C}}$ itself.
- We let $q_{G_{\mathbb{C}}} \in \mathrm{Hom}(1, G_{\mathbb{C}} \times G_{\mathbb{C}})$ and $p_{G_{\mathbb{C}}} \in \mathrm{Hom}(G_{\mathbb{C}} \times G_{\mathbb{C}}, 1)$ both be $T^*G_{\mathbb{C}}$, but with the action of $(h, h') \in G_{\mathbb{C}} \times G_{\mathbb{C}}$ on $g \in G_{\mathbb{C}}$ (considered as the zero section) is now given by $g \mapsto o(h)g(h')^{-1}$ where o is a non-inner order-2 automorphism when $\mathfrak{g}$ is of type A, D or E_6, and the identity automorphism otherwise.[2] $p_{G_{\mathbb{C}}}$, $q_{G_{\mathbb{C}}}$ and $\mathrm{id}_{G_{\mathbb{C}}}$ satisfy the zig-zag identities.

[2] Let $\mu_1 \oplus \mu_2 : T^*G_{\mathbb{C}} \to \mathfrak{g}^* \oplus \mathfrak{g}^*$ be the moment map of this $G_{\mathbb{C}} \otimes G_{\mathbb{C}}$ action. With o, $\mu_1(p)$ and $\mu_2(p)$ conjugate, $\mu_1(p) \sim \mu_2(p)$ at every point $p \in T^*G_{\mathbb{C}}$. Without o, we would have $\mu_1(p) \sim -\mu_2(p)$, ruining functoriality later.

3.2. Functor. Choose a simple, simply-connected algebraic group $G_{\mathbb{C}}$. We want to define a functor

$$\eta_{G_{\mathbb{C}}} : \mathrm{Bo}_2 \to \mathrm{HS} \tag{3.3}$$

between symmetric monoidal categories with duality. The image of the object S^1 is $G_{\mathbb{C}}$:

$$\eta_{G_{\mathbb{C}}}(S^1) = G_{\mathbb{C}}. \tag{3.4}$$

Next, because our categories have duality, it suffices to specify the image of the basic corbordisms with one, two and three incoming circles:

(3.5) $U =$  $\in \mathrm{Hom}(S^1, \varnothing)$,

(3.6) $V =$  $\in \mathrm{Hom}(S^1 \sqcup S^1, \varnothing)$,

(3.7) $W =$ 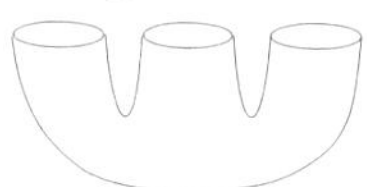 $\in \mathrm{Hom}(S^1 \sqcup S^1 \sqcup S^1, \varnothing)$.

Our convention is that in all the figures all boundary components are incoming. Other morphisms are then easily obtained by using p_{S^1} and q_{S^1}.

Let us define varieties $U_{G_{\mathbb{C}}} := \eta_{G_{\mathbb{C}}}(U)$ and $W_{G_{\mathbb{C}}} := \eta_{G_{\mathbb{C}}}(W)$. In order for $\eta_{G_{\mathbb{C}}}$ to be a functor, we must have $\eta_{G_{\mathbb{C}}}(V) = T^*G_{\mathbb{C}}$, and moreover, $U_{G_{\mathbb{C}}}$ and $W_{G_{\mathbb{C}}}$ have to satisfy basic sewing axioms of two-dimensional TFT (see e.g. [**35**]). In our case these axioms translate into the following statements:

- (Capping) $U_{G_{\mathbb{C}}} \circ W_{G_{\mathbb{C}}} = T^*G_{\mathbb{C}}$. This comes from the diagram:

(3.8) $\eta_{G_{\mathbb{C}}}($ 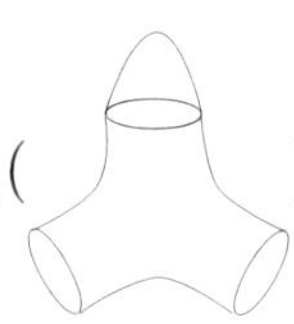 $) = \eta_{G_{\mathbb{C}}}($ 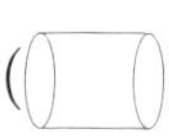 $)$.

- (Commutativity) $W_{G_{\mathbb{C}}}$ has an action of $\mathfrak{S}_3$ permuting the Hamiltonian ${G_{\mathbb{C}}}^3$ action, i.e. it has a holomorphic symplectic action of $\mathfrak{S}_3 \ltimes {G_{\mathbb{C}}}^3$. This comes from the diffeomorphism exchanging three boundaries of the pair of pants.
- (Associativity) $W_{G_{\mathbb{C}}} \circ W_{G_{\mathbb{C}}}$ has an action of $\mathfrak{S}_4$ permuting the Hamiltonian ${G_{\mathbb{C}}}^4$ action, i.e. it has a holomorphic symplectic

action of $\mathfrak{S}_4 \ltimes {G_\mathbb{C}}^4$. This comes from the following diagram:

(3.9) $\eta_{G_\mathbb{C}}($ 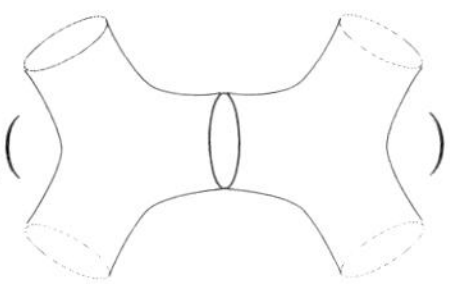 $) = \eta_{G_\mathbb{C}}($ 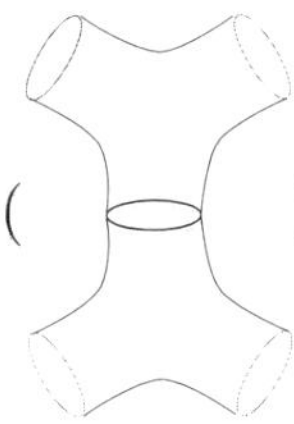 $).$

We can summarize as follows: there is a one-to-one correspondence between a 2d topological field theory $\eta_{G_\mathbb{C}}$ valued in HS and a pair of holomorphic symplectic varieties $(U_{G_\mathbb{C}}, W_{G_\mathbb{C}})$ satisfying the above three properties. We close with two remarks:

(1) The axioms imply that ${W_{G_\mathbb{C}}}^{n-2}$ has a holomorphic symplectic action of $\mathfrak{S}_n \ltimes {G_\mathbb{C}}^n$ for all positive n. Here we formally take ${W_{G_\mathbb{C}}}^{-1} = U_{G_\mathbb{C}}$ and ${W_{G_\mathbb{C}}}^{0} = V_{G_\mathbb{C}} = T^*G_\mathbb{C}$.

(2) $\eta_G(\text{torus})$ is the symplectic quotient of $T^*G_\mathbb{C}$ by the adjoint action of $G_\mathbb{C}$. This shows that the holomorphic varieties are in general singular.

3.3. Physical data. In the context of theories of class S, physical arguments show that $U_{G_\mathbb{C}}$ is a moduli space of Nahm's equation on a segment. These spaces have been studied by Kronheimer [**31**] and Bielawski [**9, 10**]. In particular $U_{G_\mathbb{C}}$ is given by

$$U_{G_\mathbb{C}} = G_\mathbb{C} \times S_n \subset G_\mathbb{C} \times \mathfrak{g}_\mathbb{C} \simeq T^*G_\mathbb{C} \tag{3.10}$$

where S_n is the Slodowy slice at a principal nilpotent element n. Recall that, by definition, the Slodowy slice $S_e \subset \mathfrak{g}_\mathbb{C}$ at a nilpotent element e is

$$S_e = \{e + v \in \mathfrak{g}_\mathbb{C} \mid [f, v] = 0\} \tag{3.11}$$

where (e, h, f) is an $\mathfrak{sl}(2)$ triple containing e, i.e. $[e, f] = h$, $[h, e] = 2e$ and $[h, f] = -2f$. Note that the principal nilpotent element n is unique up to conjugacy so the equivalence class $[U_{G_\mathbb{C}}]$ doesn't depend on the choice of n. For example, for $G_\mathbb{C} = \mathrm{SL}(N, \mathbb{C})$, an example of a principal nilpotent element n is a Jordan block of size $N \times N$, and S_n is an affine space of dimension $N - 1$. For example, for $N = 4$, we have

$$n = \begin{pmatrix} 0&1&0&0 \\ 0&0&1&0 \\ 0&0&0&1 \\ 0&0&0&0 \end{pmatrix} \quad \text{and} \quad S_n = \left\{ \begin{pmatrix} 0&1&0&0 \\ a&0&1&0 \\ b&a&0&1 \\ c&b&a&0 \end{pmatrix} \right\}. \tag{3.12}$$

Now the physical theories of class S predict the existence of a variety $W_{G_\mathbb{C}}$ satisfying the properties above needed to define a topological field theory $\eta_{G_\mathbb{C}}$. We will give some explicit examples of $W_{G_\mathbb{C}}$ in the next section but for general $G_\mathbb{C}$, $W_{G_\mathbb{C}}$ does not appear to be known, and we urge mathematicians to construct it.

We remark that the dimension of $W_{G_\mathbb{C}}$ is easily computed. The dimension of $U_{G_\mathbb{C}}$ is given by

$$\dim_\mathbb{C} U_{G_\mathbb{C}} = \dim_\mathbb{C} G_\mathbb{C} + \operatorname{rank} G_\mathbb{C}. \tag{3.13}$$

Now, since $G_\mathbb{C}$ acts effectively, the capping axiom implies that

$$\dim_\mathbb{C} W_{G_\mathbb{C}} = 3\dim_\mathbb{C} G_\mathbb{C} - \operatorname{rank} G_\mathbb{C}, \tag{3.14}$$

It would be interesting to know if there is a unique holomorphic symplectic manifold satisfying the above criteria.

4. Examples

4.1. The case $\mathfrak{g} = A_1$. For $\mathfrak{g} = A_1$, $W_{G_\mathbb{C}}$ is given by the flat symplectic space $W_{G_\mathbb{C}} = \mathbb{C}^2 \otimes \mathbb{C}^2 \otimes \mathbb{C}^2$ [**19**]. It satisfies all the properties listed in the above under the natural $\mathrm{SL}(2,\mathbb{C})^3$ action, together with the $\mathfrak{S}_3$ action permuting the three $\mathbb{C}^2$ factors. The associativity is somewhat nontrivial. It turns out that

(4.1) $\eta_{A_1}($ 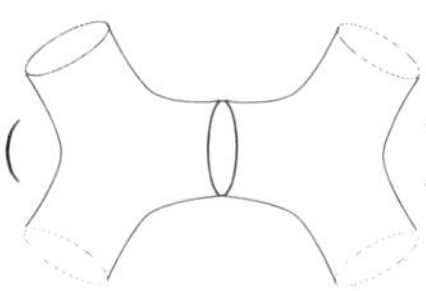 $) = (\mathbb{C}^2 \otimes \mathbb{C}^2 \otimes \mathbb{C}^2) \times (\mathbb{C}^2 \otimes \mathbb{C}^2 \otimes \mathbb{C}^2) // \mathrm{SL}(2,\mathbb{C})$

is the Atiyah-Drinfeld-Hitchin-Manin construction of the (closure of the) minimal nilpotent orbit of $\mathrm{SO}(8,\mathbb{C})$, or equivalently the framed centered 1-instanton moduli space of $\mathrm{SO}(8)$ gauge fields on $\mathbb{R}^4$ [**14**]. $\mathrm{SO}(8,\mathbb{C})$ has $\mathrm{SL}(2,\mathbb{C})^4$ subgroup, as shown in Fig. 1. The outer automorphism of $\mathrm{SO}(8)$ then provides the action of $\mathfrak{S}_4$, see Sec. 10 of [**40**]

4.2. The case $\mathfrak{g} = A_2$. For $\mathfrak{g} = A_2$, $W_{G_\mathbb{C}}$ is believed to be the (closure of the) minimal nilpotent orbit of $E_6(\mathbb{C})$ [**3, 22, 19**]. The group $E_6(\mathbb{C})$ has a special maximal proper subgroup $\mathrm{SL}(3)^3$, determined from the extended Dynkin diagram shown in Fig. 1. Outer automorphisms of $E_6(\mathbb{C})$ provide the action of $\mathfrak{S}_3$ on $W_{G_\mathbb{C}}$ permuting the three $\mathrm{SL}(3)$ actions. The capping and the associativity have not yet been checked. We believe it is a straightforward calculation.

4.3. A general conjectural property of $W_{G_\mathbb{C}}$. $W_{G_\mathbb{C}}$ has moment maps of the G^3 action

$$\mu_i : W_{G_\mathbb{C}} \to \mathfrak{g}_\mathbb{C}^* \tag{4.2}$$

for $i = 1, 2, 3$. From string theory analysis, it is believed (see e.g. Appendix C of [**7**]) that for any element of $P \in \mathbb{C}[\mathfrak{g}_\mathbb{C}]^{G_\mathbb{C}}$ we have

$$P(\mu_1) = P(\mu_2) = P(\mu_3) \tag{4.3}$$

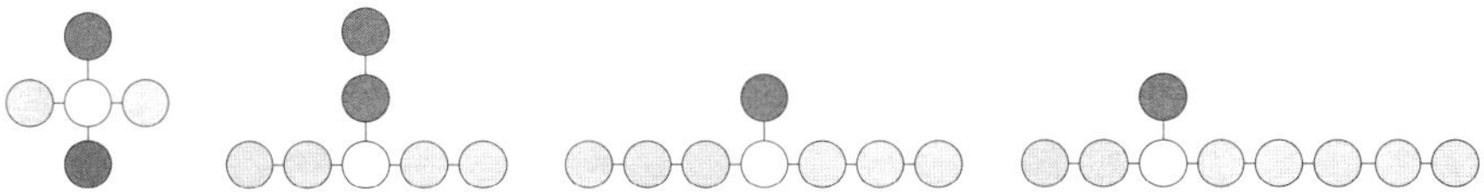

FIGURE 1. Extended Dynkin diagrams of D_4, E_6, E_7 and E_8, together with subdiagrams corresponding to $\mathrm{SL}(2)^4$, $\mathrm{SL}(3)^3$, $\mathrm{SL}(4)^2 \times \mathrm{SL}(2)$, and $\mathrm{SL}(6) \times \mathrm{SL}(3) \times \mathrm{SL}(2)$ subgroups

where we regarded P as a function on $\mathfrak{g}_{\mathbb{C}}^*$. This equality for $W_{A_1} = \mathbb{C}^2 \otimes \mathbb{C}^2 \otimes \mathbb{C}^2$ for the quadratic P reduces to the uniqueness of Cayley's hyperdeterminant. This equality for W_{A_2}, which is the minimal nilpotent orbit of $E_6(\mathbb{C})$, can be checked using its defining equations due to Joseph [**28**].

4.4. Some more conjectural properties for $\mathfrak{g} = A_n$. There are a few more results about W_{A_n} deduced from string theory analysis for general n. To describe them, we need to generalize the construction slightly. We enlarge the morphisms of the source category to be two-dimensional surfaces with marked points, with additional data at each marked point given by a homomorphism $\rho : \mathfrak{sl}(2) \to \mathfrak{g}_{\mathbb{C}}$. Correspondingly, we introduce holomorphic symplectic manifolds

(4.4) $\eta_{G_{\mathbb{C}}}($ 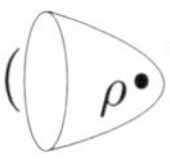 $) = G_{\mathbb{C}} \times S_{\rho(e)} \subset G_{\mathbb{C}} \times \mathfrak{g} \simeq T^* G_{\mathbb{C}}$

which are also the moduli spaces of Nahm's equation on a segment with appropriate boundary conditions [**9**]. Note that this variety has a Hamiltonian action of $G_{\mathbb{C}} \times Z(\rho)$, where $Z(\rho)$ is the centralizer of $\rho(\mathrm{SL}(2))$ inside $G_{\mathbb{C}}$. Note also that for $\rho = 0$ this manifold is $T^* G_{\mathbb{C}}$ itself.

Then we can associate to a sphere with three marked points a holomorphic symplectic variety using the holomorphic symplectic quotient specified by the following figure:

(4.5) $\eta_{G_{\mathbb{C}}}($ 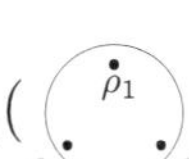 $) = \eta_{G_{\mathbb{C}}}($ 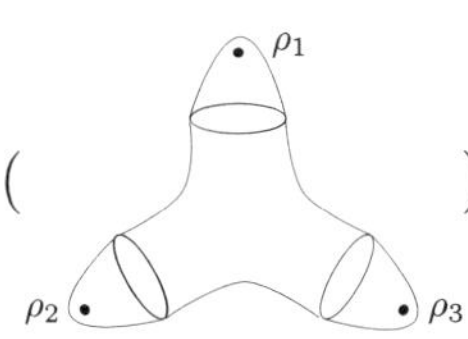 $).$

We denote this variety by $\eta_{G_{\mathbb{C}}}(\rho_1, \rho_2, \rho_3)$. Similarly, we can define the holomorphic symplectic manifolds $\eta_{G_{\mathbb{C}}}(\rho_1, \rho_2, \rho_3, \rho_4)$ associated to a sphere with four punctures marked by $\rho_{1,2,3,4}$.

For $G_{\mathbb{C}} = SL(N, \mathbb{C})$, the homomorphism $\rho : \mathfrak{sl}(2) \to \mathfrak{sl}(N)$ can be identified with a partition of N, which we denote by $[i^{e_i}, j^{e_j}, \ldots]$ for $N = \sum_i e_i i$. In this notation $\rho = 0$ corresponds to $[1^N]$. Now we can start listing the known properties of $\eta_{A_{N-1}}(\rho_1, \rho_2, \rho_3)$.

4.4.1. *Symplectic vector spaces.* With suitable choices of $\rho_{1,2,3}$, we can realize symplectic vector spaces in various representation of $\mathrm{SL}(N)$. For example,

$$\eta_{A_{N-1}}([1^N], [N-1, 1], [1^N]) = V \otimes V^* \oplus V^* \otimes V, \tag{4.6}$$

$$\begin{aligned}\eta_{A_{N-1}}([1^N], [\lfloor \tfrac{N+1}{2} \rfloor, \lfloor \tfrac{N}{2} \rfloor], [\lfloor \tfrac{N}{2} \rfloor, \lfloor \tfrac{N-1}{2} \rfloor, 1]) \\ = \Lambda^2 V \oplus \Lambda^2 V^* \oplus V \otimes \mathbb{C}^2 \oplus V^* \otimes \mathbb{C}^2\end{aligned} \tag{4.7}$$

where $V = \mathbb{C}^N$ [**19, 37, 11**]. More examples for A_3, A_4 and D_4 can be found in [**11, 12**].

From (4.6), we can deduce that

$$\eta_{A_{N-1}}([1^N], \underbrace{[N-1, 1], \ldots, [N-1, 1]}_{k}, [1^N]) \tag{4.8}$$

$$= \left[\bigoplus_{i=1}^{k} V_i \otimes V_{i+1}^* \oplus V_i^* \otimes V_{i+1} \right] // \mathrm{SL}(V_2) \times \mathrm{SL}(V_3) \times \cdots \times \mathrm{SL}(V_k)$$

where $V_i \simeq \mathbb{C}^N$. This is an SL-version of Nakajima's quiver varieties [**36**].

4.4.2. *Instanton moduli spaces.* When the choices of $\rho_{1,2,3}$ are related to the structure of an extended Dynkin diagram, $\eta_{A_n}(\rho_1, \rho_2, \rho_3)$ can sometimes be identified with instanton moduli spaces [**6, 8**]:

- $\eta_{A_{2k-1}}([k^2], [k^2], [k^2], [k^2])$ is the framed centered k-instanton moduli space of D_4 on $\mathbb{R}^4$ of dimension $12k - 2$. Note again the special maximal proper subgroup $\mathrm{SL}(2)^4$ of D_4, as shown in Fig. 1. Note that this reduces to the statement on $W_{A_1} \circ W_{A_1}$ in Section 4.2 when $k = 1$.
- $\eta_{A_{3k-1}}([k^3], [k^3], [k^3])$ is the framed centered k-instanton moduli space of E_6 on $\mathbb{R}^4$ of dimension $24k - 2$. Note that $Z([k^3]) = \mathrm{SL}(3)$ and $\mathrm{SL}(3)^3$ is a special maximal proper subgroup of E_6. Note also that this reduces to the statement on W_{A_2} in Section 4.1 when $k = 1$.
- $\eta_{A_{4k-1}}([k^4], [k^4], [2k^2])$ is the framed centered k-instanton moduli space of E_7 on $\mathbb{R}^4$ of dimension $36k-2$. Note that $Z([k^4]) = \mathrm{SL}(4)$ and $Z([2k^2]) = \mathrm{SL}(2)$. $\mathrm{SL}(4)^2 \times \mathrm{SL}(2)$ is a special maximal proper subgroup of E_7 as shown in Fig. 1.

- $\eta_{A_{6k-1}}([k^6],[2k^3],[3k^2])$ is the framed centered k-instanton moduli space of E_8 on $\mathbb{R}^4$ of dimension $60k-2$. Note again the special maximal proper subgroup $\mathrm{SL}(6)\times\mathrm{SL}(3)\times\mathrm{SL}(2)$ of E_8, as shown in Fig. 1.

The analysis of the case $k=1$ goes back to [**5, 33, 34, 26**].

5. Functor to the category of hyper-Kähler manifolds

Before ending this note, let us briefly discuss why we chose the holomorphic symplectic varieties as the target category. The Higgs branch of a supersymmetric theory not only comes with a holomorphic symplectic structure, but is equipped with hyper-Kähler structure. The category HK of hyper-Kähler manifolds with triholomorphic action can be defined naturally by saying that elements of Obj(HK) are compact semi-simple groups and that $\mathrm{Hom}(G,G')$ consists of hyper-Kähler manifolds with triholomorphic action of $G\times G'$. It seems more natural to take HK as the target category, but it turns out that η_G in the hyper-Kähler sense is not quite a topological quantum field theory.

The subtlety can be understood by considering $T^*G_{\mathbb{C}}\in\mathrm{Hom}(G,G)$, whose hyper-Kähler structure was originally constructed by Kronheimer using the Nahm equation [**31**]. The hyper-Kähler metric g can be replaced by g/a where $a\in\mathbb{R}_+$ is a positive real number without destroying the $G\times G$ invariance or hyper-Kähler structure; let us denote the resulting rescaled hyper-Kähler space by $T^*{G_{\mathbb{C}}}^a$. It can be checked that

$$T^*{G_{\mathbb{C}}}^a\times T^*{G_{\mathbb{C}}}^{a'}///G=T^*{G_{\mathbb{C}}}^{a+a'},\tag{5.1}$$

i.e. the hyper-Kähler quotient changes the overall factor of the metric. This fact follows naturally by the fact that $T^*{G_{\mathbb{C}}}^a$ is the moduli space of Nahm's equation on a segment of length a. Then, to have an identity in $\mathrm{Hom}(G,G)$, we need to take the $a\to 0$ limit of the Riemannian manifold $T^*{G_{\mathbb{C}}}^a$, which does not exist in the usual sense.

It seems likely that η_G becomes a functor to HK if we change the source to be the category without identity of 2-bordisms with the area, so that a morphism is a pair (C,a) where C is an orientable 2-manifold with boundaries and a is a positive real number which can be thought of as the area of C. The composition of two morphisms then adds the area. Then, for example, $\eta_{A_1}(W,a)$ is a hyper-Kähler manifold which is equivalent to $\mathbb{C}^2\otimes\mathbb{C}^2\otimes\mathbb{C}^2$ as a holomorphic symplectic manifold for any a; there is a sense in which the $a\to 0$ limit of $\eta_{A_1}(W,a)$ is the flat hyper-Kähler metric on $\mathbb{C}^2\otimes\mathbb{C}^2\otimes\mathbb{C}^2$. Similarly, $\eta_{A_2}(W,a)$ is equivalent to the minimal nilpotent orbit of $E_6(\mathbb{C})$ as the hyper-Kähler manifold only in the $a \to 0$ limit; at finite a, the hyper-Kähler metric is only invariant under $SU(3)^3$, not under the full E_6. These points need to be studied more carefully.

6. Further Extensions

To conclude we would like to mention two further directions in which this work could be extended. Both extensions appear to us to be nontrivial open problems.

First, quite generally, when the target category has an action of a discrete group Γ a natural generalization of a topological field theory is to the equivariant case, where the cobordism category is a category of principal Γ-bundles over a surface, where Γ is a discrete group. The sewing axioms are known in this case and can be found in [**35**]. In our case, we can pick as Γ a discrete subgroup of $\mathrm{Aut}(G)$ if G is simply-laced. Physics predicts the existence of a Γ-equivariant topological field theory extending the one we have described [**41**]. In particular, the principal Γ-bundle over S^1 with holonomy $x \in \Gamma$ is mapped to $(G^x)^\vee$, where $^\vee$ is the Langlands dual.

Second, it is natural to ask whether the topological field theory we have described fits nicely into the structure of an extended topological field theory in the sense described in [**38, 39, 32, 15, 29**]. Given the relation of theories of class S to conformal field theory it is natural to expect a 0-1-2-3 theory. The extension to level 3 should involve Lagrangian subvarieties of the holomorphic symplectic manifolds and be related to the physics of domain walls in theories of class S.

Acknowledgements

The authors thank Francesco Benini, Dan Freed, Ron Donagi, Davide Gaiotto, Hiraku Nakajima, Andy Neitzke and Graeme Segal for discussions. We also thank Daniel Freed and Daniel Roggenkamp for remarks on the draft. The work of YT is supported in part by NSF grant PHY-0969448 and by the Marvin L. Goldberger membership through the Institute for Advanced Study. The work of YT is also supported in part by World Premier International Research Center Initiative (WPI Initiative), MEXT, Japan through the Institute for the Physics and Mathematics of the Universe, the University of Tokyo. The work of GM is supported by the DOE under grant DE-FG02-96ER40959.

References

[1] Luis F. Alday, Davide Gaiotto, and Yuji Tachikawa, *Liouville Correlation Functions from Four-Dimensional Gauge Theories*, Lett. Math. Phys. **91** (2010), 167–197, `arXiv:0906.3219 [hep-th]`.

[2] Philip C. Argyres, Anton Kapustin, and Nathan Seiberg, *On S-Duality for Non-Simply-Laced Gauge Groups*, JHEP **06** (2006), 043, `arXiv:hep-th/0603048`.

[3] Philip C. Argyres and Nathan Seiberg, *S-Duality in $\mathcal{N} = 2$ Supersymmetric Gauge Theories*, JHEP **12** (2007), 088, `arXiv:0711.0054 [hep-th]`.

[4] Philip C. Argyres and John R. Wittig, *Infinite Coupling Duals of $\mathcal{N} = 2$ Gauge Theories and New Rank 1 Superconformal Field Theories*, JHEP **01** (2008), 074, `arXiv:0712.2028 [hep-th]`.

[5] Tom Banks, Michael R. Douglas, and Nathan Seiberg, *Probing F-Theory with Branes*, Phys. Lett. **B387** (1996), 278–281, `arXiv:hep-th/9605199`.

[6] Francesco Benini, Sergio Benvenuti, and Yuji Tachikawa, *Webs of Five-Branes and $\mathcal{N} = 2$ Superconformal Field Theories*, JHEP **09** (2009), 052, `arXiv:0906.0359 [hep-th]`.

[7] Francesco Benini, Yuji Tachikawa, and Brian Wecht, *Sicilian Gauge Theories and $\mathcal{N} = 1$ Dualities*, JHEP **01** (2010), 088, `arXiv:0909.1327 [hep-th]`.

[8] Francesco Benini, Yuji Tachikawa, and Dan Xie, *Mirrors of 3D Sicilian Theories*, JHEP **09** (2010), 063, `arXiv:1007.0992 [hep-th]`.

[9] Roger Bielawski, *Hyperkähler structures and group actions*, J. London Math. Soc. **55** (1997), 400. `http://www.maths.leeds.ac.uk/~rb/ga.pdf`.

[10] ———, *Lie groups, Nahm's equations and hyperkähler manifolds*, Algebraic Groups (Yuri Tschinkel, ed.), Universitätsdrucke Göttingen, 2005. `arXiv:math.DG/0509515`.

[11] Oscar Chacaltana and Jacques Distler, *Tinkertoys for Gaiotto Duality*, JHEP **11** (2010), 099, `arXiv:1008.5203 [hep-th]`.

[12] ———, *Tinkertoys for the D_N Series*, `arXiv:1106.5410 [hep-th]`.

[13] Ron Y. Donagi, *Seiberg-Witten Integrable Systems*, Algebraic Geometry Santa Cruz 1995, Proceedings of Symposia in Pure Mathematics, vol. 62/2, 1997, pp. 3–43. `arXiv:alg-geom/9705010`.

[14] Michael R. Douglas and Gregory W. Moore, *D-Branes, Quivers, and ALE Instantons*, `arXiv:hep-th/9603167`.

[15] Daniel S. Freed, Michael J. Hopkins, Jacob Lurie, and Constantin Teleman, *Topological Quantum Field Theories from Compact Lie Groups*, CRM Proceedings Lecture Notes, vol. 50, AMS, 2009, pp. 367–403. `arXiv:0905.0731 [math.AT]`.

[16] Abhijit Gadde, Elli Pomoni, Leonardo Rastelli, and Shlomo S. Razamat, *S-Duality and 2D Topological QFT*, JHEP **03** (2010), 032, `arXiv:0910.2225 [hep-th]`.

[17] Abhijit Gadde, Leonardo Rastelli, Shlomo S. Razamat, and Wenbin Yan, *The Superconformal Index of the E_6 SCFT*, JHEP **1008** (2010), 107, `arXiv:1003.4244 [hep-th]`.

[18] ———, *The 4D Superconformal Index from q-deformed 2D Yang- Mills*, `arXiv:1104.3850 [hep-th]`.

[19] Davide Gaiotto, *$\mathcal{N} = 2$ Dualities*, `arXiv:0904.2715 [hep-th]`.

[20] Davide Gaiotto and Juan Maldacena, *The Gravity Duals of $\mathcal{N} = 2$ Superconformal Field Theories*, `arXiv:0904.4466 [hep-th]`.

[21] Davide Gaiotto, Gregory W. Moore, and Andrew Neitzke, *Wall-Crossing, Hitchin Systems, and the WKB Approximation*, `arXiv:0907.3987 [hep-th]`.

[22] Davide Gaiotto, Andrew Neitzke, and Yuji Tachikawa, *Argyres-Seiberg Duality and the Higgs Branch*, Commun. Math. Phys. **294** (2010), 389–410, `arXiv:0810.4541 [hep-th]`.

[23] Davide Gaiotto and Edward Witten, *S-Duality of Boundary Conditions in $\mathcal{N} = 4$ Super Yang-Mills Theory*, J. Stat. Phys. **135** (2009), 367–403, `arXiv:0807.3720 [hep-th]`.

[24] ———, *Supersymmetric Boundary Conditions in $\mathcal{N} = 4$ Super Yang-Mills Theory*, Adv Theor. Math. Phys. **13** (2009), 721–896, `arXiv:0804.2902 [hep-th]`.

[25] Amihay Hanany and Noppadol Mekareeya, *Tri-Vertices and $SU(2)$'s*, JHEP **02** (2011), 069, `arXiv:1012.2119 [hep-th]`.

[26] Kenneth A. Intriligator and N. Seiberg, *Mirror Symmetry in Three Dimensional Gauge Theories*, Phys. Lett. **B387** (1996), 513–519, `arXiv:hep-th/9607207`.

[27] H. Itoyama and A. Morozov, *Integrability and Seiberg-Witten Theory: Curves and Periods*, Nucl. Phys. **B477** (1996), 855–877, `arXiv:hep-th/9511126`.

[28] B. Joseph, *The minimal orbit in a simple Lie algebra and its associated maximal ideal*, Ann. Sci. École Norm. Sup. Sér. 4 **9** (1976), 1. `http://www.numdam.org/item?id=ASENS_1976_4_9_1_1_0`.

[29] Anton Kapustin, *Topological Field Theory, Higher Categories, and Their Applications*, Proceedings of the ICM, 2010, pp. 2021–2043. `arXiv:1004.2307 [math.QA]`.

[30] Albrecht Klemm, Wolfgang Lerche, Peter Mayr, Cumrun Vafa, and Nicholas P. Warner, *Self-Dual Strings and $\mathcal{N} = 2$ Supersymmetric Field Theory*, Nucl. Phys. **B477** (1996), 746–766, `arXiv:hep-th/9604034`.

[31] Peter B. Kronheimer, *A hyperkähler structure on the cotangent bundle of a complex Lie group*, MSRI preprint (1988), `arXiv:math.DG/0409253`.

[32] Jacob Lurie, *On the classification of topological field theories*, Current developments in mathematics, 2008, Int. Press, Somerville, MA, 2009, pp. 129–280.

[33] Joseph A. Minahan and Dennis Nemeschansky, *An $\mathcal{N} = 2$ Superconformal Fixed Point with E_6 Global Symmetry*, Nucl. Phys. **B482** (1996), 142–152, `arXiv:hep-th/9608047`.

[34] ———, *Superconformal Fixed Points with E_N Global Symmetry*, Nucl. Phys. **B489** (1997), 24–46, `arXiv:hep-th/9610076`.

[35] Gregory W. Moore and Graeme Segal, *D-Branes and K-Theory in 2D Topological Field Theory*, `arXiv:hep-th/0609042`.

[36] Hiraku Nakajima, *Instantons on ALE Spaces, Quiver Varieties and Kac-Moody Algebras*, Duke Math. Journal **76** (1994), 365.

[37] Dimitri Nanopoulos and Dan Xie, *$\mathcal{N} = 2$ SU Quiver with USp ends or SU ends with antisymmetric matter*, JHEP **08** (2009), 108, `arXiv:0907.1651 [hep-th]`.

[38] Graeme Segal, *What is an elliptic object?*, Elliptic Cohomology (Haynes R. Miller and Douglas C Ravenel, eds.), London Mathematical Society Lecture Note, vol. 342, Cambridge University Press, 2005, pp. 306–317.

[39] ———, *Locality of holomorphic bundles, and locality in quantum field theory*, The many facets of geometry (Oscar García-Prada, Jean-Pieere Bouguignon, and Simon Salamon, eds.), Oxford University Press, 2010, pp. 164–176.

[40] N. Seiberg and Edward Witten, *Monopoles, Duality and Chiral Symmetry Breaking in $\mathcal{N}=2$ Supersymmetric QCD*, Nucl. Phys. **B431** (1994), 484–550, `arXiv:hep-th/9408099`.

[41] Yuji Tachikawa, *$\mathcal{N} = 2$ S-Duality via Outer-Automorphism Twists*, J. Phys. **A** (2011), Math. Theor.44:182001, `arXiv:1009.0339 [hep-th]`.

[42] Cumrun Vafa, *Geometric Origin of Montonen-Olive Duality*, Adv. Theor. Math. Phys. **1** (1998), 158–166, `arXiv:hep-th/9707131`.

[43] Edward Witten, *Solutions of four-dimensional field theories via M theory*, Nucl.Phys. **B500** (1997), 3–42, `arXiv:hep-th/9703166 [hep-th]`.

[44] ———, *Geometric Langlands and the Equations of Nahm and Bogomolny*, Proc. Roy. Soc. Edinburgh, Section A **140** (2009), 857–895, `arXiv:0905.4795 [hep-th]`.

[45] ———, *Geometric Langlands from Six Dimensions*, CRM Proceedings Lecture Notes, vol. 50, AMS, 2009, pp. 281–310. `arXiv:0905.2720 [hep-th]`.

[46] Niclas Wyllard, *A_{N-1} Conformal Toda Field Theory Correlation Functions from Conformal $\mathcal{N}=2$ $SU(N)$ Quiver Gauge Theories*, JHEP **11** (2009), 002, `arXiv:0907.2189 [hep-th]`.

NHETC and Department of Physics and Astronomy, Rutgers University, Piscataway, NJ 088550849, USA
E-mail address: gmoore@physics.rutgers.edu

School of Natural Sciences, Institute for Advanced Study Princeton, New Jersey 08504, USA
Current address: IPMU, University of Tokyo, Chiba 277-8583 Japan
E-mail address: yuji.tachikawa@ipmu.jp

Proceedings of Symposia in Pure Mathematics
Volume **85**, 2012

The Witten equation and the geometry of the Landau-Ginzburg model

Yongbin Ruan

Abstract. Twenty years ago, Witten proposed a first order nonlinear PDE (Witten equation) as the replacement of Cauchy-Riemann equation in the Landau-Ginzburg setting. Several years ago, a comprehensive theory was developed by Fan-Jarvis-Ruan to solve Witten equation in the case of pure Landau-Ginzburg model. In this article, we describe several possible generalizations into toric setting.

1. Introduction

In 1991, a celebrated conjecture of Witten [**Wi1**] asserted that the intersection theory of Deligne-Mumford moduli space is governed by the KdV hierarchy. His conjecture was soon proved by Kontsevich [**Ko**]. Since then, the Witten-Kontsevich theorem has introduced the seemingly alien concept of integrable hierarchies into the geometry. The KdV hierarchy is the first of a family of integrable hierarchies (Drinfeld-Sokolov/Kac-Wakimoto hierarchies) associated to integrable representations of affine Kac-Moody algebras. Immediately after Kontsevich's solution of Witten's conjecture, a great deal of effort was spent in investigating other integrable hierarchies in Gromov-Witten theory. In fact, this question was very much in Witten's mind when he proposed his famous conjecture in the first place. Around the same time, he also proposed a sweeping generalization of his conjecture [**Wi2, Wi3**]. The core of his generalization is a remarkable first order nonlinear elliptic PDE associated to an arbitrary quasihomogeneous singularity. It has the simple form

$$\bar{\partial} u_i + \overline{\frac{\partial W}{\partial u_i}} = 0,$$

where W is a quasihomogeneous polynomial, and u_i is interpreted as a section of an appropriate orbifold line bundle on a Riemann surface $\mathbb{C}$. Some simple examples are

(A_r-case): $\bar{\partial} u + \bar{u}^r = 0$
(D_n-case): $\bar{\partial} u_1 + n\bar{u}_1^{n-1} + \bar{u}_2^2 = 0, \bar{\partial} u_2 + 2\bar{u}_1\bar{u}_2 = 0.$
(E_7-case): $\bar{\partial} u_1 + 3\bar{u}_1^2 + \bar{u}_2^3 = 0, \bar{\partial} u_2 + 3\bar{u}_1\bar{u}_2^2 = 0.$

During the last decade, a comprehensive treatment of the Witten equation has been carried and a new Gromov-Witten type theory has been constructed by

2000 *Mathematics Subject Classification.* Primary .

Fan-Jarvis-Ruan [**FJR1, FJR2, FJR3**]. In particular, Witten's conjecture for ADE-integrable hierarchies has been verified.

More importantly, the theory of Fan-Jarvis-Ruan-Witten has found new applications to the Landau-Ginzburg counterpart of Gromov-Witten theory in the famous Landau-Ginzburg/Calabi-Yau correspondence, a far-reaching duality from physics. The question there is to solve the Gromov-Witten theory of a Calabi-Yau hypersurface $X_W = \{W = 0\}$ for a quasihomogenous polynomial $W : \mathbb{C}^N \to \mathbb{C}$. Naively, Gromov–Witten theory can be thought of as solving the Cauchy–Riemann equation $\overline{\partial} f = 0$ for the map $f \colon \Sigma \to X_W$, where Σ is a compact Riemann surface and X_W is the weighted projective hypersurface $\{W = 0\}$. In many ways, the difficulty and the interest of the computation of Gromov–Witten invariants come from the fact that X_W is a nonlinear space. On the other hand, Witten's equation works on a linear target $\mathbb{C}^n$. The nonlinearity is reflected in the zero order term $\overline{\partial W}$. Namely, we trade a linear equation on a nonlinear target with a nonlinear equation on a linear target! If we compare the Cauchy-Riemann equation on a Calabi-Yau manifold to the anti-self-dual equation in Donaldson theory, the Witten equation plays the role of the Seiberg-Witten equation. Indeed, it exhibits some of characteristics of Seiberg-Witten theory: (1) the moduli space is essentially compact and there is no bubbling; (2) the equation is not conformally invariant. It is generally believed that the theory of Fan-Jarvis-Ruan-Witten is much easier to compute than Gromov-Witten theory in similar fashion as Seiberg-Witten invariants versus Donaldson invariants. Three years ago, Chiodo and the author launched a program to establish the equivalence of Gromov-Witten theory and the theory of Fan-Jarvis-Ruan-Witten. We have achieved a great deal of success.

After FJRW-theory was established, it was the author's intention to write a survey article to expose our new theory and its applications to more general audience. Its title was mentioned already in several papers [**ChiR10, CIR, ChiR11, KS92**]. Over the last three years, I wrote two other surveys focusing on the applications of FJRW-theory to integrable hierarchies and the Landau-Ginzburg/Calabi-Yau correspondence. In these two surveys, the role of PDE is minimized. In a sense, part of the goal of the original survey has already been accomplished. However, the original approach of FJRW-theory from the Witten equation has its merit. Furthermore, the Witten equation treated by Fan-Jarvis-Ruan-Witten is the simplest case of a family of PDE describing Witten's so-called gauged linear sigma model. Witten's gauged linear sigma model equation is a combination of the original Witten equation with the symplectic vortex equation studied by several authors [**CRRS, RT, GW**]. There is ample opportunity for young people to work in this area. It seems to be worthwhile to present this very important prespective. In this article, we will try to do this. In particular, I will NOT address the issues of applications. Readers are referred to my two other surveys in which the applications are the main focus.

I will start from the original Witten equation for the Landau-Ginzburg model where a complete treatment was done by Fan-Jarvis-Ruan. I will discuss various fundamental issues regarding this equation such as (1) set-up; (2) possible boundary value and state space; (3) compactness; (4) wall crossing and (5) algebraic treatment. It will serve as a model for the more general equation. Then, we will discuss Witten's gauged linear sigma model equation. Here, the theory is much less developed and many fundamental issues are still unresolved. I hope that the current survey will point out some useful directions. We omit two important directions, the LG-mirror of the toric variety and a non-abelian generalization of the Witten equation. The theory we described here works best in the quasihomogeneous case. Usually, the LG-mirror of toric variety is not quasihomogeneous. However, we can still consider the Witten equation for the case of disc and torus where the tangent bundle is trivial. I believe that it still gives useful information. The non-abelian generalization is indeed very exciting and at the frontier of physics right now. It is perhaps more suitable for another survey in the future.

2. Theory of Fan-Jarvis-Ruan-Witten

Let's take a careful look at Witten equation

$$\bar{\partial} u_i + \overline{\frac{\partial W}{\partial u_i}} = 0,$$

where W is a quasihomogeneous polynomial, and u_i is interpreted as a section of an appropriate orbifold line bundle on a Riemann surface $\mathbb{C}$. A simple computation shows that

$$\bar{\partial} u_i \in \Omega^{0,1}(L_i), \overline{\frac{\partial W}{\partial u_i}} \in \Omega^0(\overline{W(L_0, \dots, L_N) \otimes L_i^{-1}}).$$

It is clear that the two terms do not live in the same space unless the L_i have some special properties. This can be summarized as the following interesting moduli problem in algebraic geometry. It can be thought of as background data for the Witten equation.

2.0.1. *Moduli space of W-structures.* Let's first set up some notation. C denotes a marked nodal stable Riemann surface. L_i is an orbifold line bundle associated to the variable z_i. Without loss of generality, we assume that all the orbifold points of L_i are marked or nodal points.

$$K_{log,C} = K_C \otimes \vartheta(x_1) \otimes \cdots \otimes \vartheta(x_k)$$

where $x_1, \dots, x_k$ are all the marked points.

DEFINITION 2.1. Suppose that $W = \sum_{j=1}^{s} c_j W_j$ where $W_j = \prod_{i=1}^{N} z_i^{b_{ij}}$ is a monomial. For any non-degenerate, quasihomogeneous polynomial $W \in \mathbb{C}[x_1, \dots, x_N]$, we define a *W-structure* to be the data of an N-tuple $(L_1, \dots, L_N)$ of orbifold line bundles on C and isomorphisms $\varphi_j : W_j(L_1, \dots, L_N) \to K_{C,log}$ for every $j \in \{1, \dots, s\}$, where

$$W_j(L_1, \dots, L_N) = L_1^{\otimes b_{1,j}} \otimes \cdots \otimes L_N^{\otimes b_{N,j}}.$$

Without loss of generality, we assume that the orbifold structure only occurs at marked or nodal points.

DEFINITION 2.2. Given any two W-structures $\mathbb{L} := (L_1, \dots, L_N, \varphi_1, \dots \varphi_s)$ and $\mathbb{L}' := (L'_1, \dots, L'_N, \varphi'_1, \dots \varphi'_s)$ on C, it is clear that a set of morphisms $\xi_j : L_j \to L'_j$ of orbifold line bundles for $j \in \{1, \dots, N\}$ will induce a morphism $\Xi_l : W_l(L_1, \dots, L_N) \to W_l(L'_1, \dots, L'_N)$ for every $l \in \{1, \dots, s\}$.

An *isomorphism of W-structures* $\Upsilon : \mathbb{L} \to \mathbb{L}'$ on C is defined to be a collection of isomorphisms $\xi_j : L_j \to L'_j$ such that for every $l \in \{1, \dots, s\}$ we have $\varphi_l = \varphi'_l \circ \Xi_l$.

DEFINITION 2.3. We denote the stack of stable W-orbicurves by $\mathcal{W}_{g,k}$.

Note that for each marked point $x_j \in C$, an orbifold line bundle L at x_j is given by the local orbifold chart

$$L_i|_{\Delta/\mathbb{Z}/m\mathbb{Z}} \cong (\Delta \times \mathbb{C})/\mathbb{Z}/m\mathbb{Z},$$

where Δ is a disc and the generator $e \in \mathbb{Z}/m\mathbb{Z}$ acts as $e(z,s) = (\exp(2\pi i/m)z, \exp(2\pi i v_i/m)s)$. We choose the same m for each L_i. Let $\gamma = diag(\exp(2\pi i v_1/m), \cdots, \exp(2\pi i v_k/m))$. We further require that $\langle\gamma\rangle$ acts on $\mathbb{C}^N$ effectively. It is rather amazing that

LEMMA 2.4. *γ defines an automorphism of W.*

The upshot of the lemma is that the orbifold structure of $\bigoplus_i L_i$ is completely parameterized by the automorphisms of W. Therefore, the group $G_{max} = Aut(W)$ enters the theory in a natural way. One may wonder if we can insist L_i be an honest line bundle. Then, there will be no W-structure in general. Even if it exists sometimes, its degeneration on a nodal Riemann surface will naturally introduce an orbifold structure at nodes. In short, there is no way one can avoid the orbifold structure. The theory of Fan-Jarvis-Ruan-Witten naturally corresponds to the orbifolded singularity (W, G_{max}) where the orbifold group is G_{max}. It is possible to generalize it to (W, G) for $\langle J\rangle \subset G$ (see the discussion of admissible groups). The current theory does not work for any group smaller than $\langle J\rangle$ in any generality.

DEFINITION 2.5. A marked point x_j of a W-curve is called *Narrow* if the fixed point locus $\mathbb{C}^N_\gamma \subseteq \mathbb{C}^N$ is just $\{0\}$. The point x_j is called *Broad* otherwise and the variable $z_i \in \mathbb{C}^N_\gamma$ is called a Broad variable.

There are several natural morphisms analogous to that of $\overline{\mathcal{M}}_{g,k}$.

Stabilization. Forgetting the W-structure and the orbifold structure gives a morphism

$$st : \mathcal{W}_{g,k} \to \overline{\mathcal{M}}_{g,k}.$$

THEOREM 2.6. *$\mathcal{W}_{g,k}$ is a compact smooth complex orbifold and st is a finite map.*

Components. $\mathcal{W}_{g,k}$ has many components. It is easy to show that the orbifold structure at each marked point remains the same for each component.

DEFINITION 2.7. For any choice $\boldsymbol{\gamma} := (\gamma_1, \ldots, \gamma_k) \in G^k_{max}$ we define $\mathcal{W}_{g,k}(\boldsymbol{\gamma})$ to be the open and closed substack of $\mathcal{W}_{g,k}$ with the given orbifold structure specified by $\boldsymbol{\gamma}$.

We define the *type* of any W-structure in $\mathcal{W}_{g,k}(\boldsymbol{\gamma})$ to be $\boldsymbol{\gamma} \in G^k_{max}$.

Another classical construction is to desingularize the orbifold structure. Namely, one can change the local trivialization of L_i by

$$(z,s) \to (z^m, z^{v_i}s).$$

Under the new local trivialization, $L_i - (L_i|_{x_j})$ extends to an orbifold line bundle $|L_i|(x_j)$ smooth over x_j. Furthermore,

$$H^0(C, L_i) = H^0(C, |L_i|(x_j)),\ H^1(C, L_i) = H^1(C, |L_i|(x_j)).$$

After we desingularize all the orbifold structure, we denote it by $|L_i|$.

The desingularization leads to an important selection rule that must be satisfied for $\mathcal{W}_{g,k}(\boldsymbol{\gamma})$ to be non-empty. Although the degree of an orbifold bundle L_i on

C may be a rational number, the degree of $|L_i|$ must be an integer, so for all $i \in \{1, \ldots, s\}$ the following equations must hold for integral values of $\deg(|L_j|)$:

$$\sum_{j=1}^{N} b_{ji} \deg(|L_j|) = 2g - 2 + k - \sum_{l=1}^{k} \sum_{j=1}^{N} b_{ji} \Theta_j^{\gamma_l}.$$

Moreover, because W is non-degenerate, the weights q_j are uniquely determined by the requirement that they satisfy the equations $\sum_{j=1}^{N} b_{ji} q_j = 1$ for all $i \in \{1, \ldots, s\}$, so we find that for every $j \in \{1, \ldots, N\}$ we have

$$\deg(|L_j|) = \left(q_j(2g - 2 + k) - \sum_{l=1}^{k} \Theta_j^{\gamma_l} \right) \in \mathbb{Z}.$$

Let's consider the condition on $|L_i|$. A simple calculation yields

$$\begin{aligned} W_i(|L_1|(x_k), \ldots, |L_N|(x_k)) &\to K_{C,log} \otimes \vartheta((-\textstyle\sum_{j=1}^{N} b_{ij} \Theta_j x_k) \\ &= K_C \otimes \vartheta(((1 - \textstyle\sum_{j=1}^{N} b_{ij} \Theta_j) x_k) \end{aligned}$$

It leads to an alternative description of W-structure as the condition

$$W_i(|L_1|, \ldots, |L_N|) = K_C \otimes \vartheta(\sum_k m_k x_k)$$

where we use weights m_k to replace the orbifold structure. This was the original description of Witten.

Forgetful morphism. As we mentioned earlier, we can desingularize the orbifold structure at a marked point x_j and obtain orbifold line bundles $|L_i|(x_j)$. But $|L_1|(x_j), \ldots, |L_N|(x_j)$ is not a W-structure in general except that the orbifold structure at x_j is J.

If $\gamma = (\gamma_1, \ldots, J)$, then we can forget the last marked point in the same way as the forgetful morphism for $\overline{\mathcal{M}}_{g,k}$. More precisely, If $(L_1, \ldots, L_N)$ is a W-structure with the above orbifold decorations. $(|L_1|(x_k), \cdots, |L_N|(x_k))$ is a W-structure on C with x_k being forgotten. The reason that this happens is the following calculation:

$$\begin{aligned} W_i(|L_1|(x_k), \ldots, |L_N|(x_k)) &\to K_{C,log} \otimes \vartheta((-\sum_{j=1}^{N} b_{ij} \Theta_j^J) x_k) \\ &= K_{C,log} \otimes \vartheta((-\sum_{j=1}^{N} b_{ij} q_j) x_k) \\ &= K_{C,log} \otimes \vartheta(-x_k) \\ &= K_{C',log}, \end{aligned}$$

where C' is the marked Riemann surface obtained from C by forgetting x_k.

Therefore, we define a morphism

$$\pi : \mathcal{W}_{g,k}(\gamma_1, \ldots, \gamma_{k-1}, J) \to \mathcal{W}_{g,k-1}(\gamma_1, \ldots, \gamma_{k-1}).$$

We call it *the forgetful morphism.*

Note that the essential property of γ_i that allows the forgetful morphism to exist is the fact that $\sum_{j=1}^{N} b_{lj} \Theta_j^{\gamma_i} = 1$ for every $l \in \{1, \ldots, s\}$. Since the weights q_j are uniquely determined by W, this means that a marked point x_i may not be forgotten unless $\gamma_i = J$.

EXAMPLE 2.8. For the three-pointed, genus-zero W-structure, the choice of orbifold line bundles $L_1, \dots, L_N$ providing the W-structure is unique, if it exists at all. If it exists, then $\mathcal{W}_{0,3}(\gamma)$ is isomorphic to BG_{max}.

There are two important gluing homomorphisms in Gromov-Witten theory.

$$\rho_{tree} : \overline{\mathcal{M}}_{g,k+1} \times \overline{\mathcal{M}}_{g_2,k_2+1} \to \overline{\mathcal{M}}_{g_1+g_2,k_1+k_2},$$

$$\rho_{loop} : \overline{\mathcal{M}}_{g,k+2} \to \overline{\mathcal{M}}_{g+1,k}.$$

Unfortunately, there is no direct lift of ρ_{tree}, ρ_{loop} to the moduli stack of W-curves because there is no canonical way to glue the fibers of the line bundles L_i on the two points that map to a node.

But we could describe a gluing process in terms of an additional structure that we call *rigidification.* Suppose the fiber of the W-structure at the marked point p is $[(L_1 \oplus L_2 \oplus \cdots \oplus L_N)/G_p]$, where G_p is the local group. The rigidification can be thought as a G_p-equivariant map $\psi : \bigoplus_i L_i \to \mathbb{C}^N$ commuting with the W-structure. For any element $g \in G_p$, the rigidification $g\psi$ is considered to be an equivalent rigidification.

Alternatively, ψ is equivalent to a choice of basis $e_i \in L_i$ such that $W_j(e_1, \dots, e_N) = dz/z$ and the basis $g(e_1), \dots, g(e_N)$ is considered to be an equivalent choice. In particular, if $L_{i_1}, \dots, L_{i_m}$ are the line bundles fixed by G_p (we call the corresponding variables z_{i_j} the *Broad variables*) then in each equivalance class of rigidification, the basis elements $e_{i_1}, \dots, e_{i_m}$ for the subspace $\bigoplus_{j=1}^m L_{i_j}\Big|_p$ are unique, but the basis elements for the terms not fixed by G_p (the *Narrow variables*) are only unique up to the action of G_p.

It is clear that the group G_{max}/G_p acts transitively on the set of rigidifications within a single orbit. Let $\mathcal{W}^{rig_p}_{g,k}$ be the equivalence classes of W-curves with a rigidification at p. The group G_{max}/G_p acts on $\mathcal{W}^{rig_p}_{g,k}$ by interchanging the rigidifications. It is clear that $\mathcal{W}^{rig_p}_{g,k}$ is a principal G_{max}/G_p-bundle over $\mathcal{W}_{g,k}$. We use $\mathcal{W}^{rig}_{g,k}$ to denote the moduli stack of rigidified W-structures at all the marked points.

Now we describe the gluing. To simplify notation, we ignore the orbifold structures at other marked points and denote the types of the marked points p_+, p_- being glued by γ_+, γ_-. Recall that the resulting orbifold structures must be balanced, which means that $\gamma_- = \gamma_+^{-1}$. Let

$$\psi_\pm : j^*_{p_\pm}(L_1 \oplus \cdots \oplus L_N) \to [\mathbb{C}^N/G_{p_\pm}]$$

be the rigidifications. Moreover, the residues at p_+, p_- are opposite to each other. The obvious identification will not preserve the rigidifications. Here, we fix once and for all an isomorphism

$$I : \mathbb{C}^N \to \mathbb{C}^N$$

such that $W(I(z)) = -W(z)$. I has been explicitly constructed previously as follows. Suppose that $q_i = n_i/d$ for the common denominator d. Choose $\xi^d = -1$. Then,

$$I(z_1, \dots, z_N) = (\xi^{n_1} z_1, \dots, \xi^{n_N} z_N).$$

The identification by I induces a W-structure on the nodal orbifold Riemann surface with a rigidification at the nodal point. Forgetting the rigidification at the node yields the lifted gluing morphisms

$$\widetilde{\rho}_{tree,\gamma} : \mathcal{W}^{rig_p}_{g_1,k_1+1}(\gamma) \times \mathcal{W}^{rig_q}_{g_2,k_2+1}(\gamma^{-1}) \to \mathcal{W}_{g_1+g_2,k_1+k_2},$$

$$\widetilde{\rho}_{loop,\gamma} : \mathcal{W}^{rig_{p,q}}_{g,k+2}(\gamma, \gamma^{-1}) \to \mathcal{W}_{g+1,k}.$$

2.0.2. *Cylindric metric, boundary condition and virtual cycle.* A choice of W-structure does not solve our problem completely. Suppose that $u_i \in \Omega^0(L_i)$ and $L_1, \ldots, L_N$ is a W-structure. Then,

$$\bar{\partial} u_i \in \Omega^{0,1}(L_i), \overline{\frac{\partial W}{\partial u_i}} \in \Omega^{0,1}_{log}(\bar{L}_i^{-1}),$$

where log means a $(0,1)$-form with possible singularities of order ≤ 1. Namely, the Witten equation has singular coefficients! This is a fundamental phenomenon for the application of the Witten equation. One of the most difficult conceptual advances in the entire theory was to generate the A-model state space from the study of the Witten equation. Now it is understood that the singularity of the Witten equation is the key. Unfortunately, the appearance of singularities makes the Witten equation very difficult to study analytically. In fact, it has taken the authors six years to construct the theory!

Another subtle issue is the fact that we need an isomorphism $\bar{L}_i^{-1} \cong L_i$ for the two terms of the Witten equation living in the same space. The required isomorphism can be obtained by a choice of metric. A nontrivial fact is that such a metric can be constructed uniformly from a metric of the underlying Riemann surface. Then the question is: which metric should we choose on the Riemann surface? We should mention that a different choice of metric often leads to a completely different looking theory, including a different dimension for its moduli space. Apparently, there is no physical guidance for the correct metric we should choose. The authors have experimented with both smooth and cylindrical metrics near marked points. Now we understand that both choices are important for the theory!

Let's briefly outline the construction. Let $\mathcal{W}_{g,k}(\gamma_1, \ldots, \gamma_k)$ be the moduli space of W-structures decorated by the orbifold structure γ_i at the marked point x_i. It can be considered as the background data to set up the Witten equation. Unfortunately, it is rather difficult to solve the Witten equation due to the fact that W is highly degenerate. It is much easier to solve a perturbed equation for $W + W_0$, where W_0 is a linear perturbation term such that $W_\gamma + W_{0\gamma}$ is a holomorphic Morse function for every γ. Here $W_\gamma, W_{0\gamma}$ are the restrictions of W, W_0 to the fixed point set $\mathbb{C}^N_\gamma$. The background data for the perturbed Witten equation is the moduli space of rigidified W-structures $\mathcal{W}^{rig}_{g,k}(\gamma_1, \ldots, \gamma_k)$.

The crucial part of the analysis is to show that a solution of the Witten equation converges to a critical point of $W_{\gamma_i} + W_{0\gamma_i}$. This enables us to construct a moduli space

$$\mathcal{W}^s_{g,k}(\kappa_{j_1}, \ldots, \kappa_{j_k})$$

of solutions of the perturbed Witten equation converging to the critical point κ_{j_i} at the marked point x_i. We call W_0 *strongly regular* if (i) $W_{\gamma_i} + W_{0\gamma_i}$ is holomorphic Morse; (ii) the critical values of $W_{\gamma_i} + W_{0\gamma_i}$ have distinct imaginary parts. The first important result is

THEOREM 2.9. *If W_0 is strongly regular, then $\mathcal{W}^s_{g,k}(\kappa_{j_1}, \ldots, \kappa_{j_k})$ is compact and has a virtual fundamental cycle $[\mathcal{W}^s_{g,k}(\kappa_{j_1}, \ldots, \kappa_{j_k})]^{vir}$ of degree*

$$2((c_W - 3)(1-g) + k - \sum_i \iota_{\gamma_i}) - \sum_i N_{\gamma_i}.$$

Here, ι_{γ_i} is certain degree shifting number.

It turns out to be convenient to map the above virtual cycle into $H_*(\mathcal{W}^{rig}_{g,k}, \mathbb{Q})$ even though it is not a subspace of the latter in any way. This is the first step of the construction.

2.0.3. *Wall crossing and state space.* We have not yet seen the state space. Then, a crucial new phenomenon comes into play when we study how the above virtual cycle changes when we vary the perturbation. It turns out that the above virtual cycle *does* depend on the perturbation. It will change when W_0 fails to be strongly regular. Recall that for a strongly regular perturbation we can construct a canonical system of HORIZONTAL paths $u_i^{\pm}$ and the associated Lefschetz thimble $\Delta_i^{\pm}$. When we perturb W_0 crossing the "wall" (where the imaginary parts of critical values become the same), we arrive at another canonical system of paths and its Lefschetz thimble $\Delta_i'^{\pm}$. The relation between $\Delta_i^{\pm}$ and $\Delta_i'^{\pm}$ is determined by the well-known Picard-Lefschetz formula. For example, the readers can find explicit formula in Arnold's book. The "wall crossing formula" for virtual fundamental cycles can be summarized in the following *quantum Picard-Lefschetz theorem*:

THEOREM 2.10. *When W_0 varies, $[\mathcal{W}^s_{g,k}(\kappa_{j_1}, \ldots, \kappa_{j_k})]^{vir}$ transforms in the same way as the Lefschetz thimble $\Delta^-_{j_i}$ attached to the critical point k_{j_i}. The later is given by Picard-Lefschetz formula.*

Δ_i^+'s transform in the opposite way as Δ_i^-'s. It is well-known that the "diagonal class" $\sum_i \Delta_i^- \otimes \Delta_i^+$ is independent of the perturbation. This suggests the following definition of an "extended virtual class". To simplify the notation, we assume that there is only one marked point with the orbifold decoration γ. Then, the wall crossing formula of $[\mathcal{W}^s_{g,1}(\kappa_i)]^{vir}$ shows precisely that

$$\sum_j [\mathcal{W}^s_{g,1}(\kappa_j)]^{vir} \otimes \Delta_j^+$$

viewed as a class in $H_*(\mathcal{W}^{rig}_{g,1}(\gamma), \mathbb{Q}) \otimes H_{N_\gamma}(\mathbb{C}^{N_\gamma}_\gamma, W^\infty_\gamma, \mathbb{Q})$ is independent of the perturbation. Now, we define

$$[\mathcal{W}^s_{g,1}(\gamma)]^{vir} = \sum_j [\mathcal{W}^s_{g,1}(\kappa_j)]^{vir} \otimes \Delta_j^+.$$

The above definition can be generalized to multiple marked points in an obvious way. It is obvious that

$$[\mathcal{W}^s_{g,k}(\gamma_1, \cdots, \gamma_k)]^{vir} \in H_*(\mathcal{W}^{rig}_{g,k}(\gamma_1, \cdots, \gamma_k), \mathbb{Q}) \otimes \prod_i H_{N_{\gamma_i}}(\mathbb{C}^{N_{\gamma_i}}, W^\infty_{\gamma_i}, \mathbb{Q})$$

of degree

$$2((c_W - 3)(1 - g) + k - \sum_i \iota_{\gamma_i}).$$

COROLLARY 2.11. $[\mathcal{W}^s_{g,k}(\gamma_1, \ldots, \gamma_k)]^{vir}$ is independent of the perturbation W_0.

W_0 is only part of the perturbation data. Eventually, we want to work on $\mathcal{W}_{g,k}$. It is known that $so : \mathcal{W}^s_{g,k} \to \mathcal{W}_{g,k}$ is a quasi-finite proper map by forgetting all the rigidifications. We can define

$$[\mathcal{W}_{g,k}(\gamma_1, \ldots, \gamma_k)]^{vir} := \frac{(-1)^\chi}{\deg(so)} (so)_* [\mathcal{W}^s_{g,k,W}(\gamma_1, \ldots, \gamma_k)]^{vir},$$

where

$$\chi = c_W(1-g) + k - \sum_i \iota_{\gamma_i}.$$

The independence of the above virtual cycle on rigidification implies that

$$[\mathcal{W}_{g,k}(\gamma_1, \dots, \gamma_k)]^{vir} \in H_*(\mathcal{W}_{g,k}(\gamma_1, \dots, \gamma_k), \mathbb{Q}) \otimes \prod_i H_{N_{\gamma_i}}(\mathbb{C}^{N_{\gamma_i}}, W^{\infty}_{\gamma_i}, \mathbb{Q})^G.$$

Namely, our effort to cancel the wall crossing leads to an additional space

$$\mathcal{H}_W = \prod_i H_{N_{\gamma_i}}(\mathbb{C}^{N_{\gamma_i}}, W^{\infty}_{\gamma_i}, \mathbb{Q})^{G_{max}}.$$

Compared with Gromov-Witten theory, $\mathcal{H}_W$ plays the role of state space or cohomology. Now, we formally define the *state space* of FJRW-theory as $\mathcal{H}_W$. In some way, our method to discover the state space is unusual! It only comes at the end of the construction. Conceptually, it is the most difficult part of the construction.

One gets a cleaner formula by pushing down $[\mathcal{W}_{g,k}(\gamma)]^{vir}$ to $\overline{\mathcal{M}}_{g,k}$.

DEFINITION 2.12. Let $\Lambda^W_{g,k} \in \hom(\mathcal{H}_W^{\otimes k}, H^*(\overline{\mathcal{M}}_{g,k}))$ be given for homogeneous elements $(\alpha_1, \dots, \alpha_k)$ with $\alpha_i \in \mathcal{H}_{\gamma_i}$ by

$$\Lambda^W_{g,k}(\alpha_1, \dots, \alpha_k) := \frac{|G_{max}|^g}{\deg(st)} PDst_* \left([\mathcal{W}_{g,k}(\gamma)]^{vir} \cap \prod_{i=1}^k \alpha_i \right),$$

and then extended linearly to general elements of $\mathcal{H}_W^{\otimes k}$.

Let 1 be the distinguished generator of $\mathcal{H}_J$, and let $\langle \, , \rangle^W$ denote the pairing on the state space $\mathcal{H}_W$.

THEOREM 2.13. *The collection* $(\mathcal{H}_{W,G_{max}}, \langle \, , \rangle^W, \{\Lambda^W_{g,k}\}, 1)$ *is a cohomological field theory in the sense of Kontsevich-Manin.*

2.0.4. *Admissible group.* So far, we are working with the group G_{max}. It is also useful to generalize it to other groups. The group $\langle J \rangle$ is particularly important in application.

DEFINITION 2.14. A subgroup $G \subset G_{max}$ is called *admissible* if there is a quasihomogeneous Laurent series Z of the same quasihomogeneity such that $G = G_{max}(W + Z)$.

Let $(\mathcal{W} + \mathcal{Z})_{g,k}$ be the moduli space of W-structures defined by $W + Z$. It is an open closed substack of $\mathcal{W}_{g,k}$. One can show that $(\mathcal{W} + \mathcal{Z})_{g,k}$ is independent of Z and we denote it by $\mathcal{W}_{g,k,G}$. It has the property that all the orbifold decorations are from G. In essence, we just use Z to single out an open closed substack of $\mathcal{W}_{g,k}$ satisfying all the formal properties of $\mathcal{W}_{g,k}$ with orbifold decoration only from G. It was shown in [**FJR2**] that $\langle J \rangle$ is admissible. Furthermore, Krawitz proved that any group G containing J is admissible [**Kr**]. Furthermore, an explicit description of $\mathcal{W}_{g,k,\langle J \rangle}$ has been worked out in [**ChiR10**].

One can simply restrict the virtual class to $\mathcal{W}_{g,k,G}$ and define a cohomological field theory $\Lambda^W_{g,k,G}$ as well as $\mathcal{F}^g_{W,G}, \mathcal{D}_{W,G}$.

2.0.5. *Narrow sector and algebraic treatment.* The general construction of Fan, Jarvis and Ruan is analytic in natural. However, there is a subsector where the Witten equation has the zero solution only. This leads to an algebraic treatment of this subsector. This subsector has the property that the markings are all narrow. Then, the Witten equation breaks into two separate equations

$$\bar{\partial} u_i = 0, \ \overline{\frac{\partial W}{\partial u_i}} = 0.$$

The first equation says that u_i is a holomorphic section. The second equation implies $u_i = 0$ by nondegeneracy of W.

In this case, the virtual cycle can be formulated in terms of the topological Euler class. Let's outline it. The W-structure $L_1, \ldots, L_N$ can be assembled into a single vector bundle

$$E = \oplus_{j=1}^N L_j.$$

Since the universal curve $\pi\mathcal{C} \to W_{g,n}(h_1, \ldots, h_n)$ is a flat morphism of relative dimension one the pushforward $R\pi_* E$ in the derived categories can be represented by a two-term complex of the form

$$\delta = \oplus \delta_j : \mathcal{E} \to \mathcal{F}.$$

Consider the direct sum of all partial derivatives $\overline{\partial_j W}$ for all $j = 1, \ldots, N$; we get $\overline{\partial W} = \oplus_j \overline{\partial W_j} \mathcal{E} \to \mathcal{F}$. Set $\mathbb{W} = \delta + \overline{\partial W}$. The nondegeneracy condition for $\mathbb{W}$ extends immediately by the nondegeneracy of the polynomial W. We can pull back $\mathcal{F}$ to $\mathcal{E}$ and view $\mathbb{W}$ as a section. Then in [**FJR2**] the virtual cycle is defined via a topological Euler class construction

$$[W_{g,n,G}(h_1, \ldots, h_n)]^{vir} = (-1)^{-\chi(E)} e(\mathbb{W} : \mathcal{E} \to \mathcal{F}) \cap [W_{g,n,G}(h_1, \ldots, h_n)]$$

We highlight two subcases.

Concavity.: Suppose that all markings are narrow and suppose that for every fibre C of the universal curve, $H^0(C, L_j) = 0$ for all j. Then the virtual cycle is given by

$$[W_{g,n,G}(h_1, \ldots, h_n)]^{\mathrm{vir}} = c_{\mathrm{top}}\left((R^1\pi_* E)^\vee\right) \cap [W_{g,n,G}(h_1, \ldots, h_n)].$$

Index zero.: Suppose that the dimension of $W_{g,n,G}(h_1, \ldots, h_n)$ is zero and that all markings are narrow. Furthermore, let us assume that the $\pi_* E$ and $R^1\pi_* E$ are both vector spaces and share the same rank. Then the virtual cycle is just the degree of $\mathbb{W}$.

The above definition is not quite algebraic because we use the complex conjugate in $\mathbb{W}$. The effort to remove it leads to several versions of algebraic treatment by Polishchuk-Vaintrob [**PV01**], Chiodo [**Ch06**] and Li-Zhang [**LZ**]. The one closest to the conventional treatment in Gromov-Witten theory is the third one. It goes as follows.

Consider $\bar{\partial} u_i = 0$ first. Let $\mathcal{E}$ be its moduli, i.e., the W-structure together with sections of L_i. Note that it is noncompact. $\mathcal{E}$ has a so-called perfect relative tangent-obstruction complex given by the two-term complex

$$\pi_* E \to R^1\pi_* E.$$

The main problem is

$$\partial W(u_1, \ldots, u_N) \in (R^1\pi_* E)^*.$$

Instead of mapping it to the right space by taking complex conjugates, Li-Zhang's observation is that it can be naturally interpreted as a map

$$\partial W(u_1, \ldots, u_N) : R^1\pi_* E \to \mathbb{C}.$$

The latter is called *a cosection.* Then, they can plug it into their general machinery to construct a virtual cycle supported at the zero locus of $\partial W(u_1, \ldots, u_N)$, which is precisely $W_{g,n,G}$.

2.1. Landau-Ginzburg/Calabi-Yau correspondence. The genus-zero invariants of both theories are encoded by two Lagrangian cones, $\mathcal{L}_{\mathrm{GW}}$ and $\mathcal{L}_{\mathrm{FJRW}}$, inside two symplectic vector spaces, $(\mathcal{V}_{\mathrm{GW}}, \Omega_{\mathrm{GW}})$ and $(\mathcal{V}_{\mathrm{FJRW}}, \Omega_{\mathrm{FJRW}})$. The two symplectic vector spaces also allow us to state the conjectural correspondence in higher genera. We recall the two settings simultaneously by using the subscript W, which can be read as GW or FJRW.

We define the vector space $\mathcal{V}_{\mathrm{W}}$ and its symplectic form Ω_{W}. The elements of $\mathcal{V}_{\mathrm{W}}$ are Laurent series with values in a state space H_{W};

$$\mathcal{V}_{\mathrm{W}} = H_{\mathrm{W}} \otimes \mathbb{C}((z^{-1})).$$

In FJRW theory the state space is normally the entire space $\mathcal{H}_{W,G}$. In GW theory the state space is $H_{CR}([X_W/\widetilde{G}]; \mathbb{C})$. We choose a basis $\phi_0, \ldots, \phi_k$ for the state space of FJRW theory and a basis $\varphi_0, \ldots, \varphi_k$ for the state space of GW theory. We label 1_{j_W} by ϕ_0 and 1_{X_W} by φ_0, respectively (these two classes play a special role in (2.6)). We express the basis of H_{W} as $\Phi_0, \ldots, \Phi_k$ and the dual basis as $\Phi^0, \ldots, \Phi^k$.

The vector space $\mathcal{V}_{\mathrm{W}}$ is equipped with the symplectic form

$$\Omega_{\mathrm{W}}(f_1, f_2) = \mathrm{Res}_{z=0}\langle f_1(-z), f_2(z)\rangle_{\mathrm{W}},$$

where $\langle\ ,\ \rangle_{\mathrm{W}}$ is the inner pairing discussed above. In this way $\mathcal{V}_{\mathrm{W}}$ is polarized as $\mathcal{V}_{\mathrm{W}} = \mathcal{V}_{\mathrm{W}}^+ \oplus \mathcal{V}_{\mathrm{W}}^-$, with $\mathcal{V}_{\mathrm{W}}^+ = H_{\mathrm{W}} \otimes \mathbb{C}[z]$ and $\mathcal{V}_{\mathrm{W}}^- = z^{-1}H_{\mathrm{W}} \otimes \mathbb{C}[[z^{-1}]]$, and can be regarded as the total cotangent space of $\mathcal{V}_{\mathrm{W}}^+$. The points of $\mathcal{V}_{\mathrm{W}}$ are parametrized by Darboux coordinates $\{q_a^h, p_{l,j}\}$ and can be written as

$$\sum_{a\geq 0}\sum_{h=0}^{k} q_a^h \Phi_h z^a + \sum_{l\geq 0}\sum_{j=0}^{k} p_{l,j}\Phi^j(-z)^{-1-l}.$$

We review the definitions of the potentials encoding the invariants of the two theories. In FJRW theory, the invariants are the intersection numbers

$$\langle \tau_{a_1}(\phi_{i_1}), \ldots, \tau_{a_n}(\phi_{i_n})\rangle_{g,n}^{\mathrm{FJRW}} = \int_{\overline{\mathcal{M}}_{g,n}} \prod_{i=1}^{n} \psi_i^{a_i} \cap \Lambda_{g,n,G}^W(\phi_{i_1}, \ldots, \phi_{i_n}), \tag{2.1}$$

with $\Lambda_{g,n,G}^W$. In GW theory, the invariants are the intersection numbers

$$\langle \tau_{a_1}(\varphi_{h_1}), \ldots, \tau_{a_n}(\varphi_{h_n})\rangle_{g,n,\delta}^{\mathrm{GW}} = \prod_{i=1}^{n} \mathrm{ev}_i^*(\varphi_{h_i})\psi_i^{a_j} \cap [X_W]_{g,n,\delta}^{\mathrm{vir}}. \tag{2.2}$$

The generating functions are respectively

$$\mathcal{F}_{\mathrm{FJRW}}^g = \sum_{\substack{a_1,\ldots,a_n \\ h_1,\ldots,h_n}} \langle \tau_{a_1}(\phi_{h_1}), \ldots, \tau_{a_n}(\phi_{h_n})\rangle_{g,n}^{\mathrm{FJRW}} \frac{t_{a_1}^{h_1}\cdots t_{a_n}^{h_n}}{n!} \tag{2.3}$$

and

$$\mathcal{F}^g_{\mathrm{GW}} = \sum_{\substack{a_1,\dots,a_n\\h_1,\dots,h_n}} \sum_{\delta\geq 0} \langle \tau_{a_1}(\varphi_{h_1}),\dots,\tau_{a_n}(\varphi_{h_n})\rangle^{\mathrm{GW}}_{g,n,\delta} \frac{t^{h_1}_{a_1}\cdots t^{h_n}_{a_n}}{n!}. \tag{2.4}$$

In this way, both theories yield a power series

$$\mathcal{F}^g_{\mathrm{W}} = \sum_{\substack{a_1,\dots,a_n\\h_1,\dots,h_n}} \sum_{\delta\geq 0} \langle \tau_{a_1}(\Phi_{h_1}),\dots,\tau_{a_n}(\Phi_{h_n})\rangle^{\mathrm{W}}_{g,n,\delta} \frac{t^{h_1}_{a_1}\cdots t^{h_n}_{a_n}}{n!}$$

in the variables t^i_a (for FJRW theory the contribution of the terms $\delta > 0$ is set to zero, whereas $\langle\ \rangle^{\mathrm{W}}_{g,n,0}$ should be read as $\langle\ \rangle^{\mathrm{FJRW}}_{g,n}$).

We can also define the partition function

$$\mathcal{D}_{\mathrm{W}} = \exp\left(\textstyle\sum_{g\geq 0}\hbar^{g-1}\mathcal{F}^g_{\mathrm{W}}\right). \tag{2.5}$$

Let us focus on the genus-zero potential $\mathcal{F}^0_{\mathrm{W}}$. The dilaton shift

$$q^h_a = \begin{cases} t^0_1 - 1 & \text{if } (a,h) = (1,0)\\ t^h_a & \text{otherwise.}\end{cases} \tag{2.6}$$

makes $\mathcal{F}^0_{\mathrm{W}}$ into a power series in the Darboux coordinates q^h_a. Now we can define $\mathcal{L}_{\mathrm{W}}$ as the cone

$$\mathcal{L}_{\mathrm{W}} := \{\boldsymbol{p} = d_{\boldsymbol{q}}\mathcal{F}^0_{\mathrm{W}}\} \subset \mathcal{V}_{\mathrm{W}}.$$

With respect to the symplectic form Ω_{W}, the subvariety $\mathcal{L}_{\mathrm{W}}$ is a Lagrangian cone whose tangent spaces satisfy the geometric condition $zT = \mathcal{L}_{\mathrm{W}} \cap T$ at any point (this happens because both potentials satisfy so called the string equation, dilaton equation and topological recursion equation).

Every point of $\mathcal{L}_{\mathrm{W}}$ can be written as follows

$$-z\Phi_0 + \sum_{\substack{0\leq h\leq k\\a\geq 0}} t^h_a\Phi_h z^a + \sum_{\substack{n\geq 0\\\delta\geq 0}} \sum_{\substack{0\leq h_1,\dots,h_n\leq k\\a_1,\dots,a_n\geq 0}} \sum_{\substack{0\leq\epsilon\leq k\\l\geq 0}} \frac{t^{h_1}_{a_1}\cdots t^{h_n}_{a_n}}{n!(-z)^{l+1}} \langle\tau_{a_1}(\Phi_{h_1}),\dots,\tau_{a_n}(\Phi_{h_n}), \tau_l(\Phi_\epsilon)\rangle^{\mathrm{W}}_{0,n+1,\delta}\Phi^\epsilon,$$

where the term $-z\Phi_0$ performs the dilaton shift.

REMARK 2.15 (J-function). Setting a and a_i to zero, we obtain the points of the form

(2.7)

$$-z\Phi_0 + \sum_{0\leq h\leq k} t^h_0\Phi_h + \sum_{\substack{n\geq 0\\\delta\geq 0}} \sum_{0\leq h_1,\dots,h_n\leq k} \sum_{\substack{0\leq\epsilon\leq k\\l\geq 0}} \frac{t^{h_1}_0\cdots t^{h_n}_0}{n!(-z)^{k+1}} \langle\tau_0(\Phi_{h_1}),\dots,\tau_0(\Phi_{h_n}), \tau_l(\Phi_\epsilon)\rangle^{\mathrm{W}}_{0,n+1,\delta}\Phi^\epsilon,$$

which uniquely determine the rest of $\mathcal{L}_{\mathrm{W}}$ (via multiplication by $\exp(\alpha/z)$ for any $\alpha \in \mathbb{C}$—*i.e.* via the string equation—and via the divisor equation in GW theory). We define the J-function

$$t = \sum_{h=0}^k t^h_0\Phi_h \mapsto J_{\mathrm{W}}(t,z)$$

from the state space H_{W} to the symplectic vector space $\mathcal{V}_{\mathrm{W}}$ so that $J_{\mathrm{W}}(t,-z)$ equals the expression (2.7).

2.1.1. *The conjecture.* The following conjecture can be regarded as a geometric version of the physical Landau–Ginzburg/Calabi–Yau correspondence [**VW89**] [**Wi3**]. The formalism is analogous to the conjecture of Coates-Corti-Iritani-Tseng/ Coates-Ruan [**CIT09, CR**] on crepant resolutions of orbifolds and uses Givental's quantization from [**Gi04**], which is naturally defined in the above symplectic spaces $\mathcal{V}_{\mathrm{FJRW}}$ and $\mathcal{V}_{\mathrm{GW}}$.

CONJECTURE 2.16 (LG/CY correspondence). Consider the Lagrangian cones $\mathcal{L}_{\mathrm{FJRW}}$ and $\mathcal{L}_{\mathrm{GW}}$.

(1) There is a bigraded space isomorphism between $\mathcal{H}_{W,G}$ and $H^*_{CR}([X_W/\tilde{G}],\mathbb{C})$.

(2) There is a degree-preserving $\mathbb{C}[z,z^{-1}]$-valued linear symplectic isomorphism
$$\mathbb{U}_{\mathrm{LG/CY}}:\mathcal{V}_{\mathrm{FJRW}}\to\mathcal{V}_{\mathrm{GW}}$$
and a choice of analytic continuation of $\mathcal{L}_{\mathrm{FJRW}}$ and $\mathcal{L}_{\mathrm{GW}}$ such that $\mathbb{U}_{\mathrm{LG/CY}}(\mathcal{L}_{\mathrm{FJRW}})=\mathcal{L}_{\mathrm{GW}}$.

(3) Up to an overall constant, the total potential functions up to a choice of analytic continuation are related by quantization of $\mathbb{U}_{\mathrm{LG/CY}}$; i.e.
$$\mathcal{D}_{\mathrm{GW}}=\widehat{\mathbb{U}}_{\mathrm{LG/CY}}(\mathcal{D}_{\mathrm{FJRW}}).$$

The first part of the conjecture is the *cohomological LG/CY correspondence* and has been proved in full generality by Chiodo-Ruan [**ChiR9**]. The second part is the genus zero LG/CY correspondence and has been verified for the quintic 3-fold with group $\langle J\rangle$ by Chiodo-Ruan and the Fermat polynomial with group $\langle J\rangle$ by Chiodo-Iritani-Ruan [**CIR**].

3. Gauged linear sigma model

In this and the next section, we describe the background of the gauged Witten equation. The part understood relatively well in physics is the abelian case corresponding to a hypersurface or complete intersection in toric varieties. There is a nonabelian version corresponding to a hypersurface or complete intersection in Grassmannian, which is very much at the frontier of physics. In this survey, we discuss the abelian cases.

3.1. Toric varieties and phases. Toric varieties are a class of orbifolds constructed as geometric invariant/symplectic quotients of $\mathbb{C}^N//(\mathbb{C}^*)^d$. More generally, we also allow an additional finite abelian group G (*admissible group*) and consider the quotient $(\mathbb{C}^N//(\mathbb{C}^*)^d)/G$. The combinatorial data of the above GIT-quotient is encoded in the weights of action. More precisely, the j-th copy of $\mathbb{C}^*$ acts on $\mathbb{C}^N$ by the weights
$$\lambda_j(z_1,\ldots,z_N)=(\lambda_j^{q_{1j}}z_1,\ldots,\lambda_j^{q_{Nj}}z_N).$$
The integral matrix (q_{ij}) is called the *weight or charge matrix*. In particular, some q_{ij} could be negative and hence the resulting toric variety could be non-compact.

Since $\mathbb{C}^*$ is non-compact, the set theoretic quotient could be non-Hausdorff. Geometric invariant theory gives a way to construct Hausdorff quotients. The key is to choose a dense open subset U consisting of *stable* orbits. The geometric invariant quotient depends on such a choice of stability. It can be conveniently formulated in terms of symplectic geometry. $\mathbb{C}^N$ is equipped with the standard Kahler form $\omega = \sum_i dz_i \wedge d\bar{z}_i$. The above action is Hamiltonian with moment map $D = (D_1, \dots, D_n)$ with

$$D_j = \sum_i q_{ij}|z_i|^2.$$

Let $\mu \in \mathbb{R}^n$ be a regular value of the moment map. The toric variety/symplectic quotient is defined as

$$X_\mu = \mathbb{C}^N//(\mathbb{C}^*)^d = D^{-1}(\mu)/U(1)^d.$$

In terms of geometric invariant theory, we choose U_μ to be the set of orbits intersecting $D^{-1}\mu$. Then,

$$X_\mu = U_\mu//(\mathbb{C}^*)^d = D^{-1}(\mu)/U(1)^d.$$

If all the weights $q_{ij} > 0$, $D_j > 0$ and any $\mu = (\mu_1, \dots, \mu_d)$ with $\mu_i > 0$ is a regular value. The toric variety X_μ is compact and smooth as an orbifold/Deligne-Mumford stack. Furthermore, a different regular value μ defines the same toric variety and we will often drop μ from the notation. X carries a collection of canonical line bundles. i-th $\mathbb{C}^*$ defines a $\mathbb{C}^*$-principal bundle P_i over X. By a abuse of the notation, we will use P_i to denote its corresponding $\mathbb{C}$-bundle as well. The action of $(\mathbb{C}^*)^d$ on the coordinate z_ρ defines a line bundle L_ρ. It is clear that $L_\rho = \otimes_i P^{q_{\rho i}}$.

When some $q_{ij} < 0$, the situation changes. The above toric variety depends on a choice of regular value μ. There is a chamber structure for $\mathbb{R}^d$ such that (i) the toric varieties will be isomorphic for the regular values in the same chamber; (ii) the toric varieties of regular values of different chambers will be birational to each other. These chambers are referred to as *phases* in the physics literature. A particular important case is of *Calabi-Yau weights* where $\sum_i q_{ij} = 0$ for all i. In this case, the toric variety is Calabi-Yau and necessarily non-compact. Let's look at some important examples studied by Witten [**Wi4**].

EXAMPLE 3.1. The easiest example is weighted projective space $WP(q_1, \dots, q_N)$. Here, we consider $\mathbb{C}^N//\mathbb{C}^*$ with the weight $(q_1, \dots, q_N)$ for $q_i > 0$. The moment map is $D = \sum_i q_i|z_i|^2$. Notice that the image of the moment map is $\mathbb{R}_{\geq 0}$. D is a quadratic function and the only critical point is zero. Hence $\mu = 0$ is a critical value and there is only one chamber $\mu > 0$. When $D = \mu > 0$, $(z_1, \dots, z_N) \neq (0, \dots, 0)$. We obtain weighted projective space $WP(q_1, \dots, q_N)$.

EXAMPLE 3.2. Suppose that $q_i > 0$ for $i = 1, \cdots, N$ and $q = \sum_i q_i$. Consider $\mathbb{C}^{N+1}//\mathbb{C}^*$ with the weights $(q, \dots, q_n, -q)$. This is a Calabi-Yau weight. The moment map

$$D = \sum_i^n q_i|z|_i^2 - q|z_{N+1}|^2.$$

D is a quadratic function with the only critical point at zero. Therefore, the only critical value is $\mu = 0$ and there are two phases $\mu > 0$ or $\mu < 0$.

For $\mu > 0$,

$$\sum_i^n q_i|z_i|^2 = q|z_{N+1}|^2 + \mu.$$

Hence, $(z_1,\dots,z_N) \neq (0,\dots,0)$. We obtain weighted projective space $X_{base} = WP(q_1,\dots,q_N)$ and a map from the toric variety to $WP(q_1,\dots,q_n)$. The toric variety for μ can be expressed as the total space of the line bundle P^{-q}, which is the canonical bundle $K_{WP(q_1,\dots,q_n)}$.

For $\mu < 0$,

$$\sum_i^n q_i|z_i|^2 - \mu = q|z_{N+1}|^2.$$

Hence, $z_{N+1} \neq 0$. We can quotient z_{N+1} by $\mathbb{C}^*$. Here, we choose the basis λ^{-1}. z_{N+1} can be considered to have positive weight q. Moreover, it has isotropy group $\mathbb{Z}_q$, the q-th roots of unity. The quotient is $WP(q) = B\mathbb{Z}_q$. Note that the canonical $\mathbb{C}^*$ bundle P is reduced to a $\mathbb{Z}_q$ bundle. Note that $\mathbb{Z}_q$ acts on the $(z_1,\dots,z_N)$ by $(\xi^{q_1}z_1,\dots,\xi^{q_N}z_N)$ for $\xi^q = 1$. We remark that this action is that of J in FJRW-theory. Again, the toric variety can be viewed as the total space of a rank N orbifold vector bundle over $pt/\mathbb{Z}_q$ which is isomorphic to $\mathbb{C}^N/\langle J\rangle$. This is a special phase where we obtain a kind of toric variety of finite group instead of $\mathbb{C}^*$. We call it *a Landau-Ginzburg phase or a pure Landau-Ginzburg phase* [**Wi4**]. This example underlies Witten's physical argument of the Landau-Ginzburg/Calabi-Yau correspondence for Calabi-Yau hypersurfaces of weighted projective spaces.

EXAMPLE 3.3. The above example admits a straight-forward generalization $\mathbb{C}^{N+M}/\mathbb{C}^*$ with the weights $(q_1,\dots,q_N,-q'_1,\dots,-q'_M)$ for $q_i, q'_i > 0$. We do not assume the Calabi-Yau condition. The moment map is $D = \sum_i q_i|z_i|^2 - \sum_j q'_j|z_{N+j}|^2$. Again, there are two phases $\mu > 0, \mu < 0$. For $\mu > 0$, $(z_1,\dots,z_N) \neq (0,\dots,0)$, we can quotient it out first to obtain $\oplus_j P^{-q'_j}$ over $WP(q_1,\dots,q_N)$. Similarly, for $\mu < 0$, the toric variety is $\oplus_i P^{-q_i}$ over $WP(q'_1,\dots,q'_M)$.

EXAMPLE 3.4. Consider the action of $\mathbb{C}^*$ on $\mathbb{C}^N$ with positive weights $(q_1,\dots,q_N)$. Its quotient is weighted projective space $WP(q_1,\dots,q_N)$. Consider another weighted projective space given by the weights $(q'_1,\dots,q'_M)$. Let $p = \sum_i q_i, p' = \sum_j q'_j$. Consider the toric variety $\mathbb{C}^{N+M} \times \mathbb{C}/(\mathbb{C}^*)^2$ with weights

$$\begin{pmatrix} q_1, & \dots, & q_N & 0, & \dots, & 0 & -p \\ 0, & \dots, & 0, & q'_1, & \dots, & q'_M & -p' \end{pmatrix}$$

The moment map is

$$D_1 = \sum_i q_i|z_i|^2 - p|z_{N+M+1}|^2; \; D_2 = \sum_j q'_j|z_{j+N}|^2 - p'|z_{N+M+1}|^2.$$

The critical points are (i) $z_1 = \dots = z_N, z_{N+M+1} = 0$; (ii) $z_{N+1} = \dots = z_{N+M} = 0, z_{N+M+1} = 0$; (iii) $z_1 = \dots = z_{N+M} = 0$. The critical values are (i) $\mu_1 = 0$; (ii) $\mu_2 = 0$; (iii)$\mu_1, \mu_2 < 0, \frac{\mu_1}{p} = \frac{\mu_2}{p'}$. It divides $\mathbb{R}^2$ into five phases.

For $\mu_1, \mu_2 > 0$, $(z_1,\dots,z_N) \neq (0,\dots,0), (z_{N+1},\dots,z_{N+M}) \neq (0,\dots,0)$. The toric variety can be expressed as the total space of the line bundle $P_1^{-p} \otimes P_2^{-p'}$ of bidegree $(-p,-p')$ or $K_{WP(q_1,\dots,q_N)} \otimes K_{WP(q'_1,\dots,q'_M)}$ over $WP(q_1,\dots,q_N) \times WP(q'_1,\dots,q'_M)$.

For $\mu_1 > 0, \mu_2 < 0$, a similar analysis implies $(z_1,\dots,z_N) \neq (0,\dots,0), z_{N+M+1} \neq 0$ and we can divide it by $(\mathbb{C}^*)^2$. Since two actions on z_{N+M+1} intertwine, we did not obtain $WP(q_1,\dots,q_N) \times BZ_p$. Instead, we obtain a nontrivial gerbe over $WP(q_1,\dots,q_N)$. Let's be more specific. Dividing it by the first copy of $\mathbb{C}^*$, we obtain $P_1^{-p} \to WP(q_1,\dots,q_N)$ where P_1 is the $\mathbb{C}^*$ bundle associated with the first

$\mathbb{C}^*$. Then, we quotient out the second $\mathbb{C}^*$. We obtain an object which is a $B\mathbb{Z}_{p'}$-bundle over $WP(q_1, \ldots, q_N)$. This is a highly stacky object and called *a gerbe* over $WP(q_1, \ldots, q_N)$. We denote it by $WP(q_1, \ldots, q_N)^{-\frac{p}{p'}}$. The toric variety is the total space of $\oplus_i P_2^{-q_i'}$. The analysis of the phase $\{\mu_1 < 0, \mu_2 > 0\}$ is similar.

For $\mu_1, \mu_2 < 0$, $\frac{\mu_1}{p} > \frac{\mu_2}{p'}$, $D_1 = \mu_1, D_2 = \mu_2$ is equivalent to

$$D_1 = \mu_1,\ \tilde{D}_2 = \frac{1}{p}\sum_j q_i|z_i|^2 - \frac{1}{p'}\sum_i q_j'|z_{j+N}|^2 = \frac{\mu_1}{p} - \frac{\mu_2}{p'} > 0.$$

Then, we have

$$z_{N+M+1} \neq 0,\ (z_1, \ldots, z_N) \neq (0, \ldots, 0).$$

By repeating the above analysis, the toric variety is the total space of $\oplus_i P_2^{-q_i'}$. The other phase can be described similarly.

EXAMPLE 3.5. Consider $\mathbb{C}^9//U(1)^2$ with weights

$$(1, 1, 1, 0, 0, 0, 0, -3, 0), (0, 0, 0, 1, 1, 1, 1, -1, -3).$$

The moment map is

$$D_1 = |z_1|^2 + |z_2|^2 + |z_3|^2 - 3|z_8|^2,$$

$$D_2 = |z_4|^2 + |z_5|^2 + |z_6|^2 + |z_7|^2 - |z_8|^2 - 3|z_9|^2.$$

The critical values are (i) $\mu_1 = 0$, (ii) $\mu_2 = 0$, (iii) $\mu_2 = \frac{\mu_1}{3}$ and $\mu_1, \mu_2 < 0$. Again we have five phases.

For the chamber $\mu_1, \mu_2 > 0$, we obtain the total space $L_1 \oplus L_2$ where $L_1 = P_1^{-3}$ has bi-degree $(-3, 0)$ and $L_2 = P_1^{-1} \otimes P_2^{-3}$ has bi-degree $(-1, -3)$.

For the phase $\mu_1 > 0, \mu_2 < 0$, $(z_1, z_2, z_3) \neq (0, 0, 0)$, $(z_8, z_9) \neq 0$. We can quotient them by $\mathbb{C}^* \times \mathbb{C}^*$ to obtain $C\mathbb{P}^2 \times WP(1, 3)$. The toric variety is the total space of the sum of 4 line bundles of bi-degrees $(0, -1)$.

The phase $\mu_1 < 0, \mu_2 > 0$ is similar.

The interesting chamber is $\mu_2 > \frac{\mu_1}{3}$. Then, $z_8 \neq 0, z_9 \neq 0$. We can quotient it by $\mathbb{C}^* \times \mathbb{C}^*$ to obtain an orbifold $WP(9) = B\mathbb{Z}_9$, where

$$\mathbb{Z}_9 = \{(\xi, \eta) \in \mathbb{C}^* \times \mathbb{C}^*; \xi^3 = \eta^{-1}, \eta^3 = 1\}.$$

The toric variety is $\mathbb{C}^7/\mathbb{Z}_7$ with action $(\xi z_1, \xi z_2, \xi z_3, \xi^{-3} z_4, \ldots, \xi^{-3} z_7)$. It is interesting that we obtain a Landau-Ginzburg phase again. A very interesting question is when the given problem has a Landau-Ginzburg phase. This problem has been studied in depth by Clarke and Guffin [**CG**].

3.2. Superpotential and gauged linear sigma model. Another piece of data for Witten's gauged linear sigma model is so-called *superpotential.* A superpotential is a polynomial $W : \mathbb{C}^N \to \mathbb{C}$ invariant under $(\mathbb{C}^*)^d$. Therefore, it descends to a holomorphic function on the toric variety $\mathbb{C}^N/(\mathbb{C}^*)^d$ in every phase. Therefore, W must be constant if $\mathbb{C}^N/(\mathbb{C}^*)^d$ is compact. For W to be interesting, the toric variety is necessarily non-compact. The cases of current interest are the Calabi-Yau toric varieties.

DEFINITION 3.6. W is called *nondegenerate* if (i) the critical points of W outside of the critical loci of D are smooth submanifolds; (ii) W defines a nondegenerate critical point in any normal direction to the critical orbifold.

The critical sub-orbifold is referred as the space of *classical vacua* in the physics literature.

EXAMPLE 3.7. Suppose that $F : \mathbb{C}^N \to \mathbb{C}$ is a nondegenerate quasihomogeneous polynomial of weights $(q_1, \ldots, q_N)$ and degree d such as in FJRW-theory. Let

$$W = pF : \mathbb{C}^{N+1} \to \mathbb{C}.$$

Here, we assign the weight $-d$ to the variable p. Then, we have a Calabi-Yau weight. Let's compute the critical set of W.

$$\partial_p W = F = 0, \partial_{z_i} W = p\partial_{z_i} F = 0$$

implies that either $p \neq 0, (z_1, \ldots, z_N) = (0, \ldots, 0)$ or $p = 0, F(z_1, \ldots, z_N) = 0$.

In the phase $\mu > 0$, the critical sub-orbifold is $p = 0, F(z_1, \ldots, z_N) = 0$, the degree d-hypersurface of zero section $WP(q_1, \ldots, q_N)$. We call this phase the *Calabi-Yau phase.* In the phase $\mu < 0$, the critical subset is $(0, \ldots, 0)$. It is clearly nondegenerate.

EXAMPLE 3.8. Suppose that we have several quasi-homogeneous polynomials $F_1, F_2, \ldots, F_k$ of degree $(d_1, \ldots, d_k)$. We first assume that they intersect transversely in $WP(q_1, \ldots, q_N)$ and define a complete intersection. Let

$$W = \sum_i p_i F_i : \mathbb{C}^{N+k} \to \mathbb{C},$$

where we assign weights $-d_i$ to p_i. The Calabi-Yau weight is $\sum_i q_i = \sum_j d_j$, where the complete intersection defined by $F_1 = \cdots = F_k = 0$ is a Calabi-Yau orbifold. One can view it as the *gauged linear sigma model for the complete intersection.*

Our previous analysis shows that there are two phases; (i) $\mu > 0$, the toric variety is $\oplus_j P^{-d_j}$ over $WP(q_1, \ldots, q_N)$ and (ii) $\mu < 0$, the toric variety is $\oplus_i P^{-q_i}$ over $WP(d_1, \ldots, d_k)$.

Let's calculate the critical set of W.

$$\partial_{p_j} W = F_j = 0, \partial_{z_i} W = \sum_j p_j \partial_{z_i} F_j = 0$$

In the chamber $\mu > 0$, $(z_1, \ldots, z_N) \neq (0, \ldots, 0)$. Since F_i intersects transversely, the dF_i is linearly independent for $(z_1, \ldots, z_N) \neq (0, \ldots, 0)$. Therefore, $p_i = 0$. Therefore, the critical set is the complete intersection in the zero section of toric variety. In the chamber $\mu < 0$, $(p_1, \ldots, p_k) \neq (0, \ldots, 0)$. This implies that $(z_1, \ldots, z_N) = (0, \ldots, 0)$. Then, the critical set is the zero section of the corresponding toric variety. However, in this case, for any $(p_1, \ldots, p_n)$, we have a pure LG-model of super-potential $\sum_i p_i F_i$. One can view it as a family of pure LG-theories.

EXAMPLE 3.9. Consider $F_1 = \sum_{i=1}^4 s_i^3, F_2 = \sum_{j=1}^3 S_j T_j^3$. They define hypersurfaces of $\mathbb{P}^3 \times \mathbb{P}^2$ of bi-degree $(3, 0), (1, 3)$. Now, we consider the gauged linear sigma model of the intersection. Namely, we define the superpotential

$$W = p_1 F_1 + p_2 F_2.$$

The critical points are given by the equation

$$F_1 = 0, F_2 = 0, p_1 \partial_{S_i} F_1 + p_2 \partial_{S_i} F_2 = 0, p_1 \partial_{T_j} F_1 + p_2 \partial_{T_j} F_2 = 0.$$

For the chamber $\mu_1 > 0, \mu_2 > 0$, the critical set is the complete intersection $F_1 = 0, F_2 = 0$ in the zero section of the corresponding toric variety.

For the chamber $\mu_2 > \frac{\mu_1}{3}$, $p_1 \neq 0, p_2 \neq 0$. The toric variety is $\mathbb{C}^7/\mathbb{Z}_9$. The super potential is $W = p_1F_1 + p_2F_2$ for any fixed $p_1 \neq 0, p_2 \neq 0$. For example, we can take $p_1 = p_2 = 1$ to obtain $W = \sum_{i=1}^3(S_i^3 + S^iT_i^3) + S_4^3$.

Now, we formalize the notion of gauged linear sigma model. Recall that if all the weights are positive, then there is only one phase and we obtain a compact toric variety. For the general case, a choice of phases pick a subset of variables $z_{i_1}, \ldots, z_{i_k}$ where $(\mathbb{C}^*)^d$ with appropriate choice of basis acts on $\mathbb{C}^k$ with positive weights. Hence, we obtain a compact toric variety $X_{base} = \mathbb{C}^k//(\mathbb{C})^d$. There is a canonical $(\mathbb{C}^*)^d$-bundle over X_{base}. The remaining variable z_j defines a toric line bundle L_j with nonpositive multi-degree given by the weights of the actions of $(\mathbb{C}^*)^d$. Let $V = \oplus_j L_j$. We call the pair (V, W) *a classical gauged linear sigma model or just a gauged linear sigma model.*

3.3. Admissible group. For the purpose of mirror symmetry, it is important to extend the above construction to include an admissible group. Suppose that G is a finite subgroup of diagonal symmetry $(\mathbb{C}^*)^N$ preserving W. We can enlarge $(\mathbb{C}^*)^d$ to the group $(\mathbb{C}^*)^d_G$ generated by $(\mathbb{C}^*)^d$ and G. Then, we can apply the same construction to obtain a gauged linear sigma model over each phase with larger isotropy group.

4. Mirror symmetry

There is a rich mirror symmetry phenomenon among the gauged linear sigma model and its phases. The simplest one is the pure LG-phase where there is a beautiful mirror symmetry construction (*LG-to-LG mirror*) of Berglund-Hübsch-Krawitz [**BH93, Kr**]. In their construction, the superpotential plays a prominent role while the geometry of the ambient weighted projective space is suppressed. On the other hand, there is famous mirror symmetry construction (*CY-to-CY mirror*) of Batyrev-Borisov for the Calabi-Yau complete intersection of a Gorenstein toric variety. The Batyrev-Borisov construction emphasize the ambient toric structure while the super-potential is suppressed. These two extremal cases are understood the best. There is some on-going effort to bridge the above two extremes [**Bo, Cl**].

We first review Berglund-Hübsch-Krawitz construction. Then, we will discuss the toric mirror symmetry of Batyrev-Borisov. Then, we make some remarks on the intermediate phases.

4.1. LG-to-LG mirror. Berglund and Hübsch [**BH93**] consider polynomials in N variables having N monomials

$$W(x_1, \ldots, x_N) = \sum_{i=1}^{N} \prod_{j=1}^{N} x_j^{m_{i,j}}. \tag{4.1}$$

Note that each of the N monomials has coefficient one; indeed, since the number of variables equals the number of monomials, even when we start from a polynomial of the form $\sum_{i=1}^N \gamma_i \prod_{j=1}^N x_j^{m_{i,j}}$, it is possible to reduce to the above expression by conveniently rescaling the N variables (this uses the nondegeneracy condition). In this way assigning a polynomial W as above amounts to specifying its exponent square matrix

$$E_W = (m_{i,j})_{1\leq i,j\leq N}.$$

The polynomials studied in [**BH93**] are called "invertible" because the matrix E_W is an invertible $N \times N$ matrix as a consequence of the uniqueness of the charges $q_1, \ldots, q_N$ (nondegeneracy of W). There is a strikingly simple classification of invertible nondegenerate singularities by Kreuzer and Skarke [**KS92**].

An invertible potential W is nondegenerate if and only if it can be written, for a suitable permutation of the variables, as a sum of invertible potentials (with disjoint sets of variables) of one of the following three types:

$$W_{\text{Fermat}} = x^a.$$
$$W_{\text{loop}} = x_1^{a_1}x_2 + x_2^{a_2}x_3 + \cdots + x_{N-1}^{a_{N-1}}x_N + x_N^{a_N}x_1.$$
$$W_{\text{chain}} = x_1^{a_1}x_2 + x_2^{a_2}x_3 + \cdots + x_{N-1}^{a_{N-1}}x_N + x_N^{a_N}.$$

One can compute the charges $q_1, \ldots, q_N$ by simply setting

$$q_i = \textstyle\sum_j m^{i,j}, \tag{4.2}$$

the sum of the entries on the ith line of $E_W^{-1} = (m^{i,j})_{1\leq i,j\leq N}$.

Each column $(m^{1,j}, \ldots, m^{N,j})$ of the matrix E_W^{-1} can be used to define the diagonal matrix

$$\rho_j = \text{Diag}(\exp(2\pi \mathtt{i} m^{1,j}), \ldots, \exp(2\pi \mathtt{i} m^{N,j})). \tag{4.3}$$

In fact these matrices satisfy the following properties: $\rho_j^* W = W$; *i.e.* W is invariant with respect to ρ_j. Furthermore the group $\text{Aut}(W)$ of diagonal matrices α such that $\alpha^* W = W$ is generated by the elements $\rho_1, \ldots, \rho_N$:

$$\text{Aut}(W) := \{\alpha = \text{Diag}(\alpha_1, \ldots, \alpha_N) \mid \alpha^* W = W\} = \langle \rho_1, \ldots, \rho_N \rangle.$$

For instance, the above-mentioned matrix j_W whose diagonal entries are $\exp(2\pi\mathtt{i}q_1), \ldots$, and $\exp(2\pi\mathtt{i}q_N)$ lies in $\text{Aut}(W)$ and is indeed the product $\rho_1 \cdots \rho_N$. Recall that

$$SL_W = \text{Aut}(W) \cap SL(\mathbb{C}^N),$$

the matrices with determinant 1; in Berglund and Hübsch's construction we consider groups G containing j_W (A-admissible) and included in SL_W (B-admissible). Let W be a nondegenerate invertible potential of charges $q_1, \ldots, q_N$ satisfying the *Calabi–Yau condition*

$$\textstyle\sum_j q_j = 1. \tag{4.4}$$

The geometrical meaning of this condition is that $X_W = \{W = 0\}$ is of Calabi–Yau type in the sense that ω is trivial (adjunction formula: $d = \sum_j w_j$).

Another important geometric condition is the Gorenstein condition of the ambient weighted projective space $\mathbb{P}(w_1, \ldots, w_N)$. The Gorenstein condition corresponds to a numerical condition $w_i | \sum_j w_j$. For weighted projective spaces, the Gorenstein condition is equivalent to another well known condition, namely that the associated toric variety is reflexive. In the Calabi–Yau case, $d = \sum_j w_j$. Then X_W is a Calabi–Yau hypersurface of Gorenstein weighted projective space if and only if $w_j | d$. If W is a Fermat polynomial, the ambient weighted projective stack is Gorenstein. Otherwise, the ambient weighted projective stack is not Gorenstein in general. Among all the known examples of Calabi–Yau hypersurfaces, the Fermat polynomials constitute only a small fraction.

4.1.1. *The polynomial* $W^\vee$. Following Berglund–Hübsch, we consider the transposed polynomial $W^\vee$ defined by the property

$$E_{W^\vee} = (E_W)^\vee.$$

Namely, the polynomial $W^\vee$ is defined by transposing the matrix $(m_{i,j})$:

$$W^\vee(x_1, \dots, x_N) = \sum_{i=1}^{N} \prod_{j=1}^{N} x_j^{m_{j,i}}. \tag{4.5}$$

This construction preserves the type of polynomial sending Fermat to Fermat, loop to loop and chain to chain. This shows that $W^\vee$ is nondegenerate if and only if W is nondegenerate. Recall that q_j is the sum of the jth column of the inverse matrix E_W^{-1}. Hence, the charges $\overline{q}_1, \dots, \overline{q}_N$ of $W^\vee$ are the sums of the rows of E_W^{-1}. Therefore,

$$\textstyle\sum_j q_j = \sum_j \overline{q}_j.$$

In this way, $W^\vee$ is of Calabi–Yau type if and only if W is of Calabi–Yau type.

The striking idea of Berglund and Hubsch is that W and $W^\vee$ should be related by mirror symmetry. Clearly this is not true in the naive way: the mirror of a Fermat quintic three-fold X_W is not the quintic itself as one would get by transposing the corresponding exponent matrix. Instead, as already discussed in the introduction, the mirror $X_W^\vee$ is the quotient of X_W by the automorphism group $(\mathbb{Z}_5)^3$. It was already understood by Berglund–Hübsch that the correct statement should read

$$(W, G) \text{ mirror to } (W^\vee, G^\vee)$$

for a conjectural dual group $G^\vee$. Many examples of dual groups have been constructed in the literature. The general construction was given only recently by Krawitz [**Kr**].

4.1.2. *The group* $G^\vee$. The group $G^\vee$ is contained in $\mathrm{Aut}(W^\vee)$. Recall that $\mathrm{Aut}(W^\vee)$ is spanned by the diagonal symmetries $\rho_1^\vee, \dots, \rho_N^\vee$ determined by the columns of $(E_W^\vee)^{-1}$ as in (4.3):

$$\mathrm{Aut}(W^\vee) = \langle \rho_1^\vee, \dots, \rho_N^\vee \rangle.$$

Then $G^\vee$ is the subgroup defined by

$$G^\vee = \left\{ \textstyle\prod_{j=1}^N (\rho_i^\vee)^{a_i} \mid \text{ if } \prod_{j=1}^N x_i^{a_i} \text{ is } G\text{-invariant} \right\}. \tag{4.6}$$

Alternatively, we express any $g \in G$ as $g = \rho_1^{k_1} \dots \rho_N^{k_N}$ and $h \in G^\vee$ as $h = (\rho_1^\vee)^{l_1} \dots (\rho_N^\vee)^{l_N}$. Then, $G^\vee$ is determined by imposing within $\mathrm{Aut}(W^\vee)$ the following conditions for all $g = \rho_1^{k_1} \dots \rho_N^{k_N} \in G$

$$\begin{bmatrix} k_1 & \dots & k_N \end{bmatrix} E_W^{-1} \begin{bmatrix} l_1 \\ \vdots \\ l_N \end{bmatrix} \in \mathbb{Z}.$$

We have the following properties: transposition is an involution $(G^\vee)^\vee = G$, it is inclusion-reversing ($H \subseteq K \Rightarrow H^\vee \supseteq K^\vee$), it sends the trivial subgroup of $\mathrm{Aut}(W^\vee)$ to the total group $\mathrm{Aut}(W)$, and it exchanges $\langle J_W \rangle$ and $SL_{W^\vee}$.

4.1.3. *Mirror symmetry conjectures between LG models.* Now, we can state the mirror symmetry conjectures between pure LG-models. Here, "mirror" means that the A model and the B model are exchanged.

CONJECTURE 4.1 (mirror symmetry $LG-LG$). Suppose that W is a nondegenerate invertible polynomial. Then the Landau–Ginzburg models (W, G) and $(W^\vee, G^\vee)$ are mirror to each other.

REMARK 4.2. Since we have not given a precise meaning to the notion of mirror, the above conjectures should be viewed as a guideline instead of a mathematical statement. Once the mathematical theories are developed on both sides, the above conjecture can be turned into more precise mathematical statements. This is the case for the state space. We have formulated the A-model state space $\mathcal{H}_{FJRW}$. The B-model state space $\mathcal{Q}_{W^\vee,G^\vee}$ was known much early to physicists Intriligator-Vafa as an orbifolded Milnor ring

$$\mathcal{Q}_{W^\vee,G^\vee} = (\oplus_{h\in G^\vee}\mathcal{Q}_{W_h^\vee})^{G^\vee},$$

where $\mathcal{Q}_{W_h^\vee}$ is the Milnor ring of $W_h^\vee$. As a vector space, $\mathcal{Q}_{W^\vee,G^\vee}$ is isomorphic to $\mathcal{H}_{W^\vee,G^\vee}$. However, they have different gradings (see [**Kr**] for the detail).

The following was proven by Krawitz.

THEOREM 4.3 (Krawitz [**Kr**]). *Suppose that W is invertible. Then, there is a bigraded vector space isomorphism*

$$\mathcal{H}_{W,G} \cong \mathcal{Q}_{W^\vee,G^\vee}.$$

REMARK 4.4. The isomorphism in the theorem is interesting in its own right. The basic idea is to exchange monomials with group elements. This isomorphism, however, does not respect the gradings. Let us express an element of $\mathcal{Q}_{W,G}$ as $\bigwedge x_i^{l_i}dx_i \mid \prod_i \rho_i^{k_i+1}\rangle$ where $\bigwedge x_i^{l_i}dx_i$ is fixed by $\prod_i \rho_i^{k_i+1}$. Here, we use the presentation of an element of $\mathrm{Aut}(W)$ in terms of the generators ρ_i. Then the mirror map in Krawitz's theorem [**Kr**] is of the form

$$\bigwedge_i x_i^{l_i}dx_i \mid \prod_i \rho_i^{k_i+1}\Big\rangle \longmapsto \bigwedge_i x_i^{k_i}dx_i \mid \prod_i \rho_i^{l_i+1}\Big\rangle.$$

The proof uses Kreuzer and Skarke's decomposition of invertible polynomials. Note that there is no analogous decomposition on the CY side. This is the main reason why the LG side is easier to work with in this case.

REMARK 4.5. Recently, Borisov has found [**Bo**] a new proof of the theorem above via vertex algebras. This approach may actually lead to a unified setup including both Berglund–Hübsch and Borisov–Batyrev duality.

On the A model side, we have rigorous theories, FJRW theory for the LG model and GW theory for the CY model. The counterpart of these theories for the B model side is incomplete. The genus-zero theory should correspond to a Frobenius manifold structure; however, unless G is a trivial group, it appears delicate to define the suitable G-orbifold version extending the state space $\mathcal{Q}_{W,G} = \bigoplus_{g\in G}(\mathcal{Q}_{W_g})^G$. Due to Kaufmann–Krawitz [**Ka06, Kr**], we can provide at least an orbifold Frobenius algebra construction; *i.e.*, a ring structure on the state space.

Krawitz checked that his vector space isomorphism for the case $G = \mathrm{Aut}(W)$

$$\mathcal{H}_{W,\mathrm{Aut}(W)} \cong \mathcal{Q}_{W^\vee,(e)}$$

respects the Frobenius algebra structure. He also provided evidence for the same statement for $G \subseteq SL_W$ and W of loop type and for other special cases related to Arnold's strange duality. We refer to [**Kr**] for precise statements.

4.2. CY-to-CY mirror. In this subsection, we review the Batyrev-Borisov mirror construction for the Calabi-Yau complete intersection of toric varieties. This is a rather classical subject with a large body of references. Our main reference is a book by Cox and Katz [**CK**]. Instead of listing all the relevant papers, we will refer the readers to the book for a complete set of references.

The Batyrev-Borisov construction is based on the notion of reflexible polytope. Let M be a lattice. A polytope $\Delta \subset M_{\mathbb{R}} = M \otimes \mathbb{R}$ is the convex hull of finitely many integral elements $v_i \in M$. Suppose that $0 \in \Delta$ and $N = Hom_{\mathbb{Z}}(M, \mathbb{Z})$ is the dual lattice. One can define its polar dual $\Delta^0 \in N_{\mathbb{R}}$ by

$$\Delta^0 = \{w \in N_{\mathbb{R}}, \langle w, v\rangle \geq -1 \text{ for any } v \in \Delta\}.$$

Δ is *reflexive* iff (i) Δ^0 is the convex hull of finitely many integral elements of N; (ii) 0 is the only interior integral point of Δ. It is easy to check that Δ^0 is also reflexive. Recall that Berglund-Hubsch-Krawitz mirror symmetry is based on the combinatorial duality of transposed exponent matrices. Batyrev-Borisov mirror symmetry is based on the combinatorial duality of reflexive polytopes $\Delta \leftrightarrow \Delta^0$.

Starting from Δ, one can construct a Fano/Gorenstein toric variety $\mathbb{P}_\Delta$ as follows. We first define the normal fan Σ_Δ as the collection of cones over the faces of Δ^0 (not Δ). The standard toric technique defines a Gorenstein toric variety $\mathbb{P}_\Delta = \mathbb{P}_\Sigma$. Let $\Sigma(1)$ be the set of one-dimensional cones. Each $v_i \in \Sigma(1)$ generates a toric divisor D_i. Any other divisor D can be expressed as a linear combination $D = \sum_i a_i D_i$. Then, D defines a polytope

$$\Delta_D = \{w \in M_{\mathbb{R}}, \langle w, v_i\rangle \geq -a_i \text{ for any } v_i \in \Sigma(1)\}.$$

Note that the anti-canonical divisor $-K_{\mathbb{P}_\Delta} = \sum_i D_i$. Then, $\Delta = \Delta_{-K_{\mathbb{P}_\Delta}}$. One can also recover the presentation of a toric variety in the previous section. There is a well-known short exact sequence

$$0 \to M \to \mathbb{Z}^{\Sigma(1)} \to A_{n-1}(\mathbb{P}_\Delta) \to 0,$$

where the first map is defined as $m \to \{\langle m, v_i\rangle\}$ and the second map can be taken as the definition of the Chow group $A_{n-1}(\mathbb{P}_\Delta)$. It induces a map

$$0 \to Hom_{\mathbb{Z}}(A_{n-1}(\mathbb{P}_\Delta), \mathbb{C}^*) \to (\mathbb{C}^*)^{\Sigma(1)}.$$

Therefore, $Hom_{\mathbb{Z}}(A_{n-1}(\mathbb{P}_\Delta, \mathbb{C}^*)$ acts on $\mathbb{C}^{\Sigma(1)}$.

$$\mathbb{P}_\Delta = (\mathbb{C}^{\Sigma(1)} - Z(\Delta))/Hom_{\mathbb{Z}}(A_{n-1}(\mathbb{P}_\Delta, \mathbb{C}^*).$$

We can also read off the moment map as well. $(\mathbb{C}^*)^{\Sigma(1)}$ acts on $\mathbb{C}^{\Sigma(1)}$ with moment map $(z_1, \ldots, z_{\Sigma(1)}) \to \frac{1}{2}(||z_1||^2, \ldots, ||z_{\Sigma(1)}||^2) \in \mathbb{R}^{\Sigma(1)}$. Our basic exact sequence induces a map

$$\mathbb{R}^{\Sigma(1)} = \mathbb{Z}^{\Sigma(1)} \otimes \mathbb{R} \to A_{n-1}(\mathbb{P}_\Delta) \otimes \mathbb{R}.$$

The moment map is the composition of two maps. We remark that if Σ is complete, i.e., the $\Sigma(1)$ generating $N_{\mathbb{R}}$, $\mathbb{P}_{\Sigma(1)}$ is compact and there is only one phase. If Σ is incomplete such as the examples in the last section, there are different phases and $\mathbb{P}_\Delta$ is not unique. But they are birational to each other.

The Calabi-Yau manifold is taken as a hypersurface $V \in |-K_{\mathbb{P}_\Delta}|$. When $\Sigma(1)$ generating N rationally (*simplicially*), $\mathbb{P}_\Delta$ is an orbifold. In this setting, we do not fix a particular representative of V. Instead, we consider that the entire family of $V \in |-K_{\mathbb{P}_\Delta}|$ is mirror to the entire family of $V^0 \in |-K_{\mathbb{P}_{\Delta^0}}|$. We can obtain a complete family by taking a further projective subdivision Σ' of Σ and considering a Calabi-Yau hypersurface of $X_{\Sigma'}$. Here, $X_{\Sigma'} \to \mathbb{P}_\Delta$ corresponds to a certain toric resolution. It was shown that all the family fits nicely together and constitutes the Calabi-Yau families of Batyrev-Borisov mirror symmetry. In this sense, LG-to-LG mirror symmetry is stronger. Namely, it define a precise correspondence between families. The main theorem of Batyrev-Borisov is

THEOREM 4.6.

$$h^{p,q}(V) = h^{n-1-p,q}(V^0).$$

Here, the Hodge number is the Hodge number of Chen-Ruan cohomology.

EXAMPLE 4.7. The fan for projective space $\mathbb{P}^n$ is generated by the standard basis $e_1, \dots e_n$ of $\mathbb{Z}^n$ and $e_0 = -\sum_i e_i$. The corresponding reflexive polytope Δ^0 are convex hull of the vertices $(-1, \dots, -1), (n+1, -1, \dots, -1), (-1, n+1, \dots, -1), \dots, (-1, \dots, -1, n+1)$. The polar dual Δ^0 are convex hull of the vertices $e_1, \dots, e_n, (-1, \dots, -1)$. The corresponding toric variety $\mathbb{P}_{\Delta^0} = \mathbb{P}^n/G$, where

$$G = \{(a_0, a_1, a_2, a_3, a_4) \in (\mathbb{Z}_5)^5; \sum_i a_i \equiv 0 \bmod 5\}/\mathbb{Z}_5,$$

where $\mathbb{Z}_5 \subset (\mathbb{Z}_5)^5$ is the diagonal subgroup. The mirror Calabi-Yau family is

$$\{x_0^5 + x_1^5 + x_2^5 + x_3^5 + x_4^5 + \psi x_0 x_1 x_2 x_3 x_4 = 0\}/G.$$

EXAMPLE 4.8. The weighted projective space $\mathbb{P}(c_0, \dots, c_n)$ (assume $c_0, \dots, c_n$ have no common factor) corresponds to the following toric variety: consider the rank n lattice $N = \mathbb{Z}e_0 + \cdots + \mathbb{Z}e_n/\mathbb{Z}(c_0 e_0 + \cdots + c_n e_n)$ i.e. the quotient of $\mathbb{Z}^{n+1}$ by the relation $c_0 e_0 + \cdots + c_n e_n = 0$. The Δ^0 is the convex hull of the primitive integral vectors $(e_0, \dots, e_n)$.

By definition, Δ^0 is reflexive if for every maximal cone C in the fan, the $\mathbb{Q}$-linear function u_C with value -1 on all primitive integer vectors of the cone takes integer values on N i.e. the linear function u_C defines an element in the dual lattice $Hom(N, \mathbb{Z})$ (which is the corresponding vertex of the polar polytope Δ). Suppose that u_i takes value -1 on the face generated by e_j for $j \neq i$. On the other hand,

$$u_i(c_0 e_0 + \cdots + c_n e_n) = 0.$$

Hence, $c_i(u_i(e_i) - 1) = -\sum_j c_j$. Hence, $c_i | \sum_j c_j$. Therefore, $WP(c_0, \dots, c_n)$ is reflexive iff $c_i | \sum_j c_j$. Clearly, this is equivalent to the existence of the Fermat Calabi-Yau hypersurface representative representative $W = \sum_i x^{d/w_i}$.

Let $d = \sum_j c_j$ and $c_i | d$. Then, Δ is the convex hull of u_i where $u_i(e_j) = -1$ for $j \neq i$ and $u_i(e_i) = -d/c_i + 1$. Again, it satisfies the relation $c_0 u_0 + \cdots + c_n u_n = 0$. A similar argument shows that

$$\mathbb{P}_{\Delta^0} = WP(c_0, \dots, c_n)/G$$

where

$$G = \{(a_0, \dots, a_n) \in \mathbb{Z}_{d/c_0} \times \cdots \times \mathbb{Z}_{d/c_n}; \sum c_i a_i = 0 \bmod \mathrm{d}\}/\mathbb{Z}_d.$$

Here, $\mathbb{Z}_d$ is the subgroup of $\prod_j \mathbb{Z}_{d/c_j}$ generated by $(1, \dots, 1)$. Moreover, the mirror family is

$$\{x^{d/c_0} + \cdots + x^{d/c_n} + \psi x_0 x_1 x_2 x_3 x_4 = 0\}/G.$$

The above construction has been generalized to the complete intersection of a reflexive toric variety. A Calabi-Yau complete intersection V^0 is the zero set of a transverse section of $K_1 \oplus \cdots \oplus K_k$ where $-K_{\mathbb{P}^\Delta} = E_1 + \cdots + E_k$. We want E_j to be toric as well. Recall that the vertices of Δ^0 define one-dimensional cones of fan and the toric divisors D_i. The decomposition $-K_{\mathbb{P}^\Delta} = E_1 + \cdots + E_k$ corresponds to a decomposition

$$\Sigma(1) = I_1 \cup \cdots \cup I_k$$

where $E_j = \sum_{i \in I_j} D_i$.

DEFINITION 4.9. The decomposition $\Sigma(1) = I_1 \cup \cdots I_k$ or $-K_{\mathbb{P}^\Delta} = E_1 + \cdots + E_k$ is called a nef-partition if for each j, E_j is a Cartier divisor spanned by its global sections. Equivalently, the $\Delta_j = \Delta_{E_j}$ is the convex hull of elements of vertices of Δ.

Clearly,

$$\Delta = \Delta_1 + \cdots + \Delta_k.$$

We also call the above decomposition a *nef partition.* Let $\nabla_j = Conv(\{0\} \cup I_j) \subset N_{\mathbb{R}}$. It is clear that

$$\Delta^0 = Conv(\nabla_1 \cup \cdots \nabla_k).$$

Batyrev-Borison showed that $\nabla = \nabla_1 + \cdots + \nabla_k$ is a reflexive polytope. It is clear that $\nabla = \nabla_1 + \cdots + \nabla_k$ is a nef partition on its own and we call it the *dual nef partition.* It induces a family of Calabi-Yau complete sections V^0 of $\mathbb{P}_\nabla$. Batyrev-Borisov proved that V, V^0 are mirror in the sense that

$$h^{p,q}(V) = h^{n-1-p,q}(V^0).$$

We want to emphasis that $\nabla \neq \Delta^0$ in general.

REMARK 4.10. Here is an important remark. The A-model theory is independent of the choice of V while the B-model theory does depends on the choice of V. To resolve this conflict, we consider a special limit corresponding to the maximal degeneration of V. This limit is called the *large complex structure limit* defined by the condition that the local monodromy is maximally unipotent. In this sense, Batyrev-Borisov mirror symmetry can be thought as the fact that the A-model of V (independent of V) is mirror to the large complex structure limit of V^0. The different families from the toric resolution will have generally different A-model theories. They will correspond to different large complex structure limits. One consequence of the above picture is that these different A-model theories should be connected by analytic continuation! This is the content of the so-called *crepant resolution conjecture.* LG-to-LG mirror symmetry also fits with the above picture nicely. Berglund-Hübsch-Krawitz construction sends FJRW-theory to a special point of the B-model. Therefore, it also connects to Calabi-Yau A-model theory by an analytic continuation. This can be thought of as the B-model description of the LG/CY-correspondence.

4.3. Intermediate/Hybrid phase. The LG-to-LG mirror symmetry and the CY-to-CY mirror symmetry represent two extremes. There are two interesting questions; (i) Are they compatible; (ii) Is there mirror symmetry between the intermediate phases. These intermediate phases are often referred as *hybrid phases.* The answers for both questions are affirmative. (i) is the subject of the LG/CY correspondence. For invertible hypersurfaces of weighted projective spaces, the cohomological LG/CY correspondence was established by Chiodo-Ruan [**ChiR9**]. We should mention that their results also imply the mirror symmetry of Hodge numbers (*cohomological mirror symmetry*) for hypersurfaces of non-reflexive weighted projective spaces, which extends the range of mirror symmetry at the *classical level* greatly. The effort of extending the Batyrev-Borisov construction to non-reflexive cases leads to some new constructions (see [**Bo**]). However, the results of [**Bo**] are not strong enough yet to cover the cohomological mirror symmetry for those non-reflexive examples. Moreover, it is a very interesting problem to find the LG-dual of complete intersections, which is explored by Chiodo-Nagel [**CN**] and the others.

The current proof of the LG/CY correspondence is abstract. It would be desirable to establish it by going through the different phases. Here, we assume the Calabi-Yau weights and hence toric varieties are Calabi-Yau in each phase. Note that in all cases, there is a special chamber in which $\mu_i > 0$. The toric variety is a bundle over a compact toric variety and the critical set is a Calabi-Yau complete intersection of zero section. Once we construct the appropriate theory from the corresponding gauged Witten equation, the new theory is expected be equivalent to Gromov-Witten theory of the complete intersection in the sense of Conjecture 2.1.2 (without the issue of analytic continuation). We call this phase the *Calabi-Yau phase*.

We can also discuss a generalized version of the LG/CY correspondence as a conjecture that the theories constructed out of the gauged Witten equation in different phases are equivalent in the sense of the conjecture in section 2. I want to emphasize that the corresponding theories are not yet constructed except for the special case of the pure LG-model. But this unsatisfactory state of affairs should not stop us from speaking about certain aspects of this conjecture. Recall that the state space of the FJRW-theory comes at the last step of the theory. It is amazing that the more general state space has already been proposed by Chiodo-Nagel.

Their key observation is that FJRW-state space can be phrased as a relative Chen-Ruan cohomology. Namely,

$$\mathcal{H}^{p,q}_{FJRW,G} = H^{p-\sum_i q_i, q-\sum_i q_i}_{CR}([\mathbb{C}^N/G], W^\infty, \mathbb{C}),$$

where W^∞ is thought as an open sub-orbifold. Recall that Chen-Ruan cohomology is defined as a sum of the ordinary cohomology of sectors. The relative Chen-Ruan cohomology is again a sum of the ordinary relative cohomology of each sector. The main difference from the FJRW-theory is the additional degree shift by $q_W = \sum_i q_i$.

The above observation motivated them to define the following state space. Recall that over any phase, the toric variety is the total space of a vector bundle V over a compact smooth toric orbifold X. The super-potential is $W : V \to \mathbb{C}$. By assumption, W_{fiber} is a nondegenerate quasi-homogenous singularity. In fact, let $(b_{i_1}, \ldots, b_{i_l})$ be homogeneous coordinate of a base point in X. W_{fiber} is obtained by setting $z_{i_j} = b_{i_j}$. Suppose that the degree of W_{fiber} is $(Q_1, \ldots, Q_d)$. Let $q_W = \sum_j \sum_i \frac{q_{ij}}{Q_j}$.

DEFINITION 4.11. (Chiodo-Nagel) Define the state space of (V, W) to be

$$\mathcal{H}^{p,q}(V, W) = H^{p-q_W, q-q_W}_{CR}(V, W^\infty, \mathbb{C}).$$

Using Chiodo-Nagel's state space, we can proposed the following generalized *Cohomological LG/CY-correspondence.*

CONJECTURE 4.12. Suppose that the weights are of Calabi-Yau type. The state spaces $\mathcal{H}(V, W)$ of different phases are isomorphic as graded vector spaces.

REMARK 4.13. On the Calabi-Yau phases, it follows from a theorem of Dolgachev-Steenbrink that $\mathcal{H}_{V,W}$ is isomorphic to the Chen-Ruan cohomology of a complete intersections.

With the above identification in the CY phase, the cohomological LG/CY correspondence was proved by Chiodo-Ruan [**ChiR9**] for the case of a Calabi-Yau hypersurface of a weighted projective space and by Chiodo-Nagel [**CN**] for the case of a Calabi-Yau complete intersection of a weighted projective space.

For the second questions, a very general mirror symmetry construction has been proposed by Clarke [**Cl**]. It would be very interesting to test Clarke's proposal using the above state space. We refer the readers to the original paper for exposition.

5. Gauged Witten equation without super-potential

After setting up the classical gauged linear sigma model, we want to quantize the theory. In the A-model setting, this is provided by solving a version of the Witten equation. The original Witten equation applies to the case of the Landau-Ginzburg phase where the symmetry/gauge group is finite. In the general case of the gauged linear sigma model, the symmetry/gauge group is higher-dimensional. We call this more general case the *gauged Witten equation.* It has the form of the original Witten equation together with a new equation of the form $F_A =$ moment map where A is a connection of certain bundle. This type of equation has appeared classically in differential geometry in the 80's. Furthermore, the Cauchy-Riemann equation ∂ in the Witten equation should be replaced by ∂_A coupled with the connection A. This family of equations is more complicated. However, it shares some of the general characteristic of ordinary Witten equation. A new ingredient from the gauge field is the *stability condition*, to which we will pay special attention. In this section, we will treat the case $W = 0$. This category can be thought of as an approach to prove the *crepant resolution conjecture.* In the next section, we treat the case in which $W \neq 0$.

Before we get into the details of PDE, let's make several remarks for motivations. The current family of PDE was introduced by Witten to quantize the gauged linear sigma model. Namely, we are looking for a Gromov-Witten type theory similar to that of FJRW-theory. With this in mind, we will pay special attention to the general elements of Gromov-Witten type theory such as *state space, gluing law and cohomological field theory.* On the other hand, a natural PDE doesn't come by so often and is interesting in its own right.

5.1. Equation. The basic ingredient of the gauged Witten equation is a connection A together with a section S of a certain bundle. Let me be more precise. Recall the basic construction of a toric variety as $X_\mu = \mathbb{C}^N/(\mathbb{C}^*)^d$. Recall the collection of canonical line bundls. P_i is associated with each $\mathbb{C}^*$-action and $L_\rho = \otimes_i P^{q_{\rho i}}$ is associated with each coordinate. Let C be a Riemann surface. Now, we associate a line bundle $\mathcal{P}_i$ to each $\mathbb{C}^*$-action and $\mathcal{L}_\rho = \otimes_i \mathcal{P}^{q_{\rho i}}$ to each coordinate. The variables of the gauged Witten equation is a $r + N$-tuple $(A_1, \ldots A_r, s_1, \ldots, s_N)$ such that A_i is a connection on $\mathcal{P}_i$ and s_ρ is a section of $\mathcal{L}_j$. A_i induces a connection on $\mathcal{L}_\rho$ and hence we can define $\bar{\partial}_A$, the Cauchy-Riemann equation coupled with the connection $A = (A_1, \dots, A_r)$. The Witten equation has the form

$$\bar{\partial}_A s_\rho = 0.$$

The gauged Witten equation consists of the above equation and another equation of the form

$$*F_A = D - \mu,$$

where F_A is the curvature viewed as a two-form on C, $*$ is the Hodge star operator, D is the moment map and μ is a regular value of the moment map. Furthermore, the above equation is invariant under gauge transformations as follows. Suppose that $g : C \to U(1)^d$ is an automorphism. Then, g acts on A by $A + g^{-1}dg$ and on s_ρ by the induced action (still denoted by gs_ρ).

Several important remarks are in order. The above equation has been studied a great deal by several authors under the name *symplectic vortex equation.* It is also important to note that the second equation depends on a choice of metric via the Hodge star operator $*$. More explicitly, it depends only on the volume form of the metric. Similar to the case of the original Witten equation, the choice of metric is very important here. For our purposes, we only care about the metric near marked points. So far, the most common choice is the smooth metric as in the case of the early treatment of the Witten equation. Similar to the Witten equation, there are some subtle issues such as the compactness of the solution space of these equation. This problem is particularly crucial when we try to vary C, which we certainly want to do. Our experience with the Witten equation told us that it is better to choose a cylindrical metric near marked points and nodes. We will make this choice.

Once we have made this choice, it was argued [**CRRS, RT**] that a finite energy solution will converge to a point x such that $D(x) - \mu = 0$. Furthermore, x could vary in an orbit of T^d by gauge transformations. Let $\mathcal{M}$ be the moduli space of solutions modulo the gauge transformations. The above observation leads to a version of the evaluation map

$$e_i : \mathcal{M} \to D^{-1}(\mu)/T^d = X_\mu.$$

Therefore, this theory will have a state space $H^*(X_\mu, \mathbb{C})$. One should be able to go ahead to compactify the moduli space, define a virtual cycle and construct a Gromov-Witten type theory. The LG/CY-correspondence between different phases corresponds to the crepant resolution conjecture. The latter has been studied by Coates-Corti-Iritani-Tseng [**CIT09**] using a different method. Presumbly, the gauged Witten equation/symplectic vortex equation will provide an alternative approach. As far as author knows, the analytic theory of the above Witten equation is still open. However, we will focus on the algebraic geometric treatment which is already available via the theory of stable quotients.

5.2. Algebraic theory. The existence of an algebraic geometric counterpart is based on the following observation in gauge theory. Instead of considering real connections and real gauge transformations $g : C \to U(1)^d$, we consider complex connection $A_\mathbb{C}$ and complex gauge transformation $g_\mathbb{C} : C \to (\mathbb{C}^*)^d$. By applying complex gauge transformations, we can achieve two properties: (i) we can make $\mathcal{L}_\rho$ holomorphic, (ii) we can always solve the equation

$$*F_A = D - \mu.$$

Namely, we change the moduli problem to that of a holomorphic bundle $\mathcal{L}_\rho$ with holomorphic section s_ρ. The original equation is transformed to a certain *stability condition* which now we describe. It is interesting that such a modular theory was first constructed in the non-abelian case of the Grassmannian under the name of *stable quotient* [**MOR**]. Then, it was generalized to toric varieties under the name *stable toric quasimaps* [**CKM**]. In this section, we will describe stable toric quasimaps.

We first assume that the action of $(\mathbb{C}^*)^d$ on $\mathbb{C}^M$ has positive weights and hence $\mathbb{C}^M/(\mathbb{C}^*)^d$ is compact. Recall that each $U(1)$-factor give arise to a holomorphic line bundle $\mathcal{P}_i$ over the toric variety. $\mathcal{P}_i$ is useful for at least two purposes. Firstly, given any effective $\beta \in H_2(X_\mu, \mathbb{Z})$, we can assign a multi-degree $(f_1, \dots, f_d)$ for $f_i = \int_\beta c_1(\mathcal{P}_i)$. Furthermore, any ample line bundle $\mathcal{L}$ can be written as $\oplus \mathcal{L}_i^{\alpha_i}$ for $\alpha_i > 0$.

DEFINITION 5.1. A stable toric quasimap of class β is given by the data

$$((C, x_1, \dots, x_k), (\mathcal{P}_1, \dots, \mathcal{P}_r), (s_1, \dots, s_N))$$

- $(C, x_1, \dots, x_k)$ is a connected, nodal, marked curve of genus g.
- $\mathcal{P}_i$ is a holomorphic line bundle on C of degree f_i.
- s_ρ is a holomorphic section of $\mathcal{L}_\rho = \otimes_i \mathcal{P}^{q_{\rho i}}$ satisfying
- (**Nondegeneracy**) There are at most finitely many points other than nodes and marked points such that the image of $S = (s_1, \dots, s_N)$ is on unstable orbit.
- (**Stability**) For any component C' of C,

$$\omega_{log}(C) \otimes (\otimes_{ij} \mathcal{P}_i^{q_{ij}})^\epsilon$$

is ample for any small ϵ.

The equivalence relation can be defined in an obvious way. We use $\mathcal{T}_{g,k}(X, \beta)$ to denote the moduli space of equivalence classes of stable toric quasimaps of genus g, k-marking and homology class β.

It was established in [**CKM**] that $\mathcal{T}_{g,k}(X, \beta)$ is a compact Deligne-Mumford stack and carries a perfect two-term obstruction theory. Hence, we obtain a virtual fundamental cycle $[\mathcal{T}_{g,k}(X, \beta)]^{vir}$. Its virtual dimension is the same as that of the moduli space of stable maps.

To construct a virtual cycle for the gauged linear sigma model $V = \oplus_{\rho'} L_{\rho'}$, we need to handle the remaining variable $z_{\rho'}$. Here is the key observation. Note that $\mathcal{P}_i$ has non-negative degree on each component and positive degree on at least one component.

LEMMA 5.2. *Suppose that the weights on $z_{\rho'}$ are $(-q_{\rho'1}, \dots, -q_{\rho'd})$ for $q_{\rho'i} > 0$. The line bundle $\mathcal{L}_{\rho'} = \otimes_i \mathcal{P}_i^{-q_{\rho' i}}$ over C has no nonzero global section.*

$\mathcal{L}_{\rho'}$ has non-positive degree on each component and a negative degree on at least one component. The lemma follows from component by component analysis starting from negative degree component.

The above lemma implies that $R^1\pi_*\mathcal{L}_{\rho'}$ forms a vector bundle over $\mathcal{T}_{g,k}(X_{base}, \beta)$. Let $\Upsilon_{\rho'} = C_{top}(R^1\pi_*(\mathcal{L}_{\rho'}))$.

DEFINITION 5.3. We define the gauged linear sigma model virtual cycle

$$[\mathcal{T}_{g,k}(V, \beta))]^{vir} = \prod_{\rho'} \Upsilon_{\rho'} \cap [\mathcal{T}_{g,k}(X_{base}, \beta)]^{vir}.$$

6. Gauged Witten equation with $W \neq 0$

In this section, we introduce the Witten equation with nontrivial W. Compared with the above case $W = 0$, the theory is significantly more complicated. The comparisons between the two cases are best illustrated by the stable map to $B\mathbb{Z}_r$ and r-spin curve theory. The latter corresponds to the case $W = x^r$. A stable map from C to $B\mathbb{Z}_r$ can be characterized as an orbifold line bundle $L \to C$ with the property

$$L^r \cong \mathcal{O}_C.$$

On the other hand, an r-spin structure is an orbifold line bundle $L \to C$ with the property

$$L^r \cong \omega_{C,log}.$$

Namely, we replace $\mathcal{O}_C$ by $\omega_{C,log}$. Recall that in the definition of toric stable quasimap, $\mathcal{L}_\rho = \prod_i \mathcal{P}_i^{i\rho}$. Let $W = \sum_j W_j$ for monomial W_j. Since W_j is invariant under the action of $(\mathbb{C}^*)^d$,

$$W_j(\mathcal{L}_1, \cdots, \mathcal{L}_N) = \mathcal{O}_C.$$

From the above comments, in our new theory we should set

$$W_j(\mathcal{L}_1, \ldots, \mathcal{L}_N) = \omega_{C,log}.$$

This is the W-structure.

6.1. Equation. As we described previously, the gauged Witten equation has the form

$$\bar{\partial}_A s_i + \overline{\partial W}(s_1, \ldots, s_N) = 0, *F_A = D(s_1, \ldots, s_N) - \mu.$$

Now, we have to interpret the variable properly. We proceed as in the case of the original Witten equation. Let s_i be a section of $\mathcal{L}_i$. The compatibility condition of the above equation is

$$W_j(\mathcal{L}_1, \ldots, \mathcal{L}_N) \cong \omega_{C,log}.$$

Furthermore, we choose a metric on C cylindrical along each marked and nodal point. Using W_j, we can fix a metric on L_ρ. In contrast to the case of the original Witten equation, W here is not nondegenerate as a singularity. We can formally define the moduli space of W-structures. It will be an Artin stack only. The nondegeneracy condition of W in the Landau-Ginzburg setting is crucial to make it a Deligne-Mumford stack. In the case of toric stable quasimaps, the additional data of a section is needed in order to have a good compactification.

An attractive approach is to utilize the toric vector bundle structure $V \to X_{base}$ where we divide the coordinates as *base coordinate* via *fiber coordinate*. Then, we view the gauged linear sigma model as a family of Landau-Ginzburg models over X_{base}. An over-simplified version is to assume that the section s_i corresponding to the base variable in fact satisfies two separate equations

$$\bar{\partial} s_i = 0, \partial_i W(S) = 0.$$

Then, for the base variable, we solve an equation with $W = 0$ to obtain the moduli space of toric stable quasimaps $\mathcal{T}_{g,k}(X_{base}, \beta)$. Here, we take the algebraic version by considering the holomorphic connection A. Let $(\mathcal{L}_{i_1}, \ldots, \mathcal{L}_{i_k}, s_{i_1}, \ldots, s_{i_k}) \in \mathcal{T}_{g,k}(X_{base}, \beta)$, where $s_{i_j} \in H^0(\mathcal{L}_{i_j})$. $(\mathcal{L}_{i_j}, s_{i_j})$ can be identified with an effective divisor $\mathcal{O}(\mathcal{L}_{i_j})$. Now, we write $W_j = W_{base,j} W_{fiber,j}$ where $W_{base,j}$ is a monomial of base coordinates and $W_{fiber,j}$ is a monomial of fiber coordinates. We can rewrite the condition

$$W_j(\mathcal{L}_1, \ldots, \mathcal{L}_N) = \omega_{C,log}$$

as

$$W_{fiber,j}(\mathcal{L}_{fibers}) = \omega_{C,log} \otimes W_{base,j}(\mathcal{L}^*_{bases}) = \omega_{C,log} \otimes \mathcal{O}(-D)$$

for an effective divisor D. The above is the alternative description of W-structure. In particular, $-D$ gives rise to the orbifold structure we are looking for. Let $\mathcal{W}_{g,k}(V, \beta)$ be the moduli space of the above W-structure. Then, there is a finite morphism

$$\pi : \mathcal{W}_{g,k}(V, \beta) \to \mathcal{T}_{g,k}(X_{base}, \beta).$$

Then, we can apply the technique from FJRW-theory to study the moduli space of fiber variable sections satisfying an ordinary Witten equation. We shall leave this to interested readers. Finally, for the base variable s_i, we still need to consider $\partial_i W$. We view it as a cosection

$$\partial_i W : H^1(\mathcal{L}_i) \to \mathbb{C}$$

and use it to modify the virtual cycle. We leave the detail to the interested readers and move into algebraic category.

6.2. Algebraic theory. In the last subsection, we sketch a construction of the moduli space of fiberwise W-structures $\mathcal{W}_{g,k}(V, \beta)$. Let's consider its virtual cycle more carefully. $\mathcal{L}_{base}, s_{base}$ has a two-term obstruction theory which we can pull back. Note that its obstruction space is a direct sum of $H^1(\mathcal{L}_{base})$. $\mathcal{L}_{fiber}$ is governed by the fiberwise W-structure. Since they satisfy a relation

$$W_{fiber,j}(\mathcal{L}_{fibers}) = \omega_{C,log} W_{base,j}(\mathcal{L}^*_{bases}),$$

its degree is not necessarily negative and it could have nonzero global section. As in the case of the LG-model, we define an extended moduli space $\mathcal{T}_{g,k}(V, \beta)$ by including the fiber variable global section. It is non-compact. But it carries a two-term obstruction theory whose obstruction space is the $\oplus_\rho H^1(\mathcal{L}_\rho)$. The Witten map

$$\oplus_\rho \partial_\rho W : \oplus_\rho H^1(\mathcal{L}_\rho) \to \mathbb{C}$$

is again a co-section. Using the general theory of co-section localization, we can construct the virtual cycle $[\mathcal{W}_{g,k}(V, \beta)]^{vir}$.

References

BH93. P. Berglund, T. Hübsch, *A Generalized Construction of Mirror Manifolds*, Nuclear Physics B, vol **393**(1993), 397–391.

BK97. P. Berglund, S. Katz, *Mirror Symmetry Constructions: A Review*, Mirror Symmetry II, AMS/IP Stud. Adv. Math 1, Amer. Math. Soc. Providence, RI (1997), 71–86, Preprint version: arXiv:hep-th/9406008.

Bo. L. Borisov, *Berglund–Hübsch mirror symmetry via vertex algebras*, Preprint version: arXiv:1007.2633v3

CIT09. T. Coates, H. Iritani, H.-H. Tseng, *Wall-Crossings in Toric Gromov Wtten Theory I: Crepant Examples.* Geometry and Topology, **13** (2009), 2675–2744.

Ch06. A. Chiodo, *The Witten top Chern class via K-theory.* J. Algebraic Geom. **15** (2006), no. 4, 681–707. Preprint version: math.AG/0210398.

CIR. A. Chiodo, H. Iritani, Y. Ruan, *LG/CY correspondence and global mirror symmetry,* preprint.

CN. A. Chiodo, J. Nagel, in preparation.

ChiR9. A. CHIODO AND Y. RUAN, *LG/CY correspondence: the state space isomorphism*, Adv. Math., 227, Issue 6 (2011), 2157-2188.

ChiR10. A. CHIODO AND Y. RUAN, *Landau–Ginzburg/Calabi–Yau correspondence for quintic three-folds via symplectic transformations*, Invent. Math. (2010) **182**, 117–165. Preprint version: arXiv:0812.4660.

ChiR11. A. CHIODO AND Y. RUAN, *A global mirror symmetry framework for the Landau–Ginzburg/Calabi–Yau correspondence*, Preprint.

Cl. P. CLARKE, *Duality for toric Landau-Ginzburg models*, Preprint: arXiv:0803.0447.

CR. T. COATES, Y, RUAN, *Quantum Cohomology and Crepant Resolutions: A Conjecture*, Preprint: arXiv:0710.5901.

CG. P. CLARKE, J. GUFFIN, *On the existence of affine Landau-Ginzburg phases in gauged linear sigma models*, Preprint version: arXiv:1004.2937

CKM. I. CIOCAN-FONTANINE, B. KIM, D. MAULIK, *Stable quasimaps to GIT quotients*, Preprint: arXiv:1106.3724

CRRS. K. CIELIEBAK, A. RITA GAIO, IGNASI MUNDET I RIERA AND D. SALAMON, *The symplectic vortex equations and invariants of Hamiltonian group actions*, J. Symplectic Geom. 1(2002) 543-645

CK. D. COX AND S. KATZ, *Mirror symmetry and algebraic geometry*, Math Survey and Monographs, Vol 68.

FJR1. H. Fan, T. Jarvis and Y. Ruan, *Geometry and analysis of spin equations*, Comm. Pure. Appl. Math. 61(2008) 745-788

FJR2. H. Fan, T. Jarvis and Y. Ruan, *The Witten Equation and Its Virtual Fundamental Cycle*, arXiv:math/0712.4025.

FJR3. H. Fan, T. Jarvis and Y. Ruan, *The Witten equation, mirror symmetry and quantum singularity theory*, arXiv:0712.4021

Gi04. A. GIVENTAL, *Gromov–Witten invariants and quantization of quadratic hamiltonians.* In "Frobenius manifolds", 91–112, Aspects Math., E36, Vieweg, Wiesbaden, 2004, Preprint version: math.AG/0108100.

GW. E. GONZALEZ AND C. WOODWARD, *Area dependence in gauged Gromov-Witten theory*, Preprint version: arXiv:0811.3358

Ka06. KAUFMANN, RALPH, *Singularities with symmetries, orbifold Frobenius algebras and mirror symmetry.* Contemp. Math., **403** (2006), 67-116

Ko. M. Kontsevich, *Intersection theory on the moduli space of curves and the matrix Airy function*, Comm. Math. Phys. **164** (1992), 1–23.

Kr. M. KRAWITZ, *FJRW rings and Landau–Ginzburg Mirror Symmetry*, Preprint: arXiv:0906.0796.

KS. M. KRAWITZ, Y. SHEN, *LG/CY correspondence of all genera for elliptic orbifold* $\mathbb{P}^1$, Preprint version: arXiv:1106.6270

KS92. M. KREUZER AND H. SKARKE, *On the classification of quasihomogeneous functions*, Comm. Math. Phys. **150** (1992), no. 1, 137–147.

KS93. M. KREUZER AND H. SKARKE, *All abelian symmetries of Landau-Ginzburg potentials*, Nucl. Phys. B **405** (1993), no 2-3, 305–325. Preprint version: hep-th/9211047.

MOR. A. MARIAN, D. OPREA, AND R. PANDHARIPANDE, *The moduli space of stable quotients*, Geom. Topol. (to appear).

LZ. J. LI, H. ZHANG, in preparation.

PV01. A. POLISHCHUK, A. VAINTROB, *Algebraic construction of Witten's top Chern class.* Advances in algebraic geometry motivated by physics (Lowell, MA, 2000), 229–249, Contemp. Math., **276**, Amer. Math. Soc., Providence, RI, 2001. Preprint version: math.AG/0011032.

RT. IGNASI MUNDET I RIERA AND G. TIAN, *A compactification of the moduli space of twisted holomorphic maps*, Acta. Math. 222 (2009) 1117-1196

VW89. C. VAFA AND N. WARNER, *Catastrophes and the classfication of conformal field theories*, Phys. Lett. 218B (1989) 51.

Wi1. E. Witten, *Two-dimensional gravity and intersection theory on the moduli space*, Surveys in Diff. Geom. **1** (1991), 243–310.

Wi2. E. Witten, *Algebraic geometry associated with matrix models of two-dimensional gravity*, Topological models in modern mathematics (Stony Brook, NY, 1991), Publish or Perish, Houston, TX, 1993, 235–269.

Wi3. E. Witten, private communication.

Wi4. E. Witten, *Phases of $N = 2$ Theories In Two Dimensions*, Nucl.Phys. B **403** (1993), 159-222.

DEPARTMENT OF MATHEMATICS, UNIVERSITY OF MICHIGAN, ANN ARBOR, MI 48109
E-mail address: ruan@umich.edu

Proceedings of Symposia in Pure Mathematics
Volume **85**, 2012

Non-Kähler Calabi-Yau Manifolds

Li-Sheng Tseng and Shing-Tung Yau

ABSTRACT. String theory has had a profound influence on research in Calabi-Yau spaces over the past twenty-five years. We first briefly mention some of the work in Kähler Calabi-Yau manifolds that was influenced by the discovery of mirror symmetry in the late 1980s. We then discuss some of the mathematical motivations behind the recent work on non-Kähler Calabi-Yau manifolds, which arise in string compactifications with fluxes. After extending mirror symmetry to non-Kähler Calabi-Yau manifolds, we show how this leads to new cohomologies and invariants of non-Kähler symplectic manifolds.

1. Introduction

String theory and mathematics have had a very close interaction over the past thirty years. Indeed, the interaction has been extremely fruitful and produced many beautiful results. As a prime example, mathematical research on Calabi-Yau spaces over the past two decades has been strongly motivated by string theory, and in particular, mirror symmetry.

Mirror symmetry started from the simple observation by Dixon [**17**] and Lerche-Vafa-Warner [**42**] around 1989, of a possible geometric realization of flipping the sign of a representation of the superconformal algebra. Geometrically, it implied that Calabi-Yaus should come in pairs with the pair of Hodge numbers, $h^{1,1}$ and $h^{2,1}$, exchanged. Shortly following this observation, Greene-Plesser [**33**] gave an explicit construction of the mirror of the Fermat quintic using an orbifold construction. And soon after, Candelas-de la Ossa-Green-Parkes [**13**] discovered as a consequence of mirror symmetry a most surprising formula for counting rational curves on a general quintic.

The identification of the topological A- and B-models by Witten [**59**] further inspired a lot of rigorous mathematical work needed to justify various definitions and relations, such as the Gromov-Witten invariants, multiple cover formula, and other related topics. More works by Witten [**60**], Kontsevich [**40, 41**] and many others led to the proofs, independently by Givental [**29**] and Lian-Liu-Yau [**46**], of

1991 *Mathematics Subject Classification.* Primary 32Q25; Secondary 32J81, 53D05, 53D37.
Key words and phrases. Differential geometry, Calabi-Yau manifolds, symplectic geometry, and mirror symmetry.

the Candelas *et al.* formula for the genus zero Gromov-Witten invariants in the mid-1990s. As for the genus one Gromov-Witten invariants, the string prediction of Bershadsky-Cecotti-Ooguri-Vafa (BCOV) [**9**] made in 1993 for the quintic was only proved by Zinger and Jun Li [**62, 44**] about five years ago.

Though we now know much about mirror symmetry, many important questions remain open and progress continues to be made. In fact, the higher genus $g \geq 2$ case is still mathematically not well-understood. In the celebrated work of BCOV [**10**], a holomorphic anomaly equation for higher genus partition functions, F_g, was written down. Yamaguchi-Yau [**61**] in 2004 were able to show that F_g for $g \geq 2$ are polynomials of just five generators: $(V_1, V_2, V_3, W_1, Y_1)$. When these generators are assigned degrees $(1, 2, 3, 1, 1)$ respectively, F_g becomes a quasi-homogeneous polynomial of degree $(3g-3)$. This result was used by Huang-Klemm-Quackenbush [**38**] to compute the partition function on the mirror quintic up to genus $g = 51$.

As noted by BCOV, the higher genus B-model partition function can come from the quantization of the Kodaira-Spencer gauge theory. Towards this aim, Costello and Si Li have recently made significant progress. They have found a prescription for quantizing the Kodaira-Spencer theory and have successfully carried it out in the elliptic curve case [**45**].

Separately, much of the work on mirror symmetry has been based on toric geometry. To go beyond the toric cases, one needs to study period integrals and the differential equations which govern them under complex structure deformations. In this regard, Lian, Song, and Yau [**47, 48**] have very recently been able to describe explicitly a Picard-Fuchs type differential system for Calabi-Yau complete intersections in a Fano variety or a homogeneous space.

And finally, from the geometric perspective, Strominger-Yau-Zaslow [**53**] gave a T-duality explanation of mirror symmetry. This viewpoint has been clarified in much detail in the work of Gross-Siebert [**34, 35, 36**] during the past decade.

As we can see from the influence of mirror symmetry, string theory has had a strong effect on the development of mathematics. But let us now turn to a more recent developing area of string-math collaboration. This is the study of non-Kähler manifolds with trivial canonical bundle. They are sometimes called non-Kähler Calabi-Yaus. For string theory, they play an important role as they appear in supersymmetric flux compactifications. However, we will begin first by describing why mathematicians were interested in them prior to string theory.

2. Non-Kähler Calabi-Yau

A large class of compact non-Kähler Calabi-Yau threefolds were already known in the mid-1980s by a construction of Clemens [**16**] and Friedman [**25**]. Their construction starts from a smooth Kähler Calabi-Yau threefold, Y.

Suppose Y contains a collection of mutually disjoint rational curves. These are curves that are isomorphic to $\mathbb{CP}^1$ and have normal bundles $\mathcal{O}(-1) \oplus \mathcal{O}(-1)$.

Following Clemens, we can contract these rational curves and obtain a *singular* Calabi-Yau threefold X_0 with ordinary double-point singularities. Friedman then gave a condition to deform X_0 into a smooth complex manifold X_t. What we have described is just the compact version of the local conifold transition which physicists are familiar with

$$Y \dashrightarrow X_0 \dashrightarrow X_t \ .$$

Here, X_t's canonical bundle is also trivial, so it is a Calabi-Yau too. But in general X_t is non-Kähler. To see this, we can certainly contract enough rational curves so that $H^2(Y)$ is killed and $b_2 = 0$. In this case, after smoothing, we end up with a complex non-Kähler complex manifold which is diffeomorphic to a k-connected sum of $S^3 \times S^3$, with $k \geq 2$.

In 1987, Reid [**50**] put forth an interesting proposal, often called Reid's fantasy. Reid wanted to make sense of the vast collection of diverse Calabi-Yau threefolds. He speculated that all (Kähler) Calabi-Yau threefolds that can be deformed to Moishezon manifolds fit into a single universal moduli space in which families of smooth Calabi-Yaus of different homotopy types are connected to one another by the Clemens-Friedman conifold transitions that we have just described.

Now if we want to test this proposal, understanding non-Kähler Calabi-Yau manifolds becomes essential. For example, a question one can ask is: what geometrical structures exist on these non-Kähler Calabi-Yau manifolds? If the metrics are no longer Kähler, do they have some other property?

2.1. Balanced Metrics. A good geometric structure to consider is the one studied by Michelsohn [**49**] in 1982. Recall that a hermitian metric, with an associated $(1,1)$-form ω, is Kähler if

$$d\omega = 0 \qquad\qquad \text{(Kähler)} \ .$$

For threefolds, Michelsohn analyzed the weaker *balanced* condition:

$$d(\omega\wedge\omega) = 2\,\omega\wedge d\omega = 0 \qquad \text{(balanced)} \ .$$

As should be clear, a Kähler metric is always balanced but a balanced metric need not be Kähler.

The balanced condition has good mathematical properties. It is preserved under proper holomorphic submersions and also under birational transformations as shown by Alessandrini-Bassanelli [**2**]. There are also simple non-Kähler compact balanced manifolds. For example:

- Calabi [**14**] showed that a non-trivial bundle of complex tori over a Riemann surface cannot be Kähler, but it does have a balanced metric [**32**].
- The natural metric on compact six-dimensional twistor spaces is balanced. As Hitchin showed, only those associated with S^4 and $\mathbb{CP}^2$ are Kähler [**37**]. One can get a non-Kähler Calabi-Yau by taking branched covers of twistor spaces. Sometimes, if the four-manifold is an orbifold, the singularities on the twistor space may be resolved to also give a non-Káhler Calabi-Yau.
- Three-dimensional Moishezon spaces are balanced.

So how about the non-Kähler Calabi-Yaus from conifold transitions? Do they admit a balanced metric? In this regard, J. Fu, J. Li, and S.-T. Yau proved the following theorem.

Theorem 2.1 (Fu-Li-Yau [**26**]). *Let Y be a smooth Kähler Calabi-Yau threefold and let $Y \to X_0$ be a contraction of mutually disjoint rational curves. Suppose X_0 can be smoothed to a family of smooth complex manifolds X_t. Then for sufficiently small t, X_t admits smooth balanced metrics.*

This constructive theorem provides us with balanced metrics on a large class of complex threefolds. In particular, for the Clemens-Friedman construction, the theorem implies

Corollary (Fu-Li-Yau [**26**]). *There exists a balanced metric on $\#_k(S^3 \times S^3)$ for any $k \geq 2$.*

Knowing that a balanced metric is present is useful. But to really understand Reid's proposal for Calabi-Yau moduli space, it is important to define some *canonical* balanced metric which would satisfy an additional condition, like the Ricci-flatness condition for the Kähler Calabi-Yau case. So we would like to have a natural condition, and here string theory gives some suggestions. As Calabi-Yau has played an important role in strings, one may ask what would be the natural setting to study compact conifold transitions and non-Kähler Calabi-Yau in physics.

Physicists have been interested in non-Kähler manifolds for more than a decade now in the context of compactifications with fluxes and model building (see e.g. [**30, 19**]). In this scenario, if one desires compact spaces without singularities from branes, then one should consider working in heterotic string theory.

2.2. Strominger's System. In the heterotic theory, the conditions for preserving $N = 1$ supersymmetry with H-fluxes were written down by Strominger [**52**] in 1986. Strominger's system of equations specifies the geometry of a complex threefold X (with a holomorphic three-form Ω) and in addition a holomorphic vector bundle E over X. The Hermitian metric ω of the manifold X and the metric h of the bundle E satisfy the following system of differential equations:

(1) $d(\| \Omega \|_\omega \, \omega \wedge \omega) = 0\,;$

(2) $F_h^{2,0} = F_h^{0,2} = 0\,, \quad F_h \wedge \omega^2 = 0\,;$

(3) $i\partial\overline{\partial}\, \omega = \frac{\alpha'}{4}\left[\mathrm{tr}\big(R_\omega \wedge R_\omega\big) - \mathrm{tr}\big(F_h \wedge F_h\big)\right]\,.$

Notice that the first equation is equivalent to the existence of a (conformally) balanced metric. The second is the Hermitian-Yang-Mills equations which is equivalent to E being a stable bundle. The third equation is the anomaly equation. When X is Kähler and E is the tangent bundle T_X, the system is then solved with $h = \omega_{CY}$, the Kähler Calabi-Yau metric.

Using a perturbation method, J. Li and S.-T. Yau [**43**] have constructed smooth solutions on a class of Kähler Calabi-Yau manifolds with *irreducible* solutions for vector bundles with gauge group $SU(4)$ and $SU(5)$. Andreas and Garcia-Fernandez [**5, 6**] have generalized our construction on Kähler Calabi-Yau manifolds for any stable bundle E that satisfies $c_2(X) = c_2(E)$. In recent years, our collaborators and other groups have also constructed solutions of the Strominger system on non-Kähler Calabi-Yaus [**28, 7, 8, 27, 21, 58**].

As is clear in heterotic string theory, understanding stable bundles on Calabi-Yau threefolds is important. In this regard, Donagi, Pantev, Bouchard and others have done nice work constructing stable bundles on Kähler Calabi-Yaus to obtain realistic heterotic models of nature [**18, 12**]. Also, Andreas and Curio [**3, 4**] have done analysis on the Chern classes of stable bundles on Calabi-Yau threefolds, verifying in a number of cases a proposal of Douglas-Reinbacher-Yau [**20**].

But returning to conifold transitions on compact Calabi-Yaus, it has been proposed by Yau to use Strominger's system to study Reid's proposal. Certainly the first condition that there exists a balanced metric can be satisfied. As we have already mentioned, Fu-Li-Yau [**26**] showed the existence of a balanced metric under conifold transitions. However, the second condition from the heterotic string adds a stable gauge bundle into the picture. So one needs to know about the stability of holomorphic bundles through a global conifold transition. In his recent PhD thesis, M.-T. Chuan [**15**] examined how to carry a stable vector bundle through a conifold transition, from a Kähler to a non-Kähler Calabi-Yau. Under the assumption that the initial stable holomorphic bundle is trivial in a neighborhood of the contracting rational curves, he proved that the resulting holomorphic bundle on the non-Kähler Calabi-Yau also has a Hermitian Yang-Mills metric, and hence is stable. This shows that two of the three conditions of Strominger's system, the existence of a balanced metric and a Hermitian-Yang-Mills metric on the bundle, can be satisfied. The last condition, the anomaly equation, which couples the two metrics, is perhaps the most demanding and difficult to analyze.

J. Fu and S.-T. Yau have analyzed carefully the anomaly equation when the manifold is a T^2 bundle over a $K3$ surface [**28**]. In this case, the anomaly equation reduces down to a Monge-Ampère type equation on the $K3$:

$$\triangle(e^u - \frac{\alpha'}{2} f e^{-u}) + 4\alpha' \frac{\det u_{i\bar{j}}}{\det g_{i\bar{j}}} + \mu = 0\,,$$

where f and μ are functions on the $K3$ satisfying $f \geq 0$ and $\int_{K3} \mu = 0$ and $u_{i\bar{j}}$ is the $\partial\bar{\partial}$ partial derivative matrix on the function u. It would be interesting to show that the anomaly equation can be satisfied throughout the non-Kähler Calabi-Yau moduli space.

2.3. Symplectic Conifold Transitions: Smith-Thomas-Yau. So far we have discussed conifold transitions between Calabi-Yaus that although non-Kähler still maintain a complex structure. The contraction of a rational curve $\mathbb{CP}^1$ (and the inverse operation of resolution) is naturally a complex operation. The smoothing of a conifold singularity by S^3 on the other hand is naturally symplectic. Friedman's condition is needed to ensure that a smoothed out Calabi-Yau contains a global complex structure.

But instead of preserving the complex structure, we can preserve the symplectic structure throughout the conifold transition. This would be the symplectic *mirror* of the Clemens-Friedman conifold transition. In this case, we would collapse disjoint Lagrangian three-spheres, and then replace them by symplectic two-spheres. Such a symplectic transition was proposed in a work of Smith-Thomas-Yau [**51**] in 2002.

Locally, of course, there is a natural symplectic form in resolving the singularity by a two-sphere. But there may be obstructions to patching the local symplectic forms to get a global one. Smith-Thomas-Yau wrote down the condition (analogous to Friedman's complex condition) that ensures a global symplectic structure. This symplectic structure however may not be compatible with the complex structure. So in general, the symplectic conifold transitions result in non-Kähler manifolds, but they all have $c_1 = 0$ and so they are called symplectic Calabi-Yaus. In fact, Smith-Thomas-Yau used conifold transitions to construct many real six-dimensional non-Kähler symplectic Calabi-Yaus.

In the symplectic conifold transition, if we can collapse all disjoint three-spheres, then such a process should result in a manifold diffeomorphic to a connected sum of $\mathbb{CP}^3$s. This mirrors the complex case, which after collapsing all disjoint rational curves gives a connected sum of $S^3 \times S^3$s. More recently, Fine-Panov [**23, 24**] have also constructed interesting simply-connected symplectic Calabi-Yaus with Betti number $b_3 = 0$, which means that they cannot be Kähler.

As mentioned above, a balanced structure can always be found in a complex conifold transition. So similarly, we can ask if there is any geometric structure present before and after a symplectic conifold transition? Here we will be looking for a condition on the globally $(3,0)$-form which in the general non-Kähler case is no longer d-closed. Again, we can turn to string theory for a suggestion. Is there a mirror dual of a complex balanced manifold in string theory that is symplectic and generally non-Kähler?

Such a symplectic mirror will not be found in heterotic string theory. All supersymmetric solutions satisfy the Strominger system in heterotic string. So the mirror dual of a complex balanced manifold with a bundle should be another complex balanced manifold with a bundle. But it turns out the answer can be found in type II string theories. As we will describe below, the equations for non-Kähler Calabi-Yaus in type II string also give us insights into the natural cohomologies on non-Kähler manifolds.

3. Type II Strings: Non-Kähler Calabi-Yau Mirrors

In type II string theory, supersymmetric compactifications preserving a $SU(3)$ structure have been studied by many authors in the last ten years. Since we are interested in non-Kähler geometries of compact manifolds, any supersymmetric solution will have orientifold sources. The type of sources helps determine the type of non-Kähler geometries. We shall describe the supersymmetric equations written in a form very similar to that of Graña-Minasian-Petrini-Tomasiello [**31**] and Tomasiello [**54**]. More details of our description here can be found in [**57**].

3.1. Complex Balanced Geometry in Type IIB. The supersymmetric equations that involve complex balanced threefolds are found in type IIB theory in the presence of orientifold 5-branes (and possibly also D5-branes). These branes are wrapped on holomorphic curves. In this case, the conditions on the Hermitian $(1,1)$-form ω and $(3,0)$-form Ω can be written as [**54**]

$$\begin{aligned} d\Omega &= 0 && \text{(complex integrability)} \\ d(\omega\wedge\omega) &= 0 && \text{(balanced)} \\ 2i\,\partial\bar{\partial}(e^{-2f}\omega) &= \rho_B && \text{(source)} \end{aligned}$$

where ρ_B is the sum of the currents Poincaré dual to the holomorphic curves that the five-brane sources wrap around, and f is a distribution that satisfies

$$i\,\Omega\wedge\overline{\Omega} = 8\,e^{2f}\,\frac{\omega^3}{3!}\ .$$

The balanced and the source equations together are noteworthy in that they share a resemblance with the Maxwell equations. With the Hodge star operator defined with respect to the compatible Hermitian metric, we can write $\omega = *(\omega^2/2)$. The equations can then be expressed (ignoring the conformal factor)

$$d(\omega^2/2) = 0\ ,$$

$$2i\,\partial\bar{\partial} * (\omega^2/2) = \rho_B\ .$$

Now this might have been somewhat expected as the five-brane sources are associated with the three-form field strength F_3 which is hidden in the source equation. These two equations however do tell us something more.

Let us recall the Maxwell case. Maxwell's equations in four dimensions are

$$d\,F_2 = 0\,,$$
$$d * F_2 = \rho_e\,,$$

where ρ_e is the Poincaré dual current of some electric charge configuration. Now, if we consider the deformation $F_2 \to F_2 + \delta F_2$ with the source fixed, that is, $\delta\rho_e = 0$, this leads to

$$d(\delta F_2) = d * (\delta F_2) = 0\ ,$$

which are the harmonic conditions for a degree two form in de Rham cohomology. Thus, the de Rham cohomology is naturally associated with Maxwell's equations.

For the type IIB complex balanced equations, we can also deform $\omega^2 \to \omega^2 + \delta\omega^2$. If this deformation is performed with the source current and the conformal factor fixed (i.e. $\delta\rho_A = \delta A = 0$), then we arrive at the conditions

$$d(\delta\omega^2) = \partial\bar{\partial} * (\delta\omega^2) = 0\ ,$$

which turn out to be precisely the harmonicity conditions for a (2,2)-element of the Bott-Chern cohomology:

$$H^{p,q}_{BC} = \frac{\ker d \cap \mathcal{A}^{p,q}}{\text{im } \partial\overline{\partial} \cap \mathcal{A}^{p,q}} \,.$$

This cohomology was introduced by Bott-Chern [**11**] and Aeppli [**1**] in the mid-1960s.

The string equations thus strongly suggest that the Bott-Chern cohomology (or the dual Aeppli cohomology, see Table 1) is the natural one to use for studying complex balanced manifolds. Let us point out that when the manifold is Kähler, the $\partial\overline{\partial}$-lemma holds. In this case, the Bott-Chern and the Dolbeault cohomology are in fact isomorphic. So the Bott-Chern cohomology is really most useful in the non-Kähler setting, and especially when the $\partial\overline{\partial}$-lemma fails to hold.

3.2. Symplectic Mirror Dual Equations in Type IIA. The mirror dual to the complex balanced manifold is found in the type IIA string. Roughly, the type IIA equations can be obtained from the IIB equations by first replacing $\omega^2/2$ with $(\text{Re}\, e^{i\,\omega})$ and then exchanging $e^{i\,\omega}$ with Ω [**22**]:

$$d(\omega^2/2) = 0 \;\Leftrightarrow\; d(\text{Re}\, e^{i\,\omega}) = 0 \qquad \longleftrightarrow \qquad d\,\text{Re}\;\Omega = 0\,.$$

Thus, $d\,\text{Re}\,\Omega = 0$ is the condition that is suggested by string theory for symplectic conifold transitions.

This condition turns out to be part of the type IIA supersymmetry conditions in the presence of orientifold (and D-) six-branes wrapping special Lagrangian submanifolds. The type IIA equations that are mirror to the type IIB complex balanced system can be written as follows:

$$\begin{aligned} d\omega &= 0\,, && \text{(symplectic)}\\ d\,\text{Re}\;\Omega &= 0\,, && \text{(almost complex)}\\ \partial_+\partial_- * (e^{-2f}\,\text{Re}\,\Omega) &= \rho_A\,, && \text{(source)} \end{aligned}$$

with

$$8\,\frac{\omega^3}{3!} = i\,e^{2f}\Omega\wedge\overline{\Omega}\,.$$

In the above system, ρ_A is the current Poincaré dual to the wrapped special Lagrangian submanifolds. The operators ∂_+ and ∂_- are linear symplectic operators that can be thought of as the symplectic analogues of the Dolbeault operators, ∂ and $\overline{\partial}$, and were recently introduced by us in [**56**]. If we naively compare the above symplectic system with the mirror complex one, a natural question arises: does the type IIA symplectic system suggest the existence of a symplectic cohomology of the form

$$\frac{\ker d}{\text{im } \partial_+\partial_-}$$

analogous to the complex Bott-Chern cohomology? Interestingly, as we found in [**55, 56**], such a cohomology is indeed natural and finite-dimensional on a compact symplectic manifold. Moreover, it provides new invariants for non-Kähler symplectic manifolds. But to discuss more about this symplectic cohomology and its relation to the IIA symplectic system, we will need certain properties of the symplectic differential operators, (∂_+, ∂_-). Since these linear operators are new, let us proceed now to give some more details.

4. Symplectic Differential Operators and Cohomologies

4.1. Linear Differential Symplectic Operators. Like their Dolbeault counterparts, (∂_+, ∂_-) can be naturally defined by an intrinsically symplectic decomposition of the exterior derivative. Recall that in the complex case, the differential forms are decomposed into (p,q) components $\mathcal{A}^{p,q}$. The exterior derivative d acting on each component gives two terms:

$$d: \mathcal{A}^{p,q} \rightarrow \mathcal{A}^{p+1,q} \oplus \mathcal{A}^{p,q+1} .$$

This then defines the Dolbeault operators ∂ and $\overline{\partial}$ as the projections of $d\mathcal{A}^{p,q}$ onto $\mathcal{A}^{p+1,q}$ and $\mathcal{A}^{p,q+1}$, respectively.

On a symplectic space (M, ω) of dimension $d = 2n$, we can do the analogous analysis. Indeed, there is also a decomposition of differential forms, specifically into representations of the $\mathfrak{sl}(2)$ lie algebra. This is well-known in the Kähler literature as the Lefschetz decomposition. Let us however emphasize that this decomposition requires only a non-degenerate two-form, which we do have here in ω. More explicitly, acting on a differential form $A \in \Omega^*(M)$, the $\mathfrak{sl}(2)$ generators take the form

$$\begin{aligned} L: &\ A \rightarrow \omega \wedge A \\ \Lambda: &\ A \rightarrow \frac{1}{2}(\omega^{-1})^{ij}\, i_{\partial_{x^i}} i_{\partial_{x^j}} A \\ H: &\ A \rightarrow (n-k)\, A \qquad \text{for } A \in \Omega^k(M) \end{aligned}$$

with commutation relations

$$[H, \Lambda] = 2\Lambda\,, \quad [H, L] = -2L\,, \quad [\Lambda, L] = H\,.$$

The $\mathfrak{sl}(2)$ irreducible modules are standardly constructed from the highest weight forms, which are commonly called primitive forms. Denoting the space of primitive forms by $\mathcal{P}^*(M)$, let us recall that a differential form is primitive, i.e. $B_s \in \mathcal{P}^s(M)$, if

$$\Lambda\, B_s = 0 \quad \text{or equivalently,} \quad L^{n+1-s}\, B_s = 0\,.$$

Hence, an irreducible $\mathfrak{sl}(2)$ module is the span of the elements

$$\left\{B_s\,,\ \omega \wedge B_s\,,\ \omega^2 \wedge B_s\,,\ \ldots\,,\ \omega^{n-s} \wedge B_s\right\}\,.$$

Since each element of this basis consists of ω raised to some power r exterior multiplied with a primitive s-form, it is natural to label basis elements of the $\mathfrak{sl}(2)$ module by the pair (r,s) and define

$$\mathcal{L}^{r,s}(M) = \left\{A \in \Omega^{2r+s}(M) \,\middle|\, A = \omega^r \wedge B_s \text{ and } \Lambda\, B_s = 0\right\}\,.$$

In a rough sense then, the space of forms $\mathcal{L}^{r,s}$ are the symplectic analogs of $\mathcal{A}^{p,q}$ of complex geometry.

Continuing the analogy with the complex case, let us act on $\mathcal{L}^{r,s}$ by the exterior derivative d. Since $d\omega = 0$, we have

$$d\mathcal{L}^{r,s} = d\,(\omega^r \wedge B_s) = \omega^r \wedge dB_s \ .$$

Clearly, the derivative only acts on the primitive forms. This is suggestive that much if not all of the data of differential forms on symplectic manifolds are encoded within primitive forms. Now as for d acting on a primitive form, it can be shown (see for example [**39**]) that

$$dB_s = B^0_{s+1} + \omega \wedge B^1_{s-1} \ .$$

Combining the above two equations, we find that

$$d : \mathcal{L}^{r,s} \ \rightarrow \ \mathcal{L}^{r,s+1} \ \oplus \ \omega \wedge \mathcal{L}^{r,s-1}$$

which has only two components on the right hand side just as in the complex case. Therefore, projecting onto each component, we can express the exterior derivative as [**56**]

$$d \ = \ \partial_+ \ + \ \omega \wedge \partial_-$$

where the first-order differential operators (∂_+, ∂_-) are defined by the derivative mapping

$$\begin{aligned} \partial_\pm : \mathcal{L}^{r,s} &\longrightarrow \mathcal{L}^{r,s\pm 1} \\ \partial_\pm : \mathcal{P}^{s} &\longrightarrow \mathcal{P}^{s\pm 1} \qquad \text{for } r = 0 \end{aligned}$$

By the above definition, ∂_+ and ∂_-, respectively, raise and decrease the degree of the forms by one. Moreover, (∂_+, ∂_-) are operators that map primitive forms to primitive forms (in the case of $r = 0$). And similarly to their complex counterparts, it follows from $d^2 = 0$ and the Lefschetz decompostion that they square to zero, i.e.

$$(\partial_+)^2 = (\partial_-)^2 = 0 \ ,$$

and anticommute: $\omega \wedge (\partial_+\partial_-) = -\,\omega \wedge (\partial_-\partial_+)\,.$

4.2. Symplectic Cohomologies and a Type IIA System. With the linear symplectic operators, (∂_+, ∂_-) and their properties at hand, we can now write down an interesting primitive symplectic elliptic complex.

Proposition (Tseng-Yau [**56**]). *On a symplectic manifold of dimension $d = 2n$, the following differential complex is elliptic.*

$$\begin{array}{ccccccccccc}
0 & \xrightarrow{\partial_+} & \mathcal{P}^0 & \xrightarrow{\partial_+} & \mathcal{P}^1 & \xrightarrow{\partial_+} & \cdots & \xrightarrow{\partial_+} & \mathcal{P}^{n-1} & \xrightarrow{\partial_+} & \mathcal{P}^n \\
 & & & & & & & & & & \big\downarrow {\scriptstyle \partial_+\partial_-} \\
0 & \xleftarrow{\partial_-} & \mathcal{P}^0 & \xleftarrow{\partial_-} & \mathcal{P}^1 & \xleftarrow{\partial_-} & \cdots & \xleftarrow{\partial_-} & \mathcal{P}^{n-1} & \xleftarrow{\partial_-} & \mathcal{P}^n
\end{array}$$

Symplectic (M,ω)		Complex (M,J)	
$PH^s_{\partial_\pm} = \dfrac{\ker\partial_\pm \cap \mathcal{P}^s}{\text{im }\partial_\pm \cap \mathcal{P}^s}$,	$s < n$	$H^{p,q}_{\bar\partial} = \dfrac{\ker\bar\partial \cap \mathcal{A}^{p,q}}{\text{im }\bar\partial \cap \mathcal{A}^{p,q}}$	(Dolbeault)
$PH^s_{\partial_+\partial_-} = \dfrac{\ker\partial_+\partial_- \cap \mathcal{P}^s}{(\text{im }\partial_+ + \text{im }\partial_-) \cap \mathcal{P}^s}$,	$s \leq n$	$H^{p,q}_{\partial\bar\partial} = \dfrac{\ker\partial\bar\partial \cap \mathcal{A}^{p,q}}{(\text{im }\partial + \text{im }\bar\partial) \cap \mathcal{A}^{p,q}}$	(Aeppli)
$PH^s_{\partial_++\partial_-} = \dfrac{\ker d \cap \mathcal{P}^s}{\text{im }\partial_+\partial_- \cap \mathcal{P}^s}$,	$s \leq n$	$H^{p,q}_{\partial+\bar\partial} = \dfrac{\ker d \cap \mathcal{A}^{p,q}}{\text{im }\partial\bar\partial \cap \mathcal{A}^{p,q}}$	(Bott-Chern)

TABLE 1. Finite-dimensional primitive symplectic cohomologies (Tseng-Yau [**55, 56**]) and their complex analogs. The complex cohomologies involving $\partial\bar\partial$ were introduced in the mid-1960s by Aeppli [**1**] and Bott-Chern [**11**].

Since the complex is elliptic, we can write down four different types of finite-dimensional primitive cohomologies associated with it.

$$\begin{aligned} PH^s_{\partial_\pm}(M) &= \frac{\ker\partial_\pm \cap \mathcal{P}^s(M)}{\text{im }\partial_\pm \cap \mathcal{P}^s(M)} \qquad \text{for } 0 \leq s < n\ , \\ PH^n_{\partial_+\partial_-}(M) &= \frac{\ker\partial_+\partial_- \cap \mathcal{P}^n(M)}{\text{im }\partial_+ \cap \mathcal{P}^n(M)}\ , \\ PH^n_{\partial_++\partial_-}(M) &= \frac{\ker\partial_- \cap \mathcal{P}^n(M)}{\text{im }\partial_+\partial_- \cap \mathcal{P}^n(M)}\ . \end{aligned}$$

Furthermore, by considering extended elliptic complexes involving non-primitive forms, it is possible to also define $PH^s_{\partial_+\partial_-}(M)$ and $PH^s_{\partial_++\partial_-}(M)$ for all degree s, $0 \leq s \leq n$. On a compact symplectic manifold, these cohomologies are also finite-dimensional [**55**]. We list the finite-dimensional symplectic cohomologies and their complex analogues in Table 1.

In [**55, 56**], we analyzed some of the basic properties of the new symplectic cohomologies. Since they are associated with an elliptic complex, each has an associated elliptic Laplacian and thus have standard nice Hodge theoretical properties. Moreover, we have calculated the cohomologies explicitly for some non-Kähler symplectic nilmanifolds and found that the cohomologies indeed lead to new symplectic invariants. Perhaps not too surprisingly, the new invariants do not contain new information when the manifold is Kähler.

Finally, returning back to the type IIA symplectic system of equations in Section 3.2. the middle-degree cohomology

$$PH^n_{\partial_++\partial_-} = \frac{\ker d \cap \mathcal{P}^n}{\text{im }\partial_+\partial_- \cap \mathcal{P}^n}$$

with $n = 3$ turns out to be particularly relevant. Consider deforming the type IIA system by $\Omega \longrightarrow \text{Re}\,\Omega + \delta\text{Re}\,\Omega$ with $\delta\rho_A = 0$ and conformal factor remaining fixed,

$\delta A = 0$. Then, $\delta\mathrm{Re}\,\Omega$ is required to satisfy (dropping the conformal factor)

$$d(\delta\mathrm{Re}\,\Omega) = 0\ , \qquad \partial_+\partial_- * (\delta\mathrm{Re}\,\Omega) = 0\ ,$$

which are the harmonicity conditions of the primitive $PH^n_{\partial_+ + \partial_-}$ cohomology. In fact, assuming $\delta\rho_A = \delta A = 0$, it can be shown [**57**] that a subspace of the linearized deformation of the type IIA symplectic system can be parametrized by the primitive cohomology with

$$\delta\mathrm{Re}\,\Omega \in P\mathcal{H}^3_{\partial_+ + \partial_-} \cap \mathrm{Re}\,\mathcal{A}^{2,1}\ .$$

5. Concluding Remarks

String theory has motivated much research in non-Kähler geometry in the past couple of years. We fully expect that investigations especially revolving around six-dimensional non-Kähler geometry will remain very active in the near future too. Of particular interest, six-dimensional non-Kähler geometries can have relations with four- and three-dimensional manifolds. As has been known for some time, one can construct many non-Kähler six-manifolds by the twistor construction. The twistor space of an anti-self dual four-manifolds has a complex structure, and the twistor space of a hyperbolic four-manifold has a symplectic structure. The S^3 bundle over a hyperbolic three-manifold is also complex. (Fine-Panov have given examples of the hyperbolic constructions [**24**].) There should also be interesting dualities relating complex and symplectic structures on non-Kähler six-manifolds. We have no doubt that string theory will continue to be a guiding influence in future work on non-Kähler geometry.

References

[1] A. Aeppli, *On the cohomology structure of Stein manifold.* In: Proc. Conf. Complex Analysis (Minneapolis, Minn., 1964), 58–70, Springer, Berlin, 1965.

[2] L. Alessandrini and G. Bassanelli, *Modifications of compact balanced manifolds.* C. R. Acad. Sci. Paris Sér. I Math 320 (1995) no. 12, 1517–1522.

[3] B. Andreas and G. Curio, *On possible Chern classes of stable bundles on Calabi-Yau threefolds.* J. Geom. Phys. 61 (2011) 1378–1384, arXiv:1010.1644.

[4] B. Andreas and G. Curio, *Spectral bundles and DRY-conjecture.* J. Geom. Phys. 62 (2012) 800–803, arXiv:1012.3858.

[5] B. Andreas and M. Garcia-Fernandez, *Solutions of the Strominger system via stable bundles on Calabi-Yau threefolds.* arXiv:1008.1018.

[6] B. Andreas and M. Garcia-Fernandez, *Heterotic non-Kähler geometries via polystable bundles on Calabi-Yau threefolds.* J. Geom. Phys. 62 (2012) 183–188, arXiv:1011.6246.

[7] K. Becker, M. Becker, J.-X. Fu, L.-S. Tseng and S.-T. Yau, *Anomaly cancellation and smooth non-Kähler solutions in heterotic string theory.* Nucl. Phys. B 751 (2006), 108–128, arXiv:hep-th/0604137.

[8] M. Becker, L.-S. Tseng and S.-T. Yau, *New heterotic non-Kähler geometries.* Adv. Theor. Math. Phys. 13 (2009), no. 6, 1815–1845, arXiv:0807.0827.

[9] M. Bershadsky, S. Cecotti, H. Ooguri and C. Vafa, *Holomorphic anomalies in topological field theories.* Nuclear Phys. B 405 (1993), no. 2-3, 279–304.

[10] M. Bershadsky, S. Cecotti, H. Ooguri and C. Vafa, *Kodaira-Spencer theory of gravity and exact results for quantum string amplitudes.* Comm. Math. Phys. 165 (1994), no. 2, 311–427.

[11] R. Bott and S.S. Chern, *Hermitian vector bundles and the equidistribution of the zeroes of their holomorphic sections.* Acta Math. 114 (1965), 71–112.

[12] V. Bouchard and R. Donagi, *An $SU(5)$ heterotic standard model.* Phys. Lett. B 633 (2006), no. 6, 783–791.

[13] P. Candelas, X. de la Ossa, P. Green and L. Parkes, *A pair of Calabi-Yau manifolds as an exactly soluble superconformal theory.* Nuclear Phys. B 359 (1991), no. 1, 21–74.
[14] E. Calabi,*Construction and properties of some 6-dimensional almost complex manifolds.* Trans. Amer. Math. Soc. 87 (1958), 407–438.
[15] M.-T. Chuan, *Existence of Hermitian-Yang-Mills metrics under conifold transitions.* arXiv:1012.3107.
[16] C.H. Clemens, *Double solids.* Adv. in Math. 47 (1983), no. 2, 107–230.
[17] L. Dixon, *Some world-sheet properties of superstring compactifications, or orbifolds and otherwise.* In: Superstrings, Unified Theories and Cosmology 1987 (Trieste, 1987), ICTP Ser. Theoret. Phys. vol. 4, World Sci., Teaneck, NJ, 1988, 67–126.
[18] R. Donagi, B. Ovrut, T. Pantev and D. Waldram, *Standard-model bundles.* Adv. Theor. Math. Phys. (2002) no. 5, 563–615, arXiv:math/0008010.
[19] M.R. Douglas and S. Kachru, *Flux compactification.* Rev. Modern Phys. 79 (2007), no. 2, 733–796.
[20] M.R. Douglas, R. Reinbacher, and S.-T. Yau, *Branes, bundles and attractors: Bogomolov and beyond.* math.AG/0604597.
[21] M. Fernández, S. Ivanov, L. Ugarte and R. Villacampa, *Non-Kähler heterotic string compactifications with non-zero fluxes and constant dilaton.* Comm. Math. Phys. 288 (2009), no. 2, 677–697.
[22] S. Fidanza, R. Minasian and A. Tomasiello, *Mirror symmetric SU(3)-structure manifolds with NS fluxes.* Commun. Math. Phys. 254 (2005), 401–423.
[23] J. Fine and D. Panov, *Symplectic Calabi-Yau manifolds, minimal surfaces and the hyperbolic geometry of the conifold.* J. Differential Geom. 82 (2009), no. 1, 155–205.
[24] J. Fine and D. Panov, *Hyperbolic geometry and non-Kähler manifolds with trivial canonical bundle.* Geom. Topol. 14 (2010), no. 3, 1723–1763.
[25] R. Friedman, *Simultaneous resolution of threefold double points.* Math. Ann. (1986), no. 4, 671–689.
[26] J.-X. Fu, J. Li and S.-T. Yau, *Constructing balanced metrics on some families of non-Kahler Calabi-Yau threefolds.* arXiv:0809.4748.
[27] J.-X. Fu, L.-S. Tseng and S.-T. Yau, *Local heterotic torsional models*, Comm. Math. Phys. 289 (2009) no. 3, 1151–1169, arXiv:0806.2392.
[28] J.-X. Fu and S.-T. Yau, *The theory of superstring with flux on non-Kähler manifolds and the complex Monge-Ampére equation.* J. Differential Geom. 78 (2008), no. 3, 369–428, arXiv:hep-th/0604063.
[29] A. Givental, *Equivariant Gromov-Witten invariants.* Internat. Math. Res. Notices (1996), no. 13, 613–663.
[30] M. Graña, *Flux compactifications in string theory: a comprehensive review.* Phys. Rept. 423 (2006), 91–158, arXiv:hep-th/0509003.
[31] M. Graña, R. Minasian, M. Petrini and A. Tomasiello, *Generalized structures of N=1 vacua.* JHEP 0511 (2005) 020, arXiv:hep-th/0505212.
[32] A. Gray, *Some examples of almost Hermitian manifolds.* Illinois J. Math. 10 (1966), 353–366.
[33] B. Greene and M. Plesser, *Duality in Calabi-Yau moduli space.* Nucl. Phys. B 338 (1990), no. 1, 15–37.
[34] M. Gross and B. Siebert, *Affine manifolds, log structures, and mirror symmetry.* Turkish J. Math. 27 (2003), no. 1, 33–60.
[35] M. Gross and B. Siebert, *Mirror symmetry via logarithmic degeneration data. I.* J. Differential Geom. 72 (2006), no. 2, 169–338.
[36] M. Gross and B. Siebert, *Mirror symmetry via logarithmic degeneration data. II.* J. Algebraic Geom. 19 (2010), no. 4, 679–780.
[37] N. Hitchin, *Kählerian twistor spaces.* Proc. London Math. Soc. (3) 43 (1981), no. 1, 133–150.
[38] M.X. Huang, A. Klemm and S. Quackenbush, *Topological string theory on compact Calabi-Yau: modularity and boundary conditions.* In: Homological Mirror Symmetry, 45–102, Lecture Notes in Phys., 757, Springer, Berlin, 2009, arXiv:hep-th/0612125.
[39] D. Huybrechts, Complex Geometry: An Introduction, Springer-Verlag, Berlin (2005).
[40] M. Kontsevich, *Intersection theory on the moduli space of curves and the matrix Airy function.* Comm. Math. Phys. 147 (1992), no. 1, 1–23.

[41] M. Kontsevich, *Enumeration of rational curves via torus actions.* In: The Moduli Space of Curves (Texel Island, 1994), 335–368, Progr. Math., 129, Birkhuser Boston, Boston, MA, 1995.

[42] W. Lerche, C. Vafa and N.P. Warner, *Chiral rings in $N = 2$ superconformal theories.* Nucl. Phys. B 324 (1989), no. 2, 427–474.

[43] J. Li and S.-T. Yau, *The existence of supersymmetric string theory with torsion.* J. Differential Geom. 70 (2005), 143–181.

[44] J. Li and A. Zinger, *On the genus-one Gromov-Witten invariants of complete intersections.* J. Differential Geom. 82 (2009), no. 3, 641–690.

[45] S. Li, PhD Thesis, Harvard University, 2011.

[46] B. Lian, K. Liu and S.-T. Yau, *Mirror principle I.* Asian J. Math. 1 (1997), no. 4, 729–763.

[47] B. Lian, R. Song, and S.-T. Yau, *Period integrals and topological systems.* arXiv:1105.2984.

[48] B. Lian and S.-T. Yau, *Period integrals of CY and general type complete intersections*, arXiv:1105.4872.

[49] M.-L. Michelsohn, *On the existence of special metrics in complex geometry.* Acta Math. 149 (1982) no. 3-4, 261–295.

[50] M. Reid, *The moduli space of 3-folds with $K = 0$ may nevertheless be irreducible.* Math. Ann. 278 (1987) 329–334.

[51] I. Smith, R.P. Thomas and S.-T. Yau, *Symplectic conifold transitions.* J. Differential Geom. 62 (2002), no. 2, 209–242.

[52] A. Strominger, *Superstring with torsion.* Nucl. Phys. B 274 (1986) 253–284.

[53] A. Strominger, S.-T. Yau and E. Zaslow, *Mirror symmetry is T-duality.* Nucl. Phys. B 479 (1996), no. 1-2, 243–259.

[54] A. Tomasiello, *Reformulating supersymmetry with a generalized Dolbeault operator*, JHEP 0802 (2008) 010, arXiv:0704.2613.

[55] L.-S. Tseng and S.-T. Yau, *Cohomology and Hodge theory on symplectic manifolds: I.* arXiv:0909.5418.

[56] L.-S. Tseng and S.-T. Yau, *Cohomology and Hodge theory on symplectic manifolds: II.* arXiv:1011.1250.

[57] L.-S. Tseng and S.-T. Yau, *Generalized cohomologies and supersymmetry.* arXiv:1111.6968.

[58] L. Ugarte and R. Villacampa, *Non-nilpotent complex geometry of nilmanifolds and heterotic supersymmetry.* arXiv:0912.5110.

[59] E. Witten, *Mirror manifolds and topological field theory.* In: Essays on Mirror Manifolds, 120–158, Int. Press, Hong Kong, 1992.

[60] E. Witten, *Two-dimensional gravity and intersection theory on moduli space.* In: Surveys in Differential Geometry (Cambridge, MA, 1990), 243–310, Lehigh Univ., Bethlehem, PA, 1991.

[61] S. Yamaguchi and S-T. Yau, *Topological string partition functions as polynomials.* J. High Energy Phys. 0407 (2004), 047.

[62] A. Zinger, *Reduced genus-one Gromov-Witten invariants.* J. Differential Geom. 83 (2009), no. 2, 407–460.

Department of Mathematics, University of California, Irvine, CA 92697 USA
E-mail address: `lstseng@math.uci.edu`

Department of Mathematics, Harvard University, Cambridge, MA 02138 USA
E-mail address: `yau@math.harvard.edu`

Proceedings of Symposia in Pure Mathematics
Volume **85**, 2012

F-theory GUTs: Global Aspects and Phenomenology

Sakura Schäfer-Nameki

ABSTRACT. F-theory provides an ideal framework for studying and realizing supersymmetric Grand Unified Theories (GUTs) within string theory. We review recent developments in this subject, giving an outline of how the GUT gauge groups, matter fields and Yukawa couplings are realized geometrically from F-theory compactifications on singular elliptic Calabi-Yau fourfolds. The main emphasis of this exposition will be on global aspects of these compactifications: we review how to resolve the singular Calabi-Yau, thereby giving a concrete picture of how the fiber structure changes in codimension 2 and 3, where matter and Yukawas are realized. Furthermore, a complete understanding of the resolved geometry allows the construction of consistently quantized G_4-flux, and we explain the connection to analogous constructions in local models. Finally, we discuss how simple anomaly considerations lead to generic constraints on the spectra of the theories that can arise in the framework of F-theory GUT realizations, and summarize their potential phenomenological implications.
This is a contribution to the proceedings of the *String Math 2011* conference, which was held at the University of Pennsylvania.

CONTENTS

1. The Basic Physics Setup

Supersymmetry is one of the main paradigms of physics beyond the Standard Model. However, at the energy scales that we observe at present, supersymmetry is broken and we have to study, in addition to supersymmetric extensions of the

1991 *Mathematics Subject Classification.* Primary 81T30, Secondary 14J32 81V22.
Key words and phrases. F-theory, singular elliptic Calabi-Yau manifolds, string phenomenology.
The author thanks Joe Marsano, Natalia Saulina and Matt Dolan for collaboration on the work presented at String Math 2011, and reported on in this contribution to the proceedings. The author would like to thank the organizers of String Math 2011 for giving her the opportunity to present this work and for organizing a very stimulting, exciting conference.

Standard Model, various ways to break supersymmetry consistently with ovservation. The parameter space and the phenomenology for such models are vast, even in the case of the minimal extension (MSSM). Although experiments put constraints on these parameters, one hope as a theorist is that some navigational help can be obtained by embedding such gauge theoretic setups into string theory.

Instead of considering the MSSM with $SU(3) \times SU(2) \times U(1)$ gauge theory embedded into string theory, we will focus on a description in terms of a grand unified gauge group (GUT). This is strongly motivated by the precision unification of the gauge couplings in the MSSM at around 10^{16}GeV, which is very suggestive of a simple gauge group at that energy scale. The minimal choice for a GUT group is $SU(5)$, and in addition to the gauge degrees of freedom, we require the following matter content

- 3 generations of

$$\mathbf{10}_M = \begin{pmatrix} Q \sim (\mathbf{3},\mathbf{2})_{+1/6} \\ U^c \sim (\bar{\mathbf{3}},\mathbf{1})_{-2/3} \\ E^c \sim (\mathbf{1},\mathbf{1})_{+1} \end{pmatrix}, \qquad \bar{\mathbf{5}}_M = \begin{pmatrix} D^c \sim (\bar{\mathbf{3}},\mathbf{1})_{+1/3} \\ L \sim (\mathbf{1},\mathbf{2})_{-1/2} \end{pmatrix}$$

- Higgs multiplets

$$\mathbf{5}_H = \begin{pmatrix} H_u \sim (\mathbf{1},\mathbf{2})_{+1/2} \\ H_u^{(3)} \sim (\mathbf{3},\mathbf{1})_{-1/3} \end{pmatrix}, \qquad \bar{\mathbf{5}}_H = \begin{pmatrix} H_d \sim (\mathbf{1},\mathbf{2})_{-1/2} \\ H_d^{(3)} \sim (\bar{\mathbf{3}},\mathbf{1})_{+1/3} \end{pmatrix}$$

- Top and bottom Yukawa couplings, i.e. superpotential couplings

$$W \sim \lambda_t \, \mathbf{5}_H \times \mathbf{10}_M \times \mathbf{10}_M + \lambda_b \, \bar{\mathbf{5}}_H \times \bar{\mathbf{5}}_M \times \mathbf{10}_M$$

In addition, we have to devise a mechanism to lift the Higgs triplets, which are not part of the MSSM, as well as break the GUT gauge group to the Standard Model one: $SU(3) \times SU(2) \times U(1)$.

In this report, we focus on the class of F-theory embeddings of supersymmetric extensions of the Standard Model, first pioneered in [**DW, BHV, HTTWY**]. In the next section we summarize the basics of F-theory compactifications and how gauge theories arise from singular elliptically fibered Calabi-Yau (CY) compactifications of F-theory. We then focus on global aspects of the geometry, study the resolution of the CY, and discuss G-fluxes. In the last section we will derive generic set of constraints arising in F-theory GUT compactifications and study their phenomenological implications. This report is a summary of the talk given by the author at *String Math 2011* mainly on the results in [**MSSN1, MSSN2, DMSS, DMS, MSSN3**]. For detailed references on the subject we refer the reader to the references in the papers, and to the review [**W**].

2. F-theory GUT Basics

2.1. F-theory. Gauge degrees of freedom in string theory arise from open strings ending on branes. In particular, non-abelian gauge theories in $p+1$ dimensions can be constructed from stacks of branes, where the non-abelian gauge bosons arise from massless open strings stretching between a set of coincident p-branes. In perturbative (more specifically, type II) string theory this can be achieved, thus providing a framework for realization of GUT models. However, not all Yukawa couplings are automatically generated at tree-level, in particular the top Yukawa coupling is perturbatively forbidden and has to be generated by instanton effects.

F-theory provides an alternative to this, where all couplings arise in the same geometric fashion.

To set the scene, recall that F-theory is an alternative name for all non-perturbative type IIB vacua, in particular the axio-dilaton $\tau = C_0 + ie^{-\phi}$ is allowed to vary. It is useful geometrize τ, consistently with the S-duality $\mathrm{SL}_2\mathbb{Z}$ action $\tau \mapsto (a\tau + b)/(c\tau + d)$. This leads us to consider elliptically fibered compactifications, where τ is the complex structure of the elliptic fiber. In particular, to generate four dimensional $N = 1$ supersymmetric vacua we consider compactifications on elliptic Calabi-Yau (CY) fourfolds Y_4.

Geometrically, the non-abelian gauge degrees of freedom arise from M2-branes that wrap the collapsed $\mathbb{P}^1$s in the elliptic fiber[1]. The abelian part has a slightly different origin: recall that in addition to the geometry, string backgrounds are specified by additional fluxes, which in the case of F-theory amounts to G_4 four-form fluxes which are of type $(2,2)$, with potential C_3. Denote by $\omega_i^{1,1}$ the $(1,1)$ forms, which have one leg on the fiber. The $U(1)$ gauge fields arise from decomposing C_3 into these

$$C_3 = A_i \wedge \omega_i^{(1,1)} . \tag{2.1}$$

As we will see in the next section, the divisors dual to $\omega^{(1,1)}$ can be explicitly constructed in the resolved CY fourfold, and will be called Cartan divisors.

G_4-flux plays an important role in F-theory model building, for instance it is crucial for generating chiral matter, i.e. lifting certain zero-modes of matter multiplets. Such G-flux would in the type IIB language correspond to brane worldvolume flux, and this G-flux has to have one leg on the elliptic fiber. This is equivalent to requiring that the divisors dual to the flux are orthogonal to all vertical and horizontal divisors (i.e. those that are either elliptic fibrations over curves in the base or are entirely inside the section of the fibration). Making sure of this requirement makes studying G_4-flux in the singular space tricky. However, resolving the CY fourfold allows us to construct surfaces dual to the G_4-flux explicitly. Finally, G-flux has to satisfy the quantization condtion

$$G_4 + c_2(Y_4) \in H^4(Y_4, \mathbb{Z}) , \tag{2.2}$$

where c_2 is the second Chern class of the fourfold. Ensuring correct quantization of the flux again requires knowledge of the resolved CY. All in all we will see that the resolved geometry will give a powerful tool to analyze F-theory compactifications. This will be discussed in section 3.

2.2. Geometric Setup. For the present purposes we focus on elliptic fourfolds Y_4 with a Weierstrass form

$$y^2 = x^3 + fx + g , \tag{2.3}$$

where f and g are global sections of $\mathcal{O}(-4K_{B_3})$ and $\mathcal{O}(-6K_{B_3})$, and where the base of the elliptic fibration is the threefold denoted by B_3. Furthermore let S_{GUT} be a component of the discriminant $\Delta = 4f^3 + 27g^2$, locally characterized by a section z. This will be the divisor S_{GUT}, on which the GUT gauge degrees of freedom are realized, and the type of singularity determines the gauge group. For concreteness

[1]This makes use of writing F-theory as a limit of M-theory compactified on a CY fourfold, where the volume of the fiber is shrunk to zero size. We refer the reader to [**W, D**] for more details.

(and from a physics point of view, minimality of the unified gauge group) we will focus on the $SU(5)$ case, i.e. the discriminat will locally have the behavior

$$\Delta = O(z^5)\,. \tag{2.4}$$

The Weierstrass model can then be put globally into Tate form for $SU(5)$ [**BIKMSV, KMSS**]

$$y^2 = x^3 + b_0 z^5 + b_2 z^3 x + b_3 z^2 y + b_4 z x^2 + b_5 xy\,, \tag{2.5}$$

where the b_m are sections of $\mathcal{O}((m-6)K_{B_3} + (m-5)S_{\rm GUT})$ and the discriminant has the expansion

$$\Delta = z^5 (b_5^4 P) + z^6 (b_5^2 (b_4 P + b_5 R)) + O(z^7)\,, \tag{2.6}$$

where $P = b_0 b_5^2 - b_2 b_3 b_5 + b_3^2 b_4$.

Locally the singularity type can enhance. For instance in codimension 2, the higher singularity type can be loosely[2] referred to as rank 1 enhancements to either $SU(6)$ or $SO(10)$ along curves inside $S_{\rm GUT}$. The physics intuition is that along matter curves other 7-branes intersect the GUT 7-branes, and result in a local higher rank gauge group. The matter is obtained from higgsing the locally enhanced gauge group to the GUT group. For instance, the adjoint of $SU(6)$ gives rise under $SU(6) \to SU(5) \times U(1)$, to **5** and $\bar{\mathbf{5}}$ matter. Likewise in codimension 3, loosely (again a precise description of the fibers will follow in the section on global structure) the singularity type can enhance further to $SU(7), SO(12), E_6$, which corresponds to the various Yukawa interactions of the GUT matter fields. In terms of the Tate model, these loci are summarized as follows:

- Codimension 1: $z = 0$ specifies $S_{\rm GUT} \subset B_3$
  $\Rightarrow SU(5)$ gauge fields
- Codimension 2: Matter curves arise from $D.S_{\rm GUT}$
  - SU(6): $z = P = b_0 b_5^2 - b_2 b_3 b_5 + b_3^2 b_4 = 0$
    $\Rightarrow$ matter in **5** and $\bar{\mathbf{5}}$
  - SO(10): $z = b_5 = 0$
    $\Rightarrow$ matter in **10**
- Codimension 3: Yukawa couplings, i.e. intersections of matter curves
  - $SO(12)$: $z = b_3 = b_5 = 0$
    $\Rightarrow$ down Yukawa
  - E_6: $z = b_4 = b_5 = 0$
    $\Rightarrow$ up Yukawa

2.3. Local versus Global Models. A huge amount of effort has been put into studying not the complete geometry of the CY, but the so-called local limit. Geometrically this corresponds to "zooming into the vicinity of $S_{\rm GUT}$" – we should give a precise definition of this limit once we have constructed the smooth CY fourfold in the next section. Physically it corresponds to considering the theory in the limit when only the gauge degrees of freedom on the 7-branes are taken into account, neglecting the precise embedding into an elliptic CY geometry. The theory can be modeled in this limit by Higgs bundles, and for most purposes by the spectral data of the Higgs bundle. This is discussed in the contribution to these proceedings [**W-StringMath**]. We will therefore keep this discussion brief, and

[2]We will discuss the precise structure of the fibers in codimension greater than 1 later on and will for the moment refer to them in terms of their naive group theoretic characterization.

only comment on the spectral cover construction for the Higgs bundles in the local limit of the resolved CY fourfold.

3. Global Aspects

Local models are sufficient to understand the structure and properties in the vicinity of the 7-branes, where a suitable description in terms of spectral covers exists. In this limit one can realize additional $U(1)$ symmetries, which are phenomenologically relevant, and G-fluxes can be constructed (on the local limit, see the contribution [**W-StringMath**] to these proceedings). However, it is far from clear, that these notions lift to the global CY fourfold. Recall that the local model arises when we consider only the geometry in the vicinity of the component of the discriminant, $S_{\rm GUT}$.

These issues can be described in complete generality by constructing the resolution of the singular CY fourfold [**EY**], see the contribution by Esole to this volume [**E-StringMath**], which was then extended to incorporate among other things G-fluxes and the local limit in [**MSN**].

3.1. Geometry of the resolution. The Tate form for an $SU(5)$ (or I_5) singularity (2.5) can be resolved by embedding it first into the 5-dimensional space that is a $\mathbb{P}^2$ bundle defined as

$$X_5 = \mathbb{P}\left(\mathcal{O} \oplus K_{B_3}^{-2} \oplus K_{B_3}^{-3}\right) . \tag{3.1}$$

Divisors are obtained by pull-back from the base B_3 of the fibration, and a new divisor arises from the hyperplane class in $\mathbb{P}^2$, with section σ. The Tate form can then be written as

$$wy^2 = x^3 + b_0 w^3 z^5 + b_2 x w^2 z^3 + b_3 y w^2 z^2 + b_4 x^2 w z + b_5 x y w\,, \tag{3.2}$$

where

(3.3)

Section	Bundle
w	$\mathcal{O}(\sigma)$
x	$\mathcal{O}(\sigma + 2c_1)$
y	$\mathcal{O}(\sigma + 3c_1)$

where $c_1 = c_1(S_{\rm GUT})$. The geometry (3.2) has a singularity at $x = y = z = 0$, which in the first step is resolved by introducing a new section ζ_1, such that

$$x = x_1\zeta_1\,, \quad y = y_1\zeta_1\,, \quad z = z_1\zeta_1\,, \tag{3.4}$$

which replaces the singular locus with the exceptional divisor $\zeta_1 = 0$, with projective coordinates $[x_1, y_1, z_1]$, resulting after a proper transform in the new equations for the CY fourfold[3]

$$\text{Tate}_1: \quad y_1(y_1 - b_5 x_1) = \zeta_1\left(b_2\zeta_1 x_1 z_1^3 + b_4 x_1^2 z_1 + b_3 y_1 z_1^2 + b_0\zeta_1^2 z_1^5 + x_1^3\right) . \tag{3.5}$$

Next blowup along $x_1 = y_1 = \zeta_1 = 0$, which yields a new exceptional divisor $\zeta_2 = 0$

$$x_1 = x_2\zeta_2\,, \quad y_1 = y_2\zeta_2\,, \quad \zeta_1 = \zeta_2\zeta_{12}\,, \tag{3.6}$$

arriving at

$$y_2\left(y_2 - b_5 x_2 - b_3\zeta_{12} z_1^2\right) = \zeta_2\zeta_{12}\left(b_2\zeta_{12} x_2 z_1^3 + b_4 x_2^2 z_1 + b_0\zeta_{12}^2 z_1^5 + \zeta_2 x_2^3\right) . \tag{3.7}$$

[3]We set $w = 1$ in what follows.

This binomial geometry can be written as

$$v_1 v_2 = u_1 u_2 u_3 \,, \tag{3.8}$$

and has 6 distinct resolutions, which are related by flops. We consider as an example excuting the blowups along

$$y_2 = \zeta_2 = 0 \qquad \text{and} \qquad \left(-y_2 + b_5 x_2 + b_3 \zeta_{12} z_1^2\right) = \zeta_{12} = 0 \,. \tag{3.9}$$

Denote the two exceptional divisors ζ_2 and ζ_3, with the blowup equations

$$\begin{aligned} y_2 &= \zeta_3 y_3 \,, & \zeta_2 &= \zeta_3 \zeta_{23} \\ \left(-y_2 + b_5 x_2 + b_3 \zeta_{12} z_1^2\right) &= \zeta_4 y_4 \,, & \zeta_{12} &= \zeta_4 \zeta_{124} \,. \end{aligned} \tag{3.10}$$

which results in the final smooth CY fourfold

$$y_3 y_4 + \zeta_{23}\zeta_{124} \left(\zeta_4 \zeta_{124} z_1^3 \left(b_2 x_2 + b_0 \zeta_4 \zeta_{124} z_1^2\right) + b_4 x_2^2 z_1 + \zeta_3 \zeta_{23} x_2^3\right) = 0 \,. \tag{3.11}$$

From the last two blowups we also have the additional relation from (3.10)

$$\left(-y_3 \zeta_2 + b_5 x_2 + b_3 \zeta_{124} \zeta_4 z_1^2\right) = \zeta_4 y_4 \tag{3.12}$$

The sections have the following classes in terms of c_1, σ and the exceptional divisors E_i given in terms of $\zeta_i = 0$:

(3.13)

Section	Bundle
y_3	$\mathcal{O}(\sigma + 3c_1 - E_1 - E_2 - E_3)$
ζ_{23}	$\mathcal{O}(E_2 - E_3)$
y_4	$\mathcal{O}(\sigma + 3c_1 - E_1 - E_2 - E_4)$
ζ_{124}	$\mathcal{O}(E_1 - E_2 - E_4)$
ζ_3	$\mathcal{O}(E_3)$
ζ_4	$\mathcal{O}(E_4)$
x_2	$\mathcal{O}(\sigma + 2c_1 - E_1 - E_2)$
z_1	$\mathcal{O}(S_{\rm GUT} - E_1)$

3.2. Alternative Resolution. Following the Tate algorithm, a slightly simpler way of resolving the geometry can be obtained [**BIKMSV**] (see also forthcoming [**LS**]). The blowup in this case proceeds after the first resolution by blowup along the exctional divisor, which inside the blown-up Tate form is split

$$\text{Tate}_1|_{\zeta_1=0} = y_1(y_1 - b_5 x_1) \,. \tag{3.14}$$

It is clear that

$$\text{Tate}_1 \qquad \Leftrightarrow \qquad y_1(y_1 - b_5 x_1) = \zeta_1(\cdots) \,, \tag{3.15}$$

and thus is singular along $y_1 = \zeta_1 = 0$. Introducing a new section ζ_2

$$y_1 = y_2 \zeta_2 \,, \qquad \zeta_1 = \zeta_2 \zeta_{12} \,, \tag{3.16}$$

the space after this resolution can be written in the form

$$-x_1 \left(\zeta_{12} x_1 \left(b_4 z_1 + x_1\right) + b_5 y_2\right) = \zeta_2 \left(\zeta_{12}^2 z_1^3 \left(b_2 x_1 + b_0 \zeta_2 \zeta_{12} z_1^2\right) + b_3 \zeta_{12} y_2 z_1^2 - y_2^2\right) \,. \tag{3.17}$$

From this we can read off that the next blowup is along $x_1 = \zeta_2 = 0$, wich $x_1 = x_3 \zeta_3$ and $\zeta_2 = \zeta_{23} \zeta_3$

$$y_2 \left(\zeta_{23} \left(y_2 - b_3 \zeta_{12} z_1^2\right) - b_5 x_3\right) = \zeta_3 \zeta_{12} \left(b_2 \zeta_{12} \zeta_{23} x_3 z_1^3 + b_4 x_3^2 z_1 + b_0 \zeta_{12}^2 \zeta_{23}^2 z_1^5 + \zeta_3 x_3^3\right) \,, \tag{3.18}$$

and finally along $y_2 = \zeta_3 = 0$ with $y_2 = y_4\zeta_4$ and $\zeta_3 = \zeta_{34}\zeta_4$ which gives rise to the smooth fourfold

(3.19)
$$\begin{aligned}&\zeta_4\zeta_{23}y_4^2 - b_5x_3y_4 - \zeta_4\zeta_{12}\zeta_{34}^2x_3^3\\ &\quad - b_2\zeta_{12}^2\zeta_{23}\zeta_{34}x_3z_1^3 - b_4\zeta_{12}\zeta_{34}x_3^2z_1 - b_3\zeta_{12}\zeta_{23}y_4z_1^2 - b_0\zeta_{12}^3\zeta_{23}^2\zeta_{34}z_1^5 = 0\,.\end{aligned}$$

This resolution has various advantages, as e.g. it does not require to keep track of the additional relation (3.12), and generalizes systematically to any other singularity.

3.3. Codimension one. The first thing to read off from the resolved geometry (3.11) are the so-called Cartan divisors, i.e. the new divisors, which in the smooth geometry arise from the blowup $\mathbb{P}^1$s in the I_5 fiber, fibered over the divisor $S_{\rm GUT}$ in the base. Concretely, the divisor $z = 0$, which in the singular geometry corresponded to the I_5 singularity fibered over $S_{\rm GUT}$ becomes reducible in the resolved geometry:

(3.20)
$$z = \zeta_3\zeta_4\zeta_{23}\zeta_{124}z_1\,.$$

The divisors are in the classes

(3.21)

Component	Class in $\tilde{Y}_4$	Notation
$(z_1 = 0)\,|_{\tilde{Y}_4}$	$(S_{\rm GUT} - E_1)$	$\mathcal{D}_{-\alpha_0}$
$(\zeta_{124} = 0)\,|_{\tilde{Y}_4}$	$(E_1 - E_2 - E_4)$	$\mathcal{D}_{-\alpha_1}$
$(\zeta_3 = 0)\,|_{\tilde{Y}_4}$	E_3	$\mathcal{D}_{-\alpha_2}$
$(\zeta_{23} = 0)\,|_{\tilde{Y}_4}$	$(E_2 - E_3)$	$\mathcal{D}_{-\alpha_3}$
$(\zeta_4 = 0)\,|_{\tilde{Y}_4}$	E_4	$\mathcal{D}_{-\alpha_4}$

The subscripts indicate to which root of $SU(5)$ the divisors correspond to, with α_0 the extended node. The intersection matrix is precisely the (extended) Cartan matrix for $SU(5)$, as expected.

3.4. Codimension two and three. More unexpected things happen in higher codimension. From the local limit, we expect that matter fields are generated in codimension two and Yukawa couplings between matter fields in codimension three. As these seem to naively be governed by simple group theoretic rules: e.g. higgsing the adjoint of a higher rank gauge group G to $SU(5)\times U(1)$ or $SU(5)\times U(1)^2$, the naive expectation may be to find Kodaira fibers of the type G at those restricted loci on $S_{\rm GUT}$. As an example, consider $SO(10)$, whose adjoint decomponses into the adjoint of $SU(5)\times U(1)$ and two **10** multiplets, thus giving rise to matter fields in the **10**. As summarized in section 2.2 this occurs at $b_5 = 0$, where we would expect the fiber structure to enhance.

To study this in the resolved CY fourfold geometry (3.11), consider how the Cartan divisors $\mathcal{D}_{-\alpha_i}$ behave along the locus $b_5 = 0$ and likewise at $P = 0$, which is the locus of $SU(6)$ enhancement and the two Yukawa points. As observed in [**EY**], the fibers are generically not fibers of the type $SU(6), SO(10)$ or $SO(12), E_6$. However, the Cartan divisors become reducible along these subloci, precisely in such a way that the physical interpretation in terms of matter fields or Yukawa couplings is consistent [**MSN**]. More concretely, in [**MSN**] we showed that the Cartan divisors corresponding to roots, $-\alpha_2$ and $-\alpha_4$, split above $b_5 = 0$, i.e. the

10 matter locus, as

$$(3.22)\quad \begin{aligned} (1,-2,1,0) &\to (1,-1,1,-1)+(0,-1,0,1)\\ -\alpha_2 &\to [-\mu_{10}+\alpha_1+\alpha_2+\alpha_3]+[\mu_{10}-\alpha_1-2\alpha_2-\alpha_3]\\ (0,0,1,-2) &\to (1,-1,1,-1)+(1,0,0,-1)+(-2,1,0,0)\\ -\alpha_4 &\to [-\mu_{10}+\alpha_1+\alpha_2+\alpha_3]+[\mu_{10}-\alpha_2-\alpha_3-\alpha_4]+[-\alpha_1]\,, \end{aligned}$$

where $\mu_{\mathbf{10}}$ is the highest weight of the **10** representation of $SU(5)$. All other Cartan divisors $\mathcal{D}_{-\alpha_i}$ remain unchanged. In particular, this splitting indicates that there is one additional, independent component of $b_5 = z = 0$, which is characterized by the weight $(1,-1,1,-1)$, so that in total we have six independent divisors. However, crucially, the splitting occurs into weights of representations of $SU(5)$ and not into the roots of $SO(10)$.

Similarly, at $P = 0$, i.e. the $SU(6)$ locus, we get

$$(3.23)\quad \begin{aligned} (1,-2,1,0) &\to (0,-1,1,0)+(1,-1,0,0)\\ -\alpha_2 &\to [\mu_5-\alpha_1-\alpha_2]+[-\mu_5+\alpha_1]\,, \end{aligned}$$

and all other Cartan divisors remain unsplit, so that there is again one new component of $z = 0$ when we move to the locus $P = 0$.

Finally, in higher codimension, we expect the matter divisors that we discussed so intersect at points. For instance, recall that the E_6 point generates the top Yukawa coupling $\mathbf{10}\times\mathbf{10}\times\mathbf{5}$, and is characterized by $b_5 = b_4 = z = 0$. For instance consider the splitting of one of the **10** divisor. It was shown in [**MSN**] that although the fibers are not E_6 the splitting is such that the **10** divisor in (3.22) splits into another **10** and a **5**

$$(3.24)\quad \begin{aligned} (1,-1,1,-1) &\to (1,0,0,-1)+(0-1,1,0)\\ [-\mu_{10}+\alpha_1+\alpha_2+\alpha_3] &\to [\mu_{10}-\alpha_2-\alpha_3-\alpha_4]+[\mu_5-\alpha_1-\alpha_2-\alpha_3]\,. \end{aligned}$$

In particular, the rank of the fiber does not change. Similarly, the **5** divisor (3.23) can be shown to split as

$$(3.25)\quad \begin{aligned} (1,-1,0,0) &\to (1,0,0,-1)+(0,-1,0,1)\\ -\mu_5+\alpha_1 &\to [\mu_{10}-\alpha_2-\alpha_3-\alpha_4]+[\mu_{10}-\alpha_1-2\alpha_2-\alpha_3]\,, \end{aligned}$$

i.e. the matter divisor becomes homologous to the sum of two **10**s. Alternatively, we can describe this as follows: we can think of the Cartan divisors as fibering the blowup $\mathbb{P}^1$s over $S_{\rm GUT}$. The splitting corresponds to the existence of a 3-chain connecting the $\mathbb{P}^1$ that corresponds to the matter divisor in the $-(1,-1,0,0)$ with the $\mathbb{P}^1$ corresponding to **10** matter divisors in $(1,0,0,-1)$ and $(0,-1,0,1)$. This 3-chain degenerates as the E_6 locus is reached. This is precisely what is required for consistency with generating of the top Yukawa coupling.

To summarize, the fibers in higher codimensions are not necessarily of Kodaira type. However, the splitting of the $z = 0$ locus along the codimension 2 and 3 loci corresponding to matter divisors, and Yukawas (and naively, to an enhancement to a higher rank gauge group) is precisely consistent with generating the right structures required for matter (new independent components along the loci $z = b_5 = 0$ etc) and the Yukawas (matter splitting consistently with the required coupling).

3.5. Fluxes. We mentioned earlier that one of the nice applications of the resolution $\tilde{Y}_4$ of Y_4 is the construction of G-fluxes. First we require the second Chern class of $\tilde{Y}_4$ for the quantization of G. Direct computation yields [**MSN**]

$$\begin{aligned} c_2(\tilde{Y}_4) &= \left(3\sigma^2 + 13\sigma c_1 + c_2 + 11c_1^2\right) \\ &\quad - E_3 \cdot E_4 - c_1 \cdot (8E_1 + 3E_2 + 6E_3 - 2E_4) + S_{\rm GUT} \cdot (2E_1 + 4E_3 - 2E_4) \end{aligned} \tag{3.26}$$

As we explained earlier, G_4-flux of relevance for our purposes has to be a $(2,2)$ form with one leg in the fiber direction. We construct these by specifying a dual holomorphic surface G in $\tilde{Y}_4$, which has to be orthogonal to all horizontal and vertical divisors, in particular, for any divisors D_i that are pullbacks of divisors in the base, we impose

$$G \cdot_{\tilde{Y}_4} \sigma \cdot_{\tilde{Y}_4} D_1 = G \cdot_{\tilde{Y}_4} D_1 \cdot_{\tilde{Y}_4} D_2 = 0\,. \tag{3.27}$$

Furthermore, requiring that the $SU(5)$ gauge symmetry remains unbroken by the flux means that intersections with the Cartans should vanish

$$G \cdot_{\tilde{Y}_4} \mathcal{D}_{-\alpha_i} \cdot_{\tilde{Y}_4} D_1 = 0\,. \tag{3.28}$$

Surfaces of this type that can be obtained from intersections of divisors in the ambient fivefold with the resolved CY fourfold were constructed in [**MSN**] and it was shown that the most general such flux is a linear combination

$$G = \frac{1}{2}\left(E_3 \cdot_{\tilde{Y}_4} E_4 + E_2 \cdot_{\tilde{Y}_4} c_1\right) + \sum_i E_i \cdot_{\tilde{Y}_4} (a_i c_1 + b_i S_{\rm GUT}) + pE_3 \cdot_{\tilde{Y}_4} E_4 + qE_2 \cdot_{\tilde{Y}_4} E_4\,. \tag{3.29}$$

Imposing the quantization condition

$$G_4 + c_2(Y_4) \in H^4(Y_4, \mathbb{Z})\,, \tag{3.30}$$

with c_2 as in (3.26) reduces this to

$$G = -\frac{1}{2}(1 + 2a_2)\left(5E_3 \cdot_{\tilde{Y}_4} E_4 + c_1 \cdot_{\tilde{Y}_4} [2E_1 - E_2 - 2E_3 - 4E_4]\right)\,. \tag{3.31}$$

Using the intersection relations in the fourfold we can compute the induced chirality on the matter surfaces. Let us denote $\mathcal{S}_{\mathbf{R}}$ the matter surface associated to the $SU(5)$ representation $\mathbf{R}$, then

$$G \cdot_{\tilde{Y}_4} \mathcal{S}_{\mathbf{10}} = G \cdot_{\tilde{Y}_4} \mathcal{S}_{\overline{\mathbf{5}}} = -\frac{1}{2}(2a_2 + 1)\, S_{\rm GUT} \cdot_{B_3} c_1 \cdot_{B_3} (6c_1 - 5S_{\rm GUT})\,. \tag{3.32}$$

Quite nicely, this result reduces to an intersection in $S_{\rm GUT}$. We can make contact with the usual local model chirality formula by rewriting this as

$$G \cdot_{\tilde{Y}_4} \mathcal{S}_{\mathbf{10}} = G \cdot_{\tilde{Y}_4} \mathcal{S}_{\overline{\mathbf{5}}} = -\frac{1}{2}(2a_2 + 1)\, \eta \cdot_{S_{\rm GUT}} (\eta - 5c_1(S_{\rm GUT}))\,, \qquad a_2 \in \mathbb{Z}\,, \tag{3.33}$$

which reproduces the expressions in the local limit.

Finally, we can also compute the D3-tadpole, which is determined from

$$n_{D3,\text{induced}} = -\frac{\chi}{24} + \frac{1}{2}G^2\,, \tag{3.34}$$

and essentially tells us whether additional D3-branes need to be introduced. The Euler characteristic for the resolved CY is

$$\chi(Y_4) = 4\int_{B_3} c_1 c_2 + 120\int_{B_3} c_1^3 = 24\left(3 + 4\int_{B_3} c_1^3\right)\,. \tag{3.35}$$

Inserting this into (3.34) yields

$$n_{D3,\text{induced}} = \frac{5}{8}(1+2a_2)^2\, c_1 \cdot_{B_3} (6c_1 - 5S_{\text{GUT}}) \cdot_{B_3} S_{\text{GUT}}\,. \tag{3.36}$$

Again we can make contact with the expressions known in the local models [**BGJW**] by expressing this in terms of the local model intersections in S_{GUT}

$$n_{D3,\text{induced}} = \frac{5}{8}(1+2a_2)^2\, \eta \cdot_{S_{\text{GUT}}} (\eta - 5c_1(S_{\text{GUT}}))\,. \tag{3.37}$$

3.6. Spectral Covers and Local Limit. To complete the picture, we should (re)derive the local spectral cover description [**W-StringMath**] from the global, resolved CY fourfold. Let us recall how we expect spectral covers to emerge [**MSSN3**]. We start with the Tate divisor, defined as the divisor in singular CY

$$\mathcal{C}_{\text{Tate}}: \qquad wy^2 = x^3 + w\left(b_0 w^2 z^5 + b_2 wxz^3 + b_3 wyz^2 + b_4 wx^2 z + b_5 wxy\right) \tag{3.38}$$

by

$$\left(b_0 w^2 z^5 + b_2 wxz^3 + b_3 wyz^2 + b_4 wx^2 z + b_5 wxy\right) = 0\,. \tag{3.39}$$

Then we note that the meromorphic section y/x is holomorphic when restricted to this divisor. We call this t and use the fact that $y^2 = x^3$ along the Tate divisor (in the $w = 1$ patch) to write

$$b_0 z^5 + b_2 z^3 t^2 + b_3 z^2 t^3 + b_4 z t^4 + b_5 t^5 = 0\,. \tag{3.40}$$

In the limit $t \to 0$, $z \to 0$, with $s = z/t$ fixed this turns into

$$t^5\left[b_0 s^5 + b_2 s^3 + b_3 s^2 + b_4 s + b_5\right] = 0\,. \tag{3.41}$$

Taking a proper transform, what remains is precisely the Higgs bundle spectral cover $\mathcal{C}_{\text{Higgs,loc}}$.

We can generalize this to the smooth CY fourfold. First note that the various sections in the resolved space are

$$\begin{aligned} x &= \zeta_3\zeta_4^2\zeta_{24}^2\zeta_{123}x_2 \\ y &= \zeta_3\zeta_4^2\zeta_{24}^2\zeta_{123}y_2 \\ z &= \zeta_3\zeta_4\zeta_{24}\zeta_{123}z_1\,, \end{aligned} \tag{3.42}$$

and $y_2 = \zeta_3 y_3$, so that

$$\frac{y}{x} = \frac{y_3\zeta_3}{x_2} \quad \text{and} \quad \frac{zx}{y} = \frac{\zeta_4\zeta_{24}\zeta_{123}x_2 z_1}{y_3}\,. \tag{3.43}$$

So to send $y/x \to 0$ while holding zx/y fixed, in the resolved geometry, we simply have to take

$$\zeta_3 \to 0\,. \tag{3.44}$$

In the limit $\zeta_3 \to 0$, the Tate divisor $\mathcal{C}_{\text{Tate}}$ in the resolved fourfold limits to the spectral cover $\mathcal{C}_{\text{Higgs,loc}}$ of the Higgs bundle in the worldvolume theory. Here we will summarize the procedure: first we need to determine the proper transform of the Tate divisor $\mathcal{C}_{\text{Tate}}$ in the smooth CY fourfold. The Tate divisor is in the class

$$\mathcal{C}_{\text{Tate}} = [3\sigma + 6c_1 - 3E_1 - E_2 - 2E_3 + E_4]_{\tilde{Y}_4}\,. \tag{3.45}$$

and intersecting with the Cartan divisors yields

$$\mathcal{C}_{\text{Tate}} \cdot_{\tilde{Y}_4} \mathcal{D}_{-\alpha_i} = (0,1,0,0)\times 5 \tag{3.46}$$

consistent with the fact that each of the five sheets of $\mathcal{C}_{\text{Tate}}$ restricted to $z = 0$ intersects the roots as $(0,1,0,0)$. Note that this is the highest weight of the **10**, consistent with claims that the 5 sheets of $\mathcal{C}_{\text{Tate}}$ locally behave like a sum of five exceptional lines of dP_9 that transform as **10**'s of $SU(5)_{\text{GUT}}$ [**MSSN3**].

The Higgs bundle spectral cover emerges by restricting $\mathcal{C}_{\text{Tate}}$ to $\zeta_3 = 0$

$$\mathcal{C}_{\text{Higgs,loc}} = \mathcal{C}_{\text{Tate}} \cdot_{\tilde{Y}_4} E_3 \,. \tag{3.47}$$

Indeed, going to a local patch with $x_2 = 1$ we reproduce the standard Higgs bundle spectral cover expression

$$b_0 {y_3}^5 + b_2 {y_3}^3 + b_3 {y_3}^2 + b_4 y_3 + b_5 = 0 \,. \tag{3.48}$$

We can check a few other points for consistency: for instance, the **10** matter surface $\mathcal{S}_{\mathbf{10}}$ of meets $\mathcal{C}_{\text{Higgs,loc}} = \mathcal{C}_{\text{Tate}} \cdot_{\tilde{Y}_4} E_3$ precisely along the curve

$$y_3 = b_5 = 0 \tag{3.49}$$

i.e. the **10** matter curve in the local model. Similarly we can show that the $\bar{\mathbf{5}}$ matter surface intersects $\mathcal{C}_{\text{Higgs,loc}} = \mathcal{C}_{\text{Tate}} \cdot_{\tilde{Y}_4} E_3$ along the standard $\bar{\mathbf{5}}$ matter curve $P = 0$.

3.7. Spectral Cover Fluxes. Finally, we can make contact with the spectral cover construction of fluxes, as described in [**MSSN3**], that are supposed to correspond to curves inside $\mathcal{C}_{\text{Higgs,loc}}$. Consider $p^*\Sigma$, where $\Sigma = D \cdot S_{\text{GUT}}$ for a divisor D in the base B_3 and p is the projection map

$$p : \mathcal{C}_{\text{Higgs,loc}} \to S_{\text{GUT}} \,. \tag{3.50}$$

We can then define a surface

$$\mathcal{S}_{p^*D} = \mathcal{C}_{\text{Tate}} \cdot D \,, \tag{3.51}$$

which however is not yet satisfying the requirements of orthogonality with respect to all vertical and horizontal divisors. For this we need to make a subtraction of those contributions, and get

$$\mathcal{S}_{p^*D} = \mathcal{C}_{\text{Tate}} \cdot D - (3\sigma + 6c_1) \cdot D \,. \tag{3.52}$$

There is another choice of flux corresponding to σ which in the local model denotes the section of the $\mathbb{P}^1$ bundle $\mathbb{P}(\mathcal{O} \oplus K_{S_{\text{GUT}}})$ in which the local model is usually embedded. The surface class of $\mathcal{S}_{\sigma \cdot \mathcal{C}}$ was determined in [**MSN**] to be

$$\mathcal{S}_{\sigma \cdot \mathcal{C}} = c_1 \cdot_{\tilde{Y}_4} (E_4 - E_1) - E_3 \cdot_{\tilde{Y}_4} E_4 \,, \tag{3.53}$$

and with these definitions for the surfaces $\mathcal{S}_{p^*D}$ and $\mathcal{S}_{\sigma \cdot \mathcal{C}}$ the intersection relations from [**MSSN3**], where these expressions had been conjectured, can be verified explicitly

$$\begin{aligned} \mathcal{S}^2_{\sigma \cdot \mathcal{C}} &= -c_1 S^2_{\text{GUT}} \\ \mathcal{S}_{\sigma \cdot \mathcal{C}} \cdot \mathcal{S}_{p^*D} &= -6 c_1 S_{\text{GUT}} D \\ \mathcal{S}^2_{p^*D} &= -30 S_{\text{GUT}} D^2 \,. \end{aligned} \tag{3.54}$$

The generic traceless flux in the local model is therefore

$$\begin{aligned} G_{\text{local model}} &= 5\mathcal{S}_\sigma - \mathcal{S}_{p^*c_1} \\ &= -\left(5E_3 \cdot_{\tilde{Y}_4} E_4 + c_1 \cdot_{\tilde{Y}_4} [2E_1 - E_2 - 2E_3 - 4E_4]\right) \,, \end{aligned} \tag{3.55}$$

which is exactly proportional to the flux that we constructed earlier in (3.29). Finally, we can reproduce the flux quantization in the local models: in local models the flux is quantized such that a curve γ in $\mathcal{C}_{\text{Higgs,loc}}$ defining the flux satisfies

$$\gamma + \frac{1}{2} r \in H_2(\mathcal{C}_{\text{Higgs,loc}}, \mathbb{Z}) , \tag{3.56}$$

where r is the ramification divisor of the five-sheeted covering $\mathcal{C}_{\text{Higgs,loc}} \to S_{\text{GUT}}$, which can be written as

$$r = \mathcal{C}_{\text{Higgs,loc}} \cdot (\mathcal{C}_{\text{Higgs, loc}} - \sigma - \sigma_\infty) . \tag{3.57}$$

Here $\sigma_\infty = \sigma + \pi^* c_1$[4]. We can now identify the surface corresponding to r

$$\mathcal{S}_r = 3S_\sigma + S_{p^*(5c_1 - 4S_{\text{GUT}})} , \tag{3.58}$$

and identify the odd part as that of c_2, as required by the quantization

$$\begin{aligned} \mathcal{S}_r^{(odd)} &= c_1 E_2 + E_3 E_4 \\ c_2(\tilde{Y}_4)^{(odd)} &= c_1 E_2 + E_3 E_4 . \end{aligned} \tag{3.59}$$

In conclusion, the resolution of the CY fourfold gives a most suitable framework to study F-theory GUTs, allowing a concise characterization of G-fluxes, quantization and connecting this back to the local model and the Higgs bundle spectral cover.

4. Survey of F-theory GUTs and Phenomenology

In this final section we will review some of the recent developments obtained in understanding the phenomenological relevance or characteristics, of GUTs that arise in F-theory [**DMS**]. The main constraint that we will discuss arises within the local models, see [**DW, MSSN2**] and [**W-StringMath**] in these proceedings.

4.1. Models with additional $U(1)$s and Constraints. Despite many advantages compared to non-unified models, many supersymmetric GUTs suffer from phenomenological problems. Usually one encounters problems with proton decay, i.e. the lifetime of the proton ends up being much less than what is required from observation. There are various sources of proton decay in an $SU(5)$ GUT. Particularly worrisome is the dimension 5 coupling

$$W_{\text{dim5}} \sim \frac{Q^3 L}{\Lambda} . \tag{4.1}$$

Another phenomenologically unfavorable coupling, which should be absent at tree-level, is the μ-term

$$W_\mu \sim \mu H_u H_d . \tag{4.2}$$

One way to forbid such couplings is to introduce additional $U(1)$ symmetries, under which all fields carry a charge q_i, and only couplings that are neutral under the $U(1)$ are allowed. This leads us to consider F-theory $SU(5)$ GUTs with an additional $U(1)$. Furthermore, the GUT breaking from $SU(5)$ to the MSSM gauge group is done, as is most common in F-theory GUTs, by hypercharge flux F_Y, which lifts the exotic XY bosons arising from the gauge fields of $SU(5)$ and restricts to the Higgs matter curves such that the zero modes of the Higgs triplets are lifted

$$F_Y|_{\Sigma_{\mathbf{5}_H}} = +1 , \qquad F_Y|_{\Sigma_{\bar{\mathbf{5}}_H}} = -1 . \tag{4.3}$$

[4]For a review of the local model notation, see [**MSSN3**] appendix D.

This setup of hypercharge flux and $U(1)$ symmetry is however not automatically consistent for any distribution of charges: $G^2_{MSSM} \times U(1)$ mixed anomaly cancellation for the additional $U(1)$ implies the following relation [**DP, M, MSN, DMSS**]

$$\sum_{\mathbf{10}\text{ matter curves},\Sigma^i_{\mathbf{10}}} q_i \int_{\Sigma^{(i)}_{\mathbf{10}}} F_Y = \sum_{\overline{\mathbf{5}}\text{ matter curves, }\Sigma^a_{\overline{\mathbf{5}}}} q_a \int_{\Sigma^{(a)}_{\overline{\mathbf{5}}}} F_Y \,. \tag{4.4}$$

The two phenomenological onsequences of these anomaly cancellation relations:

COROLLARY 4.1. *In an F-theory $SU(5)$ GUT with an additional $U(1)$ gauge symmetry, the relation (4.4) imply the following:*

(1) *In a minimal $SU(5)$ GUT, i.e. no exotics in the spectrum, then the only $U(1)$ compatible with the MSSM couplings is $U(1)_\chi$*

	$\mathbf{10}_M$	$\overline{\mathbf{5}}_M$	$\mathbf{5}_H$	$\overline{\mathbf{5}}_H$
$U(1)_\chi$	-1	3	2	-2

(2) *If the $U(1)$ is a PQ symmetry, i.e. $q_{H_u} + q_{H_d} \neq 0$, which forbids $W_{\rm dim5}$ and W_μ, then there are always non-GUT exotics in the spectrum.*

PROOF. (1) Consider a minimal $SU(5)$ GUT with $U(1)$. What $U(1)$s are consistent with anomalies? GUT breaking and absence of exotics requires that F_Y restricts as:

$$\int_{\Sigma_{H_u}} F_Y = +1\,, \quad \int_{\Sigma_{H_d}} F_Y = -1\,, \quad \int_{\text{other }\Sigma'} F_Y = 0\,,. \tag{4.5}$$

Inserting this into (4.4)

$$\begin{aligned} \sum_{\mathbf{10}\text{ matter curves},\Sigma^i_{\mathbf{10}}} q_i \int_{\Sigma^{(i)}_{\mathbf{10}}} F_Y &= \sum_{\overline{\mathbf{5}}\text{ matter curves, }\Sigma^a_{\overline{\mathbf{5}}}} q_a \int_{\Sigma^{(a)}_{\overline{\mathbf{5}}}} F_Y \\ \Rightarrow \quad q_{H_u} + q_{H_d} &= 0 \end{aligned} \tag{4.6}$$

The only possible $U(1)$ compatible with MSSM interactions and $SU(5)$ is $U(1)_\chi$. But $U(1)_\chi$ does not forbid $W_{\rm dim\ 5}$!

(2) Phenomenologically we require absence of:

$$W \sim \mu H_u H_d \qquad \text{and} \qquad W_{\rm dim5} \sim \frac{1}{\Lambda} Q^3 L\,,$$

which requires the $U(1)$ to be a *PQ* symmetry, i.e.

$$q_{H_u} + q_{H_d} \neq 0\,. \tag{4.7}$$

In particular, this implies from (4.4) that in the presence of a PQ symmetry, the spectrum always has exotics that do not fall into complete GUT group representations. More precisely, these are vector-like pairs with respect to the MSSM gauge group:

$$\sum_{\mathbf{10}\text{ matter curves, }i} \int_{\Sigma^{(i)}_{\mathbf{10}}} F_Y = 0\,, \qquad \sum_{\overline{\mathbf{5}}\text{ matter curves, }a} \int_{\Sigma^{(a)}_{\overline{\mathbf{5}}}} F_Y = 0\,. \tag{4.8}$$

□

4.2. Survey. In the last subsection we showed that anomaly cancellation has radical consequences for the types of models that are consistent. To summarize: the setup we consider is

- F-theory realization of supersymmetric $SU(5)$ GUT
- Matter fields and Higgs fields arise from codimension 2 singularities and all MSSM Yukawa couplings from codimension 3
- The GUT group is broken by hypercharge flux F_Y
- Higgs triplets are lifted by hypercharge flux F_Y
- Require an additional $U(1)_{PQ}$ symmetry, which forbids the dimension 5 proton decay operator and the μ-term, whose charges satisfy $q_{H_u}+q_{H_d} \neq 0$

Then the anomaly relation (4.4) implies that such models will always contain multiplets that do not fall into complete GUT group representations. This result is generic in the sense that it does not rely on the specifics of the geometric details, for instance the elliptic CY fourfold base.

Since these non-GUT exotics are vector-like pairs in terms of the MSSM gauge group, a natural way to give them a mass (and therefore make them at least not immediately inconsistent with observation) is to couple them to a singlet field X, that also carries a non-trivial $U(1)_{PQ}$ charge

$$W_{\text{ex}} = \lambda X f_{\text{ex}} \bar{f}_{\text{ex}} \,. \tag{4.9}$$

We can characterize the complete set of possible exotics in terms of four integers K, L, M, N

(4.10)

$SU(5)$ origin	Exotic Multiplet	Degeneracy
$\mathbf{10} \oplus \overline{\mathbf{10}}$	$(\mathbf{1},\mathbf{1})_{+1} \oplus (\mathbf{1},\mathbf{1})_{-1}$	$M+N$
	$(\mathbf{3},\mathbf{2})_{+1/6} \oplus (\overline{\mathbf{3}},\mathbf{2})_{-1/6}$	M
	$(\overline{\mathbf{3}},\mathbf{1})_{-2/3} \oplus (\mathbf{3},\mathbf{1})_{+2/3}$	$M-N$
$\overline{\mathbf{5}} \oplus \mathbf{5}$	$(\overline{\mathbf{3}},\mathbf{1})_{+1/3} \oplus (\mathbf{3},\mathbf{1})_{-1/3}$	K
	$(\mathbf{1},\mathbf{2})_{-1/2} \oplus (\mathbf{1},\mathbf{2})_{+1/2}$	$K-L$

with

$$M \geq |N| \qquad K \geq 0 \qquad K-L \geq 0 \,. \tag{4.11}$$

In particular: the PQ-charges of the Higgs multiplets are:

$$q_{H_u} + q_{H_d} = q_X \Delta \,, \qquad \Delta \equiv N - L \,.$$

4.3. Gauge Mediation. A natural purpose for these exotics, given their vector-like nature with respect to the MSSM gauge group, is to use them as messenger fields in a gauge mediated supersymmetry breaking setup.

In a nutshell: a standard paradigm to break supersymmetry (which is observationally necessary, given the non-observation of the superpartners of the Standard Model particles at the same energies as the Standard Model particles), is to introduce a sector in the theory that breaks supersymmetry, but is not necessarily directly coupled to the visible sector, i.e. MSSM. The particulars of how the supersymmetry breaking is mediated (communicated) to the visible sector characterize the phenomenological properties of a given model. Two common ways to mediated supersymmetry breaking is either by gravitational interactions, or by gauge interactions.

In gauge mediated supersymmetry breaking, new fields are introduced, which are vector-like, and are charged under the MSSM gauge group. E.g. common choices are $f = \mathbf{5}$ and $\bar{f} = \bar{\mathbf{5}}$ pairs, or $\mathbf{10}$s. Unification requires these to be, at least in effective field theory models, complete GUT multiplets, in order not to disrupt the gauge coupling unification. The messenger fields couple to the supersymmetry breaking sector, which is usually modeled by a so-called SUSY spurion field X, which develops an F-term that breaks supersymmetry. Schematically the superpotential takes the form

$$W \supset F_X X + X f \bar{f} + W_{MSSM}\,, \tag{4.12}$$

where W_{MSSM} contains all the standard MSSM superpotential couplings, as discussed before. When X picks up a scalar vacuum expectation value, this effectively gives mass to the messenger fields

$$\langle X \rangle \sim M_{\text{Mess}}. \tag{4.13}$$

The supersymmetry breaking is then communicated to the visible sector by messenger field loops. In particular, general expressions for the masses of the superpartners of the Standard Model fermions and gauge bosons can be obtained: the one-loop gaugino masses are

$$M_{1/2}(M_{\text{Mess}}) = \delta b_i \left(\frac{\alpha_i(M_{\text{Mess}})}{4\pi} \right) \left(\frac{F_X}{M_{\text{Mess}}} \right), \tag{4.14}$$

and the squared two-loop scalar masses are

$$M_{\tilde{Q}}^2(M_{\text{Mess}}) = \sum_{\text{relevant } i} \frac{c_i\, \delta b_i\, \alpha_i(M_{\text{Mess}})^2}{8\pi^2} \left| \frac{F_X}{M_{\text{Mess}}} \right|^2 . \tag{4.15}$$

Here α_i are the gauge couplings at the given energy scale, which is M_{Mess}, c_i the quadratic Casimirs of the gauge group factor i and δb_i are the beta-function coefficients that are induced by the messenger fields. For K pairs of standard messengers $f = \mathbf{5}$ and $\bar{f} = \bar{\mathbf{5}}$ these are e.g. $\delta b_i = K$, for $i = 1, 2, 3$ and are universal.

4.4. Gauge Mediation with F-theory Exotic Messengers. We showed that in an $SU(5)$ GUT with an additional $U(1)_{PQ}$ symmetry, realized in F-theory in terms of matter and Higgses arising on curves inside a surface S_{GUT} and hypercharge flux to break the GUT group to the MSSM gauge group, there are always non-GUT exotics that are vectorlike with respect to the MSSM, but charged differently under the $U(1)_{PQ}$ symmetry. It is very suggestive therefore to embed this into a gauge mediated supersymmetry breaking setup, where the exotics play the role of messenger fields. The singlet SUSY spurion X then has to be charged under the $U(1)_{PQ}$ in order to couple to the exotics as in (4.9). In fact, the Higgs fields can couple via a Kaehler potential coupling to the singlet

$$\int d^4\theta \frac{X^\dagger H_u H_d}{M_{\text{Mess}}} \tag{4.16}$$

which generates a μ term dynamically from the F-term of X

$$\mu \sim \frac{F_X}{M_{\text{Mess}}}\,. \tag{4.17}$$

The gaugino and scalar mass equations are identical to (4.14) and (4.15), with the only change due in the beta-function shifts, which are of course dependent on the particular representation content of the messenger fields (aka exotics). In terms of the integers K, L, M, N these now generalize to

$$\begin{aligned}\delta b_1 &= 3M - \frac{2}{5}N + K - \frac{3}{5}L \\ \delta b_2 &= 3M + K - L \\ \delta b_3 &= 3M - N + K\,.\end{aligned} \tag{4.18}$$

One immediate potential problem is that these expressions cease to be universal, unlike in the case of GUT multiplet messengers. In a detailed analysis of all contributions to the renormalization of the gauge couplings (including low-scale and high-scale thresholds, as well as F-theory specific corrections, that are absent in a generic effective field theory setup), we can show that there is a range of parameters (i.e. choices for the stringy corrections, scales M_{Mess} and F_X) such that unification remains intact [**DMS**].

4.5. LHC Phenomenology. The model proposed in the last section is not something one would consider without being induced to do so, in this case, by arguments of UV completion within the framework of F-theory GUTs with additional $U(1)_{PQ}$ symmetry. In this framework we have argued that this is the generic type of model that arises – irrespective of the concrete Calabi-Yau. So, once put into this position, one can ask whether there are distinct features of it, that could be a signature setting this class of models apart from standard gauge mediation or GUT models.

Indeed, from the gaugino mass expression it is clear that the usual gaugino mass ratios, which hold for all models with GUT multiplet messengers (as well as for minimal supergravity mediated supersymmetry breaking models)

$$M_1 : M_2 : M_3 \simeq 1 : 2 : 6 = \alpha_1 : \alpha_2 : \alpha_3\,, \tag{4.19}$$

are modified to the ratios

$$M_1 : M_2 : M_3 \simeq 1 : 2\left(\frac{3M + K - L}{3M - 2N/5 + K - 3L/5}\right) : 6\left(\frac{3M + K - N}{3M + K - 2N/5 - 3L/5}\right)\,, \tag{4.20}$$

that explicitly depend on the non-GUT messenger content of the model.

In [**DMS**] we considered various choices for (K, L, M, N), which yield interesting non-universalities in the gaugino spectrum, and chose a small set of benchmark models, which we then analyzed in more detail.

(4.21)

L	Benchmark Model	$(K,\ L,\ M,\ N)$	M_{Mess}	$\Lambda = F_X/M_{\text{Mess}}$	$m_{\tilde{g}}$
-2	$\text{NonUniv}_{L=-2}$	$(0,-2,1,0)$	10^{14}GeV	8×10^4GeV	1686GeV
-1	$\text{NonUniv}_{L=-1}^{\text{SmallFlux}}$	$(2,-1,0,0)$	10^{14}GeV	9×10^4GeV	1307GeV
$+2$	$\text{NonUniv}_{L=2}$	$(3,2,0,0)$	10^{14}GeV	7×10^4GeV	1489GeV
-1	$\text{NonUniv}_{L=-1}^{\text{LargeFlux}}$	$(6,-1,7,0)$	10^{14}GeV	1.3×10^4GeV	2337GeV
0	mGMSB5	$(3,0,0,0)$	10^{10}GeV	7×10^4GeV	1503GeV

The most interesting aspect to understand in this class of theories is, whether these could be distinguished from very similar gauge mediation models with GUT messengers, such as the last in the previous table denoted by mGMSB5. In the first

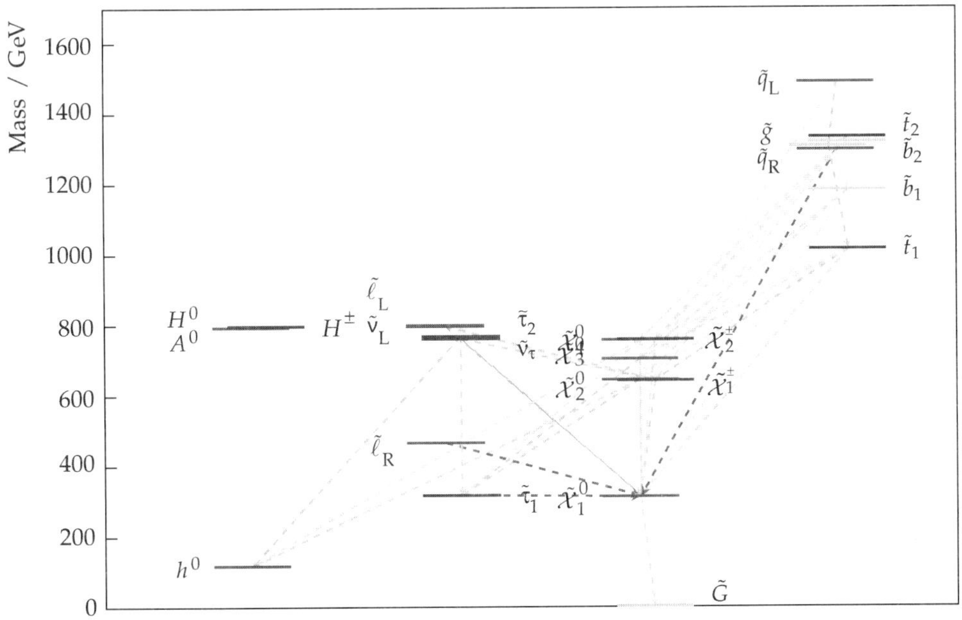

FIGURE 1. This figure shows the spectrum of the benchmark model $\text{NonUniv}_{L=-1}^{\text{SmallFlux}}$ with $M_{\text{Mess}} = 10^{14}$ GeV and $\Lambda = 9 \times 10^4$ GeV.

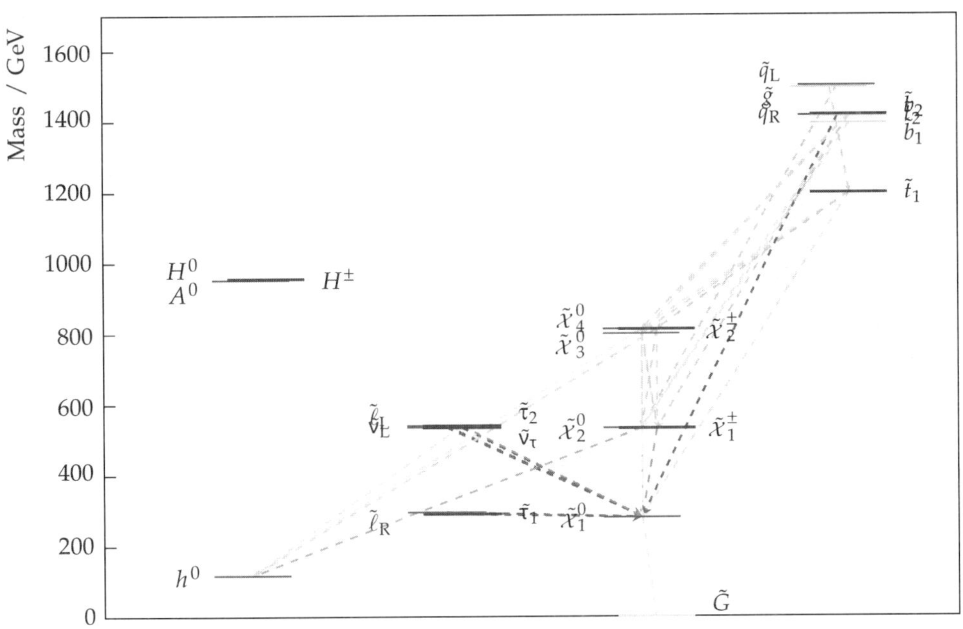

FIGURE 2. This figure shows the spectrum of an mGMSB point with similar phenomenology to $\text{NonUniv}_{L=-1}^{\text{SmallFlux}}$.

part of figure 1 the spectrum of one benchmark is shown. A very similar looking spectrum for a model with GUT messengers is shown in the second part of table 1. The shifts in the beta-functions for this example are $(\delta b_1, \delta b_2, \delta b_3) = (2.6, 3, 2)$. This choice interpolates between minimal gauge mediation with two and three sets of $\mathbf{5} \oplus \bar{\mathbf{5}}$ messengers. It can be fit in the weakly interacting sector with the three sets of GUT messengers, whereas the strongly interacting part of the spectrum is more like in the case of two sets of GUT messengers.

Model	m_{ll}	m_{llq}^{edge}	m_{llq}^{thr}	m_{lq}^{high}	m_{lq}^{low}
GUT Gauge Mediation	138.7	1126	306	1102	396
$\text{NonUniv}_{L=-1}^{\text{SmallFlux}}$	330.2	1011	550	856	688

This behavior can be made quantitative by studying kinetic edges corresponding to the decay channel

$$\tilde{q}_L \to \chi_2^0 q \to \tilde{l}_R^{\pm} l^{\mp} q \to \chi_1^0 l^{\pm} l^{\mp} q \tag{4.22}$$

The most useful quantity for this comparison is

$$m_{ll} = (m^2_{\tilde{\chi}^1_0} - m^2_{\tilde{\ell}_R})(m^2_{\tilde{\ell}_R} - m^2_{\tilde{\chi}^2_0})/m^2_{\tilde{\ell}_R} \tag{4.23}$$

which is the kinetic invariant for the dilepton channel $\chi_2^0 \to \tilde{l}^{\pm} l^{\mp} \to \chi_1^0 l^{\pm} l^{\mp}$. The values of this kinetic invariant for the comparison of the benchmark model $\text{NonUniv}_{L=-1}^{\text{SmallFlux}}$ versus a gauge mediation model with standard GUT messengers, as listed in the table below, seem to be large enough to distinguish these models at the LHC.

References

[BHV] C. Beasley, J. J. Heckman and C. Vafa, "GUTs and Exceptional Branes in F-theory - I," JHEP **0901**, 058 (2009) [arXiv:0802.3391 [hep-th]] • "GUTs and Exceptional Branes in F-theory - II: Experimental Predictions," JHEP **0901**, 059 (2009) [arXiv:0806.0102 [hep-th]].

[BIKMSV] M. Bershadsky, K. A. Intriligator, S. Kachru, D. R. Morrison, V. Sadov and C. Vafa, "Geometric singularities and enhanced gauge symmetries," Nucl. Phys. B **481**, 215 (1996) [hep-th/9605200].

[BGJW] R. Blumenhagen, T. W. Grimm, B. Jurke and T. Weigand, Nucl. Phys. B **829**, 325 (2010) [arXiv:0908.1784 [hep-th]].

[D] F. Denef, "Les Houches Lectures on Constructing String Vacua," arXiv:0803.1194 [hep-th].

[DMSS] M. J. Dolan, J. Marsano, N. Saulina and S. Schafer-Nameki, "F-theory GUTs with U(1) Symmetries: Generalities and Survey," Phys. Rev. D **84**, 066008 (2011) [arXiv:1102.0290 [hep-th]].

[DMS] M. J. Dolan, J. Marsano and S. Schafer-Nameki, "Unification and Phenomenology of F-Theory GUTs with $U(1)_{PQ}$," JHEP **1112**, 032 (2011) [arXiv:1109.4958 [hep-ph]].

[DW] R. Donagi and M. Wijnholt, "Model Building with F-Theory," arXiv:0802.2969 [hep-th] • "Breaking GUT Groups in F-Theory," arXiv:0808.2223 [hep-th] • "Higgs Bundles and UV Completion in F-Theory," arXiv:0904.1218 [hep-th].

[DP] E. Dudas and E. Palti, "On hypercharge flux and exotics in F-theory GUTs," JHEP **1009**, 013 (2010) [arXiv:1007.1297 [hep-ph]].

[E-StringMath] M. Esole, String Math 2011 Proceedings.

[EY] M. Esole and S. -T. Yau, "Small resolutions of SU(5)-models in F-theory," arXiv:1107.0733 [hep-th].

[HTTWY] H. Hayashi, R. Tatar, Y. Toda, T. Watari and M. Yamazaki, "New Aspects of Heterotic–F Theory Duality," Nucl. Phys. B **806** (2009) 224 [arXiv:0805.1057 [hep-th]].

[KMSS] S. Katz, D. R. Morrison, S. Schafer-Nameki and J. Sully, "Tate's algorithm and F-theory," JHEP **1108**, 094 (2011) [arXiv:1106.3854 [hep-th]].

[LS] C. Lawrie and S. Schafer-Nameki, to appear.

[MSSN1] J. Marsano, N. Saulina and S. Schafer-Nameki, JHEP **1004**, 095 (2010) [arXiv:0912.0272 [hep-th]].

[MSSN2] J. Marsano, N. Saulina and S. Schafer-Nameki, "Monodromies, Fluxes, and Compact Three-Generation F-theory GUTs," JHEP **0908**, 046 (2009) [arXiv:0906.4672 [hep-th]].

[MSSN3] J. Marsano, N. Saulina and S. Schafer-Nameki, "A Note on G-Fluxes for F-theory Model Building," JHEP **1011**, 088 (2010) [arXiv:1006.0483 [hep-th]] • "On G-flux, M5 instantons, and U(1)s in F-theory," arXiv:1107.1718 [hep-th]

[M] J. Marsano, Phys. Rev. Lett. **106**, 081601 (2011) [arXiv:1011.2212 [hep-th]].

[MSN] J. Marsano and S. Schafer-Nameki, "Yukawas, G-flux, and Spectral Covers from Resolved Calabi-Yau's," JHEP **1111**, 098 (2011) [arXiv:1108.1794 [hep-th]].

[W] T. Weigand, "Lectures on F-theory compactifications and model building," Class. Quant. Grav. **27**, 214004 (2010) [arXiv:1009.3497 [hep-th]].

[W-StringMath] M. Wijnholt, String Math 2011 Proceedings.

DEPARTMENT OF MATHEMATICS, KING'S COLLEGE, THE STRAND, WC2R 2LS, LONDON, ENGLAND

E-mail address: `ss299 theory caltech edu, gmail: sakura.schafer.nameki`

Proceedings of Symposia in Pure Mathematics
Volume **85**, 2012

Higgs Bundles and String Phenomenology

Martijn Wijnholt

ABSTRACT. String phenomenology is the branch of string theory concerned with making contact with particle physics. The original models involved compactifying ten-dimensional supersymmetric Yang-Mills theory on an internal Calabi-Yau three-fold. In recent years, this picture has been extended to compactifications of supersymmetric Yang-Mills theory in seven, eight or nine dimensions. We review some of this progress and explain the fundamental role played by Higgs bundles in this story.

CONTENTS

1. String Phenomenology

1.1. Particle physics and Grand Unification.

To date the most successful description of particle physics is the Standard Model. It is a gauge theory with gauge group $SU(3)_c \times SU(2)_w \times U(1)_Y$ and with three fermionic generations of matter:

$$\begin{array}{lll} Q = (\mathbf{3},\mathbf{2})_{1/6} & u^c = (\bar{\mathbf{3}},\mathbf{1})_{-2/3} & d^c = (\bar{\mathbf{3}},\mathbf{1})_{1/3} \\ L = (\mathbf{1},\mathbf{2})_{-1/2} & e^c = (\mathbf{1},\mathbf{1})_1 & \end{array} \tag{1.1}$$

2010 *Mathematics Subject Classification.* Primary: 81T30, Secondary: 14D21, 14J81, 81V22.
Key words and phrases. Differential geometry, algebraic geometry, string theory, unification.

In addition it includes a bosonic Higgs field $h = (\mathbf{1}, \mathbf{2})_{-1/2}$. Three of its components are swallowed by gauge bosons through the Higgs mechanism, and the remaining scalar (called the Higgs boson) may soon be discovered at the LHC.

At first sight, these groups and matter representations may look rather random. However it has been known for some time that these representations fit together very nicely if we embed the gauge group into a larger semi-simple group [**1**]. The simplest choice is

$$SU(3) \times SU(2) \times U(1) \;\subset\; SU(5) \tag{1.2}$$

In this case, the matter representations are combined into

$$\mathbf{10} \;=\; (Q, u^c, e^c), \qquad \bar{\mathbf{5}} \;=\; (L, d^c) \tag{1.3}$$

where $\mathbf{10}$ denotes the two-index anti-symmetric representation of $SU(5)$, and $\bar{\mathbf{5}}$ denotes the anti-fundamental representation of $SU(5)$.

This 'unification' of the representations can be continued. The next step is $SU(5) \subset SO(10)$, which unifies the matter representations as well as the neutrino into a single spinor representation of $SO(10)$:

$$\mathbf{16} \;=\; \mathbf{10} + \bar{\mathbf{5}} + \mathbf{1} \tag{1.4}$$

where the singlet corresponds to the right-handed neutrino. One may continue with $SO(10) \subset E_6$, which unifies matter with the Higgs field.

This remarkable bit of group theory could be a simple mathematical curiosity, explained to some extent by anomaly cancellation. However, there is important dynamical evidence that nature makes use of this idea. One of the most compelling pieces of evidence is obtained by adding supersymmetric partners around a TeV, and extrapolating the one-loop running of the gauge couplings upwards. One arrives at the following remarkable picture of the couplings as a function of $\log(E)$ [**2, 3**]:

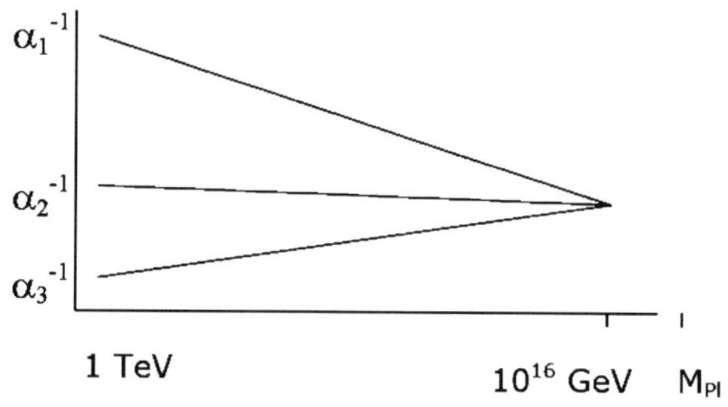

FIGURE 1. Important scales in particle physics. The weak scale $E \sim 1$ TeV is associated with electro-weak symmetry breaking and will be explored by the LHC. Extrapolating the gauge couplings of the strong force (α_3), weak force (α_2), and hypercharge (α_1), we see that they meet at an energy scale $E \sim 10^{16}$ GeV which is called the GUT scale. At the Planck scale $E \sim 10^{19}$ GeV gravity becomes strong and we need a full theory of quantum gravity.

The unification of couplings gives strong evidence for the idea that the three forces of the Standard Model merge into a single force at high energy scales [**1**]. Such gauge theories go under the name of Grand Unified Gauge Theories or simply GUTs. The convergence of couplings isn't just an accident of one number coming out right. The fact that the couplings meet just below the Planck scale is also very suggestive. And the picture is fairly robust and in principle independent of supersymmetry; for example it is not materially affected by adding complete $SU(5)$ multiplets to the theory at intermediate energy scales. Without supersymmetry though, the couplings don't meet with the same impressive accuracy.

Furthermore, these models not only provide an explanatory framework for some known patterns, but they also yield a new and very characteristic prediction. As was emphasized by Georgi and Glashow, GUT models collect particles with different baryon number in the same multiplet and therefore lead to proton decay, so looking for such decay is one of the best ways to test the hypothesis. In fact, proton decay is one of the few ways we have to directly probe energy scales as large as 10^{16} GeV. It is one of the outstanding questions of particle physics and remains on the experimental agenda. For constraints on Grand Unified models from proton decay, both with and without supersymmetry, see eg. [**4, 5, 6**].

1.2. The heterotic string.

String theory gives a further generalization of this picture by also adding quantum gravity into the mix. We would like to understand how to recover Grand Unification from string theory. This was first achieved by [**7**] in '85.

We start with the ten-dimensional $E_8 \times E_8$ heterotic string. Of course its symmetry group is much too big. We want to break this to four-dimensional $N = 1$ super-Poincaré invariance and $SU(5)$ gauge group. To break the Poincaré symmetry, we take the space-time of the form

$$\mathbf{R}^{1,3} \times X_6 \tag{1.5}$$

It is possible to make a more general Ansatz with a warp factor, but we will not consider it as it is still poorly understood. Preserving $4d$ $N = 1$ super-Poincaré then implies that X_6 must be a Calabi-Yau three-fold. To break the gauge symmetry we turn on a non-trivial gauge field along X_6. Four-dimensional supersymmetry then implies that the $E_8 \times E_8$ connection must satisfy the Hermitian Yang-Mills equations:

$$F^{0,2} = 0, \quad g^{i\bar{j}} F_{i\bar{j}} = 0 \tag{1.6}$$

Given a solution (X_6, V) to these equations, we may then calculate the effective four-dimensional gauge theory by computing certain Dolbeault cohomology groups. The Standard Model gauge group and the charged matter fields (the 'visible sector') all come from Kaluza-Klein reduction of the $10d$ gauge theory, using just one of the two E_8 groups. They are given by the Dolbeault cohomology groups

$$H^p(X_6, V_{E_8}) \tag{1.7}$$

where V_{E_8} is the adjoint bundle for one of the E_8 gauge groups. The second E_8 is said to yield a hidden sector. For $p = 0$ this cohomology group counts generators of the $4d$ gauge group, and for $p = 1$ this counts $4d$ matter fields. So we want to find pairs (X_6, V_{E_8}) such that

$$H^0(X_6, V_{E_8}) = su(3) \oplus su(2) \oplus u(1) \tag{1.8}$$

$$H^1(X_6, V_{E_8}) = \sum_{i=1}^{3}(\mathbf{10}_i \oplus \bar{\mathbf{5}}_i) \oplus \text{Higgs} \tag{1.9}$$

We used $SU(5)$ notation for the quarks and leptons to emphasize that they come in complete $SU(5)$ multiplets, even though the GUT group is broken. Furthermore, there is a Yoneda product

$$H^1(X_6, V_{E_8}) \times H^1(X_6, V_{E_8}) \times H^1(X_6, V_{E_8}) \rightarrow \mathbf{C} \tag{1.10}$$

which computes the $4d$ Yukawa couplings. Here we used the Calabi-Yau condition and the obvious anti-symmetric three-index tensor for E_8 to get a number. The Kähler potential (kinetic terms) may not be computed exactly, however there are numerical techniques that allow one to approximate it [**8**].

Note that even ignoring gravity, string theory has added another level of unification to the story of Grand Unification. In conventional $4d$ GUTs, matter and gauge fields still appeared as separate entities. But by adding extra dimensions we have managed to unify gauge and matter fields into a single structure: the unique $10d$ supersymmetric gauge theory with E_8 gauge group.

1.3. The landscape.

A solution (X, V) gives only a first (tree level) approximation to the $4d$ physics. Simple solutions tend to have a large number of moduli, and there are no-go theorems that say that suitable classical solutions without moduli do not exist. The presence of light moduli contradicts known experimental facts, such as the equivalence principle or Big-Bang nucleosynthesis. Furthermore the correct theory of particle physics below a TeV (i.e. the Standard Model) is not supersymmetric. Finding a stable weakly coupled solution is a complicated dynamical problem, and unfortunately many predictions depend on it.

It is by now appreciated that many of the moduli of (X, V) (in particular complex structure moduli, vector bundle moduli and even some Kähler moduli) can be lifted at tree level, leaving only very few moduli S_i to be stabilized by quantum effects. However the number of such classical solutions (X, V) which we would like to use as a first approximation to a true solution, keeping S_i fixed, appears to be very large. The existence of an exponentially large number of solutions is well-advertized in the context of F-theory flux vacua [**9, 10**]. It is less well-known that there is an equally large heterotic landscape. This was overlooked for many years, but should have been expected on the basis of heterotic/F-theory duality. In particular, the construction of heterotic bundles on elliptic Calabi-Yau manifolds was related in [**11**] to a Noether-Lefschetz problem, which leads to an exponentially large number of solutions, in the same manner as the F-theory landscape.

Let us consider the construction of the MSSM in the heterotic string, on a fixed elliptically fibered Calabi-Yau three-fold X. Apart from the Calabi-Yau, the main ingredient is a stable rank five bundle V with $c_1(V) = 0$ and $\frac{1}{2}c_3(V) = 3$. There is a bound on $c_2(V)$ from tadpole cancellation, which will show up as a cut-off Λ below.

dim	*stringy realization*
10d	$E_8 \times E_8$ heterotic string
9d	strongly coupled type I'
8d	F-theory on ALE
7d	M-theory on ALE

Table 1: *Branes with exceptional gauge symmetry in string theory.*

Using the results of [**11**] we can make the following rough, Bousso/Polchinski-like estimate for the number of isolated solutions:

$$\#\,\text{solutions} \;\simeq\; \frac{1}{(r/2)!}\,(2\pi\Lambda)^{\mathrm{r}/2} \;\simeq\; \left(\frac{4\pi e\Lambda}{r}\right)^{r/2} \;\simeq\; 10^{1000} \tag{1.11}$$

Here Λ is a cut-off due to tadpole cancellation, which we took conservatively of order $\Lambda \sim r/24$. The number $r \sim 10^3$ to 10^5 comes from the rank of the H_2 homology lattice of a degree five spectral cover over $\mathbf{CP}^2$. This version of the landscape arises solely from compactifying the $10d$ E_8 gauge theory, i.e. from the choice of solution to the Hermitian-Yang-Mills equation on a rank five bundle, holding everything else (in particular the background X) fixed. As such, it directly affects the parameters in the visible sector.

These numbers are so large that is simply pointless to try to enumerate the set of MSSMs, even on a fixed X. However, generic solutions are expected to have qualitatively similar particle physics. This lends support to the traditional idea of naturalness in model building: unless there is extra well-motivated structure, dimensionless parameters should be assumed to be of order unity.

1.4. Extension to seven, eight and nine dimensions.

In '85 our picture of string theory was rather limited, and the only known realization of E_8 gauge theory was in the context of the heterotic string. However, during the duality revolution in the '90s we learned that lower dimensional E_8 gauge theories are also realized in string theory. Thus our excuse for only constructing GUT models from ten-dimensional gauge theory has disappeared. These constructions were carried out in a series of papers in the last three years (some yet to appear).

The basic summary for the realization of E_8 gauge theories in string theory is given in table 1. There are no supersymmetric gauge theories above ten dimensions, which is why the table ends at the upper end. It is possible to realize E_8 gauge theory below 7 dimensions, however then there is not enough room to get chiral matter in $4d$. This is why the table ends at the lower end.

The entries for $d = 7, 8, 9$ look slightly mysterious: they do not correspond to weakly coupled string theories and the way the non-abelian gauge symmetries arise is not completely obvious, as it involves non-perturbative physics. Obviously we cannot do justice to it here, but let us at least give a very quick summary.

The starting point is M-theory, the conjectural non-perturbative completion of eleven-dimensional supergravity. M-theory on a smooth space-time does not give rise to non-abelian gauge symmetry, but M-theory on a singular ALE space of type ADE gives rise to seven-dimensional super-Yang-Mills theory localized at the singularity, with the corresponding ADE gauge group [**12**]. An important consistency check is that when we resolve the singularity, quantized $M2$-membranes wrapping the vanishing cycles reproduce the off-diagonal components of the Yang-Mills multiplet (i.e. those not proportional to the Cartan generators). M-theory on an elliptically fibered space-time with section can be mapped to ten-dimensional type IIb supergravity on the section times a circle. In the limit that the elliptic fiber shrinks to zero, the circle decompactifies. The modular parameter of the elliptic fiber is identified with a varying axio-dilaton of the IIb supergravity, $\tau = a + ie^{-\phi}$, and the mechanism of non-abelian gauge enhancement is similar to M-theory. This is called F-theory [**13**]. M-theory on an S^1/Z_2 orbifold gives rise to $10d$ E_8 gauge theory on each of the two walls. This is the Hořava-Witten picture [**14**]. By shrinking the interval, we recover the weakly coupled heterotic string. By instead compactifying on S^1 and shrinking it, we get the type I' theory [**15**]. In each of these contexts, when the non-abelian gauge symmetry is of type $SU(n)$, $SO(n)$ or $USp(n)$, we can often extrapolate to a weakly coupled D-brane configuration in perturbative string theory, but for the exceptional cases this is not possible. These issues were only understood after the non-perturbative behaviour of string theory became clearer in the '90s.

In this review we will be firmly focused on the Yang-Mills theory itself, without much consideration of its origins. Part of the reason for this is that if we want to consider exceptional gauge groups in higher dimensions, then the string point of view involves singular geometries and strongly coupled physics, whereas the higher dimensional Yang-Mills theory is weakly coupled in the derivative expansion. We will briefly explain how to connect with string theory in section 2.4. The upshot will be that we can reconstruct the local geometry from the Yang-Mills theory. The type I' story requires a separate discussion, which will not be attempted here.

Unification with eight-dimensional Yang-Mills theory was initiated in [**16, 17, 18**] and is currently the most developed and best understood of the new cases. The main reason for this is that in even dimensions one can use the powerful techniques of complex and algebraic geometry. The seven-dimensional story was initiated in [**19**] and the nine-dimensional story will appear in [**20**].

The strategy that we will use is similar to the heterotic string. The main new idea is that instead of Hermitian-Yang-Mills bundles, we have to consider their close cousins: Higgs bundles. This consists of a bundle E together with a map

$$\phi : E \to E \otimes N \tag{1.12}$$

where N is a suitable twisting bundle. The connection A_μ on E and the 'Higgs field' ϕ have to satisfy certain versions of Hitchin's equations, which we discuss below. (The terminology is standard and the field ϕ has no direct relation to the Higgs boson h of the Standard Model).

2. Higgs bundles

2.1. Dimensional reduction.

The maximally supersymmetric Yang-Mills theory in p dimensions is uniquely determined by the choice of gauge group. Its Lagrangian can be obtained by dimensional reduction from the $10d$ supersymmetric Yang-Mills theory. Therefore we would expect that the BPS conditions can be obtained from dimensional reduction of the $6d$ Hermitian Yang-Mills equations (1.6).

There is a well-known analogous story involving reduction of the $4d$ Hermitian Yang-Mills equations, aka the ASD equations. Reducing to three dimensions yields the Bogomolnyi equations:

$$F = *D_A\phi \tag{2.1}$$

The reduction to two dimensions is known as Hitchin's equations:

$$\bar{\partial}_A\phi = 0, \quad \Lambda F + [\phi, \phi^*] = 0 \tag{2.2}$$

where Λ denotes contraction with the Kähler form. These equations turn out to have many applications, and several variants have been considered. We will be interested in some of its higher dimensional variants.

Indeed, let us now consider the dimensional reduction of the $6d$ Hermitian Yang-Mills equations (1.6), relevant for compactifications of $10d$ Yang-Mills down to four dimensions. For $8d$ Yang-Mills on $\mathbf{R}^{1,3} \times S_4$ we get $F^{0,2} = 0$, as well as equations (2.2) above on S_4. For $7d$ Yang-Mills on $\mathbf{R}^{1,3} \times Q_3$ we get

$$F - [\phi, \phi] = 0, \quad D_A\phi = 0, \quad D_A^\dagger\phi = 0 \tag{2.3}$$

This is the real version of Hitchin's equations, defined on a real three-manifold Q_3. The gauge and Higgs field can be assembled into a complex gauge field $A + i\phi$, and the first two equations above assert that this is a complex flat connection on Q_3. For $9d$ Yang-Mills we compactify on $\mathbf{R}^{1,3} \times X_5$, where X_5 is a five-manifold given by an S^1-fibration over a Kähler surface. Denote complex coordinates on the base by z and along the S^1 by y. Then we get

$$F_\perp^{0,2} = 0, \quad F_{\bar{z}y} = iD_{\bar{z}}\phi, \quad \Lambda F_\perp + iD_y\phi = 0 \tag{2.4}$$

where $\perp$ denotes the components orthogonal to the S^1. This is a higher dimensional generalization of the Bogomolnyi equations.

2.2. Compactification of eight-dimensional SYM.

We can now summarize the main results of the Kaluza-Klein reduction of the higher dimensional Yang-Mills theory to four dimensions. We take the $8d$ supersymmetric Yang-Mills theory as our main example [**16, 17, 11, 21**]. The Higgs bundle will be defined on a compact Kähler surface S. As we have seen, the bosonic fields are given by a connection A_μ on a bundle $ad(E)$ on S, and a complex Higgs field Φ valued in $ad(E) \otimes K_S$. The twisting by the canonical bundle is required to preserve the $4d$ super-Poincaré invariance. The equations are given by

$$F^{0,2} = 0, \quad \bar{\partial}_A\Phi = 0, \quad \Lambda F + [\Phi, \Phi^*] = 0 \tag{2.5}$$

In practice, we actually need meromorphic Higgs bundles, or in recent physics language, we need to insert certain surface operators along a curve Σ_∞ in S. The simplest such defects lead to parabolic Higgs bundles [**22**]. They may be thought of as arising from integrating out hypermultiplets living on Σ_∞, which become dynamical when we embed the gauge theory into a compact model. We will not discuss these defects explicitly here.

Given a solution to these equations, we would like to know the effective gauge theory in four dimensions obtained from the Kaluza-Klein reduction of the eight-dimensional gauge theory on $\mathbf{R}^{1,3} \times S$. We define a two-term complex $\mathcal{E}^\bullet$:

$$\mathcal{E}^\bullet \;=\; ad(E) \stackrel{\Phi}{\to} ad(E) \otimes K_S \tag{2.6}$$

The $4d$ gauge theory is derived by computing the hypercohomology groups

$$\mathbb{H}^p(S, \mathcal{E}^\bullet) \tag{2.7}$$

For $p = 0$ this counts generators of the $4d$ gauge group, and for $p = 1$ it counts $4d$ matter fields. We can compute the Yukawa couplings from a Yoneda product

$$\mathbb{H}^1(S, \mathcal{E}^\bullet) \times \mathbb{H}^1(S, \mathcal{E}^\bullet) \;\to\; \mathbb{H}^2(S, \mathcal{E}^\bullet) \tag{2.8}$$

and conjecturally we can even numerically approximate the Hermitian-Einstein metric and therefore the kinetic terms [**22**]. Of course this parallels the analogous statements for the heterotic string we discussed earlier.

These statements perhaps seem somewhat abstract. It is often possible to give more intuitive pictures for the wave functions of the massless modes. The main picture in this regard is that a generic zero of Φ can be interpreted as a vortex string on S, and as is well known one tends to get charged zero modes localized on such a vortex. Thus, while the wave functions of the four-dimensional gauge fields are spread out over S, the wave functions of charged matter such as the $\mathbf{10}$ or $\bar{\mathbf{5}}$ of an $SU(5)$ GUT tends to be localized on some Riemann surface in S, see figure 2. For $7d$ super-Yang-Mills theory compactified on Q_3 we get a similar picture, with gauge fields spread over Q_3 but chiral matter in the $\mathbf{10}$ or $\bar{\mathbf{5}}$ generically localized at points on Q_3. Such localization leads to interesting possibilities for phenomenology. But for non-generic configurations the intuition can be misleading, and the advantage of the above formulation is that it is precise and general.

2.3. Higgs/spectral correspondence.

Now that we have the general statements, we need a good method for constructing solutions and computing the hypercohomology groups. Hitchin's equations are highly non-linear, so trying to solve them in closed form is bound to fail. Instead we will make use of two standard techniques. The first is the technique of splitting up the equations in "F-terms" and "D-terms", i.e.:

- a pair of complex equations: $F^{0,2} = \bar{\partial}_A \Phi = 0$
- a moment map: $\Lambda F + [\Phi, \Phi^*] = 0$

The strategy is then to temporarily ignore the D-terms, and focus on the F-terms, which we can solve exactly up to *complexified* gauge transformations. This amounts to focussing on the holomorphic structure and ignoring the Hermitian metric. Once we have done that, we can go back and try to apply some powerful results in geometric invariant theory regarding D-flatness.

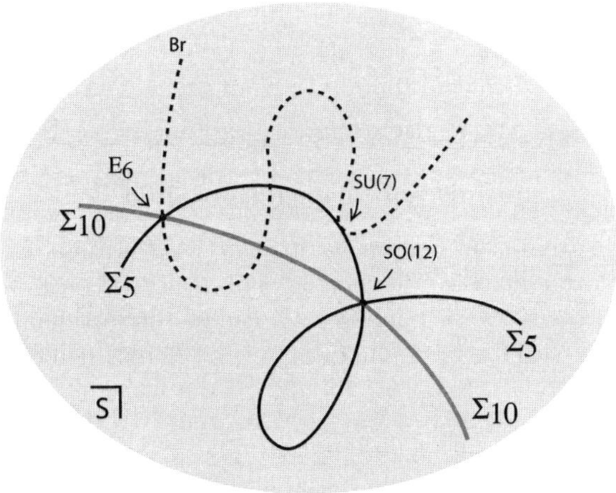

FIGURE 2. *Schematic picture of the matter curves, branch locus and their intersections in an* $SU(5)$ *GUT model, for generic values of the complex structure moduli.*

To solve the F-term equations, we use the second essential tool, namely the Higgs/spectral correspondence. Essentially it turns the problem of constructing solutions to the F-terms into an 'abelianized' version which is much simpler to solve. To avoid the somewhat intricate group theory of E_8, we will take E to be a rank n vector bundle transforming in the fundamental representation of $U(n)$.

The Higgs/spectral correspondence states that there is an equivalence

$$\text{Higgs bundle } (E, \Phi) \leftrightarrow \text{ spectral sheaf } \mathcal{L} \tag{2.9}$$

mapping the (holomorphic data of the) Higgs bundle to its spectral data. That is, we interpret Φ as a map $E \to E \otimes K_S$, and then we fiberwise replace Φ by the eigenvalues and E by the eigenvectors. The relation can concisely expressed through a short exact sequence

$$0 \to \pi^* E \xrightarrow{\Psi} \pi^*(E \otimes K_S) \to \mathcal{L} \to 0 \tag{2.10}$$

where $\Psi \equiv \Phi - \lambda I_{n\times n}$, π^* is pull-back to the total space of the canonical bundle, and λ is the tautological section of $\pi^* K_S$. For future use, we also denote the total space of K_S by X. The similarity to our notation for the heterotic Calabi-Yau three-fold is not entirely coincidental. There are versions of this correspondence also for meromorphic Higgs bundles.

Such spectral sheaves are fairly simple to write down explicitly, certainly compared to solving non-abelian equations. As an example relevant for phenomenological models, let us consider breaking an E_8 gauge group to $SU(5)_G$. This requires writing down an $SL(5, \mathbf{C})_H$ Higgs bundle over S, where the structure group H is the commutant of G in the (complexified) E_8. By the Higgs/spectral correspondence, the data of such an $SL(5, \mathbf{C})_H$ Higgs bundle is given by

- the spectral cover C in X (the support of $\mathcal{L}$), given explicitly by a degree five polynomial;

- the spectral line bundle L in $\mathrm{Pic}(C)$, which may be specified by writing an explicit divisor D on C and putting $L = \mathcal{O}(D)$. We then define $\mathcal{L} = i_* L$, where i is the inclusion $C \hookrightarrow X$ and i_* is the push-forward.

Here we can see the Noether-Lefschetz problem rearing its head. We need to choose a curve in a surface in a three-fold. Simple examples constructed by hand will have many moduli. But if the curve is sufficiently 'generic' then deforming the surface or the three-fold will destroy the curve, in other words the corresponding deformation moduli are stabilized. However, finding the most rigid solutions requires us to solve a complicated Noether-Lefschetz problem, and the number of solutions grows exponentially with the rank of $H^2(C)$. This brings us back to the landscape problem. If these vacua are really stable after including quantum corrections (and most are probably not), then we'd have to ask who lives in the other mathematically consistent universes, and how was our universe able to solve the complicated computational problem of finding the most rigid, stable solutions. It is probably best to focus on general features of the whole class of solutions, rather than on individual solutions.

Leaving such questions aside, another important point about the separation into F-terms and D-terms is that changing the Hermitian metric on E does not affect the hypercohomology groups $\mathbb{H}^p(S, \mathcal{E}^\bullet)$. Therefore the low energy spectrum and Yukawa couplings are independent of the Hermitian metric on E, and in fact they may be computed directly in terms of the spectral sheaf $\mathcal{L}$ using Ext groups. Needless to say, this simplifies one's life tremendously.

This takes care of the F-terms. The D-terms (i.e. the moment map equation) may now be interpreted as an equation for the Hermitian metric on E, and a solution (if it exists) is called a Hermitian-Einstein metric. As mentioned already, solving this equation in closed form is virtually impossible. Instead one would like to appeal to a Higgs bundle version of the Donaldson-Uhlenbeck-Yau theorem:

A unique HE metric exists $\leftrightarrow$ the Higgs bundle is poly-stable.

A Higgs bundle is stable if for every Higgs sub-bundle $F \subset E$, we have

$$\mu(F) < \mu(E) \tag{2.11}$$

The slope $\mu(E)$ is defined to be the ratio of the degree of E (with respect to some choice of Kähler form) and the rank of E. A Higgs bundle is said to be poly-stable if it is the sum of stable Higgs bundles with the same slope.

The beauty of this type of statement is that existence of the Hermitian-Einstein metric is a differential-geometric criterion, whereas poly-stability is an algebro-geometric criterion. In particular, one can easily translate the requirement of stability for the Higgs bundle (E, Φ) to stability for the spectral sheaf $\mathcal{L}$. Unfortunately the proven versions of this correspondence do not suffice in our context, because we really need certain meromorphic incarnations of Higgs bundles. In particular, the stability condition depends on boundary data associated to the defect/surface operator, as is well-known for parabolic Higgs bundles. Modulo that issue however, we see that the Higgs/spectral correspondence answers the problem posed at the beginning of this section: constructing solutions and understanding the $4d$ effective theory boils down almost entirely to constructing and studying suitable spectral sheaves $\mathcal{L}$.

In other dimensions, the story is in principle similar, but less understood. In M-theory we were dealing with (2.3), which describes complex flat connections on a real manifold Q_3. Up to complexified gauge transformations, we can exchange such a complex flat connection for spectral data. In this case, the spectral cover corresponds to a Lagrangian A-brane in the cotangent bundle T^*Q_3 [**19**], and the $4d$ spectrum is computed by Floer cohomology groups of the A-brane.

2.4. Spectral/ALE correspondence.

Although we have claimed that our constructions are realized in string theory, the reader may be puzzled that the descriptions we have obtained seem to look very different from the traditional descriptions of F-theory, M-theory and type I', which we briefly touched on in section 1.4. Four-dimensional compactifications of F-theory are usually said to involve elliptically fibered Calabi-Yau manifolds with a configuration for a certain three-form (mathematically, a 2-gerbe with some additional properties). Four-dimensional compactifications of M-theory are usually said to involve manifolds with G_2-holonomy and a flat three-form (mathematically, a flat 2-gerbe). So far these did not yet appear in our story. In order to see the relation, we need another dictionary: the spectral/ALE correspondence [**23, 11, 19**].

In order to avoid the somewhat complicated representation theory of E_8, we will illustrate the correspondence by focussing on $U(n)$ Higgs bundles and A_{n-1}-type ALE fibrations. Consider the ALE fibration $Y \to S$ given by

$$y^2 \;=\; x^2 + a_0 z^n + \ldots + a_n \tag{2.12}$$

The a_i are sections of certain line bundles on S. Let us define a fibration $R \to S$ whose fibers over a point on S are the n lines defined by

$$y \;=\; x, \quad a_0 z^n + \ldots + a_n \;=\; 0 \tag{2.13}$$

The second equation specifies n points in the z-plane, hence defines an n-fold covering $C \to S$, which we identify with the spectral cover. We also get a natural projection $p : R \to C$ by replacing each line with a point. We further have a natural inclusion $i : R \hookrightarrow Y$. Then the spectral/ALE correspondence is given by the 'cylinder' maps i_*p^* and p_*i^*:

$$H^{i,j}(\bar{C}) \;\leftrightarrow\; H_p^{i+1,j+1}(\bar{Y}) \tag{2.14}$$

where H_p denotes a certain primitive part of the cohomology, and $\bar{C}$ and $\bar{Y}$ denote certain compactifications. In particular this maps the class of the spectral line bundle in Pic($\bar{C}$) to the Deligne cohomology class of a 2-gerbe on $\bar{Y}$. The more sophisticated version for exceptional gauge groups was described in [**23, 11**]. Essentially this mapping establishes a isomorphism of certain Hodge structures associated to the spectral cover and ALE sides.

The spectral/ALE correspondence exists for any gauge group, but the group theory tends to get more involved when we go away from the unitary groups. It so happens that for $SU(5)_{GUT}$ model building, which is the most relevant case phenomenologically, the group theoretic aspects simplify somewhat. A configuration in an E_8 gauge theory with an unbroken $SU(5)_{GUT}$ group corresponds to the following ALE fibration over S [**11**]:

$$y^2 \;=\; x^3 + a_0 z^5 + a_2 z^3 x + a_3 z^2 y + a_4 z x^2 + a_5 x y \tag{2.15}$$

Here the a_i are sections of certain line bundles over S. Note that we can easily put this in the form of a Weierstrass model by making some coordinate redefinitions. Under the spectral/ALE correspondence this gets mapped to an E_8 spectral cover which decomposes into various pieces. One of these pieces is a five-fold cover C_5 of S, given by a degree five equation in the total space of $K_S \to S$:

$$0 = a_0 s^5 + a_2 s^3 + a_3 s^2 + a_4 s + a_5 \tag{2.16}$$

Here s is a local coordinate on the fiber of K_S. This is precisely the spectral cover for the fundamental representation of the $SL(5,\mathbf{C})_H$ Higgs bundle mentioned in section 2.3. The intersection of C_5 with the zero section yields a curve $a_5 = 0$ where the A_4 singularity of the generic ALE fiber (2.15) further degenerates to type D_5, as one can easily check. This is the curve $\Sigma_{\mathbf{10}}$ where matter fields on the $\mathbf{10}$ and $\overline{\mathbf{10}}$ of $SU(5)_{GUT}$ propagate, see figure 2. One also gets a ten-fold covering C_{10}, the spectral cover in the anti-symmetric representation of our $Sl(5,\mathbf{C})_H$ Higgs bundle in section 2.3. Its intersection with the zero section yields a curve $a_0 a_5^2 - a_2 a_3 a_5 + a_3^2 a_4 = 0$ where the A_4 singularity of the ALE fiber degenerates to A_5. This is the curve $\Sigma_{\mathbf{5}}$ where the $\mathbf{5}$ and $\bar{\mathbf{5}}$ matter of $SU(5)_{GUT}$ propagate, again see figure 2. One gets various further degenerations in higher codimension. For details of this construction, see [**11**].

Using the spectral/ALE correspondence for E_8 we obtain an elliptic Calabi-Yau four-fold Y with boundary together with a Deligne cohomology class. We can then proceed to glue Y into a compact Calabi-Yau four-fold [**11**]. In this way we recover the traditional description of F-theory vacua.

In the M-theory context, the spectral data was given by a Lagrangian A-brane in T^*Q_3. Under the spectral/ALE correspondence, it gets mapped to an ALE fibration over Q_3 with a flat 2-gerbe, and with singularities at the location of non-abelian gauge groups and charged chiral matter. We have argued that this seven-dimensional non-compact manifold should admit a metric with G_2 holonomy [**19**]. This should then be glued into a compact G_2 manifold.

Note that in the traditional descriptions, we are dealing with singular spaces. For example to get an $SU(5)$ gauge theory in four dimensions from F-theory, we needed A_4 type quotient singularities along a section isomorphic with S which further degenerate in higher codimension. Of course, the problem with singularities is that they are singular. In order to do physics, we need some way to 'smooth' these singularities. The traditional way to do this is by making a crepant resolution (in physics terms, using M/F-duality and moving out on the Coulomb branch). Then one can quantize solitons wrapped on the exceptional cycles to deduce some of the physics. Unfortunately this extrapolation is not valid at the level of D-terms and it obscures many aspects of the physics which are relevant for phenomenology. For example a proper definition of the F-theory 2-gerbe and the analogue of its hypercohomology groups has not yet been given, in part because the 2-gerbe may obstruct such resolutions. Furthermore it is not clear how to extend this approach to other setting like M-theory on G_2 manifolds with singularities, where no natural resolution is available. The Higgs bundle/Yang-Mills theory approach yields another way to smooth the singularities, which has proven more useful for the questions discussed here.

2.5. Some open problems.

Here we collect some open mathematical problems. They were mostly already pointed out in the text, but we group them here for convenience. It is not meant to be a complete list, and we only list questions that fit in the scope of this review.

- **Explicit constructions.** As we reviewed, in even dimensions there are good techniques for constructing solutions to the F-term part of the Yang-Mills-Higgs equations. (We discuss the D-terms separately below). But it is an open problem to find non-abelian solutions to the equations in odd dimensions, given in (2.3) and (2.4). Furthermore, given a solution, one would like an effective method for computing hypercohomology groups.

  For solving (2.3) one could appeal to Donaldson-Corlette, which says that there exists a unique solution (called the harmonic metric) if the closure of the monodromy representation is a reductive subgroup of G. However, this theorem assumes that Q_3 is smooth and compact. As we mentioned, for phenomenological applications one should add source terms to (2.3) and (2.4) along a submanifold. An interesting sub-problem is to classify the most general supersymmetric boundary conditions (source terms).

  A class of non-abelian solutions for the type I' case will appear in [**20**]. The strategy there is to assume an S^1 symmetry and make use of the Fourier-Mukai transform to turn the question into a holomorphic problem. For the M-theory models we could use a similar strategy, by using mirror symmetry for T^*Q_3. Presumably there are other approaches. Is it possible to give a more general construction?

  The equations encountered in this review often reappear in other contexts, see eg. [**24**] or the literature on $Sl(N)$ knot invariants. The source terms are said to be associated to 'defect operators.' Thus a positive answer could also be helpful in other contexts.

- **D-flatness.** Once we can construct solutions to the F-term equations, we still have to solve the moment map equation. This is called proving D-flatness in physics terminology. The standard strategy is to formulate some type of stability condition, and then to try and prove an analogue of the Donaldson-Uhlenbeck-Yau theorem for Hermitian-Yang-Mills connections.

  In the holomorphic case one can often make natural conjectures for the stability criterion (eg. for parabolic Higgs bundles), but a proof of D-flatness is missing. Mochizuki has given proofs for certain classes of parabolic Higgs bundles, which unfortunately are slightly different from the Higgs bundles considered here. For the odd dimensional cases, it is not clear to us what the correct stability condition should be.

  It would also be interesting to develop numerical schemes for approximating the solutions. For Higgs bundles one can presumably use a modified version of balanced metrics [**22**].

- **Deformation theory.** To find the low energy effective action, we needed to understand the infinitesimal deformations of the Higgs bundle. They were classified by certain hypercohomology groups of the Higgs bundle. In the spectral cover picture, one is asking for the infinitesimal deformations of a coherent sheaf or a Lagrangian brane with flat connection. These are classified by Ext groups or Floer cohomology groups respectively.

But we also had a third picture: ALE fibrations with a 2-gerbe, satisfying some additional conditions. (The conditions in F-theory are somewhat complicated to state; see appendix C of [**16**] where a compactified version of ALE fibrations is considered). So we have a natural question: what classifies the first order infinitesimal deformations in the ALE fibration picture? And what are the analogues of the Yoneda pairing and the higher Massey-like products, which compute Yukawa couplings and higher order interactions?

It is not clear to us how to answer this question, and maybe a good answer is not possible. There are many well-understood spectral cover configurations that get mapped to poorly understood ALE-fibrations. As a simple example, one may consider a non-abelian bundle on a degenerate non-reduced cover. This should get mapped to an ALE fibration with singularities, and some kind of non-abelian 2-gerbe along the singularities, whatever that means exactly. Such configurations cannot be lifted to the resolution, but should be included in the general formulation. More complicated examples can be found in [**22**]. Thus understanding the 2-gerbe is an important part of the problem. One could further ask about stability conditions in this picture. This is also poorly understood.

- **Local versus global.** ALE fibrations are said to be local because they are non-compact. For string phenomenology, we think of them as a local piece of a compact manifold, and eventually we would like to embed them in a concrete global model, i.e. in a compact manifold. In the context of F-theory there has been a lot of progress on this question. Questions remain, but this is outside the scope of this review. In the context of M-theory, we would like to embed our ALE fibration in a compact G_2 holonomy manifold. The problem here is obvious: there are very few constructions of compact G_2 manifolds and none appears to be suitable for compactifying our ALE fibrations.

- **Finite or infinite landscape.** For the equations we have considered, the expected dimension of the moduli space is zero. That is, generic solutions should be isolated, and it makes sense to try to count them, in an appropriate sense. It is an interesting problem to make the counting more precise.

  We discussed some crude estimates for the expected number of solutions of the Hermitian-Yang-Mills or Hitchin equations which reproduce the supersymmetric standard model or a unified extension thereof. In toy models of flux vacua it appears that the true number of solutions can be much larger. In fact, there is a basic question if the true number of solutions is even finite. The prevailing opinion is that it should be finite, but we recently constructed an infinite sequence which appears to be stable and evades the known no-go theorems [**25**]. (It should be remembered that proving stability is of course the hardest part). Such infinite sequences would address some concerns raised in [**26**].

There are of course also many interesting physical questions. In particular one would like to get an interesting physical prediction that distinguishes the extra-dimensional models from conventional four-dimensional models. Once we have done the Kaluza-Klein reduction, we are working within four-dimensional effective field theory. So are there processes where we might see the extra dimensions in the foreseeable future? The answer appears to be yes, at least in principle. Proton decay probes the GUT scale and could provide such characteristic signatures [**27, 28, 29**]. The extra dimensions become important at the GUT scale and the proton feels the higher dimensional interactions. For further questions and developments, see the talk by S. Schäfer-Nameki at this conference [**30**], and the additional reviews [**31, 32, 33, 34, 35**].

Acknowledgements: I would like to thank the organisers of String-Math 2011 for an inspiring conference. I would also like to thank R. Donagi and T. Pantev for collaboration on the material presented here, and R. Donagi and the referee for comments on the manuscript.

References

[1] H. Georgi and S. L. Glashow, "Unity Of All Elementary Particle Forces," Phys. Rev. Lett. **32**, 438 (1974).

[2] H. Georgi, H. R. Quinn and S. Weinberg, "Hierarchy Of Interactions In Unified Gauge Theories," Phys. Rev. Lett. **33**, 451 (1974).

[3] S. Dimopoulos, S. Raby and F. Wilczek, "Supersymmetry And The Scale Of Unification," Phys. Rev. D **24**, 1681 (1981).

[4] S. Raby, "Proton decay," hep-ph/0211024.

[5] P. Nath and P. Fileviez Perez, "Proton stability in grand unified theories, in strings and in branes," Phys. Rept. **441**, 191 (2007) [arXiv:hep-ph/0601023].

[6] G. Senjanovic, "Proton decay and grand unification," AIP Conf. Proc. **1200**, 131 (2010) [arXiv:0912.5375 [hep-ph]].

[7] P. Candelas, G. T. Horowitz, A. Strominger and E. Witten, "Vacuum Configurations for Superstrings," Nucl. Phys. B **258**, 46 (1985).

[8] M. R. Douglas, R. L. Karp, S. Lukic and R. Reinbacher, "Numerical solution to the hermitian Yang-Mills equation on the Fermat quintic," JHEP **0712**, 083 (2007) [arXiv:hep-th/0606261].

[9] R. Bousso and J. Polchinski, "Quantization of four form fluxes and dynamical neutralization of the cosmological constant," JHEP **0006**, 006 (2000) [arXiv:hep-th/0004134].

[10] F. Denef and M. R. Douglas, "Distributions of flux vacua," JHEP **0405**, 072 (2004) [arXiv:hep-th/0404116].

[11] R. Donagi and M. Wijnholt, "Higgs Bundles and UV Completion in F-Theory," arXiv:0904.1218 [hep-th].

[12] E. Witten, "String theory dynamics in various dimensions," Nucl. Phys. B **443**, 85 (1995) [arXiv:hep-th/9503124].

[13] C. Vafa, "Evidence for F-Theory," Nucl. Phys. B **469**, 403 (1996) [arXiv:hep-th/9602022].

[14] P. Horava and E. Witten, "Heterotic and type I string dynamics from eleven dimensions," Nucl. Phys. B **460**, 506 (1996) [arXiv:hep-th/9510209].

[15] J. Polchinski and E. Witten, "Evidence for Heterotic - Type I String Duality," Nucl. Phys. B **460**, 525 (1996) [arXiv:hep-th/9510169].

[16] R. Donagi and M. Wijnholt, "Model Building with F-Theory," arXiv:0802.2969 [hep-th], to appear in ATMP.

[17] C. Beasley, J. J. Heckman and C. Vafa, "GUTs and Exceptional Branes in F-theory - I," JHEP **0901**, 058 (2009) [arXiv:0802.3391 [hep-th]].

[18] H. Hayashi, R. Tatar, Y. Toda, T. Watari and M. Yamazaki, "New Aspects of Heterotic–F Theory Duality," Nucl. Phys. B **806**, 224 (2009) [arXiv:0805.1057 [hep-th]].

[19] T. Pantev and M. Wijnholt, "Hitchin's Equations and M-Theory Phenomenology," J. Geom. Phys. **61**, 1223 (2011) [arXiv:0905.1968 [hep-th]].

[20] T. Pantev and M. Wijnholt, "Unification and type I'." For an overview, see http://www.math.unm.edu/ vassilev/goldensands/tpantev.pdf.

[21] H. Hayashi, T. Kawano, R. Tatar and T. Watari, "Codimension-3 Singularities and Yukawa Couplings in F-theory," Nucl. Phys. B **823**, 47 (2009) [arXiv:0901.4941 [hep-th]].

[22] R. Donagi and M. Wijnholt, "Gluing Branes, I," arXiv:1104.2610 [hep-th].

[23] G. Curio and R. Y. Donagi, "Moduli in N = 1 heterotic/F-theory duality," Nucl. Phys. B **518**, 603 (1998) [arXiv:hep-th/9801057].

[24] A. Kapustin and E. Witten, "Electric-magnetic duality and the geometric Langlands program," Commun. Number Theory and Physics **1** (1997), 1-236, arXiv:hep-th/0604151.

[25] R. Donagi, M. Wijnholt.

[26] M. Dine and N. Seiberg, "Is The Superstring Weakly Coupled?," Phys. Lett. B **162**, 299 (1985).

[27] T. Friedmann and E. Witten, "Unification scale, proton decay, and manifolds of G(2) holonomy," Adv. Theor. Math. Phys. **7**, 577 (2003) [arXiv:hep-th/0211269].

[28] I. R. Klebanov and E. Witten, "Proton decay in intersecting D-brane models," Nucl. Phys. B **664**, 3 (2003) [arXiv:hep-th/0304079].

[29] R. Donagi and M. Wijnholt, "Breaking GUT Groups in F-Theory," arXiv:0808.2223 [hep-th], to appear in ATMP.

[30] S. Schäfer-Nameki, "F-theory: global aspects and phenomenology," talk at this conference.

[31] J. J. Heckman and C. Vafa, "From F-theory GUTs to the LHC," arXiv:0809.3452 [hep-ph].

[32] M. Wijnholt, "F-Theory, GUTs and Chiral Matter," arXiv:0809.3878 [hep-th].

[33] J. J. Heckman, "Particle Physics Implications of F-theory," Submitted to: Ann.Rev.Nucl.Part.Sci. [arXiv:1001.0577 [hep-th]].

[34] M. Wijnholt, "F-theory and unification," Fortsch. Phys. **58**, 846 (2010).

[35] T. Weigand, "Lectures on F-theory compactifications and model building," Class. Quant. Grav. **27**, 214004 (2010) [arXiv:1009.3497 [hep-th]].

Ludwig-Maximilians-Universität, Arnold-Sommerfeld Center, Theresienstrasse 37, D-80333 Munich, Germany

Contributed Talks

Proceedings of Symposia in Pure Mathematics
Volume **85**, 2012

Topological T-duality with Monodromy

David Baraglia

ABSTRACT. We extend topological T-duality from principal torus bundles to the more general class of affine torus bundles allowing for arbitrary monodromy representations. The conditions under which an affine torus bundle admits a T-dual are determined. We construct isomorphisms in twisted cohomology and twisted K-theory for T-duals.

1. Introduction

T-duality is a duality arising from string theory between spaces $X, \hat{X}$ which are torus fibrations equipped with degree 3 integral cohomology classes $h, \hat{h}$. The dual spaces can be thought of as target spaces for string theories and the duality is an equivalence between these two theories. The degree 3 class $h \in H^3(X, \mathbb{Z})$ called the *H-flux* can be interpreted as the Diximier-Douady class classifying a gerbe $\mathcal{G}$ on X. Aside from the presence of gerbes the other ingredient that makes up T-duality is that the spaces involved $X, \hat{X}$ are torus fibrations over a common base M. Roughly speaking the non-singular fibers of $X, \hat{X}$ are understood as being dual tori. A local expression for T-duality given by the Buscher rules [**7**] makes more precise the duality between the fibers of X and $\hat{X}$. However the Buscher rules apply only in local charts where both the torus bundle and the gerbe admit trivializations. This leaves open the question of how to understand the global nature of T-duality.

In [**4**] a global topological description of T-duality was formulated which was shown to be consistent with the Buscher rules [**5**]. This *topological T-duality* expresses a purely topological statement of T-duality that does not involve metrics or other geometric structures. The standard starting point for topological T-duality is a principal torus bundle $X \to M$ and a degree 3 integral cohomology class h (or gerbe) on X. In this paper we describe an extension of topological T-duality to non-principal torus bundles. In [**1**] we studied the case of general circle bundles and in [**2**] we extended these results to arbitrary rank. In this paper we discuss the main results of [**2**] leaving aside some of the more technical details.

The main distinguishing feature of non-principal torus bundles is that they can have a non-trivial monodromy. One motivation for developing a topological T-duality with non-trivial monodromy comes from topological mirror symmetry

---

1991 *Mathematics Subject Classification.* Primary 19L50; Secondary 53C08; 55N25.

[**11**]. Typically a singular torus fibration has non-trivial monodromy around the singular fibers. We would like T-duality to incorporate this feature in the hope of developing a topological mirror symmetry with flux, expanding on [**14**].

The main result of this paper is that topological T-duality works just as well for the class of affine torus bundles, defined in Section 2 as it does for principal torus bundles. We will see that there are still some novelties compared with the principal case, such as the need to use *graded gerbes.* The relevance of graded gerbes in string theory has been observed for instance in [**9**] in connection with certain discrete symmetries of type II string theory. We formulate a definition of topological T-duality for affine torus bundles extending the principal case. We are able to prove existence under conditions analogous to the principal case and prove isomorphisms in twisted K-theory and twisted cohomology.

2. Affine torus bundles

2.1. Classification. Let $V = \mathbb{R}^n$ be the standard real n-dimensional vector space, $\Lambda = \mathbb{Z}^n$ the standard rank n lattice and $T^n = V/\Lambda$ the corresponding torus. We define $\mathrm{Aff}(T^n)$ to be the semi-direct product $\mathrm{Aff}(T^n) = \mathrm{GL}(n,\mathbb{Z}) \ltimes T^n$ which acts on T^n by affine transformations.

DEFINITION 2.1. An *affine torus bundle* is a torus bundle $X \to M$ with structure group $\mathrm{Aff}(T^n)$. Thus there is a principal $\mathrm{Aff}(T^n)$-bundle $P \to M$ such that X is the associated bundle $X = P \times_{\mathrm{Aff}(T^n)} T^n$.

Classifying affine torus bundles goes a long way towards a full classification of torus bundles. As explained in [**1**] it is known that every torus bundle of rank $n \leq 3$ has an affine structure.

We assume throughout that the base manifold M is connected. Let $P \to M$ be a principal $\mathrm{Aff}(T^n)$-bundle over M. Using the homomorphism $\mathrm{Aff}(T^n) \to \mathrm{GL}(n,\mathbb{Z})$ we obtain from P a principal $\mathrm{GL}(n,\mathbb{Z})$-bundle. Such bundles correspond to conjugacy classes of homomorphisms $\rho : \pi_1(M) \to \mathrm{GL}(n,\mathbb{Z})$. Any such representative ρ will be called the *monodromy* of P. The monodromy ρ acts on the lattice Λ and thus defines a local system Λ_ρ with coefficient group Λ. Let $\pi : X \to M$ be the associated affine torus bundle $X = P \times_{\mathrm{Aff}(T^n)} T^n$. The projection $\pi : X \to M$ determines a local system $R^1\pi_*\mathbb{Z}$ which describes the degree 1 cohomology of the fibers. We have that $R^1\pi_*\mathbb{Z}$ is isomorphic to Λ^*_ρ, the dual local system to Λ_ρ.

In addition to the monodromy there is another piece of topological data attached to an affine torus bundle $X \to M$. It is a class $c \in H^2(M, \Lambda_\rho)$ which represents the translational part of the $\mathrm{Aff}(T^n)$-valued transition functions. We call c the *twisted Chern class* of X. The explicit construction of c is explained in [**1**]. Strictly speaking c is valued in a cohomology group $H^2(M, S)$, where S is the local system dual to $R^1\pi_*\mathbb{Z}$. A choice of monodromy representation $\rho : \pi_1(M) \to \mathrm{GL}(n,\mathbb{Z})$ allows us to identify S with Λ_ρ and thus identify c with an element of $H^2(M, \Lambda_\rho)$, but this depends on the choice of isomorphism $S \simeq \Lambda_\rho$. To rephrase this slightly, given an affine torus bundle $X \to M$ we can find an associated pair (ρ, c) consisting of a monodromy representation ρ and twisted Chern class $c \in H^2(M, \Lambda_\rho)$, but the pair (ρ, c) is only defined up to the natural action of $\mathrm{GL}(n,\mathbb{Z})$.

THEOREM 2.2 ([**1**]). *To every pair (ρ, c) where ρ is a homomorphism $\rho : \pi_1(M) \to \mathrm{GL}(n,\mathbb{Z})$ and c is a class $c \in H^2(M, \Lambda_\rho)$ there is an affine T^n-bundle with monodromy ρ and twisted Chern class c. Two pairs (ρ, c),(ρ', c') determine the same T^n-bundle if and only if they are related by the natural action of* $\mathrm{GL}(n,\mathbb{Z})$.

Implicit in this theorem is the non-trivial fact that two principal $\mathrm{Aff}(T^n)$-bundles P_1, P_2 determine the same T^n-bundle (up to bundle isomorphism that covers the identity on M) if and only if P_1, P_2 are isomorphic as principal bundles.

2.2. Leray-Serre spectral sequence. Let $\pi : X \to M$ be a rank n affine torus bundle with monodromy ρ and twisted Chern class $c \in H^2(M, \Lambda_\rho)$. Let $\{(E_r^{p,q}, d_r)\}$ denote the Leray-Serre spectral sequence associated to $\pi : X \to M$, using integral coefficients in cohomology. The E_2 stage is given by $E_2^{p,q} = H^p(M, \wedge^q \Lambda_\rho^*)$. Recall also that the map $\pi : X \to M$ induces a filtration on the cohomology groups of X which we write as

$$0 = F^{m+1,m} \subseteq F^{m,m} \subseteq F^{m-1,m} \subseteq \cdots \subseteq F^{0,m} = H^m(X, \mathbb{Z}).$$

The spectral sequence converges to the associated graded ring; specifically we have isomorphisms $E_\infty^{p,q} = F^{p,p+q}/F^{p+1,p+q}$. The pull-back $\pi^* : H^p(M,\mathbb{Z}) \to H^p(X,\mathbb{Z})$ can be described in terms of the spectral sequence as the composition $H^p(M,\mathbb{Z}) = E_2^{p,0} \to E_\infty^{p,0} = F^{p,p} \to H^p(X,\mathbb{Z})$.

We need a description of the differential d_2. For $c \in \Lambda$ and $d \in \wedge^q \Lambda^*$ let $i_c d \in \wedge^{q-1}\Lambda^*$ denote the contraction of d by c. This gives a corresponding contraction map $\Lambda_\rho \otimes \wedge^q \Lambda_\rho^* \to \wedge^{q-1}\Lambda_\rho^*$ of local systems. Combining with the cup product we get a natural contraction map $H^r(M, \Lambda_\rho) \otimes H^s(M, \wedge^q \Lambda_\rho^*) \to H^{s+r}(M, \wedge^{q-1}\Lambda_\rho^*)$.

THEOREM 2.3 ([**1**]). *The differential $d_2 : H^p(M, \wedge^q \Lambda_\rho^*) \to H^{p+2}(M, \wedge^{q-1}\Lambda_\rho^*)$ is given by contraction with the twisted Chern class: $d_2(a) = c \smile a$.*

3. Graded gerbes

We have found that it is natural to formulate T-duality in terms of graded gerbes. Here we briefly survey the main features of graded gerbes and their application to twisted K-theory. Let X be a smooth manifold. By a *cover* of X we mean a smooth manifold Y and a surjective submersion $f : Y \to X$. This generalizes the notion of a cover of X by open subsets $\{U_i\}$ for we can take Y to be the disjoint union of the open sets. For a cover $Y \to X$ we write $Y^{[k]}$ for the k-fold fiber product. We can think of $Y^{[2]}$ as a topological groupoid such that two pairs $(x, y), (y', z) \in Y^{[2]}$ are composable if and only if $y = y'$ and their composition is taken to be $(x, y)(y, z) = (x, z)$. If $L \to Y^{[2]}$ is a line bundle on $Y^{[2]}$ then we write $L_{x,y}$ for the fiber of L over a point $(x, y) \in Y^{[2]}$.

DEFINITION 3.1. A *bundle gerbe* [**12**] on X consists of a cover $f : Y \to X$, a Hermitian line bundle $L \to Y^{[2]}$ on $Y^{[2]}$ and for each pair of composable pairs $(x, y), (y, z) \in Y^{[2]}$ an isomorphism $\theta_{x,y,z} : L_{x,y} \otimes L_{y,z} \to L_{x,z}$ of Hermitian line bundles such that $\theta_{x,y,z}$ depends smoothly on the triple $(x, y, z) \in Y^{[3]}$. In addition θ is required to satisfy an associativity condition: for each quadruple $(x, y, z, w) \in Y^{[4]}$ there are two ways to use θ to go from $L_{x,y} \otimes L_{y,z} \otimes L_{z,w}$ to $L_{x,w}$ depending on the order of multiplication. These two ways are required to be equal.

DEFINITION 3.2. A *graded bundle gerbe* [**10**] is a bundle gerbe $(f : Y \to X, L \to Y^{[2]}, \theta)$ together with an assignment of a (continuously varying) $\mathbb{Z}_2$-grading on L such that θ has degree 0. That is, for every point $(x,y) \in Y^{[2]}$ the line bundle $L_{x,y}$ is assigned a degree $\alpha_{x,y} \in \mathbb{Z}_2$ that varies continuously with (x,y) and such that for every triple $(x,y,z) \in Y^{[3]}$ we have $\alpha_{x,y} + \alpha_{y,z} = \alpha_{x,z}$.

We can construct a category of graded bundle gerbes on X, actually a 2-category [**13, 15, 10**] but for the purposes of twisted K-theory it is sufficient to take the 1-category quotient. We let $\mathrm{GrGrb}(X)$ denote the 1-category quotient of graded bundle gerbes on X. The most important features of $\mathrm{GrGrb}(X)$ that we need for T-duality and twisted K-theory are as follows [**13, 15, 10**]:

- The only morphisms in $\mathrm{GrGrb}(X)$ are isomorphisms.
- $\mathrm{GrGrb}(X)$ is contravariant with respect to X in the sense that for any smooth map $\phi : Y \to X$ there is a pull-back functor $\phi^* : \mathrm{GrGrb}(X) \to \mathrm{GrGrb}(Y)$ such that $(\phi \circ \psi)^* = \psi^* \circ \phi^*$.
- If $\mathcal{G}, \mathcal{H} \in \mathrm{GrGrb}(X)$ are isomorphic then $\mathrm{Hom}(\mathcal{G}, \mathcal{H})$ is a torsor for the group $H^0(X, \mathbb{Z}_2) \oplus H^2(X, \mathbb{Z})$ of graded Hermitian line bundles on X.
- Isomorphism classes of elements in $\mathrm{GrGrb}(X)$ are in bijection with the set $H^1(X, \mathbb{Z}_2) \times H^3(X, \mathbb{Z})$. For a graded gerbe $\mathcal{G}$ the pair $[\mathcal{G}] = (\alpha, h) \in H^1(X, \mathbb{Z}_2) \times H^3(X, \mathbb{Z})$ will be called the graded Dixmier-Douady class of $\mathcal{G}$. We may also call α the *grading class* and h the *Dixmier-Douady class*.

Given graded bundle gerbes $\mathcal{G}, \mathcal{H}$ on X it is possible to define their product $\mathcal{G} \otimes \mathcal{H}$. For simplicity we assume $\mathcal{G}, \mathcal{H}$ are defined on the same cover $f : Y \to X$, otherwise we can pullback to a common refinement. Let $\mathcal{G}, \mathcal{H}$ correspond to graded line bundles $L, M \to Y^{[2]}$ and multiplications $\theta_{x,y,z}, \tau_{x,y,z}$. To define $\mathcal{G} \otimes \mathcal{H}$ we define a multiplication on the graded tensor product $L \otimes M$ as follows. For any triple $(x,y,z) \in Y^{[3]}$ the isomorphism $(L_{x,y} \otimes M_{x,y}) \otimes (L_{y,z} \otimes M_{y,z}) \to L_{x,z} \otimes M_{x,z}$ is obtained by first swapping the inner two factors and then applying $\theta_{x,y,z} \otimes \tau_{x,y,z}$. Note that when we swap $L_{y,z}$ and $M_{x,y}$ we use the graded sign convention so if that if both these spaces are odd we must introduce a minus sign in the isomorphism $M_{x,y} \otimes L_{y,z} \to L_{y,z} \otimes M_{x,y}$. The multiplication descends to isomorphism classes and gives $H^1(X, \mathbb{Z}_2) \times H^3(X, \mathbb{Z})$ an abelian group structure. Note that it is not the product structure but rather a (generally) non-trivial extension of $H^1(X, \mathbb{Z}_2)$ by $H^3(X, \mathbb{Z})$. The product is given by [**10**] $(a,h)(b,k) = (a+b, h+k+\beta(ab))$ where $\beta : H^2(X, \mathbb{Z}_2) \to H^3(X, \mathbb{Z})$ is the Bockstein homomorphism.

For any orthogonal vector bundle $V \to X$ on X we can define a graded gerbe $L(V)$ called the *lifting gerbe of* V [**13**] which captures the obstructions to V admitting a Spin^c-structure. Let $P \to X$ be the orthogonal frame bundle of V. There is a natural map $\phi : P^{[2]} \to \mathrm{O}(n)$ where n is the rank of V. The map ϕ is defined by letting $\phi(pg, p) = g$ for any $p \in P$, $g \in \mathrm{O}(n)$. There is a natural graded line bundle A on $\mathrm{O}(n)$ associated to the principal circle bundle $\mathrm{Pin}^c(n) \to \mathrm{O}(n)$. The grading on A comes from the determinant map $\mathrm{O}(n) \to \mathbb{Z}_2$. Setting $L = \phi^*(A)$ gives a graded line bundle $L \to P^{[2]}$ which will define the lifting gerbe. The multiplication on L is induced by the group composition on $\mathrm{Spin}^c(n)$. One checks that this defines a graded gerbe $L(V)$, the lifting gerbe of V. The isomorphism class of $L(V)$ is given by $[L(V)] = (w_1(V), W_3(V)) \in H^1(X, \mathbb{Z}_2) \oplus H^3(X, \mathbb{Z})$. If V is a rank n vector bundle without metric we can still define a lifting gerbe $L(V, h)$ by choosing a metric h on V. Although $L(V, h)$ depends on h we have as a consequence of the contractibility of the space of metrics on V that for any two metrics h_1, h_2 there exists a canonical isomorphism $L(V, h_1) \to L(V, h_2)$ as elements of $\mathrm{GrGrb}(X)$.

One of the main reasons for working with graded bundles gerbes is their application to twisted K-theory. For any graded gerbe $\mathcal{G} \in \mathrm{GrGrb}(X)$ one may define a *twisted K-theory group* [**3, 10**] $K^*(X, \mathcal{G})$ which is a $\mathbb{Z}_2$-graded abelian group. The most important properties of twisted K-theory that we need for T-duality are:

- $K^*(X, \cdot\,) : \mathrm{GrGrb}(X) \to \mathrm{Gr}_{\mathbb{Z}_2}\mathrm{Ab}$ is a functor from graded gerbes on X to $\mathbb{Z}_2$-graded abelian groups.
- There is a product structure $K^i(X, \mathcal{G}) \otimes K^j(X, \mathcal{H}) \to K^{i+j}(X, \mathcal{G} \otimes \mathcal{H})$ which is graded commutative and associative in the evident sense.
- For any proper map $\phi : Y \to X$ there are pullback maps $\phi^* : K^*(X, \mathcal{G}) \to K^*(Y, \phi^*(\mathcal{G}))$ which are contravariant in ϕ.
- For any smooth map $\phi : Y \to X$ there is a pushforward [**8, 10**] $\phi_* : K^i(Y, \phi^*(\mathcal{G}) \otimes L(\phi)) \to K^{i-n}(Y, \mathcal{G})$ where $L(\phi)$ is the lifting gerbe of $TY \oplus \phi^*(TX)$ and n is the rank of $TY \oplus \phi^*(TX)$. In the case where $\phi : Y \to X$ is a fiber bundle we can replace $L(\phi)$ by $L(V)$ where $V = \mathrm{Ker}(\phi_*)$ is the vertical tangent bundle.

4. Topological T-duality

Given torus bundles $\pi : X \to M$, $\hat{\pi} : \hat{X} \to M$ we denote by C the fiber product $C = X \times_M \hat{X}$. We call C the *correspondence space.* Let $p, \hat{p}$ denote the projections $p : C \to X$, $\hat{p} : C \to \hat{X}$ and $q : C \to M$ the projection $q = \pi\, p = \hat{\pi}\, \hat{p}$. Associated to each of the above maps is the Leray-Serre spectral sequence and the corresponding filtration in cohomology. To distinguish these we label by the corresponding map, thus for example $E_r^{p,q}(\pi)$ is the spectral sequence associated to $\pi : X \to M$ and $\cdots \subseteq F^{p+1,n}(\pi) \subseteq F^{p,n}(\pi) \subseteq \cdots H^n(X, \mathbb{Z})$ the filtration on cohomology.

DEFINITION 4.1. Let $\pi : X \to M$, $\hat{\pi} : \hat{X} \to M$ be rank n affine torus bundles over a common base M. Let $\mathcal{G}$, $\hat{\mathcal{G}}$ be graded gerbes on X, $\hat{X}$ respectively. Suppose that the graded Dixmier-Douady class of $\mathcal{G}$ is $[\mathcal{G}] = (\pi^*(\alpha), h)$, where $\alpha \in H^1(M, \mathbb{Z}_2)$, $h \in H^3(X, \mathbb{Z})$ and similarly $[\hat{\mathcal{G}}] = (\hat{\pi}^*(\hat{\alpha}), \hat{h})$ for $\hat{\alpha} \in H^1(M, \mathbb{Z}_2)$, $\hat{h} \in H^3(\hat{X}, \mathbb{Z})$. Let ρ be the monodromy of $\pi : X \to M$ and $c \in H^2(M, \Lambda_\rho)$ the twisted Chern class. Similarly define $\hat{\rho}$, $\hat{c}$. We say that $(X, \mathcal{G})$ and $(\hat{X}, \mathcal{G})$ are *T-dual* if

(T1) The monodromies $\rho, \hat{\rho}$ are dual representations.
(T2) The flux h lies in the subspace $F^{2,3}(\pi) \subseteq H^3(E, \mathbb{Z})$ and likewise $\hat{h}$ lies in $F^{2,3}(\hat{\pi})$.
(T3) The twisted Chern class $\hat{c} \in H^2(M, \Lambda_{\hat{\rho}})$ is a representative in $E_2^{2,1}(\pi)$ for the image of h in $E_\infty^{2,1}(\pi) \simeq F^{2,3}(\pi)/F^{3,3}(\pi)$.
(T4) Similarly $c \in H^2(M, \Lambda_\rho)$ is a representative in $E_2^{2,1}(\hat{\pi})$ for the image of $\hat{h}$ in $E_\infty^{2,1}(\hat{\pi}) \simeq F^{2,3}(\hat{\pi})/F^{3,3}(\hat{\pi})$.
(T5) There is an isomorphism of graded gerbes $\theta : p^*(\mathcal{G}) \to \hat{p}^*(\hat{\mathcal{G}}) \otimes q^*(L_\rho)$ where L_ρ is the lifting gerbe for the vector bundle $V_\rho = \Lambda_\rho \otimes \mathbb{R}$ associated to ρ.
(T6) The isomorphism θ must satisfy the Poincaré property explained below.

We now define the Poincaré property, based on [**6**, (2,7)]. The rough idea is that θ looks locally like the Poincaré line bundle on the fibers. Let $u \in M$ be a point in the base M and set $T_u = \pi^{-1}(u)$, $\hat{T}_u = \hat{\pi}^{-1}(u)$. From the above properties we find that $\mathcal{G}$ restricted to the fiber T_u is trivializable. Let $\tau_u : 0 \to \mathcal{G}|_{T_u}$ be such a trivialization and similarly let $\hat{\tau}_u : 0 \to \hat{\mathcal{G}}|_{\hat{T}_u}$ be a trivialization of $\hat{\mathcal{G}}$ restricted to the fiber $\hat{T}_u$. Obviously one can find a trivialization $\psi_u : 0 \to L_\rho|_{\{u\}}$ of the lifting gerbe L_ρ restricted to the singleton set $\{u\}$. Now we can pullback the various trivializations to $q^{-1}(u) = T_u \times \hat{T}_u$ and obtain an isomorphism

$$\iota_u = (\hat{p}^*(\hat{\tau}_u) \otimes q^*(\psi_u)) \circ p^*(\tau_u^{-1}) : p^*(\mathcal{G})|_{T_u \times \hat{T}_u} \to \hat{p}^*(\hat{\mathcal{G}})|_{T_u \times \hat{T}_u} \otimes q^*(L_\rho)|_{T_u \times \hat{T}_u}.$$

On the other hand we can simply restrict θ over u to get another such isomorphism. The set of isomorphisms between two isomorphic gerbes in $\mathrm{GrGrb}(T_u \times \hat{T}_u)$ is a torsor for $H^0(T_u \times \hat{T}_u, \mathbb{Z}_2) \times H^2(T_u \times \hat{T}_u, \mathbb{Z})$, so there is a class $t \in H^0(T_u \times \hat{T}_u, \mathbb{Z}_2) \times H^2(T_u \times \hat{T}_u, \mathbb{Z})$ such that $\theta|_{T_u \times \hat{T}_u}$ is obtained from ι_u through translation by t. Note however that the class t depends on the choice of trivializations $\tau_u, \hat{\tau}_u, \psi_u$. Changes in trivializations make it so that t is only well-defined modulo the span of $p^*(H^2(T_u, \mathbb{Z})), \hat{p}^*(H^2(\hat{T}_u, \mathbb{Z})), q^*(H^0(\{u\}, \mathbb{Z}_2))$. The quotient space can be identified with $H^1(T_u, \mathbb{Z}) \otimes H^1(\hat{T}_u, \mathbb{Z}) \simeq (\mathbb{Z}^n) \otimes (\mathbb{Z}^n)$. Letting u vary we obtain a section of the local system $\Lambda_\rho \otimes \Lambda_{\hat{\rho}} = \mathrm{Hom}(\Lambda_\rho, \Lambda_\rho)$. The Poincaré property, axiom (T6), is the requirement that this class is the identity section of $\mathrm{Hom}(\Lambda_\rho, \Lambda_\rho)$.

Condition (T5) of Definition 4.1 immediately implies some further topological conditions for T-duality by taking graded Dixmier-Douady classes:

$$\alpha = \hat{\alpha} + w_1(V_\rho), \tag{4.1}$$

$$p^*(h) = \hat{p}^*(\hat{h}) + q^*(W_3(V_\rho) + \beta(w_1(V_\rho)\alpha)). \tag{4.2}$$

To get Equation (4.1) we used the fact that $q^* : H^1(M, \mathbb{Z}_2) \to H^1(C, \mathbb{Z}_2)$ is injective. Equation (4.2) is perhaps surprising since in the case of principal torus bundles we simply have an equality $p^*(h) = \hat{p}^*(\hat{h})$ but when ρ is non-trivial we have found that this equality needs to be modified for twisted K-theory. Yet another topological condition is obtained from (T3) (the same condition also follows from (T4)). Since $\hat{c}$ represents a class in $E_\infty^{2,1}(\pi)$ we must have $d_2\hat{c} = 0$, where d_2 is the differential $d_2 : E_2^{2,1}(\pi) \to E_2^{4,0}(\pi)$. The condition can be expressed as

$$\langle c, \hat{c} \rangle = 0$$

where we use $\langle\, , \,\rangle$ to express the duality pairing $\Lambda_\rho \otimes \Lambda_{\hat{\rho}} \to \mathbb{Z}$ of local systems. We remark also that condition (T6) imposes some topological conditions on $h, \hat{h}$ beyond (4.2), but it is less straightforward to express this cohomologically.

THEOREM 4.2 ([**2**]). *Let $(X, \mathcal{G})$ be a pair consisting of a rank n affine torus bundle $\pi : X \to M$ and a graded gerbe $\mathcal{G}$ on X. Suppose that the graded Dixmier-Douady class of $\mathcal{G}$ has the form $[\mathcal{G}] = (\pi^*(\alpha), h)$ where $\alpha \in H^1(M, \mathbb{Z}_2)$ and $h \in F^{2,3}(\pi) \subseteq H^3(X, \mathbb{Z})$. Then there exists a T-dual pair $(\hat{X}, \hat{\mathcal{G}})$.*

For rank $n > 1$ it turns out that the T-dual of a pair $(X, \mathcal{G})$ is not uniquely determined up to isomorphism. In fact even the topology of $\hat{X}$ is generally not uniquely determined. To see this one simply has to note that the image of h in $F^{2,3}(\pi)/F^{3,3}(\pi) \simeq E_\infty^{2,1}(\pi)$ generally has more than one lift to a class $\hat{c} \in E_2^{2,1}(\pi) \simeq$

$H^2(M, \Lambda_{\hat{\rho}})$. In fact if $\hat{c}$ is one such lift then any other lift has the form $\hat{c} + d_2 B$ where d_2 is the differential $d_2 : E_2^{0,2}(\pi) \to E_2^{2,1}(\pi)$ and B is an element of $E_2^{0,2}(\pi) \simeq H^0(M, \wedge^2 \Lambda_\rho^*)$. On the other hand for circle bundles where $n = 1$ we can show that the T-dual pair $(\hat{X}, \hat{\mathcal{G}})$ is unique up to isomorphism [**1**].

5. Differential form description

Let $\pi : X \to M$ be a rank n affine torus bundle. Although X is generally not a principal bundle we can still introduce notions of invariant forms and connections on X using the affine structure. Since X is affine there exists a principal $\mathrm{Aff}(T^n)$-bundle $\tilde{\pi} : \tilde{X} \to M$ such that X is the quotient $X \simeq \tilde{X}/\mathrm{GL}(n, \mathbb{Z})$. Let $\tilde{p} : \tilde{X} \to X$ be the projection.

DEFINITION 5.1. We say that a differential form $\omega \in \Omega^k(X)$ is *invariant* if $\tilde{p}^*\omega$ is invariant under the $\mathrm{Aff}(T^n)$-action. More generally if A is a flat vector bundle on M then the $\mathrm{Aff}(T^n)$-action on $\tilde{X}$ lifts to an action on $\tilde{\pi}^*(A)$. Thus if $\omega \in \Omega^k(X, \pi^*(A))$ is a $\pi^*(A)$-valued differential form on X we may say that ω is *invariant* if $\tilde{p}^*\omega \in \Omega^k(\tilde{X}, \tilde{\pi}^*(A))$ is $\mathrm{Aff}(T^n)$-invariant.

Let $\rho : \pi_1(M) \to \mathrm{GL}(n, \mathbb{Z})$ be the monodromy of $\pi : X \to M$. We let V_ρ denote the corresponding rank n flat vector bundle on M. The affine structure on X gives the vertical bundle $\mathrm{Ker}(\pi_* : TX \to TM)$ the structure of a flat vector bundle on X. In fact $\mathrm{Ker}(\pi_*)$ and $\pi^*(V_\rho)$ are isomorphic as flat vector bundles.

DEFINITION 5.2. A *twisted connection* on X is a $\pi^*(V_\rho)$-valued invariant 1-form $A \in \Omega^1(X, \pi^*(V_\rho))$ such that the restriction of A to the vertical bundle $A : \mathrm{Ker}(\pi_*) \to \pi^*(V_\rho)$ is an isomorphism of flat vector bundles.

It is not hard to see that twisted connections on X correspond to connections on $\tilde{X}$ in the ordinary sense. Thus the notion of a twisted connection is not really a new concept, but we prefer to work with forms on X rather than the $\mathrm{GL}(n, \mathbb{Z})$-cover $\tilde{X}$ and twisted connections allow us to do this. Suppose now that A is a twisted connection on X. There is a unique V_ρ-valued 2-form F on M such that $d_\nabla A = \pi^*(F)$, where ∇ denotes the flat connection on V_ρ (and its pullback to X). We call F the *(twisted) curvature* of A. It is almost immediate that the cohomology class of F in $H^2(M, V_\rho)$ is the image of the twisted Chern class $c \in H^2(M, \Lambda_\rho)$ under the change of coefficient morphism $\Lambda_\rho \to \Lambda_\rho \otimes \mathbb{R} \simeq V_\rho$.

Let us apply these notions to a T-dual pair $(X, \mathcal{G})$, $(\hat{X}, \mathcal{G})$ of affine torus bundles over M equipped with graded gerbes. Let $h \in H^3(X, \mathbb{Z})$ and $\hat{h} \in H^3(\hat{X}, \mathbb{Z})$ be the corresponding Dixmier-Douady classes and $h_\mathbb{R}, \hat{h}_\mathbb{R}$ their images in real cohomology.

THEOREM 5.3 ([**2**]). *There exist invariant 3-form representatives $H \in \Omega^3(X)$, $\hat{H} \in \Omega^3(\hat{X})$ for $h_\mathbb{R}, \hat{h}_\mathbb{R}$, twisted connections $A \in \Omega^1(X, \pi^*(V_\rho))$, $\hat{A} \in \Omega^1(\hat{X}, \hat{\pi}^*(V_\rho^*))$ with twisted curvatures $F, \hat{F}$ and a 3-form $H_3 \in \Omega^3(M)$ such that*

$$H = H_3 + A \wedge \pi^*(\hat{F}), \tag{5.1}$$

$$\hat{H} = H_3 + \hat{A} \wedge \hat{\pi}^*(F). \tag{5.2}$$

In particular it follows that

$$\hat{p}^*(\hat{H}) - p^*(H) = d\mathcal{B} \tag{5.3}$$

where letting $\langle\, ,\, \rangle$ *denote the pairing of* V_ρ *and* V_ρ^* *we have*

$$\mathcal{B} = \langle p^*(A) \wedge \hat{p}^*(\hat{A}) \rangle. \tag{5.4}$$

Crucial to the proof is the existence of an isomorphism $\theta : p^*(\mathcal{G}) \to \hat{p}^*(\hat{\mathcal{G}}) \otimes q^*(L_\rho)$ of graded gerbes satisfying the Poincaré property. Theorem 5.3 can also be promoted to the level of gerbes with connections and curvings.

6. T-duality transformations

6.1. Twisted cohomology. Let (A, ∇) be a flat vector bundle on X and $H \in \Omega^3(X)$ a closed 3-form. We define a $\mathbb{Z}_2$-graded complex $(S^i_A(X), d_{A,H})$, $i \in \mathbb{Z}_2$ by setting $S^i_A(X) = \bigoplus_{k \equiv i \bmod (2)} \Omega^k(X, A)$ and $d_{A,H} : S^i_A(X) \to S^{i+1}_A(X)$ by $d_{A,H}\omega = d_\nabla \omega + H \wedge \omega$. The corresponding cohomology groups will be denoted by $H^i(X, (A, H))$ and called the *twisted cohomology* for (A, H). Alternatively if A is the flat vector bundle associated to a representation $\rho : \pi_1(M) \to \mathrm{GL}(n, \mathbb{R})$ we may write $H^i(X, (\rho, H))$. Observe that $H^i(X, (A, H))$ depends up to isomorphism only on H through the cohomology class of H in $H^3(X, \mathbb{R})$.

Suppose that $\pi : X \to M$ is an affine torus bundle, that A is a flat vector bundle on M and H is an invariant 3-form on M. The subspaces $\iota : S^i_{\pi^*(A),\mathrm{inv}}(X) \to S^i_{\pi^*(A)}(X)$ of invariant forms are preserved by the differential $d_{A,H}$, so we get invariant cohomology groups $H^i_{\mathrm{inv}}(X, (\pi^*(A), H))$ and the inclusion ι induces a linear map $\iota_* : H^i_{\mathrm{inv}}(X, (\pi^*(A), H)) \to H^i(X, (\pi^*(A), H))$.

THEOREM 6.1. *The map* $\iota_* : H^i_{\mathrm{inv}}(X, (\pi^*(A), H)) \to H^i(X, (\pi^*(A), H))$ *is an isomorphism.*

PROOF. We have a map of chain complexes $\iota : S^*_{\pi^*(A),\mathrm{inv}}(X) \to S^*_{\pi^*(A)}(X)$. In addition the fibration $\pi : X \to B$ induces a natural filtration on both complexes, since we can filter the space of differential forms by how many indices they have in the fiber direction. It is straightforward that ι preserves the filtrations and thus induces a morphism $\iota_r : E^{p,q}_{r,\pi^*(A),\mathrm{inv}}(X) \to E^{p,q}_{r,\pi^*(A)}(X)$ between the associated spectral sequences. Moreover one sees that already at the $r = 1$ stage ι_1 is an isomorphisms essentially because every cohomology class on the n-torus is represented by a unique invariant form. □

Now suppose that $(X, \mathcal{G})$, $(\hat{X}, \hat{\mathcal{G}})$ are T-dual affine torus bundles with graded gerbes over M. Let $\mathcal{G}, \hat{\mathcal{G}}$ be classified by pairs $(\pi^*(\alpha), h) \in H^1(X, \mathbb{Z}_2) \oplus H^3(X, \mathbb{Z})$, $(\hat{\pi}^*(\hat{\alpha}), \hat{h}) \in H^1(X, \mathbb{Z}_2) \oplus H^3(X, \mathbb{Z})$, for some classes $\alpha, \hat{\alpha} \in H^1(M, \mathbb{Z}_2)$. Choose twisted connections $A, \hat{A}$ and forms $H, \hat{H}, H_3$ as in Theorem 5.3, so in particular Equations (5.1),(5.2),(5.3) hold where $\mathcal{B}$ is defined as in (5.4). From this data we will construct a T-duality map $T : H^i(X, (\pi^*(\alpha), H)) \to H^{i-n}(\hat{X}, (\hat{\pi}^*(\hat{\alpha}), \hat{H}))$ generalizing [**4**]. Note that we think of $\alpha, \hat{\alpha}$ as determining flat real line bundles $\mathbb{R}_\alpha, \mathbb{R}_{\hat{\alpha}}$. The map T is defined analogous to the Fourier-Mukai transform, that is we define a map $T : S^i_{\pi^*(\alpha)}(X) \to S^{i-n}_{\hat{\pi}^*(\hat{\alpha})}(\hat{X})$ as follows

$$T(\omega) = \int_{C/\hat{X}} e^{-\mathcal{B}} p^*(\omega) \tag{6.1}$$

where $\int_{C/\hat{X}}$ denotes integration over the fibers of $\hat{p} : C \to \hat{X}$. Note that we are not assuming that $\hat{p} : C \to \hat{X}$ is oriented, instead the integration picks up sign factors due to change of fiber orientation that can be accounted for by the local system $\det(V_\rho)$. Thus if ω is $\pi^*(\mathbb{R}_\alpha)$-valued then $T(\omega)$ has values in $\hat{\pi}^*(\mathbb{R}_\alpha \otimes \det(V_\rho))$ which by Equation (4.1) is isomorphic to $\hat{\pi}^*(\mathbb{R}_{\hat{\alpha}})$. One immediately observes that T is a chain map, that is $T \circ d_{\pi^*(\alpha),H} = d_{\hat{\pi}^*(\hat{\alpha}),\hat{H}} \circ T$ so we get a morphism $T : H^i(X,(\pi^*(\alpha),H)) \to H^{i-n}(\hat{X},(\hat{\pi}^*(\hat{\alpha}),\hat{H}))$ of twisted cohomologies.

THEOREM 6.2. *The map* $T : H^i(X,(\pi^*(\alpha),H)) \to H^{i-n}(\hat{X},(\hat{\pi}^*(\hat{\alpha}),\hat{H}))$ *is an isomorphism.*

PROOF. Since T sends invariant forms to invariant forms it defines a map $T : H^i_{\text{inv}}(X,(\pi^*(\alpha),H)) \to H^{i-n}_{\text{inv}}(\hat{X},(\hat{\pi}^*(\hat{\alpha}),\hat{H}))$. By Theorem 6.1 it suffices to show this map is an isomorphism. Using the connections $A, \hat{A}$ to decompose invariant forms one sees that $T : S^i_{\pi^*(\alpha),\text{inv}}(X) \to S^{i-n}_{\hat{\pi}^*(\hat{\alpha}),\text{inv}}(\hat{X})$ is an isomorphism at the level of chain complexes. □

6.2. Twisted K-theory. Let $(X,\mathcal{G}),(\hat{X},\hat{\mathcal{G}})$ be a T-dual pair of rank n affine torus bundles over a base M as in Definition 4.1 and in addition fix an isomorphism $\theta : p^*(\mathcal{G}) \to \hat{p}^*(\hat{\mathcal{G}}) \otimes q^*(L_\rho)$ satisfying axiom (T6). From this data we can immediately construct a map $T : K^*(X,\mathcal{G}) \to K^{*-n}(\hat{X},\hat{\mathcal{G}})$ as follows:

$$T(x) = \hat{p}_*(\theta(p^*(x))). \tag{6.2}$$

Starting with the class $x \in K^*(X,\mathcal{G})$ one takes the pullback $p^*(x)$, applies the isomorphism in twisted K-theory induced by θ and finally we perform the pushforward under the map $\hat{p} : C \to \hat{X}$.

THEOREM 6.3. *If M is a smooth manifold with a finite good cover (for instance if M is compact) then the map* $T : K^*(X,\mathcal{G}) \to K^{*-n}(\hat{X},\hat{\mathcal{G}})$ *defined in (6.2) is an isomorphism.*

PROOF. The proof follows [**6**]. Choosing a finite good open cover on the base and pulling back to X or $\hat{X}$ we get Mayer-Vietoris sequences for $K^*(X,\mathcal{G})$ and $K^{*-n}(\hat{X},\hat{\mathcal{G}})$. By restriction T yields a chain map between these sequences. By the 5-lemma and induction on the number of elements in the cover we reduce to the case where the base is contractible. In this case using axiom (T6) we find T reduces to a K-theoretic Fourier-Mukai transform and is thus an isomorphism. □

References

1. D. Baraglia, Topological T-duality for general circle bundles, arXiv:1105.0290v2 (2011).
2. D. Baraglia, Topological T-duality for torus bundles with monodromy, arXiv:1201.1731v1, (2012).
3. P. Bouwknegt, A. Carey, V. Mathai, M. Murray, D. Stevenson, Twisted K-theory and K-theory of bundle gerbes. *Comm. Math. Phys.* **228** no. 1, 17-45, (2002).
4. P. Bouwknegt, J. Evslin, V. Mathai, T-duality: topology change from H-flux. *Comm. Math. Phys.* **249** no. 2, 383-415, (2004).
5. P. Bouwknegt, K. Hannabuss, V. Mathai, T-duality for principal torus bundles and dimensionally reduced Gysin sequences. *Adv. Theor. Math. Phys.* **9** no. 5, 749-773, (2005).
6. U. Bunke, P. Rumpf, T. Schick, The topology of T-duality for T^n-bundles. *Rev. Math. Phys.* **18** no. 10, 1103-1154, (2006).
7. T. H. Buscher, Path-integral derivation of quantum duality in nonlinear sigma-models. *Phys. Lett. B* **201** no. 4, 466-472, (1988).
8. A. L. Carey, B.-L. Wang, Thom isomorphism and push-forward map in twisted K-theory. *J. K-Theory* **1** no. 2, 357-393, (2008).

9. J. Distler, D. S. Freed, G. W. Moore, Spin structures and superstrings, Surveys in differential geometry. Volume XV. Perspectives in mathematics and physics, 99-130, Surv. Differ. Geom., 15, Int. Press, Somerville, MA, (2011).
10. D. S. Freed, M. J. Hopkins, C. Teleman, Loop groups and twisted K-theory I. *J. Topol.* **4**, no. 4, 737-798 (2011).
11. M. Gross, Topological mirror symmetry. *Invent. Math.* **144** no. 1, 75-137, (2001).
12. M. K. Murray, Bundle gerbes. *J. London Math. Soc.* (2) **54** no. 2, 403-416, (1996).
13. D. Stevenson, The Geometry of Bundle Gerbes, PhD thesis, University of Adelaide, math.DG/0004117 (2000).
14. A. Tomasiello, Topological mirror symmetry with fluxes. *J. High Energy Phys.* no. 6, **067**, 29 pp, (2005).
15. K. Waldorf, More morphisms between bundle gerbes. *Theory Appl. Categ.* **18** no. 9, 240-273, (2007).

Mathematical sciences institute, The Australian National University, Canberra ACT 0200, Australia

E-mail address: `david.baraglia@anu.edu.au`

Proceedings of Symposia in Pure Mathematics
Volume **85**, 2012

Variable transformation defects

Nicolas Behr and Stefan Fredenhagen

ABSTRACT. We investigate defects between supersymmetric Landau-Ginzburg models whose superpotentials are related by a variable transformation. It turns out that there is one natural defect, which can then be used to relate boundary conditions and defects in the different models. In particular this defect can be used to relate Grassmannian Kazama-Suzuki models and minimal models, and one can generate rational boundary conditions in the Kazama-Suzuki models from those in minimal models. The defects that appear here are closely related to the defects that are used in Khovanov-Rozansky link homology.

1. Introduction

Matrix factorisations are a beautiful mathematical subject in the sense that they are easy to define and still have a lot of interesting structures. Furthermore they can be used and applied in physics, where they describe boundary conditions and defects in $N = 2$ supersymmetric Landau-Ginzburg (LG) models (see e.g. [**18**] for an overview).

In the simplest setting, a matrix factorisation consists of two square matrices p^0 and p^1 of the same size with polynomial entries whose product is the identity matrix multiplied by a given potential $W \in \mathbb{C}[x_1, \ldots, x_n]$,

$$p^0 \cdot p^1 = W \cdot \mathbb{1} \ , \quad p^1 \cdot p^0 = W \cdot \mathbb{1} \ . \tag{1.1}$$

An example of that is given by

$$x^{\ell} \cdot x^{k-\ell} = x^k \ , \tag{1.2}$$

where p^0 and p^1 are just polynomials (1×1 matrices). This example describes B-type boundary conditions in $N = 2$ minimal models [**19, 6**].

In addition to boundary conditions one can also consider B-type defects between Landau-Ginzburg models with superpotentials W and W'. They can be described by matrix factorisations of the difference $W - W'$ [**20, 21, 8**]. Defects are very important and useful objects in two-dimensional field theory: one of their most crucial properties is that they can be fused by bringing them on top of each other to produce a new defect [**24, 8**]. In this way, defects define an interesting algebraic structure that turns out to be useful in analysing symmetries and dualities (see e.g. [**15**]), and bulk and boundary renormalisation group flows (see e.g. [**17, 3, 9, 14**])

2010 *Mathematics Subject Classification.* Primary 81T40; Secondary 81T45, 13C14.

in such models. As defects can also be fused onto boundaries, they may be used to relate or to generate boundary conditions. In particular, if we know defects between different theories, we can generate boundary conditions in one model from boundary conditions in the other model by fusion of the defect.

In this work we will analyse defects between LG models with potentials W and W' that are related by a variable transformation. If these transformations are non-linear, the two physical theories will be different. We will see that in such a situation there is one natural defect that acts in a simple, but non-trivial way on matrix factorisations. After analysing its properties we will apply it in a number of examples. In particular we demonstrate how it can be used to generate matrix factorisations in Kazama-Suzuki models from those in minimal models.

2. Variable transformations via defects

A B-type defect separating two $N=2$ supersymmetric Landau-Ginzburg models with superpotentials W and W', respectively, can be described by a matrix factorisation of the difference $W-W'$ of the potentials [**8, 10**]. To be more precise, let R and R' be polynomial algebras over $\mathbb{C}$, and $W\in R$, $W'\in R'$. A (W,W')-defect matrix factorisation is then a pair $({}_RM_{R'},Q)$ where ${}_RM_{R'}={}_RM^0_{R'}\oplus {}_RM^1_{R'}$ is a free, $\mathbb{Z}/2\mathbb{Z}$-graded R-R'-bimodule, and Q is an odd bimodule map,

$$Q=\begin{pmatrix}0 & p^1\\ p^0 & 0\end{pmatrix}\,, \tag{2.1}$$

such that $Q^2=W\cdot\mathrm{id}_M-\mathrm{id}_M\cdot W'$. As M is assumed to be free, Q can be written as a matrix with polynomial entries. A B-type boundary condition is a special defect, for which one side is trivial, e.g. $R'=\mathbb{C}$, $W'=0$.

Morphisms between defects $({}_RM_{R'},Q)$ and $({}_R\tilde{M}_{R'},\tilde{Q})$ are bimodule maps $\varphi: M\to\tilde{M}$ with $\tilde{Q}\circ\varphi=\varphi\circ Q$ modulo exact maps of the form $\tilde{Q}\circ\psi+\psi\circ Q$. Matrix factorisations are considered to be equivalent if there exist two morphisms $\phi: M\to\tilde{M}$ and $\psi:\tilde{M}\to M$ such that $\phi\circ\psi$ and $\psi\circ\phi$ equal the identity map up to exact terms. Consider e.g. $({}_RM_R,Q)$ and $({}_RM_R,S\circ Q\circ S^{-1})$ for an even isomorphism $S:M\to M$. These factorisations are then equivalent, with the morphisms being $\phi=S$ and $\psi=S^{-1}$. When we write the $\mathbb{Z}/2\mathbb{Z}$-gradation explicitly, the action of $S=\left(\begin{smallmatrix}s^0 & 0\\ 0 & s^1\end{smallmatrix}\right)$ on p^0 and p^1 amounts to similarity transformations,

$$p^0\mapsto s^1p^0(s^0)^{-1}\,,\quad p^1\mapsto s^0p^1(s^1)^{-1}\,. \tag{2.2}$$

One of the most interesting properties of defects is that they can be fused. Physically this means that two defects can be put on top of each other producing a new defect [**24, 8**]. Mathematically this amounts to defining the tensor product [**25, 21**] of two matrix factorisations $({}_RM_{R'},Q)$ and $({}_{R'}\tilde{M}_{R''},\tilde{Q})$. As a module this is simply the graded tensor product

$$M\otimes\tilde{M}=\left(M^0\otimes_{R'}\tilde{M}^0\oplus M^1\otimes_{R'}\tilde{M}^1\right)\oplus\left(M^1\otimes_{R'}\tilde{M}^0\oplus M^0\otimes_{R'}\tilde{M}^1\right)\,, \tag{2.3}$$

and the associated module map is

$$Q\hat{\otimes}\tilde{Q} := \begin{pmatrix} & 0 & p^1\otimes\mathrm{id} & \mathrm{id}\otimes\tilde{p}^1 \\ & & -\mathrm{id}\otimes\tilde{p}^0 & p^0\otimes\mathrm{id} \\ p^0\otimes\mathrm{id} & -\mathrm{id}\otimes\tilde{p}^1 & & 0 \\ \mathrm{id}\otimes\tilde{p}^0 & p^1\otimes\mathrm{id} & & \end{pmatrix} . \tag{2.4}$$

For $R' = R$ and $W' = W$, there is a special defect called the identity defect, which we denote by $({}_RI_R, {}_W\mathcal{I}_W)$. Fusing the identity defect onto some defect reproduces the original defect; it serves therefore as a unit object with respect to the tensor product. Its precise construction can be found in [**21, 20, 10**].

For different superpotentials $W \in R$ and $W' \in S$ there is in general no natural defect factorisation. On the other hand, if there exists a ring homomorphism

$$\phi : R \to S \ , \quad \text{such that } \phi(W) = W' \ , \tag{2.5}$$

then we can naturally map R-modules to S-modules and vice versa by extension or restriction of scalars: via the homomorphism ϕ the ring S has a natural R-S-bimodule structure, ${}_RS_S$, where the multiplication from the left is defined via the homomorphism ϕ. Given a right R-module M_R we can then map it to a right S-module by

$$\phi^* : M_R \mapsto (M_R) \otimes_R ({}_RS_S) \ , \tag{2.6}$$

which describes the extension of scalars from R to S. On the other hand, a left S-module ${}_S\tilde{M}$ has a natural R-module structure using the homomorphism ϕ. This restriction of scalars from S to R can be written as the map

$$\phi_* : {}_S\tilde{M} \mapsto ({}_RS_S) \otimes_S ({}_S\tilde{M}) \ . \tag{2.7}$$

ϕ^* and ϕ_* act also on module homomorphisms in an obvious way, so they define functors on the categories of R- and S-modules. Notice that ϕ^* maps free modules to free modules, whereas this is not guaranteed for ϕ_*. We assume in the following that the R-module ${}_RS$ is free, such that ϕ_* maps free modules to free modules.

We can apply these functors also to matrix factorisations. In particular we can apply them to the identity factorisations $({}_RI_R, {}_W\mathcal{I}_W)$ and $({}_SI_S, {}_{W'}\mathcal{I}_{W'})$ to obtain two (W, W')-defects with $W' = \phi(W)$,

$$({}_RI_S^A, {}_W\mathcal{I}_{W'}^A) = (\phi^*({}_RI_R), \phi^*({}_W\mathcal{I}_W)) \ , \quad ({}_RI_S^B, {}_W\mathcal{I}_{W'}^B) = (\phi_*({}_SI_S), \phi_*({}_{W'}\mathcal{I}_{W'})) \ . \tag{2.8}$$

We now claim that these two defects are actually equivalent. To show this we take the first defect and fuse the identity defect ${}_SI_S$ from the right, and compare it to the second defect onto which we fuse the identity defect ${}_RI_R$ from the left. As a module we obtain

$$({}_RI_S^A) \otimes_S ({}_SI_S) \cong ({}_RI_R) \otimes_R ({}_RS_S) \otimes_S ({}_SI_S) \cong ({}_RI_R) \otimes_R ({}_RI_S^B) \ . \tag{2.9}$$

Since

$$({}_W\mathcal{I}_W \otimes_R \mathrm{id}_S) \,\hat{\otimes}\, ({}_{W'}\mathcal{I}_{W'}) = ({}_W\mathcal{I}_W) \,\hat{\otimes}\, (\mathrm{id}_S \otimes_S {}_{W'}\mathcal{I}_{W'}) \ , \tag{2.10}$$

also the factorisations agree, so that we indeed find that these two defects are equivalent. We call them $({}_RI_S, {}_W\mathcal{I}_{W'})$.

By a similar consideration as above we see that when we fuse $({}_RI_S, {}_W\mathcal{I}_{W'})$ to the left, it acts by the functor ϕ^*, whereas it acts by the functor ϕ_* when we fuse it to defects to the right. Thus we have a very simple description for the fusion result for this defect. Analogously we can construct the defect $({}_SI_R, {}_{W'}\mathcal{I}_W)$.

Let us explicitly describe how the defect $({}_R I_S, {}_W\mathcal{I}_{W'})$ acts by fusion. First consider the (simpler) fusion to the left on a defect $({}_{R'}M_R, Q)$. For a rank $2m$ free R'-R-bimodule ${}_{R'}M_R$ we can think of Q as a $2m \times 2m$ matrix with entries Q_{ij} in $R' \otimes_{\mathbb{C}} R$. Fusing $({}_R I_S, {}_W\mathcal{I}_{W'})$ onto this defect from the right, we obtain a free R'-S-module of rank $2m$, and a matrix $\tilde{Q}$ with entries $\tilde{Q}_{ij} = (\mathrm{id} \otimes \phi)(Q_{ij})$, i.e. we just replace the variables of R by the variables of S via the map ϕ.

We now assume that ${}_R S$ is a finite rank free R-module,

$$\rho : R^{\oplus n} \xrightarrow{\sim} {}_R S \ . \tag{2.11}$$

With the help of the R-module isomorphism ρ we can then explicitly describe how the defect $({}_R I_S, {}_W\mathcal{I}_{W'})$ acts by fusion to the right on a defect $({}_S M_{S'}, Q)$. If ${}_S M_{S'}$ is free of rank $2m$, then Q can be represented as a $2m \times 2m$ matrix with entries $Q_{ij} \in S \otimes_{\mathbb{C}} S'$. After the fusion we have an R-S'-module of rank $2mn$, and each entry Q_{ij} is replaced by the $n \times n$ matrix that represents the map $\rho^{-1} \circ Q_{ij} \circ \rho$ (where we tacitly extend ρ to mean $\rho \otimes \mathrm{id}_{S'}$).

A particular situation occurs when all Q_{ij} are of the form $\phi(\tilde{Q}_{ij})$. As ρ is an R-module map, the map $\rho^{-1} \circ Q_{ij} \circ \rho$ can then be represented by the $n \times n$ matrix $\tilde{Q}_{ij} \cdot \mathbb{1}_{n\times n}$. The resulting defect is therefore a direct sum of n identical defects. As an example, consider the fusion of $({}_R I_S, {}_W\mathcal{I}_{W'})$ on $({}_S I_R, {}_{W'}\mathcal{I}_W)$. By the arguments above this fusion results in a direct sum of n identity defects,

$$({}_R I_S, {}_W\mathcal{I}_{W'}) \otimes ({}_S I_R, {}_{W'}\mathcal{I}_W) \cong ({}_R I_R, {}_W\mathcal{I}_W)^{\oplus n} \ . \tag{2.12}$$

In the special case that ϕ is a ring isomorphism, $\phi : R \to R$, and $W' = \phi(W) = W$, the construction above leads to symmetry or group-like defects,

$$G^{\phi} = ({}_R M_R, (\mathrm{id} \otimes \phi)({}_W\mathcal{I}_W)) \ , \tag{2.13}$$

which have been discussed in [**15, 8**]. The fusion of such defects is particularly simple,

$$G^{\phi} \otimes G^{\psi} \cong G^{\psi\circ\phi} \ , \tag{2.14}$$

and G^{ϕ} is invertible with inverse $G^{\phi^{-1}}$. These defects therefore form a group.

3. Examples and applications

In this section we want to apply the formalism of the foregoing section to physically interesting examples.

Minimal models. Let us first look at the one-variable case, $R = \mathbb{C}[y]$, and choose the potential to be $W(y) = y^k$. The corresponding Landau-Ginzburg model describes a minimal model at level $k-2$. Consider now the ring homomorphism

$$\phi_1 : p(y) \mapsto p(x^d) \tag{3.1}$$

that maps polynomials in R to those in $S = \mathbb{C}[x]$. The transformed potential is $W'(x) = x^{kd}$. We observe that ${}_R S$ is a free R-module of rank d,

$$\begin{array}{rccl} \rho : & R^{\oplus d} & \to & S \\ & (p_1(y), \ldots, p_d(y)) & \mapsto & \sum_{j=1}^{d} x^{j-1} p_j(x^d) \ . \end{array} \tag{3.2}$$

Let us now look at the corresponding defect between these two minimal models. We consider the explicit construction $({}_R I^B_S, {}_W\mathcal{I}^B_{W'})$ via ϕ_*. We start with the identity defect $({}_S I_S, {}_{W'}\mathcal{I}_{W'})$ that is given by a rank 2 matrix ${}_{W'}\mathcal{I}_{W'} = \left(\begin{smallmatrix} 0 & \imath_0 \\ \imath_1 & 0 \end{smallmatrix}\right)$ with $\imath_0 = (x - x')$ and $\imath_1 = (W(x) - W(x'))/(x - x')$. Here we denoted by x' the variable corresponding to the right S-module structure. Under the map ϕ_* acting on the left S-module structure the entry $\imath_0$ is then replaced by

$$\tilde{\imath}_0 = \begin{pmatrix} -x' & & & y \\ 1 & -x' & & \\ & \ddots & \ddots & \\ & & 1 & -x' \end{pmatrix} \xrightarrow[\text{transformation}]{\text{similarity}} \begin{pmatrix} y - x'^d & & & \\ & 1 & & \\ & & \ddots & \\ & & & 1 \end{pmatrix} . \tag{3.3}$$

We therefore explicitly see that this defect is equivalent to $({}_R I_S^A, {}_W \mathcal{I}_{W'}^A)$ that we obtain from the identity defect in the y-variables by expressing one of the variables in terms of x. This defect is related to the generalised permutation boundary conditions in two minimal models [**12, 13**] by the folding trick.

We now want to apply this defect to matrix factorisations $({}_S M, Q)$ that describe boundary conditions. The elementary factorisation $x^\ell \cdot x^{kd-\ell}$ will be called $Q_\ell^{(x)}$, and correspondingly $Q_\ell^{(y)}$ refers to the y-factorisation $y^\ell \cdot y^{k-\ell}$. Fusing the defect $({}_R I_S, {}_W \mathcal{I}_{W'})$ to $(S^{\oplus 2}, Q_{rd+\ell}^{(x)})$ results in a superposition $\big(R^{\oplus d}, (Q_r^{(y)})^{\oplus d-\ell} \oplus (Q_{r+1}^{(y)})^{\oplus \ell}\big)$ (for $0 \leq \ell \leq d-1$). The factorisation $Q_0^{(y)}$ is trivial, so we see that the basic factorisation $Q_1^{(x)}$ is just mapped to the basic factorisation $Q_1^{(y)}$.

We can also consider defects in minimal models. Of particular interest are the group-like defects $G_{(y)}^n$ [**8**] that induce the map $y \mapsto \eta^n y$. Here $\eta = \exp \frac{2\pi i}{k}$ such that the potential y^k is invariant. Obviously we have $G^n \cong G^{n+k}$, and the group law is just $G_{(y)}^m \otimes G_{(y)}^n = G_{(y)}^{m+n}$. As a (W,W)-defect matrix factorisation, $G_{(y)}^n$ corresponds to $(y - \eta^n y') \cdot \frac{W(y)-W(y')}{y-\eta^n y'}$. Similarly, $G_{(x)}^n$ denotes the group-like defect corresponding to the map $x \mapsto \exp \frac{2\pi i n}{kd} x$. Given such a defect $G_{(x)}^n$ one can ask what happens to it when we sandwich it between the defects ${}_R I_S$ and ${}_S I_R$. Surprisingly the result can again be expressed in terms of group-like defects, namely

$$({}_R I_S, {}_W \mathcal{I}_{W'}) \otimes G_{(x)}^n \otimes ({}_S I_R, {}_{W'} \mathcal{I}_W) \cong \left(G_{(y)}^n\right)^{\oplus d} . \tag{3.4}$$

***SU*(3)/*U*(2) Kazama-Suzuki model.** As a more interesting example we look at a defect between an $SU(3)/U(2)$ Kazama-Suzuki model and a product of two minimal models. Consider the two-variable polynomial rings $R = \mathbb{C}[y_1, y_2]$ and $S = \mathbb{C}[x_1, x_2]$, and the ring homomorphism

$$\phi : p(y_1, y_2) \mapsto p(x_1 + x_2, x_1 x_2) , \tag{3.5}$$

which replaces the y_i by the elementary symmetric polynomials in the x_j. The potential in the x-variables is that of two minimal models,

$$W'(x_1, x_2) = x_1^k + x_2^k \qquad (k \geq 4) . \tag{3.6}$$

It is symmetric in x_1 and x_2 and thus it can be expressed in terms of the elementary symmetric polynomials leading to the potential $W(y_1, y_2)$ in the y-variables such that $\phi(W) = W'$. This then describes the $SU(3)/U(2)$ Kazama-Suzuki model (see e.g. [**16, 5**]).

The R-module ${}_R S$ is free of rank 2 with the explicit R-module isomorphism

(3.7)
$$\begin{array}{llcl} \rho : & R \oplus R & \to & {}_R S \\ & (p_1(y_1,y_2), p_2(y_1,y_2)) & \mapsto & p_1(x_1+x_2, x_1x_2) + (x_1 - x_2) p_2(x_1+x_2, x_1x_2) , \end{array}$$

with inverse

$$\rho^{-1}: \quad {}_RS \to R\oplus R \qquad (3.8)$$
$$p(x_1,x_2) \mapsto \left(p_s(x_1,x_2)\Big|_{y_i}, \tfrac{p_a(x_1,x_2)}{x_1-x_2}\Big|_{y_i}\right),$$

where $p_{s/a}(x_1,x_2) = \frac{1}{2}(p(x_1,x_2) \pm p(x_2,x_1))$, and $|_{y_i}$ means to replace in a symmetric polynomial in x_j the elementary symmetric polynomials by the y_i.

The (W,W')-defect between the Kazama-Suzuki model (y-variables) and the minimal models (x-variables) acts on y-factorisations simply by replacing variables. However, given an x-factorisation with matrix Q, a matrix element Q_{ij} is replaced by a 2×2 matrix,

$$Q_{ij} \mapsto \begin{pmatrix} (Q_{ij})_s & (x_1-x_2)(Q_{ij})_a \\ \frac{(Q_{ij})_a}{x_1-x_2} & (Q_{ij})_s \end{pmatrix}\Bigg|_{y_1,y_2}. \qquad (3.9)$$

As an example consider the boundary condition based on the factorisation

$$(x_1-\xi x_2)\cdot\frac{W'(x_1,x_2)}{x_1-\xi x_2} = W'(x_1,x_2) \qquad (3.10)$$

with $\xi = \exp\frac{\pi i}{k}$ (these are the so-called permutation factorisations [**2, 7**]). By the map (3.9) the factor $(x_1-\xi x_2)$ is mapped to

(3.11)
$$(x_1-\xi x_2)\mapsto \begin{pmatrix} \frac{1-\xi}{2}y_1 & \frac{1+\xi}{2}(y_1^2-4y_2) \\ \frac{1+\xi}{2} & \frac{1-\xi}{2}y_1 \end{pmatrix} \xrightarrow[\text{transf.}]{\text{similarity}} \begin{pmatrix} y_1^2-2(1+\cos\frac{\pi}{k+2})y_2 & 0 \\ 0 & 1\end{pmatrix}.$$

This means that the linear polynomial factorisation in x is mapped to a polynomial factorisation in the y-variables. The interesting fact is now that both factorisations describe rational boundary states in the corresponding conformal field theories [**5**]. One can go further and consider the x-factorisation

$$\big((x_1-\xi x_2)(x_1-\xi^3x_2)\big)\cdot\frac{W'(x_1,x_2)}{(x_1-\xi x_2)(x_1-\xi^3 x_2)} = W'(x_1,x_2)\,. \qquad (3.12)$$

The quadratic factor is mapped to

(3.13) $(x_1-\xi x_2)(x_1-\xi^3x_2)$
$$\mapsto \begin{pmatrix} \frac{1-\xi^4}{2}(y_1^2-2y_2)-(\xi+\xi^3)y_2 & \frac{1+\xi^4}{2}(y_1^2-4y_2)y_1 \\ \frac{1+\xi^4}{2}y_1 & \frac{1-\xi^4}{2}(y_1^2-2y_2)-(\xi+\xi^3)y_2 \end{pmatrix}$$
$$\xrightarrow[\text{transf.}]{\text{similarity}} \begin{pmatrix} y_1^2-2(1+\cos\frac{\pi}{k})y_2 & 0 \\ y_1 & y_1^2-2(1+\cos\frac{3\pi}{k})y_2\end{pmatrix}.$$

Again this factorisation has been identified with a rational boundary condition in the Kazama-Suzuki model in [**5**]. This example shows that the variable transformation defect is indeed very useful in generating interesting matrix factorisations. In a subsequent publication we will show that, with the help of this variable transformation defect, one can also generate rational defects in Kazama-Suzuki models which then allow one to generate in principle all factorisations corresponding to rational boundary conditions in these models.

The defect considered here actually also appears in the link homology of Khovanov and Rozansky [**21**], namely the diagram on the right in figure 1 corresponds in our language to the defect $({}_{S'}I_R)\otimes({}_RI_S)$ (where $S'=\mathbb{C}[x_3,x_4]$). The diagram on the left of figure 1 simply corresponds to the identity defect in x-variables. One

FIGURE 1. Basic building blocks that appear in the resolution of crossings [**21**, figure 9]: the identity defect ${}_{S'}I_S$ in the x-variables to the left, and the basic wide-edge graph on the right corresponding to $({}_{S'}I_R) \otimes ({}_R I_S)$ (with $S = \mathbb{C}[x_1, x_2]$, $S' = \mathbb{C}[x_3, x_4]$).

of the fundamental equivalences in the link homology displayed in figure 2 would read in our notation (with $S'' = \mathbb{C}[x_5, x_6]$)

$$
\big(({}_{S''}I_R) \otimes ({}_R I_{S'})\big) \otimes \big(({}_{S'}I_R) \otimes ({}_R I_S)\big) \cong \big(({}_{S''}I_R) \otimes ({}_R I_S)\big)^{\oplus 2} \,, \tag{3.14}
$$

which follows immediately from (2.12). It would be very interesting to also consider

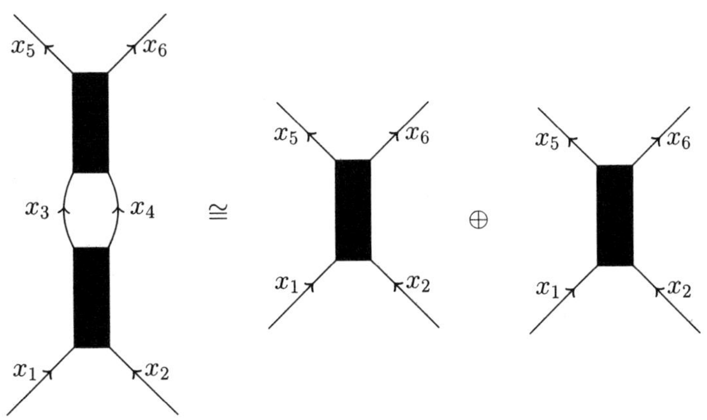

FIGURE 2. One of the fundamental diagram equivalences of [**21**, figure 35 and Prop. 30] (up to grading).

the morphisms between the defects in figure 1 that are needed to formulate the complex of defects assigned to crossings (see [**21**, figure 46]) in our framework, but we leave this for future work.

***SU*(*n* + 1)/*U*(*n*) Kazama-Suzuki models.** The last example has a beautiful generalisation to a defect between an $SU(n+1)/U(n)$ Kazama-Suzuki model and n copies of minimal models. We consider the polynomial rings $R = \mathbb{C}[y_1, \ldots, y_n]$ and $S = \mathbb{C}[x_1, \ldots, x_n]$, and the potential

$$
W'(x_1, \ldots, x_n) = x_1^k + \cdots + x_n^k \,. \tag{3.15}
$$

The ring homomorphism is defined by

$$\phi(y_j) = \sum_{i_1<\cdots<i_j} x_{i_1}\cdots x_{i_j} \,, \tag{3.16}$$

and it maps the y_j to the elementary symmetric polynomials in the x_i. It is an old result in invariant theory [**1**, section II.G] that ${}_RS$ is a free R-module of rank $n!$. To get an explicit R-module isomorphism between ${}_RS$ and $R^{\oplus n!}$, one needs to choose a good basis in S. The simplest choice [**1**] is to take the $n!$ polynomials given by

$$x_1^{\nu_1}x_2^{\nu_2}\cdots x_n^{\nu_n} \,, \quad \text{where } \nu_i \leq i-1 \,. \tag{3.17}$$

Another possibility with some computational advantages is provided by the Schubert polynomials $X_\sigma(x_1,\ldots,x_n)$, for which there is one for each permutation σ in the symmetric group S_n. It was shown in [**22**] that any polynomial $p(x_1,\ldots,x_n)$ has a unique expansion

$$p(x_1,\ldots,x_n) = \sum_{\sigma\in S_n} p_\sigma(x_1,\ldots,x_n)X_\sigma(x_1,\ldots,x_n) \,, \tag{3.18}$$

where the p_σ are totally symmetric. The map ρ is then given by

$$\begin{aligned}\rho: \quad R^{\oplus n!} &\to S \\ (p_\sigma(y_1,\ldots,y_n))_{\sigma\in S_n} &\mapsto \sum_{\sigma\in S_n} p_\sigma(x_1+\cdots+x_n,\ldots,x_1\cdots x_n)X_\sigma(x_1,\ldots,x_n) \,.\end{aligned} \tag{3.19}$$

4. Conclusion

We have seen that there is a natural defect between Landau-Ginzburg theories whose superpotentials are related by a variable transformation. The fusion of this defect onto other factorisations has an explicit and simple description via the functors ϕ^* and ϕ_* corresponding to extension and restriction of scalars.

The examples have shown that these defects can be used to relate boundary conditions or defects in different LG models. In particular, one can use such defects between minimal models and Grassmannian Kazama-Suzuki models to put into use the knowledge that is already available for minimal models to obtain factorisations for the Kazama-Suzuki models. In this way one can for example generate all factorisations corresponding to rational boundary conditions in the $SU(3)/U(2)$ model, as we will show in a subsequent publication.

In the $SU(3)/U(2)$ example it turns out that the defects discussed here are crucial in constructing factorisations for rational topological defects. These finitely many elementary defects and their superpositions form a closed semiring that is isomorphic to the fusion semiring. Realising such a finite-dimensional semiring (in the sense that as a semigroup it is isomorphic to a direct product of finitely many copies of $\mathbb{N}_0$) in terms of defect matrix factorisations reflects the rational structure of the conformal field theory that is otherwise hard to see in the LG formulation. It would be interesting to investigate whether the existence of such an algebraic structure automatically signals an enhanced symmetry in the CFT.

Finally, we have seen that our defects generate the building blocks of Khovanov-Rozansky homology, except for the morphisms between defect building blocks. In other words, one could say that our formulation provides a physical setup of the Khovanov-Rozansky factorisations as a sequence of Kazama-Suzuki models separated by defects. By generalisation to $SU(n+1)/U(n)$ models, this is also true for the higher graphs appearing in the MOY calculus [**23**]. Our analysis can therefore be seen as a physical supplement to the recent results in [**4, 11**].

Acknowledgements

We thank Nils Carqueville, Dan Murfet and Ingo Runkel for useful discussions.

References

[1] Emil Artin, *Galois theory*, 2nd ed., Notre Dame Mathematical Lectures, no. 2, University of Notre Dame Press, 1944.

[2] Sujay K. Ashok, Eleonora Dell'Aquila, and Duiliu-Emanuel Diaconescu, *Fractional branes in Landau-Ginzburg orbifolds*, Adv. Theor. Math. Phys. **8** (2004), 461–513, [hep-th/0401135].

[3] Constantin Bachas and Matthias Gaberdiel, *Loop operators and the Kondo problem*, JHEP **0411** (2004), 065, [hep-th/0411067].

[4] Hanno Becker, *Khovanov-Rozansky homology via Cohen-Macaulay approximations and Soergel bimodules*, Diploma thesis, University of Bonn, 2011, arXiv:1105.0702.

[5] Nicolas Behr and Stefan Fredenhagen, *D-branes and matrix factorisations in supersymmetric coset models*, JHEP **11** (2010), 136, [1005.2117].

[6] Ilka Brunner, Manfred Herbst, Wolfgang Lerche and Bernhard Scheuner, *Landau-Ginzburg realization of open string TFT*, JHEP **11** (2006), 043, [hep-th/0305133].

[7] Ilka Brunner and Matthias R. Gaberdiel, *Matrix factorisations and permutation branes*, JHEP **07** (2005), 012, [hep-th/0503207].

[8] Ilka Brunner and Daniel Roggenkamp, *B-type defects in Landau-Ginzburg models*, JHEP **08** (2007), 093, [0707.0922].

[9] ———, *Defects and bulk perturbations of boundary Landau-Ginzburg orbifolds*, JHEP **04** (2008), 001, [0712.0188].

[10] Nils Carqueville and Ingo Runkel, *On the monoidal structure of matrix bi-factorisations*, J. Phys. A **43** (2010), 275401, [0909.4381].

[11] Nils Carqueville and Daniel Murfet, *Computing Khovanov-Rozansky homology and defect fusion*, (2011), [1108.1081].

[12] Claudio Caviezel, Stefan Fredenhagen, and Matthias R. Gaberdiel, *The RR charges of A-type Gepner models*, JHEP **01** (2006), 111, [hep-th/0511078].

[13] Stefan Fredenhagen and Matthias R. Gaberdiel, *Generalised $N = 2$ permutation branes*, JHEP **11** (2006), 041, [hep-th/0607095].

[14] Stefan Fredenhagen, Matthias R. Gaberdiel, and Cornelius Schmidt-Colinet, *Bulk flows in Virasoro minimal models with boundaries*, J.Phys.A **A42** (2009), 495403, [0907.2560].

[15] Jürg Fröhlich, Jürgen Fuchs, Ingo Runkel, and Christoph Schweigert, *Duality and defects in rational conformal field theory*, Nucl.Phys. **B763** (2007), 354–430, [hep-th/0607247].

[16] Doron Gepner, *Scalar field theory and string compactification*, Nucl. Phys. **B322** (1989), 65.

[17] Kevin Graham and Gerard M.T. Watts, *Defect lines and boundary flows*, JHEP **04** (2004), 019, [hep-th/0306167].

[18] Hans Jockers and Wolfgang Lerche, *Matrix Factorizations, D-Branes and their Deformations*, Nucl. Phys. Proc. Suppl. **171** (2007), 196–214, [0708.0157].

[19] Anton Kapustin and Yi Li, *D-branes in Landau-Ginzburg models and algebraic geometry*, JHEP **12** (2003), 005, [hep-th/0210296].

[20] Anton Kapustin and Lev Rozansky, *On the relation between open and closed topological strings*, Commun.Math.Phys. **252** (2004), 393–414, [hep-th/0405232].

[21] Mikhail Khovanov and Lev Rozansky, *Matrix factorizations and link homology*, Fund. Math. **199** (2008), no. 1, 1, [math/0401268].

[22] Ian G. Macdonald, *Schubert polynomials*, Surveys in Combinatorics (A.D. Keedwell, ed.), London Mathematical Society Lecture Note Series, vol. 166, Cambridge University Press, Cambridge, 1991, p. 73.

[23] Hitoshi Murakami, Tomotada Ohtsuki, and Shuji Yamada, *HOMFLY polynomial via an invariant of colored plane graphs*, Enseign. Math. (2) **44** (1998), no. 3-4, 325.

[24] V. B. Petkova and J. B. Zuber, *Generalised twisted partition functions*, Phys. Lett. **B504** (2001), 157–164, [hep-th/0011021].

[25] Yuji Yoshino, *Tensor products of matrix factorizations*, Nagoya Math. J. **152** (1998), 39.

Max Planck Institute for Gravitational Physics, Potsdam, Germany
E-mail address: Nicolas.Behr@aei.mpg.de

Max Planck Institute for Gravitational Physics, Potsdam, Germany
E-mail address: Stefan.Fredenhagen@aei.mpg.de

Proceedings of Symposia in Pure Mathematics
Volume **85**, 2012

The D-brane U-scan

Eric A. Bergshoeff and Fabio Riccioni

ABSTRACT. We consider the D-branes that occur in IIA/IIB string theory compactified on a torus. We review how a general expression for the Wess-Zumino term of such branes is derived. We also review the method to determine the D-brane Wess-Zumino term in a U-duality covariant way, and we apply it to derive all the branes obtained by transforming the D-branes under U-duality in any dimension above five. We finally determine all the supersymmetric branes supporting worldvolume tensor multiplets that occur in these theories in any dimension.

1. Introduction

D-branes, that are defined as the endpoints of open strings, are a crucial ingredient in the study of non-perturbative string theories. The fact that string theories possess non-perturbative U-duality symmetries implies that D-branes are mapped by these symmetries to other non-perturbative branes. In order to study how D-branes transform under U-duality, it is sufficient to consider the low-energy limit, in which this duality can be seen as a continuous symmetry which is the global symmetry of the corresponding supergravity theory. We are interested in theories with maximal supersymmetry, that is, IIA/IIB string theory compactified on a torus. For these theories the continuous U-duality symmetries are given in Table 1.

In general, p-branes couple electrically, via a Wess-Zumino (WZ) term, to the $(p+1)$-form potentials of the low-energy supergravity theory. It turns out that the supersymmetry algebra of maximal supergravity theories closes not only on the propagating gauge fields and their magnetic duals, but also on $(D-2)$-forms (dual to scalars), $(D-1)$-forms (dual to constants) and D-forms (objects with vanishing field-strength). The allowed U-duality representations of these potentials have been classified in [**1, 2, 3**]. The branes associated to these potentials have codimension equal to or lower than 2, and as such there are consistency issues that one has to consider in introducing them, most notably the fact that in order to have a finite energy solution one has to introduce orientifolds.[1] Nonetheless, one can still determine the number of these branes by only imposing the weaker requirement

Key words and phrases. Duality, supersymmetry. 2010 Mathematics subject classification: Primary 81T30.

[1]The U-duality representations of branes with codimension higher than two have been extensively studied in the past, see e.g. [**4**].

dimension D	duality group G
11	1
10A	$\mathbb{R}^+$
10B	$\mathrm{SL}(2,\mathbb{R})$
9	$\mathrm{GL}(2,\mathbb{R})$
8	$\mathrm{SL}(3,\mathbb{R}) \times \mathrm{SL}(2,\mathbb{R})$
7	$\mathrm{SL}(5,\mathbb{R})$
6	SO(5,5)
5	$\mathrm{E}_{6(6)}$
4	$\mathrm{E}_{7(7)}$
3	$\mathrm{E}_{8(8)}$

TABLE 1. *The U-duality groups for all maximal supergravities in dimensions $3 \leq D \leq 11$. The group is always over the real numbers and of split real form. In $D = 10$ we distinguish between IIA and IIB supergravity.*

of local existence, which amounts to requiring the existence of a supersymmetric effective action. Remarkably, this analysis reveals that these "non-standard" branes are fewer than the corresponding potentials.

The counting of non-standard branes was performed in the IIB case in [**5**]. Given the $\mathrm{SL}(2,\mathbb{R})$ covariant forms of the theory [**6**]

$$A_{2,\alpha}\ (\mathbf{2}) \quad A_4\ (\mathbf{1}) \quad A_{6,\alpha}\ (\mathbf{2}) \quad A_{8,\alpha\beta}\ (\mathbf{3}) \quad A_{10,\alpha\beta\gamma}\ (\mathbf{4}) \quad A_{10,\alpha}\ (\mathbf{2}) \tag{1.1}$$

one can project on the RR sector, which is associated to D-branes, introducing the charges q^α and $\tilde{q}^\alpha$ such that

$$B_2 = q^\alpha A_{2,\alpha} \qquad C_2 = \tilde{q}^\alpha A_{2,\alpha} \tag{1.2}$$

are the NS-NS and RR 2-forms. Performing an $\mathrm{SL}(2,\mathbb{R})$ transformation therefore corresponds to swapping q^α and $\tilde{q}^\alpha$. The charge that projects on the RR 8-form is

$$\tilde{q}^\alpha q^\beta q^\gamma q^\delta \epsilon_{\alpha\beta} \quad , \tag{1.3}$$

and denoting by $n(q)$ and $n(\tilde{q})$ the number of q's and $\tilde{q}$'s that occur in a given expression for the charges, we label the charge in eq. (1.3) with $(n(q), n(\tilde{q})) = (3,1)$. By swapping q and $\tilde{q}$ this charge can only be mapped to the charge (1,3). Consequently, there are two supersymmetric branes inside the triplet of 8-forms $A_{8,\alpha\beta}$.[2] Similarly, the D9-brane corresponds to projecting the quadruplet of 10-forms $A_{10,\alpha\beta\gamma}$ via the charge (4,1), which can only be rotated to the charge (1,4) giving in total two 9-branes inside the quadruplet. This analysis also reveals that the way the tension scales with the string coupling,

$$\text{Tension} \sim (g_S)^\alpha \quad , \tag{1.4}$$

is given by

$$\alpha = -n(\tilde{q}) \quad . \tag{1.5}$$

The aim of this contribution is to review the results of [**8**], where the same techniques were applied to maximal supergravity theories in lower dimensions, leading

[2]The third potential does not correspond to a single supersymmetric brane, but can be associated to a bound state of two supersymmetric branes [**7**].

to a universal gauge-invariant expression for the WZ terms of the D-branes. We will then apply these methods to determine in any dimension above five the number of non-standard branes that are obtained by performing U-duality rotations on the D-branes. Finally, we will also determine in any dimension the number of 5- branes whose worldvolume degrees of freedom are described by a tensor multiplet. These branes, that we call "tensor branes", always belong to different U-duality multiplets with respect to the D-branes.

2. D-branes in various dimensions

We wish to repeat the IIB analysis of [**5**] in any dimension D for any maximal supergravity theory. We list the n-forms on which the supersymmetry algebra closes as

$$A_{1,M_1}\,, \quad A_{2,M_2}\,, \quad A_{3,M_3}\,, \quad \cdots \quad , A_{D,M_D}\,, \tag{2.1}$$

where $M_1, M_2, ..., M_D$ denote the representations of U-duality to which the various forms belong. For all the spacetime-filling D-forms and for all forms of rank higher than five these representations turn out to be reducible.

Under gauge transformations, in general each form in (2.1) transforms with respect to the gauge parameters of the lower rank forms according to the gauge algebra of the given theory, whose generalized structure constants are invariant tensors of the U-duality group. For instance, the 2-form A_{2,M_2} transforms as

$$\delta A_{2,M_2} = d\Lambda_{1,M_2} - f^{M_1 N_1}{}_{M_2} \Lambda_{0,M_1} F_{2,N_1} \quad , \tag{2.2}$$

where Λ_{0,M_1} and F_{2,M_1} are the gauge parameter and field strength of the 1-form A_{1,M_1} and $f^{M_1 N_1}{}_{M_2}$ is an invariant tensor. This leads to the gauge-invariant field strength

$$F_{3,M_2} = dA_{2,M_2} + f^{M_1 N_1}{}_{M_2} A_{1,M_1} F_{2,N_1} \quad . \tag{2.3}$$

Similarly, one can define generalized structure constants for higher rank fields, and the closure of the gauge algebra determines the constraints that these invariant tensors have to consistently satisfy [**9, 8**].

In any dimension D, the U-duality group contains the T-duality subgroup that has the universal expression SO(d, d), with $d = 10 - D$. Given that T-duality is a perturbative symmetry of string theory, we expect the D-branes to form representations of the T-duality group. It is therefore useful to decompose the U-duality representations with respect to T-duality according to

$$\text{U-duality} \supset \text{SO}(d,d) \times \mathbb{R}^+ \quad . \tag{2.4}$$

In general, for each supersymmetric brane occurring in the theory, the scaling of its tension with respect to the string coupling, determined by the parameter α in (1.4), is related to the $\mathbb{R}^+$-weight of the corresponding potential.

Direct inspection in all dimensions [**8, 10**] reveals a universal structure, at least for the highest values of α, which we summarize here:

- Neglecting the scalars, the highest possible value of α is $\alpha = 0$, which corresponds to the fundamental fields $B_{1,A}$ and B_2 in the vector and singlet representations of T-duality respectively. These fields are associated to the fundamental particles and the fundamental string.

- The RR fields, that is, the fields associated to D-branes, that have $\alpha = -1$, are spinors of T-duality of alternating chirality $C_{2n,\dot{a}}$ and $C_{2n+1,a}$ [**8**]. For standard branes this has been discussed in e.g. [**11**].
- The solitonic fields, associated to branes with $\alpha = -2$, are antisymmetric tensors of T-duality [**10**].
- The fields associated to branes with $\alpha = -3$ also have a universal structure. It turns out that they belong to tensor-spinor representations of T-duality [**12**].
- In the case of forms belonging to reducible representations, the RR fields (associated to D-branes) always belong to the highest dimensional representation.

We are now in the position to generalize the IIB analysis of [**5**] that was reviewed in the introduction. We introduce two charges $q_A^{M_1}$ and $\tilde{q}_a^{M_1}$ that project the 1-form A_{1,M_1} on the fundamental and RR 1-forms:

$$B_{1,A} = q_A^{M_1} A_{1,M_1} \qquad C_{1,a} = \tilde{q}_a^{M_1} A_{1,M_1} \quad . \tag{2.5}$$

It follows that the charges that project the U-duality covariant fields (2.1) on the fundamental 2-form B_2 and all the RR fields of rank higher than 1 are all given in terms of these charges [**8**], and using the properties of the invariant tensors of the various U-duality groups it can be shown that these expressions are always unique. As an example, we write here the expression for the fundamental and RR 2-forms:

$$B_2 = q_A^{M_1} q_B^{N_1} f_{M_1N_1}{}^{M_2} \eta^{AB} A_{2,M_2} \qquad C_{2,\dot{a}} = q_A^{M_1} \tilde{q}_a^{N_1} f_{M_1N_1}{}^{M_2} \Gamma^{Aa}_{\dot{a}} A_{2,M_2} \quad . \tag{2.6}$$

Exactly as in the IIB case, the charges that project on the fundamental fields have $n(\tilde{q}) = 0$, while the charges that project on the RR fields have $n(\tilde{q}) = 1$. The general expression for the charge corresponding to the RR fields of rank higher than 2 was derived in [**8**].

In order to write down a gauge invariant WZ term for a D-brane, one introduces the world-volume fields $b_{0,A}$ and b_1, such that the corresponding gauge invariant field strengths are

$$\mathcal{F}_{1,A} = db_{0,A} + B_{1,A} \qquad \mathcal{F}_2 = db_1 + B_2 \quad . \tag{2.7}$$

It follows that the WZ terms have the universal and elegant expression

$$e^{\mathcal{F}_2} e^{\mathcal{F}_{1,A}\Gamma^A} C \quad , \tag{2.8}$$

where Γ_A are the gamma matrices of the T-duality group SO(d,d) acting on the various RR fields, which are spinors of T-duality. The above formula thus provides an expression for the WZ term of a D-brane for each component of the T-duality spinor. Using the properties of these gamma matrices, one can show that the world-volume fields contained in the expression (2.8), together with the transverse scalars, form the bosonic sector of a world-volume vector multiplet as required [**8, 10**].

3. The D-brane U-scan

In this section we determine the numbers of different branes obtained by performing U-duality transformations on the D-branes discussed in the previous section. Since for standard branes the result of this analysis is straightforward, because the number of these branes is equal to the dimension of the representation of the corresponding field, we will concentrate on the non-standard branes. The strategy is the same as in the IIB case reviewed in the introduction: given the expression for

brane	U repr	$\alpha = -1$	$\alpha = -2$	$\alpha = -3$	$\alpha = -4$
6-brane	$2 \subset \mathbf{3}_3$	1	–	1	
7-brane	$2 \subset \mathbf{3}_4$	1	–	1	
8-brane	$2 \subset \mathbf{4}_4$	1	–	–	1

TABLE 2. *The non-standard branes in nine dimensions. The U-duality group is $SL(2,\mathbb{R}) \times \mathbb{R}^+$. The subscript indicates the $\mathbb{R}^+$-weight.*

the charge of the D-brane, which for a RR n-form is an expression with $n-1$ q's and one $\tilde{q}$, we perform all possible rotations of the q's and $\tilde{q}$'s. In each expression, we determine the value of α of the corresponding brane according to

$$\alpha = -n(\tilde{q}) \quad . \tag{3.1}$$

We perform this analysis in any dimension above five. The lower-dimensional cases are more involved and we leave them as an open problem.

3.1. D=9. Given that the U-duality symmetry in nine dimensions is $\mathrm{SL}(2,\mathbb{R})\times\mathbb{R}^+$, the nine-dimensional analysis is similar to the IIB case. The 1-forms are $A_{1,\alpha}$, A_1 in the $\mathbf{2}\oplus\mathbf{1}$ of $\mathrm{SL}(2,\mathbb{R})$. We introduce the charges $\tilde{q}^\alpha$, q^α and q such that

$$B_1 = q^\alpha A_{1,\alpha} \qquad B_1' = qA_1 \qquad C_1 = \tilde{q}^\alpha A_{1,\alpha} \quad , \tag{3.2}$$

where B_1 and B_1' are the selfdual and anti-selfdual parts of a vector of the T-duality group $\mathrm{SO}(1,1)$.

The charges that project on the RR fields of higher rank are obtained in terms of these charges according to the following rule: going from even to odd rank corresponds to adding a charge q^α, and going from odd to even corresponds to adding a charge q. For example, the charge of a D1-brane is $\tilde{q}^\alpha q$, while the charge of a D2-brane is $\epsilon_{\alpha\beta}\tilde{q}^\alpha q^\beta q$, which shows that while the first can be related to a fundamental string of charge $q^\alpha q$, the second is a U-duality singlet. Proceeding this way, one can determine all the charges that project onto the RR fields. For the D6-brane one obtains

$$\tilde{q}^\alpha q^\beta q^\gamma q^\delta q^3 \epsilon_{\alpha\beta} \quad , \tag{3.3}$$

for the D7-brane one gets

$$\tilde{q}^\alpha q^\beta q^\gamma q^\delta q^4 \epsilon_{\alpha\beta} \tag{3.4}$$

and finally for the D8-brane one gets

$$\tilde{q}^\alpha q^\beta q^\gamma q^\delta q^\epsilon q^4 \epsilon_{\alpha\beta} \quad . \tag{3.5}$$

It is instructive to label these charges as $(n(q), n(\tilde{q}))$ in terms of the number of q^α and $\tilde{q}^\alpha$ that occur (the number of q charges is not relevant because it is invariant under U-duality). In particular, the charges of the D6-brane and D7-brane are labelled by (3,1), while the charge of the D8-brane is labelled by (4,1). Performing a U-duality transformation corresponds to swapping q^α and $\tilde{q}^\alpha$, which takes (3,1) to (1,3) and (4,1) to (1,4). This implies that for instance one cannot get a brane with $n(\tilde{q}) = 2$. The result, giving the full set of non-standard branes in nine dimensions, is summarized in Table 2.

brane	U repr	$\alpha = -1$	$\alpha = -2$	$\alpha = -3$	$\alpha = -4$
5-brane	$6 \subset (\mathbf{8}, \mathbf{1})$	$2 \subset (\mathbf{2}, \mathbf{1})$	$2 \subset (\mathbf{3}, \mathbf{1})$	$2 \subset (\mathbf{2}, \mathbf{1})$	
6-brane	$6 \subset (\mathbf{6}, \mathbf{2})$	$2 \subset (\mathbf{1}, \mathbf{2})$	–	$4 \subset (\mathbf{3}, \mathbf{2})$	
7-brane	$6 \subset (\mathbf{15}, \mathbf{1})$	$2 \subset (\mathbf{2}, \mathbf{1})$	–	$2 \subset (\mathbf{4}, \mathbf{1})$	$2 \subset (\mathbf{3}, \mathbf{1})$

TABLE 3. *The non-standard branes of $D = 8$ maximal supergravity. The U-duality symmetry is $SL(3, \mathbb{R}) \times SL(2, \mathbb{R})$ and the T-duality is $SL(2, \mathbb{R}) \times SL(2, \mathbb{R})$.*

3.2. D=8. In eight dimensions the U-duality group is $\mathrm{SL}(3, \mathbb{R}) \times \mathrm{SL}(2, \mathbb{R})$ and the T-duality group is $\mathrm{SL}(2, \mathbb{R}) \times \mathrm{SL}(2, \mathbb{R})$ which is indeed isomorphic to $\mathrm{SO}(2, 2)$. The 1-form $A_{1,Ma}$ is in the $(\overline{\mathbf{3}}, \mathbf{2})$ of the U-duality group, and is projected on the fundamental and RR fields through the charges

$$q_{\dot{a}}^{M} \quad \text{and} \quad \tilde{q}^{M} \quad , \tag{3.6}$$

where $\dot{a}$ denotes the $\mathbf{2}$ of the $\mathrm{SL}(2, \mathbb{R})$ inside $\mathrm{SL}(3, \mathbb{R})$. The fundamental and RR 1-forms are

$$B_{1,a\dot{a}} = q_{\dot{a}}^{M} A_{1,Ma} \qquad C_{1,a} = \tilde{q}^{M} A_{1,Ma} \quad . \tag{3.7}$$

The charges projecting on the fundamental string and the D-string are $\epsilon_{MNP} q_{\dot{a}}^{N} q_{\dot{b}}^{P} \epsilon^{\dot{a}\dot{b}}$ and $\epsilon_{MNP} \tilde{q}^{N} q_{\dot{a}}^{P}$ respectively, and one can similarly construct the charges for the higher rank forms using invariant tensors of the U-duality group. Given that there is always one and only one independent expression that gives the right charges, we can neglect the U-duality indices, so that for instance the charges for the fundamental and D-string can be schematically written as

$$q_{\dot{a}} q_{\dot{b}} \epsilon^{\dot{a}\dot{b}} \quad \text{and} \quad \tilde{q} q_{\dot{a}} \quad . \tag{3.8}$$

Moreover, we can label the charge in terms of the three numbers $(n(q_{\dot{1}}), n(q_{\dot{2}}), n(\tilde{q}))$, so that for instance the charge of the fundamental string is $(1, 1, 0)$ while the charge of the D-string is the doublet $(1, 0, 1)$ and $(0, 1, 1)$. One can easily see that performing $\mathrm{SL}(3, \mathbb{R})$ transformations these charges are mapped into each other.

The charge projecting on the RR 3-form is

$$\tilde{q} q_{\dot{a}} q_{\dot{b}} \epsilon^{\dot{a}\dot{b}} \tag{3.9}$$

which we label as $(1, 1, 1)$ and is clearly a singlet. This corresponds to a doublet of D2-branes, because the field possesses an a index with respect to the $\mathrm{SL}(2, \mathbb{R})$ not inside $\mathrm{SL}(3, \mathbb{R})$ which is left understood in our notation.

Proceeding in this way one can determine the charge of all the branes in the theory. We now concentrate on the non-standard branes. The D5-brane charge is

$$\tilde{q} q_{\dot{a}} q_{\dot{b}} q_{\dot{c}} q_{\dot{d}} q_{\dot{e}} \epsilon^{\dot{b}\dot{c}} \epsilon^{\dot{d}\dot{e}} \tag{3.10}$$

which is a doublet of charges $(3, 2, 1)$ and $(2, 3, 1)$. This can be rotated to the two solitonic branes $(1, 3, 2)$ and $(3, 1, 2)$, as well as to the doublet of $\alpha = -3$ branes with charges $(1, 2, 3)$ and $(2, 1, 3)$.

The D6-brane charge is

$$\tilde{q} q_{\dot{a}} q_{\dot{b}} q_{\dot{c}} q_{\dot{d}} q_{\dot{e}} q_{\dot{f}} \epsilon^{\dot{a}\dot{b}} \epsilon^{\dot{c}\dot{d}} \epsilon^{\dot{e}\dot{f}} \tag{3.11}$$

brane	U repr	$\alpha = -1$	$\alpha = -2$	$\alpha = -3$	$\alpha = -4$
4-brane	$20 \subset \mathbf{24}$	$4 \subset \overline{\mathbf{4}}$	$12 \subset \mathbf{15}$	$4 \subset \mathbf{4}$	
5-brane	$20 \subset \overline{\mathbf{40}}$	$4 \subset \mathbf{4}$	$4 \subset \mathbf{10}$	$12 \subset \overline{\mathbf{20}}$	
6-brane	$20 \subset \mathbf{70}$	$4 \subset \overline{\mathbf{4}}$	–	$12 \subset \mathbf{36}$	$4 \subset \mathbf{10}$

TABLE 4. *The number of non-standard branes in $D = 7$ maximal supergravity. The U-duality group is $SL(5,\mathbb{R})$ and the T-duality group is $SL(4,\mathbb{R})$.*

which we label by $(3,3,1)$. This is a doublet of branes because the 7-form field possesses an extra a index. The fact that there are no 2's in this partition of charges immediately implies that there cannot be any solitonic brane because the charge can never be rotated to something with two $\tilde{q}$'s. One can instead rotate to the two charges $(1,3,3)$ and $(3,1,3)$, corresponding to $\alpha = -3$ branes, which makes a total of four branes due to the extra a index of the field.

We finally consider the 8-forms. The D7-brane charge is

$$\tilde{q} q_{\dot{a}} q_{\dot{b}} q_{\dot{c}} q_{\dot{d}} q_{\dot{e}} q_{\dot{f}} q_{\dot{g}} \epsilon^{\dot{b}\dot{c}} \epsilon^{\dot{d}\dot{e}} \epsilon^{\dot{f}\dot{g}} \tag{3.12}$$

which corresponds to the doublet of charges $(4,3,1)$ and $(3,4,1)$. Again, this cannot be rotated to a solitonic brane, while there are two branes with $\alpha = -3$, that is, $(4,1,3)$ and $(1,4,3)$, and two branes with $\alpha = -4$, that is $(1,3,4)$ and $(3,1,4)$. The overall result is summarized in Table 3.

3.3. D=7. We now consider the branes in seven dimensions, where the U-duality is $\mathrm{SL}(5,\mathbb{R})$ and the T-duality is $\mathrm{SL}(4,\mathbb{R})$, which is isomorphic to $\mathrm{SO}(3,3)$. As in the previous subsection we do not write explictly the U-duality indices, and we project the 1-form $A_{1,MN}$ in the $\overline{\mathbf{10}}$ of $\mathrm{SL}(5,\mathbb{R})$ on the $\overline{\mathbf{4}}$ and $\mathbf{6}$ of $\mathrm{SL}(4,\mathbb{R})$ through the charges

$$\tilde{q}_{a5} \quad \text{and} \quad q_{ab} \quad , \tag{3.13}$$

where the index 5 has been left explicit to make the embedding of the $\mathbf{4}$ of $\mathrm{SL}(4,\mathbb{R})$ inside the $\mathbf{5}$ of $\mathrm{SL}(5,\mathbb{R})$ more transparent.

The charges of the 2-forms are

$$Q^{d}_{(1,1)} = \tilde{q}_{5a} q_{bc} \epsilon^{5abcd} \tag{3.14}$$

and

$$Q^{5}_{(2,0)} = q_{ab} q_{cd} \epsilon^{abcd5} \tag{3.15}$$

which correspond to the D1-brane and the fundamental string respectively. Here we denote with $Q_{(n,m)}$ the charge of an $(n+m)$-form with $n(q) = n$ and $n(\tilde{q}) = m$. Since one can rotate any d index to the index 5, all the 1-branes are connected by U-duality. We can also derive a general rule to determine how $n(\tilde{q})$ changes when we perform such a rotation. The rule is the following: if we rotate an upstairs index from a to 5, $n(\tilde{q})$ decreases by one unit and viceversa. If we rotate a downstairs index from a to 5, $n(\tilde{q})$ increases by one unit and viceversa.

We now move to the 3-forms. The charge for the D2-brane reads

$$Q_{(2,1)e} = \tilde{q}_{5a} q_{bc} q_{de} \epsilon^{5abcd}. \tag{3.16}$$

brane	U repr	$\alpha=-1$	$\alpha=-2$	$\alpha=-3$	$\alpha=-4$	$\alpha=-5$
3-brane	$40\subset\mathbf{45}$	$8\subset\mathbf{8}_V$	$24\subset\mathbf{28}$	$8\subset\mathbf{8}_V$		
4-brane	$80\subset\mathbf{144}$	$8\subset\mathbf{8}_S$	$32\subset\mathbf{56}_C$	$32\subset\mathbf{56}_S$	$8\subset\mathbf{8}_C$	
5-brane	$80\subset\mathbf{320}$	$8\subset\mathbf{8}_V$	$8\subset\mathbf{35}_V$	$48\subset\mathbf{160}_V$	$8\subset\mathbf{35}_V$	$8\subset\mathbf{8}_V$

TABLE 5. *The non-standard branes of $D=6$ maximal supergravity. The U-duality group is $SO(5,5)$ and the T-duality group is $SO(4,4)$.*

This charge can be rotated to $Q_{(1,2)5}$ which is the charge for a solitonic brane, in agreement with our rule.

We can now write down the expression for the charges of the non-standard D-branes. For the D4-brane one gets

$$Q_{(4,1)i}{}^{5}=\tilde{q}_{5a}q_{bc}q_{de}q_{fg}q_{hi}\epsilon^{5abcd}\epsilon^{efgh5}\quad . \tag{3.17}$$

This charge can be rotated to the one with $\alpha=-2$, $Q_{(3,2)i}{}^{j}$ with $i\neq j$, which corresponds to 12 branes, and to the $\alpha=-3$ charge $Q_{(2,3)5}{}^{j}$ in the **4** of SL$(4,\mathbb{R})$.

Next we move to the 5-branes. The D5-branes have charge

$$Q_{(5,1)}{}^{5,l5}=\tilde{q}_{5a}q_{bc}q_{de}q_{fg}q_{hi}q_{jk}\epsilon^{5abcd}\epsilon^{efgh5}\epsilon^{ijkl5}\quad , \tag{3.18}$$

which can be rotated to the $\alpha=-2$ charge $Q_{(4,2)}{}^{l,l5}$ corresponding to 4 branes inside the **10** of SL$(4,\mathbb{R})$, and to the $\alpha=-3$ charge $Q_{(3,3)}{}^{l,lm}$, with $l\neq m$, corresponding to 12 branes inside the $\overline{\mathbf{20}}$ of SL(4).

Finally, we consider the 6-branes. The charge of the D6-branes is

$$Q_{(6,1)m}{}^{55}=\tilde{q}_{5a}q_{bc}q_{de}q_{fg}q_{hi}q_{jk}q_{lm}\epsilon^{5abcd}\epsilon^{efgh5}\epsilon^{ijkl5} \tag{3.19}$$

in the $\overline{\mathbf{4}}$ of SL$(4,\mathbb{R})$. This cannot be rotated to a charge with $\alpha=-2$, while one can rotate to the $\alpha=-3$ charge $Q_{(4,3)m}{}^{nn}$, with $m\neq n$, which gives 12 branes inside the **36** of SL$(4,\mathbb{R})$, and to the $\alpha=-4$ charge $Q_{(3,4)5}{}^{nn}$ which gives four 6-branes inside the **10** of SL$(4,\mathbb{R})$. This concludes the seven-dimensional analysis, which is summarised in Table 4.

3.4. D=6. The six-dimensional case can in principle be treated precisely as the higher dimensional ones, but as we will see the analysis is considerably simplified because the U-duality is the orthogonal group SO(5,5). Given the 1-form $A_{1,\alpha}$ in the **16** of SO(5,5), one introduces the charges $q^{\alpha}_{\dot a}$ and $\tilde{q}^{\alpha}_{a}$ such that [3]

$$B_{1,\dot a}=q^{\alpha}_{\dot a}A_{1,\alpha}\qquad C_{1,a}=\tilde{q}^{\alpha}_{a}A_{1,\alpha}\quad . \tag{3.20}$$

Therefore, swapping q and $\tilde{q}$ simply corresponds to swapping the chirality of the T-duality spinors. For every $(m+n)$-form, given the charge denoted by $(n(q),n(\tilde{q}))$, one can always construct the charge in which $n(q)$ and $n(\tilde{q})$ are interchanged. This, together with the fact that the number of D-branes, solitonic branes and $\alpha=-3$ branes is known in all dimensions [**10, 12**], automatically determines all the other branes. For the 4-forms, one has 8 D3-branes, corresponding to the charge $(3,1)$,

[3] Note that in this six-dimensional case we have denoted the fundamental 1-form with a T-duality spinor index instead of a vector index as in all other cases. This is consistent due to triality of SO(4,4).

dimension	U repr	$\alpha = -2$	$\alpha = -3$	$\alpha = -4$
$D = 6$	$16 \subset \overline{\mathbf{126}}$	$8 \subset \mathbf{35}_{\rm S}$	–	$8 \subset \mathbf{35}_{\rm C}$
$D = 7$	$5 \subset \overline{\mathbf{15}}$	$4 \subset \overline{\mathbf{10}}$	–	$1 \subset \mathbf{1}$
$D = 8$	$2 \subset (\mathbf{1}, \mathbf{3})$	$2 \subset (\mathbf{1}, \mathbf{3})$		
$D = 9$	$1 \subset \mathbf{1}$	1		

TABLE 6. *The tensor branes in any dimension. These branes are 5-branes, and thus they can only exist in dimension six and above.*

24 solitonic branes with charge $(2, 2)$, and 8 $\alpha = -3$ branes, corresponding to the charge $(1, 3)$. The 5-forms give rise to 8 D4-branes (charge $(4, 1)$), 32 solitonic branes (charge (3,2)), 32 $\alpha = -3$ branes (charge (2,3)) and 8 $\alpha = -4$ branes (charge (1,4)). The 6-forms in the **320** give rise to 8 D5-branes (charge (5,1)), 8 solitons (charge (4,2)), 48 $\alpha = -3$ branes (charge (3,3)), 8 $\alpha = -4$ branes (charge (2,4)) and finally 8 $\alpha = -5$ branes (charge (5,1)). The full result is summarized in Table 5.

4. Tensor branes

The analysis we have performed leaves aside another class of branes that occur in string theory, namely the supersymmetric branes whose worldvolume degrees of freedom are described by a tensor multiplet. We call these branes "tensor branes", and the fact that tensor multiplets only occur in six dimensions implies that tensor branes must have a six-dimensional worldvolume, and thus only occur in supergravity theories in six dimensions or higher.

By looking at the way the 6-form potentials decompose under T-duality in various dimensions, one can deduce that in dimensions higher than seven such branes can only be solitonic. The counting of solitonic tensor branes in any dimension was performed in [**10**], and gives one tensor brane in nine dimensions, two in eight dimensions, four in seven dimensions and eight in six dimensions. The solitonic branes reveal the rule to determine the charge of these branes in general. The rule is that one has to take the square of the charge of the 3-form. This rule implies that the number of tensor branes equals in all cases the number of 2-branes, and that the only allowed tensor branes have either $\alpha = -2$ or $\alpha = -4$.

In seven dimensions, taking the square of the charge (3.16) one obtains the charge $Q_{(3,1)a}Q_{(3,1)a}$, describing 4 branes inside the $\overline{\mathbf{10}}$ of $\mathrm{SL}(4, \mathbb{R})$. This can be rotated to the charge $Q_{(3,2)5}Q_{(3,2)5}$, which is a singlet tensor brane with $\alpha = -4$. The six-dimenisonal case works in exactly the same way, giving 8 solitonic tensor branes from the square of the D2-brane charge and other 8 $\alpha = -4$ tensor branes from the square of the solitonic 2-brane charge. The results for all dimensions are summarized in Table 6.

5. Conclusions

We have reviewed the analysis of [**8**], giving a completely general gauge-invariant expression for the WZ terms of all D-branes in any dimensions and a method to construct a U-duality covariant expression for the RR fields of any maximal supergravity theory. We have applied this method to determine the number of branes that are obtained by performing U-duality transformations on the D-branes.

It is important to emphasize that the number of solitonic (i.e. $\alpha = -2$) branes and the $\alpha = -3$ branes have been derived recently by the authors using other independent methods [**10, 13, 12**], and they coincide in all cases with the ones derived here.

Acknowledgements

E.B. would like to thank the organizers of the String-Math 2011 conference for the opportunity to present this work.

References

[1] F. Riccioni and P. C. West, "The E(11) origin of all maximal supergravities," JHEP **0707** (2007) 063 [arXiv:0705.0752 [hep-th]].

[2] E. A. Bergshoeff, I. De Baetselier and T. A. Nutma, "E(11) and the embedding tensor," JHEP **0709** (2007) 047 [arXiv:0705.1304 [hep-th]].

[3] B. de Wit, H. Nicolai and H. Samtleben, "Gauged Supergravities, Tensor Hierarchies, and M-Theory," JHEP **0802** (2008) 044 [arXiv:0801.1294 [hep-th]].

[4] N. A. Obers and B. Pioline, "U duality and M theory, an algebraic approach," proceedings of the second international conference *Quantum aspects of gauge theories, supersymmetry and unification*, Corfu (Greece), 20-26 September 1998, p. 424-433, Springer (1999), arXiv:hep-th/9812139.

[5] E. A. Bergshoeff, M. de Roo, S. F. Kerstan, T. Ortín and F. Riccioni, "SL(2,R)-invariant IIB Brane Actions," JHEP **0702** (2007) 007 [arXiv:hep-th/0611036].

[6] E. A. Bergshoeff, M. de Roo, S. F. Kerstan and F. Riccioni, "IIB Supergravity Revisited," JHEP **0508** (2005) 098 [arXiv:hep-th/0506013]; E. A. Bergshoeff, J. Hartong, P. S. Howe, T. Ortín and F. Riccioni, "IIA/IIB Supergravity and Ten-forms," JHEP **1005** (2010) 061 [arXiv:1004.1348 [hep-th]].

[7] E. A. Bergshoeff, J. Hartong, T. Ortín and D. Roest, "Seven-branes and Supersymmetry," JHEP **0702** (2007) 003 [arXiv:hep-th/0612072].

[8] E. A. Bergshoeff, F. Riccioni, "D-Brane Wess-Zumino Terms and U-Duality," JHEP **1011** (2010) 139. [arXiv:1009.4657 [hep-th]].

[9] F. Riccioni, D. Steele and P. West, "The E(11) origin of all maximal supergravities: The Hierarchy of field-strengths," JHEP **0909** (2009) 095 [arXiv:0906.1177 [hep-th]].

[10] E. A. Bergshoeff, F. Riccioni, "String Solitons and T-duality," JHEPA,1105,131. 2011 **1105** (2011) 131. [arXiv:1102.0934 [hep-th]].

[11] M. Fukuma, T. Oota and H. Tanaka, "Comments on T dualities of Ramond-Ramond potentials on tori," Prog. Theor. Phys. **103** (2000) 425 [arXiv:hep-th/9907132].

[12] E. A. Bergshoeff and F. Riccioni, "Branes and wrapping rules," Phys. Lett. B **704** (2011) 367 [arXiv:1108.5067 [hep-th]].

[13] E. A. Bergshoeff and F. Riccioni, "Dual doubled geometry," Phys. Lett. B **702** (2011) 281 [arXiv:1106.0212 [hep-th]].

Centre for Theoretical Physics, University of Groningen, Nijenborgh 4, 9747 AG Groningen, The Netherlands

E-mail address: `E.A.Bergshoeff@rug.nl`

INFN Sezione di Roma, Dipartimento di Fisica, Università di Roma "La Sapienza", Piazzale Aldo Moro 2, 00185 Roma, Italy

E-mail address: `Fabio.Riccioni@roma1.infn.it`

Proceedings of Symposia in Pure Mathematics
Volume **85**, 2012

An invitation to algebraic topological string theory

Nils Carqueville and Michael M. Kay

ABSTRACT. The purpose of this note is to provide a short invitation to the universal algebraic approach to topological string theory. In the first section we make an attempt to explain the origin of this approach and how it fits into the bigger picture of full string theory, while in the second half of this note we will introduce the relevant notions in more detail and discuss some of our main results on bulk-deformed open topological string amplitudes.

1. Introduction

We start our discussion with the bulk sector. In the low energy limit this amounts to the study of $\mathcal{N} = 1$ supergravity in ten dimensions and its solutions, viewed as vacua of closed string theory. We restrict our attention to such solutions that have a six-dimensional compact factor M. Then the Einstein equations imply that this manifold is a Calabi-Yau space.

In the framework of closed perturbative string theory these solutions M are realised as the possible targets of two-dimensional sigma models with $\mathcal{N} = (2, 2)$ superconformal symmetry. More precisely, solutions of classical gravity appear as the low energy limit of the worldsheet description. There is a distinguished set of *marginal* fields of these $\mathcal{N} = (2, 2)$ sigma models that implement infinitesimal deformations which may transform the metric of the corresponding target Calabi-Yau manifold to another nearby Calabi-Yau space. In general the superconformal field theory (CFT) $\mathcal{C}$ associated to M is deformed to another $\mathcal{N} = (2, 2)$ CFT, which may be more "stringy" in nature in the sense that it does not need to have a geometric interpretation. The marginal fields can be viewed as defining sections of the cotangent bundle of the moduli space of $\mathcal{N} = (2, 2)$ CFTs containing our initial $\mathcal{C}$, namely the space of vacua continuously connected to $\mathcal{C}$.

More generally one can consider all *chiral primary* fields that deform away from $\mathcal{C}$ in the much larger moduli space of $\mathcal{N} = (2, 2)$ supersymmetric field theories. Locally (anti-) chiral primaries define Riemann normal coordinates on this moduli space for a neighbourhood of the CFT we started with. We will denote these coordinates by t_i and refer to them as closed moduli.

The chiral primaries can be divided into left- and right-moving zero modes of either the (c, c) (chiral, chiral) or the (a, c) (antichiral, chiral) sectors. The

2010 *Mathematics Subject Classification.* 81T45, 18G55, 81T30.

first sector corresponds to type IIB string theory, while the second corresponds to type IIA. From the point of view of the superconformal algebra, the two types are related by an involutive outer automorphism. Thus, for every type IIA theory there should be a corresponding type IIB theory – this is precisely the mirror symmetry conjecture. Of the chiral primaries that deform the metric of the Calabi-Yau, those in the type IIB theory correspond to complex structure deformations, while those in the type IIA theory govern deformations of the complexified Kähler structure. Thus by Yau's theorem the type IIA and type IIB theories together contain complete information about the moduli space of metrics.

Both the (c,c) and (a,c) fields, together with their operator product expansion, naturally form an algebra. In fact these are Frobenius algebras that define two topological field theories (TFTs) called the B-model and A-model, respectively. More generally one can construct two such TFTs from any $\mathcal{N}=(2,2)$ CFT by a procedure known as topological twisting where the chiral primary fields appear as the cohomology of a BRST operator. The subsector of full string theory that builds on these TFTs is the one that we are interested in here. It is precisely the sector of all chiral primaries, and from it one can compute quantities of the *full theory* like the effective superpotential that we discuss below.

We will now consider either one of these topological twists. The associated algebra of fields ϕ_i is encoded in the structure constants $\langle\phi_i(0),\phi_j(1)\phi_k(\infty)\rangle_{\text{bulk}}$ where $\langle\,\cdot\,,\,\cdot\,\rangle_{\text{bulk}}$ is the topological metric and $0,1,\infty$ is the standard choice of punctures on the genus zero worldsheet. Note that the set of structure constants represents only a single point in our moduli space of $\mathcal{N}=(2,2)$ field theories. What we would like to describe is a (possibly only infinitesimal) patch around that point. The local coordinates of this patch are the closed moduli t_i, which allow us to deform the TFT *correlators* $\langle\phi_i(0),\phi_j(1)\phi_k(\infty)\rangle_{\text{bulk}}$ to the *amplitudes*

$$\left\langle\phi_i(0),\phi_j(1)\phi_k(\infty)\,\mathrm{e}^{\sum_l t_l\int\phi_l^{(2)}}\right\rangle_{\text{bulk}} \tag{1.1}$$

where $\phi_l^{(2)}$ are the two-form descendants of the chiral primaries ϕ_l.

Let us expand the exponential in the amplitudes and define higher maps ℓ_n which act on the fields ϕ_i as follows:

$$\left\langle\phi_{i_0}(0),\phi_{i_1}(1)\phi_{i_2}(\infty)\int\phi_{i_3}^{(2)}\cdots\int\phi_{i_n}^{(2)}\right\rangle_{\text{bulk}}=\left\langle\phi_{i_0},\ell_n(\phi_{i_1},\ldots,\phi_{i_n})\right\rangle_{\text{bulk}}.$$

The topological metric $\langle\,\cdot\,,\,\cdot\,\rangle_{\text{bulk}}$ together with the maps ℓ_n define what is called a Calabi-Yau L_∞-structure [**25, 26, 11**] (see the next section for more details). This type of algebra completely encodes the full structure of classical closed topological string theory: knowing the maps ℓ_n and the topological metric one can compute all amplitudes.

Now we turn to the boundary sector. There are several reasons for introducing open strings, ranging from phenomenological to mathematical. For example, from a phenomenological perspective open strings and branes allow for nonabelian gauge symmetries in four-dimensional effective low energy field theories, and their presence also reduces the amount of spacetime supersymmetry. Another compelling reason for introducing open strings appears if one leaves the classical description of gravity by turning on the string coupling constant. It was in fact shown in [**24**] that the string coupling constant can be interpreted as the deformation parameter that quantises the moduli space of Calabi-Yau manifolds attached to a given M, hence providing a first step to the quantisation of gravity. In this framework open strings are necessary to describe excited states of the quantum theory [**20**].

In the following we will restrict to the genus zero case, which already has a rich structure that demands deeper understanding. We are mostly interested in boundary conditions that describe BPS branes. In the topologically twisted theory they descend to branes whose only open string states are chiral primaries that have an operator product that is strictly associative. Hence these branes and open strings naturally form the objects and morphisms of a category. Together with sewing relations and the boundary topological metric $\langle\,\cdot\,,\,\cdot\,\rangle_{\mathrm{bdry}}$, this data combined with the closed TFT structure defines an open-closed TFT as in [**14, 19**]. For notational simplicity we will mostly consider the case of only one brane in this note.

The open and closed structure at one point of our moduli space of $\mathcal{N}=(2,2)$ theories is encoded in the deformed open string three-point correlator

$$\Big\langle \psi_{a_0}(p_0), \psi_{a_1}(p_1)\psi_{a_2}(p_2)\, \mathcal{P}\, \mathrm{e}^{\sum_i u_{a_i} \int \psi_{a_i}^{(1)}} \Big\rangle_{\mathrm{bdry}}$$

which is the boundary version of the bulk sector expression (1.1). Here the ψ_a denote the open string chiral primaries inserted at generic points p_0, p_1, p_2 on the boundary of the disk, and the $\psi_a^{(1)}$ are their descendants implementing deformations of the purely open theory. The u_a are free moduli which are to be viewed as local coordinates of a non-commutative space; setting them to zero we obtain the TFT structure constants.

Just like in the bulk sector we can expand the exponential and consider individual open string amplitudes which are now given in terms of *higher products* $\widetilde{r}_n$:

$$\Big\langle \psi_{a_0}(p_0), \psi_{a_1}(p_1)\psi_{a_2}(p_2) \int \psi_{a_3}^{(1)} \cdots \int \psi_{a_n}^{(1)} \Big\rangle_{\mathrm{bdry}} = \Big\langle \psi_{a_0}, \widetilde{r}_n(\psi_{a_1} \otimes \cdots \otimes \psi_{a_n}) \Big\rangle_{\mathrm{bdry}}.$$

The topological metric $\langle\,\cdot\,,\,\cdot\,\rangle_{\mathrm{bdry}}$ and the maps $\widetilde{r}_n$ define a *Calabi-Yau A_∞-algebra* which arises from the symmetries of amplitudes and which completely encodes the full structure of open topological string theory [**9, 7**] (see the next section for the definition). Similarly, for arbitrarily many branes we are led to a Calabi-Yau A_∞-category. In the case of the B-model with target M this category is the bounded derived category of coherent sheaves $\mathbb{D}^b(M)$, while the boundary sector of the A-model is the derived Fukaya category $\mathrm{Fuk}(M)$.

Recall that in the purely closed sector the mirror symmetry conjecture says that there is an involution $\mathcal{I}$ acting on the set of moduli spaces of Calabi-Yau manifolds, exchanging complex structure deformations with complexified Kähler structure deformations. In the boundary sector the conjecture extends to homological mirror symmetry which in particular states that $\mathbb{D}^b(M)$ and $\mathrm{Fuk}(\mathcal{I}(M))$ are equivalent as Calabi-Yau A_∞-categories. Thus we have identified a second important reason to study A_∞-algebras.

So far we have sketched how the structure of the full moduli space of an open-closed topological string theory at genus zero consists of A_∞-categories fibred over the commutative moduli space of closed TFTs. In the next section we will give a prescription for how to glue these fibres in a neighbourhood of a chosen point in moduli space. In particular we will arrive at an explicit recursive formula for the *bulk-deformed open string amplitudes*

$$\Big\langle \psi_{a_0}(p_0), \psi_{a_1}(p_1)\psi_{a_2}(p_2) \int \psi_{a_3}^{(1)} \cdots \int \psi_{a_n}^{(1)}\, \mathrm{e}^{\sum_i t_i \int \phi_i^{(2)}} \Big\rangle_{\mathrm{bdry}}$$

which we will express as $\langle\psi_{a_0}, \widetilde{r}^t_n(\psi_{a_1}\otimes\cdots\otimes\psi_{a_n})\rangle_{\text{bdry}}$ in terms of the higher products $\widetilde{r}^t_n$ of a curved A_∞-algebra. Once these amplitudes are computed one immediately obtains their generating function

$$\mathcal{W}_{\text{eff}}(t,u) = \sum_{n\geqslant 2}\frac{1}{n+1}\Big\langle\psi_{a_0}, \widetilde{r}^t_n(\psi_{a_1}\otimes\cdots\otimes\psi_{a_n})\Big\rangle_{\text{bdry}} u_{a_0}u_{a_1}\cdots u_{a_n} \tag{1.2}$$

which is also the F-term D-brane superpotential of the effective four-dimensional low energy field theory.

Our construction splits into two parts, the first of which is restricted to the case of B-twisted affine Landau-Ginzburg models (with arbitrary potential W), while the second part is completely general. Let us mention some of the reasons why Landau-Ginzburg models are interesting theories to study in this context. Firstly, the simple description of their boundary conditions in terms of matrix factorisations allows both for very explicit calculations and for a direct analysis of the underlying structure, unfettered by unnecessary complications. Secondly, by the CY/LG correspondence there is an equivalence between orbifolded Landau-Ginzburg models with quasi-homogeneous potential W and B-models whose compact targets are hypersurfaces $\{W = 0\}$ in projective space, both in the bulk [**23**] and boundary sector [**8**]. Finally, Landau-Ginzburg models are important since via RG flow they are expected to be related to full CFTs. Several quantities of interest are invariants under the flow and can thus be studied on the often more accessible Landau-Ginzburg side of this CFT/LG correspondence. Again we stress that the CFTs describable in this way can but need not have a geometric interpretation, thus covering a larger class of possible string vacua.

Finally we comment on the relation between the algebraic approach advocated here and other approaches to topological string theory. These are more geometric in nature, both in the sense that they apply only to sigma models with a certain class of target manifolds M and that they derive the effective superpotential $\mathcal{W}_{\text{eff}}$ as a geometric quantity. In fact $\mathcal{W}_{\text{eff}}$ is treated as a commutative function of marginal fields only, and the focus is on computing this function. There are two related geometric approaches which we briefly discuss in turn.

The most widely used approach aims to generalise the special geometry of the closed string sector [**3**] to the open and closed B-model, initiated in [**17, 22**]. While a general proof is lacking, in this approach $\mathcal{W}_{\text{eff}}$ is computed as a linear combination of relative period integrals over 3-cycles ending on D2-branes. The computation of these period integrals involves equations of Picard-Fuchs type, and once these are solved the challenge is to find the initial conditions that will yield the correct linear combination of relative periods. This approach works for general toric targets, and it has been applied successfully to several examples with compact Calabi-Yau manifolds.

A second approach [**1, 2**] uses the CY/LG correspondence and is restricted to a subclass of compact Calabi-Yau manifolds. It assumes knowledge of an explicit family (parametrised by open moduli) of BPS D2-branes in the associated B-model. This family is then transported to the Landau-Ginzburg side where $\mathcal{W}_{\text{eff}}$ can be computed to first order in the bulk moduli simply by computing TFT three-point correlators. By iterating this procedure one may obtain a complete non-perturbative description of the open string moduli space attached to the initial family of D2-branes.

The geometric approaches to computing $\mathcal{W}_{\text{eff}}$ are very successful and efficient for the classes of models they are applicable to. This level of efficiency has not yet been achieved in the approach via homotopy algebras, which is however mostly due to the fact that it has received much less attention so far. On the other hand, the algebraic approach is founded on the symmetries of amplitudes and thus universally applies to any topological string theory. As we will see below the A_∞-structure encoding the amplitudes is derived directly from a Chern-Simons-esque string field theory. In this sense it is also more conceptual, and the effective superpotentials (1.2) are obtained as a byproduct.

2. Algebraic approach to bulk-deformed open topological string theory

In this section we start with a very short introduction to A_∞- and L_∞-algebras. Then we discuss B-twisted Landau-Ginzburg models and explain how such algebras describe their boundary and bulk sectors, respectively. Finally, we review the construction of bulk-induced deformations of open topological string theory for such models. In an attempt to hide less significant technical details from this exposition, we shall treat signs, degrees, shifts etc. rather negligently. For a full account we refer to our paper [**6**].

2.1. A_∞- and L_∞-basics. Let us start by recalling the necessary background on *algebras with higher structures*. Most importantly, a *curved A_∞-algebra* is a graded vector space A together with an operator ∂ on the space $T_A = \bigoplus_{n\geqslant 0} A^{\otimes n}$ satisfying (i) $\Delta\partial = (\partial\otimes 1 + 1\otimes\partial)\Delta$ and (ii) $\partial^2 = 0$. Here Δ is the comultiplication defined by $\Delta(a_1\otimes\cdots\otimes a_n) = \sum_{j=0}^{n}(a_1\otimes\cdots\otimes a_j)\otimes(a_{j+1}\otimes\cdots\otimes a_n)$, and property (i) tells us that ∂ is a *coderivation*, i. e. the dual notion of a derivation. One can show that it is completely determined by the family of *higher products*

$$r_n = \pi_A \partial\big|_{A^{\otimes n}} : A^{\otimes n} \longrightarrow A\,, \quad n \geqslant 0\,,$$

where π_A is the projection $T_A \to A$. Accordingly we will use both (A,∂) and (A, r_n) to denote A_∞-algebras. The maps r_n are constrained by quadratic relations coming from property (ii). In terms of $C = r_0(1)$, $d = r_1$ and $a\cdot b = \pm r_2(a\otimes b)$ the constraint $\partial^2 = 0$ in particular implies $d(C) = 0$, $d^2(a) = a\cdot C - C\cdot a$ and the product rule $d(a\cdot b) = d(a)\cdot b \pm a\cdot d(b)$. In the special case when $r_n = 0$ for all $n \geqslant 3$ the only additional constraint is that the product r_2 is associative, and hence the data $(A, C, d, \cdot)$ define a *curved differential graded (DG) algebra*.

From the above we see that for any curved A_∞-algebra r_1 is a differential (to be interpreted as the BRST operator in topological string theory) whenever the *curvature* r_0 is central or vanishes. We call the A_∞-algebra *minimal* if the differential vanishes too. Furthermore, (A, r_n) is *cyclic* with respect to a pairing $\langle\,\cdot\,,\,\cdot\,\rangle$ on A if $\langle a_0, r_n(a_1\otimes\cdots\otimes a_n)\rangle = \pm\langle a_n, r_n(a_2\otimes\cdots\otimes a_n\otimes a_0)\rangle$ for all $n\geqslant 0$. Finally, an A_∞-algebra is *Calabi-Yau* if it is minimal, cyclic with respect to a non-degenerate pairing, and has a unit compatible with the higher products. As recalled in the previous section, every open topological string theory is described by a Calabi-Yau A_∞-algebra.

We also need the notion of a *morphism* between curved A_∞-algebras (A,∂) and (A',∂'). This is a map F from T_A to $T_{A'}$ such that $\Delta' F = (F\otimes F)\Delta$ and $F\partial = \partial' F$. Again, F is determined by maps $F_n = \pi_{A'}F|_{A^{\otimes n}}$ subject to certain compatibility conditions with the higher products coming from $F\partial = \partial' F$. We call F an *A_∞-isomorphism* if F_1 is an isomorphism $A \to A'$. In the case of (non-curved) A_∞-algebras F is called an *A_∞-quasi-isomorphism* if F_1 induces an isomorphism between the cohomologies of the differentials r_1 and r_1'.

The most fundamental result on (non-curved) A_∞-algebras is the *minimal model theorem* [**10, 18**]. It states that any A_∞-algebra (A, r_n) is A_∞-quasi-isomorphic to a minimal A_∞-algebra $(H = H_{r_1}(A), \widetilde{r}_n)$ which is unique up to A_∞-isomorphism.

To see this explicitly, let us restrict to the case where (A, r_n) is a DG algebra and decompose $A \cong H \oplus B \oplus L$ where $B = \mathrm{Im}(r_1)$ and L is the complement of $\mathrm{Ker}(r_1)$. Such a decomposition provides us with a map $G = (r_1|_L)^{-1}\pi_B : A \to A$ which together with $\lambda_2 = r_2$ allows us to recursively define for $n \geqslant 3$:

$$\lambda_n = -r_2(G \otimes 1)(\lambda_{n-1} \otimes 1) - r_2(1 \otimes G)(1 \otimes \lambda_{n-1}) - \sum_{\substack{i,j \geqslant 2, \\ i+j=n}} r_2(G \otimes G)(\lambda_i \otimes \lambda_j) .$$

Then the higher products on H are given by $\widetilde{r}_n = \pi_H \lambda_n$, and the components of the A_∞-quasi-isomorphism $F : (H, \widetilde{r}_n) \to (A, r_n)$ are the inclusion $F_1 : H \hookrightarrow A$ and $F_n = G\lambda_n F_1$ for $n \geqslant 2$. Note that these formulas precisely arise from the Feynman diagrams computed in the topological string field theory (A, r_n) where G plays the role of the propagator [**15**].

Just like A_∞-algebras are generalisations of associative algebras A with higher products $r_n : A^{\otimes n} \to A$, L_∞-algebras are generalisations of Lie algebras V with higher brackets $\ell_n : V^{\wedge n} \to V$. Similarly, there are appropriate notions of L_∞-morphisms which again can be presented as maps $L_n : V^{\wedge n} \to V'$, subject to certain compatibility conditions. The detailed definitions will not matter to us as the only L_∞-algebras relevant in this note are *DG Lie algebras*, i. e. vector spaces endowed with a graded anti-symmetric bracket that satisfies the super Jacobi identity, and a differential compatible with the bracket. It is however crucial to view them as special L_∞-algebras since L_∞-morphisms (with higher components) between them will be important in what follows. The root of this fact lies in the relation between deformations and solutions to Maurer-Cartan equations that we discuss next.

Our aim is to study bulk-deformed open topological string theory. Hence we must explain what a *deformation* of an A_∞-algebra (A, ∂) is. By definition it is a map $\delta \in \mathrm{End}(T_A)$ such that $(A, \partial + \delta)$ is a curved A_∞-algebra. This means that δ is a coderivation, $\delta \in \mathrm{Coder}(T_A)$, and $(\partial + \delta)^2$ must vanish. Thus if we write $[\,\cdot\,,\,\cdot\,]$ for the graded commutator and use $\partial^2 = 0$, we see that δ must solve

$$[\partial, \delta] + \frac{1}{2}[\delta, \delta] = 0 . \tag{2.1}$$

Let us recall that for any DG Lie algebra $(V, d, [\,\cdot\,,\,\cdot\,])$ its associated *Maurer-Cartan equation* reads $d(\delta) + \frac{1}{2}[\delta, \delta] = 0$. We denote its space of formal power series solutions modulo gauge transformations $\delta \mapsto \delta + d(\delta) + [\varphi, \delta]$ as $\mathcal{MC}(V, d, [\,\cdot\,,\,\cdot\,])$. It is an important result [**13**] that given an L_∞-morphism $L : (V, d, [\,\cdot\,,\,\cdot\,]) \to (V', d', [\,\cdot\,,\,\cdot\,]')$, the map

$$\delta \longmapsto \sum_{n \geqslant 1} \frac{1}{n!} L_n(\delta^{\wedge n}) \tag{2.2}$$

induces a map $\mathcal{MC}(V, d, [\,\cdot\,,\,\cdot\,]) \to \mathcal{MC}(V', d', [\,\cdot\,,\,\cdot\,]')$. Furthermore, this is an isomorphism if L is an L_∞-quasi-isomorphism.

We now make the obvious yet crucial observation that the deformation condition (2.1) is precisely the Maurer-Cartan equation of the DG Lie algebra $\mathrm{Coder}(T_A)$

with differential $[\partial, \cdot]$ and bracket $[\cdot, \cdot]$. In addition, solving (2.1) to first order (up to gauge transformations) is identical to computing *Hochschild cohomology* $\mathrm{HH}^\bullet(A, \partial) = H_{[\partial, \cdot]}(\mathrm{Coder}(T_A))$. More importantly, the above result on transporting Maurer-Cartan solutions will allow us to construct all A_∞-deformations if we can find a quasi-isomorphic DG Lie algebra whose Maurer-Cartan solutions are known. In the remainder of this note we will review how to do this for the case of Landau-Ginzburg models, where the known Maurer-Cartan solutions will be precisely the space of bulk fields.

2.2. Landau-Ginzburg models. We shall consider the topological B-twist of Landau-Ginzburg models with affine target $X = \mathbb{C}^N$ and potential $W \in R = \mathbb{C}[x_1, \dots, x_N]$. The *on-shell* space of states in the bulk sector of such two-dimensional topological field theories is the Jacobian $\mathrm{Jac}(W) = R/(\partial_1 W, \dots, \partial_N W)$ [**21**]. This is obtained from the *off-shell* bulk space of polyvector fields $T_{\mathrm{poly}} = \Gamma(X, \bigwedge T^{(1,0)}X)$ by taking cohomology with respect to the BRST operator $[-W, \cdot]_{\mathrm{SN}}$. Here we denote by $[\cdot, \cdot]_{\mathrm{SN}}$ the Schouten-Nijenhuis bracket, the extension of the Lie bracket to polyvector fields. Note that T_{poly} together with the BRST differential and the bracket has the structure of a DG Lie algebra. A direct computation shows that its Maurer-Cartan solutions are precisely the on-shell bulk space $\mathrm{Jac}(W)$.

The boundary sector is defined by matrix factorisations of W, i. e. odd supermatrices D with polynomial entries such that D squares to $W \cdot \mathrm{id}$ [**12, 4, 16**]. For simplicity we will only consider one such boundary condition. Then the *off-shell* space A is simply the space of all polynomial matrices of the same size as D. The boundary BRST operator is given by the graded commutator $[D, \cdot]$, and together with matrix multiplication this makes A a DG algebra (A, ∂). By definition its cohomology is the *on-shell* space H.

Since the off-shell algebra (A, ∂) is an A_∞-algebra (describing open topological string field theory for Landau-Ginzburg models), we can apply the minimal model theorem to produce an A_∞-structure $\widetilde{\partial}$ on H together with an A_∞-quasi-isomorphism $F : (H, \widetilde{\partial}) \to (A, \partial)$. For generic choices of propagators $(H, \widetilde{\partial})$ will not be Calabi-Yau, but this can be corrected using methods of non-commutative geometry as explained in [**5**]. Hence we can assume that for any Landau-Ginzburg model and all their branes we can construct a Calabi-Yau A_∞-algebra $(H, \widetilde{\partial})$ that encodes the full structure of open topological string theory.

2.3. Bulk-deformed Landau-Ginzburg models. Now we discuss bulk deformations of $(H, \widetilde{\partial})$. We saw that any deformation must solve the Maurer-Cartan equation of $(\mathrm{Coder}(T_H), [\widetilde{\partial}, \cdot], [\cdot, \cdot])$, but finding such solutions directly is hard. However, we are only interested in *bulk-induced deformations*, which we define as the image of on-shell bulk fields under an L_∞-morphism

$$(T_{\mathrm{poly}}, [-W, \cdot]_{\mathrm{SN}}, [\cdot, \cdot]_{\mathrm{SN}}) \longrightarrow (\mathrm{Coder}(T_H), [\widetilde{\partial}, \cdot], [\cdot, \cdot]) . \tag{2.3}$$

The left-hand side is the off-shell bulk algebra, and we already know that its Maurer-Cartan solutions precisely form the on-shell bulk space $\mathrm{Jac}(W)$. Hence our remaining task is to construct the L_∞-map (2.3). We shall do this in two main steps which are Theorems 1 and 2 below.

To set the stage, let us write the A_∞-structure of the off-shell open string algebra (A, ∂) as $\partial = \partial_1 + \partial_2$ to emphasise that it is a DG algebra with differential

$r_1 = [D, \cdot]$ and matrix multiplication r_2. There is another natural curved A_∞-structure on A which we write as $(A, \partial_0 + \partial_2)$, where the coderivation ∂_0 corresponds to the curvature $r_0 = -W \cdot \mathrm{id}$. Similarly, the polynomial ring R is also a curved algebra $(R, \widehat{\partial}_0 + \widehat{\partial}_2)$, where $\widehat{\partial}_0$ corresponds to multiplication with $-W$ and $\widehat{\partial}_2$ to the usual product. Now we can take the first step:

THEOREM 1 ([**6**]). *There is a sequence of explicit L_∞-quasi-isomorphisms*

$$\begin{aligned}(T_{\mathrm{poly}}, [-W, \cdot]_{\mathrm{SN}}, [\cdot\,, \cdot]_{\mathrm{SN}}) &\xrightarrow{\varphi_1} (\mathrm{Coder}(T_R), [\widehat{\partial}_0 + \widehat{\partial}_2, \cdot], [\cdot\,, \cdot]) \\ &\xrightarrow{\varphi_2} (\mathrm{Coder}(T_A), [\partial_0 + \partial_2, \cdot], [\cdot\,, \cdot]) \\ &\xrightarrow{\varphi_3} (\mathrm{Coder}(T_A), [\partial_1 + \partial_2, \cdot], [\cdot\,, \cdot])\,.\end{aligned}$$

The map φ_3 is simply the adjoint action of the A_∞-isomorphism T (which "cancels" the "tadpole" r_0) whose non-vanishing components are $T_0 = D$ and $T_1 = 1_A$, and φ_2 is Morita equivalence for curved DG-algebras.

It turns out that the map φ_1 deserves more attention; it is an interesting variant of deformation quantisation. We recall that the latter is a method to quantise the algebra of classical observables $C^\infty(M, \mathbb{R})$ on a phase space M, which we take to be $\mathbb{R}^d$. The idea is to deform the commutative, associative multiplication $\widehat{r}_2$ on $C^\infty(M, \mathbb{R})$ to an associative but non-commutative $\star$-product of quantum observables. This deformation again leads to a Maurer-Cartan equation. Kontsevich's solution [**13**] was to explicitly construct an L_∞-quasi-isomorphism

$$K : (\Gamma(M, \textstyle\bigwedge TM), [\cdot\,, \cdot]_{\mathrm{SN}}) \longrightarrow (\mathrm{Coder}(T_{C^\infty(M,\mathbb{R})}), [\widehat{\partial}_2, \cdot], [\cdot\,, \cdot])\,,$$

thus providing a one-to-one correspondence between Poisson structures on M and perturbative $\star$-products of quantum observables.

Note that the map K is precisely the special case of our map φ_1 if $W = 0$. Indeed, we can show that K continues to be an L_∞-map after "turning on" W. Furthermore, it also remains a quasi-isomorphism as a direct computation yields $\mathrm{HH}^\bullet(R, \widehat{\partial}_0 + \widehat{\partial}_2) \cong \mathrm{Jac}(W)$. Hence we arrive at a generalisation of deformation quantisation, namely that the DG Lie algebra of polyvector fields with differential $[-W, \cdot]$ governs deformations of $(R, \widehat{\partial}_0 + \widehat{\partial}_2)$.

With Theorem 1 we have explicitly classified all deformations of the off-shell open string algebra (A, ∂), and we found that all such deformations are bulk-induced. We stress that this off-shell, string field theoretic result is important by itself as it contains information on descendants of ground states. However, one is also interested in computing amplitudes and effective superpotentials in on-shell open topological string theory. Accordingly, we will discuss how to transport deformations of (A, ∂) to those of $(H, \widetilde{\partial})$. Again, this will be achieved by an L_∞-morphism that maps off-shell deformations on-shell via (2.2):

THEOREM 2 ([**6**]). *There is an explicit L_∞-map*

$$(\mathrm{Coder}(T_A), [\partial, \cdot], [\cdot\,, \cdot]) \longrightarrow (\mathrm{Coder}(T_H), [\widetilde{\partial}, \cdot], [\cdot\,, \cdot])\,.$$

The main tool that allows us to construct this map is an L_∞-version of the *homological perturbation lemma*, which is another structure transfer result. Given two complexes (C_1, d_1), (C_2, d_2) with morphisms $i : (C_2, d_2) \to (C_1, d_1)$, $p : (C_1, d_1) \to (C_2, d_2)$ and $h \in \mathrm{End}(C_1)$, we call these data a deformation retraction if $pi = 1_{C_2}$

and $1_{C_1} - ip = d_1 h + h d_1$. In this situation the perturbation lemma states that

$$\delta \longmapsto \sum_{n\geqslant 1} p(\delta h)^n \delta i \tag{2.4}$$

maps a deformation δ of (C_1, d_1) to a deformation of (C_2, d_2), and this can be understood in terms of an L_∞-quasi-isomorphism between the endomorphism algebras of the two complexes.

We now use this method in our setting. It was shown in [**6**] that for any A_∞-algebra (A, ∂) with minimal model $(H, \widetilde{\partial})$ there is a deformation retraction

$$(T_H, \widetilde{\partial}) \underset{\bar{F}}{\overset{F}{\rightleftarrows}} (T_A, \partial) \circlearrowleft U \,.$$

Here F is our usual minimal model quasi-isomorphism, and $\bar{F}, U$ are defined recursively by

$$U_n = -\frac{1}{2} G r_2 \Big(\sum_{i=1}^{n-1} (U_i \otimes (1_{T_A} + F\bar{F})_{n-i} + (1_{T_A} + F\bar{F})_{n-i} \otimes U_i \Big) \,,$$

$$\bar{F}_n = -\frac{1}{2} \pi_H r_2 \Big(\sum_{i=1}^{n-1} (U_i \otimes (1_{T_A} + F\bar{F})_{n-i} + (1_{T_A} + F\bar{F})_{n-i} \otimes U_i \Big) \,.$$

Furthermore, these maps are such that (2.4) maps coderivations to coderivations.

Now we have everything in place to apply the above to the case where (A, ∂) and $(H, \widetilde{\partial})$ are the off-shell and on-shell open string algebras. It follows from Theorem 1 that all deformations δ of (A, ∂) are coderivations determined by $1 \mapsto \sum_i t_i \phi_i$ where ϕ_i are on-shell bulk fields in $\mathrm{Jac}(W)$ and t_i are the closed moduli. Then Theorem 2 tells us that the associated deformations $\widetilde{\delta}$ of the on-shell algebra $(H, \widetilde{\partial})$ are given by

$$\widetilde{\delta} = \sum_{n\geqslant 1} \bar{F} (\delta U)^n \delta F \,.$$

All the ingredients $F, \bar{F}, U$ can be algorithmically computed, thus leading to completely explicit expressions for the bulk-deformed open string amplitudes

$$\Big\langle \psi_{a_0}, \widetilde{r}^t_n (\psi_{a_1} \otimes \cdots \otimes \psi_{a_n}) \Big\rangle_{\mathrm{bdry}} = \Big\langle \psi_{a_0}(p_0), \psi_{a_1}(p_1) \psi_{a_2}(p_2) \int \psi^{(1)}_{a_3} \cdots \int \psi^{(1)}_{a_n} \, \mathrm{e}^{\sum_i t_i \int \phi^{(2)}_i} \Big\rangle_{\mathrm{bdry}}$$

where the higher products $\widetilde{r}^t_n$ are the components of $\widetilde{\partial} + \widetilde{\delta}$.

Acknowledgements. We thank I. Brunner, D. Murfet and A. Recknagel for helpful comments on the manuscript.

References

[1] M. Baumgartl, I. Brunner, and M. R. Gaberdiel, *D-brane superpotentials and RG flows on the quintic*, JHEP **0707** (2007), 061, [arXiv:0704.2666].

[2] M. Baumgartl, I. Brunner, and M. Soroush, *D-brane Superpotentials: Geometric and Worldsheet Approaches*, Nucl. Phys. B **843** (2011), 602–637, [arXiv:1007.2447].

[3] M. Bershadsky, S. Cecotti, H. Ooguri, and C.Vafa, *Kodaira–Spencer Theory of Gravity and Exact Results for Quantum String Amplitudes*, Commun.Math. Phys **165** (1994), 311–428, [hep-th/9309140].

[4] I. Brunner, M. Herbst, W. Lerche, and B. Scheuner, *Landau-Ginzburg Realization of Open String TFT*, JHEP **0611** (2003), 043, [hep-th/0305133].

[5] N. Carqueville, *Matrix factorisations and open topological string theory*, JHEP **0907** (2009), 005, [arXiv:0904.0862].

[6] N. Carqueville and M. M. Kay, *Bulk deformations of open topological string theory*, [arXiv:1104.5438].

[7] K. J. Costello, *Topological conformal field theories and Calabi-Yau categories*, Adv. in Math. **210** (2007), [math.QA/0412149].

[8] M. Herbst, K. Hori, and D. Page, *Phases Of $N = 2$ Theories In $1 + 1$ Dimensions With Boundary*, [arXiv:0803.2045].

[9] M. Herbst, C. I. Lazaroiu, and W. Lerche, *Superpotentials, A_∞ Relations and WDVV Equations for Open Topological Strings*, JHEP **0502** (2005), 071, [hep-th/0402110].

[10] T. Kadeishvili, *On the homology theory of fibre spaces*, Uspekhi Mat. Nauk **35:3** (1980), 183–188, English translation in Russian Math. Surveys, 35:3 (1980), 231–238, [math.AT/0504437].

[11] H. Kajiura and J. Stasheff, *Homotopy algebras inspired by classical open-closed string field theory*, Commun. Math. Phys. **263** (2006), 553–581, [math.QA/0410291].

[12] A. Kapustin and Y. Li, *D-branes in Landau-Ginzburg Models and Algebraic Geometry*, JHEP **0312** (2003), 005, [hep-th/0210296].

[13] M. Kontsevich, *Deformation quantization of Poisson manifolds*, Lett. Math. Phys. **66** (2003), 157–216, [math.QA/9709040].

[14] C. I. Lazaroiu, *On the structure of open-closed topological field theory in two dimensions*, Nucl. Phys. B **603** (2001), 497–530, [hep-th/0010269].

[15] C. I. Lazaroiu, *String field theory and brane superpotentials*, JHEP **0110** (2001), 018, [hep-th/0107162].

[16] C. I. Lazaroiu, *On the boundary coupling of topological Landau-Ginzburg models*, JHEP **0505** (2005), 037, [hep-th/0312286].

[17] W. Lerche, P. Mayr, and N. P. Warner, *Holomorphic $N = 1$ Special Geometry of Open–Closed Type II Strings*, [hep-th/0207259].

[18] S. A. Merkulov, *Strongly homotopy algebras of a Kähler manifold*, Internat. Math. Res. Notices **3** (1999), 153–164, [math.AG/9809172].

[19] G. W. Moore and G. Segal, *D-branes and K-theory in 2D topological field theory*, [hep-th/0609042].

[20] A. Neitzke and J. Walcher, *Background Independence and the Open Topological String Wavefunction*, Nucl. Phys. B **843** (2011), 602–637, [arXiv:0709.2390].

[21] C. Vafa, *Topological Landau-Ginzburg Models*, Mod. Phys. Lett. A **6** (1991), 337–346.

[22] J. Walcher, *Opening Mirror Symmetry on the Quintic*, Commun. Math. Phys. **276** (2007), 671–689, [hep-th/0605162].

[23] E. Witten, *Phases of $N = 2$ Theories in Two Dimensions*, Nucl. Phys. B **403** (1993), 159–222, [hep-th/9301042].

[24] E. Witten, *Quantum Background Independence In String Theory*, [hep-th/9306122].

[25] B. Zwiebach, *Closed string field theory: Quantum action and the BV master equation*, Nucl. Phys. B **390** (1993), 33–152, [hep-th/9206084].

[26] B. Zwiebach, *Oriented open-closed string theory revisited*, Annals Phys. **267** (1998), 193–248, [hep-th/9705241].

Arnold Sommerfeld Center for Theoretical Physics, LMU München & Excellence Cluster Universe

E-mail address: `nils.carqueville@physik.uni-muenchen.de`

Arnold Sommerfeld Center for Theoretical Physics, LMU München & Excellence Cluster Universe

E-mail address: `michael.kay@physik.uni-muenchen.de`

Proceedings of Symposia in Pure Mathematics
Volume **85**, 2012

Landau-Ginzburg Mirror Symmetry for Orbifolded Frobenius Algebras

Amanda Francis, Tyler Jarvis, Drew Johnson, and Rachel Suggs

Abstract. We prove the Landau-Ginzburg Mirror Symmetry Conjecture at the level of (orbifolded) Frobenius algebras for a large class of invertible singularities, including arbitrary sums of loops and Fermats with arbitrary symmetry groups. Specifically, we show that for a quasi-homogeneous polynomial W and an admissible group G within the class, the Frobenius algebra arising in the FJRW theory [**FJR07b**] of $[W/G]$ is isomorphic (as a Frobenius algebra) to the orbifolded Milnor ring [**Kra10, Kau06, IV90**] of $[W^T/G^T]$, associated to the dual polynomial W^T and dual group G^T.

1. Introduction

1.1. Background. The Landau-Ginzburg mirror symmetry conjecture says that the A-model of a given singularity with a given symmetry group should match the B-model of an appropriately chosen *transpose singularity* and *transpose group*. The full conjecture is that there should be cohomological field theories for both the A- and B-models and that these should match in all genera.

The A-model has been constructed by Fan, Jarvis, and Ruan [**FJR08**], [**FJR07a, FJR07b**], following ideas of Witten. For any non-degenerate, quasi-homogenous (weighted homogeneous) polynomial W, and for any admissible group of diagonal symmetries G, they constructed a cohomological field theory which is often called *FJRW theory*. A restriction of this theory to genus zero with three marked points gives a Frobenius algebra for the A-model which we denote $\mathscr{A}_{W,G}$.

For the B-model, in the case that the orbifold group is trivial, the Frobenius algebra is given by the Milnor ring (or local algebra) of the singularity. For cases orbifolded by non-trivial groups, the orbifold B-model, as a vector space, was given by Intriligator and Vafa [**IV90**]; however, until more recently the orbifold B-model was lacking a definition of the product structure. In [**Kra10**], Krawitz, following ideas of Kaufmann [**Kau02, Kau03, Kau06**], wrote down a multiplication for the orbifold Milnor ring, which we call $\mathscr{B}_{W,G}$.

The construction of the mirror dual of a given invertible polynomial was first described by Berglund-Hübsch [**BH93**]. It is defined for so-called *invertible singularities*, that is, polynomials with the same number of monomial as variables which define an isolated singularity at the origin. The general construction of the dual

2000 *Mathematics Subject Classification.* Primary: 14B05, 32S25, 14J81, 57R56. Secondary: 14J81, 14N35, 53D45, 32S40, 32S55.

This research partially supported by NSA grant H98230-10-1-0181.

group and the general definition of the mirror map (at the level of vector spaces) is due to Krawitz [**Kra10**], based in part on ideas from [**Kre94**].

Krawitz proved that the LG Mirror Symmetry Conjecture holds on the level of bi-graded vector spaces for all invertible polynomials and all admissible groups. At the level of Frobenius algebras, he only proved the conjecture for the case when the B-model has trivial orbifold group. This generalized some previous work of [**KPA$^+$10**], [**Aco09**], and [**FS09**]. A few special cases of the conjecture with non-trivial orbifolded B-model have also been verified [**Kra10, Ber09**].

This paper proves the LG Mirror Symmetry Conjecture for arbitrary admissible groups of symmetries. We show that for a wide class of polynomials and admissible orbifold groups, the A-model is isomorphic, as a Frobenius algebra, to the orbifold B-model of the dual singularity and dual group. The precise conditions on the polynomials and groups is given in Property ($*$). In terms of the classification [**KS92**] of invertible polynomials into Fermats, chains, and loops, our result applies many cases, including arbitrary sums of Fermats and loops, with arbitrary choice of admissible group.

The product structure of the FJRW theory is determined by the genus-zero, three-point correlators, which are $\mathbb{C}$-valued functions on $\mathscr{A}_{W,G}^{\otimes 3}$. The FJRW theory satisfies the axioms of a cohomological field theory as well as several additional axioms that facilitate computation. When orbifolding by a trivial symmetry group on the B-side, and thus by the maximal group on the A-side, the relevant insertions are mostly what we call *narrow* [**FJR07b**]. In these cases, the axioms of [**FJR07b**] provide straightforward ways to compute the correlators.

When we orbifold by non-trivial symmetry groups on the B-side (and thus smaller groups on the A-side) the situation is more difficult, more of the insertions may be *broad*, and are much more difficult to compute. The general case of the problem is a difficult PDE-problem, not yet explicitly solved in most cases.

Orbifolding by a smaller group on the A-side also introduces another difficulty. When the orbifold group is a (decoupled) product of groups acting on these sums, as in the case of the maximal symmetry group, the FJRW ring is the tensor product of the pieces, and the B-model can similarly be broken up as a tensor product (see Axiom 4 and Proposition 2.3.2). Thus, previous attention has been focused on these atomic types. However, if we consider more arbitrary groups are not necessarily decoupled products of groups, this method does not apply directly. This paper introduces a new strategy to take advantage of the "breaking up into tensor products" technique for more general orbifold groups. This also allows us to avoid computation of some of the difficult correlators. Essentially, for each product, we take a subalgebra containing the factors that can also be thought of as a subalgebra of a theory with a group that does break up as a direct product in a useful way.

Consider a non-degenerate, invertible singularity $W \in \mathbb{C}[x_1, \ldots, x_N]$ and an admissible group G of symmetries of W. We prove that LG mirror symmetry holds for all (W, G) that have the following property:

PROPERTY ($*$). *Let W be a non-degenerate, invertible singularity, and let G be an admissible group of symmetries of W. We say that* the pair (W, G) has Property ($*$) *if,*

(1) *W can be decomposed as $W = \sum_{i=1}^{M} W_i$, where the W_i are themselves invertible polynomials having no variables in common with any other W_j.*

(2) *For any element g of G whose associated sector $\mathscr{A}_g \subseteq \mathscr{A}_{W,G}$ is nonempty, and for each $i \in \{1, \ldots, M\}$ the action of g fixes either all of the variables in W_i or none of them.*
(3) *For any element g' of G^T whose associated sector of $\mathscr{B}_{g'} \subseteq \mathscr{B}_{W^\mathsf{T},G^\mathsf{T}}$ is non-empty, and for each $i \in \{1, \ldots, M\}$ the action of g' fixes either all of the variables in W_i^T or none of them.*

REMARK 1.1.1. If W is a two-variable chain, or if W is a sum of Fermat and loop type singularities (see Proposition 2.1.1 of this paper), then Property $(*)$ is satisfied for *any* admissible group G, and Theorem 3.0.3 will hold.

Let $\mathscr{A}_{W,G}$ and $\mathscr{B}_{W^\mathsf{T},G^\mathsf{T}}$ be the FJRW ring (LG A-model Frobenius algebra) and orbifold Milnor ring (LG B-model Frobenius algebra), respectively. Our main theorem is the following.

THEOREM 3.0.3. *If W and G satisfy Property $(*)$, then there is an isomorphism of Frobenius algebras: $\mathscr{A}_{W,G} \cong \mathscr{B}_{W^\mathsf{T},G^\mathsf{T}}$.*

Borisov [**Bor10**] has constructed both an A-model and a B-model Frobenius algebra, and he has proved mirror symmetry for his constructions in the case that the singularity is Calabi-Yau and the symmetry group is both admissible and a subgroup of SL_N. Borisov's B-model agrees with the Krawitz/Kaufmann $\mathscr{B}_{W^\mathsf{T},G^\mathsf{T}}$, but without our mirror symmetry result, it is not at all clear that Borisov's A-model agrees with the FJRW construction $\mathscr{A}_{W,G}$. Furthermore, our result applies to polynomials that are not necessarily Calabi-Yau, and to admissible groups that are not necessarily subgroups of SL_N.

2. Review of the Constructions

2.1. Quasihomogeneous Polynomials. We call a polynomial *invertible* if it has the same number of variables as monomials. We start with a quasi-homogeneous, invertible polynomial in variables $X_1, \ldots X_N$:

$$W = \sum_{i=1}^{N} c_i \prod_{j=1}^{N} X_j^{a_{ij}} \in \mathbb{C}[X_1, \ldots, X_N].$$

The matrix $A = (a_{ij})$ encodes the exponents of the polynomial. We require that the weights q_i of X_i be uniquely determined by the condition that W have weighted degree 1. We can write the q_i as

$$q_i = \frac{w_i}{d} \quad \text{with} \quad \gcd(d, w_1, \ldots, w_N) = 1. \tag{1}$$

In such a polynomial the c_i can be absorbed by rescaling the variables, so we will always assume from now on that $c_i = 1$.

We also require that the polynomial define an isolated singularity at the origin; i.e., the system of equations $\left\{\frac{\partial W}{\partial X_i} = 0\right\}$ has a unique solution at the origin. A polynomial satisfying these two conditions we call *non-degenerate*. Invertible, non-degenerate, quasi-homogeneous polynomials are completely classified.

PROPOSITION 2.1.1 ([**KS92**]). *Any invertible, non-degenerate, quasi-homogeneous polynomial is the decoupled sum of polynomials of one of three* atomic types*:*

$$\begin{aligned} \textit{Fermat} \text{ type: } W &= X^a, \\ \textit{loop} \text{ type: } W &= X_1^{a_1}X_2 + X_2^{a_2}X_3 + \cdots + X_N^{a_N}X_1, \\ \textit{chain} \text{ type: } W &= X_1^{a_1}X_2 + X_2^{a_2}X_3 + \cdots + X_N^{a_N}. \end{aligned}$$

We also assume that $a_i \geq 2$, so that there are no terms of the form X_iX_j.

REMARK 2.1.2. Assuming that the variables are listed in order as above, an atomic polynomial is determined by its type and by the vector of non-one exponents $\mathbf{a} = (a_1, \dots, a_N)^\mathsf{T}$.

Throughout this paper we will use boldface type to represent a column vector, and regular italic type to represent entries in the vector. Thus, by $\mathbf{g}$ we mean the vector $(g_1, \dots, g_N)^\mathsf{T}$, where the N must be understood from context. We also write $\mathbf{1}$ for the vector $(1, \dots, 1)^\mathsf{T}$.

We form the *Berglund-Hübsch dual* or *transpose singularity* W^T by taking the polynomial with exponent matrix A^T. If W is invertible and non-degenerate, then W^T is also invertible and non-degenerate. We will generally denote the variables of W^T by Y_i. Taking the transpose of any of the atomic types preserves the type. The variables, however, will be in the reverse order.

2.1.1. *Diagonal symmetry groups.* There is an action of $(\mathbb{C}^*)^N$ on $\mathbb{C}[X_1, \dots, X_N]$, where the tuple $(\lambda_1, \dots, \lambda_N) \in (\mathbb{C}^*)^N$ acts on X_j by multiplication by λ_j.

DEFINITION 2.1.3. The *maximal group of diagonal symmetries* $G_W^{\max}$, or simply $G^{\max}$, if W is clear from context, is the maximal subgroup of $(\mathbb{C}^*)^N$ which fixes the polynomial W.

We prefer to think of symmetries $G_W^{\max}$ as a subgroup of $(\mathbb{Q}/\mathbb{Z})^N$, where the element $[\mathbf{g}]$ corresponding to the class of the vector $\mathbf{g} \in \mathbb{Q}^N$ acts on X_j by multiplication by $\exp(2\pi i g_j)$. We can find a special set of generators for $G_W^{\max}$ and a special element J, as follows.

DEFINITION 2.1.4. The group element in $(\mathbb{Q}/\mathbb{Z})^N$ corresponding to the ith column of A^{-1} we denote by ρ_i.

The element $J := \sum_{j=1}^N \rho_j = [A^{-1}\mathbf{1}] = [\mathbf{q}] \in (\mathbb{Q}/\mathbb{Z})^N$ is called the *exponential grading element.*

PROPOSITION 2.1.5 (Krawitz). *The ρ_i generate the maximal symmetry group $G^{\max}$. Similarly, the rows $\bar{\rho}_i$ of A^{-1} generate the maximal symmetry group of W^T.*

DEFINITION 2.1.6 (Krawitz). The dual group G^T is a group of symmetries of W^T defined by

$$G^\mathsf{T} := \left\{ [\mathbf{g}] \mid \mathbf{g}^\mathsf{T} A \mathbf{b} \in \mathbb{Z}^N \text{ for all } [\mathbf{b}] \in G \right\}. \tag{2}$$

One can check that G^T is a group and that the definition is independent of the presentation of the elements of G. Additionally, the transpose group has the following properties, which are verified in [**Kra10**]:

- If G contains J, then G^T is contained in SL_N, and vice-versa.
- $(G^\mathsf{T})^\mathsf{T} = G$.
- If $G' \leq G$, then $G^\mathsf{T} \leq (G')^\mathsf{T}$.

2.1.2. *Milnor Rings.*

DEFINITION 2.1.7. The *Jacobian ideal* $\mathrm{Jac}(W)$ of a polynomial W is the ideal generated by the partial derivatives:

$$\mathrm{Jac}(W) := \left(\frac{\partial W}{\partial X_1}, \ldots, \frac{\partial W}{\partial X_N}\right)$$

and the *Milnor ring* $\mathscr{Q}_W$ (also called the *local algebra*) is

$$\mathscr{Q}_W := \mathbb{C}[X_1, \ldots, X_N]/\mathrm{Jac}(W).$$

The Milnor ring is a finite-dimensional $\mathbb{C}$-vector space, graded by the weighted degree of the monomials. The subspace of highest weighted degree is one-dimensional, is spanned by the Hessian determinant $\mathrm{hess}(W)$, and has weighted degree equal to the *central charge* $\hat{c} = \sum_{j=1}^{N}(1 - 2q_j)$.

The residue pairing $\langle\ ,\ \rangle$ makes $\mathscr{Q}_W$ into a Frobenius algebra; that is, for every $a, b, c \in \mathscr{Q}_W$ we have

$$\langle a \cdot b, c\rangle = \langle a, b \cdot c\rangle.$$

The residue pairing in $\mathscr{Q}_W$ can be computed as $\langle a, b\rangle = \mu \frac{ab}{\mathrm{hess}(W)}$, by which we mean

$$ab = \mu\, \mathrm{hess}(W)\, \langle a, b\rangle + \text{lower-degree terms}, \tag{3}$$

where μ is the dimension of the vector space $\mathscr{Q}_W$.

2.2. The A-model. The Landau-Ginzburg A-model Frobenius algebra is the so-called *FJRW ring*. We give a description in terms of Milnor rings which is more elementary than that used in [**FJR07b**] and is sufficient for our computations.

DEFINITION 2.2.1. Any subgroup of G of $G^{\max}$ containing the exponential grading element J is called *admissible*. For each $g \in G^{\max}$, we define W_g to be the restriction of the polynomial W to the subspace of $\mathbb{C}^N$ fixed by g.

2.2.1. *The A-model state space.* The state space of the FJRW ring is the underlying vector space of the ring. This is constructed by taking a direct sum of *sectors*—one for each $g \in G$.

DEFINITION 2.2.2. Let W be a non-degenerate, invertible singularity and G be admissible group of symmetries of W. For any $g \in G$, the *unprojected g-sector* $\mathscr{H}_g$ is the G-module

$$\mathscr{H}_g := \mathscr{Q}_{W_g} \cdot dX_{i_1} \wedge \cdots \wedge dX_{i_{N_g}},$$

where N_g is the number of variables fixed by g, and $i_1, \ldots i_{N_g}$ are the indices of the fixed coordinates.

The g-sector is called *narrow* if $N_g = 0$, and *broad* otherwise. Each $g = (\Theta_1, \ldots, \Theta_N) \in G$ acts on both the variable X_j and on the one-form dX_j by multiplication by $\exp(2\pi i \Theta_j)$.

The state space of the A-model, or underlying vector space of the FJRW ring, is given by taking the direct sum of the un-projected sectors and taking G invariants:

$$\mathscr{A}_{W,G} := \left(\bigoplus_{g \in G} \mathscr{H}_g\right)^G \quad \text{and} \quad \mathscr{A}_g := \mathscr{H}_g^G. \tag{4}$$

We call $\mathscr{A}_g$ the *projected g-sector*, or *g-sector of the FJRW ring.*

We denote an element of $\mathscr{H}_g$ or $\mathscr{A}_g$ by $\lceil m\,; g\rfloor$, where m is an element of the ring $\mathscr{Q}_{W_g}$. We do not explicitly write the volume form, since it is determined by the group element. This notation is not standard in the subject, but we find it clearer than other notations, especially when we are discussing the mirror map.

2.2.2. *The pairing and three-point correlators.* The Milnor ring always has a basis of monomials, and since the G-action is diagonal, the G-invariants $\mathscr{A}_g$ also have a basis of monomials. Thus, we can write a basis for $\mathscr{H}_g$ or of $\mathscr{A}_g$ with elements of the form $\lceil \mathbf{X}^{\mathbf{r}}\,; g\rfloor$, where by $\mathbf{X}^{\mathbf{r}}$ we mean the monomial $\prod_{j=1}^{N_g} X_{i_j}^{r_j} \in \mathscr{Q}_g$.

DEFINITION 2.2.3. We endow $\mathscr{A}_{W,G}$ with a pairing as follows. Since g and $-g$ fix the same coordinates, there is a natural isomorphism

$$I : (\mathscr{A}_g)^G \to (\mathscr{A}_{-g})^G \quad \text{given by} \quad \lceil \mathbf{X}^{\mathbf{r}}\,; g\rfloor \mapsto \lceil \mathbf{X}^{\mathbf{r}}\,; -g\rfloor .$$

The pairing on the Milnor ring $\mathscr{Q}_g$ induces a pairing

$$\langle \cdot,\cdot\rangle_g : (\mathscr{A}_g)^G \otimes (\mathscr{A}_{-g})^G \to \mathbb{C}$$

$$\langle \lceil \mathbf{X}^{\mathbf{r}}\,; g\rfloor , \lceil \mathbf{X}^{\mathbf{s}}\,; -g\rfloor\rangle_g = \langle \lceil \mathbf{X}^{\mathbf{r}}\,; g\rfloor , I^{-1}(\lceil \mathbf{X}^{\mathbf{s}}\,; -g\rfloor)\rangle_{\mathscr{Q}_{W_g}} .$$

The pairing on the full state space $\mathscr{A}_{W,G}$ is the "direct sum" of these pairings—that is, if two basis elements are from sectors g and $-g$, we use the pairing $\langle\cdot,\cdot\rangle_g$ above, and otherwise, the pairing is 0.

The FJRW theory provides a full cohomological field theory, but for the purposes of this paper, we are only concerned with the Frobenius algebra structure, arising from the genus-zero, three-point correlators, which we denote by

$$\langle \cdot,\cdot,\cdot\rangle^{W,G} : \mathscr{A}_{W,G}^{\otimes 3} \to \mathbb{C}.$$

We omit the superscripts W, G when they are clear from context. We discuss the computation of these correlators in Section 2.2.3. The product in the FJRW ring is given by

$$\alpha \star \beta = \sum_{\tau,\sigma} \langle \alpha,\beta,\tau\rangle\, \eta^{\tau,\sigma}\sigma, \tag{5}$$

where the sum is over all pairs of elements σ,τ from a fixed basis of $\mathscr{A}_{W,G}$, and where $\eta^{\tau,\sigma}$ is the inverse of the pairing matrix with respect to this basis. The FJRW ring $\mathscr{A}_{W,G}$ is a graded Frobenius algebra with a $\mathbb{Q}$-grading that we call the *W-degree.*

2.2.3. *A-model axioms.* The genus-zero, three-point correlators may be difficult to compute in general. However, they satisfy some axioms that allow us to compute them in many cases. These axioms all follow immediately, by restriction to the genus-zero, three-point case, from the axioms for the FJRW virtual cycle described in [**FJR07b**].

Throughout this section, for $i \in \{1,2,3\}$ we assume that $\gamma_i \in \mathscr{A}_{g_i} \subseteq \mathscr{A}_{W,G}$ are elements of the FJRW ring, with $g_i = \mathbf{\Theta}^i = (\Theta^i_1,\ldots,\Theta^i_N)^{\mathsf{T}}$ and $0 \le \Theta^i_j < 1$ for every i, j. We will only list here the axioms we need for this paper. The remaining axioms can be found in [**FJR07b**, Thm 4.1.8] and [**KPA$^+$10**, §1.2].

AXIOM 1 (Symmetry). *Let $\sigma \in S_3$. Then*

$$\langle \gamma_1,\gamma_2,\gamma_3\rangle = \langle \gamma_{\sigma(1)},\gamma_{\sigma(2)},\gamma_{\sigma(3)}\rangle .$$

The next axioms relate to the degree of certain line bundles. For genus zero, these degrees are given by

$$l_j = q_j - \sum_{i=1}^{3} \Theta_j^i \text{ for } j \in \{1, 2, 3\} .$$

AXIOM 2 (Integer Line Bundle Degrees). *The correlator* $\langle \gamma_1, \gamma_2, \gamma_3 \rangle$ *vanishes unless* $l_j \in \mathbb{Z}$ *for* $j = 1, \ldots, N$.

The following observation, due to Krawitz, follows from Axiom 2.

PROPOSITION 2.2.4. *Suppose* $\langle \gamma_1, \gamma_2, \gamma_3 \rangle$ *does not vanish. Then* $g_3 = J - g_1 - g_2$. *Thus* $\gamma_1 \star \gamma_2 \in \mathscr{A}_{g_1+g_2-J}$, *and* $\mathscr{A}_{g_1+J} \star \mathscr{A}_{g_2+J} \subset \mathscr{A}_{g_1+g_2+J}$.

AXIOM 3 (Pairing). *Let* $\mathbb{1} = \lceil 1 \,; J \rfloor$. *Then* $\langle \gamma_1, \gamma_2, \mathbb{1} \rangle = \langle \gamma_1, \gamma_2 \rangle$.

This axiom implies that $\mathbb{1}$ is the identity element with respect to the multiplication (5) in the FJRW ring. Do not confuse the identity element of the ring with the element $\lceil 1 \,; 0 \rfloor$ in the identity sector.

AXIOM 4 (Decoupled Sums). *Suppose* W_1 *and* W_2 *are non-degenerate, quasi-homogeneous polynomials with no variables in common. Suppose* G_1 *and* G_2 *are admissible groups of diagonal symmetries for* W_1 *and* W_2 *respectively. Then* $G_1 \times G_2$ *is an admissible group of diagonal symmetries for* $W_1 + W_2$. *Suppose*

$$\lceil m_i n_i \,; g_i + h_i \rfloor \in \mathscr{A}_{W_1+W_2, G_1 \times G_2}$$

for $i = 1, 2, 3$, *with* $m_i \in \mathscr{Q}_{W_1}$, $n_i \in \mathscr{Q}_{W_2}$, $g_i \in G_1$, *and* $h_i \in G_2$. *Then the three-point correlator*

$$\langle \lceil m_1 n_1 \,; g_1 + h_1 \rfloor, \lceil m_2 n_2 \,; g_2 + h_2 \rfloor, \lceil m_3 n_3 \,; g_3 + h_3 \rfloor \rangle^{W_1+W_2, G_1 \times G_2}$$

is equal to the product

$$\langle \lceil m_1 \,; g_1 \rfloor, \lceil m_2 \,; g_2 \rfloor, \lceil m_3 \,; g_3 \rfloor \rangle^{W_1, G_1} \cdot \langle \lceil n_1 \,; h_1 \rfloor, \lceil n_2 \,; h_2 \rfloor, \lceil n_3 \,; h_3 \rfloor \rangle^{W_2, G_2} .$$

This gives an isomorphism

$$\mathscr{A}_{W_1, G_1} \otimes \mathscr{A}_{W_2, G_2} \cong \mathscr{A}_{W_1+W_2, G_1 \times G_2}.$$

AXIOM 5. *The three-point correlator is invariant under the action of* $G^{\max}$, *i.e.*,

$$\langle h\gamma_1, h\gamma_2, h\gamma_3 \rangle = \langle \gamma_1, \gamma_2, \gamma_3 \rangle$$

for all $h \in G^{\max}$.

REMARK 2.2.5. Axiom 5 gives another valuable selection rule as follows. If $\gamma_i = \lceil \mathbf{X}^{\mathbf{r}_i} \,; g_i \rfloor$ with $g_i \in G_W$ for $i \in \{1, 2, 3\}$, then for every $h = [\mathbf{h}] \in G_W^{\max}$, denote by $\mathbf{h}|_{\mathrm{Fix}(g_i)}$ the restriction of $\mathbf{h}$ to the fixed locus of g_i; that is, the jth coordinate of $\mathbf{h}$ is set to zero in $\mathbf{h}|_{\mathrm{Fix}(g_i)}$ if the jth coordinate of g_i is non-trivial. The correlator $\langle \lceil \mathbf{X}^{\mathbf{r}_1} \,; g_1 \rfloor, \lceil \mathbf{X}^{\mathbf{r}_2} \,; g_2 \rfloor, \lceil \mathbf{X}^{\mathbf{r}_3} \,; g_3 \rfloor \rangle^{W,G}$ vanishes unless

$$\sum_{i=1}^{3} (\mathbf{r}_i + \mathbf{1})^{\mathsf{T}} \mathbf{h}|_{\mathrm{Fix}(g_i)} \in \mathbb{Z}. \tag{6}$$

2.3. The B-model. The Frobenius algebra for the B-model we call the *orbifolded Milnor ring*. In this paper, on the B-side we will only consider diagonal automorphism groups which are subgroups of SL_N. Here, we are thinking of the elements of G as linear transformations of $\mathbb{C}^N$. In our notation, $g = [\mathbf{\Theta}] \in \mathrm{SL}_N$ is equivalent to $\sum_i \Theta_i \in \mathbb{Z}$. As mentioned above, this condition is dual to the condition that the A-model orbifold group contain J.

The construction of the state space (underlying vector space) is identical to the construction of the FJRW state space. We again take the un-projected g sector

$$\mathscr{H}_g = \mathscr{Q}_{W_g} \cdot dX_{i_1} \wedge \cdots \wedge dX_{i_{N_g}}$$

with the same G-action as before, and again let the (projected) g-sector $\mathscr{B}_g$ of our B-model be the G-invariants:

$$\mathscr{B}_g := \mathscr{H}_g^G \quad \text{and} \quad \mathscr{B}_{W,G} = \left(\bigoplus_g \mathscr{H}_g \right)^G .$$

The pairing is also defined in the same way as for the A-model.

The main difference between the A- and B-models is in the product. The B-model product makes $\mathscr{B}_{W,G}$ into a G-graded Frobenius algebra. In fact the product does more—it can be defined on the entire unprojected state space $\mathscr{H}_{W,G} := \bigoplus_{g \in G} \mathscr{H}_g$ and there it defines a *G-Frobenius algebra* (cf. [**Kau06, JKK05, Tur10**]).

The product was first written down explicitly in [**Kra10**], following ideas from [**Kau06**]. To define the product on $\mathscr{B}_{W,G}$, first note that for all g there is a surjective homomorphism of Milnor rings $\mathscr{Q}_e \to \mathscr{Q}_g$ given by setting to zero all variables not fixed by g. Thus, $\mathscr{Q}_g$ may be thought of as a cyclic $\mathscr{Q}_e$-module with generator $\lceil 1\,; g \rfloor$. For any two elements $g, h \in G$ we define $W_{g,h}$ to be the restriction of the polynomial W to the subspace of $\mathbb{C}^N$ fixed by both g and h. Define

$$\mathscr{Q}_{g,h} := \mathscr{Q}_{W_{g,h}}$$

to be the Milnor ring of $W_{g,h}$. There are three surjective ring homomorphisms:

$$e_g : \mathscr{Q}_g \to \mathscr{Q}_{g,h}, \qquad e_h : \mathscr{Q}_h \to \mathscr{Q}_{g,h} \quad \text{and} \quad e_{g+h} : \mathscr{Q}_{g+h} \to \mathscr{Q}_{g,h}.$$

The residue pairing η is non-degenerate for each of the Milnor rings, and hence we can use it to construct vector-space isomorphisms $\eta^\flat_{g,h} : \mathscr{Q}_{g,h} \to \mathscr{Q}^\vee_{g,h}$ and $\eta^\sharp_{g+h} : \mathscr{Q}^\vee_{g+h} \to \mathscr{Q}_{g+h}$, where $^\vee$ indicates the dual vector space. We can also can dualize the homomorphism e_{g+h} to construct an injective linear map $e^\vee_{g+h} : \mathscr{Q}^\vee_{g,h} \to \mathscr{Q}^\vee_{g+h}$. Combining these we define

$$e^*_{g+h} = \eta^\sharp_{g+h} \circ e^\vee_{g+h} \circ \eta^\flat_{g,h} : \mathscr{Q}_{g,h} \to \mathscr{Q}_{g+h}.$$

Finally, we define I_g, I_h, I_{g+h} to be the sets of indices fixed by the group elements g, h, and $g+h$, respectively.

DEFINITION 2.3.1. If $\lceil m\,; g \rfloor \in \mathscr{B}_g$ and $\lceil n\,; h \rfloor \in \mathscr{B}_h$, then $m \in \mathscr{Q}_g$ and $n \in \mathscr{Q}_h$. We define the product $\lceil m\,; g \rfloor \star_B \lceil n\,; h \rfloor$ to be

$$\lceil m\,; g \rfloor \star_B \lceil n\,; h \rfloor := \begin{cases} \left\lceil e^*_{g+h}\left(e_g(m) \cdot e_h(n)\right)\,; g+h \right\rfloor & \text{if } I_g \cup I_h \cup I_{g+h} = \{1, 2, \ldots, N\} \\ 0 & \text{otherwise,} \end{cases} \tag{7}$$

and extend the product linearly to the rest of $\mathscr{B}_{W,G}$.

Using Equation (3), a straightforward computation shows that Definition 2.3.1 is equivalent to the formulation given in [**Kra10**]:

$$\mu_{g+h}\operatorname{hess}(W_{g,h})\lceil 1\,;g\rfloor \star_B \lceil 1\,;h\rfloor = \mu_{g,h}\operatorname{hess}(W_{g+h})\lceil 1\,;g+h\rfloor,$$

where $\operatorname{hess}(W_{g,h})$ and $\operatorname{hess}(W_{g+h})$ denote the Hessian determinants of $W_{g,h}$ and W_{g+h}, respectively, and $\mu_{g,h}$ and μ_{g+h} denote the dimensions of the vector spaces $\mathscr{Q}_{g,h}$ and $\mathscr{Q}_{g+h}$, respectively. [**Kra10**] shows this product is associative for all invertible W and all $G \leq \mathrm{SL}_N$.

The orbifold Milnor ring also has a $\mathbb{Q}$-grading that matches the W-degree in our mirror symmetry.

The orbifold Milnor ring also has a tensor product property analogous to Axiom 4. This is easy to check from the definitions.

PROPOSITION 2.3.2. *Suppose W_1 and W_2 are non-degenerate, quasi-homogeneous polynomials with no variables in common. Suppose $G_1 \leq \mathrm{SL}_{N_1}$ and $G_2 \leq \mathrm{SL}_{N_2}$ are groups of diagonal symmetries. Then $G_1 \times G_2$ is contained in $\mathrm{SL}_{N_1+N_2}$ and is a group of diagonal symmetries for $W_1 + W_2$. Moreover, there is an isomorphism*

$$\mathscr{B}_{W_1,G_1} \otimes \mathscr{B}_{W_2,G_2} \cong \mathscr{B}_{W_1+W_2,G_1\times G_2}.$$

3. The mirror map

Consider the case of a non-degenerate, invertible, quasi-homogeneous polynomial W and an admissible group G of symmetries of W. Our main result is

THEOREM 3.0.3. *If (W, G) satisfies Property $(*)$, then there is an isomorphism of Frobenius algebras: $\mathscr{A}_{W,G} \cong \mathscr{B}_{W^\intercal,G^\intercal}$.*

Marc Krawitz defined two mirror maps in [**Kra10**]. The first is an isomorphism of graded vector spaces $\mathscr{B}_{W^\intercal,G^\intercal} \to \mathscr{A}_{W,G}$, defined for any (W,G), but it is not necessarily a ring homomorphism. The second is an algebra isomorphism $\mathscr{B}_{W^\intercal,0} \to \mathscr{A}_{W,G^{\max}}$, but it is not defined for general groups. In this section, we will show how to combine these two maps to get a mirror map defined for all pairs (W, G) satisfying Property $(*)$.

3.1. Basic Setup. The fact that (W, G) satisfies Property $(*)$ means that W has a decomposition

$$W = \sum_{i=1}^{M} W_i, \tag{8}$$

where each W_i is itself a non-degenerate, invertible polynomial, having no variables in common with any of the other W_j, and such that Property $(*)$ holds for this decomposition. Unless otherwise stated, we will assume hereafter that this decomposition has been fixed, once and for all.

NOTATION 3.1.1. Notice that $\mathscr{A}_{W,G}$ has a basis of element of the form $\lceil m\,;g\rfloor$ with $m = \prod_{i=1}^{M} m_i$, where m_i is a monomial in $\mathscr{Q}_{W_i}$. For any $g \in G$ we can also write $g = \sum_i g_i$, where g_i acts trivially except on W_i. Hereafter, when any element of $\mathscr{A}_{W,G}$ is written in this form, it will mean that the element is as described above. Similar remarks apply to the B-model.

3.1.1. *The maximal unprojected state space.* For any quasi-homogeneous, non-degenerate, invertible polynomial W and any group $H \leq \mathrm{SL}_N$ of diagonal symmetries of W, the projected state space $\mathscr{B}_{W^\intercal,H}$ is a subspace of the unprojected vector space $\mathscr{H}_{W^\intercal,G^{\max}_{W^\intercal}}$. Moreover, since the definition of B-model multiplication makes no reference to the orbifold group, but only to the group elements involved, the multiplication makes sense on all of $\mathscr{H}_{W^\intercal,G^{\max}_{W^\intercal}\cap \mathrm{SL}_N}$. This gives us the following lemma.

LEMMA 3.1.2. *For any diagonal symmetry group $H \leq \mathrm{SL}_N$ of $W^\intercal$, the subspace $\mathscr{B}_{W^\intercal,H}$ is a subalgebra of $\mathscr{H}_{W^\intercal,G^{\max}_{W^\intercal}\cap \mathrm{SL}_N}$.*

If B is a subalgebra of $\mathscr{B}_{W^\intercal,H}$ that is invariant under another symmetry group $H' \leq \mathrm{SL}_N$, and if B contains only sectors with group elements in H', then B is also a subalgebra of $\mathscr{B}_{W^\intercal,H'}$.

REMARK 3.1.3. On the A-side, we have a similar situation at the level of vector spaces: for any non-degenerate W and admissible G, the projected state space $\mathscr{A}_{W,G}$ is a subspace of the unprojected vector space $\mathscr{H}_{W,G^{\max}}$. There is, however, no obvious algebra structure on the unprojected space $\mathscr{H}_{W,G^{\max}}$ that would project to give the algebra structure on every $\mathscr{A}_{W,G}$.

3.1.2. *Properties of Atomic Types.* Assume that W is one of the atomic types with vector of exponents $\mathbf{a}$, as in Proposition 2.1.1. To define the mirror map we need to define the *Chain Property.*

PROPERTY (†). *If W is a chain we say that a vector $\boldsymbol{\beta}$ has the* Chain Property *for W if $\boldsymbol{\beta} = (a_1 - 1, 0, a_3 - 1, 0, \ldots, a_{2k-1} - 1, 0, \beta_{2k+1}, \beta_{2k+2}, \ldots, \beta_N)$, where $\beta_{2k+1} \neq a_{2k+1} - 1$, and $\beta_i \leq a_i - 1$ for all $i > 2k$. In other words, if $s \in \{1, 2, \ldots, N\}$ is the smallest index for which $\beta_s \neq \delta^s_{odd}(a_s - 1)$, then s is odd.*

The following lemma from [**Kre94**] gives a description of the Milnor ring of an invertible, non-degenerate polynomial.

LEMMA 3.1.4.

(1) *The Milnor ring of a loop type polynomial is generated over $\mathbb{C}$ by the basis $\{\prod_{i=1}^N X_i^{\beta_i} : 0 \leq \beta_i \leq a_i - 1\}$ and has dimension $\mu_W = \prod_{i=1}^N a_i$.*
(2) *The Milnor ring of a chain type polynomial is generated over $\mathbb{C}$ by the basis $\{\prod_{i=1}^n X_i^{\beta_i} : 0 \leq \beta_i \leq a_i - 1\}$, where the vector $\boldsymbol{\beta}$ has the Chain Property (†). This Milnor ring has dimension $\mu_W = \sum_{i=odd}(a_i - 1)\prod_{j=i+1}^n a_j$.*
(3) *The Milnor ring of a Fermat type polynomial is generated over $\mathbb{C}$ by the basis $\{X^\beta : 0 \leq \beta \leq a_i - 2\}$ and has dimension $\mu_W = a - 1$.*

In each case, define $\mathbf{h}_W = A^\intercal \mathbf{1} - \mathbf{1}$, then the monomial $\mathbf{X}^{\mathbf{h}_W - \mathbf{1}}$ is of top degree, so the Hessian is a scalar multiple of $\mathbf{X}^{\mathbf{h}_W - \mathbf{1}}$.

To define the mirror map, we will also need the following lemma which follows trivially from [**Kre94**].

LEMMA 3.1.5. *Any non-trivial diagonal symmetry of an atomic invertible singularity W can be written uniquely in the form*

$$J + \sum_{i=1}^N r_i \rho_i, \qquad \text{or equivalently} \qquad A_W^{-1}(\mathbf{r} + \mathbf{1}), \tag{9}$$

where the vector $\mathbf{r}$ *satisfies the following constraints:*

(1) *If* W *is a loop polynomial, then* $0 \leq r_i \leq a_i - 1$. *Furthermore, if* N *is even, then the identity element can be written as*

$$0 = J + \sum_{i \text{ even}} (a_i - 1)\rho_i = J + \sum_{i \text{ odd}} (a_i - 1)\rho_i.$$

(If N *is odd, then the identity cannot be written in this form.)*

(2) *If* W *is a chain polynomial, then* $\mathbf{r}$ *satisfies the Chain Property* (†). *The identity element can be written*

$$0 = J + \sum_{n-i \text{ even}} (a_i - 1)\rho_i.$$

(3) *If* W *is a Fermat polynomial, then* $N = 1$ *and* $\mathbf{r} = (r)$ *with* $0 \leq r \leq a - 2$.

Hereafter, all elements of the Milnor rings and groups associated to Fermat, loop, and chain type polynomials will be written in the forms described above.

3.2. The mirror map. In this section we will define the mirror map. We do this first on some basic pieces we call *fundamental factors* and then use these to define it in general.

3.2.1. *Fundamental factors.* If a polynomial is atomic, it cannot be written as a disconnected sum of other polynomials, and Property (∗) implies a stricter property, which we call *fundamental.*

DEFINITION 3.2.1. We say that a pair (W, G) is *fundamental* if

(1) For every non-trivial $g \in G^{\mathsf{T}}$ the sector $\mathscr{B}_g$ of $\mathscr{B}_{W^{\mathsf{T}},G^{\mathsf{T}}}$ is either narrow (see Definition 2.2.2) or is empty.
(2) For every non-trivial $h \in G$ the sector $\mathscr{A}_h$ of $\mathscr{A}_{W,G}$ is either narrow or is empty.

We construct the mirror map $\varphi : \mathscr{B}_{W^{\mathsf{T}},G^{\mathsf{T}}} \to \mathscr{A}_{W,G}$ for all fundamental pairs (W, G) in two parts, as follows.

Firstly, the untwisted sector $\mathscr{B}_0 \subseteq \mathscr{B}_{W^{\mathsf{T}},G^{\mathsf{T}}}$ is a subalgebra of the untwisted sector of the unprojected algebra $\mathscr{H}_0 \cong \mathscr{Q}_{W^{\mathsf{T}}}$, and Krawitz's unorbifolded algebra mirror map [**Kra10**, §3.1] gives an isomorphism of this sector to the sum of the A-model twisted sectors (all narrow), $\bigoplus_{h\in G_W^{\max}} \mathscr{A}_h \subseteq \mathscr{A}_{W,G_W^{\max}}$. Our mirror map is a rescaling of that isomorphism, given on the algebra generators $\lceil Y_j\,; 0 \rfloor$ by

$$\lceil Y_j\,; 0 \rfloor \mapsto \ell^{w'_j} \lceil 1\,; \rho_j + J \rfloor, \tag{10}$$

where w'_j is the integer weight of Y_j as defined in Equation (1), and where $\ell \in \mathbb{C}$ is a constant chosen such that

$$\ell^{\mathbf{w}'\cdot\mathbf{h}_{W^{\mathsf{T}}}} = \left\langle \lceil \mathbf{Y}^{\mathbf{h}_{W^{\mathsf{T}}}}\,; 0 \rfloor, \lceil 1\,; 0 \rfloor \right\rangle = \frac{\mu_{W^{\mathsf{T}}} \mathbf{Y}^{\mathbf{h}_{W^{\mathsf{T}}}}}{\operatorname{hess}(W^{\mathsf{T}})} \tag{11}$$

Also, φ takes the identity $\lceil 1\,; 0 \rfloor \in \mathscr{B}_{W^{\mathsf{T}},G^{\mathsf{T}}}$ to the identity $\mathbb{1} = \lceil 1\,; J \rfloor \in \mathscr{A}_{W,G}$.

Secondly, when $g \in G^{\mathsf{T}}$ is non-trivial, the sector must be narrow or empty. If we write W as a decoupled sum of atomic types, then by Lemma 3.1.5 we may write g uniquely as $g = (A^{\mathsf{T}})^{-1}(\mathbf{r} + \mathbf{1})$, with the components of $\mathbf{r}$ satisfying the constraints of that lemma, as determined by the corresponding atomic types. The direct sum of all the twisted sectors $\bigoplus_{g\neq 0\in G^{\mathsf{T}}} \mathscr{B}_g$ is spanned by elements of the form $\lceil 1\,; g\rfloor$. We define the mirror map by

$$\lceil 1\,; (A_W^{\mathsf{T}})^{-1}(\mathbf{r}+\mathbf{1})\rfloor \mapsto k^{-\mathbf{w}\cdot\mathbf{r}}\lceil \mathbf{X}^{\mathbf{r}}\,; 0\rfloor, \tag{12}$$

where w_j is the integer weight of X_j as defined in Equation (1), and where $k\in\mathbb{C}$ is a constant chosen such that

$$k^{\mathbf{w}\cdot\mathbf{h}_W} = \langle \lceil \mathbf{X}^{\mathbf{h}_W}\,; 0\rfloor, \lceil 1\,; 0\rfloor\rangle = \frac{\mu_W \mathbf{X}^{\mathbf{h}_W}}{\operatorname{hess}(W)} \tag{13}$$

REMARK 3.2.2. The mirror map on the twisted sectors is a rescaling of Krawitz's vector space mirror map. The lemmas in [**Kra10**] show that the map (10) on the untwisted sector agrees with the vector space mirror map, except possibly in the case of an even variable loop. In this case, there is some ambiguity in the definition of the vector space mirror map on a certain two dimensional subspace. Krawitz made an arbitrary choice to prove the vector space isomorphism, and the algebra mirror map (10) should, in principle, resolve this ambiguity. However, (10) is given in terms of algebra generators, and we have not been able to compute the image of the vector space basis elements in question. It is possible to check, however, that the map is surjective. Our map agrees with Krawitz's vector space map, up to a possibly different choice on the ambiguous subspace, so it is still a bijection. Thus, it only remains to check that φ is a ring homomorphism, which we do in Section 4.

3.2.2. *Decoupled sums.* Let $W=\sum_{i=1}^M W_i$ be a decoupled sum of non-degenerate, invertible polynomials, and let $G=\bigoplus_{i=1}^M G_i$, where each G_i is an admissible group of symmetries of W_i. If the mirror map is defined for each of the pairs (W_i, G_i)

$$\varphi_i : \mathscr{B}_{W_i^{\mathsf{T}}, G_i^{\mathsf{T}}} \to \mathscr{A}_{W_i, G_i},$$

then the mirror map for the decoupled sum is just given by the tensor product:

$$\varphi : \mathscr{B}_{W,G} \cong \bigotimes_i \mathscr{B}_{W_i^{\mathsf{T}}, G_i^{\mathsf{T}}} \xrightarrow{\otimes_i \varphi_i} \bigotimes_i \mathscr{A}_{W_i, G_i} \cong \mathscr{A}_{W,G}.$$

3.2.3. *The general case.* We now wish to define the mirror map φ for the general case of (W, G) satisfying Property $(*)$, where G is not necessarily a direct sum of groups.

NOTATION 3.2.3. Let I a subset of $\{1,\ldots,M\}$. Then we let

$$g_I = \sum_{i\in I} g_i \quad\text{and}\quad m_I = \prod_{i\in I} m_i \quad\text{and}\quad W_I^{\mathsf{T}} = \sum_{i\in I} W_i^{\mathsf{T}}. \tag{14}$$

The algebra $\mathscr{B}_{W^{\mathsf{T}}, G^{\mathsf{T}}}$ is spanned by elements of the form $\lceil \prod m_i\,; \sum g_i\rfloor$. For each such element, we will find a partition $\mathscr{I}$ of $\{1,2,\ldots,M\}$ and for each $I\in\mathscr{I}$ we will find groups G_I of symmetries of W_I so that each pair (W_I, G_I) is fundamental and $\lceil \prod m_i\,; \sum g_i\rfloor \in \bigoplus_{I\in\mathscr{I}} \mathscr{B}_{W_I^{\mathsf{T}}, G_I^{\mathsf{T}}}$ (and which satisfy a few other properties). The previous results allow us to define the mirror map on the decoupled sum $\bigoplus_{I\in\mathscr{I}} \mathscr{B}_{W_I^{\mathsf{T}}, G_I^{\mathsf{T}}}$, and we can show that the image actually lies in $\mathscr{A}_{W,G}$. We will show that the mirror map thus defined is independent of the choices and is a homomorphism of Frobenius algebras.

DEFINITION 3.2.4. Let $\mathscr{I}$ be a partition of $\{1,2,\ldots,M\}$ and let $\{G_I\}_{I\in\mathscr{I}}$ be admissible groups of symmetries of W_I.

For any $\lceil \prod m_i\,; \sum g_i\rfloor \in \mathscr{B}_{W^{\mathsf{T}}, G^{\mathsf{T}}}$ we say that $\lceil \prod m_i\,; \sum g_i\rfloor$ *splits nicely* with respect to $\mathscr{I}$ and $\{G_I^{\mathsf{T}}\}$ if the following properties are satisfied:

(1) For every $I \in \mathscr{I}$ the pair (W_I, G_I) is fundamental.
(2) $\lceil m_I\, ; g_I \rfloor \in \mathscr{B}_{W_I^\mathsf{T}, G_I^\mathsf{T}}$.
(3) The element g_I is either trivial, or acts non-trivially on all W_i for $i \in I$.
(4) If $g_I = 0$ and $m_I \neq 1$, then $|I| = 1$, i.e., $W_I = W_j$ for some j.

LEMMA 3.2.5. *For any element* $\gamma = \lceil \prod m_i\, ; \sum g_i \rfloor \in \mathscr{B}_{W^\mathsf{T}, G^\mathsf{T}}$ *there exists a partition* $\mathscr{I}$ *and groups* $\{G_I\}$ *so that* γ *splits nicely with respect to* $\mathscr{I}, \{G_I\}$.

PROOF. Let I_0 be the set of i so that $g_i = 0$. Let I_g be the set of i with $g_i \neq 0$. Then we choose the partition

$$\mathscr{I} = \{I_g\} \cup \{\{i\}\}_{i \in I_0}.$$

We choose the group $G_{I_g}^\mathsf{T}$ for $W_{I_g}^\mathsf{T}$ to be the group generated by g_{I_g}, and we choose the trivial groups for the others. In light of Property $(*)$, it is clear that these groups satisfy the properties described in Definition 3.2.4. $\square$

For each $\mathscr{I}, \{G_I\}$ with each (W_I, G_I) fundamental, define the mirror map

$$\varphi_\mathscr{I} : \mathscr{B}_{\sum_{I \in \mathscr{I}} W_I^\mathsf{T}, \bigoplus_{I \in II} G_I^\mathsf{T}} \cong \bigotimes_{I \in \mathscr{I}} \mathscr{B}_{W_I^\mathsf{T}, G_I^\mathsf{T}} \to \bigotimes_{I \in \mathscr{I}} \mathscr{A}_{W_I^\mathsf{T}, G_I^\mathsf{T}} \cong \mathscr{A}_{\sum_{I \in \mathscr{I}} W_I, \bigoplus_{I \in \mathscr{I}} G_I}.$$

Given a $\mathscr{I}, \{G_I\}$ such that γ splits nicely, define the sub-algebra

$$B_\mathscr{I} := \mathscr{B}_{W^\mathsf{T}, G^\mathsf{T}} \cap \mathscr{B}_{\sum_{I \in \mathscr{I}} W_I^\mathsf{T}, \bigoplus_{I \in \mathscr{I}} G_I^\mathsf{T}}$$

and apply the map $\varphi_\mathscr{I}$ to $B_\mathscr{I}$. We observe from the definition of φ that the image $\varphi_\mathscr{I}(B_\mathscr{I})$ lies in the vector space $\mathscr{A}_{W,G}$.

Proposition 4.1.1 states that the algebra structure induced on the vector space $\mathscr{A}_{W,G} \cap \mathscr{A}_{\sum_{I \in \mathscr{I}} W_I, \bigoplus_{I \in \mathscr{I}} G_I}$ from $\mathscr{A}_{\sum_{I \in \mathscr{I}} W_I, \bigoplus_{I \in \mathscr{I}} G_I}$ is the same as the algebra structure induced from $\mathscr{A}_{W,G}$. Thus the subalgebra $\varphi_\mathscr{I}(B_\mathscr{I})$ of $\mathscr{A}_{W, \bigoplus_{I \in \mathscr{I}} G_I}$ is a subalgebra of $\mathscr{A}_{W,G}$. Furthermore, we have the following lemma.

LEMMA 3.2.6. *The image* $\varphi_\mathscr{I}(\gamma)$ *is independent of the choice of* $\mathscr{I}$ *and* $\{G_I\}$.

This follows from the fact that Equation (12) respects the splitting, and both (12) and (10) refer only to the group elements, and not to the subgroup.

REMARK 3.2.7. By Remark 3.2.2 our map for fundamental pairs agrees with Krawitz's vector space isomorphism (up to a minor ambiguity for even-variable loops). Since these maps all respect splitting of decoupled sums, our φ also preserves the grading and is a vector space isomorphism.

Combining all these results we have the following.

PROPOSITION 3.2.8. *The map* $\varphi : \mathscr{B}_{W^\mathsf{T}, G^\mathsf{T}} \to \mathscr{A}_{W,G}$ *is a well-defined, degree-preserving isomorphism of vector spaces.*

4. Computations in $\mathscr{A}_{W,G}$

In this section we prove two results about computations of products or correlators in $\mathscr{A}_{W,G}$. The first shows that we can change groups without changing the products and the second gives a selection rule that allows us to prove that many correlators vanish.

4.1. Changing groups. As mentioned in Remark 3.1.3, for any admissible group G of symmetries of W, the projected state space $\mathscr{A}_{W,G}$ is a vector subspace of the maximal unprojected state space $\mathscr{H}_{W,G^{\max}}$, but we have no algebra structure on $\mathscr{H}_{W,G^{\max}}$. In this section we address this problem with the following proposition.

PROPOSITION 4.1.1. *If H and H' are two admissible groups of symmetries of W, then for any $\lceil m\,;g\rfloor$ and $\lceil n\,;h\rfloor$ in $\mathscr{A}_{P,H} \cap \mathscr{A}_{P,H'}$, the product $\lceil m\,;g\rfloor \star \lceil n\,;h\rfloor$ is the same whether is it computed in $\mathscr{A}_{P,H}$ or $\mathscr{A}_{P,H'}$.*

The key observation needed to prove this proposition is the following lemma.

LEMMA 4.1.2. *Let P be a quasi-homogeneous, non-degenerate, invertible polynomial and H and H' be admissible subgroups of $G_P^{\max}$. Suppose we have $\lceil m\,;g\rfloor$, $\lceil n\,;h\rfloor$, and $\lceil p\,;k\rfloor$ in the vector space $\mathscr{A}_{P,H} \cap \mathscr{A}_{P,H'}$. We have*

$$\langle \lceil m\,;g\rfloor, \lceil n\,;h\rfloor, \lceil p\,;k\rfloor\rangle^{P,H} = \langle \lceil m\,;g\rfloor, \lceil n\,;h\rfloor, \lceil p\,;k\rfloor\rangle^{P,H'}. \tag{15}$$

That is, we may compute the three point correlator in either FJRW ring.

PROOF. It suffices to prove the result in the case that $H \leq H'$. The correlator $\langle \lceil m\,;g\rfloor, \lceil n\,;h\rfloor, \lceil p\,;k\rfloor\rangle^{P,H}$ is defined (see [**FJR07b**, Def 4.6.2]) as

$$\langle \lceil m\,;g\rfloor, \lceil n\,;h\rfloor, \lceil p\,;k\rfloor\rangle^{P,H} := \int_{\overline{\mathscr{M}}_{0,3}} \Lambda_{0,3}^{P,H}(m,n,p),$$

where $\overline{\mathscr{M}}_{0,3}$ is the stack of three-pointed, genus-zero stable curves, and $\Lambda_{0,3}^{P,H}(m,n,p)$ is defined (see [**FJR07b**, Def 4.2.1]) to be the Poincaré dual of the pushforward of the virtual cycle, capped with the classes m, n, and p:

$$\Lambda_{0,3}^{P,H}(m,n,p) := \frac{1}{\deg(\mathrm{st}^{P,H})} PD\ \mathrm{st}_*^{P,H}\left(\left[\overline{\mathscr{W}}_{0,3,H}(P;g,h,k)\right]^{vir} \cap (m \cup n \cup p)\right).$$

Here $\mathrm{st}^{P,H} : \overline{\mathscr{W}}_{0,3,H}(P;g,h,k) \to \overline{\mathscr{M}}_{0,3}$ is the forgetful map taking the stack of genus-zero, stable P-curves with admissible group H to the stack of genus-zero stable curves with three marked points, defined simply by forgetting the P-structure on the curve. The correlator $\langle \lceil m\,;g\rfloor, \lceil n\,;h\rfloor, \lceil p\,;k\rfloor\rangle^{P,H'}$ is defined similarly.

According to [**FJR07b**, 2.3.1], there is a finite morphism of stacks $a : \overline{\mathscr{W}}_{0,3,H}(P;g,h,k) \to \overline{\mathscr{W}}_{0,3,H'}(P;g,h,k)$, surjective onto an open and closed substack of $\overline{\mathscr{W}}_{0,3,H'}(P)$. Moreover, $\overline{\mathscr{W}}_{0,3,H'}(P)$ actually has only a single geometric point, corresponding to the unique genus-zero, three-pointed P-curve with markings g, h, and k, respectively. Therefore, in this case, the morphism a is surjective and finite.

Theorem 6.3.5 of [**FJR07a**] shows that the virtual class $\left[\overline{\mathscr{W}}_{0,3,H'}(P;g,h,k)\right]^{vir}$ is the pullback along a of the virtual cycle $\left[\overline{\mathscr{W}}_{0,3,H'}(P;g,h,k)\right]^{vir}$ on $\overline{\mathscr{W}}_{0,3,H'}(P)$, which gives

$$\begin{aligned}
&\langle\lceil m\,;g\rfloor\,,\lceil n\,;h\rfloor\,,\lceil p\,;k\rfloor\rangle^{P,H}\\
&\quad=\int_{\overline{\mathscr{M}}_{0,3}}\frac{PD\ \mathrm{st}_*^{P,H'}\ a_*\left(a^*\left[\mathscr{W}_{0,3,H'}(P;g,h,k)\right]^{vir}\cap(m\cup n\cup p)\right)}{\deg(\mathrm{st}^{P,H})}\\
&\quad=\int_{\overline{\mathscr{M}}_{0,3}}\frac{\deg(a)PD\ \mathrm{st}_*^{P,H'}\left(\left[\mathscr{W}_{0,3,H'}(P;g,h,k)\right]^{vir}\cap(m\cup n\cup p)\right)}{\deg(\mathrm{st}^{P,H})}\\
&\quad=\langle\lceil m\,;g\rfloor\,,\lceil n\,;h\rfloor\,,\lceil p\,;k\rfloor\rangle^{P,H'}.
\end{aligned}$$

□

PROOF OF PROPOSITION 4.1.1. By definition of multiplication, we have

$$\lceil m\,;g\rfloor\star_{P,H}\lceil n\,;h\rfloor=\sum_{\sigma,\tau}\langle\lceil m\,;g\rfloor\,,\lceil n\,;h\rfloor\,,\sigma\rangle\,\eta^{\sigma,\tau}\tau,$$

where σ and τ range over a basis of $\mathscr{A}_{P,H}$. The product is defined similarly for $\mathscr{A}_{P,H'}$. For elements $\mathscr{A}_{P,H}\cap\mathscr{A}_{P,H'}$, Lemma 4.1.2 tells us that we can compute the correlators (and thus also the pairing) in either. It suffices then to show that if we have a basis element $\lceil p\,;k\rfloor$ of $\mathscr{A}_{P,H}$ that is not in $\mathscr{A}_{P,H'}$, the correlator $\langle\lceil m\,;g\rfloor\,,\lceil n\,;h\rfloor\,,\lceil p\,;k\rfloor\rangle^{P,H}$ vanishes. This is a straightforward consequence of Axiom 5 ($G^{\max}$-invariance) and Proposition 2.2.4. □

4.2. A selection rule. We isolate the following fact since it will be used multiple times in the proof of Lemma 5.2.1.

PROPOSITION 4.2.1. *Suppose P is a decoupled sum P_1+P_2, and following Notation 3.1.1, we have three basis elements $\lceil m_1\cdot m_2\,;g_1+g_2\rfloor$, $\lceil n_1\cdot n_2\,;h_1+h_2\rfloor$, and $\lceil p_1\cdot p_2\,;k_1+k_2\rfloor$ of $\mathscr{A}_{P,H}$. Suppose $\lceil m_1\,;g_1\rfloor$ and $\lceil n_1\,;h_1\rfloor$ are invariant under $G^{\max}_{P_1}$. Then the correlator*

$$\langle\lceil m_1\cdot m_2\,;g_1+g_2\rfloor\,,\lceil n_1\cdot n_2\,;h_1+h_2\rfloor\,,\lceil p_1\cdot p_2\,;k_1+k_2\rfloor\rangle \tag{16}$$

vanishes unless both

(1) *$\lceil p_1\,;k_1\rfloor$ is also invariant under $G^{\max}_{P_1}$.*
(2) *The correlator*

$$\langle\lceil m_1\,;g_1\rfloor\,,\lceil n_1\,;h_1\rfloor\,,\lceil p_1\,;k_1\rfloor\rangle^{P_1,G^{\max}_{P_1}} \tag{17}$$

is non-vanishing.

PROOF. Condition 1 follows from $G^{\max}$-invariance (Axiom 5). If this condition is satisfied, then we can pick groups $H_1=G^{\max}_{P_1}$ and $H_2=\pi_2(H)$, where π_2 is projection onto the second factor of $G^{\max}_P\cong G^{\max}_{P_1}\times G^{\max}_{P_2}$. We can then check that Lemma 4.1.2 applies, and we can compute (16) in the ring $\mathscr{A}_{W_1+W_2,H_1\times H_2}$. Axiom 4 now shows (16) is the product of (17) and $\langle\lceil m_2\,;g_2\rfloor\,,\lceil n_2\,;h_2\rfloor\,,\lceil p_2\,;k_2\rfloor\rangle^{P_2,H_2}$. □

5. Algebra homomorphism

Although we have a well-defined mirror map, we must still prove that it respects multiplication. We do this first for fundamental factors and then in the general case. This requires several technical lemmas about loops and chains, which are provided in an appendix to this paper.

5.1. Fundamental factors.

PROPOSITION 5.1.1. *If (W, G) is a fundamental pair, then the mirror map*

$$\varphi : \mathscr{B}_{W^\intercal, G^\intercal} \to \mathscr{A}_{W,G}$$

defined in Section 3.2.1 is a ring homomorphism

PROOF. Assume that $\lceil m\,; g\rfloor$ and $\lceil m'\,; g'\rfloor$ are elements of $\mathscr{B}_{W^\intercal, G^\intercal}$. In order to check that $\varphi(\lceil m\,; g\rfloor \star_B \lceil m'\,; g'\rfloor) = \varphi(\lceil m\,; g\rfloor) \star_A \varphi(\lceil m'\,; g'\rfloor)$, we must consider three distinct cases:

(1) $g = g' = 0$
(2) $g = 0$ but $g' \neq 0$ and
(3) $g \neq 0$ and $g' \neq 0$.

In case (1) restrict the map φ to the untwisted sector $\mathscr{B}_0$, where it is a rescaling of the ring homomorphism defined in [**Kra10**].

Case (2) breaks into two parts. Firstly, in the case of $\lceil 1\,; 0\rfloor \star_B \lceil 1\,; g'\rfloor$, the element $\lceil 1\,; 0\rfloor$ is the identity element of $\mathscr{B}_{W^\intercal, G^\intercal}$ and it maps to the identity element $\mathbb{1} = \lceil 1\,; J\rfloor$ of $\mathscr{A}_{W,G}$, so the product is preserved in this case.

Secondly, in the case of $\lceil m\,; 0\rfloor \star_B \lceil 1\,; g'\rfloor$, where m is not constant, Lemma 5.2.1 (proved in the next section) shows that both products vanish.

In case (3) write W as a decoupled sum of atomic types. By Lemma 3.1.5 we may write g and g' uniquely as $g = (A^\intercal)^{-1}(\mathbf{r} + \mathbf{1})$ and $g' = (A^\intercal)^{-1}(\mathbf{s} + \mathbf{1})$, respectively, with the components of $\mathbf{r}$ and $\mathbf{s}$ satisfying the constraints of that lemma, as determined by the corresponding atomic types.

If $[A^{-1}(\mathbf{r}+\mathbf{s}+\mathbf{2})] \neq 0$ then the B-side product vanishes by definition, and the A-side product vanishes by Corollary 7.0.11.

If $[A^{-1}(\mathbf{r} + \mathbf{s} + \mathbf{2})] = 0$, then $\mathbf{r} + \mathbf{s} = \mathbf{h}_W - \mathbf{1}$. For loops this follows from Remark 7.0.9, since if either (1b) or (1c) were true the B-side product would not be a product of non-identity sectors. Similarly for Fermats, $r_i + s_i = a_i - 1$. For chains, it is easy to check that if $\mathbf{r} + \mathbf{s} \neq \mathbf{h}_W - \mathbf{1}$ then we get a contradiction to Property $(*)$. By definition, the B-side product is

$$\lceil 1\,; (A^\intercal)^{-1}(\mathbf{r} + \mathbf{1})\rfloor \star_B \lceil 1\,; (A^\intercal)^{-1}(\mathbf{s} + \mathbf{1})\rfloor = \frac{1}{\mu_{W^\intercal}} \lceil \mathrm{hess}(W^\intercal)\,; 0\rfloor . \tag{18}$$

To compute the image $\varphi(\lceil \mathrm{hess}(W^\intercal)\,; 0\rfloor^B)$ we write[1]

$$\varphi\left(\frac{1}{\mu_{W^\intercal}} \lceil \mathrm{hess}(W^\intercal)\,; 0\rfloor^B\right) = \ell^{-\mathbf{w}'\cdot(\boldsymbol{\beta}+\boldsymbol{\gamma})}\varphi\left(\lceil \mathbf{Y}^{\boldsymbol{\beta}}\,; 0\rfloor^B\right) \star_B \left(\lceil \mathbf{Y}^{\boldsymbol{\gamma}}\,; 0\rfloor^B\right)$$

with $\boldsymbol{\beta} + \boldsymbol{\gamma} = \mathbf{h}_{W^\intercal}$. We may apply [**Kra10**, Lem 3.2 and 3.5], which states that

$$\lceil 1\,; [A^{-1}(\boldsymbol{\beta} + \mathbf{1})]\rfloor \star_A \lceil 1\,; [A^{-1}(\boldsymbol{\gamma} + \mathbf{1})]\rfloor = \lceil 1\,; [A^{-1}(\boldsymbol{\beta} + \boldsymbol{\gamma} + \mathbf{1})]\rfloor \tag{19}$$

as long as $\boldsymbol{\beta} + \boldsymbol{\gamma} \preceq \mathbf{a} - \mathbf{1}$ (componentwise) and $[A^{-1}(\boldsymbol{\beta} + \boldsymbol{\gamma} + \mathbf{1})] \neq 0$. Combining with Lemma 7.0.6 gives

$$\varphi\left(\frac{1}{\mu_{W^\intercal}} \lceil \mathrm{hess}(W^\intercal)\,; 0\rfloor^B\right) = \lceil 1\,; [(A)^{-1}\mathbf{h}_{W^\intercal}]\rfloor = \lceil 1\,; -J\rfloor^A . \tag{20}$$

On the other hand, the A-side product is

---

[1] Recall that φ agrees with Krawitz' algebra homomorphism for these correlators.

$$k^{-\mathbf{w}\cdot(\mathbf{r}+\mathbf{s})}\langle\lceil\mathbf{X}^{\mathbf{r}}\,;\mathbf{0}\rfloor\,,\,\lceil\mathbf{X}^{\mathbf{s}}\,;\mathbf{0}\rfloor\rangle\lceil 1\,;-J\rfloor^{A}=\lceil 1\,;-J\rfloor^{A}.$$

□

5.2. Mixed products. The following lemma finishes the proof of Proposition 5.1.1. We must prove this lemma in more generality than just for fundamental pairs, since we need to apply it to pairs of elements which cannot be split nicely by a common partition.

We will continue to use the notation of Section 3.1 and we will write the vector of variables of W_j^{T} as $\mathbf{Y}_j$ and similarly the vector of variables of W_j as $\mathbf{X}_j$.

LEMMA 5.2.1. *Suppose we have a pair* $\lceil\prod m_i\,;\sum g_i\rfloor^B$ *and* $\lceil\prod n_i\,;\sum h_i\rfloor^B$ *and for some* j, *we have* $m_j=1$, $g_j=[(A_j^{\mathsf{T}})^{-1}(\mathbf{s}+1)]\neq 0$, *and* $n_j=\mathbf{Y}_j^{\boldsymbol{\beta}}\neq 1$, $h_j=0$. *Then the following products both vanish.*

$$\left\lceil\prod m_i\,;\sum g_i\right\rfloor\star_B\left\lceil\prod n_i\,;\sum h_i\right\rfloor\quad\text{and}\quad\varphi(\left\lceil\prod m_i\,;\sum g_i\right\rfloor)\star_A\varphi(\left\lceil\prod n_i\,;\sum h_i\right\rfloor)$$

PROOF OF LEMMA 5.2.1. Since $g_j+h_j\neq 0$ but $\mathbf{Y}_j^{\boldsymbol{\beta}}\neq 1$, the product $\lceil\prod m_i\,;\sum g_i\rfloor^B\star_B\lceil\prod n_i\,;\sum h_i\rfloor$ will vanish. It remains to see that $\varphi(\lceil\prod m_i\,;\sum g_i\rfloor)\star_A\varphi(\lceil\prod n_i\,;\sum h_i\rfloor)$ vanishes as well. The image of $\lceil\prod m_i\,;\sum g_i\rfloor$ will be a linear combination of things of the form $\lceil\prod\hat{m}_i\,;\sum\hat{g}_i\rfloor$, where $\hat{m}_j=\mathbf{X}_j^{\mathbf{s}}$, $\hat{g}_j=0$.

We can see that the image of $\lceil\prod n_i\,;\sum h_i\rfloor$ will be a linear combination of things of the form $\left\lceil\prod\hat{n}_i\,;\hat{h}_i\right\rfloor$ Lemma 3.1.5 shows exactly when $\hat{h}_j=0$. If W_j is an even variable loop, $\hat{h}_j=0$ only when $\beta_k=\delta^k_{\text{odd}}(a_k-1)$ or $\beta_k=\delta^k_{\text{even}}(a_k-1)$, which implies that n_j is $\prod_{k\text{ even/odd}}X_{j,k}^{a_{j,k}-1}$. For the potentially non-vanishing correlators, we can apply Proposition 4.2.1 and examine pieces of the form

$$\left\langle\lceil\mathbf{X}_j^{\mathbf{s}}\,;0\rfloor\,,\left\lceil\prod_{k\text{ odd/even}}X_{j,k}^{a_{j,k}-1}\,;0\right\rfloor\,,\lceil 1\,;J_{W_j}\rfloor\right\rangle,\tag{21}$$

where we filled in the third spot using Proposition 2.2.4. We see that this is non-vanishing only if $\mathbf{X}_j^{\mathbf{s}}$ pairs with $\prod_{k\text{ odd/even}}X_{j,k}^{a_{j,k}-1}$, which will only happen if $\mathbf{s}$ is of the form $s_k=\delta^k_{\text{odd/even}}(a_k-1)$. But in that case we have $[(A_j^{\mathsf{T}})^{-1}(\mathbf{s}+\mathbf{1})]=0$, contradicting our assumption. For the case that W_j is a chain, a similar argument produces a contradiction to the Chain Property (†); thus the product vanishes.

Suppose now that $\boldsymbol{\beta}$ is not of these special forms. Then $\hat{n}_j=1$, $\hat{h}_j=[(A_j^{\mathsf{T}})^{-1}(\boldsymbol{\beta}+\mathbf{1})]\neq 0$. Again using Proposition 4.2.1 and Proposition 2.2.4, we can examine pieces of the form

$$\langle\lceil\mathbf{X}_j^{\mathbf{s}}\,;0\rfloor\,,\lceil 1\,;[(A_j)^{-1}(\boldsymbol{\beta}+\mathbf{1})]\rfloor\,,\lceil 1\,;[-A_j{}^{-1}\boldsymbol{\beta}]\rfloor\rangle.\tag{22}$$

It follows from Lemma 3.1.5 (1) that $(A_j)^{-1}\boldsymbol{\beta}\neq 0$. Applying Remark 2.2.5 to $\bar{\rho}_{j,k}$ this correlator vanishes unless $\bar{\rho}_{i,k}^{\mathsf{T}}(\mathbf{s}+\mathbf{1})\in\mathbb{Z}$. But this would imply that $[(A_j^{\mathsf{T}})^{-1}(\mathbf{s}+\mathbf{1})]=0$, again contradicting the hypothesis. □

5.3. The general case. We can now prove the homomorphism property in the general case.

THEOREM 5.3.1. *If* (W,G) *satisfies Property* $(*)$ *then then the mirror map* $\varphi:\mathscr{B}_{W^{\mathsf{T}},G^{\mathsf{T}}}\to\mathscr{A}_{W,G}$ *is a ring homomorphism*

Given a general pair of elements $\lceil\prod m_i\,;\sum g_i\rfloor$ and $\lceil\prod n_i\,;\sum h_i\rfloor$ in $\mathscr{B}_{W^\intercal,G^\intercal}$, if they are not of the form described in Lemma 5.2.1, then we will show that we can pick a partition $\mathscr{I}$ and groups $\{G_I\}$ such that both elements split nicely with respect to $\mathscr{I}$ and $\{G_I\}$. Since the algebra $\mathscr{B}_{\sum_{I\in\mathscr{I}} W_I, \bigoplus_{I\in\mathscr{I}} G_I^\intercal}$ is a tensor product of fundamental factors, the desired result will follow.

DEFINITION 5.3.2. Excluding the cases described in Lemma 5.2.1, consider a pair of elements $\lceil\prod m_i\,;\sum g_i\rfloor$ and $\lceil\prod n_i\,;\sum h_i\rfloor$ in $\mathscr{B}_{W^\intercal,G^\intercal}$. Define the partition $\mathscr{I} = \{I_g, I_h, I_{g,h}\} \cup \{\{i\}\}_{i\in I_0}$. as follows.

- Let I_h be the set of indices such that $g_i = 0$ and $h_i \neq 0$ (then by assumption $m_i = 1$).
- Let I_g be the set of indices such that $h_i = 0$ and $g_i \neq 0$ (then $n_i = 1$).
- Let $I_{g,h}$ be the set of indices where $g_i, h_i \neq 0$.
- Let I_0 be the set of indices such that $g_i = h_i = 0$.

Also define groups as follows:

- Let $G^\intercal_{I_h}$ be the group of symmetries of W_{I_h} generated by h_{I_h}.
- Let $G^\intercal_{I_g}$ be the group of symmetries of W_{I_g} generated by g_{I_g}.
- Let $G^\intercal_{I_{g,h}}$ be the group of symmetries of $W_{I_{g,h}}$ generated by both $h_{I_{g,h}}$ and $g_{I_{g,h}}$.
- Let $G^\intercal_i$ be the trivial group of symmetries of W_i for $i \in I_0$.

LEMMA 5.3.3. *The groups $\{G^\intercal_I\}$ are each contained in* SL. *Both $\lceil\prod m_i\,;\sum g_i\rfloor$ and $\lceil\prod n_i\,;\sum h_i\rfloor$ split nicely (see Definition 3.2.4) with respect to $\mathscr{I}$ and $\{G^\intercal_I\}$.*

This follows from the from the definitions, using the the fact that $\sum g_i$ preserves $\lceil\prod n_i\,;\sum h_i\rfloor$ and vice versa.

6. The Pairing

Having established the algebra isomorphism, to complete the proof of Theorem 3.0.3 we need only check that the pairing is preserved by φ.

PROPOSITION 6.0.4. *If (W, G) satisfy Property $(*)$, then the mirror map $\varphi : \mathscr{B}_{W^\intercal,G^\intercal} \to \mathscr{A}_{W,G}$ preserves the pairing and therefore is an isomorphism of Frobenius algebras.*

PROOF. Assume that $\lceil m\,;g\rfloor$ and $\lceil m'\,;g'\rfloor$ are elements of $\mathscr{B}_{W^\intercal,G^\intercal}$. We must check that

$$\langle\varphi(\lceil m\,;g\rfloor\,, \lceil m'\,;g'\rfloor)\rangle_B = \langle\varphi(\lceil m\,;g\rfloor), \varphi(\lceil m'\,;g'\rfloor)\rangle_A\,. \tag{23}$$

LEMMA 6.0.5. *For any mixed pair, as described in the hypothesis of Lemma* 5.2.1, *both pairings in Equation* (23) *vanish.*

PROOF. On the B-side this is immediate, since $g + g' \neq [0]$. The form of their images on the A-side was computed explicitly in the proof of Lemma 5.2.1, and we saw in (21) that if the pairing were non-trivial, it would violate the hypothesis. In (22), we can see that the images do not come from inverse sectors. □

Now, by using the same partitions we used for the verification of the homomorphism property, we can reduce to the case of a fundamental pair (W, G). Furthermore, it suffices to check for each factor in the tensor product. We again have three cases to check: (1) $g = g' = 0$, (2) $g = 0$ but $g' \neq 0$ and (3) $g \neq 0$ and $g' \neq 0$.

Case (2) follows immediately from Lemma 6.0.5. Cases (1) and (3) are straightforward computations, very similar to the computations done in the proof of the third case of Proposition 5.1.1. □

7. Appendix: Some loop and chain lemmas

In the following proofs, unless otherwise indicated, we assume that W is a atomic polynomial with variables ordered as in Proposition 2.1.1 and with exponent matrix A. We take $\mathbf{h}_W$ as defined in Lemma 3.1.4. When discussing loop polynomials, we take the indices modulo N.

The first lemma follows immediately from the definitions.

LEMMA 7.0.6. *Let J be the corresponding exponential grading element corresponding to W (see Definition 2.1.4). Then*

$$-J = \left[A^{-1}\mathbf{h}_{W^{\mathsf{T}}}\right] \in (\mathbb{Q}/\mathbb{Z})^N.$$

LEMMA 7.0.7. *If $\mathbf{X}^{\mathbf{t}} \in \mathscr{Q}_W$ is a scalar multiple of the Hessian, then $[(A^{\mathsf{T}})^{-1}(\mathbf{t}+\mathbf{2})] = 0$.*

PROOF. If $\mathbf{t} = \mathbf{h}_W - \mathbf{1}$, then the result follows from Lemma 7.0.6. The more general case is a straightforward computation once one translates the Jacobian relations into an operation on exponent vectors. □

LEMMA 7.0.8. *Suppose that $\mathbf{v} = A^{\mathsf{T}}\mathbf{n}$, where $\mathbf{n}$ has integer entries and the entries of $\mathbf{v}$ satisfy $2 \leq v_i \leq 2a_i$.*

(1) *If W is a loop type polynomial, then one of the following is true:*
   (a) $\mathbf{v} = \mathbf{h}_W + \mathbf{1}$.
   (b) $v_i = \delta^i_{odd}(2a_i - 2) + 2$.
   (c) $v_i = \delta^i_{even}(2a_i - 2) + 2$.
   The latter two cases can occur only if N is even.
(2) *If W is a chain type polynomial, then there is some $s \in \{1, \ldots, N\}$ where n_s is the first coordinate of $\mathbf{n}$ equal to 1, and*

$$v_i = \begin{cases} \delta^i_{odd}(2a_i - 2) + 2 & i < s \\ a_s + 2\delta^s_{even} & i = s \\ 1 + a_i & i > s \end{cases}$$

PROOF. Consider first W a loop. We have the set of inequalities

$$2 \leq a_i n_i + n_{i-1} \leq 2a_i. \tag{24}$$

If $n_j \leq 0$ for some j, then we see that $n_{j-1} \geq 2$. Then $a_{j-1}2 + n_{j-2} \leq 2a_{j-1}$ which implies that $n_{j-2} \leq 0$. Thus we see that if any n_i is not 1, then the entries of the vector $\mathbf{n}$ must alternate being less than or equal to zero and being greater than or equal to 2. This is impossible if N is odd.

Suppose now that for some j, $n_j \geq m$, where $m \geq 3$ is an integer. Using the inequality $a_j m + n_{j-1} \leq 2a_i$ repeatedly, one shows that we can find an entry of $\mathbf{n}$ larger than any natural number, a contradiction. Thus $n_i \leq 2$ for all i, and it then follows that $n_i \geq 0$. Thus, either $\mathbf{n} = \mathbf{1}$, $\mathbf{n} = [2, 0, 2, 0, \ldots, 2, 0]^{\mathsf{T}}$, or $\mathbf{n} = [0, 2, 0, 2, \ldots, 0, 2]^{\mathsf{T}}$, and the latter two cases can occur only when n is even. Then the result follows from computing $A^{\mathsf{T}}\mathbf{n}$.

To prove the statement about chain polynomials, start by considering the possibilities for n_1, and it is straightforward to check that $\mathbf{n}$ is of the form $[2, 0, \ldots, 2, 0, 1, \ldots 1]$ or $[2, 0, \ldots, 0, 2, 1, \ldots, 1]$, from which the result follows. □

REMARK 7.0.9.

(1) Suppose that for a loop polynomial $0 \le r_i, s_i \le a_i - 1$ (as in Lemma 3.1.5 (1)) and $[(A^\mathsf{T})^{-1}(\mathbf{r}+\mathbf{s}+\mathbf{2})] = 0$. The vector $\mathbf{v} = \mathbf{r}+\mathbf{s}+\mathbf{2}$ satisfies the hypothesis of Lemma 7.0.8 and one of the following is true:
   (a) $\mathbf{r}+\mathbf{s} = \mathbf{h}_W - \mathbf{1}$
   (b) $r_i = s_i = \delta^i_{\text{odd}}(a_i - 1)$
   (c) $r_i = s_i = \delta^i_{\text{even}}(a_i - 1)$

(2) Suppose that for a chain polynomial $\mathbf{r}$ and $\mathbf{s}$ satisfy the Chain Property (†) and $[(A^\mathsf{T})^{-1}(\mathbf{r}+\mathbf{s}+\mathbf{2})] = 0$. By Lemma 7.0.8 either $\mathbf{r}+\mathbf{s} = \mathbf{h}_W - 1$ or

$$(\mathbf{r}+\mathbf{s})_i = \begin{cases} \delta^i_{\text{odd}}(2a_i - 2) & i < s \\ a_s - 2\delta^s_{\text{odd}} & i = s \\ a_i - 1 & i > s \end{cases}$$

LEMMA 7.0.10. *If W is a loop and has an even number of variables, then in $\mathscr{Q}_W$ we have*

$$\prod_{i \text{ odd}} X_i^{2a_i-2} = \prod_{i \text{ even}} (-a_i) \prod_{i=1}^{N} X_i^{a_i-1} \quad \textit{and} \quad \prod_{i \text{ even}} X_i^{2a_i-2} = \prod_{i \text{ odd}} (-a_i) \prod_{i=1}^{N} X_i^{a_i-1}.$$

(Thus these are also multiples of $\text{hess}(W)$*.)*

PROOF. The Jacobian relations for a loop polynomial are $X_i^{a_i} = -a_{i+1}X_{i+1}^{a_{i+1}-1}X_{i+2}$. Repeated application of this relation gives the result. □

COROLLARY 7.0.11. *Suppose that $W = \sum W_i$ is a sum of Fermat, loop, and chain type polynomials with exponent matrix A, and that $a_{i,j}$ are the defining exponents of W_i. Suppose $\mathbf{X}^\mathbf{r}$ and $\mathbf{X}^\mathbf{s}$ are representatives of elements of $\mathscr{Q}_W$ with the exponent vectors* $\mathbf{r}$, $\mathbf{s}$ *satisfying the conditions of Lemma* 3.1.5*). Then the following are equivalent:*

(1) $[(A^\mathsf{T})^{-1}(\mathbf{r}+\mathbf{s}+\mathbf{2})] = 0$
(2) $\mathbf{X}^{\mathbf{r}+\mathbf{s}}$ *is a scalar multiple of the Hessian.*

PROOF. It suffices to prove this for atomic types. For the Fermat type, it is obvious. For the loop and chain types, (2) implies (1) is the content of Lemma 7.0.7. To see that (1) implies (2), first use Remark 7.0.9. Then, for loops, apply Lemma 7.0.10, if necessary, to show that $\mathbf{X}^{\mathbf{r}+\mathbf{s}}$ is a scalar multiple of $\mathbf{X}^{\mathbf{h}_W-\mathbf{1}}$. For chains, if $\mathbf{r}+\mathbf{s} \neq \mathbf{h}_W - 1$, and s is odd, then apply the Jacobian relation $X_i^{a_i} = -a_{i+1}X_{i+1}^{a_{i+1}-1}X_{i+2}$ for $i = s-2, s-4, \ldots, 1$ to show that $\mathbf{X}^{\mathbf{r}+\mathbf{s}} = \prod_{\substack{i \text{ even} \\ i<s}}(-a_i)\mathbf{X}^{\mathbf{h}_W-\mathbf{1}}$. If s is even, apply the Jacobian relation $X_i^{a_i-1}X_{i+1} = -\frac{1}{a_i}X_{i-1}^{a_{i-1}}$ for $i = s-1, s-3, \ldots, 3$ to show that $\mathbf{X}^{\mathbf{r}+\mathbf{s}} = \prod_{\substack{i \text{ odd} \\ i<s}}(-\frac{1}{a_i})\mathbf{X}^{\mathbf{h}_W-\mathbf{1}} \cdot X_1^{a_1}X_2^{a_2} = 0$. Either way, at least one of $\mathbf{r}$ or $\mathbf{s}$ violates the Chain Property (†). □

References

[Aco09] Pedro Acosta, *FJRW-Rings and Landau-Ginzburg Mirror Symmetry in Two Dimensions*, ArXiv e-prints (2009), 0906.0970v1.

[Ber09] Natalie Bergin, *The Orbifold Landau-Ginzburg Conjecture for Unimodal and Bimodal Singularities*, Master's thesis, Brigham Young University, 2009.

[BH93] Per Berglund and Tristan Hübsch, *A generalized construction of mirror manifolds*, Nuclear Phys. B **393** (1993), no. 1-2, 377–391. MR 1214325 (94k:14031)

[Bor10] Lev A. Borisov, *Berglund-Hübsch mirror symmetry via vertex algebras*, ArXiv e-prints (2010), 1007.2633v2.

[FJR07a] Huijun Fan, Tyler J. Jarvis, and Yongbin Ruan, *The Witten equation and its virtual fundamental cycle*, ArXiv e-prints (2007), 0712.4025v3.

[FJR07b] ———, *The Witten equation, mirror symmetry and quantum singularity theory*, ArXiv e-prints (2007), 0712.4021v3.

[FJR08] ———, *Geometry and analysis of spin equations*, Comm. Pure Appl. Math. **61** (2008), no. 6, 745–788. MR 2400605 (2011c:32016)

[FS09] Huijun Fan and Yefeng Shen, *Quantum ring of singularity* $X^p + XY^q$, ArXiv e-prints (2009), 0902.2327v1.

[IV90] Kenneth Intriligator and Cumrun Vafa, *Landau-Ginzburg orbifolds*, Nuclear Phys. B **339** (1990), no. 1, 95–120. MR 1061738 (91k:81131)

[JKK05] Tyler J. Jarvis, Ralph Kaufmann, and Takashi Kimura, *Pointed admissible G-covers and G-equivariant cohomological field theories*, Compos. Math. **141** (2005), no. 4, 926–978. MR 2148194 (2006d:14065)

[Kau02] Ralph M. Kaufmann, *Orbifold Frobenius algebras, cobordisms and monodromies*, Orbifolds in mathematics and physics (Madison, WI, 2001), Contemp. Math., vol. 310, Amer. Math. Soc., Providence, RI, 2002, pp. 135–161. MR 1950945 (2004f:57037)

[Kau03] ———, *Orbifolding Frobenius algebras*, Internat. J. Math. **14** (2003), no. 6, 573–617. MR 1997832 (2005b:57057)

[Kau06] ———, *Singularities with symmetries, orbifold Frobenius algebras and mirror symmetry*, Gromov-Witten theory of spin curves and orbifolds, Contemp. Math., vol. 403, Amer. Math. Soc., Providence, RI, 2006, pp. 67–116. MR 2234885 (2007d:14076)

[KPA+10] Marc Krawitz, Nathan Priddis, Pedro Acosta, Natalie Bergin, and Himal Rathnakumara, *FJRW-rings and mirror symmetry*, Comm. Math. Phys. **296** (2010), no. 1, 145–174. MR 2606631 (2011g:53188)

[Kra10] Marc Krawitz, *FJRW rings and Landau-Ginzburg Mirror Symmetry*, Ph.D. thesis, University of Michigan, 2010.

[Kre94] Maximilian Kreuzer, *The mirror map for invertible LG models*, Phys. Lett. B **328** (1994), no. 3-4, 312–318. MR 1279367 (95e:32021)

[KS92] Maximilian Kreuzer and Harald Skarke, *On the classification of quasihomogeneous functions*, Comm. Math. Phys. **150** (1992), no. 1, 137–147. MR 1188500 (93k:32075)

[Tur10] Vladimir Turaev, *Homotopy quantum field theory*, EMS Tracts in Mathematics, vol. 10, European Mathematical Society (EMS), Zürich, 2010, Appendix 5 by Michael Müger and Appendices 6 and 7 by Alexis Virelizier. MR 2674592 (2011k:57039)

Department of Mathematics, 275 TMCB, Brigham Young University, Provo, UT 84602

E-mail address: `amandaefrancis@gmail.com`

E-mail address: `jarvis@math.byu.edu`

E-mail address: `jjohnson@math.utah.edu`

Current address, Drew Johnson: University of Utah Math Department, 155 S 1400 E Room 233, Salt Lake City, UT 84112-0090

E-mail address: `rachel.suggs@byu.edu`

Proceedings of Symposia in Pure Mathematics
Volume **85**, 2012

On Hirzebruch invariants of elliptic fibrations

James Fullwood and Mark van Hoeij

Abstract. We compute all Hirzebruch invariants χ_q for D_5, E_6, E_7 and E_8 elliptic fibrations of every dimension. A single generating series $\chi(t,y)$ is produced for each family of fibrations such that the coefficient of $t^k y^q$ encodes χ_q over a base of dimension k, solely in terms of invariants of the base of the fibration.

Contents

1. Introduction

The prospect of realizing realistic particle physics (such as the Standard Model) in a regime of string theory coined 'F-theory' by its originator Cumrun Vafa [**1**][**2**] has provided a source of attraction for string theorists and their mathematician counterparts to the study of elliptic fibrations, i.e., a proper surjective morphism $\varphi : Y \to B$ such that the generic fiber is an elliptic curve. F-theory was first formulated as a non-perturbative description of Type-IIB string theory with D-branes on a complex n-fold B endowed with a multi-valued complex scalar field

$$\tau = C_0 + i/g_s \tag{1.1}$$

known to physicists as the axio-dilaton field (g_s is the string coupling constant). The multi-valued images of τ differ by elements of $SL_2(\mathbb{Z})$, thus forming a striking resemblance with the complex structure modulus of a torus $E_\tau : \mathbb{C}/(\mathbb{Z}+\tau\mathbb{Z})$. Exercising a string theorist's natural penchant for algebro-geometric descriptions of nature, Vafa formulated a geometrization of the $SL_2(\mathbb{Z})$ symmetry of the axio-dilaton

The second author was supported by NSF Grant number 1017880.

via a Calabi-Yau elliptic fibration over the type-IIB n-fold B (which describes a physical theory in 10-$2n$ real space-time dimensions), identifying the axio-dilaton with the complex structure modulus of the elliptic fiber. Not only is this formulation of non-perturbative type-IIB string theory aesthetically pleasing from a purely geometric perspective, the physical theory has attractive features such as providing promising avenues for moduli-stabilization and potential realization of GUT gauge groups which project by definition to the Standard Model gauge group at lower non-supersymmetric energy levels. To realize the elliptic fibration explicitly, physicists have primarily focused on a Weierstrass fibration, i.e., a hypersurface in a $\mathbb{P}^2$-bundle over the Type-IIB base B which in its reduced form is defined as the zero-scheme associated with the locus

$$y^2z = x^3 + fxz^2 + gz^3, \tag{1.2}$$

where f and g are sections of appropriate tensor powers of the anti-canonical bundle $\mathscr{O}(-K_B) \to B$. As in the theory of curves, every smooth elliptic fibration admitting a section is birational to a (possibly singular) fibration in Weierstrass form [**3**], often referred to in the physics literature as an 'E_8' elliptic fibration. But crucial to the physical theory associated with an elliptic fibration are the singular fibers of the fibration, as the singular fibers encode the structure of gauge theories associated with D-branes wrapping components of the discriminant locus[1] over which they appear [**4**][**5**][**6**]. And since singular fibers of a fibration are in general not preserved under a birational transformation, elliptic fibrations not in Weierstrass form enjoy their own physical relevance.

Motivated by tadpole cancellation in F-theory, in [**7**] Sethi,Vafa and Witten derived a formula for the Euler characteristic of an elliptically fibered Calabi-Yau fourfold in Weierstrass form solely in terms of the Chern classes of the base of the fibration. Similar formulas for fibrations not in Weierstrass form were derived by Klemm, Lian, Roan and Yau in [**8**]. It was later shown by Aluffi and Esole in [**9**][**10**] that these formulas are all numerical avatars of more general Chern class identities which hold not only without any Calabi-Yau hypothesis but over a base of arbitrary dimension. In this note we consider four families of fibrations $\varphi : Y \to B$ which are known to the physics community as D_5, E_6, E_7 and E_8 elliptic fibrations from a purely mathematical perspective (i.e., with no Calabi-Yau hypothesis or restrictions on the dimension of the base), and pursue similar formulas not for the Chern classes of a given fibration Y but for its Hirzebruch invariants (or arithmetic genera)

$$\chi_q(Y) := \int \mathrm{ch}(\Omega_Y^q)\mathrm{td}(Y), \tag{1.3}$$

where $\mathrm{ch}(\Omega_Y^q)$ denotes the Chern character of the qth exterior power of the cotangent bundle of Y and $\mathrm{td}(Y) := \mathrm{td}(TY) \cap [Y]$, i.e., the Todd class of the tangent bundle of Y acting on the fundamental class of Y.

As integrals are invariant under proper pushforwards of the integrand, we relate $\chi_q(Y)$ to invariants of the base by pushing forward $\mathrm{ch}(\Omega_Y^q)\mathrm{td}(Y)$ via φ_* (the

---

[1] The *discriminant* of a Weierstrass fibration is a hypersurface in the base determined by the locus $\Delta : 4f^3 + 27g^2 = 0$.

pushforward map associated with $\varphi : Y \to B$). By the celebrated Hirzebruch-Riemann-Roch theorem (later generalized by Grothendieck),

$$\chi_q(Y) = \sum_i (-1)^i \dim H^i(Y, \Omega_Y^q) = h^{q,0} - h^{q,1} + \cdots + (-1)^{\dim(Y)} h^{q,\dim(Y)}, \tag{1.4}$$

thus Hirzebruch invariants yield linear relations among the Hodge numbers of Y. For a general smooth complex projective variety X of fixed dimension, the standard approach to computing Hirzebruch invariants of X is to encode them in a generating series

$$\chi(y) = \sum_q \chi_q y^q = \int_X \prod_{i=1}^{\dim(X)} (1 + ye^{-\lambda_i}) \frac{\lambda_i}{1 - e^{-\lambda_i}}, \tag{1.5}$$

where the λ_is are the Chern roots of the tangent bundle of X. Given an elliptic fibration $\varphi : Y \to B$ of type D_5, E_6, E_7 or E_8 over a base B of *arbitrary dimension*, what we achieve in this note is a single generating series $\chi(t, y)$ for each family where the coefficient of $t^k y^q$ encodes χ_q for the given family of elliptic fibrations over a base of dimension k, solely in terms of Chern classes of the base and the first Chern class of a line bundle $\mathscr{L} \to B$ (i.e., one can see the Hirzebruch invariants of the fibration as *functions* of invariants of the base).

Let B be a smooth compact complex projective variety of arbitrary dimension endowed with a line bundle $\mathscr{L} \to B$. The elliptic fibrations we consider will all be subvarieties of an ambient projective bundle $\mathbb{P}(\mathscr{E}) \to B$ (each fibration will be precisely defined in §2), where $\mathscr{E}$ is a vector bundle over B that is constructed by taking direct sums of tensor powers of $\mathscr{L}$. Before stating the main result of this note, let us make the following definitions:

Let X be a smooth variety. We define the *Hirzebruch series* of X to be

$$\mathscr{H}_y(X) := \mathscr{H}_0(X) + \mathscr{H}_1(X)y + \mathscr{H}_2(X)y^2 + \cdots, \tag{1.6}$$

where $\mathscr{H}_q(X) := \mathrm{ch}(\Omega_X^q)\mathrm{td}(X)$ is the qth *Hirzebruch characteristic class* of X. Then given a proper morphism $\varphi : X \to B$ we define $\varphi_* \mathscr{H}_y(X)$ in the obvious way:

$$\varphi_* \mathscr{H}_y(X) := \varphi_* \mathscr{H}_0(X) + \varphi_* \mathscr{H}_1(X)y + \varphi_* \mathscr{H}_2(X)y^2 + \cdots, \tag{1.7}$$

where φ_* is the proper pushforward associated with the morphism φ. Our main result is the following

THEOREM 1.1. *Let $\varphi : Y \to B$ be an elliptic fibration of type D_5, E_6, E_7 or E_8 and let $U = e^{-c_1(\mathscr{L})}$. Then*

$$\varphi_* \mathscr{H}_y(Y) = Q \cdot \mathscr{H}_y(B), \tag{1.8}$$

where

$$(1.9)\qquad Q=\begin{cases}4-y+\frac{(y+1)(yU-3)}{(yU^2+1)}-\frac{U(y+1)^2}{(yU^2+1)^2} & \textit{for } Y \textit{ a } D_5 \textit{ fibration}\\ 3-y+\frac{(y+1)(yU^2-U-2)}{(yU^3+1)} & \textit{for } Y \textit{ an } E_6 \textit{ fibration}\\ 2-y+\frac{(y+1)(yU^3-U-1)}{(yU^4+1)} & \textit{for } Y \textit{ an } E_7 \textit{ fibration}\\ 1-y+\frac{(y+1)(yU^5-U-0)}{(yU^6+1)} & \textit{for } Y \textit{ an } E_8 \textit{ fibration}\end{cases}$$

The proof of this result is considerably streamlined via the use of the *Chern-ext character*, which we first introduce and define in §4. We note that the numbers 4, 3, 2 and 1 in the expressions for Q coincide with the number of sections of the given fibration. Reading off the coefficient of y^q in $Q\cdot\mathscr{H}_y(B)$ immediately yields

$$(1.10)\qquad \chi_q(Y)=\int_B\left(P_0\mathscr{H}_q(B)+P_1\mathscr{H}_{q-1}(B)+\cdots+P_q\mathrm{td}(B)\right),$$

where the P_is are polynomials in $U=e^{-c_1(\mathscr{L})}$ (we list all P_is in §5.1). Moreover, from Theorem 1.1 we derive generating series $\chi(t,y)$ for each family of fibrations, where the coefficient of t^ky^q encodes χ_q for the given family of fibrations over a base of dimension k. Before unveiling the $\chi(t,y)$ we need the following definitions:

Let R be a commutative ring with unity. For two series $f(t)=a_0+a_1t+a_2t^2+\cdots$ and $g(t)=b_0+b_1t+b_2t^2+\cdots$ in $R[[t]]$, we recall the *Hadamard product* of f and g is defined to be

$$(1.11)\qquad f\odot g:=a_0b_0+a_1b_1t+a_2b_2t^2+\cdots.$$

Furthermore, let $[t^d]:R[[t]]\to R$ denote the map given by $[t^d][f]:=a_d$. Now let X be a smooth projective variety of dimension d, let $g=(1+ye^{-t})\frac{t}{1-e^{-t}}$ and let $f=\ln(g)$. As a consequence of Lemma 5.1 we show

$$(1.12)\qquad \mathscr{H}_y(X)=(1+y)^d\cdot[t^d][\exp\left(f\odot(-tC'/C)\right)],$$

where $C=1-c_1t+c_2t^2-c_3t^3+\cdots\in R[[t]]$ with $R=\mathbb{Z}[c_1,c_2,\ldots,]$ [2]. As $\varphi_*(\mathscr{H}_y(Y))=Q\cdot\mathscr{H}_y(B)$ by Theorem 1.1, the generating series $\chi(t,y)$ is constructed by replacing $U=e^{-c_1(\mathscr{L})}$ by $U_t=e^{-c_1(\mathscr{L})t}$ in Q and then constructing a series in $R[[t]]$ with $R=\mathbb{Z}[c_1,c_2,\ldots]$ such that the coefficient of t^d is precisely $(1+y)^d\cdot[t^d]\exp\left(f\odot(-tC'/C)\right)$, which we interpret as the Hirzebruch series for a base of dimension d. All Hirzebruch invariants χ_q for D_5, E_6, E_7 and E_8 fibrations of all dimensions are then contained in the following

COROLLARY 1.2. *Let $Q_t=Q(e^{-c_1(\mathscr{L})t})$, where Q is defined as in Theorem 1.1 and let*

$$(1.13)\qquad \widetilde{\chi}(t,y)=Q_t\cdot\exp\left(\ln\left((1+ye^{-t})\frac{t}{1-e^{-t}}\right)\odot\frac{-tC'}{C}\right),$$

[2] As $f\odot(-tC'/C)$ is independent of d, the c_is appearing in the definition of C are countably many formal variables which acquire their familiar meaning as Chern classes of the tangent bundle of X in concrete examples.

where $C = 1 - c_1 t + c_2 t^2 - \cdots \in R[[t]]$ *with* $R = \mathbb{Z}[c_1, c_2, \ldots]$ *and* $C' = \frac{d}{dt}C$. *Then*

$$\chi(t, y) := \widetilde{\chi}(t(1+y), y) \tag{1.14}$$

is a generating series for Hirzebruch invariants of D_5, E_6, E_7 *and* E_8 *fibrations as the definition of* Q *varies according to Theorem* 1.1, *i.e., the integral of the coefficient of* $t^k y^q$ *over a base of dimension* k *is precisely* χ_q *for the given family of fibrations.*

As an illustration, the coefficient of t^4y^2 in $\chi(t,y)$ in the E_6 case is $-\frac{L}{12}(1729L^3 - 524c_1L^2 + (-17c_1^2 + 193c_2)L + 5c_1c_2 - 66c_3)$, thus χ_2 of an E_6 fibration over a base B of dimension four is (a computer implementation is located at the file 'findQ' at [**15**])

$$\int_B -\frac{L}{12}(1729L^3 - 524c_1(B)L^2 + (-17c_1(B)^2 + 193c_2(B))L + 5c_1(B)c_2(B) - 66c_3(B)). \tag{1.15}$$

2. The fibrations under consideration

We now formally define the objects under consideration, namely D_5, E_6, E_7 and E_8 elliptic fibrations (the names and definitions we use are all lifted from the physics literature [**8**]). We work over $\mathbb{C}$ though everything we say is equally valid over an algebraically closed field of characteristic zero. All fibrations are constructed by taking equations of classical elliptic curves and promoting their coefficients from scalars (or sections of line bundles over a point) to sections of line bundles over some smooth positive dimensional base variety B. As such, let B be some smooth compact complex projective variety of arbitrary dimension endowed with (a suitably ample) line bundle $\mathscr{L} \to B$. This will be the base assumption in each of the D_5, E_6, E_7 and E_8 cases.

Now let $\mathscr{E} = \mathscr{O} \oplus \mathscr{L} \oplus \mathscr{L} \oplus \mathscr{L}$. A D_5 *elliptic fibration* Y_{D_5} is defined to be a smooth complete intersection in $\mathbb{P}(\mathscr{E})$ (here we take the projective bundle of *lines* in $\mathbb{P}(\mathscr{E})$) associated with the locus

$$Y_{D_5} : \begin{cases} x^2 - y^2 - z(az + cw) = 0 \\ w^2 - x^2 - z(dz + ex + fy) = 0, \end{cases} \tag{2.1}$$

where z is a section of $\mathscr{O}(1)$ (the dual of the tautological line bundle on $\mathbb{P}(\mathscr{E})$), and x, y and w are sections of $\mathscr{O}(1) \otimes \pi^*\mathscr{L}$, where π is the projection $\pi : \mathbb{P}(\mathscr{E}) \to B$. Then we take a, c, d, e and f to be suitably generic sections of (minimal) appropriate tensor powers of $\pi^*\mathscr{L}$ such that Y_{D_5} is a smooth complete intersection in $\mathbb{P}(\mathscr{E})$ and each of the defining equations for Y_{D_5} is a well defined section of a line bundle on $\mathbb{P}(\mathscr{E})$. Taking a and d to be sections of $\pi^*\mathscr{L}^2$ and c, e and f to be sections of $\pi^*\mathscr{L}$ then defines Y_{D_5} as a variety of class $(2H + 2L)^2 \in A^*\mathbb{P}(\mathscr{E})$, where $H := c_1(\mathscr{O}(1))$ and $L := c_1(\pi^*\mathscr{L})$. Such a locus naturally determines an elliptic fibration $\varphi : Y_{D_5} \to B$, with generic fiber an elliptic curve in $\mathbb{P}^3$. Such fibrations contain fibers not on the list of Kodaira[3] and were studied extensively

[3]More precisely, over the locus $a = c = d = e = f = 0$ in the base the fibers consist of four $\mathbb{P}^1$s meeting at a point.

in [**11**] from both a mathematical and physical perspective. We note that as our results are topological in nature, they depend only on the class $[Y] \in A^*\mathbb{P}(\mathscr{E})$ of the given fibration, i.e., the explicit equations which define the fibration are essentially irrelevant. The equations are included for concreteness and to not completely sever ourselves from the physical theories with which they are associated. The definitions of E_6, E_7 and E_8 fibrations are summarized in Table 1.

	equation	ambient projective bundle	class in $A^*\mathbb{P}(\mathscr{E})$
E_6	$x^3 + y^3 = dxyz + exz^2 + fyz^2 + gz^3$	$\mathbb{P}(\mathscr{O} \oplus \pi^*\mathscr{L} \oplus \pi^*\mathscr{L})$	$3H + 3L$
E_7	$y^2 = x^4 + ex^2z^2 + fxz^3 + gz^4$	$\mathbb{P}_{1,1,2}(\mathscr{O} \oplus \pi^*\mathscr{L} \oplus \pi^*\mathscr{L}^2)$	$4H + 4L$
E_8	$y^2z = x^3 + fxz^2 + gz^3$	$\mathbb{P}(\mathscr{O} \oplus \pi^*\mathscr{L}^2 \oplus \pi^*\mathscr{L}^3)$	$3H + 6L$

TABLE 1. A summary of the pertinent data defining the smooth fibrations under consideration.

As in the D_5 case, the coefficients of E_6, E_7 and E_8 fibration are chosen to be suitably generic sections of tensor powers of $\pi^*\mathscr{L}$ such that the total space of the fibration is a smooth divisor of the indicated class in $A^*\mathbb{P}(\mathscr{E})$. We point out that the total space of an E_7 fibration is defined as a hypersurface in a *weighted* projective bundle, which is isomorphic to a bundle of quadric cones. To avoid dealing with any singularities of the ambient projective bundle while performing intersection theoretic computations, we embed $\mathbb{P}_{1,1,2}(\mathscr{E})$ in a $\mathbb{P}^3$-bundle and then realize the total space of the E_7 fibration as a complete intersection of the image of $\mathbb{P}_{1,1,2}(\mathscr{E})$ via its embedding in the $\mathbb{P}^3$-bundle with another hypersurface.

3. A motivating example

Let B be a non-singular compact complex algebraic variety of arbitrary dimension endowed with a (suitably ample) line bundle $\mathscr{L}$. We recall the definition of an E_8 elliptic fibration, i.e., a surjective proper morphism $\varphi : Y \to B$, whose total space Y is realized as a hypersurface of class $3H + 6L$ in the the Chow ring of a projective bundle $\mathbb{P}(\mathscr{E}) \xrightarrow{\pi} B$, where $\mathscr{E} = \mathscr{O} \oplus \mathscr{L}^2 \oplus \mathscr{L}^3$, $H := c_1(\mathscr{O}(1))$ and L we non-reluctantly use to denote both $c_1(\mathscr{L})$ and $\pi^*c_1(\mathscr{L})$. As one can show that $\varphi_*c(Y) = \frac{12L}{1+6L} \cdot c(B)$ [**9**], we exploit the fact that $\int_Y c(Y) = \int_B \varphi_*c(Y)$ and compute the topological Euler characteristic $\chi(Y)$ by integrating the coefficient of $t^{dim(B)}$ in the formal series $\chi^{dim(B)}(t)$, where we define $\chi^N(t)$ for general $N \in \mathbb{N}$ to be [4]

$$\chi^N(t) := \frac{12Lt}{1+6Lt} \cdot (1 + c_1t + c_2t^2 + \cdots + c_Nt^N). \tag{3.1}$$

As the series $\chi^n(t)$ and $\chi^{n+1}(t)$ are identical up to order n (for any n), we notice that the formal series

$$\chi^{E_8}(t) = \frac{12Lt}{1+6Lt} \cdot (1 + c_1t + c_2t^2 + \cdots + c_mt^m + \cdots) \tag{3.2}$$

[4]Here and throughout, terms not expanded in a series such as $\frac{12Lt}{1+6Lt}$ in the series above are a shorthand for their associated series expansions about t.

serves as a generating series for the topological Euler characteristic for Y of all possible dimensions (i.e., the coefficient of t^k encodes the Euler characteristic of Y over a base of dimension k, solely in terms of L and Chern classes of B). The c_is are then temporarily formal objects (as their subscripts tend towards infinity), acquiring their familiar meaning whence integrated upon. For example, over a base of dimension 3 the coefficient of t^3 in $\chi^{E_8}(t)$ is $12L(c_2 - 6Lc_1 + 36L^2)$, so $\chi(Y)$ over a base B of dimension 3 is

$$\int_B 12L(c_2(B) - 6Lc_1(B) + 36L^2). \tag{3.3}$$

Though admittedly these observations are all rather trivial, what is striking is that the pushforward $\varphi_* c(Y)$ is manifestly independent of the base of the fibration, i.e., over a base of dimension k the actual class in A^*B associated with $\varphi_* c(Y)$ is obtained by truncating a formal power series at order k. As such, we deem this a 'motivating example' as it exhibits general features which can be abstracted to cases of other invariants of elliptic fibrations of the form $\int \alpha$. In particular, the first step at arriving at such base independent expressions for $\varphi_* c(Y)$ is deriving a factorization of $c(Y)$ (which plays the role of α in our more general considerations) as $g(L, H) \cdot \pi^* c(B)$, where g is a rational expression depending only on L and H (as defined above). Once we have such a factorization, we get that

$$\varphi_* c(Y) = \pi_*(g(L, H))c(B) = \frac{12L}{1+6L} c(B), \tag{3.4}$$

an expression which depends in no way on the dimension of B. Essential to the base independence of the formula above is a pushforward formula which computes $\pi_*(g(L, H))$ in terms of a rational expression in L whose associated series is truncated at the dimension of the base to obtain the given class. Such a pushforward formula was recently obtained in [**14**]. More generally, for a given invariant of the form $I_Y = \int \alpha$ for some subvariety Y of a projective bundle, we seek an analogous factorization $\alpha = g(L, H)\pi^*(I_B)$, where the class associated with g is obtained by truncating a formal series in L and H at the dimension of Y. Then by applying the pushforward formula of [**14**] to $g(L, H)$ we immediately arrive at a base independent expression for the pushforward of α, which lends itself naturally to a generating series which encodes the invariant I_Y for Y over bases of arbitrary dimension. In what follows we successfully carry out this program for Hirzebruch invariants of D_5, E_6, E_7 and E_8 elliptic fibrations.

4. The Chern-ext character

We now define a series which plays a key role in our analysis:

DEFINITION 4.1. Let $\mathscr{E}$ be a vector bundle, then we define the *Chern-ext character* of $\mathscr{E}$ to be

$$\mathrm{ch}_{ext}(\mathscr{E}) := 1 + \mathrm{ch}(\mathscr{E})y + \mathrm{ch}(\Lambda^2\mathscr{E})y^2 + \cdots. \tag{4.1}$$

We note that for any commutative ring R with unity, any element of the form $f(y) = 1 + a_1 y + a_2 y^2 + \cdots$ is a unit in $R[[y]]$, thus $\frac{1}{f(y)}$ is well defined. If $\mathscr{E}$ is of rank r with Chern roots $(\lambda_1, \lambda_2, \cdots, \lambda_r)$, then $\mathrm{ch}_{ext}(\mathscr{E})$ is a polynomial in y of degree r which factors as

$$\mathrm{ch}_{ext}(\mathscr{E}) = \prod_{i=1}^{r}(1 + y \cdot \exp(\lambda_i)). \tag{4.2}$$

However, due to the dimension-independent nature of our results we prefer to think of ch_{ext} as a series, which also makes more evident the invertability of the Chern-ext character. Furthermore, given a smooth projective variety X we note that the Hirzebruch series of X is simply

$$\mathscr{H}_y(X) = \mathrm{ch}_{ext}(\Omega_X)\mathrm{td}(X). \tag{4.3}$$

Armed with such a series, we now prove the following

LEMMA 4.2. *Let*

$$0 \to \mathscr{A} \to \mathscr{B} \to \mathscr{C} \to 0 \tag{4.4}$$

be an exact sequence of vector bundles. Then

$$\mathrm{ch}_{ext}(\mathscr{B}) = \mathrm{ch}_{ext}(\mathscr{A}) \cdot \mathrm{ch}_{ext}(\mathscr{C}) \tag{4.5}$$

PROOF. ¿From the λ-ring identity ([**12**], pg.2)

$$\lambda^p(x+y) = \sum_{i=0}^{p} \lambda^i(x)\lambda^{p-i}(y), \tag{4.6}$$

along with nice properties of the Chern character ([**13**], example 3.2.3), we get

$$\mathrm{ch}(\Lambda^p\mathscr{B}) = \sum_{i=0}^{p} \mathrm{ch}(\Lambda^i\mathscr{A}) \cdot \mathrm{ch}(\Lambda^{p-i}\mathscr{C}). \tag{4.7}$$

The lemma immediately follows. □

5. The proof

5.1. Proof of main result. Let $\varphi : Y \to B$ be a D_5, E_6, E_7 or E_8 elliptic fibration as defined in §2 and let N denote both the normal bundle of Y in $\mathbb{P}(\mathscr{E})$ and the bundle on $\mathbb{P}(\mathscr{E})$ which restricts to it. Using the exact sequences (we use a superscript '$\vee$' to denote duals)

$$\begin{gathered} 0 \to N^\vee \to i^*\Omega_{\mathbb{P}(\mathscr{E})} \to \Omega_Y \to 0 \\ 0 \to \pi^*\Omega_B \to \Omega_{\mathbb{P}(\mathscr{E})} \to \Omega_{(\mathscr{E})/B} \to 0 \\ 0 \to \Omega_{\mathbb{P}(\mathscr{E})/B} \to (\pi^*\mathscr{E} \otimes \mathscr{O}(1))^\vee \to \mathscr{O}_{\mathbb{P}(\mathscr{E})} \to 0, \end{gathered}$$

along with Lemma 4.2 we get

$$i_*\left(\mathrm{ch}_{ext}(\Omega_Y)\mathrm{td}(Y)\right) = \left(\frac{\mathrm{ch}_{ext}(\mathscr{F}^\vee)}{\mathrm{ch}_{ext}(N^\vee)(1+y)} \cdot \alpha\right)\pi^*\left(\mathrm{ch}_{ext}(\Omega_B)\mathrm{td}(B)\right), \tag{5.1}$$

where $\mathscr{F}$ we use to denote $\pi^*\mathscr{E}\otimes\mathscr{O}(1)$, $\alpha=\frac{\mathscr{F}}{\mathrm{td}(N)}\cap[Y]$, $i:Y\hookrightarrow\mathbb{P}(\mathscr{E})$ is the inclusion and we use the fact that $\mathrm{ch}_{ext}(\mathscr{O}_{\mathbb{P}(\mathscr{E})})=1+y$. Furthermore, π^* (and π_*) act on Chern-ext characters in the obvious manner. Now we apply π_* to the equation above yielding

$$\varphi_*\left(\mathrm{ch}_{ext}(\Omega_Y)\mathrm{td}(Y)\right)=\pi_*\left(\frac{\mathrm{ch}_{ext}(\mathscr{F}^\vee)}{\mathrm{ch}_{ext}(N^\vee)(1+y)}\cdot\alpha\right)\mathrm{ch}_{ext}(\Omega_B)\mathrm{td}(B) \tag{5.2}$$

by the projection formula. Thus computing $\varphi_*\left(\mathrm{ch}_{ext}(\Omega_Y)\mathrm{td}(Y)\right)$ amounts to computing

$$\pi_*\left(\frac{\mathrm{ch}_{ext}(\mathscr{F}^\vee)}{\mathrm{ch}_{ext}(N^\vee)(1+y)}\cdot\alpha\right). \tag{5.3}$$

We spell out the details of this computation in the case of D_5 only, as the other cases differ inasmuch as the Chern roots of $\mathscr{F}$ and N vary from case to case. For D_5 the Chern roots of $\mathscr{F}$ and N are $(H,H+L,H+L,H+L)$ and $(2H+2L,2H+2L)$ respectively, where $H:=c_1(\mathscr{O}(1))$ and L we will use to denote both $c_1(\mathscr{L})$ and $\pi^*c_1(\mathscr{L})$. Putting this all together we get

(5.4)
$$\frac{\mathrm{ch}_{ext}(\mathscr{F}^\vee)}{\mathrm{ch}_{ext}(N^\vee)(1+y)}\cdot\alpha=\frac{(1+y\cdot e^{-H})(1+y\cdot e^{-H-L})^3}{(1+y\cdot e^{-2H-2L})^2(1+y)}\cdot\frac{H(H+L)^3(1-e^{-2H-2L})^2}{(1-e^{-H})(1-e^{-H-L})^3},$$

where we have cancelled a factor of $(2H+2L)^2$ from the numerator and denominator of α. Now let $D=\frac{(1+y\cdot e^{-H})(1+y\cdot e^{-H-L})^3}{(1+y\cdot e^{-2H-2L})^2(1+y)}\cdot\frac{H(H+L)^3(1-e^{-2H-2L})^2}{(1-e^{-H})(1-e^{-H-L})^3}$. By the pushforward formula of [**14**] we get

$$\begin{aligned}\pi_*(D) &= \frac{1}{2}\frac{d^2}{dH^2}\left(\frac{D-(a_0+a_1H+a_2H^2)}{H}\right)\Bigg|_{H=-L}\\ &= 4-y+\frac{(y+1)(yU-3)}{(yU^2+1)}-\frac{U(y+1)^2}{(yU^2+1)^2},\end{aligned}$$

where $U=e^{-L}$ and the a_is are expressions in L and y obtained by expanding D as a series in H. Identifying $\mathrm{ch}_{ext}(\Omega_Y)\mathrm{td}(Y)$ and $\mathrm{ch}_{ext}(\Omega_B)\mathrm{td}(B)$ with $\mathscr{H}_y(Y)$ and $\mathscr{H}_y(B)$ respectively yields the conclusion of Theorem 1.1. We note that reading off the coefficient of y^q in the series $\varphi_*\mathscr{H}_y(Y)$ gives us

$$\varphi_*\mathscr{H}_q(Y)=\sum_{i=0}^{q}P_{q-i}\mathscr{H}_i(B), \tag{5.5}$$

where we recall that $\mathscr{H}_q(X):=\mathrm{ch}(\Omega_X^q)\mathrm{td}(X)$ denotes the qth Hirzebruch characteristic class of a smooth variety X and the P_is are polynomials in $U=e^{-L}$. We list the explicit form of the P_is for each case in Table 2.

The fact that P_0 is the same in all cases is a consequence of the fact that $K_Y=\varphi^*(c_1(\mathscr{L})-c_1(B))$ for all the fibrations considered (see the appendix of [**11**]). With the exception of the D_5 case, all roots of P_n for $n>1$ lie on $(S^1\cup 0)\subset\mathbb{C}$

	P_0	P_1	P_n for $n>1$
D_5	$1-U$	$2U^3+3U^2-U-4$	$-U((n+1)U-n+2)(U-1)(U+1)^2(-U^2)^{n-2}$
E_6	$1-U$	$U^4+2U^3+U^2-U-3$	$-U^2(U^3-1)(U+1)^2(-U^3)^{n-2}$
E_7	$1-U$	$U^5+U^4+U^3-U-2$	$-U^3(U^4-1)(U^2+U+1)(-U^4)^{n-2}$
E_8	$1-U$	U^7+U^5-U-1	$-U^5(U^6-1)(U^2+1)(-U^6)^{n-2}$

TABLE 2. Polynomial coefficients of $\varphi_*\mathscr{H}_q(Y)$

(for the D_5 case an 'anomalous' root of $\frac{n-2}{n+1}$ appears). We note that the length of our original proof was substantially greater, as we computed each χ_q individually and then proved a recursive relation between them. As it turned out, it was much easier to compute all of the χ_qs at once, which we were easily able to do once armed with the Chern-ext character and a computer implementation of π_* [**15**].

5.2. Proof of the corollary. We first need some definitions. Let R be a commutative ring with 1, let $f = a_0+a_1t+a_2t^2+\cdots \in R[[t]]$ and let $[t^d] : R[[t]] \to R$ be defined as in §1. If $\lambda_1,\ldots,\lambda_d \in R$, we use the notation $p_i := \lambda_1^i + \cdots \lambda_d^i$ and we let

$$C := \prod_{i=1}^{d}(1-\lambda_i t) = 1 - c_1t + c_2t^2 - c_3t^3 + \cdots$$

Then c_i is the ith *symmetric polynomial*, and p_i is the ith *power polynomial* of $\lambda_1,\ldots,\lambda_d$.

LEMMA 5.1. *Let $f(t) = a_0 + a_1t + \cdots \in R[[t]]$. Then*

$$\sum_{i=1}^{d} f(\lambda_i t) = f \odot (d + p_1t + p_2t^2 + \cdots) = da_0 + f \odot (-tC'/C),$$

where $\odot$ denotes the Hadamard product as defined in §1.

PROOF. The first equality follows from the definition of the p_i. For the second, note that $-tC'/C$ is well defined because the polynomial C has a constant term of 1. The equation $-tC'/C = p_1t + p_2t^2 + \cdots$ is obvious for $d = 1$ (geometric series). For $d > 1$, recall that logarithmic derivatives turn products into sums: $(CD)'/(CD) = C'/C + D'/D$. □

Now let Y be a D_5, E_6, E_7 or E_8 elliptic fibration over a base B of some fixed dimension d. Then by Theorem 1.1 and equation (1.5) in §1,

$$\varphi_*\mathscr{H}_y(Y) = Q\cdot\mathscr{H}_y(B) = Q\cdot\prod_{i=1}^{d} g(\lambda_i), \tag{5.6}$$

where the λ_is are the Chern roots of the tangent bundle of B and $g = (1 + ye^{-t})\frac{t}{1-e^{-t}}$. Now let $f = \ln(g) = a_0 + a_1t + \cdots$ (note that $a_0 = \ln(1+y)$) . Lemma 5.1 then yields

$$\begin{aligned}\mathscr{H}_y(B) = \prod_{i=1}^{d} g(\lambda_i) &= [t^d][\prod_{i=1}^{d} g(\lambda_i t)] \\ &= [t^d][\exp(\sum_{i=1}^{d} f(\lambda_i t))] \\ &= [t^d][\exp(da_0 + f \odot (-tC'/C))] \\ &= (1+y)^d \cdot [t^d][\exp(f \odot (-tC'/C))].\end{aligned}$$

The conclusion of Theorem 1.1 then states that over a base B of dimension d we have

$$\varphi_* \mathscr{H}_y(Y) = Q \cdot (1+y)^d \cdot [t^d][\exp(f \odot (-tC'/C))]. \tag{5.7}$$

But the right hand side of (5.7) is just $[t^d]\widetilde{\chi}(t,y)\mid_{t=t(1+y)}$, with $\widetilde{\chi}(t,y)$ as defined in Corollary 1.2. The corollary then follows.

6. Discussion

After going through the proof of the main result, one immediately notices that the only data needed from the elliptic fibrations under consideration were the Chern roots of its normal bundle in $\mathbb{P}(\mathscr{E})$ along with the Chern roots of the relative tangent bundle $\mathrm{T}_{\mathbb{P}(\mathscr{E})/B}$ of the ambient projective bundle $\mathbb{P}(\mathscr{E})$. As such, our program may be carried out verbatim for any smooth subvariety of $\mathbb{P}(\mathscr{E})$, long as $\mathscr{E}$ is a direct summand of tensor powers of a fixed line bundle on the base (this assumption is needed to apply the pushforward formula from [**14**]). More precisely, take any smooth complete intersection X in some projective space $\mathbb{P}^n$ given by equations $X : (F_1 = F_2 = \cdots = F_m = 0)$, promote the coefficients of the F_i to appropriate sections of tensor powers of a fixed line bundle on some smooth base variety B and we will have then constructed a fibration $\varphi : Y \to B$ such that the generic fiber is a complete intersection which is rationally equivalent to X. Then substituting the $n+1$ Chern roots of the relative tangent bundle of the ambient $\mathbb{P}^n$-bundle where Y resides along with the m Chern roots of the normal bundle to Y into our calculations above will yield analogous results for the 'X fibration' $\varphi : Y \to B$. Our results are thus genuinely more general than the title of this note suggests (see [**15**] for more details).

We conclude by noting that a true culmination of these results will not be achieved without a Lefschetz hyperplane type theorem for varieties in projective bundles, as hypersurfaces in projective bundles are almost never ample divisors (which is the key assumption of the Lefschetz hyperplane theorem). Once such a theorem is unveiled, only the middle cohomology will be unique to a hypersurface in a projective bundle thus rendering only $\lceil \frac{d}{2} \rceil$ of its Hodge numbers as non-trivial (where d is the dimension of the hypersurface). The Hirzebruch invariants could then be used for the determination of the non-trivial Hodge numbers. As the cohomology of a projective bundle can be related to its base via the projective bundle theorem, a Lefschetz hyperplane type theorem along with the results in this paper would then relate all Hodge numbers of a hypersurface (and so complete intersections) in a projective bundle to invariants of the base.

References

[1] C. Vafa, "Evidence for F theory," Nucl. Phys. **B469**, 403-418 (1996). [hep-th/9602022].

[2] F. Denef, "Les Houches Lectures on Constructing String Vacua," [arXiv:0803.1194 [hep-th]].

[3] P. Deligne, Courbes elliptiques: formulaire d'apr'es J. Tate, Modular functions of one variable, IV (Proc. Internat. Summer School, Univ. Antwerp, Antwerp, 1972), Springer, Berlin, 1975, 53-73. Lecture Notes in Math., Vol. 476.

[4] S. H. Katz, C. Vafa, "Matter from geometry," Nucl. Phys. **B497**, 146-154 (1997). [hep-th/9606086].

[5] D. R. Morrison, W. Taylor, "Matter and singularities," [arXiv:1106.3563 [hep-th]].

[6] C. Vafa, "Geometry of Grand Unification," [arXiv:0911.3008 [math-ph]].

[7] S. Sethi, C. Vafa, and E. Witten. Constraints on low-dimensional string compactifications. *Nuclear Phys. B*, 480(1-2):213–224, 1996.

[8] A. Klemm, B. Lian, S. S. Roan, S. -T. Yau, "Calabi-Yau fourfolds for M theory and F theory compactifications," Nucl. Phys. **B518**, 515-574 (1998). [hep-th/9701023].

[9] P. Aluffi, M. Esole, "Chern class identities from tadpole matching in type IIB and F-theory,"JHEP **0903**, 032 (2009). [arXiv:0710.2544 [hep-th]].

[10] P. Aluffi, M. Esole, "New Orientifold Weak Coupling Limits in F-theory," JHEP **1002**, 020 (2010). [arXiv:0908.1572 [hep-th]].

[11] M. Esole, J. Fullwood, S.T. Yau. "D_5 elliptic fibrations: Non-Kodaira fibers and new orientifold limits of F-theory" [arXiv:1110.6177 [hep-th]]

[12] W. Fulton, S. Lang. *Riemann-Roch algebra*, volume 277 of *Grundlehren der Mathematischen Wissenschaften [Fundamental Principles of Mathematical Sciences]*. Springer-Verlag, New York, 1985.

[13] W. Fulton. *Intersection theory*, volume 2 of *Ergebnisse der Mathematik und ihrer Grenzgebiete (3) [Results in Mathematics and Related Areas (3)]*. Springer-Verlag, Berlin, 1984.

[14] J. Fullwood, "On generalized Sethi-Vafa-Witten formulas," [arXiv:1103.6066 [math.AG]].

[15] www.math.fsu.edu/~hoeij/files/Hirzebruch

Mathematics Department, Florida State University, Tallahassee, FL 32306, USA
E-mail address: jfullwoo@math.fsu.edu

Mathematics Department, Florida State University, Tallahassee, FL 32306, USA
E-mail address: hoeij@math.fsu.edu

Proceedings of Symposia in Pure Mathematics
Volume **85**, 2012

G_2-structure deformations and warped products

Sergey Grigorian

ABSTRACT. We overview the properties of non-infinitesimal deformations of G_2-structures on seven-manifolds, and in particular, focus on deformations that lie in the seven-dimensional representation of G_2 and are thus defined by a vector. We then consider deformations from G_2-structures with the torsion class having one-dimensional and seven-dimensional components (so-called conformally nearly parallel G_2-manifolds) to G_2-structures with just a one-dimensional torsion component (nearly parallel G_2-manifolds). We find that deformations between such structures exist if and only if the metric is a particular warped product metric.

1. Introduction

One of the most general geometric structures that can be constructed on a 7-dimensional manifold is a G_2-structure. A G_2-structure can be considered as a generalization of the vector cross product on $\mathbb{R}^7$ [**8**]. It is well-known that a 7-manifold admits a G_2-structure if and only if it is orientable and admits a spin structure, or equivalently, if the first two Stiefel-Whitney classes vanish [**6, 7**]. A very important special case of a G_2-structure is a torsion-free G_2-structure. This implies that the holonomy group lies in G_2. In Section 2 we give a more precise definition and an overview of the properties of G_2-structures. The concept of G_2-structures also has important applications in physics - as shown in [**16**], the most general backgrounds for M-theory compactifications with fluxes are indeed 7-manifolds with G_2-structures with some particular torsion.

Given a 7-manifold with a G_2-structure defined by the 3-form φ with torsion T, a natural question to ask is whether we can modify this 3-form to get a new G_2-structure with torsion that lies a strictly lower torsion class. If φ is deformed by a 3-form lying in the 7-dimensional component of Λ^3, it is easy to see that such a deformation will always yield a new G_2-structure and in my paper [**10**], I have explicitly calculated the new torsion in terms of the old one, and the derived the equation that v must satisfy to take a torsion T to torsion $\tilde{T}$. It was moreover shown that on closed, compact manifolds there are no such deformations from strict torsion classes W_1, W_7, $W_1 \oplus W_7$ to the vanishing torsion class W_0, and vice versa.

In this paper we use the general results from [**10**] and apply them to the situation when we want a transition from the torsion class $W_1 \oplus W_7$ to W_1. The

---

2010 *Mathematics Subject Classification.* Primary 53C10, 53C29; Secondary 53C80.

torsion class $W_1 \oplus W_7$ is known as *conformally nearly parallel* because a conformal transformation takes $W_1 \oplus W_7$ to W_1. Here, however, we show that there exists a deformation of φ in Λ^3_7 that takes a torsion in $W_1 \oplus W_7$ to W_1 if and only if the metric is a particular warped product. However G_2-structures with such torsion have been constructed by Cleyton and Ivanov in [**5**] as a warped product of an interval over a nearly Kähler manifold, so these examples fit as solutions.

2. G_2-structures

The 14-dimensional group G_2 is the smallest of the five exceptional Lie groups and is closely related to the octonions. In particular, G_2 can be defined as the automorphism group of the octonion algebra. Taking the imaginary part of octonion multiplication of the imaginary octonions defines a vector cross product on $V = \mathbb{R}^7$ and the group that preserves the vector cross product is precisely G_2. A more detailed account of the relationship between octonions and G_2 can be found in [**1, 9**].The structure constants of the vector cross product define a 3-form on $\mathbb{R}^7$, hence G_2 can alternatively be defined as the subgroup of $GL(7, \mathbb{R})$ that preserves a particular 3-form [**13**]. In general, given a n-dimensional manifold M, a G-structure on M for some Lie subgroup G of $GL(n, \mathbb{R})$ is a reduction of the frame bundle F over M to a principal subbundle P with fibre G. A G_2-structure is then a reduction of the frame bundle on a 7-dimensional manifold M to a G_2 principal subbundle. It turns out that there is a 1-1 correspondence between G_2-structures on a 7-manifold and smooth 3-forms φ for which the 7-form-valued bilinear form B_φ as defined by (2.1) is positive definite (for more details, see [**2**] and the arXiv version of [**12**]).

$$B_\varphi(u, v) = \frac{1}{6}(u \lrcorner \varphi) \wedge (v \lrcorner \varphi) \wedge \varphi \tag{2.1}$$

Here the symbol $\lrcorner$ denotes contraction of a vector with the differential form:

$$(u \lrcorner \varphi)_{mn} = u^a \varphi_{amn}.$$

Note that we will also use this symbol for contractions of differential forms using the metric.

A smooth 3-form φ is said to be *positive* if B_φ is the tensor product of a positive-definite bilinear form and a nowhere-vanishing 7-form. In this case, it defines a unique metric g_φ and volume form vol such that for vectors u and v, the following holds

$$g_\varphi(u, v)\,\mathrm{vol} = \frac{1}{6}(u \lrcorner \varphi) \wedge (v \lrcorner \varphi) \wedge \varphi \tag{2.2}$$

In components we can rewrite this as

$$(g_\varphi)_{ab} = (\det s)^{-\frac{1}{9}} s_{ab} \text{ where } s_{ab} = \frac{1}{144}\varphi_{amn}\varphi_{bpq}\varphi_{rst}\hat{\varepsilon}^{mnpqrst}. \tag{2.3}$$

Here $\hat{\varepsilon}^{mnpqrst}$ is the alternating symbol with $\hat{\varepsilon}^{12\ldots7} = +1$. Following Joyce ([**13**]), we will adopt the following definition

Definition 2.1. Let M be an oriented 7-manifold. The pair (φ, g) for a positive 3-form φ and corresponding metric g defined by (2.2) will be referred to as a G_2-structure.

Since a G_2-structure defines a metric and an orientation, it also defines a Hodge star. Thus we can construct another G_2-invariant object - the 4-form $*\varphi$. Since the Hodge star is defined by the metric, which in turn is defined by φ, the 4-form $*\varphi$ depends non-linearly on φ. For convenience we will usually denote $*\varphi$ by ψ.

For a general G-structure, the spaces of p-forms decompose according to irreducible representations of G. Given a G_2-structure, 2-forms split as $\Lambda^2 = \Lambda^2_7 \oplus \Lambda^2_{14}$, where $\Lambda^2_7 = \{\alpha \lrcorner \varphi\colon$ for a vector field $\alpha\}$ and

$$\Lambda^2_{14} = \left\{\omega \in \Lambda^2\colon\ (\omega_{ab}) \in \mathfrak{g}_2\right\} = \left\{\omega \in \Lambda^2\colon\ \omega \lrcorner \varphi = 0\right\}.$$

The 3-forms split as $\Lambda^3 = \Lambda^3_1 \oplus \Lambda^3_7 \oplus \Lambda^3_{27}$, where the one-dimensional component consists of forms proportional to φ, forms in the 7-dimensional component are defined by a vector field $\Lambda^3_7 = \{\alpha \lrcorner \psi\colon$ for a vector field $\alpha\}$, and forms in the 27-dimensional component are defined by traceless, symmetric matrices:

$$\Lambda^3_{27} = \left\{\chi \in \Lambda^3 : \chi_{abc} = h^d_{[a}\varphi_{bc]d} \text{ for } h_{ab} \text{ traceless, symmetric}\right\}. \tag{2.4}$$

By Hodge duality, similar decompositions exist for Λ^4 and Λ^5. A detailed description of these representations is given in [**2, 3**]. Also, formulae for projections of differential forms onto the various components are derived in detail in [**10, 11, 15**].

The *intrinsic torsion* of a G_2-structure is defined by $\nabla\varphi$, where ∇ is the Levi-Civita connection for the metric g that is defined by φ. Following [**15**], it is easy to see

$$\nabla\varphi \in \Lambda^1_7 \otimes \Lambda^3_7 \cong W. \tag{2.5}$$

Here we define W as the space $\Lambda^1_7 \otimes \Lambda^3_7$. Given (2.5), we can write

$$\nabla_a\varphi_{bcd} = T_a{}^e\psi_{ebcd} \tag{2.6}$$

where T_{ab} is the *full torsion tensor*. From this we can also write

$$T_a{}^m = \frac{1}{24}\left(\nabla_a\varphi_{bcd}\right)\psi^{mbcd}. \tag{2.7}$$

This 2-tensor fully defines $\nabla\varphi$ since pointwise, it has 49 components and the space W is also 49-dimensional (pointwise). In general we can split T_{ab} according to representations of G_2 into *torsion components*:

$$T = \tau_1 g + \tau_7 \lrcorner \varphi + \tau_{14} + \tau_{27} \tag{2.8}$$

where τ_1 is a function, and gives the **1** component of T. We also have τ_7, which is a 1-form and hence gives the **7** component, and, $\tau_{14} \in \Lambda^2_{14}$ gives the **14** component and τ_{27} is traceless symmetric, giving the **27** component. Hence we can split W as

$$W = W_1 \oplus W_7 \oplus W_{14} \oplus W_{27}. \tag{2.9}$$

As it was originally shown by Fernández and Gray [**6**], there are in fact a total of 16 torsion classes of G_2-structures that arise as the subsets of W to which $\nabla\varphi$ belongs. Moreover, as shown in [**15**], the torsion components τ_i relate directly to the expression for $d\varphi$ and $d\psi$. In fact, in our notation,

$$d\varphi = 4\tau_1\psi - 3\tau_7 \wedge \varphi - *\tau_{27} \tag{2.10a}$$

$$d\psi = -4\tau_7 \wedge \psi - 2 * \tau_{14}. \tag{2.10b}$$

Note that in the literature ([**3, 5**], for example) a slightly different convention for torsion components is sometimes used. Our τ_1 then corresponds to $\frac{1}{4}\tau_0$, τ_7 corresponds to $-\tau_1$ in their notation, τ_{27} corresponds to $-\tau_3$ and τ_{14} corresponds to $-\frac{1}{2}\tau_2$. Similarly, our torsion classes $W_1 \oplus W_7 \oplus W_{14} \oplus W_{27}$ correspond to $W_0 \oplus W_1 \oplus W_2 \oplus W_3$.

An important special case is when the G_2-structure is said to be torsion-free, that is, $T = 0$. This is equivalent to $\nabla\varphi = 0$ and also equivalent, by Fernández and Gray, to $d\varphi = d\psi = 0$. Moreover, a G_2-structure is torsion-free if and only if the holonomy of the corresponding metric is contained in G_2 [**13**]. The holonomy group is then precisely equal to G_2 if and only if the fundamental group π_1 is finite.

The torsion tensor T_{ab} and hence the individual components τ_1,τ_7,τ_{14} and τ_{27} must also satisfy certain differential conditions. For the exterior derivative d, $d^2 = 0$, so from (2.10), must have

$$d\left(4\tau_1\psi - 3\tau_7 \wedge \varphi - *\tau_{27}\right) = 0 \tag{2.11a}$$

$$d\left(4\tau_7 \wedge \psi + 2 * \tau_{14}\right) = 0 \tag{2.11b}$$

These conditions are explored in more detail in [**10**]. However for some simple torsion classes, the conditions on torsion components simplify. In particular, if the torsion is in the class W_1, so that only the τ_1 component is non-vanishing, we just get the condition $d\tau_1 = 0$. Similarly, for the W_7 class the condition is $d\tau_7 = 0$. For the class $W_1 \oplus W_7$, both τ_1 and τ_7 are non-vanishing, and the condition is $d\tau_1 = \tau_1\tau_7$. So if τ_1 is nowhere zero, we have

$$\tau_7 = d\left(\log \tau_1\right). \tag{2.12}$$

However if τ_1 does vanish somewhere, it was shown in [**5**] that it must in fact vanish identically, and so the torsion class reduces to W_7.

3. Deformations of G_2-structures

Suppose we have a G_2-structure on M defined by the 3-form φ, and we want to obtain a new G_2-structure $\tilde{\varphi}$ by adding another 3-form χ

$$\varphi \longrightarrow \tilde{\varphi} = \varphi + \chi \tag{3.1}$$

There are a number of challenges associated with this. Firstly, for a generic 3-form χ, the 3-form $\tilde{\varphi}$ may not even define a G_2-structure. In order for $\tilde{\varphi}$ to define a G_2-structure it has to be a positive 3-form. In this case, as shown in ([**11**]), $\tilde{\varphi}$ defines a Riemannian metric $\tilde{g}$ given by

$$\tilde{g}_{ab} = \left(\frac{\det g}{\det \tilde{g}}\right)^{\frac{1}{2}} \tilde{s}_{ab} \tag{3.2}$$

for

$$\tilde{s}_{ab} = g_{ab} + \frac{1}{2}\chi_{mn(a}\varphi_{b)}{}^{mn} + \frac{1}{8}\chi_{amn}\chi_{bpq}\psi^{mnpq} + \frac{1}{24}\chi_{amn}\chi_{bpq}\left(*\chi\right)^{mnpq} \tag{3.3}$$

In fact, $\tilde{\varphi}$ is positive if and only if $\tilde{s}$ is positive-definite, so in general this gives some algebraic constraints on χ. If we assume that $\tilde{\varphi}$ does in fact define a G_2-structure, the next question is the torsion class of the new G_2-structure. The metric $\tilde{g}$ defines a Levi-Civita connection $\tilde{\nabla}$, so the new torsion (with lowered indices) is

$$\tilde{T}_{am} = \frac{1}{24}\left(\tilde{\nabla}_a\tilde{\varphi}_{bcd}\right)\tilde{\psi}_m{}^{\tilde{b}\tilde{c}\tilde{d}} \tag{3.4}$$

Here $\tilde{\psi} = \tilde{*}\tilde{\varphi}$, the Hodge dual of $\tilde{\varphi}$ with the Hodge star $\tilde{*}$ being defined by the metric $\tilde{g}$. The tilded raised indices on $\tilde{\psi}$ denote indices raised by $\tilde{g}$. In [**10**], I derived an explicit expression for $\tilde{T}$ in terms of the old torsion T and the 3-form χ:

Proposition 3.1 ([10]). *Given a deformation of φ as in (3.1), the full torsion $\tilde{T}$ of the new G_2-structure $\tilde{\varphi}$ is given by*

$$\begin{aligned}\tilde{T}_{an} &= \frac{1}{24}\left(\frac{\det g}{\det \tilde{g}}\right)\Big(\Big(24T_a{}^m + T_a{}^e\psi_{ebcd}\,(*\chi)^{mbcd} + \\ &\quad +\psi^{mbcd}\nabla_a\chi_{bcd} + \nabla_a\chi_{bcd}\,(*\chi)^{mbcd}\Big)\,\tilde{s}_{mn} \\ &\quad -3\left(4\varphi_c{}^{bd} + \varphi_{cpq} * \chi^{pqbd} + \chi_{cpq}\psi^{pqbd} + \chi_{cpq}\,(*\chi)^{pqbd}\right)\times \\ &\quad \times\left(\delta^c_n\nabla_b s_{ad} - \frac{1}{9}\delta^c_a\tilde{g}_{bn}\tilde{g}^{\tilde{p}\tilde{q}}\nabla_d\tilde{s}_{pq}\right)\Big).\end{aligned} \tag{3.5}$$

where $\tilde{s}_{ab}$ is given by (3.3).

Using (3.5), it is then possible to extract the components of $\tilde{T}$ in $\tilde{W}_1 \oplus \tilde{W}_7 \oplus \tilde{W}_{14} \oplus \tilde{W}_{27}$ and hence determine the new torsion class. Note that since the G_2-structure is different, the decomposition $\tilde{W}_1 \oplus \tilde{W}_7 \oplus \tilde{W}_{14} \oplus \tilde{W}_{27}$ differs from $W_1 \oplus W_7 \oplus W_{14} \oplus W_{27}$. An interesting question is whether, given a G_2-structure in a specific torsion class, we can find a 3-form χ such that the new G_2-structure is in a strictly smaller torsion class. From (3.5), this obviously involves solving a non-linear differential equation for χ, subject to algebraic constraints that (3.3) is positive-definite. One way to simplify the problem is to restrict the choice of χ. Using the original G_2-structure φ we can decompose the 3-form χ according to representations of G_2. So in general it has a Λ^3_1 component that is proportional to φ, a Λ^3_7 component that is of the form $v\lrcorner\psi$ for some vector v and a more complicated Λ^3_{27} component. For a generic χ, many of the difficulties come from the Λ^3_{27} component. These of course can be avoided if we only consider deformations by 3-forms that have components only in either Λ^3_1 or Λ^3_7.

A deformation by a 3-form in Λ^3_1 is equivalent to a conformal transformation. So let $\chi = (f^3 - 1)\,\varphi$, so that $\tilde{\varphi} = f^3\varphi$. Clearly $\tilde{\varphi}$ still defines a G_2-structure. Then from (3.3) and (3.2) we get

$$s_{ab} = f^9 g_{ab} \text{ and thus, } \tilde{g}_{ab} = f^2 g_{ab}. \tag{3.6}$$

Substituting into (3.5), we find that

$$\tilde{T} = fT - df\lrcorner\varphi \tag{3.7}$$

In particular, we see that such a transformation only affects the W_7 component of the torsion. Moreover, if T has a W_7 component τ_7 that is an exact form, then we can always find a function f so that a conformal transformation will remove this torsion component.

As an example, suppose φ has torsion in the strict class $W_1 \oplus W_7$. Then from (2.12), we know that $\tau_7 = d(\log \tau_1)$. Hence if we take $f = \frac{\tau_1}{C}$ for any non-zero constant C, the new torsion will be

$$\tilde{T} = \frac{\tau_1}{C}\tau_1 g = C\left(\frac{\tau_1}{C}\right)^2 g = C\tilde{g} \tag{3.8}$$

and thus in the $\tilde{W}_1$ class. Therefore, the conformal transformation $\tilde{\varphi} = \left(\frac{\tau_1}{C}\right)^3\varphi$ reduces the class $W_1 \oplus W_7$ to W_1. Conversely, a conformal transformation of the W_1 class will result in $W_1 \oplus W_7$. Since G_2-structures in the W_1 class are sometimes called *nearly parallel*, the G_2-structures in the strict $W_1 \oplus W_7$ class are referred to as *conformally nearly parallel*. If $W_1 = 0$, then we just have the W_7 class. In this case, we know that τ_7 is closed. So by the Poincaré Lemma, we can at least locally find a function h such that $dh = \tau_7$. By taking a conformal transformation with $f = e^h$, we can thus remove the torsion locally. Hence the W_7 class is sometimes called *locally conformally parallel.*

Now suppose we look at deformations where $\chi_{bcd} = v^e\psi_{bcde} \in \Lambda^3_7$. It was shown by Karigiannis in [**14**], that in this case, if we let $|v|^2 = M$, with respect to the old metric g,

$$s_{ab} = (1+M)\, g_{ab} - v_a v_b \tag{3.9a}$$

$$\left(\frac{\det \tilde{g}}{\det g}\right)^{\frac{1}{2}} = (1+M)^{\frac{2}{3}} \tag{3.9b}$$

$$\tilde{g}_{ab} = (1+M)^{-\frac{2}{3}} \left((1+M)\, g_{ab} - v_a v_b\right) \tag{3.9c}$$

$$\tilde{g}^{\tilde{a}\tilde{b}} = (1+M)^{-\frac{1}{3}} \left(g^{mu} + v^m v^u\right) \tag{3.9d}$$

Note that the deformed metric defined above is always positive definite. To see this, suppose ξ^a is some vector, then

$$\tilde{g}_{ab}\xi^a\xi^b = \left(1+|v|^2\right)^{-\frac{2}{3}} \left(|\xi|^2 + |v|^2 |\xi|^2 - (v_a\xi^a)^2\right) \geq 0 \tag{3.10}$$

since $(v_a\xi^a)^2 \leq |v|^2 |\xi|^2$. Therefore, under such a deformation, the 3-form $\tilde{\varphi}$ is always a positive 3-form, and thus indeed defines a G_2-structure.

In [**10**], the expression for the new torsion was derived using (3.9) and (3.5). This is a very long and messy expression, which we will not reproduce here, but it gives the new torsion in terms of the old torsion components and ∇v, which was also decomposed according to G_2-representations as

$$\nabla v = v_1 g + v_7 \lrcorner \varphi + v_{14} + v_{27} \tag{3.11}$$

The expression for $\tilde{T}$ was then inverted to obtain equations for v_1, v_7, v_{14} and v_{27} in terms of the old and new torsion components. By analyzing the equations for ∇v in the case when the original torsion lies in the class $W_1 \oplus W_7$, it was shown that the new torsion vanishes if and only if the original G_2-structure was also torsion free, and moreover $\nabla v = 0$. Similarly it was shown that there are no deformations of this type which preserve the strict W_1 torsion class. Here we will attempt something different - what if the torsion is in the class $W_1 \oplus W_7$ and we want to obtain the class W_1. From the expression (3.7) we already know that this is possible to do with a conformal transformation. However if it were possible to go from $W_1 \oplus W_7$ to W_1 using $\chi \in \Lambda^3_7$, then a composition of the two types of deformation would actually give a much more complicated and interesting deformation that preserves the class W_1.

4. Conformally nearly parallel G_2-structures

Suppose now we have a G_2-structure (φ, g) with torsion lying in the strict class $W_1 \oplus W_7$ -that is, both τ_1 and τ_7 are non-zero. We then deform φ to $\tilde{\varphi}$ given by

$$\tilde{\varphi} = \varphi + v^e\psi_{bcde} \ . \tag{4.1}$$

As we know from (3.10), the metric defined by $\tilde{\varphi}$ is positive definite, so $\tilde{\varphi}$ does indeed define a G_2-structure. From [**10**], we can also write down the torsion components of $\tilde{\varphi}$. As before, $M = |v|^2$ with respect to the old metric g, and ∇v is decomposed into components as in (3.11). Here we show the expression for $\tilde{\tau}_1$ and $\tilde{\tau}_7$, the 1- and 7-dimensional components of the new torsion $\tilde{T}_{ab}$:

$$\begin{aligned}
\tilde{\tau}_1 &= \frac{\left(\left(1+\frac{1}{7}M\right)\tau_1 - v_1 - \frac{6}{7}\left(\tau_7\right)^a v_a + \frac{3}{7}\left(v_7\right)^a v_a\right)}{\left(1+M\right)^{\frac{2}{3}}} \\
\left(\tilde{\tau}_7\right)_c &= \left(\tau_7\right)_c - \frac{1}{6}\varphi_c{}^{ab}\left(\tau_7\right)_a v_b + \frac{v_c\left(6\tau_1 - 6\left(\tau_7\right)_a v^a - 8v_1 + 3\left(v_7\right)_a v^a\right)}{6\left(1+M\right)} \\
&\quad - \frac{\left(3\left(M+2\right)\left(v_7\right)_c + v^a\left(v_{27}\right)_{ac} + \varphi_{ca}{}^b v^a\left(v_{27}\right)_{bd} v^d + 3\varphi_{cab} v^a\left(v_7\right)^b\right)}{6\left(1+M\right)}
\end{aligned}$$

The expressions for $\tilde{\tau}_{14}$ and $\tilde{\tau}_{27}$ can similarly be written down in terms of τ_1, τ_7 and the components of ∇v, using the general results in [**10**], however they are rather long and not very enlightening. Also, as shown in [**10**], the linear equations for the torsion components can be solved for the components of ∇v in terms of the old torsion and the new torsion. Hence if we require the new torsion to be in a specific torsion class, this would give us a differential equation that v has to satisfy. We will try to find conditions that will have the G_2-structure in $W_1 \oplus W_7$ class to a G_2-structure that only has a 1-dimensional torsion component. Using the general expression for ∇v in [**10**], and setting $\tau_{14} = \tau_{27} = 0$ and $\tilde{\tau}_7 = \tilde{\tau}_{14} = \tilde{\tau}_{27} = 0$, we thus have:

Proposition 4.1. *Suppose (φ, g) is G_2-structure with the only non-vanishing torsion component τ_1 and τ_7. The new G_2-structure $\tilde{\varphi}$ obtained via the deformation (4.1) then has torsion in class $\tilde{W}_1$ with the non-vanishing component $\tilde{\tau}_1$ if and only if v satisfies:*

$$\begin{aligned}
(4.2)\quad \nabla_a v_b &= \left(\tau_1 - \left(1+M\right)^{\frac{2}{3}}\tilde{\tau}_1 - \left(\tau_7\right)_c v^c\right) g_{ab} \\
&\quad + 4\left(1+M\right)^{-\frac{1}{3}}\tilde{\tau}_1 v_a v_b + \frac{1}{\left(M+9\right)}\left(-3\left(M-3\right)\left(\tau_7\right)_c\varphi^c{}_{ab}\right. \\
&\quad - \left(M+33\right) v_a\left(\tau_7\right)_b + 3\left(1+M\right)\left(\tau_7\right)_a v_b \\
&\quad - \frac{1}{3}v^c\varphi_{cab}\left(9\tau_1 - 4\tilde{\tau}_1\left(M+9\right)\left(1+M\right)^{-\frac{1}{3}} + \tau_1 M - 12\left(\tau_7\right)_d v^d\right) \\
&\quad \left. + 12 v_a\varphi^{cd}{}_b\left(\tau_7\right)_c v_d - 12 v_b\varphi^{cd}{}_a\left(\tau_7\right)_c v_d + 12\left(\tau_7\right)_c v_d\psi^{cd}{}_{ab}\right)
\end{aligned}$$

Note that $\tilde{\tau}_1$ has to be a constant due to the conditions on the torsion (2.11). While it is too difficult to solve equation (4.2) directly, we can obtain conditions under which the equation is at least consistent. In this equation v has lowered indices, so we can consider this as a 1-form $v^\flat$. Then

$$\left(dv^\flat\right)_{ab} = 2\nabla_{[a}v_{b]}$$

However $d^2 = 0$, and thus the exterior derivative applied to the anti-symmetrization of (4.2) must give zero. From these considerations we get the consistency conditions in Proposition 4.2 below. The extra equations which we get from the consistency conditions are very important to simplify the equation (4.2).

Proposition 4.2. *The equation (4.2) is consistent with the necessary condition $d^2 v^\flat = 0$ if and only if all of the following conditions are satisfied:*

(1) *For some smooth function V,*

$$v^{\flat} = V\tau_7 \tag{4.3}$$

(2) *The 7-dimensional component τ_7 of the original torsion satisfies*

$$\nabla\tau_7 = -\frac{1}{4}\frac{\left(V^2\left|\tau_7\right|^2 - 3\right)(V\tau_1 + 1)}{V^2}g + \frac{1}{6}\left(V^2\tau_1^2 + 6V\tau_1 + 3\right)\tau_7 \otimes \tau_7 \tag{4.4}$$

(3) *The 1-dimensional component $\tilde{\tau}_1$ of the new torsion satisfies*

$$\tilde{\tau}_1 = \frac{1}{4V}\left(1 + V^2\left|\tau_7\right|^2\right)^{\frac{1}{3}}(V\tau_1 - 3) \tag{4.5}$$

PROOF. As outlined above, we apply the condition $\nabla_{[a}\nabla_b v_{c]} = 0$ to (4.2). During the simplification process we apply (4.2) again, and moreover use the conditions $d\tau_7 = 0$, $d\tau_1 = \tau_1\tau_7$ and $d\tilde{\tau}_1 = 0$. In the end we obtain an expression for the 3-form $d^2v^{\flat}$ in terms of v, τ_1, $\tilde{\tau}_1$, τ_7 and $\nabla\tau_7$. Since the whole 3-form must vanish, so must the components of the 3-form in Λ^3_1, Λ^3_7 and Λ^3_{27}. So let ξ_1 be the scalar corresponding to the Λ^3_1 component and let ξ_7 and ξ_{27} be the vector and the antisymmetric symmetric tensor corresponding to the Λ^3_7 and Λ^3_{27} components of $d^2v^{\flat}$, respectively.

Then by considering the equations $\xi_1 = 0$, $(\xi_7)^a v_a = 0$ and $(\xi_{27})_{mn} v^m v^n = 0$, we can express $(\nabla_a(\tau_7)_b)v^a v^b$, $\nabla^a(\tau_7)_a$ and $|\tau_7|^2$ in terms of M, $\tau_1, \tilde{\tau}_1$ and $\langle\tau_7, v\rangle$. In particular, we find that

$$\begin{aligned}
\left|\tau_7\right|^2 \quad = \quad & \frac{3\langle\tau_7, v\rangle^2\left(3M^2 - 10M + 51\right)}{\left(7M^2 - 66M - 9\right)M} \\
& -\frac{4}{3}\frac{\langle\tau_7, v\rangle(M+9)^2\left(\tau_1(1+M)^{\frac{1}{3}} - 4\tilde{\tau}_1\right)}{(1+M)^{\frac{1}{3}}\left(7M^2 - 66M - 9\right)} \\
& +\frac{2}{9}\frac{\tau_1^2 M(M+9)^2}{7M^2 - 66M - 9} - \frac{16}{9}\frac{\tau_1\tilde{\tau}_1 M(M+9)^2}{(1+M)^{\frac{1}{3}}\left(7M^2 - 66M - 9\right)} \\
& +\frac{32}{9}\frac{\tilde{\tau}_1^2 M(M+9)^2}{(1+M)^{\frac{2}{3}}\left(7M^2 - 66M - 9\right)}
\end{aligned} \tag{4.6}$$

Further, we can consider the vector equations $\xi_7^d = 0$, $\varphi_{abc}v^b\xi_7^c = 0$, $(\xi_{27})_{mn} v^n = 0$ and $\varphi_{abc}(\xi_{27})^b{}_n v^n v^c = 0$. From these, in particular, we find

$$\tau_7 = \frac{\langle\tau_7, v\rangle}{M}v \text{ and } \left|\tau_7\right|^2 = \frac{\langle\tau_7, v\rangle^2}{M}. \tag{4.7}$$

Equating (4.6) and (4.7), and solving for $\tilde{\tau}_1^2$, we obtain an expression for $\tilde{\tau}_1$ in terms of τ_1, τ_7 and v.

$$\tilde{\tau}_1 = \frac{1}{4}\frac{(1+M)^{\frac{1}{3}}(M\tau_1 - 3\langle\tau_7, v\rangle)}{M} \tag{4.8}$$

It can be checked that this expression for $\tilde{\tau}_1$ is in fact consistent with the assumption $d\tilde{\tau}_1 = 0$.

Next, from equations $(\xi_{27})_{ab} = 0$, $\varphi^{cd}{}_{(a}(\xi_{27})_{b)d}v_c = 0$ and $\varphi_a{}^{cd}\varphi_b{}^{ef}v_c v_e(\xi_{27})_{df}$, we finally obtain an expression for $\nabla_a(\tau_7)_b$. Using (4.7) and (4.8) to eliminate $|\tau_7|^2$

and $\tilde{\tau}_1$ from the resulting expression, we overall get:
(4.9)

$$\nabla\tau_7 = -\frac{\langle\tau_7, v\rangle (M-3)(M\tau_1 + \langle\tau_7, v\rangle)}{4M^2} g + \frac{\left(\tau_1^2 M^2 + 6\langle\tau_7, v\rangle M\tau_1 + 3\langle\tau_7, v\rangle^2\right)}{6\langle\tau_7, v\rangle^2}\tau_7^2$$

Now since v is proportional to τ_7, let us write $v = V\tau_7$ for some smooth function V. Then

$$M = |v|^2 = V^2|\tau_7|^2 \quad \text{and} \quad \langle\tau_7, v\rangle = V|\tau_7|^2 \tag{4.10}$$

Thus we get the expressions (4.4) and (4.5) for $\nabla\tau_7$ and $\tilde{\tau}_1$ in terms of V. □

Now it is easy to see that if v is proportional to τ_7 and $\tilde{\tau}_1$ satisfies (4.5), then the equation (4.2) for v is equivalent to the equation (4.4) for τ_7. However since these conditions are required for the consistency of (4.2), the equation (4.2) is in fact equivalent to (4.4) together with conditions (4.3) and (4.5). This is now something that we can solve, however for that we will need the following lemma.

Lemma 4.3 ([4]). *Let M be a n-dimensional Riemannian manifold. Then the metric g satisfies*

$$\nabla_a\nabla_b h = \lambda g_{ab} \tag{4.11}$$

for functions h and λ if and only if the underlying smooth manifold is $(a,b)\times N$, for a $(n-1)$-dimensional manifold N, with a warped product metric g given by

$$g = \frac{dh^2}{|\nabla h|^2} + |\nabla h|^2\hat{g} \tag{4.12}$$

where $\hat{g}$ is the induced metric on the $(n-1)$-dimensional slices.

Theorem 4.4. *Consider a deformation of (φ, g) with T_{ab} lying in the strict class $W_1\oplus W_7$ to $(\tilde{\varphi},\tilde{g})$ with $\tilde{T}_{ab}$ lying in the class $\tilde{W}_1$. Then, such a deformation exists if and only if M is a warped product manifold $I\times_f N$ for some interval I and 6-dimensional manifold N. There are three cases:*

(1) *If $v^\flat = \frac{3}{\tau_1}\tau_7$, then for a 6-dimensional metric $\hat{g}$, the original metric g and new torsion $\tilde{\tau}_1$ must be given by*

$$g = \frac{\tau_7^2}{|\tau_7|^2} + \tau_1^{-10}|\tau_7|^2\hat{g} \tag{4.13}$$

$$\tilde{\tau}_1 = 0 \tag{4.14}$$

(2) *If $v^\flat = -\frac{3}{\tau_1}\tau_7$, then*

$$g = \frac{\tau_7^2}{|\tau_7|^2} + \tau_1^2|\tau_7|^2\hat{g} \tag{4.15}$$

$$\tilde{\tau}_1^3 = \frac{\tau_1}{8}\left(\tau_1^2 + 9|\tau_7|^2\right). \tag{4.16}$$

(3) *If* $v^\flat = \frac{f}{\tau_1}\tau_7$ *where* $f^2 = \frac{9A\tau_1^3}{A\tau_1^3 - 1}$ *for an arbitrary constant* A, *then*

$$g = \frac{\tau_7^2}{|\tau_7|^2} + \frac{(f-3)^{\frac{10}{3}}}{f^{\frac{2}{3}}(f+3)^{\frac{2}{3}}}|\tau_7|^2\hat{g} \tag{4.17}$$

$$\tilde{\tau}_1^3 = \frac{1}{64}\frac{(f-3)^2}{Af(f^2-9)}\left(1 + \frac{9A\tau_1|\tau_7|^2}{A\tau_1^3 - 1}\right) \tag{4.18}$$

PROOF. From the expression for ∇v (4.2), we find that

$$dM = \frac{3}{2}\left(V^3|\tau_7|^2\tau_1 + V^2|\tau_7|^2 + V\tau_1 + 1\right)\tau_7 \tag{4.19}$$

However, from (4.10),

$$dM = 2V|\tau_7|^2 dV + 2V^2\tau_7 \lrcorner (\nabla\tau_7) \tag{4.20}$$

So equating (4.19) and (4.20), and using (4.4), we get an expression for dV:

$$dV = \frac{1}{6}V\left(3 - V^2\tau_1^2\right)\tau_7 \tag{4.21}$$

Now consider $d(V\tau_1)$. Using (4.21) and the fact that that $d\tau_1 = \tau_1\tau_7$ we find that

$$d(V\tau_1) = \frac{1}{6}\left(9V\tau_1 - V^3\tau_1^3\right)\tau_7 \tag{4.22}$$

Let $f = V\tau_1$, so we have the equation

$$df = \frac{1}{6}\left(9f - f^3\right)d(\log\tau_1). \tag{4.23}$$

Consider the constant solutions first. Since τ_1 is non-zero, the solution $f = 0$ implies that $V = 0$, and hence $v = 0$, so this is a degenerate solution. The non-trivial constant solutions are $f \equiv \pm 3$. If $f = V\tau_1 \equiv 3$, then in (4.5), $\tilde{\tau}_1 = 0$. In this case, from (4.4), we find

$$\nabla\tau_7 = \left(\frac{\tau_1^2}{3} - |\tau_7|^2\right)g + 5\tau_7 \otimes \tau_7 \tag{4.24}$$

Now using (4.24) and $d\tau_1 = \tau_1\tau_7$ we can relate the metric to the Hessian of a function via the following expression

$$\nabla_a\nabla_b\left(\tau_1^{-5}\right) = -5\tau_1^{-5}\left(\frac{\tau_1^2}{3} - |\tau_7|^2\right)g_{ab}. \tag{4.25}$$

Hence by Lemma 4.3 we get (4.13). Now consider the solution $f = V\tau_1 \equiv -3$. In this case, from (4.5) we find

$$V^2|\tau_7|^2 = -\frac{8\tilde{\tau}_1^3V^3}{27} - 1. \tag{4.26}$$

Using the fact that $V = -\frac{3}{\tau_1}$ and (4.26) in (4.4) we have

$$\nabla\tau_7 = \frac{2}{9\tau_1}\left(2\tilde{\tau}_1^3 - \tau_1^3\right)g - \tau_7 \otimes \tau_7 \tag{4.27}$$

However, in the $W_1 \oplus W_7$ class, $\tau_7 = d\log\tau_1$, so we rewrite (4.27) in terms of τ_1:

$$\nabla_a\nabla_b\tau_1 = \frac{2}{9}\left(2\tilde{\tau}_1^3 - \tau_1^3\right)g_{ab} \tag{4.28}$$

Hence, by Lemma 4.3, the metric must be (4.15). Using (4.26), we also find that $\tilde{\tau}_1$ satisfies (4.16).

Suppose now f is non-constant. Then whenever the right-hand side of (4.23) is non-zero - that is, $f \neq \pm 3$, we can separate variables in (4.23). Integrating, we obtain

$$f^2 = \frac{9A\tau_1^3}{A\tau_1^3 - 1} \tag{4.29}$$

for some positive constant A. Suppose now $f = \pm 3$ at some point, then if f is non-constant, we must have $f^2 \longrightarrow 9$ in (4.29) for some values of A and τ_1. However, we can see from (4.29) that this happens if and only if $|\tau_1| \longrightarrow \infty$, but τ_1 is smooth, so in fact, either $f \equiv \pm 3$ (so these are singular solutions of 4.23) or f is nowhere equal to ± 3 and is given by (4.29) everywhere. We have already covered the constant cases above, so now we can assume that (4.29) holds everywhere. In particular from (4.29) we also get the relations

$$V^2 = \frac{9A\tau_1}{A\tau_1^3 - 1} \text{ and } V^3 = Af\left(f^2 - 9\right) \tag{4.30}$$

From (4.5), find can find $|\tau_7|^2$ in terms of f and $\tilde{\tau}_1^3$. Also, note that from (4.23) we get

$$\nabla_a \nabla_b f = -\frac{1}{6} f\left(f^2 - 9\right) \nabla_a \left(\tau_7\right)_b + \frac{3\left(f^2 - 3\right)}{f\left(f^2 - 9\right)} \nabla_a f \nabla_b f.$$

So overall, we can rewrite (4.4) as an equation for f in the form

$$\nabla_a \nabla_b f = P(f) g_{ab} + Q\left(f\right) \nabla_a f \nabla_b f \tag{4.31}$$

for some functions $P\left(f\right)$ and $Q\left(f\right)$. The exact form of $P\left(f\right)$ is not very important, but $Q\left(f\right)$ is given by

$$Q\left(f\right) = \frac{2\left(f^2 - 3f - 6\right)}{f\left(f^2 - 9\right)}.$$

In order to reduce (4.31) to the form (4.11) we need to find a function $F\left(f\right)$ that satisfies

$$\frac{d^2 F}{df^2} + \frac{dF}{df} Q\left(f\right) = 0. \tag{4.32}$$

For such an F, the Hessian would be proportional to the metric. Let $G = \frac{dF}{df}$, then by separation of variables, we solve (4.32) for G and get the solution

$$G = \frac{6\left(f-3\right)^{\frac{2}{3}}}{f^{\frac{4}{3}}\left(f+3\right)^{\frac{4}{3}}} \tag{4.33}$$

Note that we have chosen a constant factor for convenience. Since f is nowhere vanishing and is nowhere equal to -3, this can be integrated further to find F. Hence for this F, $\nabla_a \nabla_b F$ is proportional to g_{ab}. Therefore by Lemma 4.3, the metric must be a warped product of the form

$$g = \frac{1}{|\nabla F|^2} dF^2 + |\nabla F|^2 \hat{g} \tag{4.34}$$

Note that $\frac{dF}{|\nabla F|} = \frac{df}{|\nabla f|} = \frac{\tau_7}{|\tau_7|}$, and therefore, using (4.23), we get

$$\begin{aligned} |\nabla F|^2 &= \left(\frac{dF}{df}\right)^2 |\nabla f|^2 = \frac{1}{36} G^2 |\tau_7|^2 f^2 \left(f^2 - 9\right)^2 \\ &= \frac{(f-3)^{\frac{10}{3}}}{f^{\frac{2}{3}} (f+3)^{\frac{2}{3}}} |\tau_7|^2 \end{aligned}$$

Thus we obtain the metric (4.17). From (4.5) we get the expression (4.18) for $\tilde{\tau}_1$ by substituting (4.30). □

It was shown in by Cleyton and Ivanov in [**5**] that by considering a warped product of an open interval over a 6-dimensional nearly Kähler manifold, it is possible to obtain 7-dimensional manifolds with G_2-structures. Moreover, it has been shown that for such a construction the τ_{14} torsion component (τ_2 in the notation of [**5**]) will always vanish, and moreover, it is possible to find parameters such that the τ_{27} torsion component (τ_3 in [**5**]) will also become zero, leaving only τ_1 and τ_7 non-zero (τ_0 and τ_1 in their notation), which are given in terms of the warp factor. In particular, if the metric is $dt^2 + h(t)^2 \hat{g}$ for warp factor $h > 0$, then, using our conventions for the torsion components,

$$\tau_1 = h^{-1} \sigma \sin\theta \quad \text{and } \tau_7 = h^{-1} \left(\sigma \cos\theta - h'\right) dt \tag{4.35}$$

where σ is a constant related to the scalar curvature of the 6-dimensional manifold and $\theta(t)$ satisfies $\theta' = h^{-1}\sigma \sin\theta$. These are precisely the kind of manifolds that appear as solutions in Theorem 4.4 - warped product 7-manifolds with torsion in $W_1 \oplus W_7$. The warp factors in Theorem 4.4 have to be consistent with the expressions (4.35), and so for each case in Theorem 4.4 we get a system of two first-order ODEs for h and θ. Given appropriate initial conditions we can say that solutions exist, however the analysis of these solutions is something to be investigated further. Therefore, we can construct examples of 7-manifolds with a conformally nearly parallel G_2-structure such that a non-infinitesimal deformation in Λ^3_7 gives a G_2-structure in a strictly smaller torsion class. Moreover, applying a conformal transformation (3.6) we can obtain a nearly parallel G_2-structure, for which a combination of a conformal transformation and a Λ^3_7 deformation lead to another nearly parallel G_2-structure. As we have seen, there are various non-trivial relationships between different G_2-structures and it will be a subject of further study whether it is possible find transformations between other G_2 torsion classes. In particular, so far we only have a grasp on deformations that lie in Λ^3_1 and Λ^3_7, however it is likely that deformations in Λ^3_{27} could yield the most interesting results.

References

[1] J. Baez, *The Octonions*, *Bull. Amer. Math. Soc. (N.S.)* **39** (2002) 145–205.

[2] R. L. Bryant, *Metrics with exceptional holonomy*, *Ann. of Math. (2)* **126** (1987), no. 3 525–576.

[3] R. L. Bryant, *Some remarks on G_2-structures*, in *Proceedings of Gökova Geometry-Topology Conference 2005*, pp. 75–109, Gökova Geometry/Topology Conference (GGT), Gökova, 2006. `math/0305124`.

[4] J. Cheeger and T. H. Colding, *Lower bounds on Ricci curvature and the almost rigidity of warped products*, *Ann. of Math. (2)* **144** (1996), no. 1 189–237.

[5] R. Cleyton and S. Ivanov, *Conformal equivalence between certain geometries in dimension 6 and 7*, *Math. Res. Lett.* **15** (2008), no. 4 631–640 [`math/0607487`].

[6] M. Fernández and A. Gray, *Riemannian manifolds with structure group G_2*, *Ann. Mat. Pura Appl. (4)* **132** (1982) 19–45 (1983).
[7] T. Friedrich, I. Kath, A. Moroianu and U. Semmelmann, *On nearly parallel G_2-structures*, *J. Geom. Phys.* **23** (1997), no. 3-4 259–286.
[8] A. Gray, *Vector cross products on manifolds*, *Trans. Amer. Math. Soc.* **141** (1969) 465–504.
[9] S. Grigorian, *Moduli spaces of G2 manifolds*, *Rev. Math. Phys.* **22** (2010), no. 9 1061–1097 [0911.2185].
[10] S. Grigorian, *Deformations of G2-structures with torsion*, 1108.2465.
[11] S. Grigorian and S.-T. Yau, *Local geometry of the G2 moduli space*, *Commun. Math. Phys.* **287** (2009) 459–488 [0802.0723].
[12] N. J. Hitchin, *The geometry of three-forms in six*, *J. Differential Geom.* **55** (2000), no. 3 547–576 [math/0010054].
[13] D. D. Joyce, *Compact manifolds with special holonomy.* Oxford Mathematical Monographs. Oxford University Press, 2000.
[14] S. Karigiannis, *Deformations of G_2 and Spin(7) Structures on Manifolds*, *Canadian Journal of Mathematics* **57** (2005) 1012 [math/0301218].
[15] S. Karigiannis, *Flows of G_2-Structures, I*, *Q. J. Math.* **60** (2009), no. 4 487–522 [math/0702077].
[16] P. Kaste, R. Minasian and A. Tomasiello, *Supersymmetric M-theory compactifications with fluxes on seven-manifolds and G-structures*, *JHEP* **0307** (2003) 004 [hep-th/0303127].

Simons Center for Geometry and Physics, Stony Brook University, Stony Brook, NY 11794, USA

Proceedings of Symposia in Pure Mathematics
Volume **85**, 2012

Non-commutative Solitons and Quasi-determinants

Masashi Hamanaka

This paper is dedicated to the memory of professor Miki Wadati.

ABSTRACT. We discuss the extension of soliton theory and integrable systems to non-commutative spaces, focusing on integrable aspects of non-commutative anti-self-dual Yang-Mills equations. We present Bäcklund transformations for the $G = U(2)$ non-commutative anti-self-dual Yang-Mills equations and give a wide class of exact solutions of them (not only instanton-type solutions with finite action). We find that one kind of non-commutative determinants, quasi-determinants, play a crucial role in the construction of non-commutative solutions. Finally we briefly present some examples of reduction of non-commutative anti-self-dual Yang-Mills equations to non-commutative KdV, NLS and Liouville equations. This is partially based on collaboration with C. Gilson and J. Nimmo (Glasgow).

1. Introduction

Non-commutative (NC) extension of field theories is not just a generalization of them but a fruitful direction in both physics and mathematics. In particular, non-commutative gauge theories are known to be equivalent to commutative gauge theories in background (magnetic) flux and have been applied to problems in the quantum Hall effect. Furthermore, in non-commutative spaces, singularities resolve in general and new physical objects could appear. For example, for the 4-dimensional non-commutative anti-self-dual Yang-Mills (ASDYM) equations, because of resolution of singularities in the moduli spaces, there appear $U(1)$ instanton solutions which are special to non-commutative spaces [**30, 9**]. Here it is important that the ADHM construction can be sensibly extended to non-commutative spaces and in this sense, integrability is also preserved. The corresponding D-brane interpretation is also clear and gives new insight. (For reviews, see [**7, 17, 24, 33**].)

After the development of non-commutative gauge theories, non-commutative extension of (scalar) integrable equations, such as the Korteweg-de Vries (KdV) equation, has been studied intensively. They do not (directly) belong to gauge theories; however, most of them have been proved to be derived from the 4-dimensional

1991 *Mathematics Subject Classification.* Primary 39K40; Secondary 37K10, 81R60.
Key words and phrases. Solitons, integrable systems, non-commutative geometry.
The author was supported in part by the Toyoaki Scholarship Foundation.

non-commutative anti-self-dual Yang-Mills equations by reduction. (The results are summarized in e.g. [**20**].) Hence, the KdV equation can belong to gauge theories and the non-commutative extension has a physical meaning, that is, the presence of background flux. In this context, the integrable equations relate to $N=2$ string theories [**32**] and can be applied to them by analysis of the exact soliton solutions.

In this article, we discuss extension of soliton theory and integrable systems to non-commutative spaces, focusing on integrable aspects of non-commutative anti-self-dual Yang-Mills equations. We present Bäcklund transformations for the G=U(2) non-commutative anti-self-dual Yang-Mills equations and give wide class of exact solutions of them (not only instanton-type solutions with finite action but also soliton-type solutions with infinite action)[**14, 15**]. We find that one kind of non-commutative determinants, quasi-determinants, play crucial roles in the construction of non-commutative solutions. Finally we briefly some examples of reduction of non-commutative anti-self-dual Yang-Mills equations to non-commutative KdV, NLS and Liouville equations [**20**].

2. Non-commutative Field Theory

Non-commutative spaces are characterized by the non-commutativity of the spatial coordinates x^μ:

$$[x^\mu, x^\nu] = i\theta^{\mu\nu}, \tag{2.1}$$

where the anti-symmetric tensor $\theta^{\mu\nu}$ is called the *non-commutativity parameter.* In this article, the parameter is assumed to be real constant.

Non-commutative field theories on flat spaces are given by the replacement of ordinary products in the commutative field theories with the *Moyal products* [**29**] and realized as theories deformed from the commutative ones. The Moyal-product is defined for ordinary fields explicitly by

$$\begin{aligned} f\star g(x) &:= \exp\left(\frac{i}{2}\theta^{\mu\nu}\partial_\mu^{(x')}\partial_\nu^{(x'')}\right) f(x')g(x'')\Big|_{x'=x''=x} \\ &= f(x)g(x) + \frac{i}{2}\theta^{\mu\nu}\partial_\mu f(x)\partial_\nu g(x) + O(\theta^2) \text{ (for small } \theta^{ij}), \end{aligned} \tag{2.2}$$

where $\partial_\mu^{(x')} := \partial/\partial x'^\mu$ and so on. The Moyal product is associative: $f\star(g\star h) = (f\star g)\star h$, and reduces to the ordinary product in the commutative limit: $\theta^{\mu\nu}\to 0$. The modification of the product makes the ordinary spatial coordinate "non-commutative" which means: $[x^\mu, x^\nu]_\star := x^\mu\star x^\nu - x^\nu\star x^\mu = i\theta^{\mu\nu}$. In this article, we do not discuss convergence and domain issues. (See e.g. [**8, 24, 31**].)

We note that the fields themselves are c-number functions and differentiation and integration for them are defined as usual. A nontrivial point is that non-commutative field equations contain an infinite number of derivatives in general. Hence the integrability of the equations is not so trivial as in commutative cases, especially for space-time non-commutativity.

Here are a few examples of non-commutative field equations:

- Non-commutative ASDYM equation in 4 dimensions ($[x^\mu, x^\nu]_\star = i\theta^{\mu\nu}$)

$$F^\star_{\mu\nu} = -* F^\star_{\mu\nu}, \tag{2.3}$$

where $*$ is the Hodge dual and $F^{\star}_{\mu\nu} := \partial_\mu A_\nu - \partial_\nu A_\mu + [A_\mu, A_\nu]_\star$. We note that even for $G = U(1)$, the commutator $[A_\mu, A_\nu]_\star$ in the field strength must be kept because of non-commutative gauge symmetry.

- Non-commutative KdV equation in $(1+1)$ dimensions ($[t,x]_\star = i\theta$)

$$\frac{\partial u}{\partial t} = \frac{1}{4}\frac{\partial^3 u}{\partial x^3} + \frac{3}{4}\left(\frac{\partial u}{\partial x} \star u + u \star \frac{\partial u}{\partial x}\right). \tag{2.4}$$

The ordering of variables in non-linear terms is crucial to preserve some special integrablity properties and determined in the Lax formalism as we will see later. For non-commutative KdV equation, the non-linear term $2u \cdot \partial_x u$ becomes symmetric: $\partial_x u \star u + u \star \partial_x u$.

From now on, we suppose that all field variables are non-commutative and omit the star symbols "$\star$" for non-commutative products.

3. A Brief Review of Quasi-determinants

In this section, we give a brief introduction to quasi-determinants introduced by Gelfand and Retakh in 1991 [**10**] and present a few properties of them which play important roles in section 4. A good survey is e.g. [**11**] and the relation between quasi-determinants and non-commutative symmetric functions is summarized in e.g. [**12**].

Quasi-determinants are not just a non-commutative generalization of the usual commutative determinants but are rather related to inverse matrices.

Let $A = (a_{ij})$ be an $n \times n$ matrix and $B = (b_{ij})$ be the inverse matrix of A. Here all matrix elements are assumed to belong to a (non-commutative) ring with an associative product. This general non-commutative situation includes the Moyal or non-commutative deformation which we discuss in the main sections.

Quasi-determinants of A are defined formally as the inverses of the elements of $B = A^{-1}$:

$$|A|_{ij} := b_{ji}^{-1}. \tag{3.1}$$

In the commutative limit, this reduces to

$$|A|_{ij} \longrightarrow (-1)^{i+j}\frac{\det A}{\det \tilde{A}^{ij}}, \tag{3.2}$$

where $\tilde{A}^{ij}$ is the matrix obtained from A by deleting the i-th row and the j-th column.

We can write down a more explicit form of quasi-determinants. In order to see it, let us recall the following formula for a square matrix:

$$\begin{pmatrix} A & B \\ C & D \end{pmatrix}^{-1} = \begin{pmatrix} (A - BD^{-1}C)^{-1} & -A^{-1}B(D - CA^{-1}B)^{-1} \\ -(D - CA^{-1}B)^{-1}CA^{-1} & (D - CA^{-1}B)^{-1} \end{pmatrix},$$

where A and D are square matrices, and all inverses are assumed to exist. We note that any matrix can be decomposed as a 2×2 matrix by block decomposition where the diagonal parts are square matrices, and the above formula can be applied to the decomposed 2×2 matrix. So the explicit forms of quasi-determinants are given iteratively by the following formula:

$$
\begin{aligned}
|A|_{ij} &= a_{ij} - \sum_{i'(\neq i),j'(\neq j)} a_{ii'}((\tilde{A}^{ij})^{-1})_{i'j'}a_{j'j} \\
&= a_{ij} - \sum_{i'(\neq i),j'(\neq j)} a_{ii'}(|\tilde{A}^{ij}|_{j'i'})^{-1}a_{j'j}.
\end{aligned}
\tag{3.3}
$$

It is sometimes convenient to represent the quasi-determinant as follows:

$$
|A|_{ij} = \begin{vmatrix} a_{11} & \cdots & a_{1j} & \cdots & a_{1n} \\ \vdots & & \vdots & & \vdots \\ a_{i1} & & \boxed{a_{ij}} & & a_{in} \\ \vdots & & \vdots & & \vdots \\ a_{n1} & \cdots & a_{nj} & \cdots & a_{nn} \end{vmatrix}. \tag{3.4}
$$

Examples of quasi-determinants are, for a 1×1 matrix $A = a$,

$$|A| = a,$$

and for a 2×2 matrix $A = (a_{ij})$,

$$
|A|_{11} = \begin{vmatrix} \boxed{a_{11}} & a_{12} \\ a_{21} & a_{22} \end{vmatrix} = a_{11} - a_{12}a_{22}^{-1}a_{21}, \quad |A|_{12} = \begin{vmatrix} a_{11} & \boxed{a_{12}} \\ a_{21} & a_{22} \end{vmatrix} = a_{12} - a_{11}a_{21}^{-1}a_{22},
$$

$$
|A|_{21} = \begin{vmatrix} a_{11} & a_{12} \\ \boxed{a_{21}} & a_{22} \end{vmatrix} = a_{21} - a_{22}a_{12}^{-1}a_{11}, \quad |A|_{22} = \begin{vmatrix} a_{11} & a_{12} \\ a_{21} & \boxed{a_{22}} \end{vmatrix} = a_{22} - a_{21}a_{11}^{-1}a_{12},
$$

and for a 3×3 matrix $A = (a_{ij})$,

$$
\begin{aligned}
|A|_{11} &= \begin{vmatrix} \boxed{a_{11}} & a_{12} & a_{13} \\ a_{21} & a_{22} & a_{23} \\ a_{31} & a_{32} & a_{33} \end{vmatrix} = a_{11} - (a_{12}, a_{13}) \begin{pmatrix} a_{22} & a_{23} \\ a_{32} & a_{33} \end{pmatrix}^{-1} \begin{pmatrix} a_{21} \\ a_{31} \end{pmatrix} \\
&= a_{11} - a_{12} \begin{vmatrix} \boxed{a_{22}} & a_{23} \\ a_{32} & a_{33} \end{vmatrix}^{-1} a_{21} - a_{12} \begin{vmatrix} a_{22} & a_{23} \\ \boxed{a_{32}} & a_{33} \end{vmatrix}^{-1} a_{31} \\
&\quad - a_{13} \begin{vmatrix} a_{22} & \boxed{a_{23}} \\ a_{32} & a_{33} \end{vmatrix}^{-1} a_{21} - a_{13} \begin{vmatrix} a_{22} & a_{23} \\ a_{32} & \boxed{a_{33}} \end{vmatrix}^{-1} a_{31},
\end{aligned}
$$

and so on.

Quasi-determinants have various interesting properties similar to those of determinants. Among them, the following ones play important roles in this article. In the block matrices given in these results, lower case letters denote single entries and upper case letters denote matrices of compatible dimensions so that the overall matrix is square. (By using boxes, it becomes easier to calculate various identities. Such calculations are fully presented in e.g. [**13**].)

- Non-commutative Jacobi identity [**10, 13**]

$$
\begin{vmatrix} A & B & C \\ D & f & g \\ E & h & \boxed{i} \end{vmatrix} = \begin{vmatrix} A & C \\ E & \boxed{i} \end{vmatrix} - \begin{vmatrix} A & B \\ E & \boxed{h} \end{vmatrix} \begin{vmatrix} A & B \\ D & \boxed{f} \end{vmatrix}^{-1} \begin{vmatrix} A & C \\ D & \boxed{g} \end{vmatrix}.
$$

- Homological relations [**10**]

$$\begin{vmatrix} A & B & C \\ D & f & g \\ E & \boxed{h} & i \end{vmatrix} = \begin{vmatrix} A & B & C \\ D & f & g \\ E & h & \boxed{i} \end{vmatrix} \begin{vmatrix} A & B & C \\ D & f & g \\ 0 & \boxed{0} & 1 \end{vmatrix},$$

$$\begin{vmatrix} A & B & C \\ D & f & \boxed{g} \\ E & h & i \end{vmatrix} = \begin{vmatrix} A & B & 0 \\ D & f & \boxed{0} \\ E & h & 1 \end{vmatrix} \begin{vmatrix} A & B & C \\ D & f & g \\ E & h & \boxed{i} \end{vmatrix}$$

- *Gilson-Nimmo's derivative formula* [**13**]

$$\begin{vmatrix} A & B \\ C & \boxed{d} \end{vmatrix}' = \begin{vmatrix} A & B' \\ C & \boxed{d'} \end{vmatrix} + \sum_{k=1}^{n} \begin{vmatrix} A & (A_k)' \\ C & \boxed{(C_k)'} \end{vmatrix} \begin{vmatrix} A & B \\ e_k^t & \boxed{0} \end{vmatrix},$$

where A_k is the kth column of a matrix A and e_k is the column n-vector (δ_{ik}) (i.e. 1 in the kth row and 0 elsewhere).

4. Bäcklund Transformations for NC ASDYM Equation

The non-commutative anti-self-dual Yang-Mills equation can be written in the Yang form:

$$\partial_z(J^{-1}\partial_{\tilde{z}}J) - \partial_w(J^{-1}\partial_{\tilde{w}}J) = 0. \tag{4.1}$$

The suffixes $z, w, \tilde{z}, \tilde{w}$ denote complex coordinates in 4-dimensional space. We note that all matrix elements are non-commutative. In this section, the gauge group is assumed to be $G = GL(2)$, which implies $J \in GL(2)$.

From a given solution J, we can obtain anti-self-dual gauge fields A_μ by decomposing J into two 2×2 matrices h and $\tilde{h}$ as $J = \tilde{h}^{-1}h$ and by putting

$$A_z = -(\partial_z h)h^{-1},\ A_w = -(\partial_w h)h^{-1},\ A_{\tilde{z}} = -(\partial_{\tilde{z}}\tilde{h})\tilde{h}^{-1},\ A_{\tilde{w}} = -(\partial_{\tilde{w}}\tilde{h})\tilde{h}^{-1}.$$

These actually satisfy the non-commutative anti-self-dual Yang-Mills equation.

In order to present Bäcklund transformations for the non-commutative Yang equation concretely, let us parametrize the matrix J as follows (without loss of generality):

$$J = \begin{pmatrix} f - gb^{-1}e & -gb^{-1} \\ b^{-1}e & b^{-1} \end{pmatrix}. \tag{4.2}$$

Then the following two type of transformations leave the non-commutative Yang equation invariant, that is, they are (auto-)Bäcklund transformations:

- β-transformation [**28, 14**]:

$$\begin{aligned} b^{\text{new}} &= f^{-1}, f^{\text{new}} = b^{-1}, \\ e_w^{\text{new}} &= f^{-1}g_{\tilde{z}}b^{-1}, e_z^{\text{new}} = f^{-1}g_{\tilde{w}}b^{-1}, \\ g_{\tilde{z}}^{\text{new}} &= b^{-1}e_w f^{-1}, g_{\tilde{w}}^{\text{new}} = b^{-1}e_z f^{-1}. \end{aligned}$$

- γ_0-transformation [**14**]:

$$\begin{aligned} b^{\text{new}} &= (f - gb^{-1}e)^{-1},\ f^{\text{new}} = (b - ef^{-1}g)^{-1}, \\ e^{\text{new}} &= (e - bg^{-1}f)^{-1},\ g^{\text{new}} = (g - fe^{-1}b)^{-1}. \end{aligned}$$

They are both involutive ($\beta \circ \beta = \text{id}$, $\gamma_0 \circ \gamma_0 = \text{id}$); however, the combination of them $\alpha = \gamma_0 \circ \beta$ or $\alpha' = \beta \circ \gamma_0$ becomes non-trivial to generate a series of non-trivial solutions from a class of trivial solutions R_0:

$$\begin{array}{ccccccccc} R_0 & \overset{\alpha}{\to} & R_1 & \overset{\alpha}{\to} & R_2 & \overset{\alpha}{\to} & R_3 & \to & \cdots \\ & \beta \searrow & \gamma_0 \updownarrow & \beta \searrow & \gamma_0 \updownarrow & \beta \searrow & \gamma_0 \updownarrow & \searrow & \cdots \\ & & R'_1 & \overset{\alpha'}{\to} & R'_2 & \overset{\alpha'}{\to} & R'_3 & \to & \cdots \end{array} \tag{4.3}$$

We note that by putting $b = e = f = g = \Delta_0^{-1}$ ($\Delta_0(x)$ is a scalar function.), then non-commutative anti-self-dual Yang-Mills equation reduces to a non-commutative linear equation $(\partial_z\partial_{\tilde{z}} - \partial_w\partial_{\tilde{w}})\Delta_0 = 0$ (in Euclidean space, the 4 dimensional Laplace equation). From this trivial solution, we can generate the following solutions by operating the Bäcklund transformation

- Non-commutative Atiyah-Ward ansatz solutions R_l

$$b_l = \begin{vmatrix} \Delta_0 & \Delta_{-1} & \cdots & \Delta_{-l} \\ \Delta_1 & \Delta_0 & \cdots & \Delta_{1-l} \\ \vdots & \vdots & \ddots & \vdots \\ \Delta_l & \Delta_{l-1} & \cdots & \boxed{\Delta_0} \end{vmatrix}^{-1}, \quad f_l = \begin{vmatrix} \boxed{\Delta_0} & \Delta_{-1} & \cdots & \Delta_{-l} \\ \Delta_1 & \Delta_0 & \cdots & \Delta_{1-l} \\ \vdots & \vdots & \ddots & \vdots \\ \Delta_l & \Delta_{l-1} & \cdots & \Delta_0 \end{vmatrix}^{-1},$$

$$e_l = \begin{vmatrix} \Delta_0 & \Delta_{-1} & \cdots & \boxed{\Delta_{-l}} \\ \Delta_1 & \Delta_0 & \cdots & \Delta_{1-l} \\ \vdots & \vdots & \ddots & \vdots \\ \Delta_l & \Delta_{l-1} & \cdots & \Delta_0 \end{vmatrix}^{-1}, \quad g_l = \begin{vmatrix} \Delta_0 & \Delta_{-1} & \cdots & \Delta_{-l} \\ \Delta_1 & \Delta_0 & \cdots & \Delta_{1-l} \\ \vdots & \vdots & \ddots & \vdots \\ \boxed{\Delta_l} & \Delta_{l-1} & \cdots & \Delta_0 \end{vmatrix}^{-1}.$$

- Non-commutative Atiyah-Ward ansatz solutions R'_l

$$b'_l = \begin{vmatrix} \boxed{\Delta_0} & \Delta_{-1} & \cdots & \Delta_{1-l} \\ \Delta_1 & \Delta_0 & \cdots & \Delta_{2-l} \\ \vdots & \vdots & \ddots & \vdots \\ \Delta_{l-1} & \Delta_{l-2} & \cdots & \Delta_0 \end{vmatrix}, \quad f'_l = \begin{vmatrix} \Delta_0 & \Delta_{-1} & \cdots & \Delta_{1-l} \\ \Delta_1 & \Delta_0 & \cdots & \Delta_{2-l} \\ \vdots & \vdots & \ddots & \vdots \\ \Delta_{l-1} & \Delta_{l-2} & \cdots & \boxed{\Delta_0} \end{vmatrix},$$

$$e'_l = \begin{vmatrix} \Delta_{-1} & \Delta_{-2} & \cdots & \boxed{\Delta_{-l}} \\ \Delta_0 & \Delta_{-1} & \cdots & \Delta_{1-l} \\ \vdots & \vdots & \ddots & \vdots \\ \Delta_{l-2} & \Delta_{l-3} & \cdots & \Delta_{-1} \end{vmatrix}, \quad g'_l = \begin{vmatrix} \Delta_1 & \Delta_0 & \cdots & \Delta_{2-l} \\ \Delta_2 & \Delta_1 & \cdots & \Delta_{3-l} \\ \vdots & \vdots & \ddots & \vdots \\ \boxed{\Delta_l} & \Delta_{l-1} & \cdots & \Delta_1 \end{vmatrix}.$$

The scalar functions $\Delta_i(x)$ in the matrix elements can be determined by the following recursive relation from Δ_0:

$$\frac{\partial \Delta_i}{\partial z} = -\frac{\partial \Delta_{i+1}}{\partial \tilde{w}}, \quad \frac{\partial \Delta_i}{\partial w} = -\frac{\partial \Delta_{i+1}}{\partial \tilde{z}}, \quad -l \leq i \leq l-1 \ (l \geq 2). \tag{4.4}$$

In the commutative limit, we find $b_l = f_l, b'_l = f'_l$ and the results coincide with those in [**4**].

The proof is carried out by using only identities of the quasi-determinant. For example, for the γ_0-transformation $b_l^{-1} = f_l' - g_l' b_l'^{-1} e_l'$, we have to show

$$\begin{vmatrix} \Delta_0 & \Delta_{-1} & \cdots & \Delta_{-l} \\ \Delta_1 & \Delta_0 & \cdots & \Delta_{1-l} \\ \vdots & \vdots & \ddots & \vdots \\ \Delta_l & \Delta_{l-1} & \cdots & \boxed{\Delta_0} \end{vmatrix} = \begin{vmatrix} \Delta_0 & \cdots & \Delta_{1-l} \\ \vdots & \ddots & \vdots \\ \Delta_{l-1} & \cdots & \boxed{\Delta_0} \end{vmatrix}$$
$$- \begin{vmatrix} \Delta_1 & \cdots & \Delta_{2-l} \\ \vdots & \ddots & \vdots \\ \boxed{\Delta_l} & \cdots & \Delta_1 \end{vmatrix} \begin{vmatrix} \boxed{\Delta_0} & \cdots & \Delta_{1-l} \\ \vdots & \ddots & \vdots \\ \Delta_{l-1} & \cdots & \Delta_0 \end{vmatrix}^{-1} \begin{vmatrix} \Delta_{-1} & \cdots & \boxed{\Delta_{-l}} \\ \vdots & \ddots & \vdots \\ \Delta_{l-2} & \cdots & \Delta_{-1} \end{vmatrix}$$

We easily find that this is just the non-commutative Jacobi identity! The proof for β-transformation can be done in similar way by using some identities of quasi-determinants [**14**]. This fact implies that non-commutative Bäcklund transformations are identities of quasi-determinants, which is a natural analog of the commutative fact that Bäcklund transformations are identities of determinants [**25**]. This might suggest the possibility of a new formulation of integrable systems in both higher and lower dimensions.

The form of J is also compact and shows clearly that this Bäcklund transformation is not just a gauge transformation.

In [**15**], the origin of the Bäcklund transformations is clarified in the framework of non-commutative twistor theory [**2, 34**] and the generated solutions are identified with the Atiyah-Ward ansatz solutions [**1**].

We can easily find some special solutions of the non-commutative linear equation $(\partial_z \partial_{\tilde{z}} - \partial_w \partial_{\tilde{w}})\Delta_0 = 0$. The fundamental solutions give non-commutative instanton solutions and the solution $\Delta_0 = \exp(\Sigma a_i x^i)$ leads to a "non-commutative non-linear plane wave." The latter corresponds to a soliton-type solution with infinite action and cannot be constructed by the ADHM procedure. It is worth studying the D-brane interpretation of such solutions.

5. Reductions of non-commutative anti-self-dual Yang-Mills equation

Here we briefly discuss reductions of the non-commutative anti-self-dual Yang-Mills equation to lower-dimensional non-commutative integrable equations. Let us summarize the strategy for reductions of the non-commutative anti-self-dual Yang-Mills equation to lower-dimension. Reductions are classified by some ingredients such as a choice of gauge group; a choice of symmetry; a choice of gauge fixing and so on. Gauge groups are in general $GL(N)$. We have to take the $U(1)$ part into account in the non-commutative case. A choice of symmetry reduces the non-commutative anti-self-dual Yang-Mills equations to a simple form. We note that non-commutativity must be eliminated in the reduced directions because of compatibility with the symmetry. Hence, within the reduced directions, discussion of the symmetry is the same as the commutative case. A choice of gauge fixing is the most important ingredient in this article which is shown explicitly in each subsection. The residual gauge symmetry sometimes shows the equivalence of a few reductions. Constants of integration in the process of reduction sometimes lead to parameter families of non-commutative reduced equations; however, in this article, we set all integration constants to zero for simplicity.

5.1. Reduction to NC KdV and NLS equations. Here, we present non-trivial reductions of non-commutative anti-self-dual Yang-Mills equation (2.3) with $G = GL(2)$ to the non-commutative KdV and non-linear Schrödinger (NLS) equations [**18**].

First, let us take a dimensional reduction by null translations:

$$X = \partial_w - \partial_{\tilde{w}},\ Y = \partial_{\tilde{z}}. \tag{5.1}$$

and identify space-time coordinates as $t \equiv z,\ x = w + \tilde{w}$.

If we put the following non-trivial reduction conditions on the gauge fields:

$$A_{\tilde{w}} = \begin{pmatrix} 0 & 0 \\ u/2 & 0 \end{pmatrix},\ A_{\tilde{z}} = \begin{pmatrix} 0 & 0 \\ 1 & 0 \end{pmatrix},\ A_w = \begin{pmatrix} 0 & -1 \\ u & 0 \end{pmatrix},$$
$$A_z = \frac{1}{4}\begin{pmatrix} u' & -2u \\ u'' + 2uu & -u' \end{pmatrix},$$

then we can see that Eq. (2.3) reduces to the non-commutative KdV equation

$$\dot{u} = \frac{1}{4}u''' + \frac{3}{4}(u'u + uu'). \tag{5.2}$$

If we put the following another reduction conditions on the gauge fields:

$$A_{\tilde{z}} = \frac{i}{2}\begin{pmatrix} 1 & 0 \\ 0 & -1 \end{pmatrix},\ A_{\tilde{w}} = 0,\ A_w = \begin{pmatrix} 0 & \psi \\ \varepsilon\bar{\psi} & 0 \end{pmatrix},\ A_z = i\varepsilon\begin{pmatrix} -\psi\bar{\psi} & -\varepsilon\psi' \\ \bar{\psi}' & \bar{\psi}\psi \end{pmatrix},$$

then we can see that Eq. (2.3) reduces to the non-commutative NLS equation:

$$i\dot{\psi} = \psi'' - 2\varepsilon\psi\bar{\psi}\psi \tag{5.3}$$

Both reduced equations possess infinitely many conserved quantities [**5, 6**].

5.2. Reduction to NC Liouville equation. The non-commutative Yang equations (4.1) can reduce to the non-commutative (affine) Toda [**27**], sine/h-Gordon [**16, 26**] and Liouville equations [**3**]. Here we present the reduction to the non-commutative Liouville equation.

In a similar way, first, let us take a dimensional reduction by null translations,

$$X = \partial_w,\ Y = \partial_{\tilde{w}} \tag{5.4}$$

and take the following ansatz for the Yang J-matrix:

$$J = e^{\sigma - \tilde{w}}\begin{pmatrix} e^{\phi_+} & 0 \\ 0 & e^{\phi_-} \end{pmatrix} e^{-\sigma + w}. \tag{5.5}$$

Then the non-commutative Yang equation (4.1) reduces to the non-commutative Liouville equation:

$$\partial_z\left(e^{-\phi_+}\partial_{\tilde{z}}e^{\phi_+}\right) = e^{-\phi_-}e^{\phi_+},\ \partial_z\left(e^{-\phi_-}\partial_{\tilde{z}}e^{\phi_-}\right) = -e^{-\phi_-}e^{\phi_+}, \tag{5.6}$$

where $\phi_+ = v + u,\ \phi_- = v - u$ [**3**]. In the commutative limit, this reduces to the Liouville equation

$$\partial_z\partial_{\tilde{z}}u = e^{2u}, \tag{5.7}$$

together with a decoupled equation $\partial_z\partial_{\tilde{z}}v = 0$.

Many other non-commutative integrable equations are in fact proved to be derived from non-commutative anti-self-dual Yang-Mills equation by reduction, which is summarized in [**20**]. Hence we can make a discussion on classification of lower-dimensional non-commutative integrable equations from the viewpoint of non-commutative twistor theory. In particular, Bäcklund transformations for non-commutative anti-self-dual Yang-Mills equations are presented [**14, 15**] and can be applied to reveal the infinite dimensional symmetry of non-commutative anti-self-dual Yang-Mills equations and the reduced equations. Furthermore, this implies that non-commutative integrable equations such as the non-commutative KdV equation can be embedded into a gauge theory and the Seiberg-Witten map can be applied to them. The Seiberg-Witten map connects the non-commutative description to the commutative one in the background of flux, where the degree of the time derivative is finite. Then we could discuss integrability of non-commutative integrable systems via the (non-perturbative) Seiberg-Witten map. These topics are discussed in more detail elsewhere.

References

[1] M. F. Atiyah and R. S. Ward, Commun. Math. Phys. **55**, 117 (1977).
[2] S. J. Brain and S. Majid, Commun. Math. Phys. **284**, 713 (2008) [math/0701893].
[3] I. Cabrera-Carnero, JHEP **0510** (2005) 071 [hep-th/0503147].
[4] E. Corrigan, D. B. Fairlie, R. G. Yates and P. Goddard, Commun. Math. Phys. **58**, 223 (1978).
[5] A. Dimakis and F. Müller-Hoissen, "A noncommutative version of the nonlinear Schroedinger equation," hep-th/0007015; Czech. J. Phys. **51** (2001) 1285.
[6] A. Dimakis and F. Müller-Hoissen, Phys. Lett. A **278** (2000) 139 [hep-th/0007074].
[7] M. R. Douglas and N. A. Nekrasov, Rev. Mod. Phys. **73** (2002) 977 [hep-th/0106048].
[8] P. Etingof, I. Gelfand and V. Retakh, Math. Res. Lett. **4**, 413 (1997) [q-alg/9701008].
[9] K. Furuuchi, Prog. Theor. Phys. **103** (2000) 1043 [hep-th/9912047];
[10] I. Gelfand and V. Retakh, Funct. Anal. Appl. **25**, 91 (1991); Funct. Anal. Appl. **26**, 231 (1992).
[11] I. Gelfand, S. Gelfand, V. Retakh and R. L. Wilson, Adv. Math. **193**, 56 (2005).
[12] I. M. Gelfand, D. Krob, A. Lascoux, B. Leclerc, V. S. Retakh and J. Y. Thibon, Adv. Math. **112** (1995) 218 [hep-th/9407124].
[13] C. R. Gilson and J. J. C. Nimmo, J. Phys. A **40**, 3839 (2007) [nlin.si/0701027].
[14] C. R. Gilson, M. Hamanaka and J. J. C. Nimmo, Glasgow Mathematical Journal **51A** (2009) 83 [arXiv:0709.2069].
[15] C. R. Gilson, M. Hamanaka and J. J. C. Nimmo, Proc. Roy. Soc. Lond. A **465** (2009) 2613 [arXiv:0812.1222].
[16] M. T. Grisaru, L. Mazzanti, S. Penati and L. Tamassia, JHEP **0404** (2004) 057 [hep-th/0310214].
[17] M. Hamanaka, "Noncommutative solitons and D-branes," Ph. D thesis [hep-th/0303256].
[18] M. Hamanaka, "Noncommutative solitons and integrable systems," Noncommutative Geometry and Physics, (World Sci Publ, 2005) 175 [hep-th/0504001].
[19] M. Hamanaka, J. Math. Phys. **46** (2005) 052701 [hep-th/0311206].
[20] M. Hamanaka, Nucl. Phys. B **741**, 368 (2006). [hep-th/0601209].
[21] M. Hamanaka, JHEP **0702**, 094 (2007) [hep-th/0610006].
[22] M. Hamanaka, "Noncommutative integrable systems and quasideterminants," Nonlinear and Modern Mathematical Physics, AIP Conf. Proc. **1212** (2010) 117 [arXiv:1012.6043].
[23] M. Hamanaka and K. Toda, Phys. Lett. A **316** (2003) 77 [hep-th/0211148].
[24] J. A. Harvey, "Komaba lectures on noncommutative solitons and D-branes," [hep-th/0102076].
[25] R. Hirota (translated by C. R. Gilson, A. Nagai and J. J. C. Nimmo), "The Direct Methods in Soliton Theory," (Cambridge UP, 2004) [ISBN/0521836603].
[26] O. Lechtenfeld, L. Mazzanti, S. Penati, A. D. Popov and L. Tamassia, Nucl. Phys. B **705** (2005) 477 [hep-th/0406065].
[27] K. M. Lee, JHEP **0408** (2004) 054 [hep-th/0405244].
[28] L. J. Mason and N. M. Woodhouse, *Integrability, self-duality, and twistor theory* (Oxford, 1996) [ISBN/0-19-853498-1].
[29] J. E. Moyal, Proc. Cambridge Phil. Soc. **45** (1949) 99; H. J. Groenewold, Physica **12** (1946) 405.

[30] N. Nekrasov and A. Schwarz, Commun. Math. Phys. **198** (1998) 689 [hep-th/9802068].
[31] H. Omori, Y. Maeda, N. Miyazaki and A. Yoshioka, Lett. Math. Phys. **82** (2007) 153.
[32] H. Ooguri and C. Vafa, Nucl. Phys. B **361** (1991) 469; Nucl. Phys. B **367** (1991) 83.
[33] R. J. Szabo, Phys. Rept. **378** (2003) 207 [hep-th/0109162]
[34] K. Takasaki, J. Geom. Phys. **37** (2001) 291 [hep-th/0005194].
[35] N. Wang and M. Wadati, J. Phys. Soc. Jap. **73** (2004) 1689.
[36] R. S. Ward, Phil. Trans. Roy. Soc. Lond. A **315** (1985) 451.

Department of Mathematics, Nagoya University, Nagoya, 4648602, JAPAN
E-mail address: `hamanaka@math.nagoya-u.ac.jp`

Proceedings of Symposia in Pure Mathematics
Volume **85**, 2012

Computing Cohomology on Toric Varieties

Benjamin Jurke

ABSTRACT. In these notes a recently developed technique for the computation of line bundle-valued sheaf cohomology group dimensions on toric varieties is reviewed. The key result is a vanishing theorem for the contributing components which depends on the structure of the Stanley-Reisner ideal generators. A particular focus is placed on the (simplicial) Alexander duality that provides a central tool for the two known proofs of the algorithm.

1. Introduction & Motivation

Cohomology groups play a central role in string model building, where they determine numerous critical properties like the (chiral) zero mode spectrum, Yukawa couplings or counting the number of moduli, for example. The availability of efficient methods to deal with this computational problem is therefore an important requirement.

The majority of geometries considered in theoretical physics and (string) model building is based on toric geometry [**1**–**3**]. Due to the requirement of $\mathcal{N} = 1$ supersymmetry in the effective fourdimensional theory, one encounters compact Calabi-Yau threefolds and fourfolds as the compactification spaces. These are typically constructed as hypersurfaces or complete intersections of hypersurfaces in toric varieties. An additional ingredient in string model building are background fluxes, which are described by vector bundles over the compactification space. Such vector bundles can be constructed via different methods, but for computational practicality three types are distinguished:

- the *monad bundle* construction, which using a short exact sequence constructs a non-trivial vector bundle $\mathcal{V}$ on the toric variety X from two other

2000 *Mathematics Subject Classification.* 14M25 (Primary); 13D45, 14Q99 (Secondary).
Key words and phrases. Cohomology of line bundle; Toric variety; Alexander duality.
Report number. MPP-2011-106.

bundles that are typically chosen to be Whitney sums of line bundles:

(1)
$$0 \longrightarrow \bigoplus_{i=1}^{n} \mathcal{O}_X(a_i) \overset{f}{\hookrightarrow} \bigoplus_{j=1}^{m} \mathcal{O}_X(b_j) \overset{g}{\twoheadrightarrow} \mathcal{V} \longrightarrow 0$$

- the *extension bundle* construction, which is rather similar but in practice computationally often much more difficult to handle:

(2)
$$0 \longrightarrow \bigoplus_{i=1}^{n} \mathcal{O}_X(a_i) \overset{f}{\hookrightarrow} \mathcal{W} \overset{g}{\twoheadrightarrow} \bigoplus_{k=1}^{l} \mathcal{O}_X(c_k) \longrightarrow 0$$

- the *spectral cover* construction, which produces stable holomorphic vector bundles of $SU(n)$ structure group on elliptically-fibered Calabi-Yau threefolds via the Fourier-Mukai transformation from line bundles on top of the so-called spectral cover.

In all three of those constructions line bundles provide the basic building block, thus the line bundle-valued cohomology is necessarily involved. In the setting of toric geometry one can relate the line bundles on a hypersurface (or complete intersection of hypersurfaces) to the line bundles of the ambient space via the Koszul sequence

(3)
$$0 \longrightarrow \mathcal{O}_X(-S) \hookrightarrow \mathcal{O}_X \twoheadrightarrow \mathcal{O}_S \longrightarrow 0.$$

Since short exact sequences of bundles induce long exact sequences of cohomology groups, in the end the critical starting point of every cohomology computation is the knowledge of line bundle-valued cohomology groups on the ambient toric variety.

In [**4**] a novel technique for the computation of the line bundle-valued cohomology group dimensions $h^i(X; \mathcal{L}_X)$ was introduced and later rigorously proven in [**5**, **6**]. Applications like the constructions above and generalizations surpassing the original scope have been in detail discussed in [**7**] and are summarized in [**8**]. A high-performance implementation called **cohomCalg** [**9**] was provided along with the original conjecture of the algorithm and has subsequently been improved and optimized. Note that various alternative approaches have been known for some time, like e.g. §3.5 of [**1**], [**10**], prop. 4.1 in [**11**] or §9.1 of [**2**].

Summary of contents. It is the goal of these notes to constructively follow the basic structures of the proofs [**5**, **6**] and highlight certain mathematical properties of the algorithm. In section 2 the basic notions of toric geometry and the Alexander duality are introduced. Section 3 summarizes the correspondence between sheaves and modules and shows how graded components of local cohomology can be identified with line bundle-valued sheaf cohomology groups. Section 4 shows how certain graded components can be grouped together and defines multiplicity factors. In section 5 the key vanishing result of the algorithm is formulated and section 6 explains an efficient method to compute the multiplicity factors from certain simplicial complexes while highlighting some subleties between the two proofs of the conjecture.

2. Simplicial Alexander Duality and Toric Geometry

Consider a finite vertex set V and an *(abstract) simplicial complex* Δ, which is a set of subsets of V such that for each $\sigma \in \Delta$ the subsets $\tau \subset \sigma$ are contained as well, i.e. $\tau \in \Delta$. In the context of toric geometry Δ can be identified with the toric fan Σ, where the $\sigma \in \Delta \cong \Sigma$ are called *cones* and the statement is then simply that all faces of a cone are themself cones of the simplicial complex. Given a set of vertices $\sigma \in V$ let $\hat{\sigma} := V \setminus \sigma$ denote the complement vertices. For each $\sigma \subset V$ the *restriction* of a simplicial complex Δ on V is defined by $\Delta|_\sigma := \{\tau \in \Delta : \tau \subseteq \sigma\}$.

The *Alexander dual simplicial complex* of Δ on V is then defined by

$$\Delta^* := \{\sigma \subseteq V : \hat{\sigma} \notin \Delta\}, \tag{4}$$

i.e. it consists of all sets of vertices whose complement are not cones of the original simplicial complex. Note that Δ^* itself defines an (abstract) simplicial complex on V. The Alexander dual is a true duality in the sense that $(\Delta^*)^* = \Delta$ and there are examples of self-dual simplices, where $\Delta^* = \Delta$ after a vertex relabeling. The *simplicial Alexander duality* then provides that for each i there exists an isomorphism such that

$$\tilde{H}_i(\Delta^*) \cong \tilde{H}^{|V|-3-i}(\Delta), \tag{5}$$

identifying the (reduced) simplicial homology of the complex with the cohomology of its Alexander dual, which can be treated as a standard combination of Poincaré duality and excision. Several detailed examples can be found in e.g. chap. 5 of [**12**] and further information in [**13**].

Given a cone $\sigma \in \Delta$, we can also define the *link of* σ *in* Δ by

$$\operatorname{link}_\Delta(\sigma) := \{\tau \in \Delta : \tau \cup \sigma \in \Delta,\ \tau \cap \sigma = \emptyset\}. \tag{6}$$

Note that this defines an actual simplicial complex on $\Delta|_{\hat{\sigma}}$. The simplicial Alexander duality (4) can then be restated as

$$\tilde{H}_i\big(\operatorname{link}_{\Delta^*}(\sigma)\big) \cong \tilde{H}^{|V|-|\sigma|-3-i}(\Delta|_{\hat{\sigma}}) \quad \text{for } \sigma \in \Delta^*. \tag{7}$$

The Alexander dual can also be formulated in terms of ideals and squarefree monomials, which in fact offers a more convenient perspective in the context of toric geometry. We consider a d-dimensional simplicial projective toric variety X and a fan Σ in the lattice $N \cong \mathbb{Z}^d$. As mentioned before, one can treat a fan Σ as a simplicial complex Δ and we will do so from this point on. Let $x_1, \dots, x_n$ with $n = |V|$ be homogeneous coordinates associated to the vertices in V that generate the rays (1d cones) of the fan and let $\boldsymbol{x}^\sigma := \prod_{i \in \sigma} x_i$ denote the associated squarefree monomial for some $\sigma \subseteq [n] := \{1, \dots, n\}$ after labeling the vertices via $[n] \cong V$. Then let

$$\mathfrak{m}^\sigma := \langle x_i : i \in \sigma \rangle \tag{8}$$

be the monomial prime ideal corresponding to the vertices σ in the so-called *Cox ring* $S := \mathbb{C}[\boldsymbol{x}] = \mathbb{C}[x_1, \dots, x_n]$ of homogeneous coordinates. Given a monomial ideal $J = \langle \boldsymbol{x}^{\sigma_1}, \dots, \boldsymbol{x}^{\sigma_r} \rangle$ in S, the *Alexander dual monomial ideal* is

$$J^* = \mathfrak{m}^{\sigma_1} \cap \dots \cap \mathfrak{m}^{\sigma_r}. \tag{9}$$

As before let $\hat{\sigma} := [n] \setminus \sigma$ denote the complement of some $\sigma \subseteq [n] \cong V$. Then

$$B_\Sigma := \langle \boldsymbol{x}^\sigma : \hat{\sigma} \in \Sigma \rangle \tag{10}$$

defines the *irrelevant ideal* of the fan Σ, and a minimal generating set for B_Σ is given by the monomials corresponding to the complements of the maximal cones of Σ. The *Stanley-Reisner ideal* on the other hand is defined by

$$I_\Sigma := \langle \boldsymbol{x}^\sigma : \sigma \notin \Sigma \rangle, \tag{11}$$

which—using appropiate identifications—is obviously Alexander dual to the irrelevant ideal by (4):

$$(B_\Sigma)^* \cong I_\Sigma, \qquad (I_\Sigma)^* \cong B_\Sigma. \tag{12}$$

For example, consider the fan of $\mathbb{P}^2$ with the three rays x, y and z. The irrelevant ideal is $B_{\mathbb{P}^2} = \langle x, y, z \rangle$ whereas the Stanley-Reisner ideal is $I_{\mathbb{P}^2} = \langle xyz \rangle$. The Stanley-Reisner ideal and indirectly the associated *Stanley-Reisner ring* S/I_Σ take a central role in our computational technique. Likewise, one can also consider the ring S/B_Σ which in an abuse of the terminology can be treated as the Alexander dual to S/I_Σ based on (12). See [**14**] for further information on the various relationships between (squarefree) monomial ideals and simplicial complexes.

3. Local Cohomology and Grading Reorganization

Consider the Weil divisor class group $\mathrm{Cl}(X) = \mathrm{Div}(X)/\mathrm{Div}_0(X)$ of X. Since we are considering a smooth variety X all Weil divisors are also Cartier,* such that the class group can be identified with the Picard group $\mathrm{Pic}(X) = \mathrm{CDiv}(X)/\mathrm{Div}_0(X)$ and therefore $\mathrm{Cl}(X) \cong \mathrm{Pic}(X) \cong \mathbb{Z}^{n-d}$.

The Cox ring $S = \mathbb{C}[\boldsymbol{x}] = \mathbb{C}[x_1, \dots, x_n]$ of homogeneous coordinates is a $\mathrm{Cl}(X)$-graded ring, i.e. it can be decomposed like

$$S = \bigoplus_{\alpha \in \mathrm{Cl}(X)} S_\alpha, \qquad \text{such that } S_\alpha \cdot S_\beta \subset S_{\alpha+\beta}, \tag{13}$$

and each of those graded decomposition spaces is naturally isomorphic to the space of sections of the line bundle $\mathcal{O}_X(\alpha)$, i.e.

$$S_\alpha \cong \Gamma\big(X; \mathcal{O}_X(\alpha)\big). \tag{14}$$

---

* *Weil divisors* $\mathrm{Div}(X)$ are formal sums of irreducible codimension-1 subvarieties of X. A Weil divisor is called a *Cartier divisor* $\mathrm{CDiv}(X)$ if it is locally principal, i.e. when it can be locally described by the vanishing order and locus of a rational function.

This identification forms the basis of a deeper connection between line bundle-valued cohomology and algebraic notions. Given an S-module M and an ideal $J \subset S$ the *J-torsion submodule* is defined by

$$\Gamma_J(M) := \{\boldsymbol{x} \in M : J^k \cdot \boldsymbol{x} = 0 \text{ for some } k \in \mathbb{N}\}. \tag{15}$$

The *i-th local cohomology* $H^i_J(M)$ of M with support on J is then defined by the i-th cohomology of the complex

$$0 \longrightarrow \Gamma_J(I^0) \longrightarrow \Gamma_J(I^1) \longrightarrow \Gamma_J(I^2) \longrightarrow \dots \tag{16}$$

that is obtained from an injective resolution $0 \longrightarrow I^0 := M \longrightarrow I^1 \longrightarrow \dots$ of the module M, see §9.5 of [**2**] for a detailed introduction of the $\Gamma_J(\,\cdot\,)$ functor.

The local cohomology also inherits any grading of M and the precise connection between line bundle cohomology and local cohomology can then be formulated as

$$H^i\big(X; \mathcal{O}_X(\alpha)\big) \cong H^{i+1}_{B_\Sigma}(S)_\alpha \tag{17}$$

for any divisor class $\alpha \in \mathrm{Cl}(X) \cong \mathrm{Pic}(X)$ and $i \geq 1$,[†] see prop. 2.3 of [**10**] or thm. 9.5.7 in [**2**] for a proof. Computing the i-th line bundle-valued cohomology group of a toric variety therefore is equivalent to computing the $(i+1)$-th cohomology group of the homogeneous coordinate ring S localized on the irrelevant ideal B_Σ, and the line bundle $\mathcal{O}_X(\alpha)$ determines which graded piece has to be considered.

The $\mathrm{Cl}(X)$-grading of S and $H^i_{B_\Sigma}(S)$ can be refined by introducing a $\mathbb{Z}^n$-grading induced from projective weights of the homogeneous coordinates x_i themselves, which in the physics literature are often referred to as the GLSM charges $Q^{(j)}_i$. More precisely, we consider the map

$$\begin{aligned} f : \mathbb{Z}^n &\longrightarrow \mathrm{Cl}(X) \cong \mathbb{Z}^{n-d} \\ \vec{e}_i &\mapsto [D_i] = (Q^{(1)}_i, \dots, Q^{(n-d)}_i), \end{aligned} \tag{18}$$

where $\vec{e}_i \in \mathbb{Z}^n$ is a basis vector associated to the coordinate x_i, such that a monomial $x_1^{k_1} \cdots x_n^{k_n}$ can be simply represented as $k_1\vec{e}_1 + \ldots + k_n\vec{e}_n$. The coordinate divisor D_i refers to the hypersurface $\{x_i = 0\} \subset X$ and $[D_i] \in \mathrm{Cl}(X) \cong \mathrm{Pic}(X)$ to its divisor class. In terms of this finer grading we can then use

$$H^i\big(X; \mathcal{O}_X(\alpha)\big) = \bigoplus_{\substack{\vec{u} \in \mathbb{Z}^n: \\ f(\vec{u}) = \alpha}} H^{i+1}_{B_\Sigma}(S)_{\vec{u}}, \tag{19}$$

which means that we split up $H^{i+1}_{B_\Sigma}(S)_\alpha$ from (17) into $f^{-1}(\alpha) \subset \mathbb{Z}^n$ pieces. In other words: For the grading we are considering all monomials $x_1^{k_1} \cdots x_n^{k_n}$ whose total projective weight/GLSM charge/degree is equal to $\alpha \in \mathrm{Cl}(X)$ that specifies the line bundle $\mathcal{O}_X(\alpha)$. An explicit way to compute this local cohomology $H^i_{B_\Sigma}(S)$ directly via generalized Čech cohomology of a free resolution of the *irrelevant ring*

[†] There seems to be a small error in [**6**] in the argumentation surrounding eqn. 10, but since the computation of $h^0(X; \mathcal{O}_X(\alpha)) = \dim_{\mathbb{C}} \Gamma(X; \mathcal{O}_X(\alpha))$ is equivalent to counting only $|(\alpha, \emptyset)|$ with multiplicity factor 1 (see (24) and (32) below) the end result remains unchanged.

S/B_Σ is summarized in §3 of [**5**], but we will only use it as an intermediate step—however, see the comments at the end of sec. 6 below.

It remains to understand the structure of the splitting (19) better and ideally to group the non-vanishing summands together. The graded pieces $H^{i+1}_{B_\Sigma}(S)_{\vec{u}}$ can also be computed in terms of an abstract simplicial complex on the vertex set/coordinates of the fan: For $\vec{u} \in \mathbb{Z}^n$ let

$$\operatorname{neg}(\vec{u}) := \{k \in V \cong [n] : u_k < 0\} \tag{20}$$

be the set of those indices where $\vec{u}$ has a negative entry and as usual define the complement by $\widehat{\operatorname{neg}}(\vec{u}) := [n] \setminus \operatorname{neg}(\vec{u}) = \{k \in [n] : u_k \geq 0\} \subseteq [n]$. Following prop. 3.1 in [**5**] it can then be shown that

$$H^{i+1}_{B_\Sigma}(S)_{\vec{u}} \cong \tilde{H}_{d-i-1}(\Sigma|_{\widehat{\operatorname{neg}}(\vec{u})}) \tag{21}$$

which in particular implies that $H^i_{B_\Sigma}(S)_{\vec{u}} \cong H^i_{B_\Sigma}(S)_{\vec{v}}$ if $\operatorname{neg}(\vec{u}) = \operatorname{neg}(\vec{v})$. The refined $\mathbb{Z}^n$-grading in (19) can therefore be simplified by collecting the pieces $H^i_{B_\Sigma}(S)_{\vec{u}}$ with the same $\operatorname{neg}(\vec{u})$.

Following the notation introduced in [**6**], given some $\sigma \subseteq [n] \cong V$ define

$$\tilde{\sigma} \in \mathbb{Z}^n \quad \text{such that} \quad \tilde{\sigma}_i := \begin{cases} 1 & \text{if } i \in \sigma, \\ 0 & \text{if } i \notin \sigma, \end{cases} \tag{22}$$

i.e. we have a tupel of 0s or 1s depending on whether or not the respective coordinate index appears in σ. For the computation of the dimension of $H^i\big(X;\mathcal{O}_X(\alpha)\big)$ based on (19) and using $H^i_{B_\Sigma}(S)_{\vec{u}} \cong H^i_{B_\Sigma}(S)_{-\widehat{\operatorname{neg}}(\vec{u})}$ we therefore arrive at

$$h^i\big(X;\mathcal{O}_X(\alpha)\big) = \sum_{\sigma\subseteq[n]} |(\alpha,\sigma)| \cdot h^{i+1}_{B_\Sigma}(S)_{-\tilde{\sigma}} \tag{23}$$

due to (21), where $|(\alpha,\sigma)|$ counts the number of identical $\mathbb{Z}^n$-pieces $H^i_{B_\Sigma}(S)_{\vec{u}}$ via

$$(\alpha,\sigma) := \{\vec{u} \in \mathbb{Z}^n : f(\vec{u}) = \alpha,\ \operatorname{neg}(\vec{u}) = \sigma\}. \tag{24}$$

The potentially infinite sum in (19) has therefore been reduced to a finite sum of 2^n terms, provided that the counting of (24) can be handled efficiently.[‡]

4. Graded Betti Numbers, Resolutions and Multiplicity Factors

In order to restrict the computational effort in (23) further another ingredient is required. We consider the minimal free resolution of a finitely generated graded S-module M, which can be written in the form

$$\mathcal{F}_\bullet : \quad 0 \longleftarrow V_0 \xleftarrow{\phi_1} V_1 \longleftarrow \ldots \longleftarrow V_{\ell-1} \xleftarrow{\phi_\ell} V_\ell \longleftarrow 0 \tag{25}$$

[‡]In the original conjecture [**4**] the factor $|(\alpha,\sigma)|$ appears as the counting of "rationoms"/rational functions which refers to the Laurent monomials. At this point it remains to understand the $h^{i+1}_{B_\Sigma}(S)_{-\tilde{\sigma}}$, which will turn out to be the secondary/remnant cohomology factors along with a further reduction in computational complexity.

and which is *acyclic*, meaning that it is exact everywhere except in the first position where $M \cong V_0/\operatorname{im}(\phi_1)$ holds. With respect to a $\mathbb{Z}^n$-grading the individual spaces V_i of the resolution can be written as

$$V_i = \bigoplus_{\vec{u}\in\mathbb{Z}^n} \left(S_{-\vec{u}}\right)^{\beta_{i,\vec{u}}}, \tag{26}$$

which also defines the *graded Betti numbers* $\beta_{i,\vec{u}}$, see e.g. [**15**] for details. A *minimal* resolution minimizes the ranks of the graded S-modules V_i and is unique up to isomorphisms. Furthermore, any S-module always has a free resolution with length ℓ—assuming that $V_\ell \neq 0$—smaller or equal n.

By treating both the Stanley-Reisner ideal I_Σ and the Stanley-Reisner ring S/I_Σ as $\mathbb{Z}^n$-graded S-modules, one can consider their minimal free resolutions. The associated graded Betti numbers can be calculated by the *Hochster formula*

$$\begin{aligned} \beta_{i-1,\tilde{\sigma}}(I_\Sigma) &= \beta_{i,\tilde{\sigma}}(S/I_\Sigma) \\ &= \dim \tilde{H}^{|\sigma|-i-1}(\Sigma|_\sigma) \overset{(7)}{=} \dim \tilde{H}_{i-2}\big(\operatorname{link}_{\Delta^*}(\hat{\sigma})\big), \end{aligned} \tag{27}$$

see e.g. cor. 5.12 in [**12**], which allows to rewrite (23) in the form

$$h^i\big(X;\mathcal{O}_X(\alpha)\big) = \sum_{\sigma\subseteq[n]} |(\alpha,\sigma)| \cdot \beta_{|\sigma|-i,\tilde{\sigma}}(S/I_\Sigma). \tag{28}$$

However, we still have to evaluate all possible subsets $\sigma \subseteq [n] \cong V$, i.e. 2^n terms, to compute $h^i\big(X;\mathcal{O}_X(\alpha)\big)$.

5. Reductions based on Stanley-Reisner Ideal Generators

The central point of our algorithm [**4**] concerns a further reduction of the number of terms in (28) by taking the Stanley-Reisner ideal I_Σ into account. More precisely, following definition (11) let $I_\Sigma = \langle S_1, \dots, S_t\rangle$ be generated by t square-free monomials $S_r = \boldsymbol{x}^{\sigma_r}$ in the Cox ring S. We basically need to consider all possible subsets of I_Σ-generators and consider the union of coordinates appearing in the generators. In more formal terms: for all $\tau \subseteq [t]$ denote the associated $\mathbb{Z}^n$-degree by

$$\vec{a}_\tau := \deg_{\boldsymbol{x}}(S_\tau) \in \mathbb{Z}^n \quad \text{where} \quad S_\tau := \operatorname{lcm}_{\boldsymbol{x}}\{S_i : i \in \tau\} \tag{29}$$

represents this "union of coordinates"—the least common multiple of several generators with respect to the homogeneous coordinates $x_1, \dots, x_n$ of the Cox ring. Note that $\vec{a}_\tau$ is also a series of 0s and 1s like (22).

Based on this we consider the set of all such $\vec{a}_\tau$, basically the set of all $\mathbb{Z}_n$-degrees that can arise from unions of Stanley-Reisner ideal generators S_i, which will be denoted by

$$\mathcal{P}(I_\Sigma) := \{\vec{a}_\tau : \tau \subseteq [t]\}. \tag{30}$$

Note that different $\tau \subseteq [t]$ may lead to the same $\mathbb{Z}^n$-degree $\vec{a}_\tau$. Coming back to the computation (28), the most important result is that we only have to consider combinations of the coordinates $\sigma \subseteq [n] \cong V$ that are found in $\mathcal{P}(I_\Sigma)$:

- All collections of coordinates that are not unions of coordinates of Stanley-Reisner ideal generators do not contribute to $h^i\big(X;\mathcal{O}_X(\alpha)\big)$, i.e.

$$\boxed{\beta_{r,\tilde{\sigma}}(S/I_\Sigma) = 0 \quad \text{for all} \quad \sigma \subseteq [n] \text{ where } \tilde{\sigma} \notin \mathcal{P}(I_\Sigma).} \tag{31}$$

This leads to the original algorithm formula of [**4**]. Due to a further "Serre duality for graded Betti numbers", that was first observed in [**6**] and proven in [**5**], the same is true for the complement $\hat{\sigma}$ such that the sum in (28) is in the end reduced to

$$\boxed{h^i\big(X;\mathcal{O}_X(\alpha)\big) = \sum_{\substack{\sigma\subseteq[n]\\ \tilde{\sigma},\tilde{\hat{\sigma}}\in\mathcal{P}(I_\Sigma)}} \overbrace{|(\alpha,\sigma)|}^{\text{elements in ``neg-group''}} \cdot \underbrace{\beta_{|\sigma|-i,\tilde{\sigma}}(S/I_\Sigma)}_{\text{multiplicity factor}}} \tag{32}$$

with (α,σ) defined as in (24). Formally (32) always represents a reduction compared to (28) that strongly depends on the form of the Stanley-Reisner ideal I_Σ.

6. Computing Multiplicity Factors

While the counting of the number of neg-group elements $|(\alpha,\sigma)|$ is the well-known task of solving a linear system over the integers, it remains to provide an efficient method for the computation of the graded Betti numbers $\beta_{|\sigma|-i,\tilde{\sigma}}(S/I_\Sigma)$ which have been dubbed multiplicity factors in [**4**–**8**]. This can be done by defining a subcomplex of the full abstract simplicial complex $\Delta_{[t]}$ by taking only those $\tau \subseteq [t]$ whose $\mathbb{Z}^n$-degree $\vec{a}_\tau$ has a certain value. More precisely, for a given $\sigma \subseteq [n]$ define a simplicial (sub)complex by

$$\Gamma^\sigma := \{\tau \subseteq [t] : \vec{a}_\tau = \tilde{\sigma}\}. \tag{33}$$

As mentioned before, different $\tau \subseteq [t]$ share the same $\mathbb{Z}^n$-degree $\vec{a}_\tau$ according to (29), such that the number of elements in τ—the cone dimension—becomes a major distinguishing aspect. Let $F_j(\Gamma^\sigma)$ denote the *j-faces of* Γ^σ, i.e. the $\mathbb{C}$-vector space whose basis vectors e_τ are labeled by the $\tau \in \Gamma^\sigma$ having $|\tau| = j+1$ elements. The complex mappings correspond to the standard boundary mappings in a simplicial complex, specifically by linearly extending

$$\begin{aligned} \phi_j : F_j(\Gamma^\sigma) &\longrightarrow F_{j-1}(\Gamma^\sigma) \\ e_\tau &\mapsto \sum_{i\in\tau} \operatorname{sign}(i,\tau)\, e_{\tau\setminus\{i\}}, \end{aligned} \tag{34}$$

where $\operatorname{sign}(i,\tau) := (-1)^{s-1}$ when i is the s-th element of $\tau \subset [t] = \{1,\ldots,t\}$ written in increasing order. The second central proposal of [**4**] can then be stated as follows:

- The graded Betti numbers $\beta_{r,\tilde{\sigma}}(S/I_\Sigma)$ appearing in (32) can be computed from the reduced homology dimension of the simplicial complex Γ^σ defined in (33), i.e.

$$\boxed{\beta_{r,\tilde{\sigma}}(S/I_\Sigma) = \beta_{r-1,\tilde{\sigma}}(I_\Sigma) = \dim_{\mathbb{C}} \tilde{H}_{r-1}(\Gamma^\sigma).} \tag{35}$$

Due to the basic structure of the simplicial complex Γ^σ the evaluation of this reduced homology is straightforward, thus providing a convenient method to determine the multiplicity factors required in (32).

In [**5**] an analogous computation of the graded Betti numbers/multiplicity factors is carried out using the abstract simplicial complex Λ_I, which is complementary to Γ^σ within the full simplicial complex $\Delta_{[m]}$ of all $\boldsymbol{x}^\sigma$-dividing I_Σ-generators $S_1, \ldots, S_m$, see eqn. 46 in [**6**] for the precise relationship. Moreover, the approach in [**5**] is based on computing the local cohomology $H^i_{B_\Sigma}(S)$ via a free resolution of S/B_Σ, whereas the proof in [**6**] is based on a free resolution of S/I_Σ. From this perspective both proofs [**6**] and [**5**] can be considered to be complementary: while [**5**] is somewhat more concise, the proof in [**6**] by considering the complex Γ^σ more closely follows the original conjecture in [**4**]. The key relation between the two approaches is obviously the Alexander duality according to (12).[§] Also compare §3 of [**16**].

Note that in actual computations one can determine both $\mathcal{P}(I_\Sigma)$ and the relevant simplicial complexes Γ^σ in a single step. However, since it is necessary to evaluate the entire powerset of Stanley-Reisner ideal generators with its 2^t elements, the algorithm complexity—aside from the counting of $|(\alpha,\sigma)|$ and actually computing the reduced cohomology of Γ^σ—grows exponentially with the number of Stanley-Reisner ideal generators. This behaviour is also seen in the reference implementation **cohomCalg** [9]. Compared to the starting point (23) one effectively exchanges an exponential growth in the number of vertices n by an exponential growth in the number of Stanley-Reisner ideal generators t.

When the number of Stanley-Reisner ideal generators becomes large, the algorithm and its implementation reach their limits. Other tools capable of computing line bundle-valued sheaf cohomology group dimensions include the *Sage* system [**17**], the *Macaulay2* [**18**] packages "`NormalToricVarities`" by G. Smith, which is based on §3 of [**16**], or "`ToricVectorBundles`" by R. Birkner, N. O. Ilten and L. Petersen.

[§]It should be mentioned that the generalized Čech complex associated to the resolution of the (Alexander dual) ring S/B_Σ in [**5**] is closely related to the toric Čech complex in eqn. 4.1 of [**11**]: A toric variety can be patched together from charts on

$$U_\sigma := \{(x_1, \ldots, x_n) : x_i = 0 \text{ only for } i \in \sigma\}, \tag{36}$$

i.e. on U_σ one can consider the ring of Laurent monomials $S\left[\frac{1}{\boldsymbol{x}^{\tilde{\sigma}}}\right]$ which allows for negative exponents in those coordinates x_i where $i \notin \sigma$. This coincides with the space of sections of the holomorphic line bundle $\mathcal{O}$ on U_σ.

Acknowledgments. The author would like to thank L. Borisov for some helpful discussions regarding the importance of the Alexander duality in the algorithm and would like to acknowledge the University of Pennsylvania for hospitality during the String Math 2011 conference. The algorithm discussed herein is based on the collaborative efforts together with R. Blumenhagen, T. Rahn and H. Roschy and the author would like to thank T. Rahn for discussions on the manuscript. This work was partially supported by the NFS-Microsoft grant NSF/CCF-1048082.

References

[1] W. Fulton, *Introduction to Toric Varieties.* Princeton University Press, 1993.
[2] D. A. Cox, J. B. Little, and H. Schenck, *Toric Varieties.* Graduate Studies in Mathematics. American Mathematical Society, 2011.
[3] M. Kreuzer, "Toric Geometry and Calabi-Yau Compactifications," *Ukr. J. Phys.* **55** (2010) 613, arXiv:hep-th/0612307.
[4] R. Blumenhagen, B. Jurke, T. Rahn, and H. Roschy, "Cohomology of Line Bundles: A Computational Algorithm," *J. Math. Phys.* **51** (2010) 103525, arXiv:1003.5217 [hep-th].
[5] S.-Y. Jow, "Cohomology of toric line bundles via simplicial Alexander duality," *J. Math. Phys.* **52** (2011) 033506, arXiv:1006.0780 [math.AG].
[6] T. Rahn and H. Roschy, "Cohomology of Line Bundles: Proof of the Algorithm," *J. Math. Phys.* **51** (2010) 103520, arXiv:1006.2392 [hep-th].
[7] R. Blumenhagen, B. Jurke, T. Rahn, and H. Roschy, "Cohomology of Line Bundles: Applications," arXiv:1010.3717 [hep-th].
[8] R. Blumenhagen, B. Jurke, and T. Rahn, "Computational Tools for Cohomology of Toric Varieties," *Adv. High Energy Phys.* **2011** (2011) 152749, arXiv:1104.1187 [hep-th].
[9] "**cohomCalg** package." Download link, 2010. http://wwwth.mppmu.mpg.de/members/blumenha/cohomcalg/. High-performance line bundle cohomology computation based on [**4**].
[10] D. Eisenbud, M. Mustaţă, and M. Stillman, "Cohomology on Toric Varieties and Local Cohomology with Monomial Supports," *J. Symbolic Comput.* **2** (2000) 583–600, arXiv:math/0001159.
[11] L. Borisov, "Towards the Mirror Symmetry for Calabi-Yau Complete intersections in Gorenstein Toric Fano Varieties," arXiv:alg-geom/9310001.
[12] E. Miller and B. Sturmfels, *Combinatorial Commutative Algebra.* Graduate Texts in Mathematics. Springer, 2005.
[13] J. M. Møller, "From singular chains to Alexander Duality," 2010. http://www.math.ku.dk/~moller. Lecture Notes available online.
[14] E. Miller, "Topological Cohen-Macaulay criteria for monomial ideals," in *Combinatorial aspects of commutative algebra*, vol. 502 of *Contemp. Math.*, pp. 137–155. Amer. Math. Soc., Providence, RI, 2009.
[15] W. Bruns, *Cohen-Macaulay Rings.* Cambridge University Press, 1993.
[16] D. Maclagan and G. G. Smith, "Multigraded Castelnuovo-Mumford Regularity," *J. reine angew. Math.* **571** (2004) 179–212, arXiv:math/0305214.
[17] W. A. Stein *et al.*, *Sage Mathematics Software.* The Sage Development Team, 2010. http://www.sagemath.org.
[18] D. R. Grayson and M. E. Stillman, "Macaulay2, a software system for research in algebraic geometry." http://www.math.uiuc.edu/Macaulay2/.

Northeastern University, College of Science, Department of Physics, Dana Research Center, 110 Forsyth Street, Boston, MA 02115, USA

Max-Planck-Institut für Physik, Föhringer Ring 6, 80805 München, Germany

E-mail address: b.jurke@neu.edu, mail@benjaminjurke.net

URL: http://benjaminjurke.net

Proceedings of Symposia in Pure Mathematics
Volume **85**, 2012

Fibrancy of Symplectic Homology in Cotangent Bundles

Thomas Kragh

Abstract. We describe how the result in [**Kra11**] extends to prove the existence of a Serre type spectral sequence converging to the symplectic homology $SH_*(M)$ of an exact Sub-Liouville domain M in a cotangent bundle T^*N. We will define a notion of a fiber-wise symplectic homology $SH_*(M, q)$ for each $q \in N$, which will define a graded local coefficient system on N. The spectral sequence will then have page two isomorphic to homology of N with coefficients in this graded local system.

1. Introduction

Let N be any closed smooth manifold. Let $T^*N \xrightarrow{\pi} N$ denote the cotangent bundle with its canonical projection. We may define the Liouville 1-form λ_N on T^*N by

$$\lambda_{N(q,p)}(v) = p(\pi_*(v)), \qquad q \in N, p \in T_q^*N, v \in T_{(q,p)}(T^*N).$$

It is well-known that $\omega = d\lambda_N$ is non-degenerate and hence a symplectic form on T^*N.

A (finite type) **Liouville domain** $M = (M, \lambda) = (M, \omega, \lambda)$ is a (compact) exact symplectic manifold ($\omega = d\lambda$) such that the symplectic dual vector field X of λ defined by

$$\omega(X, -) = \lambda$$

points outwards at the boundary ∂M. We will only be concerned with compact Liouville domains, so without further warning all such will be of finite type. Note that this condition implies that λ pulled back to the boundary ∂M is a contact form.

Example 1.1. Given a Riemannian structure on N the disc bundle

$$D_R T^*N = \{(q, p) \in T^*N \mid \|p\| \leq R\}$$

of any radius $R > 0$ is a Liouville domain with the 1-form λ_N defined above.

1991 *Mathematics Subject Classification.* Primary 53D05, 53D35; Secondary 70G40.
Key words and phrases. Symplectic geometry, Liouville domains, symplectic homology.
The author was supported by the Carlsberg Foundation under grant numbers 2009_01_0679 and 2010_01_0496.

An **Exact Liouville embedding** $(M', \omega', \lambda') \subset (M, \omega, \lambda)$ is an embedding $M' \subset M$ of concurrent dimensions, such that $\omega' = \omega_{|M'}$ and there exists a smooth function

$$f \colon M' \to \mathbb{R}$$

such that $\lambda_{|M'} - \lambda' = df$.

EXAMPLE 1.2. Any closed exact Lagrangian sub-manifold $L \subset D_R T^*N$ not intersecting the boundary of $D_R T^*N$ has by the Darboux-Weinstein theorem an extension to an exact Liouville embedding $D_\epsilon T^*L \subset D_R T^*N$ for some small ϵ and some Riemannian structure on L.

In Section 2 we will define symplectic homology $SH_*(M)$ of a Liouville domain M. In Section 3 We will describe a fiber-wise version of this. I.e. a symplectic homology depending on $q \in N$ denoted $SH_*(M, q)$. These will define a local coefficient system $SH_*(M, \bullet)$ on N and the following theorem may be thought of as a type of Serre spectral sequence for symplectic homology.

THEOREM 1.3. *For any exact Liouville embedding $(M, \lambda) \subset (D_R T^*N, \lambda_N)$ there is a spectral sequence $(E^r_{*,*}, d_r)_{r\geq 1}$ such that*

- *it strongly converges to a filtered quotient of $SH_*(M)$ and*
- *the second page term $E^2_{n,m}$ is isomorphic to $H_n(N; SH_m(M, \bullet))$.*

As will be evident from the construction in Section 3 the local system $SH_*(M, \bullet)$ is trivial if $M \xrightarrow{\pi} N$ is not surjective. This implies that $SH_*(M)$ is trivial, and thus generalizes the fact that any exact Lagrangian $L \subset T^*N$ surjects to N (see [**LS91**]). Indeed, in the case of $D_\epsilon T^*L$ the symplectic homology never vanishes because it is loop space homology (see [**Vit97**]) with possibly twisted coefficients (see [**Kra11**]). In [**Kra11**] we also considered products in the case of $D_\epsilon T^*L$ to get the strong result that exact Lagrangians are up to a finite covering space lift a homology equivalence. However, for simplicity we will not consider these extra structures here.

2. Symplectic Homology of $M \subset T^*N$

In this section we describe the symplectic homology of a Liouville domain M exact embedded in T^*N. The precise setup used in this section is important for understanding the construction in Section 3. One can define symplectic homology independent of the fact that M is in T^*N, but for the purpose of proving Theorem 1.3 the following explicit construction is convenient. The fact that this actually defines symplectic homology is known as localization (see e.g. [**Vit99**]).

We will need to consider Floer homology (infinite dimensional Morse homology) of restrictions of the action integral

$$A(\gamma) = \int_\gamma \lambda_N - H dt, \tag{2.1}$$

where $H \colon T^*N \to \mathbb{R}$ is a smooth Hamiltonian and $\gamma \colon I \to T^*N$ is a path. In this section, we will only consider closed loops $\gamma \colon S^1 \to T^*N$, and we denote the action restricted to these by A^Λ. It is well-known that the critical points of A^Λ are in 1-1 correspondence with closed time-1 periodic orbits of the Hamiltonian flow.

The symplectic homology will be defined as a limit of Floer homologies of A^Λ_s associated to a sequence of Hamiltonians H^s. We will define these rather explicitly.

So let

$$W = (1-\epsilon, 1] \times \partial M \subset M \subset T^*N$$

denote a symplectic collar neighborhood of $\partial M \subset M$. I.e. the 1-form λ is given by $r\lambda_{|\partial M}$, where $\lambda_{|\partial M}$ denotes the pull back of λ to the boundary, and r is the coordinate in $(1-\epsilon, 1]$. We will only need to consider Hamiltonians H^s which are locally constant away from W and which only depend on r on W.

Choose any smooth function $f\colon (1-\epsilon, \infty) \to \mathbb{R}$ which is concave such that $f(r)$ tends to $-\infty$ as r tends to $1-\epsilon$ and $f(r)$ is 0 for $r \geq 1$. We then define a smooth family of smooth functions $f_s\colon \mathbb{R}_{\geq 0} \to \mathbb{R}$ for $s > 1$ such that

- $f_s(r) = f(r) + s$ when $f(r) \geq -s/2$,
- f_s'' has a unique 0 in $(1-\epsilon, 1)$, and
- $f_s(r) = 0$ when $0 \leq r < (1-\epsilon)$.

Notice that these imply that f_s is convex on the interval from 0 to the unique 0 (in

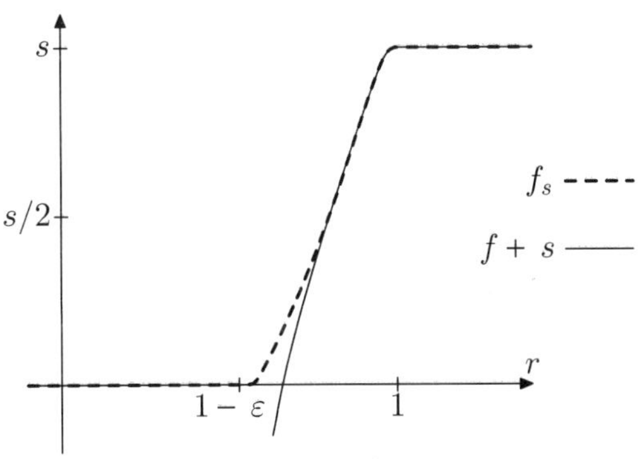

FIGURE 1. Functions f_s and $f + s$.

the second bullet point) and concave on the complement. As illustrated in Figure 1 this describes a controlled way to "cap of" $f+s$ making it smooth, bounded, non-negative, and more we will need later. We then define the smooth family of smooth Hamiltonians H^s by

$$H^s(z) = \begin{cases} f_s(r) & z = (r, x) \in W \\ 0 & z \in M - W \\ s & z \in T^*N - M \end{cases}$$

By construction the Hamiltonian flow of H^s has only constant periodic orbits outside of W and hence the action on closed loops not in W only has critical values 0 and $-s$.

There is a geometric interpretation of the action of other periodic orbits. Indeed, since we only consider loops in this section and the embedding $M \subset T^*N$ is exact we may as well integrate over λ and not λ_N in Equation (2.1). Also, if $\gamma\colon S^1 \to \mathbb{R}$ is a Hamiltonian flow curve it has to have constant r-factor. Indeed, the Hamiltonian is preserved under Hamiltonian flow. Considering variations of this r-factor (knowing that γ is a critical point for A^Λ) yields the relation

$$\int_\gamma \lambda_{|\partial M} = f_s'(r), \tag{2.2}$$

and thus one calculates that

$$A_s^\Lambda(\gamma) = rf_s'(r) - f_s(r) = -(f_s(r) - rf_s'(r)),$$

which is minus the intersection of the 2. axis with the tangent of f_s at the point $(r, f_s(r))$.

Now pick $s_0 >> 1$ large enough such that; the tangent of $f + s_0$ at the point where $f + s_0 = s_0/2$ intersects the 2. axis below zero. This and the assumptions on f_s imply that for $s > s_0$ we thus have; any tangent of f_s intersecting the 2. axis above 0 must be a tangent on the part where $f_s = f + s$ (and $f > -s/2$).

This means that any 1-periodic orbit with negative action must lie on an r-level for which $f_s(r) = f(r) + s$ (and also for r's close to it). So when increasing s we simply increase f_s by the same value in a neighborhood. This implies that the 1-periodic orbit is unchanged and the action is simply translated down with the same speed as s is translated up. Thus the critical set of A_s^Λ has its critical values *below* zero pushed downwards with the same speed as s increases.

This means that for any $a < 0$ we in fact have continuation maps to higher and higher s and may in fact define the limit

$$SH_*(M) = \operatorname*{colim}_{s\to\infty} FH_*^a(A_s^\Lambda), \tag{2.3}$$

where $FH_*^a(A_s^\Lambda)$ denotes Floer homology associated to critical values of A_s^Λ with critical value greater than a. Indeed, sliding a critical value below a corresponds to collapsing a generator in the Floer chain complex, which is always a chain map and this is why the continuation maps exist. Defining this colimit in the standard way requires picking a sequence of s's tending to infinity and perturbing the associated actions (and some compactness proofs). However, to prove Theorem 1.3 we will not use standard Floer theory. We will instead use finite dimensional approximations. The advantages of this are addressed in Section 4. Note that the definition of this limit does not depend on $a < 0$ nor the choice of sequence s going to infinity. Indeed, all critical points with critical values in $]-\infty, 0[$ for some s will eventually get their critical values translated down below any given value $a < 0$.

3. Fiber-wise Symplectic Homology

In this section we elaborate and extend the calculation of the action of 1-periodic orbits to open flow curves. We then use this to define a fiber-wise version of the symplectic homology.

Let $\gamma\colon I \to T^*N$ be any time-1 flow path for the Hamiltonian flow of H^s. The argument proving Equation (2.2) can be extended to this case using bump functions in the variable r. I.e. the equation is still valid. However, when using this to calculate A_s we get a correction term due to the fact that integrating λ and λ_N on an open path does not give the same result. Indeed, let $f\colon M \to \mathbb{R}$ be such that $df = \lambda_N - \lambda$ then

$$A_s(\gamma) = \int_\gamma \lambda_N - H^s det = \int_\gamma \lambda - H^s dt + f(\gamma(1)) - f(\gamma(0)).$$

We conclude that the geometric interpretation using intersections of tangents of f_s with the 2. axis is still valid except we have to add the term $f(\gamma(1)) - f(\gamma(0))$.

Now fix a $q \in N$ and look at the action integrals A_s^q defined on paths γ starting and ending on the Lagrangian fiber T_q^*N. It is well-known that the critical points of A_s^q are time-1 Hamiltonian flow curves starting and ending on T_q^*N. This means that we can in fact conclude that if $a < \max_{z,z' \in M} |f(z) - f(z')|$ then the critical set of A^q_s below a slides down when s is increased. Indeed, as in the previous section the time-1 flow curves do not change but their actions are translated along with s. So we may define

$$SH_*(M,q) = \operatorname*{colim}_{s\to\infty} FH^a_*(A^q_s). \tag{3.1}$$

This is what we will consider the fiber-wise symplectic homology of M. By definition we see that if $M \to N$ does not surject there must be a fiber on which H^s is constantly equal to s. This implies that for this fiber the fiber-wise symplectic homology is trivial, and thus by Theorem 1.3 all of them are trivial including the global symplectic homology $SH_*(M)$.

4. Fibrancy and Sketch of Proof of Theorem 1.3

The proof of Theorem 1.3 relies on the fact that the family of fiber-wise symplectic homologies actual looks and behaves like a Serre fibration with total space giving the total symplectic homology $SH_*(M)$. To prove this using homological algebra and infinite dimensional Floer theory seem very delicate and complicated - especially in light of transversality issues. So this is not the approach taken in [**Kra11**]. Indeed, there we use finite dimensional approximations.

However, to see why the fiber-wise symplectic homologies form a local system and understand the construction it is convenient to discuss why a smooth path $\gamma\colon [0,l] \to N$ (assumed to be parametrized by arc-length) can induce a "parallel transport" in the fiber-wise homologies. So in the following we describe why such a path gives rise to a map between the fiber-wise symplectic homologies

$$SH_*(M,\gamma(0)) \to SH_*(M,\gamma(l)). \tag{4.1}$$

First consider any $s >> 1$ fixed. Then for each $v \in [0,l]$ we have the fiber-wise action functional $A^{\gamma(v)}_s$ defined using the fiber $T^*_{\gamma(v)}N$ and one may consider the "graph" or **bifurcation diagram** of the critical set of each as a multi-valued function of v. Examples are illustrated in Figure 2. We know from the discussion

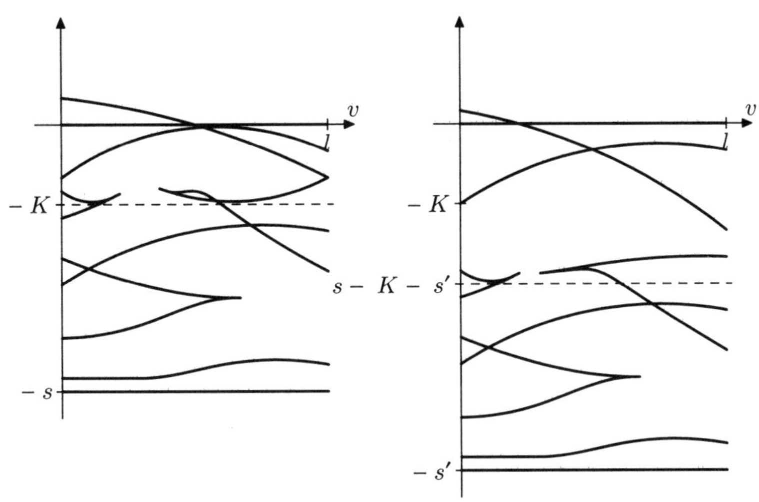

FIGURE 2. Bifurcation diagrams for same path but different s.

in the previous section that the dependence on s of this diagram is such that everything below some $-K$ is simply translated downwards with the same speed as s is translated upwards. Figure 2 shows two possible bifurcation diagrams for the same system and path γ, but for different s's. The part above $-K$ can behave arbitrarily as s changes except there is always an upper bound. This made the colimits of Floer homologies in the previous sections well-defined.

The main point is that the slopes of the "graph" pieces in the bifurcation diagrams are numerically bounded by a certain number. Indeed, Lemma 9.2 in [**Kra11**] gives this bound to be 2 if $M \subset DT^*N$, which can always be arranged by scaling M at the very beginning. Note that the proof there depends on having a Hamiltonian which have a slight slope at infinity, but in this heuristical argument we don't care about the constant critical points outside of M since we simply assume s very large so that they do not contribute to the Floer homology (i.e. the critical value $-s$ is much smaller than any $a < -K$ chosen as the energy cut-off).

The effect of such a bound is that if we make s actually depend on v and grow faster than $2v$, say $s(v) = s_0 + 3v$, then the bifurcation diagram depending on v becomes slanted. Indeed, the slopes of any critical values with value less than $-K$ have to lie in the interval $[-5, -1] = [-2, 2] - 3$. The upshot is that whenever a critical value of $A_{s(v)}^{\gamma(v)}$ co-insides with a it must go down (as a function of v) with positive speed. We are thus again collapsing generators in the Floer chain complex and have maps similar to the continuation maps described in the previous sections. I.e. we get maps

$$FH_*^a(A_{s_0}^{\gamma(0)}) \to FH_*^a(A_{s_0+3l}^{\gamma(l)}),$$

which when taking the limit as s_0 tends to infinity defines the wanted map from Equation (4.1). Usual zig-zagging arguments can be employed to prove that this is an isomorphism, but there are of course many things to check here, and proceeding using standard Floer theory seems very cumbersome.

The proof of Theorem 1.3 is carried out in [**Kra11**] using the theory of Conley indices to produce spaces instead of homology theories. I.e. the spaces have the wanted homologies (with twisted coefficients - but one can always use the appropriate twisted coefficients to make it coincide with the symplectic homology). This makes it possible to define these fiber-wise symplectic homologies even when a from Equation (3.1) is not regular for a particular fiber T_q^*N and some s. Indeed, this is needed because there is no guarantee that we can find a sequence of s's tending to infinity having a regular for all fibers simultaneously. In fact, in many cases such an a can be proven not to exist.

Working with spaces also has the advantage that it is possible to prove a Serre type fibration property in a more conventional manner, and thus lifting more than simply a path parametrized by arc length. Indeed, one may lift any compact family of smooth paths. It is important here to mention a technical but significant complication; when increasing s these finite dimensional approximations get more and more complicated and suspensions of the Conley indices need to be introduced. This effectively means that the colimits analogous to those in Equation (2.3) and Equation (3.1) are not taken in the category of spaces, but in the category of spectra - and in the fiber-wise case even in a category of parametrized spectra over N. However, the notions of Serre fibrancy, taking fiber-wise homology, and taking global homology still exist. So we, indeed, get a Serre type spectral sequence as in Theorem 1.3. We should note, however, that in [**Kra11**] only the case of $M = DT^*L$ is considered, but all the ideas generalize, and using the Hamiltonians defined here all the methods to get the spectral sequence immediately apply. However, to incorporate the products, which are very important in [**Kra11**] one needs to construct the Hamiltonian family H^s more carefully.

References

[Kra11] Thomas Kragh, *Parametrized Ring-Spectra and the Nearby Lagrangian Conjecture*, arXiv:math/1107.4674 (2011), Submitted to Geometry & Topology.

[LS91] François Lalonde and Jean-Claude Sikorav, *Sous-variétés lagrangiennes et lagrangiennes exactes des fibrés cotangents*, Comment. Math. Helv. **66** (1991), no. 1, 18–33. MR MR1090163 (92f:58060)

[Vit97] Claude Viterbo, *Exact Lagrange submanifolds, periodic orbits and the cohomology of free loop spaces*, J. Differential Geom. **47** (1997), no. 3, 420–468. MR MR1617648 (99e:58080)

[Vit99] C. Viterbo, *Functors and computations in Floer homology with applications. I*, Geom. Funct. Anal. **9** (1999), no. 5, 985–1033. MR 1726235 (2000j:53115)

Department of Mathematics, Massachusetts Institute of Technology, Cambridge, Massachusetts 02139

E-mail address: tkragh@math.mit.edu

Proceedings of Symposia in Pure Mathematics
Volume **85**, 2012

Curved String topology and Tangential Fukaya Categories

Daniel Pomerleano

1. Introduction

In this paper, we construct new examples of two-dimensional TQFTs over the prop $C_*(\overline{\mathcal{M}_{g,n}})$. Our primary methods are algebraic: we make use of the well known theorem of Kontsevich and Soibelman [**KonSoi**] that

Given a compact and smooth $\mathbb{Z}/2\mathbb{Z}$ graded Calabi-Yau A_∞ algebra B for which the Hodge to De-Rham spectral sequence degenerates, a choice of splitting for this spectral sequence gives rise to a TQFT

For a compact, smooth, Calabi-Yau variety, a (dg-version of) the derived category of quasicoherent sheaves $QCoh(\mathcal{X})$ satisfes all of the above conditions. Homological Mirror Symmetry [**Kon**] predicts that the associated TQFT is expected to be equivalent to Gromov-Witten TQFT on the mirror CY variety $\mathcal{X}^\vee$. Now consider $\mathcal{Y}$ to be a smooth but non-compact Calabi-Yau variety. Then $QCoh(\mathcal{Y})$ is a non-compact Calabi-Yau category, and by a modified version of the theorem of Kontsevich and Soibelman, we can get a so-called *positive-output* TQFT. The Landau-Ginzburg model uses deformation theory to compactify these theories by deforming the above category by a superpotential w, which is an algebraic function with a proper critical set. Recent work [**Pre, LinPom**] shows that this gives rise to a TQFT.

Similarly, there is a positive output TQFT called *string topology* for a compact oriented manifold $\mathcal{Q}$ associated to the dg-category of dg-modules $D(C_*(\Omega\mathcal{Q}))$ over the dg algebra $C_*(\Omega\mathcal{Q})$ [**Lur**], where $\Omega\mathcal{Q}$ denotes the based loop space of $\mathcal{Q}$ at some arbitrary point. Throughout this paper, all coefficients are taken to be $\mathbb{C}$, the field of complex numbers. As we explain below, this category is a smooth but not compact category. The relationship with string topology is revealed by the following calculation for the Hochschild homology:

$$\mathbf{HH}_*(C_*(\Omega\mathcal{Q})) \cong C_*(\mathcal{L}\mathcal{Q})$$

There is a natural compact CY category associated to such a manifold, the category of perfect modules over $C^*(\mathcal{Q})$, which however is not smooth. Such categories give rise to TQFT's with *positive-input.* When $\mathcal{Q}$ is simply connected, these two algebras are related via Koszul duality. Namely, the following isomorphisms hold:

$$\mathbb{R}Hom_{C^*(\mathcal{Q})}(\mathbb{C},\mathbb{C}) \cong C_*(\Omega\mathcal{Q})$$
$$C^*(\mathcal{Q}) \cong \mathbb{R}Hom_{C_*(\Omega\mathcal{Q})}(\mathbb{C},\mathbb{C})$$

and in fact this gives rise to fully faithful functors:

$$perf(C_*(\Omega\mathcal{Q}) \to D(C^*(\mathcal{Q}))$$

and

$$perf(C^*(\mathcal{Q})^{op}) \to D(C_*(\Omega\mathcal{Q})^{op})$$

Here $perf(C_*(\Omega\mathcal{Q})$ or $perf(C^*(\mathcal{Q})^{op})$ denotes the subcategory of perfect modules, which is defined for the reader below. Nevertheless, $\mathbb{C}$ is not a compact generator in the category $D(C_*(\Omega\mathcal{Q}))$ which means that Koszul duality does not give rise to an equivalence of the full derived categories. The starting point for this work is that if $\mathcal{Q}$ is $T^n = S^1 \times S^1 \times \cdots \times S^1$, Dyckerhoff [**Dyc**] proved the following theorem:

THEOREM 1.1. *Let w be a function on $\mathbb{C}[[x_1, x_2, \ldots, x_n]]$ with isolated singularities. The object $\mathbb{C}$ is a compact generator for $MF(\mathbb{C}[[x_1, x_2, \ldots, x_n]], w)$. Otherwise stated, $Hom_{MF(\mathbb{C}[[x_1,x_2,\ldots,x_n]],w)}(\mathbb{C}, -)$ defines an equivalence of categories:*

$$MF(\mathbb{C}[[x_1, x_2, \ldots, x_n]], w) \to D(Hom_{MF(\mathbb{C}[[x_1,x_2,\ldots,x_n]],w)}(\mathbb{C},\mathbb{C}))$$

Here MF denotes the category of matrix factorizations, whose definition occupies much of section 2. The relationship between this theorem and the previous discussion is that $C_*(\Omega T^n)$ is isomorphic to $\mathbb{C}[z_1, z_1^{-1}, z_2, z_2^{-1}, \ldots z_n, z_n^{-1}]$, the Laurent polynomial ring in several variables. As $T^n = S^1 \times S^1 \ldots \times S^1$ is not simply connected, we complete at the augmentation ideal of this ring to obtain $\mathbb{C}[[x_1, x_2, \ldots, x_n]]$. In such cases, $MF(\mathbb{C}[[x_1, x_2, \ldots, x_n]], w)$ defines a quantum field theory. This result can be viewed as a deformed Koszul duality in the sense that $Hom_{MF(\mathbb{C}[[x_1,x_2,\ldots,x_n]],w)}(\mathbb{C},\mathbb{C}) \cong H^*(T^n)$ with a deformed A_∞ structure m.

In this paper, we will consider simply connected manifolds $\mathcal{Q}$ whose minimal models are pure Sullivan algebras (again we will review this terminology). The first part of our paper makes precise and then gives an answer to the following question:

QUESTION 1.2. *If $C^*(\mathcal{Q})$ is a pure Sullivan algebra and given an element $w \in Z(C_*(\Omega\mathcal{Q}))$, when is $\mathbb{C}$ a compact generator of $MF(C_*(\Omega\mathcal{Q}), w)$ defining an equivalence with $D(H^*(C^*(\mathcal{Q})), m)$?*

We will examine our condition in the special case that the differential of our pure Sullivan algebra is quadratic. As mentioned earlier, morally, one can think of a potential w as "compactifying" the field theory. In the final section, inspired by a program of [**Sei**], we explain how the simplest of our theories, such as when $\mathcal{Q} = \mathbb{C}P^n$ can be interpreted as geometric compactifications of the cotangent bundle T^*CP^n inside of a certain root stack. For the latest update in the relationship between Fukaya categories of T^*CP^n and string topology, the reader should consult [**Abou**].

This paper is a summary of a short talk given in June 2011 at the String-Math conference. A forthcoming paper [**Pom**] will develop further the ideas discussed herein while developing some singularity theory that is suggested by analogy with the commutative case. The author would like to thank his advisor Constantin Teleman for suggesting the possibility of transporting ideas from the Landau-Ginzburg model to String topology as well as for his support and guidance throughout this project. The author would also like to thank Mohammmed Abouzaid, Denis Auroux for help with the symplectic geometry and Toly Preygel for teaching me about curved algebras. I have learned a lot of what I know so far about this subject from them.

2. Background and Algebraic Setup

Recall that a dg-module (or A_∞-module) N over a dg-algebra A (or A_∞-algebra) is *perfect* if it is contained in the smallest idempotent-closed triangulated subcategory of $Ho(A)$ generated by A.

DEFINITION 2.1. A dg-algebra A over $\mathbb{C}$ is *compact* if A is perfect as a $\mathbb{C}$ module (in this special case this simply says that A is equivalent to a finite dimensional vector space). A dg-algebra A is *smooth* if A is perfect as an $A - A$ bimodule.

A very useful criterion for smoothness is given by the notion of finite-type of Toen and Vaquie [**ToëVaq**].

DEFINITION 2.2. A dg-algebra A is of finite type if it is a homotopy retract in the homotopy category of dg-algebras of a free algebra $(\mathbb{C}\langle v_1, v_2, \ldots, v_n\rangle, d)$ with $dv_j \in \mathbb{C}\langle v_1, v_2, \ldots, v_{j-1}\rangle$

LEMMA 2.3. *If A is of finite type then A is smooth. The converse is also true if A is assumed to be compact.*

LEMMA 2.4. *With the notation of the previous section, the dg-algebra $C_*(\Omega \mathcal{Q})$ is smooth.*

In the simply connected case, this follows from the classical Adams-Hilton construction [**AdaHil**] and the above theorem of Toën-Vaquie.

3. Pure Sullivan algebras and Curved algebras

We consider Pure Sullivan dg-algebras $\mathcal{B}$ of the form:

$$(\wedge V, d) = (\mathbb{C}[x_1, ...x_n] \otimes \bigwedge(\beta_1, ...\beta_m), d(\beta_i) = f_i(x_1, \ldots, x_n), d(x_j) = 0)$$

where the $deg(x_i)$ are even and negative, the functions f_i have no linear term, and the $deg(\beta_i)$ are odd > 1. We further assume that $dim(H^*(\mathcal{B})) < \infty$.

From a field-theoretic point of view, it is important to note that $\mathcal{B}$ is in particular elliptic and hence $H^*(\mathcal{B})$ is a Poincare duality algebra [**FelHalTho**]. Because the deformation theory of C_∞ algebras and Frobenius C_∞ algebras is known to coincide, $\mathcal{B}$ has a natural Calabi-Yau structure. The general theory of Koszul duality in turn implies that its Koszul dual $\mathcal{A}$ has a non-compact Calabi-Yau structure.

The above cochain algebras determines canonically an L_∞ model, $\mathfrak{g} = \pi_*(\Omega(\mathcal{M}))\otimes \mathbb{C}$ for the space $\mathcal{Q}$. Using the homological perturbation lemma, we have an explicit A_∞ model for $\mathcal{A} = C_*(\Omega \mathcal{Q})$ of the form

$$(Sym(\mathfrak{g}_{even}) \otimes \Lambda(\mathfrak{g}_{odd}), m)$$

A formula for the higher multiplications appears in [**Bar**], but the key facts are as follows. First, the strict morphism of the abelian Lie algebra $\pi_{even}(\Omega(Q)) \to \mathfrak{g}$ corresponds to the inclusion of $Sym(\mathfrak{g}_{even}) \cong \mathbb{C}[u_1, \ldots, u_m] \to \mathcal{A}$. The higher multiplications m_n are multi-linear in these variables for $n \geq 3$. Finally, the A_∞ algebra is strictly unital and the augmentation $U\mathfrak{g} \to \mathbb{C}$ is also a strict morphism.

The reader should be warned that in the presence of quadratic terms in the f_i, the above identification with $Sym(\mathfrak{g}_{even})\otimes\Lambda(\mathfrak{g}_{odd})$ is only an identification of vector spaces. In other words, there can be a non-trivial Lie bracket $B : \mathfrak{g}_{odd} \otimes \mathfrak{g}_{odd} \to \mathfrak{g}_{even}$, which means that forgetting higher products, $U\mathfrak{g}$ is a Clifford algebra over $Sym(\mathfrak{g}_{even})$. It also seems worth pointing out that the even variables u_i can be thought of as being Koszul dual to the odd variables β_i. Meanwhile the variables in $\mathfrak{g}_{odd}$, from here on denoted as e_j, are dual to the even variables x_j above.

For example, if $\mathcal{Q} = \mathbb{C}P^n$, we have the following specific model:

$$U(\mathfrak{g}) = \mathbb{C}[u] \otimes \Lambda(e), m_{n+1}(e, e, e, \ldots, e) = u$$

We can then consider potentials of the form $w = u^d$.

Next, we discuss how to define an appropriate category of matrix factorizations. This section adopts the ideas of the foundational work [**Pre**] to our non-commutative context. For concreteness, let us consider as before the above A_∞-algebra $\mathcal{A}$, and an element $w \in \mathbb{C}[u_1, \dots, u_m]$ of degree $2j-2$. We define a variable x of degree $2j-2$. The element w defines a mapping from

$$w : \mathbb{C}[x] \to \mathcal{A}$$

, and we can consider the A_∞ algebra $\mathcal{A}_0 = (\mathcal{A}[e], de = w)$, where e now has degree $2j-1$.

DEFINITION 3.1. We define $Pre(MF(\mathcal{A}, w))$, to be the full subcategory of $mod(\mathcal{A}_0)$ consisting of modules which are perfect over $\mathcal{A}$.

This category has a natural $\mathbb{C}[[t]]$(degree $t = -2n$) linear structure because it is acted on by the $\mathbb{C}$-finite modules $D_{fin}(\mathbb{C}[e]/e^2)$. By Koszul duality, this latter category is equivalent to the category $perf(C[[t]])$.

There is also a deformation theoretic interpretation of the above action. We note that the element tw also defines a Maurer-Cartan element in $\mathbf{HH}^*(\mathcal{A}, \mathcal{A})[[t]]$. Such a Maurer-Cartan solution allows us to twist the differential on

$$(\bigoplus_n \mathcal{A}^{\otimes n}[[t]], d_{\mathcal{A}})$$

by the differential determined by the formula:

$$td_w : a_0 \otimes a_1 \otimes \cdots \otimes a_n \mapsto \sum_{i=0}^{n-1} (-1)^{i+1} t a_0 \otimes a_1 \otimes \cdots \otimes a_i \otimes W \otimes a_{i+1} \otimes \cdots \otimes a_n.$$

giving rise to a topological coalgebra

$$\mathbf{C} = (\bigoplus \mathcal{A}^{\otimes n}[[t]], d_{\mathcal{A}} + td_w)$$

.

LEMMA 3.2. *The functor* $M \to ((\bigoplus B^i(\mathcal{A}) \otimes M)[[t]], d_{M/A} + te\wedge)$ *defines a fully faithful functor:*

$$Pre(MF(\mathcal{A}, w)) \to D(\mathbf{C} - comod)$$

Finally, we define

$$\begin{aligned} MF(\mathcal{A}, w) = Pre(MF(\mathcal{A}, w)) \otimes_{\mathbb{C}[[t]]} \mathbb{C}((t)) &\cong Pre(MF(\mathcal{A}, w))/Perf(\mathcal{A}_0) \\ &\cong D(\mathbf{C} - comod) \otimes_{\mathbb{C}[[t]]} \mathbb{C}((t)) \end{aligned}$$

It is often convenient to work with the formal Ind-completion $Ind(MF(\mathcal{A}, w))$ which we shall denote by $MF^{\infty}(\mathcal{A}, w)$.

We have constructed a category of curved modules for a curved A_{∞} algebra which arises as a deformation of an uncurved A_{∞} algebra. It is worth pointing out that there is a more general notion of a curved A_{∞} algebra, a notion which is most developed in the case of dg-algebras.

DEFINITION 3.3. A triple $B = (A, w, d)$ consisting of a $\mathbb{Z}/2\mathbb{Z}$ graded algebra A, a function of even degree w, and an derivation d of odd degree is called a graded curved dg-algebra if $d^2 = [w, a]$

DEFINITION 3.4. A (left) curved module over a curved dg-algebra is a $\mathbb{Z}/2\mathbb{Z}$ graded (left) module over A together with an odd derivation d such that $d^2 = w$.

There is a $\mathbb{Z}/2\mathbb{Z}$ graded dg-category of modules, which we denote by $B - mod$. Positselski has studied curved Koszul duality extensively and in particular defined various versions of the derived category of curved modules over a curved dg-algebra. In particular, he considers:

DEFINITION 3.5. We denote by $B - proj$ the $\mathbb{Z}/2\mathbb{Z}$ graded dg-subcategory of $B - mod$ consisting of modules M whose underlying graded modules $M^{\sharp}$ are projective. We define $Ho(B - proj)$ to be its homotopy category.

In many cases of interest, $B - proj$ coincides with the categories defined previously and is sometimes convenient to work with. In the case that A is a commutative ring and w is a non-zero function, $B - proj$ is nothing but the usual category of matrix factorizations.

For the case of a general curved A_{∞} algebra, it is a bit unclear how to construct an interesting triangulated category of modules. One possible definition is to consider topological modules over the completed bar coalgebra $\prod B^i(A)$. This amounts to considering only those modules such that $l_n : A^{\otimes n} \otimes M \to M$ vanish in sufficiently high degree and further imposing the analogous condition for morphisms between two modules [**Pos**].

4. The criterion for properness and coformal $\mathcal{Q}$

In this section we discuss a criterion for smoothness and properness of the category $MF(\mathcal{A}, w)$. To state the criterion, we must consider the category of curved bimodules

$$MF(\mathcal{A} \otimes \mathcal{A}^{op}, w \otimes 1 - 1 \otimes w)$$

and we define $\mathbf{HH}^*(MF(\mathcal{A},w))$ to be $Hom_{MF(\mathcal{A}\otimes\mathcal{A}^{op},w\otimes 1-1\otimes w)}(\mathcal{A},\mathcal{A})$. Using either description of our category, this can be computed explicitly as:

$$\mathbf{HH}^*(MF(\mathcal{A},w)) \cong (\mathbf{HH}^*(\mathcal{A},\mathcal{A})((t)), d_{\mathcal{A}} + [tw,])$$

The following is the analogue of Dyckerhoff's theorem for our situation:

THEOREM 4.1. *If* $(\mathbf{HH}^*(\mathcal{A},\mathcal{A})((t)), d_A + [tw,])$ *is finite over* $\mathbb{C}((t))$*, then* $\mathbb{C}$ *generates the category* $MF(\mathcal{A},w)$.

(Sketch of proof) We have an action of $\mathbf{D} = \mathbb{C}[u_1,\ldots,u_m]$ on $MF(\mathcal{A},w)$ which factors through the complex $\mathbf{HH}^*(MF(\mathcal{A},w))$. For any u in $\mathbf{D}$, we let $K_u(\mathbf{D})$ be the diagram

$$\mathbf{D} \xrightarrow{u} \mathbf{D}$$

For the sequence $\bar{u} = (u_1,\ldots,u_m)$ we define

$$K_{\bar{u}}(\mathbf{D}) = \otimes K_{u_i}(\mathbf{D})$$

and we consider the colimit of the obvious diagram:

$$K_{\bar{u}}(\mathbf{D}) \to K_{\bar{u}^2}(\mathbf{D}) \to K_{\bar{u}^3}(\mathbf{D})\ldots$$

which we denote by E. For any object $\mathbf{O}$ in $MF(\mathcal{A},w)$, we have an augmentation

$$E\otimes\mathbf{O}\to\mathbf{O}\to cone(e)$$

Because the action of $\mathbf{D}$ factors as above, we can conclude that $cone(e)$ is zero and that this map is an isomorphism. Now the objects $K_{\bar{u}^i}(\mathbf{D})\otimes\mathbf{O}$ are in the triangulated subcategory generated by $\mathbb{C}$ because their cohomologies are finite. Because $\mathbf{O}$ is compact and can be expressed as a colimit of $K_{\bar{u}^i}(\mathbf{D})\otimes\mathbf{O}$, we have that $\mathbf{O}$ is a direct summand of one of the $K_{\bar{u}^i}(\mathbf{D})\otimes\mathbf{O}$ generated by $\mathbb{C}$ as well.

We denote by $RHom_c(MF^\infty(\mathcal{A},w), MF^\infty(\mathcal{A},w))$ the category of continuous endofunctors in the sense of [**Toë**]. Similarly to the works [**Pre, LinPom**], we can apply our generation result to the category to prove the following fact which implies smoothness for $MF(\mathcal{A},w)$:

THEOREM 4.2. $RHom_c(MF^\infty(\mathcal{A},w), MF^\infty(\mathcal{A},w)) \cong MF^\infty(\mathcal{A}\otimes\mathcal{A}^{op}, w\otimes 1 - 1\otimes w)$

We can make this condition more tractable by considering the deformation theory of the pure Sullivan algebra $A^!$ itself. As noted in the introduction, for any simply connected space of finite type, we have fully faithful functors induced by the $C^*(\mathcal{Q}) - C_*(\Omega\mathcal{Q})$ bimodule $\mathbb{C}$. It then follows from a result of Keller [**Kel**] that for such a fully faithful functor there is a canonical equivalence in the homotopy category of $B(\infty)$ algebras:

$$\mathbf{HH}^*(C^*(\mathcal{Q}), C^*(\mathcal{Q})) \cong \mathbf{HH}^*(C_*(\Omega\mathcal{Q}), C_*(\Omega\mathcal{Q}))$$

In particular these two Koszul dual algebras have equivalent deformation theories. Suppose a commutative dga has a free-commutative model $(\bigwedge V, d)$ where V is a finite dimensional vector space. There is a very explicit complex quasi-isomorphic as a dg-Lie algebra to $\mathbf{HH}^*((\bigwedge V, d), (\bigwedge V, d))$. Recall that $T^{poly}(V)$ is the Lie-algebra of polyvector fields on $\bigwedge V$ with Schouten bracket. Part of Kontsevich's formality theorem says that the HKR map:

$$T^{poly}(V) \to \mathbf{HH}^*(\bigwedge V)$$

is the first Taylor coefficient in an L_∞ quasi-isomorphism between the two.

We can think of the derivation d as corresponding to a vector-field v. It follows from a spectral sequence argument that the HKR map gives a quasi-isomorphism:

$$(T^{poly}(V), [v, -]) \to \mathbf{HH}^*((\bigwedge V, d), (\bigwedge V, d))$$

LEMMA 4.3. *This map can be corrected to an L_∞ quasi-isomorphism. In the case of a pure Sullivan algebra, the first Taylor coefficient agrees with the HKR map.*

In the pure Sullivan case, potentials tw in $\mathbf{HH}^*(\mathcal{A}, \mathcal{A})[[t]]$ correspond to odd-polyvector fields $tw(d/de_1, d/de_2, \ldots d/de_m) \in T^{poly}(\mathcal{B})[[t]]$. After passing to the generic fiber, the Hochschild cohomology is given by :

$$(T^{poly}(V)((t)), [v + tw(d/de_1, \ldots, d/de_m),])$$

DEFINITION 4.4. By analogy with the case of ordinary matrix factorizations, we will say that w has an isolated singularity if the homology of this complex is finite dimensional.

Let $\mathcal{B}$ be a pure Sullivan algebra, whose Lie model $\mathfrak{g}$ is formal. We have that $U\mathfrak{g}$ is a graded Clifford algebra over $\mathbb{C}[u_1, \ldots, u_m]$. We let D_k be the closed subvariety of $\mathbb{C}[u_1, \ldots u_m]$ for which $rank(B) \leq k$ and assume further that the $D_k - D_{k-1}$ is smooth. Let R denote $U\mathfrak{g}/(w)$. In this setting we can relate our notion of isolated singularity to another possible notion of isolated singularity in non-commutative geometry.

THEOREM 4.5. *Let $\mathcal{B}$ be a pure Sullivan algebra, whose Lie model $\mathfrak{g}$ is formal and as above. Let w be a potential which intersects the varieties D_k transversally at every point. Then:*

(a) w has isolated singularities
(b) $Proj(R)$ has finite homological dimension as an abelian category.

The first statement is a calculation, so we explain the second one. The exact functor between derived categories

$$\pi : D^b(Gr - R) \to D^b(Proj(R))$$

has a right adjoint $R\omega$. Thus we will show that for any $M, N \in D^b(Gr - R)$, $\mathrm{Ext}^i(M, R\omega \circ \pi(N))$ vanishes for large i.

Suppose that Q is a graded prime ideal different from the maximal ideal and lying in a component of D_k, but not D_{k-1}. Now denote by P the prime ideal corresponding to the irreducible component of D_k which Q is in. One can prove that the correspondence $P \mapsto rad(PR)$ gives a bijection between (graded) prime ideals in $\mathbb{C}[u_1, \ldots, u_m]$ and (graded) prime ideals of $U\mathfrak{g}$. We have a short exact sequence

$$0 \to S \to R/(rad(PR), Q) \to R/rad(QR) \to 0$$

where S is $R/rad(QR)$ torsion by the assumption that the prime Q lie in a component of D_k but not D_{k-1}. Now we know by our condition, that $\mathbb{C}[u_1, \ldots, u_m]/Q[l]$ has a finite resolution as a $\mathbb{C}[u_1, \ldots, u_m]/P$ module and thus so does $R/(rad(PR), Q)[l]$ as a $R/rad(PR)$ module.

The above exact sequence reveals that $\mathrm{Ext}^i_{R/(rad(PR))}(R/rad(QR)[l], M)$ is $R/rad(QR)$ torsion for sufficiently large i. It is also easy to show from the transversality hypothesis that $R/rad(PR)[l]$ has finite homological dimension over R. Now the following lemma [**Bro**] and the change of ring spectral sequence enable us to conclude the result:

LEMMA 4.6. *Let R be a graded FBN ring. Given a bounded complex C in $D(Gr - R)$ if $\mathrm{Ext}^i(R/P[l], C)$ is R/P torsion for $i >> j$ for every two-sided prime ideal P then $\mathrm{Ext}^i(M, C)$ vanishes for $i >> 0$.*

The proof is as in the stated reference provided that we note the Gabriel correspondence between minimal injectives and graded prime ideals for graded FBN rings and that every bounded complex in $Gr - R$ is equivalent to a minimal complex of injectives.

EXAMPLE 4.7. For $\prod S^{2n_j}$ the condition that w has an isolated singularity is similar to the usual Jacobian condition and states that $\mathbb{C}[u_1, \ldots, u_m]/(u_i dw/du_i)$ be finite dimensional.

The field theory assigned to S^2 can be computed explicitly. Working with $\mathbb{Z}/2\mathbb{Z}$ gradings for ease of notation, we have the following calculations:

LEMMA 4.8. *The endomorphisms algebra End*$(\mathbb{C}, \mathbb{C})$*, in the category of curved modules over* $(\mathbb{C}[x], x^{2d})$*, deg(x)=1, is given by* $\mathbb{C}[e]/e^2$*, with one higher multiplication* $m_{2d}(e, e...e) = 1$

This calculation is very similar to [**Dyc**] theorem 4.7. The deformed algebra has an obvious cyclically symmetric inner product given by Poincare duality.

The calculations below are tedious but straightforward for the patient reader.

LEMMA 4.9. *The Hochschild homology in this case is given by even elements* $e_0, e_1, ...e_{2d-1}$*. The pairing on* $\mathbf{HH}_*$ *given by the above TQFT is given by* $\langle e_i, e_j\rangle$ *is non-zero if i+j=2d-1 and zero otherwise.*

LEMMA 4.10. *The Hochschild cohomology of this TQFT is concentrated in even degree and equals*

$$\mathbb{C}[e]/e^2 = 1, \textit{if } d{=}1$$

$$\mathbb{C}[e, v]/(e^2 = 0, ev = 0, v^d = 0), \textit{otherwise}$$

5. Tangential Fukaya categories

Given the close connection between TQFTs and Floer theory, in this section we aim to give a Floer theoretic interpretation of the previous sections in some special cases. For motivation, let us consider the easiest case of a symplectic mirror to a Landau-Ginzburg model, that of S^2. We think of a sphere as being the (open) disk bundle of the cotangent bundle, $D^*(S^1)$, compactified by the points at 0 and ∞. This is then mirror to $(C_*(\Omega(S^1)) \cong \mathbb{C}[z, z^{-1}], w = z + 1/z)$.

If we wish to understand the mirror to the Landau-Ginzburg model $(\mathbb{C}[z, z^{-1}], w = z^d + 1/z^d)$ we can either consider the Fukaya category of the orbifold $S^2//(\mathbb{Z}/d\mathbb{Z})$, where $\mathbb{Z}/d\mathbb{Z}$ acts by rotations that fix the two points, or more concretely a Fukaya category where we require disks to intersect the compactifying divisor with ramification of order d. If one wants to generalize this to higher dimensional projective spaces, the mirror of $\mathbb{C}P^n$ is well known to be the Landau-Ginzburg model:

$$(z_0 + z_1 + \ldots z_n,\ z_0 z_1 \ldots z_n = 1)$$

Mirror symmetry predicts that to obtain the mirror manifold to:

$$(z_0^{d_0} + z_1^{d_1} + \ldots z_n^{d_n}, z_0 z_1 \ldots z_n = 1)$$

one performs the root stack construction, to be defined below, along the toric divisors $z_i = 0$.

Given a variety $\mathcal{X}$ and a collection of effective Cartier divisor $\mathcal{D}_i$, and d_i a collection of positive integers. The Cartier divisors define a natural morphism

$$\mathcal{X} \to [\mathbb{A}^n/(\mathbb{C}^*)^n]$$

The root stack $\mathcal{X}_{(\mathcal{D}_i,d_i)}$ is defined to be the fibre product

$$\mathcal{X} \times_{[\mathbb{A}^n/(\mathbb{C}^*)^n]} [\mathbb{A}^n/(\mathbb{C}^*)^n]$$

where the map $[\mathbb{A}^n/(\mathbb{C}^*)^n] \to [\mathbb{A}^n/(\mathbb{C}^*)^n]$ is the d_i-power map. The root stack defines an orbifold, which has non-trivial orbifold stabilizers along the divisors (the coarse moduli space is exactly $\mathcal{X}$ and away from $\mathcal{D}_i$ the map $\mathcal{X}_{(\mathcal{D}_i,d_i)} \to \mathcal{X}$ is an isomorphism). The main property is that to give a map into the root stack is equivalent to giving a map into $\mathcal{X}$ which is ramified to order d_i along the divisors $\mathcal{D}_i$. The formulas in the previous paragraph easily generalize to produce mirrors to toric Fano manifolds with root constructions performed along toric divisors.

The is also a clear symplectic interpretation of the TQFT associated in section 3 to $\mathcal{Q} = \mathbb{C}P^n$ when $w = u$. Namely, we have an anti-holomorphic involution of $I : \mathbb{C}P^n \times \mathbb{C}P^n \to \mathbb{C}P^n \times \mathbb{C}P^n$ given by $(z, w) \to (\bar{w}, \bar{z})$. Its fixed point set: $\mathcal{L} : \mathbb{C}P^n \to \mathbb{C}P^n \times \mathbb{C}P^n$, is a Lagrangian submanifold. We have that $HF^*(\mathcal{L}, \mathcal{L}) \cong \mathbb{C}((t))[e]/(e^{n+1} = t)$ and in this case, the category $MF(\mathcal{A}, w)$ is isomorphic to the full subcategory of the Fukaya category of $\mathbb{C}P^n \times \mathbb{C}P^n$ split-generated by $\mathcal{L}$.

In what follows it will be important to think of $\mathbb{C}P^n \times \mathbb{C}P^n$ as a symplectic cut. $T^*\mathcal{Q} - \mathcal{Q}$ acquires a Hamiltonian S^1 action by rotating the geodesics (which then give rise to Reeb orbits when restricted to the unit cotangent bundle). This induces a natural Hamiltonian action on $(T^*\mathcal{Q} - \mathcal{Q}) \times \mathbb{C}$. The moment map got this Hamiltonian S^1 action

$$(T^*\mathcal{Q} - \mathcal{Q}) \times \mathbb{C} \to \mathbb{R}$$

is given by

$$(x, z) \mapsto H(x) + 1/2|z|^2$$

Where $H(x) = |x|$ is the Hamiltonian associated to the Hamiltonian action on $T^*\mathcal{Q} - \mathcal{Q}$. We then take the reduced space, that is the preimage of a regular value quotiented out by the S^1 action. Finally, we glue back in the zero section to obtain a manifold $\mathcal{X}$ which is a symplectic compactification of the open disk bundle $D^*(\mathcal{Q})$ by the smooth divisor $\mathcal{D}$. Concretely, the divisor is defined by the equation

$$\sum z_i w_i = 0$$

The Lagrangian $\mathcal{L}$ corresponds to the zero section. The above discussion suggests that our other deformations should be realized by performing the d-th root stack construction on $\mathbb{C}P^n \times \mathbb{C}P^n$ along $\mathcal{D}$ or in this case equivalently counting holomorphic disks with a prescribed tangency to the divisor $\mathcal{D}$. It is worth noting that the above construction can be applied to other manifolds with periodic geodesics e.g., S^n, $n > 1$, or $\mathbb{H}P^n$. In the case of S^n, one obtains S^n as a Lagrangian submanifold of the projective quadric Q_n. The symplectic compactifying divisor is isomorphic to Q_{n-1}. One can repeat the calculations below and a similar picture develops to that described in this section.

We now define the Floer homology that we wish to consider. The tangency Floer theory we define here will agree with the Floer theory in the root stack because curves with components which live entirely in the divisor $\mathcal{D}$ occurs in real codimension ≥ 2.

We want to consider the moduli space of holomorphic disks

$$f : (D^2, S^1) \to (\mathbb{C}P^n \times \mathbb{C}P^n, \mathcal{L})$$

$$\text{if } f(p) \in \mathcal{D}, \text{then } m(p) = d$$

Here $m(p)$ denotes the intersection multiplicity, which we require to be exactly d at each point of intersection. We consider the [**FOOO**] compactification of this moduli space and compactify our moduli space as a subspace of $\mathfrak{M}_k(\mathbb{C}P^n \times \mathbb{C}P^n, \mathcal{L})$ in the obvious way. We will denote the moduli space by $\mathfrak{M}_{j,d,k}$, where j denotes the number of intersection points with $\mathcal{D}$ and d the multiplicity. As in [**FOOO**], we consider some model for chains on $C_*(\mathbb{C}P^n, \mathbb{C}((t)))$ and using the evaluation maps

$$ev_i : \mathfrak{M}_{j,d,k+1} \to C_*(\mathbb{C}P^n)$$

to define a sequence of higher products

$$m_k(\alpha_1, ...\alpha_k) = \sum ev_{0,*}(\prod ev_i^*(\alpha_i))t^n$$

One can show that the standard complex structure J can be perturbed in the complement of a neighborhood of $\mathcal{D}$ to be regular for this moduli problem. We wish to prove:

THEOREM 5.1. *$HF^*(\mathcal{L}, \mathcal{L}) \cong End_{MF(\mathcal{A},w)}(\mathbb{C}, \mathbb{C})$, where $(\mathcal{A}, w)$ is the curved algebra associated in section three to $\mathcal{Q} = \mathbb{C}P^n$, with potential $w = u^d$.*

The key lemma is:

LEMMA 5.2. *The Hochschild class the infinitesimal deformation class defined on $H^*(\mathbb{C}P^n, \mathbb{C}[t]/t^2)$ using the moduli space $\mathfrak{M}_{1,d,k}$ is gauge equivalent to the d-fold cup product of the Hochschild class determined by $\mathfrak{M}_{1,1,k}$*

To prove this result, we proceed by induction and consider a certain codimension one submanifold of the moduli space of disks with two points of intersection with the divisor, one of multiplicity $d-1$ and one simple intersection. Analysis of the the boundary of this submanifold proves the desired equation. The proof of this follows the same line of reasoning as the proof in [**FOOO**] that bulk deformation :

$$H^*(X) \to \mathbf{HH}^*(Fuk(X))$$

is a ring homomorphism.

REMARK 5.3. More generally, in the formalism above, if $\mathcal{X}$ is a projective variety and $\mathcal{D}$ is a smooth ample divisor, one could consider chains S in the divisor $\mathcal{D}$ which represent classes of $H_*(\mathcal{D})$ and require that the point of tangency simultaneously lie in S. We will explore these deformations of $Fuk(\mathcal{X}-\mathcal{D})$ in our forthcoming paper.

To finish the theorem, we have a "finite determinacy" lemma:

LEMMA 5.4. *The $A(\infty)$ structure on $HF^*_{\mathcal{X}_{(\mathcal{D},d)}}(\mathcal{L},\mathcal{L}) \cong \mathbb{C}[e]/e^{n+1}((t))$ is determined by the fact that $m_j = 0$, $2 < j < 2d$ and $m_{2d}(e^{a_1}, e^{a_2}, \ldots, e^{a_{2d}}) = t$, if $\sum(a_i) = (n+1)d$*

By a Kunneth theorem, we can get similar results for manifolds of the form $\mathcal{Q} = \prod \mathbb{C}P^{n_j}$.

It seems interesting to make a closer connection between the symplectic geometry in this section and the rational homotopy theory/deformation theory of the previous section. In view of this, it is useful to note the following strong result due to McLean from a recent paper [**McL**].

THEOREM 5.5. *If $T^*\mathcal{Q}$ is symplectomorphic to an affine variety A, then $\mathcal{Q}$ is (rationally) elliptic.*

References

[AdaHil] J.F. Adams, P.J. Hilton. On the chain algebra of a loop space, Commentarii Mathematici Helvetici (1955), 305-330

[Abou] M. Abouzaid. A cotangent fiber generates the Fukaya Category, 2010,`arxiv:math/1003.4449`

[Bar] V. Baranovsky. BGG correspondence for projective complete intersections, Internat. Math. Res. Not., 45 (2005), 2759-2774.

[Bro] K. Brown. Fully Bounded Noetherian Rings of Finite Injective Dimension. *The Quarterly Journal of Mathematics*,1990,

[Dyc] T. Dyckerhoff. Compact generators in categories of matrix factorizations, 2009, `arxiv:math/0904.4713`

[FelHalTho] Y.Felix,S. Halperin, J.Thomas. *Rational Homotopy Theory.* Springer-Verlag, 2001.

[FOOO] K. Fukaya, Y. Oh, H. Oh, K. Ono. *Lagrangian Intersection Floer Theory: Anomaly and Obstruction.* AMS/IP Studies in Advanced Mathematics, 2009.

[Kel] B. Keller. Derived Invariance of Higher Structures on the Hochschild complex,2003, http://www.math.jussieu.fr/ keller/publ/dihabs.html.

[Kon] M. Kontsevich. Homological algebra of mirror symmetry. *Proceedings of the International Congress of Mathematicians, Zürich 1994*, vol. I, Birkhauser 1995, 120–139.

[KonSoi] M. Kontsevich, Y. Soibelman. Notes on A-infinity algebras, A-infinity categories, and non-commutative geometry, I, 2006, arxiv:math/0606241

[LinPom] K. Lin, D.Pomerleano. Global Matrix Factorizations.arxiv:math/1101.5847

[Lur] J. Lurie. On the Classification of Topological Field Theories.*Current Developments in Mathematics Volume 2008* (2009), 129-280

[McL] M. McLean. The growth rate of symplectic homology and affine varieties.arxiv:math/1011.2542

[Pom] D. Pomerleano. Curved String Topology and Fukaya categories. In preparation.

[Pos] L. Positselski. Two kinds of derived categories, Koszul duality, and comodule-contramodule correspondence, 2009, arxiv:math/0905.2621

[Pre] A. Preygel. Thom–Sebastiani and Duality for Matrix Factorizations, 2011, arxiv:math/1101.5834

[Sei] P. Seidel. Fukaya categories and Deformations, 2002, arxiv:math/0010032

[Toë] B. Toën. The homotopy theory of dg-categories and derived Morita theory. *Invent. Math.* 167 (2007), no. 3, 615–667.

[ToëVaq] B. Toën, M. Vaquie. Moduli of objects in dg-categories, 2005, arxiv:math/0503269

Proceedings of Symposia in Pure Mathematics
Volume **85**, 2012

Target Space Dualities of Heterotic Grand Unified Theories

Thorsten Rahn

ABSTRACT. In this article we summarize and extend the ideas and investigations on so-called target space dualities of heterotic models with $(0,2)$ world-sheet supersymmetry as they were partly presented on the String-Math 2011 conference. After the generic description of the duality, we give some novel examples involving vector bundles that are not deformations of the tangent bundle but more generic ones corresponding to $SO(10)$ and $SU(5)$ gauge theories in four dimensions. We show explicitly that the necessary conditions for a duality also hold for compactifications of this kind. Finally we will present the results of the large landscape scan of E_6 models.

1. Introduction

Heterotic string theory provides a way to build grand unified theories very naturally. A priori it arises with an $E_8 \times E_8$ gauge group which can then be broken down to some GUT group in four dimensions. In such a model, on the one hand, we have bosonic degrees of freedom that are valued in the Calabi-Yau manifold, denoted by $\mathcal{M}$ throughout this paper, and on the other hand their fermionic superpartners that are coupled to the pullback of the tangent bundle. Furthermore we have some left-moving fermions that do not necessarily couple to the pullback of the tangent bundle but more generically to the pullback of some vector bundle $\mathcal{V}$ of rank $n = 3, 4$ or 5. The only conditions on this vector bundle is that it is holomorphic, stable and that the first and second Chern classes equal those of the tangent bundle. The structure group of $\mathcal{V}$ is then $SU(3)$, $SU(4)$ or $SU(5)$ for a rank 3, 4 or 5 bundle respectively and breaks one of the E_8 factors down to the commutant of $SU(n)$ in E_8 which gives an effective four-dimensional theory with gauge group E_6, $SO(10)$ or $SU(5)$ respectively, whereas the other E_8 factor can be hidden. In the geometric picture the massless matter spectrum can be obtained by computing certain vector bundle valued cohomology classes. In case of an E_6, $SO(10)$ or $SU(5)$ gauge group the actual cohomologies that need to be computed can be found in table 1. A review on the computational tools for this kind of calculation can be found e.g. in [**1**] and references therein.

2010 *Mathematics Subject Classification.* Primary 81T30. Secondary 14J81, 81V22.

# zero modes in reps of $H \times G$	1	$h^1_{\mathcal{M}}(\mathcal{V})$	$h^1_{\mathcal{M}}(\mathcal{V}^*)$	$h^1_{\mathcal{M}}(\Lambda^2\mathcal{V})$	$h^1_{\mathcal{M}}(\Lambda^2\mathcal{V}^*)$	$h^1_{\mathcal{M}}(\mathcal{V}\otimes\mathcal{V}^*)$
E_8 $\downarrow$				248 $\downarrow$		
$SU(3)\times E_6$	$(1,78)$	$\oplus(3,27)$	$\oplus(\overline{3},\overline{27})$			$\oplus(8,1)$
$SU(4)\times SO(10)$	$(1,45)$	$\oplus(4,16)$	$\oplus(\overline{4},\overline{16})$	$\oplus(6,10)$		$\oplus(15,1)$
$SU(5)\times SU(5)$	$(1,24)$	$\oplus(5,\overline{10})$	$\oplus(\overline{5},10)$	$\oplus(10,5)$	$\oplus(\overline{10},\overline{5})$	$\oplus(24,1)$

TABLE 1. Matter zero modes in representations of the GUT group

One nice way to see how holomorphic vector bundles arise in string-theory is using the $(0,2)$ gauged linear sigma model (GLSM) [**2**]. Here one starts with a two-dimensional field theory containing certain chiral as well as Fermi superfields. The most general superpotential of such a theory can be written as

$$S=\int d^2z d\theta\left[\sum_j \Gamma^j\, G_j(X_i)+\sum_{l,a} P_l\,\Lambda^a\, F_a{}^l(X_i)\right], \tag{1}$$

where the X_i, P_l and Γ^j, Λ^a correspond to chiral and Fermi superfields. They are all charged under a certain number of $U(1)$ gauge groups in such a way that the superpotential is gauge invariant. If we denote the bosonic components of the chiral superfields X_i and P_l by x_i and p_l we will find a bosonic potential for them consisting of an F-term potential

$$V_F=\sum_j\Big|G_j(x_i)\Big|^2+\sum_a\Big|\sum_l p_l\,F_a{}^l(x_i)\Big|^2 \tag{2}$$

as well as a D-term scalar potential

$$V_D=\sum_{\alpha=1}^{r}\left(\sum_{i=1}^{d} Q_i^{(\alpha)}|x_i|^2-\sum_{l=1}^{\gamma} M_l^{(\alpha)}|p_l|^2-\xi^{(\alpha)}\right)^2, \tag{3}$$

where the $\xi^{(\alpha)}\in\mathbb{R}$ denote the Fayet-Iliopoulos (FI) parameters for each $U(1)$ gauge group. One can now go ahead and analyze the vacuum structure of such a theory and certainly this structure will crucially depend on the actual value of the FI parameters. Putting more weight on the mathematical point of view, one also calls them Kähler parameters. Different choices of these parameters can be interpreted as different triangulations of some polytope and each triangulation describes one kind of vacuum or phase of the underlying GLSM [**3**]. If one chooses them in such a way that the triangulation is maximal, i.e. it has more or the same number of maximal dimensional cones than every other possible triangulation, then our vacuum may correspond to a holomorphic vector bundle over a Calabi-Yau manifold. In the low energy effective action this will be the corresponding target space of the non-linear sigma model.

Here the idea of target space dualities comes into play. It basically works with the phases of the GLSM that do not have a completely geometric interpretation and uses some freedom in them to change the GLSM without spoiling the phase itself. This basically means that the moduli spaces of two GLSMs can be connected and that there exists a locus inside a specific phase where they coincide. After the redefinition of the data, we can just pretend that we came from the new GLSM and go back to the geometric phase there. The question is now whether we actually undergo a transition of one geometry to a different one by doing this or if we just walk around in one and the same moduli space and are dealing with a target space duality between those two geometries. Distler and Kachru first pointed out that such a thing might actually exist [**4**] and further work [**5–7**] supported this idea with specific examples by explicitly comparing the dimensions of the moduli spaces as well as the spectra of the corresponding models and found agreement:

$$\begin{aligned} h^{1,1}(\mathcal{M}) + h^{2,1}(\mathcal{M}) + h^1_{\mathcal{M}}(\mathrm{End}(\mathcal{V})) &= h^{1,1}(\widetilde{\mathcal{M}}) + h^{2,1}(\widetilde{\mathcal{M}}) + h^1_{\widetilde{\mathcal{M}}}(\mathrm{End}(\widetilde{\mathcal{V}}))\,, \\ h^i_{\mathcal{M}}(\Lambda^k\,\mathcal{V}) &= h^i_{\widetilde{\mathcal{M}}}(\Lambda^k\,\widetilde{\mathcal{V}}), \qquad \text{for } i = 0,\ldots,3\,. \end{aligned}$$

2. The Duality

In this section we will review the general procedure that can be employed to produce dual models from a given one. In order to do that we first explain how this can be done schematically and then turn to three explicit and novel examples. Here we will capture all possible structure groups for the bundle, i.e. we consider an $SU(3)$, $SU(4)$ and finally an $SU(5)$ bundle. That the duality works for such models was mentioned in [**8**] but not shown in an explicit example. We want to use the opportunity to rectify that.

2.1. General procedure. In the setting of the GLSM we mentioned above we can specify the vector bundle as the cohomology of a complex containing direct sums of line bundles

$$0 \to \mathcal{O}_{\mathcal{M}}^{\oplus r_{\mathcal{V}}} \xrightarrow{\otimes E_i{}^a} \bigoplus_{a=1}^{\delta} \mathcal{O}_{\mathcal{M}}(N_a) \xrightarrow{\otimes F_a{}^l} \bigoplus_{l=1}^{\gamma} \mathcal{O}_{\mathcal{M}}(M_l) \to 0\,. \tag{4}$$

This means that the vector bundle is given by $\mathcal{V} = \frac{\ker(F_a{}^l)}{\mathrm{im}(E_i{}^a)}$. The chiral and Fermi fields of the corresponding GLSM are charged under the various $\mathbb{C}^*$ actions of the toric variety that contains $\mathcal{M}$ as complete intersection of hypersurfaces. In the following we will denote the defining data of the configuration by two tables

x_i			Γ^j		
Q_1	$\cdots$	Q_N	$-S_1$	$\cdots$	$-S_c$

Λ^a			p_l		
N_1	$\cdots$	N_δ	$-M_1$	$\cdots$	$-M_\gamma$

.

In this scenario we can turn the requirement of matching first and second Chern classes of $\mathcal{V}$ and $T_{\mathcal{M}}$ into a combinatorial relation of the charges of the fields:

$$\begin{aligned} &\sum_{a=1}^{\delta} N_a^{(\alpha)} = \sum_{l=1}^{\gamma} M_l^{(\alpha)}, \qquad \sum_{i=1}^{d} Q_i^{(\alpha)} = \sum_{j=1}^{c} S_j^{(\alpha)}, \\ &\sum_{l=1}^{\gamma} M_l^{(\alpha)} M_l^{(\beta)} - \sum_{a=1}^{\delta} N_a^{(\alpha)} N_a^{(\beta)} = \sum_{j=1}^{c} S_j^{(\alpha)} S_j^{(\beta)} - \sum_{i=1}^{d} Q_i^{(\alpha)} Q_i^{(\beta)}\,. \end{aligned} \tag{5}$$

Here the Greek index in parenthesis labels the corresponding $\mathbb{C}^*$ action of the toric variety and the equations (5) have to be satisfied for all α and β. Now after fixing the notations we can go on to describe the explicit procedure, which consists of different steps:

The procedure:

(1) Construct the GLSM phases of a smooth $(0,2)$ model $(\mathcal{M},\mathcal{V})$.
(2) Go to a phase where one of the p_l, say p_1, is not allowed to vanish and hence obtains a vev $\langle p_1\rangle$.
(3) Perform a rescaling of k Fermi superfields by the constant vev $\langle p_1\rangle$ and exchange the role of some Λ^a and Γ^j

$$\tilde{\Lambda}^{a_i} := \frac{\Gamma^{j_i}}{\langle p_1\rangle}, \quad \tilde{\Gamma}^{j_i} := \langle p_1\rangle\, \Lambda^{a_i}\,,\ \forall i = 1,...,k\,,$$

with $\sum_i ||G_{j_i}|| = \sum_i ||F_{a_i}{}^1||$ for anomaly cancellation.
(4) Move to a region in the bundle moduli space where the Λ^{a_i} only appear in terms with P_1 for all i. This means that we choose the coefficients in the bundle defining polynomials $F_a{}^l$ such that

$$F_{a_i}{}^l = 0\,, \qquad \forall\, l \neq 1,\ i = 1,...,k\,.$$

(5) Leave the non-geometric phase and define the Fermi superfields of the new GLSM such that each term in the superpotential is $U(1)^r$ gauge invariant. This means

$$||\tilde{\Lambda}^{a_i}|| = ||\Gamma^{j_i}|| - ||P_1|| \quad \text{and} \quad ||\tilde{\Gamma}^{j_i}|| = ||\Lambda^{a_i}|| + ||P_1||\,.$$

(6) Returning to a generic point in moduli space defines a new dual $(0,2)$ GLSM which in a geometric phase corresponds to a different Calabi-Yau/vector bundle configuration $(\widetilde{\mathcal{M}},\widetilde{\mathcal{V}})$.

2.2. Examples of models with structure group $SU(n)$. We now want to be a bit more explicit and show some specific examples of the duality. We will show one example for each type of structure group $SU(3)$, $SU(4)$ and $SU(5)$. In contrast to earlier work [**8**] where we mostly focused on models that were deformations of the tangent bundle and hence given by the cohomology of the Euler sequence, here we want to give different examples that do arise from an exact monad

$$0 \to \mathcal{V} \xrightarrow{f} \bigoplus_{a=1}^{\delta} \mathcal{O}_{\mathcal{M}}(N_a) \xrightarrow{\otimes F_a{}^l} \bigoplus_{l=1}^{\gamma} \mathcal{O}_{\mathcal{M}}(M_l) \to 0\,, \tag{6}$$

and hence are given by the kernel of the map $F_a{}^l$, $\mathcal{V} = \text{Ker}\left(F_a{}^l\right)$. The way to generate the dual models of such a monad remains the same. In [**8**] we assumed that the dual bundle was stable which we also did in the examples here. While we also had to assume another property there, in the examples presented here we did not do any such thing.

Example for an $SU(3)$-model: We start with an $SU(3)$ example which consists of a holomorphic vector bundle over a codimension two complete intersection Calabi-Yau space. In [**8**] we investigated mostly $SU(3)$ models which are given by a deformation of the tangent bundle. As described there the base space undergoes usually a conifold transition. In this example here we are not dealing with a $(0,2)$ model which is a deformation of the tangent bundle but a completely independent monad. Furthermore we will see that the base will not transform via a conifold transition. Rather, in the beginning the ambient variety will remain untouched and only a different set of hypersurfaces will be chosen, also resulting in a topology change of the base. Finally, through the exchange of those specific hypersurfaces we will see that in fact the ambient space topology will be changed after all. The model data is given by

(7)

x_i							Γ^j	
0	0	0	1	1	1	1	−2	−2
1	1	1	2	2	2	0	−4	−5

Λ^a				p_l
1	0	0	2	−3
0	1	1	6	−8

As was explained in [**9**] to compute the number of chiral matter zero modes and the massless singlets, we need nothing but line bundle cohomologies. Furthermore the efficient algorithm suggested in [**10**] and proven in [**11**] and [**12**] allows one to calculate such cohomologies quite rapidly. Employing our implementation **cohomCalg Koszul** extension [**13**] for this matter we find

$$(8)\quad \begin{aligned} h^\bullet_{\mathcal{M}}(\mathcal{V}) &= (0, 120, 0, 0)\,, \\ h^{1,1}_{\mathcal{M}} + h^{2,1}_{\mathcal{M}} + h^1_{\mathcal{M}}(\mathrm{End}(\mathcal{V})) &= 2 + 68 + 322 = 392\,. \end{aligned}$$

In order to see our freedom of consistently exchanging hypersurface equations with bundle maps in the monad as described in the last section we explicitly write down the multi-degrees of the corresponding generic homogeneous functions. Using that

$$(9)\quad ||{F_a}^l|| = -||p_l|| - ||\Lambda^a||\,,$$

for the only choice $l = 1$ they read

$$(10)\quad ||G_1|| = \begin{pmatrix} 2 \\ 4 \end{pmatrix}, ||G_2|| = \begin{pmatrix} 2 \\ 5 \end{pmatrix},$$

$$(11)\quad ||{F_1}^1|| = \begin{pmatrix} 2 \\ 8 \end{pmatrix}, ||{F_2}^1|| = \begin{pmatrix} 3 \\ 7 \end{pmatrix}, ||{F_3}^1|| = \begin{pmatrix} 3 \\ 7 \end{pmatrix}, ||{F_4}^1|| = \begin{pmatrix} 1 \\ 2 \end{pmatrix}.$$

Here we can already see that the sum of the degrees of the two hypersurfaces equals the sum of the degree of the third and the fourth F. From the last section, we know how to exchange these functions and how to redefine the Λ's and Γ's in order to obtain a sensible new monad. Namely, we perform the rescalings

$$(12)\quad \begin{aligned} \tilde{\Gamma}^1 &:= \langle p_1\rangle\,\Lambda^3, & \tilde{\Gamma}^B &:= \langle p_1\rangle\,\Lambda^4, & \tilde{\Lambda}^3 &:= \frac{\Gamma^1}{\langle p_1\rangle}, & \tilde{\Lambda}^4 &:= \frac{\Gamma^B}{\langle p_1\rangle}, \\ \tilde{G}_1 &:= {F_3}^1, & \tilde{G}_2 &:= {F_4}^1, & \tilde{F}_3{}^1 &:= G_1, & \tilde{F}_4{}^1 &:= G_2\,, \end{aligned}$$

yielding the effective superpotential

$$(13)\quad \mathcal{W} = \tilde{\Gamma}^1\tilde{G}_1 + \tilde{\Gamma}^2\tilde{G}_2 + \langle p_1\rangle\left(\tilde{\Lambda}^3\tilde{F}_3{}^1 + \tilde{\Lambda}^4\tilde{F}_4{}^1 + \Lambda^1{F_1}^1 + \Lambda^2{F_2}^1\right).$$

The new charges of the constructed model read

$$(14)\quad \begin{aligned} ||\tilde{\Gamma}^1|| &= \begin{pmatrix} -3 \\ -7 \end{pmatrix}, & ||\tilde{\Gamma}^2|| &= \begin{pmatrix} -1 \\ -2 \end{pmatrix}, & ||\tilde{\Lambda}^3|| &= \begin{pmatrix} 1 \\ 4 \end{pmatrix}, & ||\tilde{\Lambda}^4|| &= \begin{pmatrix} 1 \\ 3 \end{pmatrix}, \\ ||\tilde{G}_1|| &= \begin{pmatrix} 3 \\ 7 \end{pmatrix}, & ||\tilde{G}_2|| &= \begin{pmatrix} 1 \\ 2 \end{pmatrix}, & ||\tilde{F}_3{}^1|| &= \begin{pmatrix} 2 \\ 4 \end{pmatrix}, & ||\tilde{F}_4{}^1|| &= \begin{pmatrix} 2 \\ 5 \end{pmatrix} \end{aligned}$$

and hence going back to the geometric phase we obtain the new base with a new vector bundle. We notice that the new configuration can be rewritten in a slightly simpler way. The new hypersurface $\tilde{G}_2$ has precisely the same degree as the divisor

$\{x_4 = 0\}$ and therefore the corresponding constraining equation to the ambient space simply removes this coordinate from the configuration and we obtain

(15)
$$\begin{array}{|cccccc||c|}\hline \multicolumn{6}{|c||}{x_i} & \Gamma^j \\ \hline 0 & 0 & 0 & 1 & 1 & 1 & -3 \\ 1 & 1 & 1 & 2 & 2 & 0 & -7 \\ \hline \end{array} \quad \begin{array}{|cccc||c|}\hline \multicolumn{4}{|c||}{\Lambda^a} & p_l \\ \hline 1 & 0 & 1 & 1 & -3 \\ 0 & 1 & 4 & 3 & -8 \\ \hline \end{array}\,.$$

As was generically shown, this configuration still satisfies the conditions (5) and we obtain the following topological data:

(16)
$$\begin{gathered} h^{\bullet}_{\widetilde{\mathcal{M}}}(\widetilde{\mathcal{V}}) = (0, 120, 0, 0)\,, \\ h^{1,1}_{\widetilde{\mathcal{M}}} + h^{2,1}_{\widetilde{\mathcal{M}}} + h^{1}_{\widetilde{\mathcal{M}}}(\mathrm{End}(\widetilde{\mathcal{V}})) = 2 + 95 + 295 = 392\,. \end{gathered}$$

If we compare this with the result we obtained in (8), we see that the number of chiral zero modes did not change and the total number of first order deformations stayed the same even though the Hodge number $h^{2,1}$ changed drastically.

Let us once more put some emphasis on the fact that we started with a base manifold that was of codimension two and due to the exchange ended up with a simpler space given by a codimension one Calabi-Yau manifold. Similarly, as we will see in the next example this can also happen the other way round, resulting in an increase of the codimension. Also the number of $\mathbb{C}^*$ actions can change which will be shown in the following examples, too.

An example for an *SU*(4)-model: Next we present an example of a dual pair of heterotic (0,2) models that give rise to gauge group $SO(10)$ in four dimensions and hence are equipped with a rank 4 vector bundle. The model is again not a deformation of the tangent bundle. The base is the complete intersection of a generic quartic and homogeneous degree hypersurface two inside $\mathbb{P}^5$. The defining data can be read off in the following table:

(17)
$$\begin{array}{|c||cc|}\hline x_i & \multicolumn{2}{c|}{\Gamma^j} \\ \hline \mathbb{P}^5 & -2 & -4 \\ \hline \end{array} \quad \begin{array}{|ccccccc||ccc|}\hline \multicolumn{7}{|c||}{\Lambda^a} & \multicolumn{3}{c|}{p_l} \\ \hline 1 & 1 & 1 & 1 & 1 & 1 & 1 & -3 & -2 & -2 \\ \hline \end{array}\,.$$

Clearly this model is anomaly-free, i.e. it satisfies (5) and one can show that the vector bundle is also stable.[1] It is sometimes also referred to as a positive monad, since all line bundles involved have positive degree. It has the following topological data:

(18)
$$\begin{gathered} h^{\bullet}_{\widetilde{\mathcal{M}}}(\widetilde{\mathcal{V}}) = (0, 48, 0, 0)\,, \\ h^{1,1}_{\widetilde{\mathcal{M}}} + h^{2,1}_{\widetilde{\mathcal{M}}} + h^{1}_{\widetilde{\mathcal{M}}}(\mathrm{End}(\widetilde{\mathcal{V}})) = 1 + 89 + 159 = 249\,. \end{gathered}$$

Before we move on, we introduce a new coordinate along with a new hypersurface to the model. Doing that at the same time does not change the model at all. In order to perform the exchange of polynomials F and G, we have to go to a certain region of the moduli space, exchange them and go back to the generic region in the dual configuration. The resulting base manifold can then be obtained as the

[1] By Hoppe's criterion for the initial bundle in this and the next example, to prove stability it suffices to show that $h^0(\mathcal{M}; \Lambda^k \mathcal{V}) = 0\ \forall k < \mathrm{rk}\,\mathcal{V}$ and is relatively straight forward in these cases.

conifold transition of the initial base space. The full model is then given by

(19)

x_i	Γ^j			Λ^a							p_l		
$\mathbb{P}^1$	-1	0	-1	0	0	0	1	0	0	0	0	-1	0
$\mathbb{P}^5$	-1	-4	-1	1	1	1	0	2	1	1	-3	-2	-2

,

and its topology satisfies the necessary duality check of coinciding spectrum and moduli space dimensions:

$$(20)\qquad \begin{aligned} h^\bullet_{\widetilde{\mathcal{M}}}(\widetilde{\mathcal{V}}) &= (0,48,0,0)\,, \\ h^{1,1}_{\widetilde{\mathcal{M}}} + h^{2,1}_{\widetilde{\mathcal{M}}} + h^1_{\widetilde{\mathcal{M}}}(\mathrm{End}(\widetilde{\mathcal{V}})) &= 2 + 86 + 161 = 249\,. \end{aligned}$$

An example for an $SU(5)$-model: Finally let us quickly state a different bundle over the same base from the last paragraph. We modify it such that it has no longer $SU(4)$ but rather $SU(5)$ structure. It is given by

(21)

x_i	Γ^j		Λ^a								p_l		
$\mathbb{P}^5$	-2	-4	1	1	1	1	1	1	1	1	-3	-3	-2

.

Since we have still three chiral fields p_l and eight Fermi fields Λ^a, we end up with a rank 5 vector bundle and hence with an $SU(5)$ gauge group in the four-dimensional theory. The spectrum and the dimension of the moduli space for this model can be calculated as

$$(22)\qquad \begin{aligned} h^\bullet_{\widetilde{\mathcal{M}}}(\widetilde{\mathcal{V}}) &= (0,72,0,0)\,, \\ h^{1,1}_{\widetilde{\mathcal{M}}} + h^{2,1}_{\widetilde{\mathcal{M}}} + h^1_{\widetilde{\mathcal{M}}}(\mathrm{End}(\widetilde{\mathcal{V}})) &= 1 + 89 + 288 = 378\,. \end{aligned}$$

The dual base is again given by the same conifold transition as in the last paragraph. Altogether we get

(23)

x_i	Γ^j			Λ^a								p_l		
$\mathbb{P}^1$	-1	0	-1	0	0	0	0	0	0	1	0	0	0	-1
$\mathbb{P}^5$	-1	-4	-1	1	1	1	1	1	1	0	2	-3	-3	-2

.

Calculating the topological data,

$$(24)\qquad \begin{aligned} h^\bullet_{\widetilde{\mathcal{M}}}(\widetilde{\mathcal{V}}) &= (0,72,0,0)\,, \\ h^{1,1}_{\widetilde{\mathcal{M}}} + h^{2,1}_{\widetilde{\mathcal{M}}} + h^1_{\widetilde{\mathcal{M}}}(\mathrm{End}(\widetilde{\mathcal{V}})) &= 2 + 86 + 290 = 378 \end{aligned}$$

we can verify that the necessary condition for a duality also holds and hence the conjecture extends to $SU(5)$ bundles as well.

3. Landscape Scan

In this section we will review the results that have been obtained in the large landscape scan in [**8**]. In contrast to the examples we just discussed, there we focused on models that arise as deformations of the tangent bundle of some Calabi-Yau subvariety in a toric geometry. Here we get the condition (5) as well as the bundle stability of the initial bundle automatically, which saves a lot of work.

The scanning algorithm: The algorithm that we used to scan through the large sets of configurations was programmed such that it runs through a given set of smooth configurations and produces combinatorially all dual configurations that one obtains by performing the procedure described in 2.1. We made sure to start with configurations that are smooth and have a stable bundle. But by constructing the dual models it may happen that new singularities are produced or also that the bundle is destabilized. We did not sufficiently check for these issues but rather made some necessary checks to test it.[2] The algorithm can be summarized by the following chart:

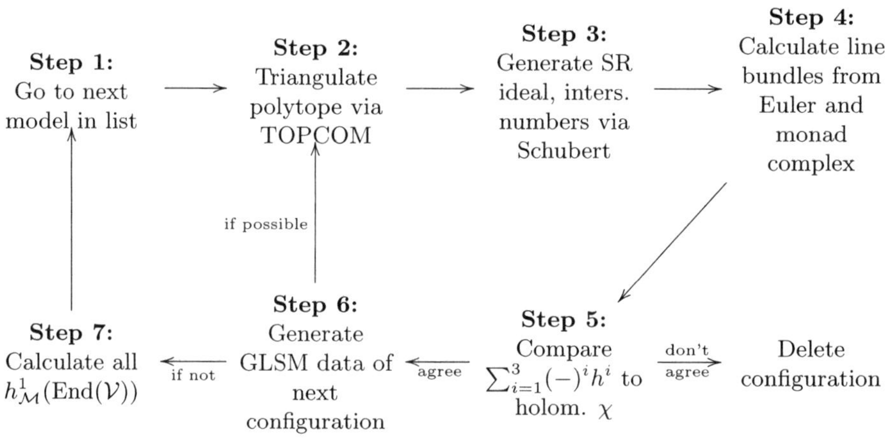

We ran through two different lists (mentioned in step 1). The first one contained Calabi-Yau manifolds defined via single hypersurfaces in toric varieties. We took the ambient spaces out of the list from [**14**] available on the website of Maximilian Kreuzer [**15**] and the second list contains codimension 2 complete intersections in weighted projective spaces which is part of the list presented in [**16**] and available at [**17**]. To resolve the ambient spaces and also to generate the set of nef partitions to obtain the codimension 2 Calabi-Yaus, we used PALP [**18**]. For the remaining steps several packages as TOPCOM [**19**] Schubert [**20**] and of course **cohomCalg Koszul** extension [**13**] along with some Mathematica routines were employed. For the interplay of TOPCOM and Schubert we use the (not published) Toric Triangulizer [**21**].

The results: Our scan ran over the list of hypersurfaces in toric varieties [**15**] where we considered all toric varieties with 7, 8 and 9 lattice points which make altogether 1,085. Additionally we scanned over a large set of complete intersections of hypersurfaces in weighted projective spaces. This list can be found online at [**17**]. For our scan we simply ran through the first 2,780 ambient spaces and chose the 16,029 possible nef partitions as starting points. All these nef partitions correspond to topologically distinct Calabi-Yau manifolds that are complete intersections of two hypersurfaces in the corresponding weighted projective space. Starting from a codimension one Calabi-Yau we performed all first duals to each of those models in the way described in 2.1. Most of the dual models of each hypersurface Calabi-Yau were given by codimension 2 complete intersections in toric varieties. Since

[2]Additionally to this assumption we also assumed for the computation of the bundle moduli that one of the maps in the involved sequences (denoted φ in (2.15) in [**8**]) is surjective.

Co-dim	Different classes	Possibly smooth models	Classes without duals	Models with matching spectrum	Models with full agreement
1	1,085	4,507	42	4,144 (100%)	1509 (94.6%)
2	16,961	79,204	718	64,332 (85%)	20,336 (91%)

TABLE 2. Some data on the landscape study: The codimension is the one of the model we started with. The percent numbers in the parentheses only cover models where these numbers could actually be calculated. In column 6, by "full agreement" we mean that the chiral spectrum of dual models as well as the sum of their complex structure, Kähler and bundle deformations agree with the initial ones.

already many of the duals are obtained by only performing the duality procedure once, we did not perform duals of duals. Similarly, most of the dual geometries coming from an intersection of two hypersurfaces were complete intersections of three hypersurfaces in some toric variety. Some details on the the results can be found in table 2.

4. Conclusions

In this letter we reviewed and extended a method to construct from a given heterotic $(0, 2)$ model, dual models that generically have the same massless spectra. Continuing earlier work we explicitly showed that this procedure actually also works for models that are not deformations of the tangent bundle and may have $SU(3)$, $SU(4)$ or $SU(5)$ structure group by testing some necessary conditions for such models to be dual.

We also presented the results of the landscape scan from earlier work which includes many configurations that actually are all deformations of the tangent bundle. They arise as codimension one and codimension two complete intersections in toric varieties and weighted projective spaces respectively. This scan provided evidence for the fact that the proposed procedure indeed generates dual configurations. Having only tested single examples for the scenario where the model is not a deformation of the tangent bundle and therefore comes generically with $SU(n)$ structure for $n = 3, 4, 5$ it remains to perform a similar larger scan for such models as well. That we did not do it so far has a reason. Namely, it is first of all not very easy to solve (5) in general for a given base geometry. The second thing is that once one has a configuration, one has to check that the bundle is non- singular and that it is furthermore stable. Since this is quite a challenge, we have not yet managed to systematically construct such models. On the other hand, if one could come up with an idea to generate all stable bundles over a given base geometry systematically, it would be no problem to check (5) for those models. This was actually already done for a subset of all bundles over specific base spaces [**22**] and one way to prove bundle stability in a somewhat systematic way for arbitrary base spaces was suggested in [**23**] and [**24**] and gives hope to enable us to overcome this challenge.

Acknowledgment. I would like to thank Ralph Blumenhagen and Benjamin Jurke for discussions and comments as well as Xin Gao for remarks on the article.

References

[1] R. Blumenhagen, B. Jurke, and T. Rahn, "Computational Tools for Cohomology of Toric Varieties," *Adv. High Energy Phys.* **2011** (2749) , arXiv:1104.1187 [hep-th].

[2] E. Witten, "Phases of N = 2 theories in two dimensions," *Nucl. Phys.* **B403** (1993) 159–222, arXiv:hep-th/9301042.

[3] P. S. Aspinwall, B. R. Greene, and D. R. Morrison, "Calabi-Yau moduli space, mirror manifolds and spacetime topology change in string theory," *Nucl. Phys.* **B416** (1994) 414–480, arXiv:hep-th/9309097.

[4] J. Distler and S. Kachru, "Duality of (0,2) string vacua," *Nucl. Phys.* **B442** (1995) 64–74, arXiv:hep-th/9501111.

[5] T.-M. Chiang, J. Distler, and B. R. Greene, "Some features of (0,2) moduli space," *Nucl. Phys.* **B496** (1997) 590–616, arXiv:hep-th/9702030.

[6] R. Blumenhagen, "Target space duality for (0,2) compactifications," *Nucl. Phys.* **B513** (1998) 573–590, arXiv:hep-th/9707198.

[7] R. Blumenhagen, "(0,2) target-space duality, CICYs and reflexive sheaves," *Nucl. Phys.* **B514** (1998) 688–704, arXiv:hep-th/9710021.

[8] R. Blumenhagen and T. Rahn, "Landscape Study of Target Space Duality of (0,2) Heterotic String Models," *JHEP* **09** (2011) 098, arXiv:1106.4998 [hep-th].

[9] R. Blumenhagen, B. Jurke, T. Rahn, and H. Roschy, "Cohomology of Line Bundles: Applications," arXiv:1010.3717 [hep-th].

[10] R. Blumenhagen, B. Jurke, T. Rahn, and H. Roschy, "Cohomology of Line Bundles: A Computational Algorithm," *J. Math. Phys.* **51** (2010) 103525, arXiv:1003.5217 [hep-th].

[11] T. Rahn and H. Roschy, "Cohomology of Line Bundles: Proof of the Algorithm," *J. Math. Phys.* **51** (2010) 103520, arXiv:1006.2392 [hep-th].

[12] S.-Y. Jow, "Cohomology of toric line bundles via simplicial Alexander duality," *Journal of Mathematical Physics* **52** no. 3, (Mar., 2011) 033506, arXiv:1006.0780 [math.AG].

[13] "**cohomCalg** package." Download link, 2010. http://wwwth.mppmu.mpg.de/members/blumenha/cohomcalg/. High-performance line bundle cohomology computation based on [**10**].

[14] M. Kreuzer and H. Skarke, "Complete classification of reflexive polyhedra in four-dimensions," *Adv.Theor.Math.Phys.* **4** (2002) 1209–1230, arXiv:hep-th/0002240.

[15] M. Kreuzer and H. Skarke. http://tph16.tuwien.ac.at/~kreuzer/CY/.

[16] A. Klemm, M. Kreuzer, E. Riegler, and E. Scheidegger, "Topological string amplitudes, complete intersection Calabi-Yau spaces and threshold corrections," *JHEP* **0505** (2005) 023, arXiv:hep-th/0410018 [hep-th].

[17] A. Klemm, M. Kreuzer, E. Riegler, and E. Scheidegger. http://hep.itp.tuwien.ac.at/~kreuzer/CY/hep-th/0410018.html.

[18] M. Kreuzer and H. Skarke, "PALP: A Package for analyzing lattice polytopes with applications to toric geometry," *Comput.Phys.Commun.* **157** (2004) 87–106, arXiv:math/0204356 [math-sc].

[19] J. Rambau, "TOPCOM: Triangulations of Point Configurations and Oriented Matroids," in *Mathematical Software—ICMS 2002*, A. M. Cohen, X.-S. Gao, and N. Takayama, eds., pp. 330–340. World Scientific, 2002. http://www.zib.de/PaperWeb/abstracts/ZR-02-17.

[20] S. Katz, S. A. Stromme, and J.-M. Økland, "Schubert." Package for intersection theory and enumerative geometry, 1992-2006.

[21] B. Jurke, "The Toric Triangulizer." Unpublished C++ wrapper for TOPCOM, Maple/SCHUBERT and associated Mathematica scripts., 2009.

[22] L. B. Anderson, J. Gray, Y.-H. He, and A. Lukas, "Exploring Positive Monad Bundles And A New Heterotic Standard Model," *JHEP* **02** (2010) 054, arXiv:0911.1569 [hep-th].

[23] L. B. Anderson, "Heterotic and M-theory Compactifications for String Phenomenology," arXiv:0808.3621 [hep-th].

[24] L. B. Anderson, J. Gray, A. Lukas, and B. Ovrut, "The Edge Of Supersymmetry: Stability Walls in Heterotic Theory," *Phys. Lett.* **B677** (2009) 190–194, arXiv:0903.5088 [hep-th].

Max-Planck-Institut für Physik, Föhringer Ring 6, 80805 München, Germany
E-mail address: rahn@mpp.mpg.de

Proceedings of Symposia in Pure Mathematics
Volume **85**, 2012

Freed-Witten anomaly and D-brane gauge theories

Fabio Ferrari Ruffino

ABSTRACT. We discuss the different nature of the gauge theories on a D-brane or a stack of D-branes, in type II superstring theory, as follows from the Freed-Witten anomaly. Usually on a D-brane world-volume there is a standard gauge theory, described by the A-field thought of as a connection on a complex vector bundle. Actually, this is a particular case, even if it is the most common one. In order to get a complete picture, within the framework provided by the geometry of gerbes with connection, it is necessary to give a joint geometrical description of the A-field and the B-field, via the language of Čech hypercohomology. The Freed-Witten anomaly, which is a global world-sheet anomaly, imposes some constraints on such fields: we will show for each case what is the nature of the corresponding gauge theory on the D-brane or stack of D-branes.

1. Introduction

In the papers [**2, 6**] we have shown how to classify the admissible configurations of the A-field and the B-field, in the presence of D-branes that are free from the Freed-Witten anomaly. We worked in the framework of type II superstring theory, and the mathematical language we used is the geometry of gerbes with connection, described via the Čech hypercohomology of sheaves. In this paper we summarize how this classification determines the nature of the gauge theory on the world-volume, starting from the definition of "non-integral vector bundles", on which the gauge theory is defined when the B-field is flat. The paper is organized as follows. In section 2 we review the definition of non-integral vector bundles, starting from the case of line bundles. In section 3 we show what are the possible geometrical natures of the gauge theory on a world-volume, within this framework. In section 4 we describe a residual gauge freedom depending on the topology of the space-time, and we show that it has a link with the twisted K-theory classification of D-brane charges. In section 5 we draw our conclusions.

2010 *Mathematics Subject Classification.* Primary 81T50; Secondary 19L50, 53C08.
Key words and phrases. Algebraic and differential topology, D-brane gauge theory.
The author was supported by FAPESP (Fundação de Amparo à Pesquisa do Estado de São Paulo).

2. Non-integral vector bundles

2.1. Line bundles. Let us consider a smooth manifold X with a good cover $\mathfrak{U} = \{U_i\}_{i\in I}$. The group of isomorphism classes of complex line bundles (with Hermitian metric) is isomorphic to the Čech cohomology group $\check{H}^1(X, \underline{U}(1))$, for $\underline{U}(1)$ the sheaf of $U(1)$-valued smooth functions on X. Via the exact sequence of sheaves

$$0 \longrightarrow \mathbb{Z} \longrightarrow \underline{\mathbb{R}} \overset{e^{2\pi i \cdot}}{\longrightarrow} \underline{U}(1) \longrightarrow 0 \tag{2.1}$$

(where $\underline{\mathbb{R}}$ is the sheaf of real smooth functions on X) we get an isomorphism $\check{H}^1(X, \underline{U}(1)) \simeq H^2(X, \mathbb{Z})$. Therefore, a line bundle is completely determined up to isomorphism by its corresponding class in $H^2(X, \mathbb{Z})$, called the first Chern class. If we consider the group of line bundles with connection, it is isomorphic to the Čech hypercohomology group [**3**]:

$$\check{H}^1(X, \underline{U}(1) \overset{\tilde{d}}{\longrightarrow} \Omega^1_{\mathbb{R}}) \tag{2.2}$$

for $\Omega^1_{\mathbb{R}}$ the sheaf of real-valued differential forms of degree 1 on X, and $\tilde{d} := \frac{1}{2\pi i} d \circ \log$. Taking the exterior differential of the connection 1-forms we can define the curvature, which is a global integral (up to a multiplicative constant) 2-form F. If (L, ∇) is a line bundle with connection, $c_1(L)$ the first Chern class and F the curvature, then $[\frac{i}{2\pi} F]_{dR} \simeq c_1(L) \otimes_{\mathbb{Z}} \mathbb{R}$, where $[F]_{dR}$ is the de-Rham cohomology class of F.

We can consider more generally a "non-integral line bundle", which is represented by transition functions $g_{\alpha\beta} : U_{\alpha\beta} \to U(1)$ such that $g_{\alpha\beta} g_{\beta\gamma} g_{\gamma\alpha}$, instead of being 1, is only required to be locally constant. Two set of transition functions are still considered equivalent when they differ by a coboundary of $\underline{U}(1)$. Such transition functions form a cocycle in the Čech cochain complex of the quotient sheaf $\underline{U}(1)/U(1)$, where $U(1)$ denotes the *constant* sheaf. Hence, we can generalize the exact sequence (2.1) to

$$0 \longrightarrow \mathbb{R} \longrightarrow \underline{\mathbb{R}} \overset{e^{2\pi i \cdot}}{\longrightarrow} \underline{U}(1)/U(1) \longrightarrow 0. \tag{2.3}$$

Since a non-integral bundle projects to a class in $\check{H}^1(X, \underline{U}(1)/U(1))$, via the Bockstein map of (2.3) we can associate to it a first Chern class in $H^2(X, \mathbb{R})$, which is therefore not necessarily integral. We can endow a non-integral line bundle with a connection, getting a class in the hypercohomology group:

$$\check{H}^1(X, \underline{U}(1)/U(1) \overset{\tilde{d}}{\longrightarrow} \Omega^1_{\mathbb{R}}). \tag{2.4}$$

We define the curvature in the same way, and we get a global 2-form F which is not necessarily integral any longer, since its de Rham cohomology class still corresponds to the (real) first Chern class of the non-integral line bundle.

2.2. Vector bundles. We can now define "non-integral bundles" of any rank. A rank-n vector bundle (with Hermitian metric) is represented by transition functions $g_{\alpha\beta} : U_{\alpha\beta} \to U(n)$, satisfying the cocycle condition. If we endow it with a connection, it is represented by the transition functions and the connection 1-forms $A_\alpha : TU_\alpha \to i\mathfrak{u}(n)$, for $\mathfrak{u}(n)$ the Lie algebra of $U(n)$, such that $A_\beta - g_{\alpha\beta}^{-1} A_\alpha g_{\alpha\beta} = \frac{1}{2\pi i} g_{\alpha\beta}^{-1} dg_{\alpha\beta}$. The curvature is a local form $F_\alpha = dA_\alpha + A_\alpha \wedge A_\alpha$, whose gauge transformations are the action of the transition functions by conjugation. Hence we can define the Chern classes via the symmetric polynomials P_i, which are invariant by conjugation, as

$$c_i[\{g_{\alpha\beta}, A_\alpha\}] = [P_i(\tfrac{i}{2\pi}F)]. \tag{2.5}$$

As above, we can define a non-integral rank-n vector bundle as a class of transition functions $g_{\alpha\beta} : U_{\alpha\beta} \to U(n)$ such that $g_{\alpha\beta}g_{\beta\gamma}g_{\gamma\alpha}$, instead of being 1, is locally constant, and its value on a triple intersection belongs to the center of $U(n)$, i.e. $U(1)\cdot I_n$ for I_n the identity matrix. This is a particular case of twisted vector bundle [**10**], since the latter is represented by transition functions such that $g_{\alpha\beta}g_{\beta\gamma}g_{\gamma\alpha}$ takes values in the center of $U(n)$, without being locally constant in general. We can endow a non-integral vector bundle with a connection, which is represented by 1-forms $A_\alpha : TU_\alpha \to i\mathfrak{u}(n)$ satisfying, as in the ordinary case, $A_\beta - g_{\alpha\beta}^{-1}A_\alpha g_{\alpha\beta} = \frac{1}{2\pi i}g_{\alpha\beta}^{-1}dg_{\alpha\beta}$. It is easy to show that the gauge transformations of the curvature are still the action by conjugation of the transition functions, contrary to the general case of twisted bundles, so that equations (2.5) still give well-defined cohomology classes in $H^{\mathrm{ev}}(X, \mathbb{R})$, which depend only on the bundle (not on the connection), but they are not necessarily integral. Thus, even the Chern character, defined as usual as

$$\mathrm{ch}[\{g_{\alpha\beta}, A_\alpha\}] = [\mathrm{Tr}\,\exp(\tfrac{i}{2\pi}F)],$$

in the case of non-integral bundles takes values in a lattice which is different from the one of ordinary vector bundles.

3. Gauge theories on a D-brane

We now show how non-integral bundles are related to the gauge theory on a D-brane or stack of D-branes world-volume, in the framework of type II superstring theory with non-trivial B-field in general.

3.1. Single D-brane. We recall that, for a single D-brane, the world-sheet path-integral measure contains the terms [**7**]:

$$e^{iS} = \mathrm{pfaff}\, D_\phi \cdot \exp\Big(2\pi i \cdot \int_\Sigma \phi^* B\Big) \cdot \exp\Big(2\pi i \cdot \int_{\partial\Sigma} \phi^* A\Big) \tag{3.1}$$

where Σ is the world-sheet, ϕ is the trajectory of the world-sheet in the space-time, $\mathrm{pfaff}\, D_\phi$ is the pfaffian of the Dirac operator on the world-sheet, coupled to the space-time via ϕ, and A and B are the A-field and the B-field. The term involving the B-field can be interpreted as the holonomy over Σ of a gerbe with connection [**3**]. We recall that the latter is an element of the hypercohomology group:

$$\check{H}^2(X, \underline{U}(1) \xrightarrow{\tilde{d}} \Omega^1_{\mathbb{R}} \xrightarrow{d} \Omega^2_{\mathbb{R}}). \tag{3.2}$$

In this case, X is the space-time, the connection 2-forms are the local representatives of the B-field, while the curvature, which is a global 3-form, is the H-flux. Actually the B-field should be described in a more refined way [**4, 5**], but this is enough for our purposes. Similarly, the pfaffian $\mathrm{pfaff}\, D_\phi$ is a section of a line bundle over the loop space of Y, which is determined [**3**] by a *flat* gerbe with connection on the world-volume Y, called the *spin^c-gerbe*, whose first Chern class is the third integral Stiefel-Whitney class $W_3(Y) \in H^3(Y, \mathbb{Z})$, and whose flat holonomy is the second

Stiefel-Whitney class $w_2(Y) \in H^2(Y, \mathbb{Z}_2)$ [**11**]. Since Σ is an open surface, such that $\phi(\partial\Sigma) \subset Y$, the product of the two terms

$$\text{pfaff}\, D_\phi \cdot \exp\left(2\pi i \int_\Sigma \phi^* B \right)$$

is in general a section of a line bundle over the space of maps from Σ to X, sending $\partial\Sigma$ to Y, and becomes a well-defined number only when the tensor product of the two gerbes on Y is trivial and trivialized. If we call $[H] \in H^3(X, \mathbb{Z})$ the integral first Chern class of the B-field gerbe, it must therefore happen that

$$W_3(Y) + [H]|_Y = 0 \tag{3.3}$$

and the term $W_3(Y) + [H]|_Y$ is called the *Freed-Witten anomaly* [**7**]. In this case, the A-field provides a reparametrization of the tensor product of the two gerbes, which trivializes it. Therefore, if we represent the B-field gerbe, restricted to Y, as $\{g_{\alpha\beta\gamma}, -\Lambda_{\alpha\beta}, B_\alpha\}$ and the spin^c gerbe as $\{\eta^{-1}_{\alpha\beta\gamma}, 0, 0\}$, for $\eta^{-1}_{\alpha\beta\gamma}$ representing $w_2(Y)$ in the cohomology of the constant sheaf $U(1)$, then it must hold that:

$$\{g_{\alpha\beta\gamma}\eta^{-1}_{\alpha\beta\gamma}, -\Lambda_{\alpha\beta}, B_\alpha\} \cdot \check{\delta}^1\{h_{\alpha\beta}, A_\alpha\} = \{1, 0, B + F\}$$

i.e.

$$\check{\delta}^1\{h_{\alpha\beta}, A_\alpha\} = \{g^{-1}_{\alpha\beta\gamma}\eta_{\alpha\beta\gamma}, \Lambda_{\alpha\beta}, B + F\} \tag{3.4}$$

for $F|_{U_\alpha} = dA_\alpha$. Since the choice of the representatives is arbitrary, we see with the language of Čech hypercohomology the well-known fact that, in general, B and F are not gauge-invariant, because, even if they are chosen to be globally defined, they are subject to large gauge transformations $B \to B + \Phi$ and $F \to F - \Phi$. The situation is different when the H-flux, restricted to the world-volume, is zero, i.e. when the B-field gerbe is flat on the world-volume. In this case, we can choose a cocycle $\{g_{\alpha\beta\gamma}, 0, 0\}$ for the B-field gerbe on the world-volume: this is clear by analogy with flat line bundles, for which, choosing *parallel* sections, the transition functions become locally constant and the connection 1-forms vanish. In this case, (3.4) becomes:

$$\check{\delta}^1\{h_{\alpha\beta}, A_\alpha\} = \{g^{-1}_{\alpha\beta\gamma}\eta_{\alpha\beta\gamma}, 0, F\} \tag{3.5}$$

which is the equation of a *non-integral* line bundle. This is why, in general, when the large gauge transformations can be canonically fixed, we do not always get a gauge theory in the usual sense, but a gauge theory on a non-integral line bundle. In particular, we miss the information about the torsion part (e.g. the Aharonov-Bohm effect in electromagnetism), and the Wilson loop of the A-field is a section of a flat line bundle over the loop space on Y, rather than a well-defined complex number of modulus 1, since it compensates in (3.1) the anomaly of the other two terms.

If it turns out that $w_2(Y) = 0$ and that the B-field gerbe, restricted to Y, has trivial holonomy, than we can choose transition functions equal to 1 for both the gerbes, so that (3.5) becomes

$$\check{\delta}^1\{h_{\alpha\beta}, A_\alpha\} = \{1, 0, F\}. \tag{3.6}$$

Only in this case, which is of course the most common one, do we get a canonical gauge theory, in the sense that it is defined on a line bundle in the usual sense, so that we recover the information even about the torsion part, and we can define the holonomy of the A-field over a loop as a number.

3.2. Stack of D-branes. For a stack of D-branes the action (3.1) becomes [9]

$$e^{iS} = \text{pfaff}\, D_\phi \cdot \exp\left(2\pi i \int_\Sigma \phi^* B \right) \cdot \text{Tr}\, \mathcal{P} \exp\left(2\pi i \int_{\partial\Sigma} \phi^* A \right) \tag{3.7}$$

where $\mathcal{P}$ is the path-ordering operator. The terms involving the B-field and the pfaffian of the Dirac operator remains unchanged, therefore we still need to trivialize the product of the spin^c-gerbe and the B-field gerbe restricted to the D-brane. Thus we still need that

$$\check{\delta}^1\{h_{\alpha\beta}, A_\alpha\} = \{g_{\alpha\beta\gamma}^{-1}\eta_{\alpha\beta\gamma}, \Lambda_{\alpha\beta}, B + \tfrac{1}{n}\text{Tr}F\}, \tag{3.8}$$

but this time the coboundary $\check{\delta}^1$ has a different meaning, since the transition functions $h_{\alpha\beta}$ take values in $U(n)$, thus they are not necessarily a trivialization of $g_{\alpha\beta\gamma}^{-1}\eta_{\alpha\beta\gamma}$ in the $\underline{U}(1)$-cohomology. They define a twisted vector bundle. Anyhow, if $h_{\alpha\beta}h_{\beta\gamma}h_{\gamma\alpha} = h_{\alpha\beta\gamma}\cdot I_n$, then the class $[\{h_{\alpha\beta\gamma}\}] \in H^2(Y, \underline{U}(1)) \simeq H^3(Y,\mathbb{Z})$ must be a *torsion class*, because, taking the determinants, we get $\det h_{\alpha\beta}\cdot\det h_{\beta\gamma}\cdot\det h_{\gamma\alpha} = (h_{\alpha\beta\gamma})^n$, and, since $\det h_{\alpha\beta}$ takes values in $U(1)$, it follows that $(h_{\alpha\beta\gamma})^n$ is a trivial class. One can prove that for any torsion class in $H^3(Y,\mathbb{Z})$ there exists a corresponding twisted vector bundle [1], therefore, if we do not fix the number of D-branes in the stack, equation (3.8) can be satisfied if and only if $[\{g_{\alpha\beta\gamma}^{-1}\eta_{\alpha\beta\gamma}\}]$ is a torsion class. Hence, if we call $[h]$ the torsion class $[\{h_{\alpha\beta\gamma}\}]$ determined by the twisted bundle, the Freed-Witten anomaly cancellation for a stack of D-branes becomes [9]

$$[H]|_Y + [h] = W_3(Y), \tag{3.9}$$

which can be satisfied by a suitable $[h]$ if and only if $[H]|_Y$ is torsion. Therefore, if we do not impose a constraint on the number of D-branes in the stack, *the only condition that the Freed-Witten anomaly imposes on a world-volume is that the H-flux, restricted to it, is an exact form.* In particular, if H is exact on the whole space-time, there are no Freed-Witten anomalous world-volumes, even if, for some of them, there is a minimum number of D-branes that must be present in the stack, which is bigger than 1.

Even for stacks of D-branes, if the H-flux is not zero when restricted to the D-brane, there are large gauge transformations $B \to B+\Phi$ and $\frac{1}{n}\text{Tr}F \to \frac{1}{n}\text{Tr}F - \Phi$, therefore we cannot fix a gauge theory on the D-brane. When the H-flux is 0 on Y, we can choose a representative $\{g_{\alpha\beta\gamma}, 0, 0\}$ for the B-field gerbe, so that equation (3.8) becomes

$$\check{\delta}^1\{h_{\alpha\beta}, A_\alpha\} = \{g_{\alpha\beta\gamma}^{-1}\eta_{\alpha\beta\gamma}, 0, \tfrac{1}{n}\text{Tr}F\}. \tag{3.10}$$

This is the equation of a *non-integral* vector bundle. Therefore, in the general case, when we can fix a gauge theory on a stack of D-branes, it is a gauge theory on a non-integral vector bundle, which therefore does not necessarily have integral Chern classes. Of course we miss the information about the torsion part, and the Wilson loop of the A-field is a section of a line bundle over the loop space of Y, whose anomaly compensates one of the other two terms of the world-sheet path-integral measure. We remark that, contrary to the case of a single D-brane, the flat holonomy of the B-field gerbe on Y can be *any class in* $H^2(Y, U(1))$, not necessarily one whose corresponding first Chern class in $H^3(X,\mathbb{Z})$ is $W_3(Y)$.

As above, if $w_2(Y) = 0$ and the B-field gerbe, restricted to Y, has trivial holonomy, than we can choose transition functions equal to 1 for both the gerbes, so that (3.10) becomes

$$\check{\delta}^1\{h_{\alpha\beta}, A_\alpha\} = \{1, 0, \tfrac{1}{n}\mathrm{Tr}F\} \tag{3.11}$$

which defines a vector bundle in the usual sense. Only in this case do we recover the information even about the torsion part, and we can define the Wilson loop of the A-field over a closed curve as a number.

4. Residual gauge freedom and twisted K-theory

We have said that, when $w_2(Y) = 0$ and the B-field gerbe, restricted to Y, has trivial holonomy, we get a gauge theory on the world-volume in the usual sense. Actually even this is not always true, unless the space-time manifold satisfies the condition $H_1(X, \mathbb{Z}) = 0$. We show that this fact has a connection with D-brane charge classification via twisted K-theory.

4.1. Single D-brane. Let us consider a space-time X with one single D-brane with world-volume Y, supposing that $w_2(Y) = 0$ and the B-field gerbe is trivial on the whole X. Then the gauge theory on Y should be defined on an ordinary line bundle. Nevertheless, let us suppose that this line bundle is flat. Then, the Wilson loop of the A-field is quantized, which means that it depends only on the homology class of the loop. Therefore, it defines an element of the cohomology group $H^1(Y, U(1))$. The loops we consider in order to compute the holonomy are the images of the boundary of the world-sheet Σ under the trajectories $\phi : \Sigma \to X$, which satisfy $\phi(\partial\Sigma) \subset Y$. This means that the loop $\phi(\partial\Sigma)$, even if it is not homologically trivial on Y in general, is trivial on X, since Σ itself trivializes it. Therefore, if the flat line bundle on Y is the restriction of a flat line bundle on X, then the holonomy over any such loop will be trivial, because it coincides with the holonomy computed on X. Since the holonomy completely determines the connection, therefore the gauge theory, this means that *the restriction to the world-volume of a flat line bundle on the whole space-time is pure gauge.* If we have different world-volumes on X, each of them hosting only one D-brane, this residual gauge freedom concerns the restriction of the *same* flat space-time line bundle to all of the world-volumes, since the strings starting from one D-brane and ending on another provide a way to measure the difference between the holonomies on the two world-volumes; therefore, if we fix this pure gauge for one of them, it is automatically fixed for all the others.

4.2. Stack of D-branes. Let us consider a space-time X with one single *stack* of D-branes with world-volume Y, still supposing that $w_2(Y) = 0$ and the B-field gerbe is trivial on the whole X. Then the gauge theory on Y should be defined on an ordinary vector bundle. Thinking as above, we should argue that the restriction to Y of a flat vector bundle on X is pure gauge. Actually this is not the case, since it must be a *line* bundle anyway. In fact, if we consider n different single D-branes very near to each other, by what we said above the residual gauge freedom is given only by a single space-time flat bundle, not one for each D-brane. When the D-branes get closer and closer until they coincide, we have a symmetry enhancement producing a stack of D-branes, but we can still think of strings starting from one D-brane of the stack and ending on another, unless this choice is not canonical. Thus, we still have the possibility of fixing the gauge up to the restriction to Y of a space-time line bundle. Mathematically, this is due to the following fact. Let us choose a stack of 2 D-branes, and a rank-2 complex vector bundle on X which splits in the direct sum of two line bundles, with a flat connection $\mathcal{A} = A_1 \oplus A_2$. Since $\mathrm{Tr}(AB) \neq \mathrm{Tr}(A)\mathrm{Tr}(B)$ in general for $A, B \in U(n)$, we also have in general that $\mathrm{Tr}(A)\mathrm{Tr}(A^{-1}) \neq n$: therefore, even if $\partial^{(1)}\Sigma$ and $\partial^{(2)}\Sigma$ are cohomologous in X, and we choose the gauge so that $\mathcal{P}e^{\int_{\partial^{(1)}\Sigma}\phi^*\mathcal{A}} \cdot \mathcal{P}e^{\int_{\partial^{(2)}\Sigma}\phi^*\mathcal{A}} = I_n$, it does not mean that the product of the traces is n as for the trivial connection. In fact, in the present example the measure is

$$\begin{aligned} e^{iS} &= \mathrm{Tr}\begin{bmatrix} e^{\int_{\partial^{(1)}\Sigma}\phi^* A} & 0 \\ 0 & e^{\int_{\partial^{(1)}\Sigma}\phi^* A'} \end{bmatrix} \cdot \mathrm{Tr}\begin{bmatrix} e^{\int_{\partial^{(2)}\Sigma}\phi^* A} & 0 \\ 0 & e^{\int_{\partial^{(2)}\Sigma}\phi^* A'} \end{bmatrix} \\ &= \left(e^{\int_{\partial^{(1)}\Sigma}\phi^* A} + e^{\int_{\partial^{(1)}\Sigma}\phi^* A'}\right)\left(e^{\int_{\partial^{(2)}\Sigma}\phi^* A} + e^{\int_{\partial^{(2)}\Sigma}\phi^* A'}\right) \end{aligned}$$

which, expanding the product, provides four terms that, when Y and Y' were different, were the four possibilities for the strings. In this case, since $\partial^{(1)}\Sigma$ and $\partial^{(2)}\Sigma$ are cohomologous, the result is

$$e^{iS} = 2 + e^{\int_{\partial^{(1)}\Sigma}\phi^* A + \int_{\partial^{(2)}\Sigma}\phi^* A'} + e^{\int_{\partial^{(1)}\Sigma}\phi^* A' + \int_{\partial^{(2)}\Sigma}\phi^* A}$$

so that, if we fix A, we can find A'.

In general, if we have more than one stack of D-branes in X, the residual gauge freedom is given by a *unique flat* space-time *line* bundle, restricted to each world-volume.

4.3. Twisted K-theory. From the discussion above, it follows that, when the B-field is trivial, the residual gauge freedom is measured by the space-time cohomology class $H^1(X, U(1))$. Actually $H^1(X, U(1)) \simeq \mathrm{Hom}(H_1(X,\mathbb{Z}), U(1))$; in fact, a flat line bundle on X is given by a representation of $\pi_1(X)$ on $U(1)$, i.e. by a group homomorphism $\rho : \pi_1(X) \to U(1)$. Since $U(1)$ is abelian, ρ uniquely projects to a representation of the abelianization of $\pi_1(X)$, which is $H_1(X,\mathbb{Z})$ [**8**]. Therefore, the group of flat line bundles on X, i.e. $H^1(X, U(1))$, is canonically isomorphic to $\mathrm{Hom}(H_1(X,\mathbb{Z}), U(1))$.[1] Therefore, $H^1(X, U(1)) = 0$ if and only if $H_1(X,\mathbb{Z}) = 0$, hence the residual gauge freedom is not present when the space-time manifold has vanishing first homology group.

Let us consider the case of a stack of n D-branes, for $n \geq 1$, such that the B-field is flat on the world-volume. We consider the twisted K-theory classification of D-brane charges, which has a meaning for $[H]$ torsion on the whole X and $H = 0$ as a form [**12**]. The gauge bundle satisfies $h_{\alpha\beta}h_{\beta\gamma}h_{\gamma\alpha} = h_{\alpha\beta\gamma} \cdot I_n$, where $h_{\alpha\beta}$ takes values in $U(n)$, while $h_{\alpha\beta\gamma}$ takes values in $U(1)$ and is locally constant. In particular, the bundle defines a twisted K-theory class on Y, whose twisting class is $[h] = [\{h_{\alpha\beta\gamma}\}] \in \check{H}^2(Y, \underline{U}(1)) \simeq H^3(Y, Z)$, which is a torsion class since $h_{\alpha\beta\gamma}$ is locally constant. Since, by equation (3.9), $[h] = W_3(Y) - [H]|_Y$, for $i : Y \to X$ the embedding of the world-volume in the space-time, since $[H]$ is torsion on the whole X, there is a well-defined Gysin map $i_! : K_{[h]}(Y) \to K_{[H]}(X)$ [**10**], and the image of the K-theory class $[\{h_{\alpha\beta}\}]$ under this map is the D-brane charge in twisted K-theory. The point is that $K_{[H]}(X)$ is defined only up to *non-canonical* isomorphism. In fact, let us choose two representing cocycles of $[H]$, the latter thought of as an element of $\check{H}^2(X, \underline{U}(1))$, which we call $\{g_{\alpha\beta\gamma}\}$ and $\{g'_{\alpha\beta\gamma}\}$. Then we can define the twisted K-theory groups $K_g(X)$ and $K_{g'}(X)$, generated by twisted bundles with twisting cocycle g and g' respectively: it is easy to see that any cochain $\{g_{\alpha\beta}\}$ such

[1] This could also be proven using the universal coefficient theorem [**8**].

that $\{g'_{\alpha\beta\gamma}\} = \{g_{\alpha\beta\gamma}\} \cdot \check{\delta}^1\{g_{\alpha\beta}\}$ realizes an isomorphism between the two groups, but two different cochains will provide different isomorphisms. Instead, since $[H]$ is torsion and $H = 0$ as a form, let us consider the twisted K-theory group $K_{\mathrm{Hol}(B)}(X)$, where $\mathrm{Hol}(B) \in H^2(X, U(1))$ is the flat holonomy of the B-field: it is defined by *non-integral bundles*, not generically twisted bundles, such that the locally constant twisting cocycle represents the class $\mathrm{Hol}(B)$. One can construct a well-defined Gysin map $i_! : K_{[h]}(Y) \to K_{\mathrm{Hol}(B)}(X)$, where this time $[h]$ is the twisting class $[\{h_{\alpha\beta\gamma}\}]$ of the gauge bundle in the constant sheaf $U(1)$ on Y. Of course, two different holonomies with the same first Chern class will produce non-canonically isomorphic groups, but we do not need to consider this. Is the group $K_{\mathrm{Hol}(B)}(X)$ well-defined? Actually this is the case if and only if $H^1(X, U(1)) = 0$, i.e. if and only if the residual gauge freedom vanishes. In fact, for two locally constant cocycles $\{g_{\alpha\beta}\}$ and $\{g'_{\alpha\beta}\}$ representing $\mathrm{Hol}(B)$, an isomorphism between $K_g(X)$ and $K_{g'}(X)$ is provided by a cochain $\{g_\alpha\}$ such that $\{g'_{\alpha\beta}\} = \{g_{\alpha\beta}\} \cdot \check{\delta}^0\{g_\alpha\}$: such a cochain is defined up to a cocycle, but, if $H^1(X, U(1)) = 0$, every cocycle is a coboundary, and a coboundary does not affect the isomorphism, since it changes the gauge choice for the transition functions but not the bundle, as is easy to see. Therefore the isomorphism is canonical. This shows that the condition $H^1(X, U(1)) = 0$ allows for a well-defined K-theory classification group, both for the ordinary case (in which otherwise we have to quotient out by the residual gauge freedom) and the twisted case.

5. Conclusions

We have seen that the gauge theory on a Freed-Witten anomaly free world-volume Y is not always a gauge theory in the canonical sense, since this can happen only when $w_2(Y) = 0$ and the B-field gerbe, restricted to Y, has trivial holonomy. When the B-field is flat on Y, the gauge theory is defined on a non-integral vector bundle, which has well-defined Chern classes, but not belonging in general to the integral lattice in the de Rham cohomology. Moreover, we miss the information about the torsion part, corresponding for example to the Aharonov-Bohm effect in electromagnetism. For a generic non-flat B-field, there are the large gauge transformations which forbid having any canonically defined gauge theory. Moreover, when the first integral homology class of the space-time is non-vanishing, there is a residual gauge freedom represented by flat line bundles on the space-time, and the condition for this residual freedom to disappear is analogous, for flat B-field, to the one allowing to canonically define the D-brane charges via twisted K-theory, using non-integral bundles instead of generic twisted bundles.

References

[1] M. Atiyah, G. Segal, *Twisted K-theory*, Ukr. Math Bull.1 (2004) 291-334, arXiv:math/0407054

[2] L. Bonora, F. Ferrari Ruffino and R.Savelli, *Classifying A-field and B-field configurations in the presence of D-branes*, JHEP 12 (2008) 078, arXiv:0810.4291

[3] J.-L. Brylinski, *Loop spaces, Characteristic Classes and Geometric Quantization*, Birkhäuser Boston 1993

[4] J. Distler, D. S. Freed, and G. W. Moore, *Orientifold précis*, Proc. Symp. Pure Math. 83 AMS 159-172, arXiv:0906.0795

[5] J. Distler, D. S. Freed, and G. W. Moore, *Spin structures and superstrings*, Surveys in Differential geometry 15, Int'l Press 99-130, arXiv:1007.4581

[6] F. Ferrari Ruffino, *Classifying A-field and B-field configurations in the presence of D-branes - Part II: Stacks of D-branes*, Nuclear Physics, Section B 858 (2012), pp. 377-404, arXiv:1104.2798
[7] D. S. Freed and E. Witten, *Anomalies in String Theory with D-branes*, Asian J. Math **3** (1999) 819, arXiv: hep-th/9907189
[8] A. Hatcher, *Algebraic topology*, Cambridge university press, 2002
[9] A. Kapustin, *D-branes in a topologically non-trivial B-field*, Adv. Theor. Math. Phys. **4** (2000) 127 arXiv:hep-th/9909089.
[10] M. Karoubi, *Twisted bundles and twisted K-theory*, arXiv:1012.2512
[11] H. B. Lawson and M. L. Michelsohn, *Spin geometry*, Princeton, UK: Univ. Pr. (1998) 427 p.
[12] E. Witten, *D-branes and K-theory*, JHEP **9812**, 019 (1998), arXiv: hep-th/9810188

ICMC - Universidade de São Paulo, Avenida Trabalhador são-carlense 400, 13566-590 - São Carlos - SP, Brasil
E-mail address: `ferrariruffino@gmail.com`

Proceedings of Symposia in Pure Mathematics
Volume **85**, 2012

Singularity structure and massless dyons of pure Seiberg-Witten theories with SU and Sp gauge groups

Jihye Seo

ABSTRACT. We study pure Seiberg-Witten (SW) theories with $SU(r+1)$ and $Sp(2r)$ gauge groups. We study the singularity structure of moduli space of the SW curve, which survives as singularity of SW-theory. Using exterior derivative and discriminant operators, we find Argyres-Douglas loci. These complex codimension two loci in the moduli space are easily captured by demanding the order of vanishing of the double discriminant to be high. The intersection locus in the moduli space is cusp-like and node-like for mutually non-local and local pairs of vanishing cycles respectively. We also write down vanishing 1-cycles of SW curves. With $Sp(4)$ curve, we observe that some vanishing 1-cycles jump as we go across Argyres-Douglas loci, giving a concrete example of Argyres-Douglas loci living inside the wall of marginal stability.

1. Introduction

1.1. Seiberg-Witten theory. We study a class of $\mathcal{N}=2$ $d=4$ gauge theories[1]. If it contains at most two derivatives and four fermions (i.e. has low energy effective action), then it enjoys holomorphicity and can be solved exactly thanks to Seiberg-Witten (SW) theory. It was studied for gauge group $SU(2)$ originally [**SW94a, SW94b**], and then other gauge groups followed: $SU(r+1)$ [**KLYT95, AF95**], $SU(r+1)$ with and without matter [**HO95**], $SO(2r)$ [**BL95**], $SO(2r+1)$ [**DS95**], and SO with matter [**Han96**][2]. We will consider gauge groups $SU(r+1)$ (Yang-Mills theories) and $Sp(2r)$. For simplicity we restrict to pure theories, i.e., without matter (flavor).

2010 *Mathematics Subject Classification.* Primary 81T60; Secondary 14H52, 14H81, 81T30.
Key words and phrases. Seiberg-Witten curve, Sp, SU, Argyres-Douglas singularity, discriminant, vanishing 1-cycles.
Most result presented here is based on [**SD**] with Keshav Dasgupta, supported in parts by NSERC grants. We thank Philip Argyres, Heng-Yu Chen, and Al Shapere for helpful correspondences and discussions. Special thanks go to Jacques Distler, Wei Li, Andy Neitzke, and Yuji Tachikawa for crucial feedback and fruitful discussions, especially during the String-Math 2011 conference at UPenn. We thank the organizers for a stimulating atmosphere and the anonymous referee for helpful feedback.

[1]By $\mathcal{N}$ we denote the amount of supersymmetry, and we have 3 spatial and 1 time dimensions ($d=4=3+1$).

[2]By r we denote the rank of the group.

When $\mathcal{N} \geq 2$, we say supersymmetry is extended, and it allows non-trivial anti-commutators between supercharges,

$$\{Q^A, Q^B\} = Z^{AB}, \qquad A, B = 1, 2, \ldots, \mathcal{N}, \tag{1.1}$$

which we call the central charges of the superalgebra. The mass of a state is given by $M \geq \|Z\|$, where equality holds for BPS states.

SW theory consists of a SW curve and a SW differential (or 1-form). The SW curve is a complex curve (or a 2-real dimensional Riemann surface) whose 1-cycles correspond to the BPS states. Writing down the 1-cycle in terms of a symplectic basis with $\mathbb{Z}$ coefficients, we can read off the quantized electro-magnetic (dyonic) charges of the corresponding BPS state.

As reviewed in [**Ler97**], the central charge of a BPS state is given by integrating the 1-form λ_{SW} over the 1-cycle ν,

$$Z = \oint_{\nu} \lambda_{\text{SW}}. \tag{1.2}$$

Assuming λ_{SW} is free of delta-function behaviour, a vanishing 1-cycle gives a massless BPS state[3]. For pure $SU(r+1)$ and $Sp(2r)$ cases, we will see that λ_{SW} does not blow up near vanishing cycles. Therefore, the study of vanishing 1-cycles can teach us about massless BPS states in the system.

Singularity locus of SW curve $\subset$ Singularity locus of SW theory

1.2. Vanishing discriminant of hyperelliptic curves. A polynomial, $f_n(x)$ of a degree n in x can be factorized in terms of its n roots e_i's as below:

$$f_n(x) = \sum_{i=1}^{n} a_i x^i = a_n \prod_{i=1}^{n} (x - c_i), \qquad a_i, c_i \subset \mathbb{C}. \tag{1.3}$$

Its discriminant $\Delta_x(f)$ is defined as

$$\Delta_x(f_n(x)) \equiv a_n^{2n-2} \prod_{i<j} (e_i - e_j)^2, \tag{1.4}$$

and it captures the nature of the roots: the discriminant vanishes if and only if the polynomial has a degenerate root[4]. Because of the square on the RHS of (1.4), it is manifest that the discriminant is symmetric in the roots: the discriminant can also be expressed, in fact, as a polynomial written in terms of the a_i's, the coefficients of the polynomial, instead of its roots.

For an algebraic curve given by $y^2 = f(x)$, the discriminant of the RHS is also the discriminant of the curve. Singularity of the curve is captured by colliding roots on the x-plane, which happens at the discriminant locus $\Delta_x f = 0$. Note that we keep a subscript for the discriminant symbol, as a reminder of which variable we take the discriminant with respect to.

---

[3]For $SU(r+1)$ theories with flavors, [**GST11**] shows that this is true only up to a subtlety related to scaling dimensions. See [**EHIY96**] and [**EH97**] for earlier works on scaling behaviour at Argyres-Douglas loci. For pure $SU(r+1)$ and $Sp(2r)$ theories of our interest here, [**SD**] will address this issue in detail.

[4]The number of repeated roots is called the multiplicity of zero or order of vanishing: later we will ask not only whether a discriminant vanishes, but also how fast it vanishes. The order of vanishing of a double discriminant can tell apart Argyres-Douglas loci.

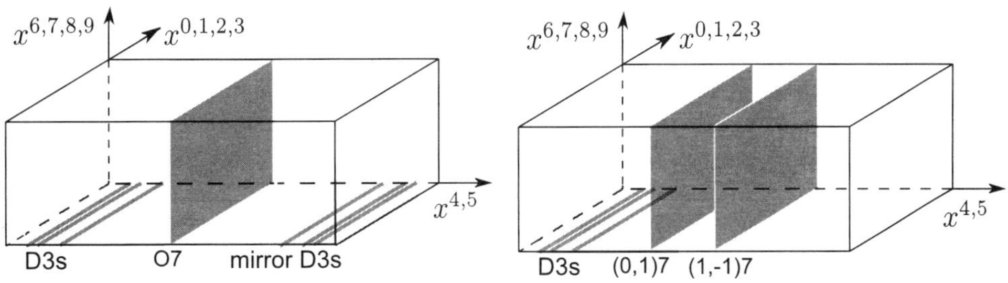

FIGURE 1. In terms of D3-branes probing 7-branes, F-theory describes the $Sp(2r)$ SW theory with antisymmetric matter.

1.3. Motivation. For $SU(n)$, exotic theories were discovered and studied in [**AD95, AF95**], where we have *mutually non-local* massless dyons. Recall from (1.2) that massless dyons are associated with vanishing 1-cycles of the SW-curve. Iff two 1-cycles have non-zero intersection number, we call them mutually non-local. In physics terms, no symplectic transformation will make them electric simultaneously. If the SW curve degenerates into a cusp form, then more than three branch points collide on the x-plane, and then multiple *mutually non-local* (i.e. having non-zero intersection) 1-cycles vanish at the same time. Argyres-Douglas points of $SU(3)$ are studied in detail in [**AF05, GR95, CH96**], and [**MM98, RSVV01**] also discuss wall crossing. In particular, Argyres-Douglas points are on the marginal stability walls [**SV99, GMN09**] and we will give concrete examples of how BPS dyon charges of vanishing cycles change across Argyres-Douglas points.

Another physical motivation to study $Sp(2r)$ curve comes from the F-theory lift of SW theories and the D3/O7 brane construction.

1.3.1. *Brane construction of pure SW $Sp(2r)$ theory.* The original $SU(2)$ SW theory has an F-theory interpretation in terms of a D3-brane probing the quantum corrected O7-plane geometry [**Sen96, BDS96**][5]. F-theory breaks the O7-plane into two (p,q)-7-branes as in figure 1.

When we have r D3-branes probing the same geometry, we obtain $Sp(2r)$ gauge theory [**DLS97, ST08**], which is captured by $Sp(2r)$ SW theory with antisymmetric matter [**AMP02**]. [**DSW11**] asks what the effects of adding extra probe D3-branes are on 7-brane dynamics, and more specifically whether the O7-plane still splits into the same two 7-branes. It will be interesting to see explicit duality and 1-1 mapping between the F-theory picture and the SW theory. As a first step toward this goal, we will study the pure SW theory for simplicity.

2. Monodromy of pure $\mathcal{N}=2$ $SU(r+1)$ theories

The SW curve and SW 1-form for pure $SU(r+1)$ [**KLYT95**] are rewritten as

$$y^2 = f_{SU(r+1)} = f_+ f_-, \quad \lambda_{SW} = -dx \log\left(-\frac{1}{2}(f_+ + f_-) - \sqrt{f_+ f_-}\right), \tag{2.1}$$

[5]Dirichlet branes and orientifold planes are boundary conditions of open string theory. The number next to D and O denotes the spatial dimensions of the object. By $SL(2,\mathbb{Z})$ transformation of a D-brane, we can obtain (p,q)-branes.

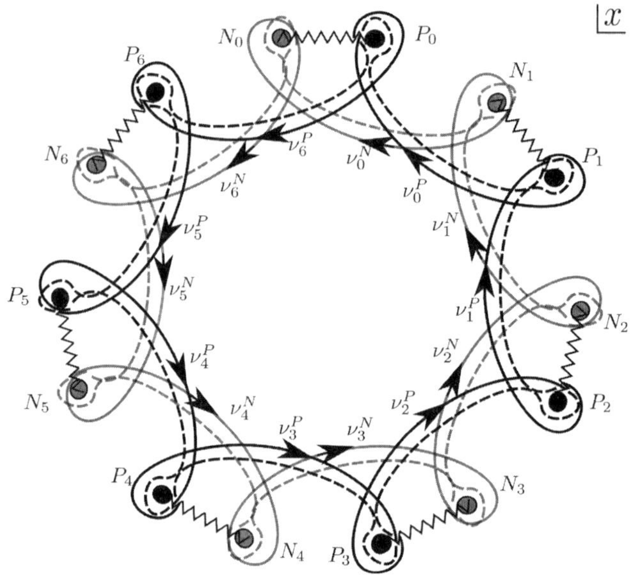

FIGURE 2. Vanishing cycles of the $SU(7)$ curve at a moduli slice (a u_5-plane) given by $u_1 = u_2 = \cdots = u_4 = 0$ and $\frac{u_6}{\Lambda^7} \in \mathbb{I}$

where $f_\pm$ are given in terms of r gauge invariant moduli $u_i \in \mathbb{C}$ $(i = 1, \ldots, r)$ as:

$$f_\pm \equiv x^{r+1} + \sum_{i=1}^{r} u_i x^{r-i} \pm \Lambda^{2r+2}, \quad f_+ = \prod_{i=0}^{r}(x - P_i), \quad f_- = \prod_{i=0}^{r}(x - N_i). \tag{2.2}$$

The discriminant of the curve factorizes as

$$\Delta_x f_{SU(r+1)} = (2\Lambda^{2r+2})^{2r+2} \Delta_x f_+ \Delta_x f_-. \tag{2.3}$$

Recall from (2.2) that $f_+ = f_- = 0$ can occur only if $\Lambda = 0$. Therefore in a quantum-corrected $(\Lambda \neq 0)$ theory, f_+ and f_- can never share a root, justifying the above factorization and binary color coding in figure 2. When $f_+ = \Delta_x f_+ = 0$ $(f_- = \Delta_x f_- = 0)$, a cycle connecting two P_i (and N_i) branch points is vanishing.

At the discriminant locus $\Delta_x f_{SU(r+1)} = 0$ and near the corresponding vanishing 1-cycle, the 1-form of (2.1) is regular

$$\lambda_{SW} = -dx \log\left(\pm\Lambda^{2r+2}\right), \qquad \text{near } f_\pm = \Delta_x f_\pm = 0, \tag{2.4}$$

confirming that the singularity of the SW curve is indeed a singularity of SW theory.

2.1. Vanishing cycles of $SU(r+1)$ curve in a hypersurface of the moduli space. In the moduli space of the $SU(r+1)$ curve, take the hypersurface given as

$$u_1 = u_2 = \cdots = u_{r-2} = 0, \qquad u_r = \text{const.} \tag{2.5}$$

The following factors of the discriminant[6]

$$\Delta_x f_\pm = (-1)^{\left[\frac{r}{2}\right]} r^r (u_{r-1})^{r+1} + (-1)^{\left[\frac{r+1}{2}\right]} (r+1)^{r+1} (u_r \pm \Lambda^{r+1})^r \tag{2.6}$$

[6]Gauss' symbol is used: $[x]$ is the greatest integer that is less than or equal to x.

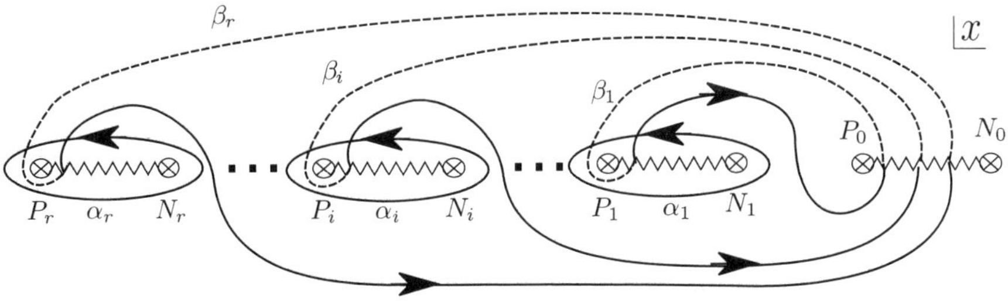

FIGURE 3. A particular choice of symplectic basis cycles for the $SU(r+1)$ curve

give us $2(r+1)$ exact solutions on the u_{r-1}-plane, and they correspond to a singularity related to vanishing 1-cycles. On the u_{r-1}-plane, choose $u_{r-1} = 0$ as a reference point: Starting from here, we make a non-contractible loop surrounding each singular point on the u_{r-1}-plane. At the reference point, P_i (and N_i) points are $\mathbb{Z}_{r+1}$ symmetric among themselves on the x-plane. Demand $\frac{u_r}{\Lambda^{r+1}}$ to be imaginary so that all P_i's are $\mathbb{Z}_2$ symmetric to (i.e. phase π rotation of) N_i's on the x-plane. Therefore all the branch points are on a circle at the reference point, as seen in figure 2. As we vary u_{r-1}, we observe the vanishing cycles on the x-plane as in figure 2. Their only non-vanishing intersection numbers are

$$\nu_i^P \cap \nu_{i+1}^P = \nu_i^N \cap \nu_{i+1}^N = 1, \quad \nu_i^N \cap \nu_i^P = \nu_i^P \cap \nu_{i+1}^N = 2. \tag{2.7}$$

In terms of the symplectic basis given in figure 3, the vanishing cycles are

$$\nu_i^P = \beta_{i+1} - \beta_i - \alpha_i, \quad \nu_i^N = \beta_{i+1} - \beta_i + \alpha_{i+1} - 2\alpha_i, \quad i = 1, \dots, r-1, \tag{2.8}$$

$$\nu_0^P = \beta_1, \quad \nu_0^N = \beta_1 + \sum_{i=1}^{r} \alpha_i + \alpha_1, \quad \nu_r^P = -\beta_r + \sum_{i=1}^{r-1} \alpha_i, \quad \nu_r^N = -\beta_r - 2\alpha_r. \tag{2.9}$$

3. Monodromy of pure $\mathcal{N} = 2$ $Sp(2r)$ theories

As explained in detail [**DSW11**], we take the no-flavor limit of [**AS96**], to obtain the SW curve and 1-form for pure $Sp(2r)$ theory,

$$y^2 = f_{Sp(2r)} = f_C f_Q, \quad \lambda = a\frac{dx}{2\sqrt{x}} \log\left(\frac{xf_C + f_Q + 2\sqrt{x}y}{xf_C + f_Q - 2\sqrt{x}y}\right), \tag{3.1}$$

$$f_C \equiv x^r + \sum_{i=1}^{r} u_i x^{r-i} = \prod_{i=1}^{r}(x - C_i), \quad f_Q \equiv xf_C + 16\Lambda^{2r+2} = \prod_{i=0}^{r}(x - Q_i). \tag{3.2}$$

The branch points of the SW curve are the r roots of f_C, the $r+1$ roots of f_Q, and a point at infinity. Transforming $x \to 1/x$, the branch point at infinity goes to the origin, and [**SD**] explains that a stable singularity does not involve the branch point at infinity. Therefore in figures 4, 7, and 8, the branch point at infinity marked with O_∞ does not participate in any of the vanishing cycles.

Like (2.3) for $SU(r+1)$, the discriminant of the $Sp(2r)$ curve factorizes as

$$\Delta_x f_{Sp(2r)} = (16\Lambda^{2r+2})^{2r} \Delta_x f_C \Delta_x f_Q. \tag{3.3}$$

Recall from (3.2) that $f_C = f_Q = 0$ can occur only if $\Lambda = 0$. Therefore in a quantum-corrected ($\Lambda \neq 0$) theory, f_C and f_Q can never share a root, justifying

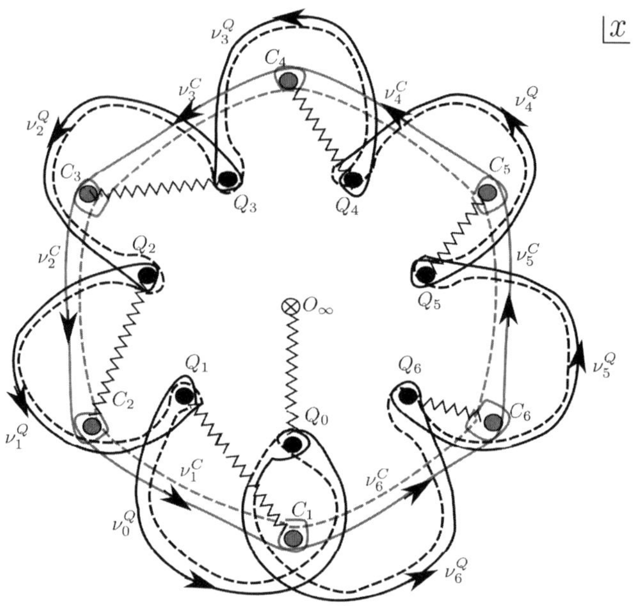

FIGURE 4. Vanishing cycles of the $Sp(12)$ curve at a moduli slice

the above factorization. On the x-plane, only C_i's (or Q_i's) can collide among themselves; from the degree of $\Delta_x f_C$ and $\Delta_x f_Q$ we can read off that each group gives r and $r+1$ vanishing cycles respectively. There is no vanishing cycle mixing these two groups of roots, and the branch point at infinity does not participate at all. The rest of the $2r+1$ branch points and the $2r+1$ vanishing cycles can all be grouped into pairs, which again allows binary color-coding in the following figures 4, 7, and 8.

At discriminant loci $\Delta_x f_{Sp(2r)} = 0$, near the corresponding vanishing 1-cycle, the 1-form of (3.1) becomes infinitesimally small [**DSW11**], far from becoming a delta function. This confirms that singularity of SW curve is indeed singularity of SW theory. At a generic point in $r_{\mathbb{C}}$-dimensional moduli space, $\Delta_x f \neq 0$ and all the branch points are separated. As we bring branch points together, we will go to a subspace with lower dimension. Maximal Argyres-Douglas theories occur at $r+1$ points computed in [**DSW11**], where all the roots of f_Q collide together[7].

3.1. Vanishing cycles of $Sp(2r)$ curve in a moduli slice. Restrict to a u_1-plane of the moduli space by fixing $u_2 = \cdots = u_{r-1} = 0$ and setting u_r to be a fixed small number. Choose $u_1 = 0$ as a reference point. Up to rank 6, if we choose u_r to be small enough[8], then branch points on x-plane are arranged such that all the Q_i's are surrounding origin O_∞, and all the C_i's are surrounding all

[7]Scaling behaviour at maximal Argyres-Douglas points for pure ABCDE SW theory were studied in [**EHIY96**] and [**EH97**], and there are two such points in moduli space for ADE groups. For B and C, number of maximal Argyres-Douglas points are $2r-1$ and $r+1$ [**Seo**], and scaling behaviour is also being studied there.

[8]For example, choosing u_r to be $-1/2$ (for rank $r=2$), $1/5$ ($r=3$), $1/7$ ($r=4,5$), and $1/9$ ($r=6$), we obtain the similar configuration on the x-plane. All with units of $16\Lambda^{2r+2}=1$.

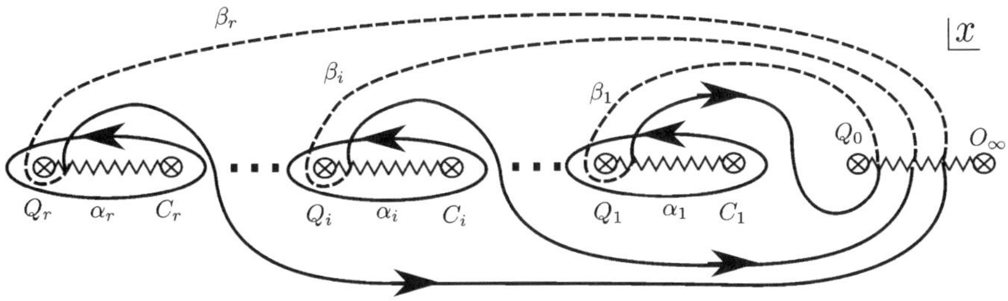

FIGURE 5. A particular choice of symplectic basis cycles for the $Sp(2r)$ curve

the Q points. Vanishing cycles have following non-zero intersection numbers [**SD**]

$$\nu_i^Q \cap \nu_{i+1}^Q = \nu_i^C \cap \nu_{i+1}^C = -1, \quad \nu_i^Q \cap \nu_i^C = \nu_i^C \cap \nu_{i-1}^Q = 2, \quad i = 1, \ldots, r, \tag{3.4}$$

$$\nu_r^Q \cap \nu_0^Q = -3, \quad \nu_0^Q \cap \nu_r^C = \nu_1^C \cap \nu_r^Q = 2. \tag{3.5}$$

Figure 4 shows such a configuration for the rank 6 case. [**SD**] conjectures that for any rank r, it is always possible to choose u_r to be small enough such that all the Q_i points are inside the ν^C cycles, such that (3.4) and (3.5) hold. Below we write down these $2r+1$ vanishing cycles in terms of the symplectic basis given in figure 5:

$$\nu_i^C = -\beta_i + \beta_{i+1} + \alpha_{i+1}, \quad \nu_i^Q = \nu_i^C - \alpha_i + \alpha_{i+1}, \quad i = 1, \ldots, r-1, \tag{3.6}$$

$$\nu_r^C = \beta_1 - \beta_r - \sum_{i=2}^{r} \alpha_i, \quad \nu_0^Q = \beta_1 + 2\alpha_1, \quad \nu_r^Q = \nu_r^C + \beta_1 + \alpha_1 - \alpha_r. \tag{3.7}$$

4. Jumping BPS spectra across Argyres-Douglas points of $Sp(4)$

So far we studied vanishing cycles within a complex codimension one hypersurface of the moduli space. Here we will see how vanishing cycles change as we vary the choice of hypersurface. We present here a detailed example of jumping BPS spectra for $Sp(4)$ SW curve. Its 2-complex dimensional moduli space is spanned by $\{u = u_1, v = u_2\} \subset \mathbb{C}$. As shown on the left of figure 6, we take a 3-real dimensional slice by restricting the phase of v to satisfy $v^3 \in \mathbb{R}$[9].

On the right of figure 6, each of the five u-planes, marked with (a) to (e), are slices of the moduli space at different magnitude of v. The complex codimension one discriminant loci of the SW curve, $\Delta_x f = 0$, are marked by Σ's, where at least one dyon becomes massless. They intersect at more singular complex codimension two loci specified by $\Delta_x f = d(\Delta_x f) = 0$: two dyons become massless simultaneously. Here the *double discriminant* also vanishes $\Delta_u \Delta_x f = 0$, as seen by colliding points on u-planes marked (b) and (d) on the right of the figure 6, which is as expected from [**ARPSW96**].

When Σ's intersect, they can intersect like nodes and cusps, as in (b) and (d) of figure 6 respectively. When this happens, the SW curve itself degenerates into either a node or a cusp, each leading to a different kind of singularity - mutually local

[9]A similar study was done in [**KLYT94**] for $SU(3)$ and in [**LPG97**] for G_2 gauge group, where they also take a slice in moduli space to show the discriminant loci and their intersection.

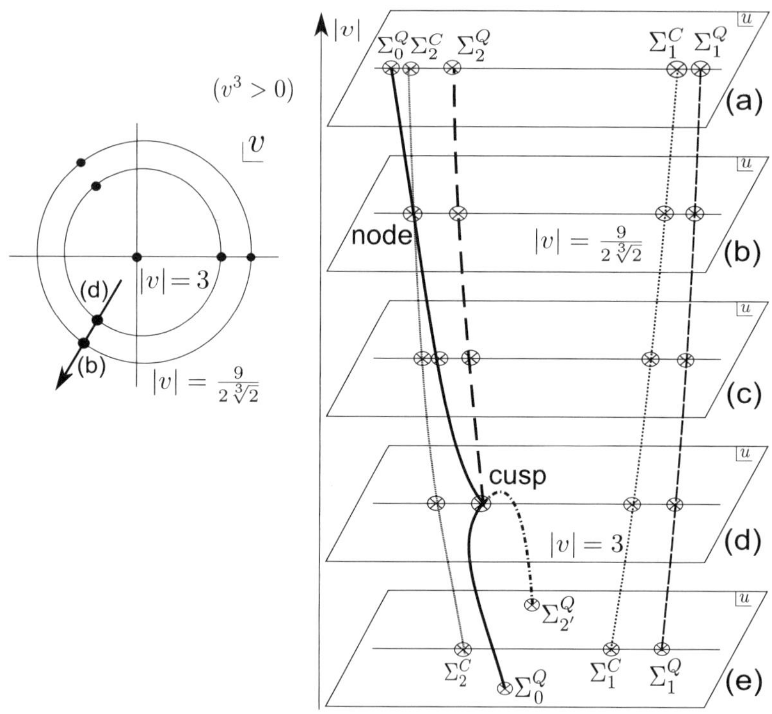

FIGURE 6. Singularity structure of pure $Sp(4)$ SW curve in a slice of moduli space

and mutually non-local (Argyres-Douglas) respectively. And this can be captured by the order of vanishing (2 and 3 respectively) of the double discriminant [**SD**][10].

For each u-plane of figures 6, we display vanishing cycles in figures 7 and 8. In (a) of figure 7, all the discriminant loci $u_i^{C,Q}$'s (given by intersection of $\Sigma_i^{C,Q}$'s with each given u-plane) are separated on the u-plane. Each of the 5 singular points on the u-plane ($u_i^{C,Q}$'s) is responsible for one vanishing 1-cycle ($\nu_i^{C,Q}$'s with corresponding choice of sub-/super scripts). By observing the relative trajectory of the branch points on x-plane, we read off the 5 vanishing cycles as below [**SD**]:

$$\nu_0^Q = \beta_1, \quad \nu_1^Q = -\beta_1 + \beta_2 + \alpha_1 + 4\alpha_2, \quad \nu_2^Q = -\beta_1 - \beta_2 - \alpha_1 - 2\alpha_2 \tag{4.1}$$

$$\nu_1^C = -\beta_1 + \beta_2 + 2\alpha_1 + 3\alpha_2, \quad \nu_2^C = -\beta_1 - \beta_2 - \alpha_2 \tag{4.2}$$

As we change the moduli $|v|$ to go from (a) to (b), singular points on the u-plane, u_0^Q and u_2^C, now collide, and the corresponding 1-cycles ν_0^Q and ν_2^C vanish simultaneously at the same place in the moduli space. However these two cycles are mutually local, and the associated branch points on the x-plane collide pairwise ($C_1 \leftrightarrow C_2$, $Q_0 \leftrightarrow Q_1$). The SW curve degenerates into a node form, $y^2 \sim (x - C)^2(x - Q)^2$.

---

[10] We obtained this result rather empirically while studying vanishing cycles, but later learned that it was used implicitly in [**ACSW05, AW05**] to classify rank 2 curves.

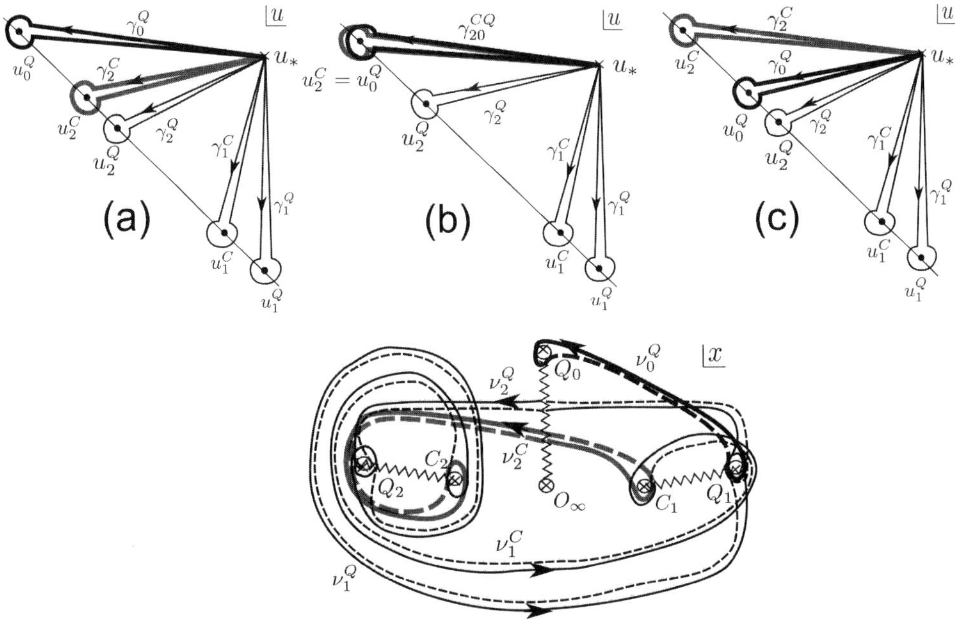

FIGURE 7. Singularity structure of pure the $Sp(4)$ SW curve as one goes through a higher singularity locus of 2 mutually local vanishing cycles

Change the moduli $|v|$ further to reach (c): u_0^Q and u_2^C are separated again as in (a). The vanishing cycles for (a), (b), and (c) remain unchanged as (4.1) and (4.2).

We will see something a little more exotic in figure 8. The discriminant loci collide at (d): two points u_0^Q and u_2^Q on the u-plane collide. Two cycles ν_0^Q and ν_2^Q become massless at the same time, but they are mutually *non*-local. Unlike the case (b) of figure 7, three branch points $Q_{0,1,2}$ collide at the same time on the x-plane. The SW curve degenerates into a cusp form $y^2 \sim (x-Q)^3$, giving Argyres-Douglas theory with $SU(2)$ singularity.

After passing the Argyres-Douglas locus, in (e), we again have 5 separated singular points on u-plane. However, the vanishing cycles are now changed from the cases of (a), (b), and (c) given in (4.1) and (4.2). Instead of ν_2^Q, we have a new vanishing cycle

$$\nu_{2'}^Q = -\beta_2 - \alpha_1 - 2\alpha_2 = \nu_0^Q + \nu_2^Q. \tag{4.3}$$

Considering that Argyres-Douglas loci live inside the wall of marginal stability [**SV99**], this jump of BPS charge is refreshing to observe.

5. Conclusion and open questions

At the discriminant loci $\Delta_x f = 0$ of the SW curve, we have a vanishing 1-cycle. In certain regions of moduli space, we identified all $2r+1$ and $2(r+1)$ vanishing cycles for pure $Sp(2r)$ and $SU(r+1)$ SW curve. With the wall crossing formula, one can in principle write down vanishing cycles everywhere in the moduli space.

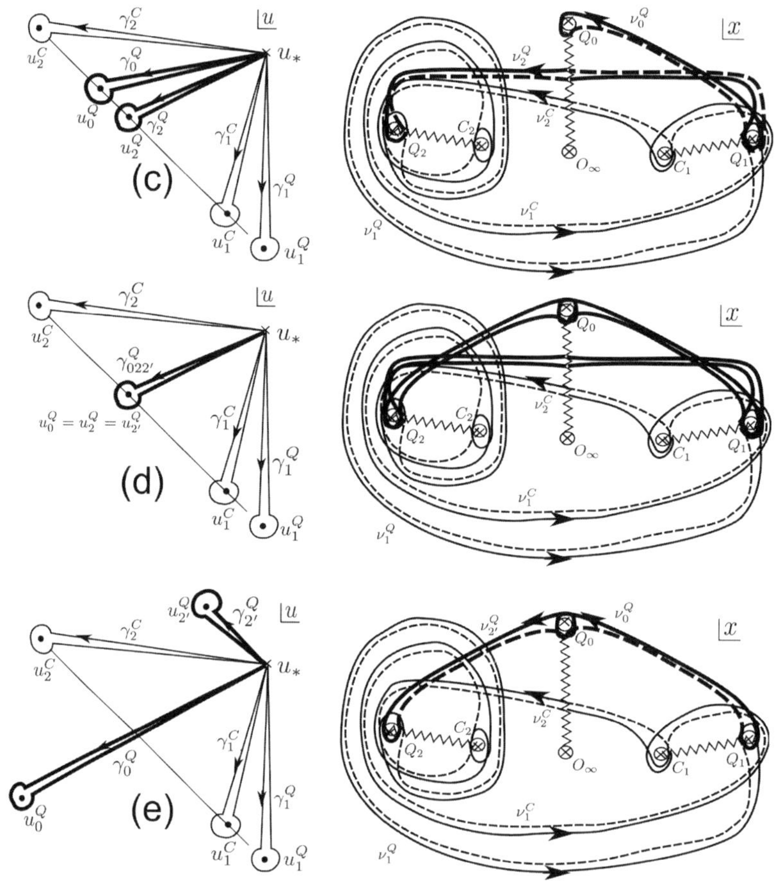

FIGURE 8. Singularity structure of pure the $Sp(4)$ SW curve near an Argyres-Douglas locus

The discriminant loci self-intersect at $d\Delta_x f = \Delta_x f = 0$, where the double discriminant also vanishes $\Delta_u \Delta_x f = 0$ (the converse does not hold [**SD**]). There multiple 1-cycles vanish simultaneously (or massless dyons coexist). If the order of vanishing of the double discriminant is 3, then we are at the Argyres-Douglas loci.

It will be interesting to study the detailed behaviour in the Argyres-Douglas neighbourhood. We are also curious to check how much the $SU(r+1)$ and $Sp(2r)$ theories resemble each other, which may happen due to their shared dual Coxeter number [**Kuc10**].

References

[ACSW05] Philip C. Argyres, Michael Crescimanno, Alfred D. Shapere, and John R. Wittig, *Classification of $N = 2$ superconformal field theories with two-dimensional Coulomb branches. I*, `arXiv:hep-th/0504070`.

[AD95] Philip C. Argyres and Michael R. Douglas, *New phenomena in SU(3) supersymmetric gauge theory*, Nucl.Phys. **B448** (1995), 93–126.

[AF95] Philip C. Argyres and Alon E. Faraggi, *The vacuum structure and spectrum of $N = 2$ supersymmetric $SU(N)$ gauge theory*, Phys. Rev. Lett. **74** (1995), 3931–3934.

[AF05] Nikolas Akerblom and Michael Flohr, *Explicit formulas for the scalar modes in Seiberg-Witten theory with an application to the Argyres-Douglas point*, JHEP **02** (2005), 057.

[AMP02] Philip C. Argyres, R. Maimon, and Sophie Pelland, *The M theory lift of two 06-planes and four D6-branes*, JHEP **0205** (2002), 008.

[ARPSW96] Philip C. Argyres, M. Ronen Plesser, Nathan Seiberg, and Edward Witten, *New N=2 Superconformal Field Theories in Four Dimensions*, Nucl. Phys. **B461** (1996), 71–84.

[AS96] Philip C. Argyres and Alfred D. Shapere, *The Vacuum Structure of N=2 SuperQCD with Classical Gauge Groups*, Nucl. Phys. **B461** (1996), 437–459.

[AW05] Philip C. Argyres and John R. Wittig, *Classification of $N = 2$ superconformal field theories with two-dimensional Coulomb branches. II*, `arXiv:hep-th/0510226`.

[BDS96] Tom Banks, Michael R. Douglas, and Nathan Seiberg, *Probing F-theory with branes*, Phys. Lett. **B387** (1996), 278–281.

[BL95] A. Brandhuber and K. Landsteiner, *On the monodromies of N=2 supersymmetric Yang-Mills theory with gauge group SO(2n)*, Phys.Lett. **B358** (1995), 73–80.

[CH96] Martin Cederwall and Magnus Holm, *Monopole and Dyon Spectra in N=2 SYM with Higher Rank Gauge Groups*, `arXiv:hep-th/9603134`.

[DLS97] Michael R. Douglas, David A. Lowe, and John H. Schwarz, *Probing F-theory with multiple branes*, Phys. Lett. **B394** (1997), 297–301.

[DS95] Ulf H. Danielsson and Bo Sundborg, *The Moduli space and monodromies of N=2 supersymmetric SO(2r+1) Yang-Mills theory*, Phys. Lett. **B358** (1995), 273–280.

[DSW11] Keshav Dasgupta, Jihye Seo, and Alisha Wissanji, *F-Theory, Seiberg-Witten Curves and $N = 2$ Dualities*, `arXiv:1107.3566 [hep-th]`.

[EH97] Tohru Eguchi and Kentaro Hori, *N=2 superconformal field theories in four-dimensions and A-D-E classification.*

[EHIY96] Tohru Eguchi, Kentaro Hori, Katsushi Ito, and Sung-Kil Yang, *Study of $N = 2$ Superconformal Field Theories in 4 Dimensions*, Nucl. Phys. **B471** (1996), 430–444.

[GMN09] Davide Gaiotto, Gregory W. Moore, and Andrew Neitzke, *Wall-crossing, Hitchin Systems, and the WKB Approximation*, `arXiv:0907.3987 [hep-th]`.

[GR95] Amit Giveon and Martin Rocek, *Effective actions and gauge field stability*, Phys. Lett. **B363** (1995), 173–179.

[GST11] Davide Gaiotto, Nathan Seiberg, and Yuji Tachikawa, *Comments on scaling limits of 4d N=2 theories*, JHEP **01** (2011), 078.

[Han96] Amihay Hanany, *On the Quantum Moduli Space of N=2 Supersymmetric Gauge Theories*, Nucl. Phys. **B466** (1996), 85–100.

[HO95] Amihay Hanany and Yaron Oz, *On the quantum moduli space of vacua of $N = 2$ supersymmetric $SU(N_c)$ gauge theories*, Nucl. Phys. **B452** (1995), 283–312.

[KLYT94] A. Klemm, W. Lerche, S. Yankielowicz, and S. Theisen, *On the monodromies of N=2 supersymmetric Yang-Mills theory*, `arXiv:hep-th/9412158 [hep-th]`.

[KLYT95] ———, *Simple singularities and N=2 supersymmetric Yang-Mills theory*, Phys. Lett. **B344** (1995), 169–175.

[Kuc10] Michael Yu. Kuchiev, *Supersymmetric N=2 gauge theory with arbitrary gauge group*, Nucl.Phys. **B838** (2010), 331–357.

[Ler97] W. Lerche, *Introduction to Seiberg-Witten theory and its stringy origin*, Nucl. Phys. Proc. Suppl. **55B** (1997), 83–117.

[LPG97] Karl Landsteiner, John M. Pierre, and Steven B. Giddings, *On the moduli space of $N = 2$ supersymmetric G(2) gauge theory*, Phys. Rev. **D55** (1997), 2367–2372.

[MM98] Marcos Mariño and Gregory W. Moore, *The Donaldson-Witten function for gauge groups of rank larger than one*, Commun. Math. Phys. **199** (1998), 25–69.

[RSVV01] Adam Ritz, Mikhail A. Shifman, Arkady I. Vainshtein, and Mikhail B. Voloshin, *Marginal stability and the metamorphosis of BPS states*, Phys. Rev. **D63** (2001), 065018.

[SD] Jihye Seo and Keshav Dasgupta, *Singularity structure and massless dyons of pure Seiberg-Witten theories with SU and Sp gauge groups (work in progress).*

[Sen96] Ashoke Sen, *F-theory and Orientifolds*, Nucl. Phys. **B475** (1996), 562–578.

[Seo] Jihye Seo, *Scaling behaviour at maximal Argyres-Douglas points of pure* $SO(2r+1)$ *and* $Sp(2r)$ *Seiberg-Witten theory (work in progress).*

[ST08] Alfred D. Shapere and Yuji Tachikawa, *Central charges of N=2 superconformal field theories in four dimensions*, JHEP **0809** (2008), 109.

[SV99] Alfred D. Shapere and Cumrun Vafa, *BPS structure of Argyres-Douglas superconformal theories*, `arXiv:hep-th/9910182`.

[SW94a] N. Seiberg and Edward Witten, *Monopole Condensation, And Confinement In N=2 Supersymmetric Yang-Mills Theory*, Nucl. Phys. **B426** (1994), 19–52.

[SW94b] ———, *Monopoles, duality and chiral symmetry breaking in* $N = 2$ *supersymmetric QCD*, Nucl. Phys. **B431** (1994), 484–550.

Physics Department, Ernest Rutherford Physics Building, McGill University, 3600 University Street, Montréal Quebec, Canada, H3A 2T8

E-mail address: `jihyeseo@hep.physics.mcgill.ca`

Proceedings of Symposia in Pure Mathematics
Volume **85**, 2012

An introduction to the theory of Higher rank stable pairs and Virtual localization

Artan Sheshmani

ABSTRACT. This article is an attempt to briefly introduce some of the results from [**She10**] on development of a higher rank analog of the Pandharipande-Thomas theory of stable pairs [**RR09**] on a Calabi-Yau threefold X. More precisely, we develop a moduli theory for highly frozen triples given by the data $\mathcal{O}_X^{\oplus r}(-n) \xrightarrow{\phi} F$ where F is a sheaf of pure dimension 1. The moduli space of such objects does not naturally determine an enumerative theory: that is, it does not naturally possess a perfect symmetric obstruction theory. Instead, we build a zero-dimensional virtual fundamental class by hand, by truncating a deformation-obstruction theory coming from the moduli of objects in the derived category of X. This yields the first deformation-theoretic construction of a higher-rank enumerative theory for Calabi-Yau threefolds. We include the results of calculations in this enumerative theory for local $\mathbb{P}^1$ using the Graber-Pandharipande [**GP99**] virtual localization technique. We emphasize that in this article we merely include the statement of our theorems and illustrate the final result of some of the computations. The proofs and more detailed calculations will appear elsewhere in a subsequent paper.

Introduction

The general philosophy behind the computation of Gromov-Witten invariants associated to a non-singular projective variety, X, is to do intersection theory over a moduli space that parametrizes curves embedded in X. In order to do intersection theory, one needs to compactify this moduli space. In 1992, Kontsevich [**Kon92**] provided the moduli of stable maps as one example of such compactification. In this case, the resulting intersection numbers are rational; however, there conjecturally exists a set of integers underlying these rational invariants which can be given an enumerative meaning. The second compactification was studied by Richard Thomas [**Tho00**] which resulted in Donaldson-Thomas theory. The invariants computed in Donaldson-Thomas theory are integers but they do not directly count curves either, due to the fact that the one-dimensional subschemes of X contain zero-dimensional subschemes which float around inside X. The third and the most recent compactification is due to Pandharipande and Thomas (PT) [**RR09**] which resulted in a new theory, known as the theory of stable pairs. Having fixed a nonsingular

1991 *Mathematics Subject Classification.* Primary 53D45, 14N35; Secondary 83E30, 81T30.
Key words and phrases. Algebraic geometry, Gromov-Witten and Donaldson-Thomas theory.
The author was supported in part by NSF Grants DMS 0244412, DMS 0555678 and DMS 08-38434 EMSW21-MCTP (R.E.G.S).

Calabi-Yau 3-fold X with $\mathrm{H}^1(\mathcal{O}_X) = 0$, the objects in PT theory of stable pairs are given as tuples (F, s), where F is a coherent sheaf with one dimensional support given by a curve embedded inside X and the section s is defined by the map $s : \mathcal{O}_X \to F$. Moreover, the stability condition for pairs means that the sheaf F is pure with respect to its support and the cokernel of the morphism s is given by a sheaf whose support is at most zero dimensional [**RR09**] (Lemma 1.3). One would like to have a higher rank enumerative theory of Calabi-Yau 3-folds, compute new invariants and possibly relate them to physics (see for example [**Mic09**] and [**Ric08**]). The present paper describes the step by step development of a higher rank analog of PT theory of stable pairs. We consider the higher rank objects in this theory called *frozen triples* given as tuples (E, F, ϕ) where $E \cong \mathcal{O}_X^{\oplus r}(-n)$ (for some r) and F is a pure sheaf with one dimensional support in X. Here $n \gg 0$ is a fixed large-enough integer which is introduced for technical deformation-theoretic reasons. We call the frozen triples of rank r "*of type* (P_F, r)" when F has fixed Hilbert polynomial P_F. Note that given a frozen triple (E, F, ϕ) of type (P_F, r) we do not make a fixed choice of isomorphism $E \cong \mathcal{O}_X^{\oplus r}(-n)$. Having fixed such choice of isomorphism leads to a different moduli of closely related objects called *highly frozen* triples given as tuples (E, F, ϕ, ψ) where (E, F, ϕ) have the same definition as before and $\psi : E \xrightarrow{\cong} \mathcal{O}_X^{\oplus r}(-n)$ is a fixed isomorphism.

Acknowledgement. I would like to thank Sheldon Katz and Tom Nevins for their help and guidance. Their influence is all over my work. I would like to thank Richard Thomas for his continuous invaluable help and sharing with me his understanding of deformation obstruction theories in the derived category. Thanks to the organizers of the String-Math-2011 conference for providing me the opportunity to present this work and to have helpful discussions with the experts interested in this subject.

1. Moduli stacks

The stability condition for frozen and highly frozen triples is compatible with PT stability of stable pairs [**RR09**] (Lemma 1.3). We call this stability condition τ'-limit-stability or in short τ'-stability. By definition [**She10**] (Lemma 4.4), a frozen (respectively highly frozen) triple (E, F, ϕ) of type (P_F, r) is τ'-limit-stable if and only if the map $E \xrightarrow{\phi} F$ has zero dimensional cokernel. We give a construction of the moduli space of τ'-stable frozen and highly frozen triples as stacks. We show that these moduli stacks are given as algebraic stacks. More precisely, the moduli stack of τ'-stable highly frozen triples of type (P_F, r) $\mathfrak{M}^{(P_F,r,n)}_{s,\mathrm{HFT}}(\tau')$ is given as a Deligne-Mumford (DM) stack while the moduli stack of τ'-stable frozen triples of type (P_F, r) $\mathfrak{M}^{(P_F,r,n)}_{s,\mathrm{FT}}(\tau')$ is given as an Artin stack. We also show that $\mathfrak{M}^{(P_F,r,n)}_{s,\mathrm{HFT}}(\tau')$ is a $\mathrm{GL}_r(\mathbb{C})$-torsor over $\mathfrak{M}^{(P_F,r,n)}_{s,\mathrm{FT}}(\tau')$. We summarize these results as follows:

THEOREM 1.1. [**She10**] *(Proposition 5.5, Corollary 6.4, Theorems 6.2, 6.5)*

(1) *Let* $\mathfrak{S}^{(P_F,r,n)}_s(\tau')$ *be the stable locus of the parametrizing scheme of highly frozen triples of type* (P_F, r) [**She10**] *(Section 6.1). Let* $\left[\frac{\mathfrak{S}^{(P_F,r,n)}_s(\tau')}{\mathrm{GL}(V_F)}\right]$ *be the stack-theoretic quotient of* $\mathfrak{S}^{(P_F,r,n)}_s(\tau')$ *by* $\mathrm{GL}(V_F)$ *where* V_F *is defined as in* [**She10**] *(Section 6.1). Then there exists an isomorphism of*

groupoids

$$\mathfrak{M}^{(P_F,r,n)}_{s,\mathrm{HFT}}(\tau') \cong \left[\frac{\mathfrak{S}^{(P_F,r,n)}_s(\tau')}{\mathrm{GL}(V_F)}\right]. \tag{1.1}$$

(2) *Let* $\left[\frac{\mathfrak{S}^{(P_F,r,n)}_s(\tau')}{\mathrm{GL}_r(\mathbb{C})\times\mathrm{GL}(V_F)}\right]$ *be the stack-theoretic quotient of* $\mathfrak{S}^{(P_F,r,n)}_s(\tau')$ *by* $\mathrm{GL}_r(\mathbb{C})\times\mathrm{GL}(V_F)$ *where* V_F. *There exists an isomorphism of groupoids:*

$$\mathfrak{M}^{(P_F,r,n)}_{s,\mathrm{FT}}(\tau') \cong \left[\frac{\mathfrak{S}^{(P_F,r,n)}_s(\tau')}{\mathrm{GL}_r(\mathbb{C})\times\mathrm{GL}(V_F)}\right]$$

(3) *The moduli stack* $\mathfrak{M}^{(P_F,r,n)}_{s,\mathrm{HFT}}(\tau')$ *is a Deligne-Mumford (DM) stack*

(4) *There exists a forgetful map* $g' : \mathfrak{M}^{(P_F,r,n)}_{s,\mathrm{FT}}(\tau') \to \left[\frac{\mathrm{Spec}(\mathbb{C})}{\mathrm{GL}_r(\mathbb{C})}\right]$ *such that the natural diagram:*

$$\begin{array}{ccc} \mathfrak{M}^{(P_F,r,n)}_{s,\mathrm{HFT}}(\tau') & \xrightarrow{\quad g \quad} & pt = \mathrm{Spec}(\mathbb{C}) \\ \pi\downarrow & & \downarrow i \\ \mathfrak{M}^{(P_F,r,n)}_{s,\mathrm{FT}}(\tau') & \xrightarrow{\quad \acute{g} \quad} & \mathcal{B}\,\mathrm{GL}_r(\mathbb{C}) = \left[\frac{\mathrm{Spec}(\mathbb{C})}{\mathrm{GL}_r(\mathbb{C})}\right] \end{array}, \tag{1.2}$$

is a fibered diagram in the category of stacks. In particular $\mathfrak{M}^{(P_F,r,n)}_{s,\mathrm{HFT}}(\tau')$ *is a* $\mathrm{GL}_r(\mathbb{C})$*-torsor over* $\mathfrak{M}^{(P_F,r,n)}_{s,\mathrm{FT}}(\tau')$.

REMARK 1.2. The notion of τ'-stability condition turns out to be a limiting GIT stability and thus one can apply the results of Wandel [**Wan10**] (Section 3) to prove that the DM stack $\mathfrak{M}^{(P_F,r,n)}_{s,\mathrm{HFT}}(\tau')$ has the stronger property of being given as a quasi-projective scheme. One can construct the higher rank theory of stable pairs over a nonsingular projective 3-fold or a noncompact 3-fold such as a toric variety. However in the instance that the base 3-fold is chosen to be non-compact, in order to get well behaved moduli spaces, we require the condition that the one dimensional support of the sheaf F appearing in a frozen or a highly frozen triple is compact. We will show later an example of this situation when X is chosen to be a toric variety given by the total space of $\mathcal{O}_{\mathbb{P}^1}(-1)\oplus\mathcal{O}_{\mathbb{P}^1}(-1)\to\mathbb{P}^1$ (local $\mathbb{P}^1$).

REMARK 1.3. If X has a torus action, then the moduli stacks $\mathfrak{M}^{(P_F,r,n)}_{s,\mathrm{HFT}}(\tau')$ and $\mathfrak{M}^{(P_F,r,n)}_{s,\mathrm{FT}}(\tau')$ inherit it (once we have chosen an equivariant structure on $\mathcal{O}_X(-n)$) [**She10**] (Section 12). Moreover, $\mathfrak{M}^{(P_F,r,n)}_{s,\mathrm{HFT}}(\tau')$ has an extra "*non-geometric*" $\mathrm{T}_0 := (\mathbb{C}^*)^r$ action by rescaling in components of $\mathcal{O}_X(-n)^{\oplus r}$ [**She10**] (Section 13). Therefore, when X is given as local $\mathbb{P}^1$, the natural "*geometric*" $\mathbf{T} := (\mathbb{C}^*)^3$ action on X induces an action on $\mathfrak{M}^{(P_F,r,n)}_{s,\mathrm{FT}}(\tau')$ while $\mathfrak{M}^{(P_F,r,n)}_{s,\mathrm{HFT}}(\tau')$ undergoes an action of $\mathbf{G} := (\mathbb{C}^*)^3\times(\mathbb{C}^*)^r$ which is due to the action of $\mathbf{T}$ together with the additional non-geometric action of T_0.

REMARK 1.4. As shown in [**She10**] (Section 6), the construction of the moduli stacks of stable frozen and highly frozen triples depends on a choice of two fixed large enough integers $n \gg 0$ and $n' \gg 0$. The first integer appears in the description of the triples $E \cong \mathcal{O}_X(-n)^{\oplus r} \to F$ and the second integer is the one for which $F(n')$ becomes globally generated and hence there exists a surjective map $V_F \otimes$

$\mathcal{O}_X(-n') \twoheadrightarrow F$ for some V_F (look at [**She10**] (Section 6.1)). The existence of such V_F is guaranteed according to Wandel [**Wan10**] (Proposition 2.4) where he shows that given a bounded family of stable triples $E \to F$ there exists an integer n' such that for every tuple (E, F) appearing in the family $E(n')$ and $F(n')$ are globally generated over X. The fact that the sheaf $F(n')$ is globally generated for large enough values of n' does not a priori imply that $\mathrm{H}^i(E_2(n)) = 0$ for all $i > 0$ and our fixed choice of n. Hence we introduce the following definition:

DEFINITION 1.5. [**She10**] (Definition 7.1). Consider $\mathfrak{M}^{(P_F,r,n)}_{s,\mathrm{HFT}}(\tau')$ and $\mathfrak{M}^{(P_F,r,n)}_{s,\mathrm{FT}}(\tau')$. Now define the open substacks $\mathfrak{H}^{(P_F,r,n)}_{s,\mathrm{HFT}}(\tau') \subset \mathfrak{M}^{(P_F,r,n)}_{s,\mathrm{HFT}}(\tau')$ and $\mathfrak{H}^{(P_F,r,n)}_{s,\mathrm{FT}}(\tau') \subset \mathfrak{M}^{(P_F,r,n)}_{s,\mathrm{FT}}(\tau')$ as follows:

(1) $\mathfrak{H}^{(P_F,r,n)}_{s,\mathrm{HFT}}(\tau') = \{(E, F, \phi, \psi) \in \mathfrak{M}^{(P_F,r,n)}_{s,\mathrm{HFT}}(\tau') \mid \mathrm{H}^1(F(n)) = 0\}$.
(2) $\mathfrak{H}^{(P_F,r,n)}_{s,\mathrm{FT}}(\tau') = \{(E, F, \phi) \in \mathfrak{M}^{(P_F,r,n)}_{s,\mathrm{FT}}(\tau') \mid \mathrm{H}^1(F(n)) = 0\}$.

From now on all our calculations are carried out over $\mathfrak{H}^{(P_F,r,n)}_{s,\mathrm{HFT}}(\tau')$ and $\mathfrak{H}^{(P_F,r,n)}_{s,\mathrm{FT}}(\tau')$ and the results in the following sections hold true for $\mathfrak{H}^{(P_F,r,n)}_{s,\mathrm{HFT}}(\tau')$ and $\mathfrak{H}^{(P_F,r,n)}_{s,\mathrm{FT}}(\tau')$ only. Also we assume that it is implicitly understood that in the following sections by the "*moduli stack of frozen or highly frozen triples*" we mean the open substack of the corresponding moduli stacks as defined in Definition 1.5. As we will show later, though we prove our results over noncompact stacks, the torus fixed loci of these moduli stacks are compact which enable us to carry out localization computations over them.

2. Deformation-obstruction theories

For a 3-fold X the natural deformation obstruction theories of stable frozen and highly frozen triples fail to provide well behaved complexes of correct amplitude over $\mathfrak{M}^{(P_F,r,n)}_{s,\mathrm{FT}}(\tau')$ and $\mathfrak{M}^{(P_F,r,n)}_{s,\mathrm{HFT}}(\tau')$ and they do not admit virtual cycles. We show that viewing the frozen and highly frozen triples as more complicated objects in $\mathcal{D}^b(X)$ given by $I^\bullet : E \to F$ and computing the fixed-determinant obstruction theory of $I^\bullet$ will be the starting step in finding a well behaved deformation obstruction theory for the moduli stacks of frozen and highly frozen triples. This method has been successfully used in [**RR09**] to obtain an alternative candidate for the obstruction theory of the moduli of stable pairs. It is important to note that in the higher rank case, despite the fact that the object $I^\bullet$ (with the fixed determinant) in the derived category does not distinguish between a frozen or a highly frozen triple, its deformation space does. In other words, it can be shown that given a frozen triple (E, F, ϕ) and a highly frozen triple (E, F, ϕ, ψ), both associated to the same object $I^\bullet \in \mathcal{D}^b(X)$, the space of flat deformations of (E, F, ϕ) and $I^\bullet$ are equally governed by the group $\mathrm{Ext}^1(I^\bullet, I^\bullet)_0$ while the space of flat deformations of (E, F, ϕ, ψ) is not equal to that of $I^\bullet$. We summarize this remark as follows:

THEOREM 2.1. [**She10**] *(Propositions 7.2, 7.4 and Theorem 7.6)*

(1) *Fix a map* $f : S \to \mathfrak{H}^{(P_F,r,n)}_{s,\mathrm{HFT}}(\tau')$. *Let* S' *be a square-zero extension of* S *with ideal* $\mathcal{I}$. *Let* $\mathcal{D}ef_S(S', \mathfrak{H}^{(P_F,r,n)}_{s,\mathrm{HFT}}(\tau'))$ *denote the deformation space of the map* f *obtained by the set of possible deformations,* $f' : S' \to$

$\mathfrak{H}_{s,\mathrm{FT}}^{(P_F,r,n)}(\tau')$ *such that* $f' \mid_S = f$. *The following statement is true:*

$$\mathcal{D}ef_S(S', \mathfrak{H}_{s,\mathrm{HFT}}^{(P_F,r,n)}(\tau')) \cong \mathrm{Hom}(I_S^\bullet, F) \otimes \mathcal{I} \tag{2.1}$$

(2) *Similarly for frozen triples let* $f : S \to \mathfrak{H}_{s,\mathrm{FT}}^{(P_F,r,n)}(\tau')$. *Let* S' *be a square-zero extension of* S *with ideal* $\mathcal{I}$. *Let* $\mathcal{D}ef_S(S', \mathfrak{H}_{s,\mathrm{FT}}^{(P_F,r,n)}(\tau'))$ *denote the deformation space of the map* f *obtained by the set of possible deformations,* $f' : S' \to \mathfrak{H}_{s,\mathrm{FT}}^{(P_F,r,n)}(\tau')$. *The following statement is true:*

$$\mathcal{D}ef_S(S', \mathfrak{H}_{s,\mathrm{FT}}^{(P_F,r,n)}(\tau')) \cong \mathrm{Ext}^1(I_S^\bullet, I_S^\bullet)_0 \otimes \mathcal{I}. \tag{2.2}$$

We show that over $\mathfrak{H}_{s,\mathrm{FT}}^{(P_F,r,n)}(\tau')$ deforming objects in the derived category leads to a 4-term deformation-obstruction complex of perfect amplitude $[-2,1]$:

THEOREM 2.2. [**She10**] *(Theorem 9.5). There exists a map in the derived category given by:*

$$R\pi_{\mathfrak{H}*}\left(R\mathcal{H}om(\mathbb{I}^\bullet, \mathbb{I}^\bullet)_0 \otimes \pi_X^*\omega_X\right)[2] \xrightarrow{ob} \mathbb{L}^\bullet_{\mathfrak{H}_{s,\mathrm{FT}}^{(P_F,r,n)}(\tau')}.$$

After suitable truncations, there exists a 4 term complex $\mathbb{E}^\bullet$ *of locally free sheaves , such that* $\mathbb{E}^{\bullet\vee}$ *is self-symmetric of amplitude* $[-2,1]$ *and there exists a map in the derived category:*

$$\mathbb{E}^{\bullet\vee} \xrightarrow{ob^t} \mathbb{L}^\bullet_{\mathfrak{H}_{s,\mathrm{FT}}^{(P_F,r,n)}(\tau')}, \tag{2.3}$$

such that $h^{-1}(ob^t)$ *is surjective, and* $h^0(ob^t)$ *and* $h^1(ob^t)$ *are isomorphisms. Here* $\mathbb{L}^\bullet_{\mathfrak{H}_{s,\mathrm{FT}}^{(P_F,r,n)}(\tau')}$ *stands for the truncated cotangent complex of the Artin stack* $\mathfrak{H}_{s,\mathrm{FT}}^{(P_F,r,n)}(\tau')$ *which is of amplitude* $[-1,1]$.

The computation of invariants over $\mathfrak{H}_{s,\mathrm{FT}}^{(P_F,r,n)}(\tau')$ requires constructing a well-behaved virtual fundamental class. Note that when X is given as local $\mathbb{P}^1$, our method of computation is to use torus-equivariant cohomology and Graber Pandharipande (GP) virtual localization technique [**GP99**]. At the moment it is not clear to us how to classify the $\mathbf{T}$-fixed loci of $\mathfrak{H}_{s,\mathrm{FT}}^{(P_F,r,n)}(\tau')$. However this obstacle does not exist for the case of highly frozen triples. The existence of the additional non-geometric T_0 action (Remark 1.3) over $\mathfrak{H}_{s,\mathrm{HFT}}^{(P_F,r,n)}(\tau')$ makes it possible to classify the $\mathbf{G}$-fixed locus of $\mathfrak{H}_{s,\mathrm{FT}}^{(P_F,r,n)}(\tau')$ as a finite union of nonsingular compact components. Hence instead of developing a higher rank theory over $\mathfrak{H}_{s,\mathrm{FT}}^{(P_F,r,n)}(\tau')$ we pull back the 4-term deformation obstruction complex in Theorem 2.2 via the forgetful map π in Theorem 1.1 (Diagram (1.2)) and try to construct a virtual fundamental class for $\mathfrak{H}_{s,\mathrm{HFT}}^{(P_F,r,n)}(\tau')$. We summarize by saying that the construction of an enumerative theory over $\mathfrak{H}_{s,\mathrm{HFT}}^{(P_F,r,n)}(\tau')$ has two advantages:

1. The construction of virtual fundamental classes and integration over DM stacks is in much more developed stage than over Artin stacks.
2. There exists a perfect classification of the torus fixed loci of highly frozen triples under the induced action of $\mathbf{G}$ on $\mathfrak{H}_{s,\mathrm{HFT}}^{(P_F,r,n)}(\tau')$ (Remark 1.3) which makes it possible to do the computations using the virtual localization technique [**GP99**].

Note that the complex $\pi^*\mathbb{E}^{\bullet\vee}$ is perfect of amplitude $[-2,1]$ and the main obstacle in constructing a well-behaved deformation obstruction theory over the DM stack $\mathfrak{H}^{(P_F,r,n)}_{s,\mathrm{HFT}}(\tau')$ is to truncate $\pi^*\mathbb{E}^{\bullet\vee}$ into a 2-term complex and define (globally) a well-behaved deformation-obstruton theory of perfect amplitude $[-1,0]$. The simplest solution to this problem is to apply a cohomological truncation operation. Doing so requires obtaining a certain *lifting* map from $g:\Omega_\pi\to\pi^*\mathbb{E}^{\bullet\vee}$ [**She10**] (Proposition 9.12 and Lemma 9.13), taking the mapping cone of this lift (and shifting by -1) and proving that the resulting complex satisfies the conditions of being a perfect obstruction theory for $\mathfrak{H}^{(P_F,r,n)}_{s,\mathrm{HFT}}(\tau')$ [**She10**] (Lemma 9.15). Here Ω_π is the relative cotangent sheaf of $\pi:\mathfrak{H}^{(P_F,r,n)}_{s,\mathrm{HFT}}(\tau')\to\mathfrak{H}^{(P_F,r,n)}_{s,\mathrm{FT}}(\tau')$. This procedure will remove the degree 1 term from the complex $\pi^*\mathbb{E}^{\bullet\vee}$. We also require to remove the degree -2 term of $\pi^*\mathbb{E}^{\bullet\vee}$ which is done by applying the same procedure to the map $g^\vee$: $\pi^*\mathbb{E}^{\bullet}\to\mathrm{T}_\pi$ [**She10**] (Diagram 9.30) obtained from dualizing the map g. We finally obtain a local truncation of $\pi^*\mathbb{E}^{\bullet\vee}$ of perfect amplitude $[-1,0]$ which we denote by $\mathbb{G}^\bullet$. Assuming that $\pi^*\mathbb{E}^{\bullet\vee}$ is given by a 4 term complex of vector bundles:

$$\pi^*E^{-2}\to\pi^*E^{-1}\to\pi^*E^{0}\to\pi^*E^{1}\tag{2.4}$$

it can be seen from our construction [**She10**] (Lemma 9.17) that locally the complex $\mathbb{G}^\bullet$ is given by

$$\pi^*E^{-2}\xrightarrow{d'}\pi^*E^{-1}\oplus T_\pi\to\pi^*E^{0}\oplus\Omega_\pi\xrightarrow{d}\pi^*E^{1}$$

which is quasi-isomorphic to a 2-term complex of vector bundles

$$\operatorname{Coker}(d')\to\operatorname{Ker}(d)\tag{2.5}$$

concentrated in degree -1 and 0.

Remark 2.3. In removing the degree -2 term of $\pi^*\mathbb{E}^{\bullet\vee}$ using the method described above, we exploit the fact that $\pi^*\mathbb{E}^{\bullet\vee}$ is a self-symmetric complex (look at [**She10**] Diagram (9.30) for more discussion).

The existence of the lifting map g is guaranteed Zariski locally over $\mathfrak{H}^{(P_F,r,n)}_{s,\mathrm{HFT}}(\tau')$ but not globally. Hence our strategy is to locally truncate $\pi^*\mathbb{E}^{\bullet\vee}$ as explained above, construct the corresponding local virtual cycles and glue the local cycles to define a globally-defined virtual fundamental class. Our main summarizing theorem of this section is as follows:

Theorem 2.4. [**She10**] *(Theorem 9.11). Consider the 4-term deformation obstruction theory* $\mathbb{E}^{\bullet\vee}$ *of perfect amplitude* $[-2,1]$ *over* $\mathfrak{H}^{(P_F,r,n)}_{s,\mathrm{FT}}(\tau')$.
1. Locally in the Zariski topology over $\mathfrak{H}^{(P_F,r,n)}_{s,\mathrm{HFT}}(\tau')$ *there exists a perfect two-term deformation obstruction theory of perfect amplitude* $[-1,0]$ *which is obtained from the suitable local truncation of the pullback* $\pi^*\mathbb{E}^{\bullet\vee}$.
2. This local theory defines a globally well-behaved virtual fundamental class over $\mathfrak{H}^{(P_F,r,n)}_{s,\mathrm{HFT}}(\tau')$.

Proving the second part of Theorem 2.4 requires an assumption which we explain next.

3. Gluing the local virtual fundamental classes

Let $\mathcal{U} = \coprod_i \mathcal{U}_i$ be an atlas of affine schemes for $\mathfrak{H}^{(P_F,r,n)}_{s,\mathrm{HFT}}(\tau')$. Fix two open charts $\mathcal{U}_\alpha$ and $\mathcal{U}_\beta$ in $\mathcal{U}$. As we mentioned above, it is guaranteed that there exists lifting maps $g_\alpha : \Omega_\pi \mid_{\mathcal{U}_\alpha} \to \pi^*\mathbb{E}^{\bullet\vee} \mid_{\mathcal{U}_\alpha}$ and $g_\beta : \Omega_\pi \mid_{\mathcal{U}_\beta} \to \pi^*\mathbb{E}^{\bullet\vee} \mid_{\mathcal{U}_\beta}$ which lead to our desired locally defined deformation obstruction complexes over $\mathcal{U}_\alpha$ and $\mathcal{U}_\beta$ respectively. However, what essentially guarantees the existence of a globally well behaved virtual fundamental class is the compatibility of these two lifts over the intersection of $\mathcal{U}_\alpha$ and $\mathcal{U}_\beta$, i.e (roughly speaking) the gluing of g_α and g_β. Having fixed the lifting maps g_α and g_β over $\mathcal{U}_\alpha$ and $\mathcal{U}_\beta$ enables one to apply the cohomological truncation operation as described in the previous section and obtain two locally truncated deformation obstruction theories of amplitude $[-1, 0]$ which we denote by $\mathbb{G}^\bullet_\alpha$ and $\mathbb{G}^\bullet_\beta$. Now let $\mathcal{U}_{\alpha\beta}$ denote the intersection of $\mathcal{U}_\alpha$ and $\mathcal{U}_\beta$. Let $\mathbb{G}^\bullet_\alpha \mid_{\mathcal{U}_{\alpha\beta}}$ and $\mathbb{G}^\bullet_\beta \mid_{\mathcal{U}_{\alpha\beta}}$ denote the pullback of $\mathbb{G}^\bullet_\alpha$ and $\mathbb{G}^\bullet_\beta$ to $\mathcal{U}_{\alpha\beta}$. It is easy to see that via the restriction to $\mathcal{U}_{\alpha\beta}$ one obtains two maps in $\mathcal{D}^b(\mathcal{U}_{\alpha\beta})$ given by $\phi_\alpha \mid_{\mathcal{U}_{\alpha\beta}} : \mathbb{G}^\bullet_\alpha \mid_{\mathcal{U}_{\alpha\beta}} \to \mathbb{L}^\bullet_{\mathfrak{H}_{\mathrm{HFT}}} \mid_{\mathcal{U}_{\alpha\beta}}$ and $\phi_\beta \mid_{\mathcal{U}_{\alpha\beta}} : \mathbb{G}^\bullet_\beta \mid_{\mathcal{U}_{\alpha\beta}} \to \mathbb{L}^\bullet_{\mathfrak{H}_{\mathrm{HFT}}} \mid_{\mathcal{U}_{\alpha\beta}}$. The gluability of the virtual cycles obtained from ϕ_α and ϕ_β depends on whether ϕ_α and ϕ_β satisfy the condition of "*giving rise to a semi-perfect obstruction theory*" over $\mathfrak{H}^{(P_F,r,n)}_{s,\mathrm{HFT}}(\tau')$ in the sense of Liang-Li [**Hua11**]:

Definition 3.1. [**Hua11**](Definition 3.1). Let X be a DM stack of finite type over an Artin stack $\mathcal{M}$. A semi perfect obstruction theory over $X \to \mathcal{M}$ consists of an étale covering $\mathcal{U} = \coprod_{\alpha\in\Lambda} \mathcal{U}_\alpha$ of X by schemes, and a truncated perfect relative obstruction theory $\phi_\alpha : \mathbb{G}^\bullet_\alpha \to \mathbb{L}^\bullet_{\mathcal{U}_\alpha/\mathcal{M}}$ for each $\alpha \in \Lambda$ such that

(1) for each α, β in Λ there is an isomorphism

$$\psi_{\alpha\beta} : \mathrm{H}^1(\mathbb{G}^\bullet_\alpha \mid_{\mathcal{U}_{\alpha\beta}}) \xrightarrow{\cong} \mathrm{H}^1(\mathbb{G}^\bullet_\beta \mid_{\mathcal{U}_{\alpha\beta}})$$

so that the collection $(\mathrm{H}^1(\mathbb{G}^\bullet_\alpha), \psi_{\alpha\beta})$ forms a descent datum of sheaves.

(2) For any pair $\alpha, \beta \in \Lambda$ the obstruction theories $\phi_\alpha \mid_{\mathcal{U}_{\alpha\beta}}$ and $\phi_\beta \mid_{\mathcal{U}_{\alpha\beta}}$ are ν-equivalent [**Hua11**] (Definition 2.9).

It is easy to see that ϕ_α and ϕ_β satisfy the condition (2) [**She10**] (Proposition 10.4). However, (1) is a technical condition whose satisfaction depends on finding a specific choice of homotopical maps $h_{\alpha\beta} : g_\alpha \to g_\beta$ and $h^\vee_{\alpha\beta} : g^\vee_\alpha \to g^\vee_\beta$ where the maps g and $g^\vee$ are the lifting maps defined in the previous section. In the present article and moreover in [**She10**] we assume that such homotopical maps are given to us and we carry out our calculation of invariants over a specific toric Calabi-Yau 3-fold. However, finding rigorously the required homotopy maps needs further effort. We suspect that when X is given as local $\mathbb{P}^1$ it *seems* possible to give a rigorous construction of such maps [**She11**].

Remark 3.2. It is important to note that essentially (as we will show later) the result of our calculations do not depend on the choice of the homotopy maps $h_{\alpha\beta}$ and $h^\vee_{\alpha\beta}$. In other words, the existence of the so-called well-defined homotopy maps will guarantee the existence of a theory of highly frozen triples for Calabi-Yau threefolds but no matter what choice of homotopy maps we make, it does not have any effect on the value of the numerical invariants which we calculate in this theory using the equivariant computations.

4. Torus-fixed loci and virtual localization computations

Let X be given as the total space of $\mathcal{O}_{\mathbb{P}^1}(-1)\oplus\mathcal{O}_{\mathbb{P}^1}(-1)\to\mathbb{P}^1$ (local $\mathbb{P}^1$). Consider the ample line bundle over X given by $\mathcal{O}_X(1)$. By our earlier notation, $\mathfrak{H}^{(P_F,r,n)}_{s,\mathrm{HFT}}(\tau')$ and $\mathfrak{H}^{(P_F,r,n)}_{s,\mathrm{FT}}(\tau')$ were defined as in Definition 1.5. It can be seen that when X is given as local $\mathbb{P}^1$ then $\mathfrak{H}^{(P_F,r,n)}_{s,\mathrm{HFT}}(\tau')=\mathfrak{M}^{(P_F,r,n)}_{s,\mathrm{HFT}}(\tau')$ and $\mathfrak{H}^{(P_F,r,n)}_{s,\mathrm{FT}}(\tau')=\mathfrak{M}^{(P_F,r,n)}_{s,\mathrm{FT}}(\tau')$, hence in this section we switch back to our earlier notation and use $\mathfrak{M}^{(P_F,r,n)}_{s,\mathrm{HFT}}(\tau')$ and $\mathfrak{M}^{(P_F,r,n)}_{s,\mathrm{FT}}(\tau')$ instead. As we described (Remark 1.3) there exists an induced action of $\mathbf{G}:=\mathbf{T}\times\mathrm{T}_0$ on $\mathfrak{M}^{(P_F,r,n)}_{s,\mathrm{HFT}}(\tau')$. It can be shown that a torus fixed point in $\mathfrak{M}^{(P_F,r,n)}_{s,\mathrm{HFT}}(\tau')$ corresponds to a $\mathbf{G}$-equivariant highly frozen triple of type (P_F,r) [**She10**] (propositions 12.2, 13.2). The key observation is that a $\mathbf{G}$-equivariant highly frozen triple of rank r is always written as a direct sum of r-copies of $\mathbf{T}$-equivariant PT stable pairs [**She10**] (Remark 13.3 and Remark 13.4):

$$[\mathcal{O}_X^{\oplus r}(-n)\to F]^{\mathbf{T}\times\mathrm{T}_0}\cong\bigoplus_{i=1}^{r}[\mathcal{O}_X(-n)\to F_i]^{\mathbf{T}}. \tag{4.1}$$

The consequence of this result is of significant importance since it enables one to immediately realize that the $\mathbf{G}$-fixed loci of $\mathfrak{M}^{(P_F,r,n)}_{s,\mathrm{HFT}}(\tau')$ are given as r-fold product of $\mathbf{T}$-fixed loci of PT moduli space of stable pairs which are conjectured by Pandharipande and Thomas in [**RR09**] (Conjecture 2) to be nonsingular and compact. Hence, though our original moduli stack is constructed as a non-compact space, its torus-fixed locus is given as a finite union of compact and non-singular components. Let $\mathbf{Q}$ denote a non-singular compact component of the torus fixed locus of $\mathfrak{M}^{(P_F,r,n)}_{s,\mathrm{HFT}}(\tau')$. Let $\mathbb{G}^{\bullet\vee}_{\mathbf{Q}}:G_{0,\mathbf{Q}}\to G_{1,\mathbf{Q}}$ be the dual of the restriction of the deformation obstruction complex in (2.5) to $\mathbf{Q}$. Using the methods described earlier, we construct the virtual fundamental class over all such $\mathbf{Q}$ and obtain the virtual localization formula [**She10**] (Equation 15.5):

$$\left[\mathfrak{M}^r_{\mathrm{s,HFT}}(\tau')\right]^{vir}=\sum_{\mathbf{Q}\subset\mathfrak{M}^r_{\mathrm{s,HFT}}(\tau')}\iota_{\mathbf{Q}*}\left(\frac{e(G_{1,\mathbf{Q}})}{e(G_{0,\mathbf{Q}})}\cdot e(T_{\mathbf{Q}})\cap[\mathbf{Q}]\right). \tag{4.2}$$

Now we compute the difference $[G_{0,\mathbf{Q}}]-[G_{1,\mathbf{Q}}]$ in the $\mathbf{G}$-equivariant $\mathcal{K}$-theory of $\mathbf{Q}$. Consider a point $p\in\mathbf{Q}$ represented by the complex $I^{\bullet\mathbf{G}}:=[\mathcal{O}_X^{\oplus r}(-n)\to F]^{\mathbf{G}}$. The difference $[G_{0,\mathbf{Q}}]-[G_{1,\mathbf{Q}}]$ over this point is the virtual tangent space at this point. We use the quasi isomorphism in (2.5) to compute the virtual tangent space:

$$\begin{aligned}&\mathcal{T}^{\mathbf{Q}}_{I^\bullet}=[\mathrm{Coker}(d')]-[\mathrm{Ker}(d)]=\\&\left([\pi^*E^1]-[\pi^*E^0]+[\pi^*E^{-1}]-[\pi^*E^{-2}]\right)+\left(\cancel{[T_\pi]}-\cancel{[\Omega_\pi]}\right),\end{aligned} \tag{4.3}$$

where E^i for $i=-1,\cdots,2$ are the corresponding terms of $\mathbb{E}^{\bullet\vee}$ in Theorem 2.2 and the cancellation in the second row is due to isomorphism of Ω_π and T_π which is seen from their triviality [**She10**] (Proposition 5.5 and exact triangle 9.21). By the construction of $\mathbb{E}^{\bullet\vee}$ and since the point $p\in\mathbf{Q}$ is represented by the complex $I^{\bullet\mathbf{G}}$

the following identity holds true:

$$\mathcal{T}^{\mathbf{Q}}_{I^{\bullet}} = \sum_{i=0}^{3}(-1)^{i+1}\cdot[\operatorname{Ext}^{i}(I^{\bullet\mathbf{G}}, I^{\bullet\mathbf{G}})_{0}] = [\chi(\mathcal{O}_X, \mathcal{O}_X)] - [\chi(I^{\bullet\mathbf{G}}, I^{\bullet\mathbf{G}})]. \tag{4.4}$$

The description of the virtual tangent space at a point plays an important role in the equivariant vertex and edge calculations [**PT09**] (Section 4.4 and Section 4.5). As it is seen, the right hand side of (4.4) is independent of choice of homotopy maps required for the existence of a globally well-behaved virtual fundamental class (Remark 3.2). Similar to computations in [**PT09**] (Section 4.3) we use Čech cohomology to compute the right hand side of (4.4) [**She10**] (Section 15.2). Moreover, we give a computation of the equivariant vertex and edge for higher rank objects [**She10**] (sections 15.2 and 15.3).

5. An Example

Assume that X is given as local $\mathbb{P}^1$. There exists two affine patches $\mathcal{U}_\alpha$ and $\mathcal{U}_\beta$ covering X. The partitions associated to the Newton polyhedron of X on each patch are given as three dimensional partitions with $\mu_1 = (1), \mu_2 = (0), \mu_3 = (0)$ [**PT09**] (Example 4.9). We compute the vertex associated to the moduli stack of highly frozen triples of rank 2 given by $I^{\bullet} := \mathcal{O}_X^{\oplus 2}(-n) \to F$. Let $\mathcal{U}_\alpha, \mathcal{U}_\beta$ denote affine open patches over the divisors $0, \infty$ on the base $\mathbb{P}^1$ respectively. Let $\mathbb{C}^*$ act on $\mathbb{C}^4$ by $t(x_0, x_1, x_2, x_3) = (tx_0, tx_1, t^{-1}x_2, t^{-1}x_3)$. We identify X as a quotient $X \cong (\mathbb{C}^4\backslash Z)/\mathbb{C}^*$ where $Z \subset \mathbb{C}^4$ is obtained by setting $x_0 = x_1 = 0$. Let $([x_0 : x_1], x_2, x_3)$ denote the coordiantes in X where $[x_0 : x_1]$ denote the homogeneous coordinates along the base $\mathbb{P}^1$ and x_2, x_3 denote the fiber coordinates. Locally in the $\mathcal{U}_\alpha$ and $\mathcal{U}_\beta$ patches the defining coordinates are given as $(\frac{x_1}{x_0}, x_2x_0, x_3x_0)$ and $(\frac{x_0}{x_1}, x_2x_1, x_3x_1)$ respectively. Consider the $\mathcal{U}_\alpha$ patch. Let us denote the local coordinates in this patch by $(\tilde{x}_1, \tilde{x}_2, \tilde{x}_3)$ where $\tilde{x}_1 = \frac{x_1}{x_0}, \tilde{x}_2 = x_2x_0, \tilde{x}_3 = x_3x_0$. Now consider the action of $\mathbf{T} = \mathbb{C}^3$ on X where locally over $\mathcal{U}_\alpha$ is given by $(\lambda_1, \lambda_2, \lambda_3)\cdot\tilde{x}_i = \lambda_i\cdot\tilde{x}_i$. We identify an action of $(\mathbb{C}^*)^2$ on X which preserves the Calabi-Yau form by considering a subtorus $\mathrm{T}' \subset \mathbf{T}$ such that

$$\mathrm{T}' = \{(\lambda_1, \lambda_2, \lambda_3) \in \mathbf{T} \mid \lambda_1\lambda_2\lambda_3 = 1\}. \tag{5.1}$$

Let $\tilde{t}_1, \cdots, \tilde{t}_3$ denote the characters corresponding to the action of λ_i. The equivariant vertex over the patch $\mathcal{U}_\alpha$ is obtained as follows:

$$\begin{aligned} tr_{R_{-\chi((\mathbb{I}^{\bullet})_\alpha, (\mathbb{I}^{\bullet})_\alpha)}} &= \mathbf{F}^{\mathbf{T}}_\alpha \cdot \frac{(w_1^{-1} + w_2^{-1})}{\tilde{t}_1^n} - \frac{\overline{\mathbf{F}^{\mathbf{T}}_\alpha}\cdot(w_1 + w_2)\cdot\tilde{t}_1^n}{\tilde{t}_1\tilde{t}_2\tilde{t}_3} \\ &+ \mathbf{F}^{\mathbf{T}}_\alpha\overline{\mathbf{F}^{\mathbf{T}}_\alpha}\frac{(1-\tilde{t}_1)(1-\tilde{t}_2)(1-\tilde{t}_3)}{\tilde{t}_1\tilde{t}_2\tilde{t}_3} + \frac{1 - \frac{(w_1+w_2)^2}{w_1w_2}}{(1-\tilde{t}_1)(1-\tilde{t}_2)(1-\tilde{t}_3)}, \end{aligned} \tag{5.2}$$

where (w_1, w_2) are defined as the weights of the action of $\mathrm{T}_0 := (\mathbb{C}^*)^{\oplus 2}$ such that w_i for $i = 1, 2$ are given by tuples $(1, 0)$ and $(0, 1)$. Here $\mathbf{F}^{\mathbf{T}}_\alpha$ denotes the equivariant character of the restriction (to the patch $\mathcal{U}_\alpha$) of the sheaf F. Similarly we compute

$tr_{R-\chi((\mathbb{I}^\bullet)_\beta,(\mathbb{I}^\bullet)_\beta)}$ **[She10]** (Equation 16.8). The edge character is obtained as follows:

$$tr_{R-\chi((\mathbb{I}^\bullet)_{\alpha\beta},(\mathbb{I}^\bullet)_{\alpha\beta})} = \Bigg(\mathbf{F}^{\mathbf{T}}_{\alpha\beta}\cdot(w_1^{-1}+w_2^{-1}) - \frac{\overline{\mathbf{F}^{\mathbf{T}}_{\alpha\beta}}\cdot(w_1+w_2)}{\tilde{t}_2\tilde{t}_3}+$$
$$\mathbf{F}^{\mathbf{T}}_{\alpha\beta}\overline{\mathbf{F}^{\mathbf{T}}_{\alpha\beta}}\frac{((1-\tilde{t}_2)(1-\tilde{t}_3)}{\tilde{t}_2\tilde{t}_3} + \frac{1-\frac{(w_1+w_2)^2}{w_1w_2}}{(1-\tilde{t}_2)(1-\tilde{t}_3)}\Bigg)\delta(\tilde{t}_1). \tag{5.3}$$

By compting the values of $\mathbf{F}^{\mathbf{T}}_{\alpha}, \mathbf{F}^{\mathbf{T}}_{\beta}, \mathbf{F}^{\mathbf{T}}_{\alpha\beta}$ we compute the vertex and edge characters in (5.2) and (5.3). The **G**-character of the virtual tangent space in (4.4) is obtained by the following equation:

$$tr_{R-\chi(I^\bullet,I^\bullet)} = tr_{R-\chi(\mathbb{I}^\bullet_\alpha,\mathbb{I}^\bullet_\alpha)} + tr_{R-\chi(\mathbb{I}^\bullet_\beta,\mathbb{I}^\bullet_\beta)} - tr_{R-\chi(\mathbb{I}^\bullet_{\alpha\beta},\mathbb{I}^\bullet_{\alpha\beta})} \tag{5.4}$$

Let $\mathbf{Q}^k$ denote the **G**-fixed component of the moduli stack of rank 2 highly frozen triples over which the highly frozen triples $\mathcal{O}_X(-n)^{\oplus 2\mathbf{G}} \xrightarrow{\phi} F^{\mathbf{G}}$ satisfy the condition that Length(Coker(ϕ)) $= k$. By the definition of the equivariant vertex ($V_{\mathbf{Q}}$) in [**She10**] (Equation 15.34) the coefficient of the degree k term in the vertex is obtained by the integral of the evaluation of the contribution of $V_{\mathbf{Q}^k}$ on $\mathbf{Q}^k$ [**She10**] (Equation 16.12):

$$\begin{aligned} w(\mathbf{Q}^k) &= \int_{\mathbf{Q}^k} e(\mathrm{T}_{\mathbf{Q}^k})e(-V_{\mathbf{Q}^k}) \\ &= \prod_{d_1+d_2=k}\left[\frac{\left(v_1(v_2-1)+\prod_{i=0}^{d_1-1}((i+n)s_1)-(s_2+s_3)\right)}{\left(v_1(1-v_2)+\prod_{i=1}^{d_1}(-1)^i\cdot(i+n)s_1\right)}\right. \\ &\quad \left.\cdot\frac{\left(v_2(v_1-1)+\prod_{i=0}^{d_2-1}((i+n)s_1)-(s_2+s_3)\right)}{\left(v_2(1-v_1)+\prod_{i=1}^{d_2}(-1)^i(i+n)s_1\right)}\right] \end{aligned} \tag{5.5}$$

where s_i denote the equivariant characters corresponding to $\tilde{t}_i$. For similar discussions look at [**PT09**] (Section 4.7) as well as the calculation in [**PT09**] (Lemma 5). By the definition of the Calabi-Yau torus T$'$ in (5.1), s_i satisfy the property that $\mathrm{s}_1+\mathrm{s}_2+\mathrm{s}_3=0$. Hence the generating series associated to the equivariant vertex of highly frozen triples of rank 2 is given by:

$$W^{\mathrm{HFT}}_{1,\emptyset,\emptyset}\mid_{r=2} = \left((1+q)^{\frac{(n+1)(s_2+s_3)}{s_1}}\right)^2 = (1+q)^{-2(n+1)}. \tag{5.6}$$

References

[GP99] T. Graber and R. Pandharipande, *Localization of virtual classes*, Invent. Math. **135** (1999), 487–518.

[Hua11] Huai-Liang Chang and Jun Li, *Semi-perfect obstruction theory and DT invariants of derived objects*, arXiv:1105.3261v1 (2011).

[Kon92] Maxim Kontsevich, *Intersection theory on the moduli space of curves and the matrix Airy function*, Commun. Math. Phys. 147 (1992).

[Mic09] Michele Cirafici, Annamaria Sinkovics, Richard J. Szabo, *Cohomological gauge theory, quiver matrix models and donaldson-thomas theory*, Nuclear physics B **809, Issue 3** (2009), 452–518.

[PT09] R. Pandharipande and R. P. Thomas, *The 3-fold vertex via stable pairs*, Geometry and Topology **13** (2009), 1835–1876.

[Ric08] Michele Cirafici, Annamaria Sinkovics, Richard J. Szabo, *Instantons and Donaldson-Thomas invariants*, Fortschritte der Physik **56, Issue 7-9** (2008), 849–855.

[RR09] R. Pandharipande and R. P. Thomas, *Curve counting via stable pairs in the derived category*, Inventiones **178** (2009), 407–447.

[She10] Artan Sheshmani, *Higher rank stable pairs and virtual localization*, arXiv1011.6342 (2010).

[She11] Artan Sheshmani, *Semi-perfect obstruction theories of higher rank stable pairs over local* $\mathbb{P}^1$, In preparation (2011).

[Tho00] R. P. Thomas, *A holomorphic Casson invariant for Calabi-Yau 3-folds, and bundles on K3-abrations*, J. Differential Geom. **54** (2000), 367–438.

[Wan10] Malte Wandel, *Moduli spaces of stable pairs in Donaldson-Thomas theory*, arXiv:1011.3328v1 (2010).

DEPARTMENT OF MATHEMATICS, UNIVERSITY OF BRITISH COLUMBIA, VANCOUVER, BC V6T 1Z2

E-mail address: artan@math.ubc.ca

URL: www.math.ubc.ca/~artan/Welcome.html

Proceedings of Symposia in Pure Mathematics
Volume **85**, 2012

HMS for punctured tori and categorical mapping class group actions

Nicolò Sibilla

ABSTRACT. Let X_n be a cycle of n projective lines, and $\mathbb{T}_n$ a symplectic torus with n punctures. In this paper we review results of [**STZ**] and [**Si**], which establish a version of homological mirror symmetry relating X_n and $\mathbb{T}_n$, and define on $D^b(Coh(X_n))$ an action of the pure mapping class group of $\mathbb{T}_n$.

1. Introduction.

As originally formulated by Kontsevich [**K**], Homological Mirror Symmetry (from now on, HMS) relates the derived category of coherent sheaves on a smooth projective Calabi-Yau manifold X, $D^b(Coh(X))$, and the Fukaya category of a compact symplectic manifold $\tilde{X}$, by stating that if X and $\tilde{X}$ are mirror partners, then $D^b(Coh(X)) \cong Fuk(\tilde{X})$. Since its proposal, much work has been done towards establishing Kontsevich's conjecture in important classes of examples, see [**PZ, S, Sh**], and references therein.

In [**STZ**], joint with Treumann and Zaslow, we address mirror symmetry in dimension 1, by proving a version of HMS which pairs singular degenerations of elliptic curves, given by cycles of projective lines, and punctured symplectic tori. This result relies on the use of a conjectural model for the Fukaya category of a punctured Riemann surface Σ, which is constructed in terms of a sheaf of dg categories, CPM($-$),[1] defined over the Lagrangian skeleton of Σ.[2] In [**Si**], using the theory of spherical objects and twist functors introduced by Seidel and Thomas in [**SeT**], we test one of the predictions of the mirror symmetry framework developed in [**STZ**], by showing that the (pure) mapping class group of a punctured torus acts by equivalences on the derived category of a cycle of projective lines.

In this paper we will review these results, by focusing on motivations and examples, and keeping the presentation of the arguments as explicit and concrete as possible. Let X_n be a cycle of projective lines, with n components, and $\mathbb{T}_n$ a

1991 *Mathematics Subject Classification.* Primary 53D37, Secondary 14F05 and 32S60.

It is a pleasure to thank David Treumann and Eric Zaslow for our collaborations, and many stimulating discussions.

[1]CPM stands for 'constructible plumbing model.' The name depends on the fact that CPM can be defined in greater generality, and supplies a dg model for the Fukaya category of *plumbings* of cotangent bundles in any dimension [**STZ2**].

[2]HMS for punctured spheres has been investigated also in [**AAEKO**].

symplectic torus with n punctures. We will review the construction of CPM for $\mathbb{T}_n$ in Section 2. The HMS statement relating the CPM model for $Fuk(\mathbb{T}_n)$ and $\mathcal{P}\mathrm{erf}(X_n)$ will be proved in Section 3. We will conclude by giving, in Section 4, a brief overview of the results in [**Si**], focusing on the case of $D^b(Coh(X_2))$.

2. A model for the Fukaya category of punctured Riemann surfaces

Starting in 2009, in various talks, Kontsevich has argued [**K1**] that the Fukaya category of a Stein manifold should have good local-to-global properties, and therefore conjecturally could be recovered as the global sections of a suitable sheaf of dg categories (note also [**S1**], and [**N1**]). This is in keeping with previous work of Nadler and Zaslow who, in [**NZ**] and [**N**], establish an equivalence between the Fukaya category of exact Lagrangians in a cotangent bundle T^*X, and the dg category of (complexes of cohomologically) constructible sheaves over X, $Sh(X)$.

Following Kontsevich's insight, in [**STZ**] we equip the Lagrangian skeleton of a punctured Riemann surface Σ with a sheaf of dg categories,[3] called CPM($-$), such that its local behavior is dictated by Nadler and Zaslow's work on cotangent bundles, while its global sections are conjecturally quasi-equivalent to the Fukaya category of exact Lagrangians in Σ, $Fuk(\Sigma)$. Before discussing the construction of CPM($-$) in Section 2.2, we collect in Section 2.1 below the necessary background on sheaf theory.

2.1. Microlocal sheaf theory in dimension 1.

In [**KS**], Kashiwara and Schapira explain how to attach to a constructible sheaf $\mathcal{F} \in Sh(X)$ a conical (i.e. invariant under fiberwise dilation) Lagrangian subset of T^*X, called *singular support*, and denoted by $SS(\mathcal{F})$. Informally, $SS(\mathcal{F})$ is an invariant encoding the co-directions along which $\mathcal{F}$ does not 'propagate.' Rather than giving the exact definition, for which we refer the reader to Section 5.1 of [**KS**], we will describe in Lemma 2.2 how the singular support works in the cases which will be relevant for us. If $\Lambda \hookrightarrow T^*X$ is a conical Lagrangian subset, denote by $Sh(X,\Lambda) \hookrightarrow Sh(X)$ the full subcategory of constructible sheaves $\mathcal{F} \in Sh(X)$ such that $SS(\mathcal{F}) \subset \Lambda$.

PROPOSITION 2.1. *Let X be a 1-dimensional manifold, let $\Lambda \hookrightarrow T^*X$ be a conical Lagrangian subset, and denote by $\pi : T^*X \to X$ the natural projection. The assigment sending a conical open subset $V \subset \Lambda$, to the dg category $Sh(\pi(V), V)$, can be extended to a sheaf of dg categories, denoted by $MSh(-)$, over Λ equipped with its natural topology.*

PROOF. The proof of the statement is discussed in Section 3.1 of [**STZ**], and depends on the microlocal theory of sheaves developed in [**KS**]. In fact, a similar statement holds in all dimensions [**STZ2**]. □

LEMMA 2.2. *Let X be a 1-dimensional manifold, and $P = \{p_1, p_2 \cdots, p_n\}$ a finite collection of points in X, then $\Lambda_P = X \cup (T^*_{p_1}X \cup \ldots \cup T^*_{p_n}X)$ is a conical Lagrangian, and $Sh(X,\Lambda_P) \subset Sh(X)$ coincides with the full subcategory of constructible sheaves which are locally constant on P, and on $X \setminus P$.*

---

[3] $Sh(X)$ is the dg enhancement of the derived category of constructible shaves of $\mathbb{C}$-vector spaces over X. From now on, we will refer to objects in $Sh(X)$ simply as 'constructible sheaves.' See [**KS**] for a comprehensive introduction to the subject.

$Sh(X, \Lambda_P)$ admits a very simple combinatorial description in terms of quiver representations. Call $\mathcal{S}$ the partition of X given by the points in P, and by the connected components of $X \setminus P$. Denote by $Q_\mathcal{S}$ the quiver whose vertices are the elements of $\mathcal{S}$, and with an arrow joining $S, S' \in \mathcal{S}$, with that orientation, if and only if S is a point, and S' is a sub-interval such that $S \in \overline{S'}$. For example, if P has cardinality 1, then $Q_\mathcal{S}$ is equal to $\bullet \leftarrow \bullet \rightarrow \bullet$ if $X = \mathbb{R}$, and to $\bullet \rightrightarrows \bullet$ if $X = S^1$.

LEMMA 2.3. $Sh(X, \Lambda_P) \cong Rep(Q_\mathcal{S})$.[4]

PROOF. If $\mathcal{F} \in Sh(X, \Lambda)$, by taking its stalks over $p_1, \ldots p_n$, and over points lying on the different components of $X \setminus P$, we obtain a complex of vector spaces for each vertex of $Q_\mathcal{S}$. Further, the restriction maps of $\mathcal{F}$ yield linear maps corresponding to the arrows of $Q_\mathcal{S}$. This prescription maps $\mathcal{F}$ to a representation of $Q_\mathcal{S}$ in a functorial way, and defines the equivalence. □

REMARK 2.4. Under the assumptions of Lemma 2.3, the sheaf $MSh(-)$ can be described explicitly. Assume, for concreteness, that $X = S^1$, and $P = \{p\}$. Also, fix an orientation on T^*S^1, and note that this allows us to decompose $T^*_pS^1$ as the union of 0 and two rays, R^+ and R^-, respectively upward and downward pointing. Below, we describe the sections of $MSh(-)$ on *contractible* open subsets $U \subset \Lambda$, and the assignment defining, on the objects, the restriction functors

$$Res_U : MSh(\Lambda) = Sh(X, \Lambda) \cong Rep(\bullet \rightrightarrows \bullet) \rightarrow MSh(U),$$

the definition on morphisms will be obvious. This is enough to reconstruct $MSh(-)$. Let $M = V_1 \underset{g}{\overset{f}{\rightrightarrows}} V_2$ be an object in $Rep(\bullet \rightrightarrows \bullet)$, then

- if $U \subset S^1$, $MSh(U) \cong \mathbb{C} - mod$, and $Res_U(M) = V_2$,
- if $U \subset R^+$, $MSh(U) \cong \mathbb{C} - mod$, and $Res_U(M) = Cone(f)$,
- if $U \subset R^-$, $Msh(U) \cong \mathbb{C} - mod$, and $Res_U(M) = Cone(g)$,
- if $p \in U$, $MSh(U) \cong Rep(\bullet \leftarrow \bullet \rightarrow \bullet)$, and $Res_U(M) = V_2 \overset{f}{\leftarrow} V_1 \overset{g}{\rightarrow} V_2$.

2.2. The construction of CPM **for** $\mathbb{T}_n$**.** A *ribbon graph* is a graph equipped with a cyclic ordering on the set of half-edges incident to each vertex. Recall that ribbon graphs label cells in the moduli space of punctured Riemann surfaces (see e.g. [**P**]). Further, if the Riemann surface Σ lies in the cell labelled by Γ_Σ, there is an embedding $\Gamma_\Sigma \hookrightarrow \Sigma$, and a nicely behaved retraction of Σ onto Γ_Σ. In the language of Stein geometry, Γ_Σ is the *skeleton* of Σ.

Bracketing issues of valency, given a pair formed by a punctured Riemann surface and its skeleton, $\Gamma_\Sigma \hookrightarrow \Sigma$, we can consider an open covering $\{U_i\}_{i \in I}$ of Γ_Σ with the property that, for all $i \in I$, there is a symplectomorphism $\phi_i : U_i \rightarrow T^*M_i$, where M_i is a 1-dimensional manifold, and $\Lambda_i := \phi_i(U_i \cap \Gamma_\Sigma) \hookrightarrow T^*M_i$ is a conical Lagrangian subset. In conformity with Kontsevich's ansatz, we should be able to recover Fuk(Σ) by first applying Nadler and Zaslow's theory [**NZ, N**] to compute the Fukaya category of each member of the covering family (this yields $Sh(M_i, \Lambda_i)$), and then exploiting the fact that the Fukaya category behaves like a sheaf over Γ_Σ.

Roughly speaking, the theory developed in [**STZ**] formalizes this heuristics by constructing a sheaf of dg categories CPM$(-)$ over Γ_Σ, which is characterized by the property that, when restricted to $U_i \cap \Gamma_\Sigma \cong \Lambda_i$, it coincides with the sheaf $MSh(-)$

[4] $Rep(Q_\mathcal{S})$ denotes the dg derived category of representations. In the proof below and everywhere in the paper, all functors, such as the stalk functor, are implictly assumed to be derived.

over Λ_i introduced in Proposition 2.1. Since $\{U_i\}_{i\in I}$ covers Γ_Σ, this prescription is actually sufficient to compute sections and restriction functors for CPM$(-)$ on arbitrary open subsets $U \subset \Gamma_\Sigma$, and therefore determines CPM$(-)$ uniquely.[5]

CONJECTURE 2.5. CPM(Γ_Σ) *is quasi-equivalent to the Fukaya category of compact exact Lagrangians in* Σ, $Fuk(\Sigma)$

For the purposes of HMS, it is important to understand how this works for a symplectic torus with n punctures, $\mathbb{T}_n$. Let Λ_i, $i \in I = \{1, \dots, n\}$, be a collection of n copies of the conical Lagrangian $\Lambda_p := S^1 \cup T^*_p S^1 \hookrightarrow T^*S^1$. Note that, if we fix an orientation on T^*S^1, Λ_p acquires a natural structure of ribbon graph. For each $i \in I$ there are open embeddings $j_i^+ : \mathbb{R}_{>0} \cong R_i^+ \hookrightarrow \Lambda_i$, and $j_i^- : \mathbb{R}_{>0} \cong R_i^- \hookrightarrow \Lambda_i$, where R_i^+ and R_i^- are defined as in Remark 2.4. Denote by Γ_n the ribbon graph constructed as the push-out of the diagram

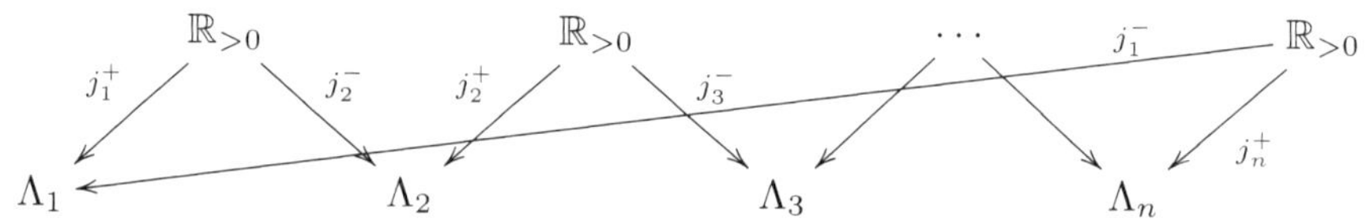

Provided that $\mathbb{T}_n$ is equipped with an appropriate complex structure, Γ_n is isomorphic to the skeleton of $\mathbb{T}_n$.[6] Also, the Λ_i supply an open covering for Γ_n. As CPM$(-)$ defines a sheaf over Γ_n, its global sections CPM(Γ_n) can be calculated in the usual way, by taking the (*homotopy*) equalizer of the Čech diagram (2.1) below, and thus, informally, picking out local sections on the Λ_i-s which agree on the overlaps,

$$\prod_{i\in I} (\mathrm{CPM}(\Lambda_i) = Sh(S^1, \Lambda_p)) \underset{Res_-}{\overset{Res_+}{\rightrightarrows}} \prod_{j\in I} (\mathrm{CPM}(\mathbb{R}_{>0}) \cong \mathbb{C} - mod). \tag{2.1}$$

Note that the functors Res_+ and Res_- can be explicitly computed, since they are products of restriction functors for $Sh(S^1, \Lambda_p) \cong Rep(\bullet \rightrightarrows \bullet)$ which were described in Remark 2.4. We will conclude this section by giving a concrete recipe for constructing homotopy equalizers in $dgCat$.

LEMMA 2.6. *Let* $\mathcal{C} \underset{G}{\overset{F}{\rightrightarrows}} \mathcal{D}$ *be a diagram in* $dgCat$, *and denote by* $\mathcal{E}$ *the dg category having,*

- *as objects, pairs* (C, u), *where* $C \in \mathcal{C}$, *and* u *is a degree zero, closed morphism* $u : F(C) \cong G(C)$, *which becomes invertible in the homotopy category,*
- *as morphisms, pairs* $(f, H) \in hom^k(C, C') \oplus hom^{k-1}(F(C), G(C'))$, *with differential given by* $d(f, H) = (df, dH - (u'F(f) - G(f)u))$. *The composition is obvious.*

Then $\mathcal{E}$, *endowed with the natural forgetful functor* $\mathcal{E} \to \mathcal{C}$, *is a homotopy equalizer for* F *and* G.

---

[5] This informal account disregards various technical aspects of the theory, for which see [**STZ**].

[6] Note that, although a different choice of complex structure on $\mathbb{T}_n$ could alter the geometry of the skeleton, this would not affect, up to quasi-equivalence, the global sections of CPM$(-)$.

PROOF. Lemma 2.6 depends on the availabilty of an explicit construction of the *path object* $P(\mathcal{D})$ for $\mathcal{D}$, which can be found in Lemma 4.1 of [**Tab1**]. This allows us to compute the homotopy equalizer in the usual way, by taking appropriate fibrant replacements. We leave the details to the reader. □

3. CPM and mirror symmetry for degenerate elliptic curves

Let X_n be a cycle of n projective lines. That is, X_n is a connected reduced curve with n nodal singularities, such that its normalization $\tilde{X}_n \xrightarrow{p} X$ is a disjoint union of n projective lines $D_1, \ldots, D_n$, with the property that the pre-image along π of the singular set interesects each D_i in exactly two points. Theorem 3.3 below yields a description of $\mathcal{P}\mathrm{erf}(X_n)$ as a suitable homotopy equalizer of dg categories. This is one of the key steps in the proof of HMS for X_n, which will be presented in Section 3.2.

3.1. Perfect complexes over a nodal curve. It will be covenient to make use of the following general result.

THEOREM 3.1 (see [**To**], Proposition 11). *Let $X = U \cup V$, where U and V are two Zariski open subschemes. Then the following square:*

$$\begin{array}{ccc} \mathcal{P}\mathrm{erf}(X) & \longrightarrow & \mathcal{P}\mathrm{erf}(U) \\ \downarrow & & \downarrow \\ \mathcal{P}\mathrm{erf}(V) & \longrightarrow & \mathcal{P}\mathrm{erf}(U \cap V) \end{array}$$

is a fiber product of dg categories.

Recall that a nodal curve is a curve having only double points as singularities.

LEMMA 3.2. *Let X be a nodal curve, with singular set Z, and normalization $\pi : \tilde{X} \to X$. Let $\sigma, \tau : Z \to X$ be two non-overlapping sections of $\pi^{-1}(Z) \to Z$. Then, for every $\tilde{\mathcal{F}}$ vector bundle on $\tilde{X}$ and isomorphism $u : \sigma^*(\tilde{\mathcal{F}}) \to \tau^*(\tilde{\mathcal{F}})$, the assignment*

$$U \subset^{open} X \mapsto \{s \in \tilde{\mathcal{F}}(\pi^{-1}(U)) | u(\sigma^*(s)) = \tau^*(s)\},$$

defines a vector bundle $\tilde{\mathcal{F}}^u$ on X such that $\pi^*(\mathcal{F}^u) \cong \tilde{\mathcal{F}}$. Conversely, if $\mathcal{F}$ is a vector bundle on X such that $\pi^*\mathcal{F} \cong \tilde{\mathcal{F}}$, then $\mathcal{F} \cong \tilde{\mathcal{F}}^u$ for some isomorphism $u : \sigma^*(\tilde{\mathcal{F}}) \to \tau^*(\tilde{\mathcal{F}})$.

PROOF. See Proposition 4.4 in [**L**]. □

THEOREM 3.3. *Let X be a nodal curve, and let Z, $\tilde{X}$, π, σ, and τ be as in Lemma 3.2; then the diagram*

$$\mathcal{P}\mathrm{erf}(X) \xrightarrow{\pi^*} D^b(Coh(\tilde{X})) \overset{\sigma^*}{\underset{\tau^*}{\rightrightarrows}} D^b(Coh(Z))$$

is an equalizer of dg categories.

PROOF. Since limits commute with limits, it is sufficient, after Theorem 3.1, to prove the claim for affine X, so we will rectrict to this case. Let E be the equalizer of the diagram

$$D^b(Coh(\tilde{X})) \overset{\sigma^*}{\underset{\tau^*}{\rightrightarrows}} D^b(Coh(Z))$$

constructed according to the prescriptions of Lemma 2.6. Recall that the objects of E are pairs $(\tilde{\mathcal{F}}, u)$, where $\tilde{\mathcal{F}}$ is an object of $D^b(Coh(\tilde{X}))$, and u is a degree zero, closed morphism $u : \sigma^*\tilde{\mathcal{F}} \cong \tau^*\tilde{\mathcal{F}}$, which becomes invertible in the homotopy category. The morphisms of E are pairs $(f, H) \in hom^k(\tilde{\mathcal{F}}, \tilde{\mathcal{G}}) \oplus hom^{k-1}(\sigma^*\pi^*\tilde{\mathcal{F}}, \tau^*\pi^*\tilde{\mathcal{G}})$, and the differential is given by $d(f, H) = (df, dH - (u'\sigma^*(f) - \tau^*(f)u))$.

Fix a natural equivalence $\alpha : \sigma^*\pi^* \cong \tau^*\pi^*$. As $\mathcal{P}\mathrm{erf}(X)$ is generated by line bundles, and E is generated by objects of the form $(\tilde{\mathcal{F}}, u)$ with $\tilde{\mathcal{F}}$ a line bundle on $\tilde{X}$, it is sufficient to define a (quasi-)equivalence ψ between these two linear subcategories. Define ψ as follows:

- if $\mathcal{F}$ is a line bundle on X, then $\psi(\mathcal{F}) = (\pi^*\mathcal{F}, \sigma^*\pi^*\mathcal{F} \xrightarrow{\alpha} \tau^*\pi^*\mathcal{F})$,
- if $\mathcal{F}, \mathcal{G}$ are line bundles on X, and $f \in hom^k(\mathcal{F}, \mathcal{G})$, then $\psi(f) = (\pi^*f, 0)$.

Consider a line bundle $\tilde{\mathcal{F}}$ over $\tilde{X}$. It follows from Lemma 3.2 that the set of isomorphism classes of line bundles $\mathcal{F}$ on X such that $\pi^*\mathcal{F} \cong \tilde{\mathcal{F}}$ carries a transitive action by $(\mathbb{C}^*)^{|Z|}$ (given by pointwise rescaling the 'compatibility' isomorphisms u, see Lemma 3.2). Further, the same is true for the set of isomorphism classes of objects of $(\tilde{\mathcal{G}}, v) \in E$, such that $(\tilde{\mathcal{G}}, v) \cong (\tilde{\mathcal{F}}, u)$ for some $u \in hom^0(\sigma^*\tilde{\mathcal{F}}, \tau^*\tilde{\mathcal{F}})$. Essential surjectivity follows from the fact that ψ defines a $(\mathbb{C}^*)^{|Z|}$-equivariant map between these two sets of isomorphism classes.

We shall prove next that ψ is quasi-fully faithful, i.e. that the map between hom-complexes defined by ψ induces an isomorphism in the homotopy category. Denote by HoE the homotopy category of E. It is sufficient to show that for all line bundles $\mathcal{F}$ on X, and for all $i \in \mathbb{N}$,

$$\psi : Hom^i_X(\mathcal{O}_X, \mathcal{F})(= H^i_X(\mathcal{F})) \xrightarrow{\cong} Hom^i_{HoE}(\psi(\mathcal{O}_X), \psi(\mathcal{F})).$$

Note that, as X and $\tilde{X}$ are affine, cohomology vanishes in positive degree. It follows that $Hom^i_{HoE}(\psi(\mathcal{O}_X), \psi(\mathcal{F})) = 0$ for all $i > 0$. [7]. Further, in degree-zero, the hom-space fits in the following short exact sequence

$$0 \to Hom^0_{HoE}(\psi(\mathcal{O}_X), \psi(\mathcal{F})) \to Hom^0_{\tilde{X}}(\mathcal{O}_{\tilde{X}}, \pi^*\mathcal{F}) \to Hom^0_Z(\sigma^*\mathcal{O}_{\tilde{X}}, \tau^*\pi^*\mathcal{F}) \to 0.$$

Thus, proving fully faithfulness boils down to showing exactness of

$$0 \to H^0_X(\mathcal{F}) \xrightarrow{\pi^*} H^0_{\tilde{X}}(\pi^*\mathcal{F}) \to H^0_Z(\tau^*\pi^*\mathcal{F}) \to 0. \tag{3.1}$$

Now, (3.1) is obtained by taking global sections of the sequence

$$0 \to \mathcal{F} \to \pi_*(\pi^*\mathcal{F}) \to \pi_*\tau_*\tau^*(\pi^*\mathcal{F}) \to 0,$$

which is exact (see the proof of Proposition 4.4 of [**L**]). Since X is affine, taking global sections is an exact operation, and this concludes the proof of Theorem 3.3. □

3.2. HMS for nodal elliptic curves. In this section we will prove that the category of perfect complexes over X_n is quasi-equivalent to $\mathrm{CPM}(\Gamma_n)$. Granting Conjecture 2.5, this result confirms well known mirror symmetry heuristics, which suggest that the mirror of X_n should be a symplectic torus with n punctures, $\mathbb{T}_n$.[8]

THEOREM 3.4 ([**Be**]). *There is an equivalence* $\beta : D^b(Coh(\mathbb{P}^1)) \xrightarrow{\cong} Rep(\bullet \rightrightarrows \bullet)$.

---

[7]Note that $Hom^1_{HoE}(\psi(\mathcal{O}_X), \psi(\mathcal{F}))$ vanishes, since it is isomorphic to the quotient of $Hom^0_Z(\sigma^*\mathcal{O}_{\tilde{X}}, \tau^*\pi^*\mathcal{F}) \cong \mathbb{C}$ by the image of the differential, which is easily seen to be surjective.

[8]Kontsevich announced related results in [**K1**]. HMS for the nodal $\mathbb{P}^1$ is also treated in [**LP**].

PROOF. For the proof, see Beilinson's famous paper [**Be**], which provides analogous descriptions of $D^b(Con(\mathbb{P}^n))$ for any n. The functor β can be defined as follows. Fix a basis x_0, x_1 for $H^0(\mathcal{O}(1))$, and set

$$\mathcal{F} \in D^b(Coh(\mathbb{P}^1)) \longmapsto \beta(\mathcal{F}) = \; R\Gamma(\mathcal{F} \otimes \mathcal{O}(-1)) \overset{\cdot x_0}{\underset{\cdot x_1}{\rightrightarrows}} R\Gamma(\mathcal{F}) \; \in Rep(\bullet \rightrightarrows \bullet),$$

with the obvious definition on morphisms. □

In view of results of Nadler and Zaslow [**NZ, N**] discussed in Section 2.2, Proposition 3.4, combined with the equivalence $Rep(\bullet \rightrightarrows \bullet) \cong Sh(S^1, \Lambda_p)$, yields a homological mirror symmetry statement pairing $D^b(Coh(\mathbb{P}^1))$, and a suitable Fukaya category of exact Lagrangians in T^*S^1. This was explained as an instance of T-duality by Fang [**F**], and fits in the framework of the *coherent-constructible correspondence* developed by Fang, Liu, Treumann and Zaslow (see [**FLTZ**]), which is one of the starting points for the project of [**STZ**].[9]

THEOREM 3.5 ([**STZ**]). *Let X_n be a cycle of n projective lines. There is a quasi-equivalence $\phi : \mathcal{P}\mathrm{erf}(X_n) \cong \mathrm{CPM}(\Gamma_n)$.*[10]

PROOF. Let $Z \hookrightarrow X_n$ be the singular set. Pick two non-overlapping sections σ, and τ of $p^{-1}(Z) \to Z$, as in Theorem 3.3, and choose an identification $D^b(Coh(Z)) \cong \prod_{j=1}^{j=n} \mathbb{C} - mod$. The proof is encoded in the following diagram:

$$\begin{array}{ccccc}
\mathcal{P}\mathrm{erf}(X_n) & \longrightarrow & \prod_{i=1}^{i=n} D^b(Coh(\mathbb{P}^1)) & \overset{\sigma^*}{\underset{\tau^*}{\rightrightarrows}} & \prod_{j=1}^{j=n} \mathbb{C} - mod \\
\downarrow{\scriptstyle \phi} & & \downarrow{\scriptstyle \beta} & & \downarrow{\scriptstyle \rho} \\
\mathrm{CPM}(\Gamma_n) & \longrightarrow & \prod_{i=1}^{i=n} Rep(\bullet \rightrightarrows \bullet) & \overset{Res^+}{\underset{Res^-}{\rightrightarrows}} & \prod_{j=1}^{j=n} \mathbb{C} - mod.
\end{array}$$

In fact, we can choose ρ in such a way that $\rho \circ \sigma^* \cong Res^+ \circ \beta$, and $\rho \circ \tau^* \cong Res^- \circ \beta$, where, abusing notation, we are denoting $\times_{i=1}^{i=n} \beta$ simply by β. This implies that the equalizer of σ^*, τ^* is quasi-equivalent to the equalizer of Res^+, Res^- (see diagram 2.1), and yields $\phi : \mathcal{P}\mathrm{erf}(X_n) \overset{\cong}{\to} \mathrm{CPM}(\Gamma_n)$. □

4. A mapping class group action on $D^b(Coh(X_n))$

Since the group of symplectic automorphisms of $\mathbb{T}_n$ acts by auto-equivalences on $Fuk(\mathbb{T}_n)$, HMS predicts the existence of a mirror action on $D^b(Coh(X_n))$.[11] This is the content of the main theorem of [**Si**], which we state below.

THEOREM 4.1 ([**Si**]). *Let $\mathrm{PM}(\mathbb{T}_n)$ be the pure mapping class group of $\mathbb{T}_n$, then, up to shift, there is an action of $\mathrm{PM}(\mathbb{T}_n)$ over $D^b(Coh(X_n))$.*

Recall that the mapping class group of an oriented surface Σ, $\mathrm{MCG}(\Sigma)$, is the group of symplectic automorphisms of Σ, up to isotopy. The pure mapping class group is the subgroup $\mathrm{PM}(\Sigma) \hookrightarrow \mathrm{MCG}(\Sigma)$ generated by Dehn twists (see

[9]The significance for mirror symmetry of the equivalence $D^b(Coh(\mathbb{P}^1)) \cong Sh(S^1, \Lambda_p)$ was first advocated by Bondal [**B**], in the context of HMS for weighted projective spaces.

[10]The result proved in [**STZ**] is actually more general, and extends to appropriate *stacky* degenerations of elliptic curves.

[11]Note in fact that, although the HMS statement of Section 3.2 involves $\mathcal{P}\mathrm{erf}(X_n)$, it is possible to show that $D^b(Coh(X_n))$ and $\mathcal{P}\mathrm{erf}(X_n)$ have the same group of auto-equivalences.

[**FM**]). Theorem 4.1 generalizes previous work of Seidel and Thomas [**SeT**], and Burban and Kreussler [**BK**], who established, respectively, the existence of an $SL(2,\mathbb{Z})$-action over $D^b(Coh(X))$, where X is a smooth elliptic curve, and over $D^b(Coh(X_1))$.[12] The proof of Theorem 4.1 depends on the availability of an explicit presentation of $\mathrm{PM}(\mathbb{T}_n)$, worked out in the first Section of [**Si**], and on Seidel and Thomas' theory of *spherical objects* and *twist functors* [**SeT**], which is designed precisely to test this aspect of mirror symmetry.

In the rest of this section, we shall briefly describe the proof of Theorem 4.1 for $n = 2$, leaving out most details, and referring the reader to [**Si**] for the general case. A presentation of $\mathrm{PM}(\mathbb{T}_2)$ can be found in [**PS**], and is reproduced below.

PROPOSITION 4.2. *The pure mapping class group* $\mathrm{PM}(\mathbb{T}_2)$ *is generated by* T_α, T_{β_1} *and* T_{β_2},[13] *with relations*

- *(Braid relations)* $T_{\beta_i}T_{\beta_j} = T_{\beta_j}T_{\beta_i}$, $T_{\beta_i}T_\alpha T_{\beta_i} = T_\alpha T_{\beta_i} T_\alpha$,
- *(G-relation)* $(T_{\beta_1}T_\alpha T_{\beta_2})^4 = 1$.

Following the discussion in Section 1 of [**SeT**], the group acting on $D^bCoh(X_2)$ is going to be a suitable central extension of $\mathrm{PM}(\mathbb{T}_2)$, whose elements should be viewed as *graded* symplectic automorphisms of the mirror of X_2, i.e. $\mathbb{T}_2$.

DEFINITION 4.3. Define $\widetilde{\mathrm{PM}}(\mathbb{T}_2)$ as the $\mathbb{Z}$-central extension of $\mathrm{PM}(\mathbb{T}_2)$, generated by T_α, $T_{\beta_i}, i = 1, 2$, and a central element t subject to the following relations

- (Braid relations), as in Proposition 4.2
- ($\tilde{G}$-relation) $(T_{\beta_1}T_\alpha T_{\beta_2})^4 = t^2$.

Let x_1 and x_2 be two smooth points lying on different components of X_2; then

THEOREM 4.4. *The assignment*

- *for all* $i = 1, 2, T_{\beta_i} \mapsto T_{\kappa(x_i)}$,
- $T_\alpha \mapsto T_{\mathcal{O}}$, *and*
- $t \mapsto [1]$,

defines an action of $\widetilde{\mathrm{PM}}(\mathbb{T}_2)$ *on* $D^b(Coh(X_2))$.

Note that the assigment described in Theorem 4.4 is compatible with mirror symmetry considerations, according to which $\mathcal{O}$ and $\kappa(x_i)$ should be mirror to Lagrangian branes whose supports are isotopic, respectively, to α and β_i. We state below two lemmas, formulated for a general cycle of projective lines X_n, which will be important for proving Theorem 4.4.

LEMMA 4.5. *Let* $F : D^b(Coh(X_n)) \to D^b(Coh(X_n))$ *be an auto-equivalence of triangulated categories. If*

- $F(\mathcal{O}) \cong \mathcal{O}$, *and*
- *for all* $i \in \{1 \dots n\}$, $F(\kappa(x_i)) \cong \kappa(x_i)$,

then there exists an isomorphisms $f : X_n \to X_n$, *such that* F *is naturally equivalent to* $f^* : D^b(Coh(X_n)) \to D^b(Coh(X_n))$.

---

[12] Recall that the mapping class groups of the torus and of the once punctured torus are both isomorphic to $SL(2,\mathbb{Z})$.

[13] The generators T_α, T_{β_i} are given by isotopy classes of Dehn twists along simple closed curves $\alpha, \beta_i \hookrightarrow \mathbb{T}_2$. Explicit representatives can be described as follows. Identify the torus $\mathbb{T}$ with $[0,1] \times [0,1]/\sim$, and set $\mathbb{T}_2 = \mathbb{T} \setminus \{p_1 = (0,0), p_2 = (\frac{1}{2}, 0)\}$, then $\alpha = [0,1] \times \{\frac{1}{2}\}$, and $\beta_i = \{\frac{i}{3}\} \times [0,1], i = 1, 2$.

PROOF. See Lemma 3.3 of [**Si**]. The key fact is that if F preserves $\mathcal{O}$ and $\kappa(x_i)$, then it will induce an automorphism of the homogeneous coordinate ring associated to the ample line bundle $\mathcal{L} = \mathcal{O}(x_1 + x_2 + \cdots + x_n)$. This implies the existence of an isomorphism $f : X_n \to X_n$ such that $F = f^*$, when restricted to the linear subcategory having as objects the tensor powers of $\mathcal{L}$. The claim then follows as in the proof of Theorem 3.1 of [**BO**]. Note that, under the assumptions of the theorem, if $n > 2$, f has to be the identity. For $n = 2$, f might be non-trivial, but has to be involutive, i.e. $f^2 = \mathrm{id}$. □

LEMMA 4.6. *Let $x \in X_n$ be a smooth point, then*

- $T_{\kappa(x)} \cong - \otimes \mathcal{O}(x)$,
- $T_{\mathcal{O}}(\kappa(x)) \cong \mathcal{O}(-x)[1]$,
- $T_{\mathcal{O}}(\mathcal{O}(x)) \cong \kappa(x)$,
- $T_{\mathcal{O}}(\mathcal{O}) \cong \mathcal{O}$.

PROOF. The first isomorphism is proved in [**SeT**], Section 3.d. For the other isomorphisms, see Lemma 2.13 in [**BK**]. □

PROOF OF THEOREM 4.4. The braid relations follow from Proposition 2.13 of [**SeT**]. It remains to check that $T_{\mathcal{O}}, T_{\kappa(x_i)}, i = 1, 2$ satisfy the $\tilde{G}$-relation. Simply by keeping track of the isomorphisms collected in Lemma 4.6, and applying the braid relations, one can see that

- $(T_{\kappa(x_1)} T_{\mathcal{O}} T_{\kappa(x_2)})^2(\mathcal{O}) \cong \mathcal{O}[1]$, and
- $(T_{\kappa(x_1)} T_{\mathcal{O}} T_{\kappa(x_2)})^2(\kappa(x_1)) \cong \kappa(x_2)[1]$, $(T_{\kappa(x_1)} T_{\mathcal{O}} T_{\kappa(x_2)})^2(\kappa(x_2)) \cong \kappa(x_1)[1]$.

Let's check this for $\kappa(x_1)$:

$$(T_{\kappa(x_1)} T_{\mathcal{O}} T_{\kappa(x_2)})(T_{\kappa(x_1)} T_{\mathcal{O}} T_{\kappa(x_2)})(\kappa(x_1)) \cong (T_{\kappa(x_1)} T_{\mathcal{O}} T_{\kappa(x_2)})(\mathcal{O}[1]) \cong \kappa(x_2)[1].$$

Consider an involution $\sigma : X_2 \to X_2$ such that $\sigma(x_1) = x_2$, and $\sigma(x_2) = x_1$. It follows from Lemma 4.5, and the comments made at the end of its proof, that there is an involution $f : X_2 \to X_2$, and a natural equivalence

$$(T_{\kappa(x_1)} T_{\mathcal{O}} T_{\kappa(x_2)})^2 \cong f^* \sigma^*[1].$$

As σ and f commute, by taking the square of this natural equivalence, one gets

$$(T_{\kappa(x_1)} T_{\mathcal{O}} T_{\kappa(x_2)})^4 \cong (f^* \sigma^*[1])(f^* \sigma^*[1]) \cong (f^*)^2 (\sigma^*)^2 [2] \cong [2].$$

This concludes the proof of the theorem. □

References

AAEKO. M. Abouzaid, D. Auroux, A. I. Efimov. L. Katzarkov, D. Orlov, "Homological mirror symmetry for punctured spheres," arXiv:1103.4322v1

Be. A. Beilinson, "Coherent sheaves on $\mathbb{P}^n$ and problems in linear algebra," Funct. An. and Appl. **12(3)** (1978) 68–69.

B. A. Bondal, "Derived categories of toric varieties," in *Convex and Algebraic geometry, Oberwolfach conference reports, EMS Publishing House* **3** (2006) 284–286.

BO. A. Bondal, D. Orlov, "Reconstruction of a variety from the derived category and groups of autoequivalences," Comp. Math. **125** (2001) 327–344.

BK. I. Burban, B. Kreussler, "Fourier-Mukai transforms and semi-stable sheaves on nodal Weierstrass cubics," J. Reine Angew. Math. **584** (2005) 45–82.

F. B. Fang, "Homological mirror symmetry is T-duality for $\mathbb{P}^n$," Comm. in Numb. Th. and Phys. **2(4)** (2008) 719–742.

FLTZ. B. Fang, D. Treumann, C.-C. Liu and E. Zaslow, "A categorification of Morelli's theorem," Invent. Math. **186(1)** (2011) 179–214.

FM. B. Farb, D. Margalit, "A primer on mapping class groups," PMS **50** Princeton University Press, 2011.

K. M. Kontsevich, "Homological algebra of mirror symmetry", Proceedings of the International Congress of Mathematicians (Zürich, 1994), 1995, 120–139.

K1. M. Kontsevich, "Symplectic Geometry of Homological Algebra," lecture at Mathematische Arbeitsgrunden 2009; notes available at `http://www.ihes.fr/ maxim/TEXTS/Symplectic_AT2009.pdf`.

KS. M. Kashiwara and P. Schapira, "Sheaves on Manifolds," Grundlehren der Mathematischen Wissenschafte **292**, Springer-Verlag, 1994.

L. J.T.A. Lang, "Relative moduli spaces of semi-stable sheaves on families of curves", Herbert Utz Verlag, 2001, 42–44.

LP. Y. Lekili, T. Perutz, "Fukaya categories of the torus and Dehn surgeries," Proc. Nat. Acad. Sci. USA 108 (2011) 8106-8113.

N. D. Nadler, "Microlocal branes are constructible sheaves", Sel. Math. New Ser. **15** (2009) 563–619.

N1. D. Nadler, "Fukaya categories as categorical Morse homology," `arXiv:1109.4848`

NZ. D. Nadler, E. Zaslow, "Constructible sheaves and the Fukaya category", J. Amer. Math. So. **22** (2009) 233–286.

P. R. Penner, "Perturbative Series and the Moduli Space of Riemann Surfaces," J. Diff. Geom. **27** (1988) 35–53.

PS. J. R. Parker, C. Series "The mapping class group of the twice punctured torus," in *Groups. Topological, Combinatorial and Geometric Aspects*, London Mathematical Society Lecture Note Series **311**, 2004, 405–486.

PZ. A. Polishchuk, E. Zaslow, "Categorical mirror symmetry: the elliptic curve," Adv. Theor. Math. Phys. **2** (1998) 443–470.

S. P. Seidel, "Homological Mirror Symmetry for the Quartic Surface," `arXiv.math:0310414`.

S1. P. Seidel, "Cotangent Bundles and their Relatives," Morse Lectures, Princeton University, 2010, currently available at `http://www-math.mit.edu/ seidel/morse-lectures-1.pdf`

Sh. N. Sheridan, "Homological mirror symmetry for Calabi-Yau hypersurfaces in projective space," `arXiv:1111.0632`

SeT. P. Seidel, R. Thomas, "Braid group actions on derived categories of coherent sheaves," Duke Math. J. **108(1)** (2001) 37–108.

Si. N. Sibilla, "A note on mapping class group actions on derived categories," `arXiv:1109.6615`.

STZ. N. Sibilla, D. Treumann, E. Zaslow, "Ribbon Graphs and Mirror Symmetry I," `arXiv:1103.2462`.

STZ2. N. Sibilla, D. Treumann, E. Zaslow, in preparation.

To. B. Toën, "Lectures on DG-categories," available at http://www.math.univ-toulouse.fr/ toen/swisk.pdf

Tab. G. Tabuada, "Une structure de catégorie de modèles de Quillen sur la catégorie des dg-catégories," Comptes Rendus de L'Académie de Sciences de Paris **340** (2005) 15–19.

Tab1. G. Tabuada, "A new Quillen model for the Morita homotopy theory of dg categories," `arXiv:0701205`.

Department of Mathematics, Northwestern University, 2033 Sheridan Road, Evanston, Il 60208

E-mail address: `sibilla@math.northwestern.edu`

Proceedings of Symposia in Pure Mathematics
Volume **85**, 2012

Vanishing Chiral Algebras and Höhn–Stolz Conjecture

Junya Yagi

ABSTRACT. Given a two-dimensional quantum field theory with $(0,2)$ supersymmetry, one can construct a chiral (or vertex) algebra. The chiral algebra of a $(0,2)$ supersymmetric sigma model is, perturbatively, the cohomology of a sheaf of chiral differential operators on a string Kähler manifold. However, it vanishes in some cases when instantons are taken into account. I will discuss the implication of this phenomenon for the Höhn–Stolz conjecture on the Witten genus.

1. Introduction

In his classic papers [**11, 12**], Edward Witten unraveled beautiful connections between elliptic genera, the geometry of loop spaces, and supersymmetric sigma models in two dimensions. Applying his idea to a special case, he was led to discover a new genus, now called the Witten genus. Let M be a closed string manifold of dimension d. (M is string if and only if it is spin and $p_1(M)/2 = 0$.) Then, the Witten genus φ_W associates to M an integral modular form $\varphi_W(M)$ of weight $d/2$.

About a decade after Witten's work, Gerald Höhn and Stephan Stolz independently arrived at the following

CONJECTURE (Höhn, Stolz [**7**]). *If M admits a Riemannian metric of positive Ricci curvature, then $\varphi_W(M) = 0$.*

The aim of this paper is to explain how recent developments in the study of supersymmetric sigma models shed new light on the Höhn–Stolz conjecture in the Kähler case. The key element in the discussion is a remarkable phenomenon exhibited by certain supersymmetric sigma models: the chiral algebras associated to these models vanish.

The emergence of a chiral algebra is a characteristic property of two-dimensional quantum field theories with $(0,2)$ supersymmetry. Although this fact had been known [**10**] for long time, it had not been paid much attention by physicists until relatively recently. During the last several years, however, there were significant advances in our understanding of the chiral algebras of $(0,2)$ supersymmetric sigma models, thanks largely to the pioneering works of Anton Kapustin [**4**] and Witten [**13**]. What inspired the physicists was the theory of chiral differential operators

2000 *Mathematics Subject Classification.* 58J26; 17B69, 81R10, 81T60.

developed earlier in mathematics [**1**–**3**, **5**, **6**]. I hope that the results presented here give some inspirations back to mathematicians.

We begin in the next section by reviewing how the Witten genus arises in the context of supersymmetric sigma models, and how it is interpreted in the language of loop space geometry. Then, in Section 3, we restrict to the Kähler case and describe the construction of the chiral algebras of $(0, 2)$ supersymmetric sigma models. We also explain some of their properties and relation to chiral differential operators. In Section 4, we outline the argument for the vanishing phenomenon and discuss its implication for the Höhn–Stolz conjecture. We end in Section 5 by raising some questions and speculating on possible answers.

2. Supersymmetric Sigma Model and the Witten Genus

Consider a sigma model on a cylinder $\mathbb{R} \times S^1$ with target space M. This is a quantum field theory describing maps $\phi\colon \mathbb{R} \times S^1 \to M$. If we add a right-moving fermion ψ_+ with values in ϕ^*TM, then the theory has $(0, 1)$ supersymmetry; that is to say, there is a Hermitian fermionic operator Q_+ satisfying

$$Q_+^2 = H - P, \qquad [Q_+, H] = [Q_+, P] = 0, \tag{1}$$

where H and P are the generators of translations in time and space, respectively. The theory we consider in this paper is the Euclidean version of this supersymmetric sigma model obtained by Wick rotating the time coordinate $t \in \mathbb{R}$.

States annihilated by the supercharge Q_+ are called supersymmetric states. Since the theory is unitary, supersymmetric states can also be characterized by the property that they have $H - P = 0$. Due to the topology of the space S^1, the momentum P is quantized and takes values in $-d/24 + \mathbb{Z}$. (The shift by $-d/24$ comes from the regularization of an infinite sum that appears in the definition of P.)

Now, cut out a finite segment of the cylinder and glue the two ends with a twist to make a torus $\mathbb{C}/2\pi(\mathbb{Z} + \tau\mathbb{Z})$. The partition function on the torus is computed by

$$Z(M) = \mathrm{Tr}\big((-1)^F q^{(H+P)/2} \bar{q}^{(H-P)/2}\big) \tag{2}$$

with $q = e^{2\pi i\tau}$. Here $(-1)^F$ is the fermion parity operator, which equals $+1$ for bosonic states and -1 for fermionic states. Since Q_+ pairs bosonic and fermionic states outside its kernel, $Z(M)$ receives contributions only from supersymmetric states. So if we let $\mathcal{H}$ be the space of supersymmetric states (that is, the kernel of Q_+) and $\mathcal{H}_n$ the subspace of $\mathcal{H}$ defined by $P = -d/24 + n$, we can write

$$Z(M) = \mathrm{Tr}_{\mathcal{H}}\big((-1)^F q^P\big) = q^{-d/24} \sum_{n=0}^{\infty} q^n \, \mathrm{Tr}_{\mathcal{H}_n}(-1)^F. \tag{3}$$

This is the equivariant index of Q_+ with respect to the natural circle action generated by P. The Witten genus of M is then given by

$$\varphi_W(M) = \eta(q)^d Z(M), \tag{4}$$

where $\eta(q) = q^{1/24} \prod_{n=1}^{\infty}(1 - q^n)$ is the Dedekind eta function. As the Witten genus and the partition function differ only by a factor, we will mainly talk about the latter, which is physically more natural to consider.

The geometric meaning of the partition function becomes clear if we view the theory as supersymmetric quantum mechanics on the free loop space $\mathcal{L}M$ of M, and canonically quantize it. After quantization, the fermion obeys the anticommutation relation

$$\{\psi^i(t,\sigma), \psi^j(t,\sigma')\} = g^{ij}\delta(\sigma - \sigma'), \tag{5}$$

where g is the metric on M. This is a loop-space version of the defining relation of Clifford algebra, with σ serving as a continuous index parametrizing the direction along the loop. Since the fermion acts on states, quantization identifies states with spinors on $\mathcal{L}M$ and $(-1)^F$ with the chirality operator. On the other hand, Q_+ is quantized as

$$Q_+ = \int d\sigma \Big(\psi^i \frac{D}{D\phi^i} + g_{ij}\psi^i \partial_\sigma \phi^j\Big), \tag{6}$$

so it is the Dirac operator $D_{\mathcal{L}M}$ on $\mathcal{L}M$ plus an extra term coupled to the Killing vector field generating rotations of loops. Taking the limit $g \to 0$ in which the extra term drops out, we learn that $Z(M)$ is the S^1-equivariant index of $D_{\mathcal{L}M}$ [**12**]. In this limit, supersymmetric states are harmonic spinors on $\mathcal{L}M$.

Taking the limit $g \to \infty$, we obtain a formula that expresses $Z(M)$ in terms of characteristic classes of M. Since Q_+^2 contains the potential

$$\int d\sigma\, g_{ij}\partial_\sigma\phi^i\partial_\sigma\phi^j, \tag{7}$$

in this limit supersymmetric states are supported mostly in the neighborhood of the configurations such that $\partial_\sigma\phi = 0$. These are the constant loops, forming a copy of M inside $\mathcal{L}M$. Such localized states can be thought of as spinors on M. More precisely, approximate supersymmetric states with $P = -d/24+n$ can be identified to leading order with spinors taking values in the vector bundle V_n, which is the coefficient of q^n in the series

$$\bigotimes_{m=1}^{\infty} S_{q^m}(TM) = 1 + qTM + q^2\big(TM + S^2(TM)\big) + \cdots. \tag{8}$$

Here $S_q(V) = 1 + qS(V) + q^2S^2(V) + \cdots$. On these localized states, Q_+ acts as the Dirac operator D_M on M. Therefore we have

$$Z(M) = q^{-d/24}\sum_{n=0}^{\infty} q^n \widehat{A}(M; V_n). \tag{9}$$

Mathematically, this formula defines the Witten genus through the relation (4).

Although the last expression of $Z(M)$ is well defined for any closed orientable manifold M, the underlying quantum field theory is not. For the theory to not suffer from quantum anomalies, M must be string. In addition, to ensure that the theory has a good ultraviolet behavior without the problem of short-distance divergences, we should require that the Ricci curvature of M is nonnegative.[1] So these conditions are necessary for $Z(M)$ to be interpreted as the S^1-equivariant index of the Dirac operator on $\mathcal{L}M$. Indeed, the former is one of the assumptions for the Höhn–Stolz conjecture. But what is the significance of the assumption that the Ricci curvature is positive?

To understand this point, recall the Lichnerowicz theorem from classical spin geometry. This theorem says that there are no harmonic spinors on a closed spin

[1]This is not a sufficient condition for the theory to have a well-defined ultraviolet limit. If the Ricci curvature is positive, then the theory is expected to be asymptotically free and well defined.

manifold with positive scalar curvature. What would be the loop-space analog of the Lichnerowicz theorem? A natural definition for the scalar curvature of $\mathcal{L}M$ at a loop $\gamma \in \mathcal{L}M$ would be the integral of the Ricci curvature of M along γ. Then, the positive Ricci curvature implies that the scalar curvature of $\mathcal{L}M$ is positive, which in turn would imply that there are no harmonic spinors on $\mathcal{L}M$. If there are no harmonic spinors, then the index of the Dirac operator vanishes and, therefore, the Witten genus vanishes. In short, the "loop-space Lichnerowicz theorem" would imply the Höhn–Stolz conjecture.

3. $(0,2)$ Supersymmetry and Chiral Algebra

From now on we assume that M is Kähler.[2] The motivation for doing so is that in this case our model has $(0,2)$ supersymmetry. This allows us to construct two important objects: the Q-cohomology of states and chiral algebra.

Recall that when M is Kähler, the Dirac operator splits into two pieces as $D_M = \mathcal{D}_M + \mathcal{D}_M^*$. The operator $\mathcal{D}_M$ squares to zero, $\mathcal{D}_M^2 = 0$, and the sections of the spinor bundle admits a $\mathbb{Z}$-grading such that $\mathcal{D}_M$ has degree 1. Thus we can define the $\mathcal{D}_M$-cohomology. Since $\{\mathcal{D}_M, \mathcal{D}_M^*\} = D_M^2$, a standard argument shows that the rth $\mathcal{D}_M$-cohomology group is isomorphic to the space of harmonic spinors of degree r.

Similarly, in the Kähler case the supercharge splits as $Q_+ = Q + Q^*$, and Q satisfies the $(0,2)$ supersymmetry algebra:

$$Q^2 = 0, \qquad \{Q, Q^*\} = H - P. \tag{10}$$

The space of states is graded by the fermion charge, which equals $(-1)^F$ modulo two and assigns Q charge 1. So, as in the case of $\mathcal{D}_M$, we can define the Q-cohomology. The fermion charge gives a $\mathbb{Z}$-grading at the level of perturbation theory, but this is reduced to a $\mathbb{Z}_{2n}$-grading nonperturbatively by instantons if $c_1(M) \neq 0$, where $2n$ is the greatest common divisor of $c_1(M)$. (Since M is spin, $c_1(M)$ is even.) By the supersymmetry algebra, the rth Q-cohomology group is isomorphic to the space of supersymmetric states of charge r. Hence, we can compute the Witten genus once we know the Q-cohomology of states.

For the application to the Höhn–Stolz conjecture, however, it turns out to be more useful to study another kind of Q-cohomology, namely the Q-cohomology in the space of local operators, with the action of Q given by the supercommutator. When the theory is conformally invariant, the Q-cohomology of states and local operators are isomorphic via the state-field correspondence. This is not true in general.

The Q-cohomology of local operators has two important properties. One is that its elements vary holomorphically: if $\mathcal{O}$ is Q-closed, then $\partial_{\bar{z}}\mathcal{O}$ is Q-exact because

$$\partial_{\bar{z}}\mathcal{O} = [H - P, \mathcal{O}] = [\{Q, Q^*\}, \mathcal{O}] = \{Q, \{Q^*, \mathcal{O}]]. \tag{11}$$

The other is that it has a natural operator product expansion (OPE) structure inherited from the underlying theory:

$$[\mathcal{O}_i(z)] \cdot [\mathcal{O}_j(z')] \sim \sum_k c_{ij}{}^k(z - z')[\mathcal{O}_k(z')]. \tag{12}$$

---

[2]This is an assumption about the classical geometry of M. Quantum corrections generally destroy the closedness of the Hermitian form ω, inducing a nonzero torsion $H = i(\bar{\partial} - \partial)\omega$. $(0,2)$ supersymmetry requires $dH = 0$. Such a geometry is called strong Kähler with torsion.

The coefficient functions ${c_{ij}}^k(z-z')$ are holomorphic except at $z = z'$ where they can have poles. The holomorphic OPE algebra generated by the Q-cohomology classes is what we call the chiral algebra of the $(0, 2)$ supersymmetric theory. The chiral algebra forms a holomorphic sector of the theory, in which correlation functions are holomorphic in the insertion points except where two points coincide. If we consider the A-model (regarding it as a $(0, 2)$ supersymmetric theory), these correlation functions generalize Gromov–Witten invariants.

The chiral algebra of $(0, 2)$ supersymmetric theory has the same structure as the chiral (or vertex) algebra of conformal field theory, except that the grading by conformal weight is missing. In fact, the former is generally not conformally invariant, certainly not for our model if $c_1(M) \neq 0$. Though classically our model has conformal invariance, quantum corrections break it if the Ricci curvature of M is nonzero. Correspondingly, though classically the energy-momentum tensor T_{zz} belongs to the Q-cohomology, there is a Q-closed local operator θ satisfying

$$[Q, T_{zz}] = \partial_z \theta \tag{13}$$

at the perturbative level. If $c_1(M) = 0$, then θ is actually Q-exact and one can find higher-order corrections to T_{zz} that make it Q-closed again. This is not possible if $c_1(M) \neq 0$. Therefore, the chiral algebra lacks the energy-momentum tensor in that case.

Despite the possible absence of conformal invariance, the chiral algebra is still graded nicely by conformal weight if we consider a slightly different variant of the theory obtained by a "twisting" procedure. After twisting, the components $T_{z\bar{z}}$, $T_{\bar{z}\bar{z}}$ of the energy-momentum tensor become Q-exact and vanish in the Q-cohomology. This implies that the generator $\overline{L}_0$ of antiholomorphic scaling vanishes, hence the holomorphic dimension L_0 is equal to the spin $L_0 - \overline{L}_0$ which takes integer values in the twisted model. Thus, L_0 provides an integral grading that is protected from small quantum corrections. Nonperturbatively, this grading is reduced to $\mathbb{Z}_n$ by instantons which carry nonzero scaling dimensions.

It was shown by Witten [**13**] that, with this grading by conformal weight, the chiral algebra of the twisted model is perturbatively isomorphic to the cohomology of a sheaf $\mathcal{D}^{ch}_M$ of chiral differential operators (or $\beta\gamma$ systems) on M. In particular, in the conformal case $c_1(M) = 0$, the character of the cohomology of $\mathcal{D}^{ch}_M$ computes the partition function[3] by the state-field correspondence:

$$\operatorname{ch} H(M; \mathcal{D}^{ch}_M) = Z(M). \tag{14}$$

By the same token, Kapustin [**4**] showed that the perturbative chiral algebra of the A-model is isomorphic to the cohomology of the chiral de Rham complex [**6**]. The case that the theory has a general "gauge bundle" E was considered by Tan [**8**] and found to be related to the construction of Gorbounov et al. [**1**]. Our model corresponds to the case $E = 0$, while the A-model is the case where E is the holomorphic tangent bundle of M.

Therefore, at the perturbative level, the chiral algebras of $(0, 2)$ supersymmetric sigma models admit a nice description which is relatively well understood mathematically. Beyond perturbation theory this is not the case any longer. Instanton

[3]On the torus (or any Riemann surface with trivial canonical bundle), the twisted and untwisted models are equivalent and have the same partition function.

effects can be violent, so much so that they can destroy many of the perturbative Q-cohomology classes—sometimes all of them.

4. Vanishing Theorem and the Höhn–Stolz Conjecture

In [**13**], Witten made a remarkable prediction. His claim was that the chiral algebras of $(0,2)$ supersymmetric sigma models, though perturbatively infinite-dimensional, could nevertheless vanish nonperturbatively in the presence of instantons. More specifically, he predicted that this phenomenon occurred for $M = \mathbb{CP}^1$ because instantons would induce the relation

$$\{Q, \theta\} \propto 1. \tag{15}$$

For $M = \mathbb{CP}^1$, nonperturbatively the chiral algebra is $\mathbb{Z}_2$-graded by fermion charge, while the grading by conformal weight is completely broken. So the operator $\{Q, \theta\}$, which has charge 2 and weight 1 perturbatively, has a chance to be proportional to the identity. Such a relation means that $1 = 0$ in the Q-cohomology, hence the chiral algebra is identically zero.

Witten's prediction was confirmed by an explicit computation in [**9**], where it was shown that the same conclusion also holds for all complete flag manifolds of semisimple Lie groups G; $M = \mathbb{CP}^1$ is the case $G = SL_2$. In fact, we have a more general

"Theorem" (Yagi [**14**]). *If M contains a rational curve with trivial normal bundle, then the chiral algebra of the $(0,2)$ supersymmetric sigma model vanishes.*

Here the normal bundle $N_{C/M}$ of a rational curve $C \subset M$ is the holomorphic vector bundle defined by the exact sequence

$$0 \longrightarrow T_C \longrightarrow T_M|_C \longrightarrow N_{C/M} \longrightarrow 0. \tag{16}$$

We put the word "Theorem" inside the quotation marks because this is not really a rigorous theorem. Mathematically, it may be better taken as a conjecture.

Let me give the outline of the "proof." Suppose that we have an instanton-induced relation $\{Q, \theta\} = \mathcal{O}$. By counting how many fermion charges and conformal weights are carried by instantons, we can show that $\mathcal{O}$ represents a Q-cohomology class of charge 0 and weight 0 perturbatively, which means that it is a function on M. (There cannot be Q-exact terms; $\mathcal{O}$ must have antiholomorphic weight 0 to leading order, but there are no perturbatively Q-exact local operators of charge 0 with that property.) On functions Q acts as the Dolbeault operator $\bar{\partial}$, so $\mathcal{O}$ is a holomorphic function. Moreover, it must be a constant since we assume that M is compact. To show that this constant is nonzero, we only need to find a correlation function that has $\{Q, \theta\}$ inside and does not vanish nonperturbatively:

$$\langle \{Q, \theta\} \cdots \rangle \neq 0. \tag{17}$$

The dots represent the insertion of additional operators. We should insert at least one operator that is not Q-closed, for otherwise the correlation function would vanish by the Q-invariance of the theory. A good choice is a function supported in the neighborhood of the rational curve C. If we insert such an operator, the correlation function receives contributions only from the instantons wrapping C and fluctuations around it. (Instantons are holomorphic maps from the worldsheet to M.) Under the assumption on the normal bundle, the fluctuations in the directions normal to C make trivial contributions and can be ignored. Then the computation reduces to the case of $M = \mathbb{CP}^1$, but we know that $\{Q, \theta\} \propto 1$ in that case, so we can find additional operators that make the correlation function nonzero. Thus, we again have the relation $\{Q, \theta\} \propto 1$, the equation $1 = 0$ holds in the Q-cohomology, and the chiral algebra vanishes nonperturbatively.

Now, with this vanishing theorem in hand, what can we say about the Höhn–Stolz conjecture? The crucial observation is the following: the Q-cohomology of states vanishes if the chiral algebra does. The argument is simple. Notice that the Q-cohomology of states is naturally a module over the chiral algebra. If the chiral algebra vanishes, then $[1] = 0$ and for any Q-closed state $|\Psi\rangle$ we have

$$[|\Psi\rangle] = [1] \cdot [|\Psi\rangle] = 0. \tag{18}$$

So any Q-closed state is Q-exact, in other words, the Q-cohomology of states is zero. Recalling that the Q-cohomology classes are in one-to-one correspondence with the supersymmetric states, we conclude that the vanishing chiral algebra implies that there are no supersymmetric states. It follows that the Witten genus vanishes.

We can actually say more. There is a Kähler structure on $\mathcal{L}M$ induced from that on M, so we can write $D_{\mathcal{L}M} = \mathcal{D}_{\mathcal{L}M} + \mathcal{D}^*_{\mathcal{L}M}$ as in the finite-dimensional case. If we canonically quantize the theory and compare the expressions of Q and $\mathcal{D}_{\mathcal{L}M}$, we realize that the two operators are related by

$$Q = e^{\mathcal{A}_0} \mathcal{D}_{\mathcal{L}M} e^{-\mathcal{A}_0}, \tag{19}$$

where the functional $\mathcal{A}_0 \colon \mathcal{L}M \to \mathbb{R}$ is the symplectic action $\mathcal{A}_H$ with $H = 0$. This shows that the Q-cohomology of states is the same as the $\mathcal{D}_{\mathcal{L}M}$-cohomology, and the supersymmetric states are in one-to-one correspondence with the harmonic spinors on $\mathcal{L}M$. Therefore, if the chiral algebra vanishes, there are no harmonic spinors on $\mathcal{L}M$ either. We should not take this statement too seriously, however. We do not even know how to construct the Dirac operator on loop space yet.

5. Questions and Speculations

A couple of questions naturally arise. In view of the Höhn–Stolz conjecture, an obvious one would be: does the chiral algebra vanish whenever $c_1(M) > 0$? If true, this implies that there are no harmonic spinors on $\mathcal{L}M$ when M admits a positive Ricci curvature, proving in the Kähler case the "loop-space Lichnerowicz theorem" and hence the conjecture. Although it is hard to tell at the moment whether this should be the case, we can hope that a deeper understanding of the physics of $(0, 2)$ supersymmetric sigma models will eventually lead to a decisive answer. As the chiral algebra is invariant under the renormalization group, it may be helpful to study the low-energy descriptions of these models.

Another important question is: how can we put all the arguments presented here on a mathematically sound footing? To this end we will need, at the very least, a rigorous construction of the chiral algebra. At the perturbative level, we know that the chiral algebra is given by the cohomology of a sheaf of chiral differential operators on M. In this approximation we only consider the neighborhood of the constant maps which make up a copy of M in the field configuration space. Beyond perturbation theory, we need to take into account the contributions from all instantons. Then, we expect that the chiral algebra is nonperturbatively given by the cohomology of some sheaf defined on the whole instanton moduli space, such that it reduces on the constant maps to the sheaf of chiral differential operators. It will be very interesting if such a sheaf is constructed and found to reproduce the vanishing theorem achieved through the physical argument.

References

[1] Vassily Gorbounov, Fyodor Malikov, and Vadim Schechtman, *Gerbes of chiral differential operators*, Math. Res. Lett. **7** (2000), no. 1, 55–66. MR1748287 (2002c:17040)

[2] ———, *Gerbes of chiral differential operators. III*, The orbit method in geometry and physics (Marseille, 2000), 2003, pp. 73–100. MR1995376 (2005a:17028)

[3] ———, *Gerbes of chiral differential operators. II. Vertex algebroids*, Invent. Math. **155** (2004), no. 3, 605–680. MR2038198 (2005e:17047)

[4] Anton Kapustin, *Chiral de Rham complex and the half-twisted sigma-model* (2005), available at `hep-th/0504074`.

[5] Fyodor Malikov and Vadim Schechtman, *Chiral de Rham complex. II*, Differential topology, infinite-dimensional Lie algebras, and applications, 1999, pp. 149–188. MR1729362 (2000j:17035b)

[6] Fyodor Malikov, Vadim Schechtman, and Arkady Vaintrob, *Chiral de Rham complex*, Comm. Math. Phys. **204** (1999), no. 2, 439–473. MR1704283 (2000j:17035a)

[7] Stephan Stolz, *A conjecture concerning positive Ricci curvature and the Witten genus*, Math. Ann. **304** (1996), no. 4, 785–800. MR1380455 (96k:58209)

[8] Meng-Chwan Tan, *Two-dimensional twisted sigma models and the theory of chiral differential operators*, Adv. Theor. Math. Phys. **10** (2006), no. 6, 759–851. MR2302273 (2009h:81267)

[9] Meng-Chwan Tan and Junya Yagi, *Chiral algebras of* $(0,2)$ *models: beyond perturbation theory*, Lett. Math. Phys. **84** (2008), no. 2-3, 257–273. MR2415553 (2009f:81196)

[10] Cumrun Vafa, *Superstring vacua*, 1989. Presented at Symp. on Fields, Strings and Quantum Gravity, Beijing, China, May 29–Jun 10, 1989.

[11] Edward Witten, *Elliptic genera and quantum field theory*, Comm. Math. Phys. **109** (1987), no. 4, 525–536. MR885560 (89i:57017)

[12] ———, *The index of the Dirac operator in loop space*, Elliptic curves and modular forms in algebraic topology (Princeton, NJ, 1986), 1988, pp. 161–181. MR970288

[13] ———, *Two-dimensional models with* $(0,2)$ *supersymmetry: perturbative aspects*, Adv. Theor. Math. Phys. **11** (2007), no. 1, 1–63. MR2320663 (2009c:17051)

[14] Junya Yagi, *Chiral Algebras of* $(0,2)$ *Models* (2010), available at `arXiv:1001.0118`.

Center for Frontier Science, Chiba University, Japan

Selected Published Titles in This Series

85 **Jonathan Block, Jacques Distler, Ron Donagi, and Eric Sharpe, Editors,** String-Math 2011, 2012

84 **Alex H. Barnett, Carolyn S. Gordon, Peter A. Perry, and Alejandro Uribe, Editors,** Spectral Geometry, 2012

83 **Hisham Sati and Urs Schreiber, Editors,** Mathematical Foundations of Quantum Field Theory and Perturbative String Theory, 2011

82 **Michael Usher, Editor,** Low-dimensional and Symplectic Topology, 2011

81 **Robert S. Doran, Greg Friedman, and Jonathan Rosenberg, Editors,** Superstrings, Geometry, Topology, and C^*-algebras, 2010

80 **D. Abramovich, A. Bertram, L. Katzarkov, R. Pandharipande, and M. Thaddeus, Editors,** Algebraic Geometry, 2009

79 **Dorina Mitrea and Marius Mitrea, Editors,** Perspectives in Partial Differential Equations, Harmonic Analysis and Applications, 2008

78 **Ron Y. Donagi and Katrin Wendland, Editors,** From Hodge Theory to Integrability and TQFT, 2008

77 **Pavel Exner, Jonathan P. Keating, Peter Kuchment, Toshikazu Sunada, and Alexander Teplyaev, Editors,** Analysis on Graphs and Its Applications, 2008

76 **Fritz Gesztesy, Percy Deift, Cherie Galvez, Peter Perry, and Wilhelm Schlag, Editors,** Spectral Theory and Mathematical Physics: A Festschrift in Honor of Barry Simon's 60th Birthday, 2007

75 **Solomon Friedberg, Daniel Bump, Dorian Goldfeld, and Jeffrey Hoffstein, Editors,** Multiple Dirichlet Series, Automorphic Forms, and Analytic Number Theory, 2006

74 **Benson Farb, Editor,** Problems on Mapping Class Groups and Related Topics, 2006

73 **Mikhail Lyubich and Leon Takhtajan, Editors,** Graphs and Patterns in Mathematics and Theoretical Physics, 2005

72 **Michel L. Lapidus and Machiel van Frankenhuijsen, Editors,** Fractal Geometry and Applications: A Jubilee of Benoît Mandelbrot, 2004

71 **Gordana Matić and Clint McCrory, Editors,** Topology and Geometry of Manifolds, 2003

70 **Michael D. Fried and Yasutaka Ihara, Editors,** Arithmetic Fundamental Groups and Noncommutative Algebra, 2002

69 **Anatole Katok, Rafael de la Llave, Yakov Pesin, and Howard Weiss, Editors,** Smooth Ergodic Theory and Its Applications, 2001

68 **Robert S. Doran and V. S. Varadarajan, Editors,** The Mathematical Legacy of Harish-Chandra, 2000

67 **Wayne Raskind and Charles Weibel, Editors,** Algebraic K-Theory, 1999

66 **Robert S. Doran, Ze-Li Dou, and George T. Gilbert, Editors,** Automorphic Forms, Automorphic Representations, and Arithmetic, 1999

65 **M. Giaquinta, J. Shatah, and S. R. S. Varadhan, Editors,** Differential Equations: La Pietra 1996, 1999

64 **G. Ferreyra, R. Gardner, H. Hermes, and H. Sussmann, Editors,** Differential Geometry and Control, 1999

63 **Alejandro Adem, Jon Carlson, Stewart Priddy, and Peter Webb, Editors,** Group Representations: Cohomology, Group Actions and Topology, 1998

62 **János Kollár, Robert Lazarsfeld, and David R. Morrison, Editors,** Algebraic Geometry Santa Cruz 1995, 1997

61 **T. N. Bailey and A. W. Knapp, Editors,** Representation Theory and Automorphic Forms, 1997

For a complete list of titles in this series, visit the
AMS Bookstore at **www.ams.org/bookstore/pspumseries/**.

ISBN 978-0-8218-7295-6

9 780821 872956

PSPUM/85